THE CONTINENTAL DRIFT CONTROVERSY
Volume IV: Evolution into Plate Tectonics

Resolution of the sixty-year debate over continental drift, culminating in the triumph of plate tectonics, changed the very fabric of Earth science. Plate tectonics can be considered alongside the theories of evolution in the life sciences and of quantum mechanics in physics in terms of its fundamental importance to our scientific understanding of the world. This four-volume treatise on *The Continental Drift Controversy* is the first complete history of the origin, debate, and gradual acceptance of this revolutionary explanation of the structure and motion of the Earth's outer surface. Based on extensive interviews, archival papers, and original works, Frankel weaves together the lives and work of the scientists involved, producing an accessible narrative for scientists and non-scientists alike.

Explanations of the curious magnetic anomalies on the seafloor and discovery and explanation of transform faults in the ocean crust in the mid-1960s led to the rapid acceptance of seafloor spreading. The birth of plate tectonics followed soon after with the geometrification of geology. Finally it was understood that the Earth's surface is divided into a small number of nearly rigid plates, most of which contain continental and oceanic parts, and whose relative motions are describable in terms of Euler's fixed point theorem. Although plate tectonics did not explain the cause or dynamic mechanism of drifting continents, it provided a convincing kinematic explanation that continues to inspire geodynamic research to the present day.

Other volumes in *The Continental Drift Controversy*:

Volume I – Wegener and the Early Debate
Volume II – Paleomagnetism and Confirmation of Drift
Volume III – Introduction of Seafloor Spreading

HENRY R. FRANKEL was awarded a Ph.D. from Ohio State University in 1974 and then took a position at the University of Missouri–Kansas City where he became Professor of Philosophy and Chair of the Philosophy Department (1999–2004). His interest in the continental drift controversy and the plate tectonics revolution began while teaching a course on conceptual issues in science during the late 1970s. The controversy provided him with an example of a recent and major scientific revolution to test philosophical accounts of scientific growth and change. Over the next thirty years, and with the support of the United States National Science Foundation, the National Endowment for the Humanities, the American Philosophical Society, and his home institution, Professor Frankel's research went on to yield new and fascinating insights into the evolution of the most important theory in the Earth Sciences.

THE CONTINENTAL DRIFT CONTROVERSY

CONTROVERSY

Volume IV: Evolution into Plate Tectonics

HENRY R. FRANKEL

University of Missouri–Kansas City

To Rosie and Maggie

CAMBRIDGE
UNIVERSITY PRESS

CAMBRIDGE
UNIVERSITY PRESS

University Printing House, Cambridge CB2 8BS, United Kingdom

Cambridge University Press is part of the University of Cambridge.

It furthers the University's mission by disseminating knowledge in the pursuit of education, learning and research at the highest international levels of excellence.

www.cambridge.org
Information on this title: www.cambridge.org/9781316616130

© Henry R. Frankel 2012

First published 2012
First paperback edition 2016

A catalogue record for this publication is available from the British Library

ISBN 978-1-107-01994-2 Hardback
ISBN 978-1-316-61613-0 Paperback

Cambridge University Press has no responsibility for the persistence or accuracy of URLs for external or third-party internet websites referred to in this publication, and does not guarantee that any content on such websites is, or will remain, accurate or appropriate.

Contents

Foreword by Steven Cande — page x

Acknowledgments — xiv

List of abbreviations — xvi

Introduction — xvii

1 Reception of competing views of seafloor evolution, 1961–1962 — 1

  1.1 Introduction — 1

  1.2 Wilson, the man — 2

  1.3 Wilson champions contractionism and continental accretion, 1949–1954 — 5

  1.4 Wilson and Scheidegger raise difficulties with continental drift and mantle convection as the cause of island and mountain arcs, 1949–1954 — 18

  1.5 Wilson continues to support contractionism and reject mobilism, 1959 — 21

  1.6 Scheidegger acknowledges paleomagnetic support for mobilism, 1958, 1963 — 31

  1.7 Wilson combines slow Earth expansion and his contractionist account of orogenic belts, 1960 — 33

  1.8 Bernal and Dietz discuss seafloor spreading, October 1961 — 35

  1.9 Wilson becomes a mobilist, 1961 — 36

  1.10 Wilson matches the Cabot and Great Glen faults in support of mobilism, 1962 — 46

  1.11 Menard and Hess correspond about seafloor spreading, 1961, 1962 — 50

  1.12 Menard defends seafloor thinning and attacks Earth expansion, 1962 — 53

  1.13 Heezen attacks seafloor spreading and seafloor thinning, 1962 — 56

  1.14 Irving cautions Hess, 1961 — 58

2 The origin of marine magnetic anomalies, 1958–1963 — 62

  2.1 Introduction — 62

  2.2 Explaining marine magnetic anomalies — 63

2.3   Interpretation of northeast Pacific marine magnetic anomalies
      prior to Vine's proposal, 1958–1962   65
2.4   Interpretation of magnetic anomalies over ridges and seamounts
      prior to Vine's proposal, 1953–1962   70
2.5   Matthews, his early life; goes to Antarctica   84
2.6   Matthews visits the Falkland Islands and favors continental drift 87
2.7   Matthews, his graduate work in the Department of Geodesy and
      Geophysics at Cambridge, 1958–1961   89
2.8   Vine's early interest in continental drift and undergraduate
      years at Cambridge, 1959–1962   91
2.9   Vine begins research, 1962   96
2.10 Mason and Raff on magnetic anomalies, 1962–1963   101
2.11 Vine reviews the literature on marine magnetic anomalies,
      October 1962 to January 1963   104
2.12 Matthews' meticulous survey over the Carlsberg Ridge,
      November 1962   109
2.13 Vine develops the Vine–Matthews hypothesis, early 1963   114
2.14 Morley's education and early work in paleomagnetism and
      aeromagnetic surveying   124
2.15 Morley accepts reversals of the geomagnetic field and continental drift 127
2.16 Morley's hypothesis   130
2.17 Morley's paper is twice rejected   136
2.18 Why Morley's paper was rejected and Vine and Matthews'
      paper was accepted   139
2.19 Two other Vine–Matthews-like hypotheses   141

3   Disagreements over continental drift, ocean floor evolution,
    and mantle convection continue, 1963–1965   148
3.1   Introduction   148
3.2   Wilson continues seeking further support for mobilism   148
3.3   The Royal Society's 1964 symposium on continental drift   162
3.4   The Everett, Bullard, and Smith fit of the continents
      surrounding the Atlantic   170
3.5   Paleomagnetism, other new evidence for continental drift,
      and mobilism's mechanism difficulty   186
3.6   Menard ends his flirtation with mobilism   193
3.7   Menard attacks seafloor spreading and Wilson's work on
      oceanic islands   198
3.8   Early responses to the Vine–Matthews hypothesis, 1964   202
3.9   Holmes on mantle convection, seafloor spreading,
      and Earth expansion   216
3.10 Rapid Earth expansion under attack, 1963–1964   227

4 Further work on the Vine–Matthews hypothesis and development
   of the idea of transform faults, 1964–1965   233
   4.1   Introduction   233
   4.2   Initial difficulties facing the Vine–Matthews hypothesis   234
   4.3   Vine, Matthews, and Cann defend and further develop the
         Vine–Matthews hypothesis, June 1964 to May 1965   236
   4.4   Mild support and criticism of the Vine–Matthews hypothesis
         during the first half of 1965   241
   4.5   Heirtzler, Le Pichon, and Talwani at Lamont Geological Observatory   245
   4.6   Lamont's view of mid-ocean ridges and rejection of the
         Vine–Matthews hypothesis: work on the Mid-Atlantic Ridge   248
   4.7   Vine, Wilson, and Hess at Madingley Rise, late 1964 to middle 1965   255
   4.8   Hess fine tunes and extends seafloor spreading, 1965   259
   4.9   Wilson develops the idea of transform faults   261
   4.10 Wilson's third trip around the world   269
   4.11 Vine independently proposes ridge-ridge transform faults   278
   4.12 Alan Coode's idea of transform faults   280

5 Continuing disagreements over the Vine–Matthews hypothesis,
   transform faults, and seafloor evolution, 1965   293
   5.1   Outline   293
   5.2   Wilson and Vine work in the northeast Pacific   294
   5.3   Lamont's view of mid-ocean ridges: work in the northeast Pacific   304
   5.4   Lamont's view of mid-ocean ridges: work on the Reykjanes Ridge   310
   5.5   Geomagnetic reversals, the dominance of remanence, and
         Vine's Ph.D. dissertation, August 1965   317
   5.6   Matthews seeks to explain the greater amplitude of the
         central anomaly   323
   5.7   The Ottawa meeting, September 1965   325
   5.8   The challenges of unraveling the Cenozoic history of the
         northeast Pacific   338

6 Resolution of the continental drift controversy   340
   6.1   Outline   340
   6.2   Lamont workers argue distribution of Atlantic Ocean floor
         sediments is incompatible with seafloor spreading: Hess disagrees   341
   6.3   Vine learns of corrections to the reversal timescale and fully
         accepts seafloor spreading: the November 1965 GSA meeting   345
   6.4   Improvements in the reversal timescale during 1966   350
   6.5   Opdyke and others at Lamont develop a reversal timescale
         based on the study of deep-sea cores   359
   6.6   Pitman's "magic" profile over the Pacific–Antarctic Ridge,
         December 1965: Pitman, Heirtzler, and Talwani accept V-M   363

6.7	Cox and Doell become mobilists	374
6.8	Vine turns the Vine–Matthews hypothesis into a difficulty-free solution	375
6.9	Sykes confirms ridge-ridge transform faults	386
6.10	Menard accepts seafloor spreading; Heezen renounces rapid Earth expansion	403
6.11	Lamont workers argue that heat flow over the Mid-Atlantic Ridge is too low for seafloor spreading; Hess disagrees	405
6.12	The Goddard conference: selling continental drift and seafloor spreading to the establishment	412
6.13	Making sense of why Le Pichon, Heirtzler, and Talwani tried so hard to prove Hess wrong	418
6.14	Maurice Ewing reluctantly accepts discontinuous seafloor spreading	428
6.15	Why seafloor spreading was rapidly accepted by most marine geologists and geophysicists	431
7	**The birth of plate tectonics**	437
7.1	Outline	437
7.2	Bryan Isacks and Jack Oliver at Lamont	438
7.3	Isacks and Oliver launch their study of deep earthquakes	443
7.4	Isacks and Oliver pin down subduction	446
7.5	McKenzie, the making of a geophysicist	456
7.6	McKenzie interprets heat flow data in terms of seafloor spreading	463
7.7	Heirtzler and company extend the reversal timescale	469
7.8	Morgan discovers plate tectonics	474
7.9	Morgan's presentations of plate tectonics: from April 1967 talk to March 1968 publication	478
7.10	Morgan's April 1967 AGU presentation	481
7.11	Morgan's 1968 paper "Rises, trenches, great faults, and crustal blocks"	485
7.12	McKenzie discovers plate tectonics	494
7.13	The keys to McKenzie's discovery	510
7.14	McKenzie and Parker's version of plate tectonics	516
7.15	Comparison of Morgan's and McKenzie and Parker's presentations of plate tectonics	527
7.16	Isacks spearheads discovery of how deep earthquakes beneath island arcs are caused	539
7.17	Le Pichon loops plate tectonics around the world	552
7.18	Isacks, Oliver, and Sykes integrate seismology and plate tectonics	562
7.19	Atwater and Menard apply plate tectonics to the great fracture zones of the northeast Pacific	576

7.20 The Great Magnetic Bight explained in terms of seafloor spreading   584
7.21 McKenzie and Morgan explore the evolution of triple junctions   591
7.22 Towards the new paleogeography: the complementarity of plate
      tectonics and APW paths and the triumph of mobilism   604

References   617
Index   656

Foreword

Graduating from Yale in June of 1970 with a major in geology and geophysics I was at a loss as to what to do with the rest of my life. It was the height of the Vietnam War, the previous months had seen disturbing anti-war protests in New Haven and around the country, and I was disillusioned with academics. I had switched majors from engineering to geology and geophysics at the start of my junior year not because of any particular interest in geology, I had not yet taken any geology courses, but because I liked the atmosphere in the geology building where I had a bursary job in my sophomore year polishing meteorites. I quickly learned that there was an ongoing revolution in the earth sciences involving continental drift and seafloor spreading. There were talks by J. Tuzo Wilson, complete with paper cut-outs showing how transform faults work, and by Teddy Bullard on the remarkable fit of the continents around the Atlantic Ocean. I read Arthur Holmes' classic textbook on physical geology which had recently been updated with a long section on paleomagnetism and continental drift. I found a discarded plot showing the location of earthquake epicenters around the globe which I scotch taped to my dorm room wall. However, my senior project involved working on a camera to take cross-sectional photos of fecal pellets on the floor of Cape Cod Bay and by the time I graduated my interest in geology was waning.

A serendipitous encounter at the end of the summer with Jim Walker, one of my undergraduate mentors (the conversation went something along the lines of "Steve, scientists at Lamont are often looking to hire recent grads to go out to sea") led to an interview the next day with Walter Pitman, who was then the head of the marine magnetics group, and a job offer. Walter gave me some papers to read about plate tectonics and suggested that I try to use the shape of marine magnetic anomalies to determine where the Pacific plate had moved over the last 80 million years. He soon sent me out to sea on the R/V Conrad for six months to look after the magnetometer and digitize the magnetic records. I have been working with marine magnetic anomalies ever since.

Upon arriving at Lamont I discovered that there had indeed been a revolution in the earth sciences and that many of the Lamont scientists, including Walter, had played pivotal roles. Walter was famous for his *"Eltanin-*19*"* profile, a magnetic anomaly record collected on the USNS *Eltanin* across the Pacific-Antarctic ridge and published in 1966 which displayed stunning mirror image symmetry about the spreading axis. Pitman's "magic profile" was a tipping point in the acceptance of seafloor spreading and continental drift by the global earth science community. I soon learned that the number of people at Lamont who were involved with the plate tectonic revolution was not only large but consisted of a fascinating coterie of characters. There was a seemingly endless set of stories about what had happened at Lamont during the heady days of the revolution, who had had flashes of brilliance, who had missed the obvious, and who had been slow to come around to accepting seafloor spreading.

At the top was Doc Ewing, the founder and director of Lamont, who was a "fixist," which meant that he was opposed to the revolution, but whose vision for collecting all types of geophysical data on two ships constantly circuiting the globe, even data whose usefulness was unclear, eventually provided Lamont scientists with an unparalleled global data base to exploit the implications of plate tectonics. He ran a tight ship. There was Bruce Heezen, originally one of Doc's protégés, who, along with Marie Tharp, had discovered the mid-ocean rift and had become a believer in an expanding earth. There had been a famous falling out between Bruce and Doc a few years earlier. There was Manik Talwani, Marc Langseth, Jim Heirtzler and Xavier LePichon, all accomplished marine geophysicists, who, in different combinations, had published several papers critical of the idea of seafloor spreading before the appearance of Walter's magic profile. There were the seismologists, Jack Oliver and his protégés Bryan Isacks and Lynn Sykes, who had instantly jumped on the implications of Walter's magic profile and published a series of classic papers outlining the fundamentals of the new global tectonics. And there was Neil Opdyke in the office next to Walter's, a paleomagnetist and one time captain of the Columbia football team, who was an early believer in continental drift, perhaps the only one at Lamont before the appearance of Walter's profile.

As I heard more stories about the events of the revolution the cast of characters quickly grew because the revolution had not started at Lamont. Harry Hess at Princeton was famous because he had written a seminal paper proposing the process of seafloor spreading in 1960. Perhaps most important from a magnetics point of view was the work of Fred Vine and Drum Matthews at Cambridge who originally proposed that marine magnetic anomalies at spreading centers contained a record of reversals of Earth's magnetic field. They had hit the bull's eye. One amazing aspect of their story is that they figured out the answer from data on the Carlsberg ridge, a slow-spreading ridge where the magnetic anomalies are much messier and harder to interpret than anomalies which form at fast-spreading centers such as the Pacific–Antarctic ridge, where Walter's magic profile was eventually collected. I also heard that there was a paper by the Canadian scientist Lawrence Morley with virtually the same idea as Vine and Matthews, which had been repeatedly rejected by journals.

And there were the magnetic data sets from the Pacific Ocean collected in the 1950s by Ron Mason, Arthur Raff and Vic Vacquier at Scripps which in retrospect contained a beautiful record of seafloor spreading but whose origin was unknown at the time they were collected. These scientists had the misfortune of trying to interpret seafloor spreading anomalies before the phenomenon of reversals of Earth's magnetic field was widely understood.

Most of the stories I had heard were incomplete. Yes, there was a famous falling out between Doc Ewing and Bruce Heezen but what exactly was it about? And yes, Talwani, Langseth, Heirtzler and LePichon at Lamont had initially rejected the Vine and Matthews interpretation of magnetic anomalies at mid-ocean ridges, but these were all very smart guys, so why did they reject it? And how was it that Fred Vine, a young, first year graduate student at Cambridge in 1963, was in a position to interpret the data set that Drum Matthews had just collected on the Carlsberg ridge? And how was it that he saw the answer? Several of Vine and Matthews' colleagues at Cambridge were also slow to accept their interpretation. These are all fascinating questions because they are at the heart of understanding how plate tectonics was discovered, and more generally, how science unfolds. This volume explores the answers to these and many more fascinating questions.

The discovery of plate tectonics took place through many small steps and some enormous leaps of scientific insight. Frankel examines all of the steps, big and small, that were involved in the development of plate tectonics. Methodically, he first outlines where the science was at a certain point, and then describes the discovery that an individual or group made, and finally shows how the broader scientific world reacts to the new discovery. At each step he examines correspondence between the principals which casts light on the thought processes going on in the scientists' heads. Frankel has directly communicated over the years with many of the principals to get them to explain the events in their own words. As he put the story together he would get back to the scientists to ask them to clarify the chain of events. He gives us short biographies of many of the individuals showing us, to the extent possible, how their backgrounds might have influenced their ability to make these discoveries.

There are many riveting points in this discourse. As a scientist, I am particularly fascinated by the Eureka moments, when the answer to a problem suddenly reveals itself to a researcher. There were many such moments in the plate tectonic revolution and they make for compelling reading. One of my favorites is when Hess, Wilson and Vine realize that there should be a symmetrical pattern of seafloor spreading anomalies on the Juan de Fuca ridge; Vine rushes to the library and brings back the publications of Mason and Raff and together they all absorb the implications of that remarkable magnetic anomaly pattern. A second inspiring Eureka moment is when Oliver, Isacks and Sykes are looking at the earthquake seismic data from the Tonga trench and realize that they are seeing the downgoing slab plunging through the mantle.

The plate tectonic revolution proceeded quickly once Pitman's magic profile was published. LePichon reversed his stance seemingly overnight and wrote the first comprehensive paper outlining Cenozoic global plate motions. Euler rotations, underpinning the theoretical basis of the motion of rigid plates, had implications for how the azimuths of fracture zones and the rate of seafloor spreading should vary along mid-ocean ridges. This was realized independently and roughly simultaneously by Dan McKenzie at Cambridge and Jason Morgan at Princeton. Morgan saw the light a few months before McKenzie did, but McKenzie was the first to publish. Frankel goes to great length to examine this interesting episode.

There are many aspects of the story behind plate tectonics of which I was unaware. For example, after reading Vine and Matthews' paper on seafloor spreading, George Backus, then a young seismologist at Scripps, realized that a test of this concept was that spreading rates should get faster the further away you got from the pivot point between two plates until you were 90° away. He wrote this idea up in a paper in 1964 and wrote a proposal to NSF requesting funds to collect magnetic profiles across the mid-ocean ridge in the South Atlantic to test this idea. It was not funded and the profiles in databases like the one amassed at Lamont eventually served the same purpose, but in retrospect Backus was way ahead of his colleagues in his thinking.

As I read through the volume I was also struck by the fact that so many of my first encounters as an undergraduate with plate tectonics in the late 1960s – J. Tuzo Wilson's demonstration of transform faults; Bullard's demonstration of the fit of the continents around the Atlantic; the second edition of Arthur Holmes's *Principles of Physical Geology*; and poster-sized charts of the global seismicity pattern – were shared by other young scientists of the era. Reading this volume kindled some fond memories from this period of my life when I was just embarking on a career in science. Frankel has done a marvelous job of capturing the wonders of a scientific revolution.

Steven Cande
Professor of Marine Geophysics, Scripps Institution of Oceanography, UCSD

Acknowledgments

I could not have undertaken and completed this book without enormous help from many Earth scientists whose work is discussed. I thank Tom Allan, Tanya Atwater, George Backus, Clem Chase, Alan Coode, Richard Doell (deceased), Jim Everett, Bob Fisher, Ron Girdler, Tony Hallam, Warren Hamilton, Jim Heirtzler, Raymond Hide, Ted Irving, Bryan Isacks, Dan Karig, Tony Laughton, Xavier Le Pichon, Ian McDougall, Michael McElhinny, Dan McKenzie, Ursula Marvin, Drum Matthews, Bill Menard, Peter Molnar, Eldridge Moores, Jason Morgan, Jason Phipps Morgan, Neil Opdyke, Bob Parker, Walter Pitman, Keith Runcorn, John Sclater, Alan Smith, Lynn Sykes, Manik Talwani, Don Tarling, Marie Tharp (deceased), David Tozer, Fred Vine, Bob White, and Tuzo Wilson (deceased) for answering questions about their work or the work of others.

I thank Ted Irving, Dan McKenzie, and Fred Vine for critically reviewing most or all chapters of this volume. I thank Tom Allan, George Backus, Alan Coode, Jim Everett, Jim Heirtzler, Bryan Isacks, Tony Laughton, Xavier Le Pichon, Dan McKenzie, Peter Molnar, Jason Morgan, Neil Opdyke, Bob Parker, Walter Pitman, Alan Smith, Lynn Sykes, and Fred Vine for critically reviewing my accounts of their work in this volume.

I thank Patricia Proctor and Susan Wilson, daughters of J. Tuzo Wilson, for giving me permission to reproduce several items from J. Tuzo Wilson's papers, and for helping me attempt to determine the whereabouts of their father while in the United Kingdom and Europe in 1965. I thank George Hess for allowing me to reproduce documents from his father's papers at Princeton University, Dick Doell (deceased) for giving me permission to reproduce several items from Alan Cox's papers, Gill Tuner for copying various documents housed within the Department of Geodesy and Geophysics at Cambridge University, and Sarah Finney, Conservator, Sedgwick Museum of Earth Sciences, for helping me locate and examine Drum Matthews' diaries,

I have greatly benefited from studying Menard's own retrospective work, *Ocean of Truth*, and from studying many of the retrospective essays in Oreskes and Le Grand's edited *Plate Tectonics*.

I should like to thank my colleagues Bill Ashworth, Bruce Bubacz, Ray Coveney, George Gale, Clancy Martin, and Dana Tulodziecki for discussions relevant to this work.

I thank Nancy V. Green and her digital imaging staff at Linda Hall Library, Kansas City, Missouri, for providing the vast majority of the images. I should also like to thank the reference librarians at Linda Hall Library, and the interlibrary staff at the Miller Nichols Library, UMKC.

I owe much to Nanette Biersmith for serving as my longtime editor and proofreader.

I thank Jeff Holzbeierlein for putting me a position to complete this volume.

I am indebted to the United States National Science Foundation, the National Endowment of the Humanities, and the American Philosophical Society for financial support. I also thank the University of Missouri Research Board and my own institution for timely grants to continue this project.

I wish to thank Susan Francis and her staff, particularly Laura Clark and Abigail Jones, at Cambridge University Press for believing in this project and for their great assistance throughout its production.

Abbreviations

AGU	American Geophysical Union
AMSOC	American Miscellaneous Society
ANU	Australian National University
APW	Apparent polar wander
FIDS	Falkland Islands Dependencies Survey
GAD	Geocentric axial dipole
GSA	Geological Society of America
GSC	Geological Survey of Canada
IGPP	Institute of Geophysics and Planetary Physics
IGY	International Geophysical Year
IUGG	International Union of Geodesy and Geophysics
JGR	*Journal of Geophysical Research*
Lamont	Lamont Geological Observatory
MIT	Massachusetts Institute of Technology
NIO	National Institute of Oceanography (UK)
NRC	National Research Council (Canada)
NRM	Natural remanent magnetization
NSF	National Science Foundation (USA)
RAS	Royal Astronomical Society (UK)
RS1	Research Strategy 1
RS2	Research Strategy 2
RS3	Research Strategy 3
Scripps	Scripps Institution of Oceanography
UCSD	University of California, San Diego
UCLA	University of California, Los Angeles
UCRN	University College of Rhodesia and Nyasaland
USCGS	United States Coast and Geodetic Survey
USGS	United States Geological Survey
V-M	Vine–Matthews hypothesis
WHOI	Woods Hole Oceanographic Institute

Introduction

Volume III described the growth of marine geophysics/geology during the 1950s and early 1960s and how it led to the development of four hypotheses to explain the origin of mid-ocean ridges: Hess's and Dietz's seafloor spreading, Menard's hypothesis of seafloor thinning, Heezen's and Carey's rapid Earth expansion, and the Ewing brothers' hypothesis that ocean ridges are produced by small-scale mantle convection. Of them, only the Ewings' was fixist. Although land-based paleomagnetism had showed (Volume II) that the Ewing brothers' fixism, and hence their hypothesis, was incorrect, most marine geophysicists and geologists failed to acknowledge paleomagnetism's confirmation of mobilism, and did not discard the Ewings' hypothesis because it did not incorporate continental drift. Moreover, during the early 1960s, none of these hypotheses warranted acceptance because the gap between what was known about ocean basins and what was implied by the hypotheses was too great to eliminate any of them.

Volume IV begins by detailing how this gap was closed by the confirmation of the Vine–Matthews hypothesis (hereafter, V-M) and the idea of ridge-ridge transform faults. Vine and Matthews proposed their hypothesis in 1963, and the idea of ridge-ridge transform faults was separately put forth by J. Tuzo Wilson and Alan Coode in 1965. Their confirmation in 1966 yielded two difficulty-free solutions. By 1967 the vast majority of marine geologists and geophysicists, and many but by no means all land-based workers, had accepted seafloor spreading and continental drift, thus ending for them the sixty year continental drift controversy. However, continued rejection of mobilism had become doubly unwarranted; unwarranted because of mobilism's support from land-based paleomagnetism and from marine geophysics/geology.

Despite the resolution of the drift controversy, plate tectonics, the eponymous crowning achievement of the revolution, had yet to be discovered, confirmed, and conceptually explicated, and it with these that the final chapter of this last volume is concerned. Plate tectonics was independently proposed in 1967 by Jason Morgan, and Dan McKenzie and Robert Parker. It combined continental drift and seafloor spreading into a precise kinematic theory of Earth's tectonics. Plate tectonics divides

Earth's surface into a small number of rigid plates whose motions relative to each other are describable with mathematical precision. Initially applicable only to Earth's present-day tectonics, plate tectonics was soon applied to the past.

There are two abiding themes. Plate tectonics is a kinematic, not a dynamic theory. Both Morgan, and McKenzie and Parker adopted a version of seafloor spreading that jettisoned any direct tie between mantle convection currents rising immediately beneath oceanic ridges and mantle convection currents sinking immediately beneath trenches. Plate tectonics became the reigning theory of the Earth sciences even though its cause remained an unknown mystery. The old-time drifters' plea that continental drift should be accepted even though its mechanism remained a mystery had gone unheeded by the vast majority of Earth scientists during the classical stage of the controversy, and then right on even into the mid-1960s, just so long as the occurrence of drifting continents remained, in the minds of most, an open question. Once drift of continents was unquestionably shown to occur, the old mechanism difficulty shrivelled away, becoming a phantom difficulty, an unsolved theoretical problem. So, ironically, mobilism gained acceptance even though its cause remained substantially unknown, although thought to involve mantle convection. Second, in the overall scheme of things, the importance of paleomagnetism did not diminish, but increased. Vine and Matthews' hypothesis is based on the paleomagnetic idea of reversals of the geomagnetic field, and that newly created seafloor acquires a thermoremanent magnetization in the direction of the geomagnetic field as it cools through its Curie point. Determining the past history of ocean basins and coordinating the formation of ocean basins relative to each other are based on matching and dating magnetic anomalies from different ocean basins, the anomalies being tied, as they were, to reversals of the geomagnetic field. In addition, paleolatitudes, be they of continents or ocean basins, cannot be determined without the use of marine magnetic anomalies and APW paths, and in the very best work APW paths and plate tectonics are used hand-in-glove to work out not only Earth's tectonic history but also Earth's climatic history, for which latitude is of overriding importance.

I should also like to underscore the important role computers began to play in the Earth sciences. This is not surprising. Vine, Morgan, McKenzie and Parker, just to mention four of the many whose work we shall examine, learned how to compute and write programs that became essential to the development and testing of V-M and plate tectonics.

As in earlier volumes, I shall describe how researchers acted in accordance with what I have identified as three standard research strategies (I, §1.13). These three research strategies are my retrospective description of how they went about their tasks, how they addressed their problems; the workers themselves did not recognize or say that they acted in this way. Research Strategy 1 (hereafter, RS1) was used by researchers to expand the problem-solving effectiveness of solutions and theories. Research Strategy 2 (hereafter, RS2) was used by them to diminish the effectiveness of competing solutions and theories; RS2 was an attacking strategy used to raise difficulties against opposing solutions, and to place all possible obstacles in their way. Workers used Research Strategy 3 (hereafter, RS3) to compare the effectiveness of competing solutions and theories, and to emphasize those aspects of a solution or theory which gave it a decided advantage over its competitors.

1

Reception of competing views of seafloor evolution, 1961–1962

1.1 Introduction

Marine geologists had the choice of four competing explanations of the origin and evolution of ocean basins: there was the fixism as espoused by the Ewing brothers (1959), the mobilistic models of seafloor spreading (Hess, 1960d, 1962; Dietz, 1961a), seafloor thinning (Menard, 1960), and rapid Earth expansion (Heezen, 1959a, 1959b; Carey, 1958). By the early 1960s, and except for expansionists, all the above invoked mantle convection.

I want now to trace the debate over ocean basin evolution through 1962. I must, however, first introduce J. Tuzo Wilson, who would about this time replace Hess and Dietz as the most voluble supporter of seafloor spreading. During the 1950s, Wilson was an ardent fixist. He and the theoretical geophysicist Adrian Scheidegger revitalized Jeffreys' contractionism and offered a sophisticated account of orogenesis. Still defending fixism through 1959, Wilson became a mobilist in 1961. Although he was not convinced that mobilism was correct because of its paleomagnetic support, it forced him to wonder if there might be something to mobilism. With this shift in attitude, he soon found ways to accommodate significant elements of Scheidegger's work within his account of orogenesis. Once converted, he never wavered; he championed mobilism with all the ardor with which he had so recently denied it.

After tracing Wilson's changing ideas through the eve of his conversion, I shall discuss the exchange about seafloor spreading between Dietz and J. D. Bernal, a crystallographer by training and a polymath by avocation, who welcomed Dietz's ideas. Returning to Wilson, I shall show that he, aware of Bernal's forthcoming welcome, joined the party and expressed his support of seafloor spreading. I shall then turn to Hess, Menard, Heezen, and Ewing. In a 1961 letter, Menard raised difficulties with Hess's seafloor spreading. Hess responded, and in 1962, Menard raised additional difficulties. Menard also continued to develop his idea of seafloor thinning. Meanwhile, in his 1962 defense of rapid Earth expansion, Heezen attacked seafloor spreading and seafloor thinning, raising difficulties about the geometry and interaction of convection currents. Ewing defended his fixist account of seafloor evolution. He argued that the paucity of terrigenous seafloor sediments, so readily explained by seafloor spreading and Earth expansion, could also be explained by fixism.

I shall conclude by examining a letter that Irving sent to Hess in 1961 after reading the famous 1960 preprint of his paper on seafloor evolution. Irving was happy to see mantle convection as mobilism's cause. However, he was concerned that not enough was yet known about Earth's interior to speculate about mobilism's mechanism, and feared that Earth scientists would again get bogged down in fruitless speculation about mechanism. He wanted efforts to be further directed at proving the reality of continental drift. As it turned out, Hess's seafloor spreading spawned Wilson's idea of transform faults and the Vine–Matthews hypothesis. As we shall see below, once both were confirmed, they provided the concrete evidence that Irving wanted. But Irving at the time was right about the dangers of a further extension of the long drawn-out quest for a mechanism. Even though many accepted seafloor spreading with the confirmation of its two key corollaries, they mistakenly thought that it provided a mechanism. With the transformation of seafloor spreading into plate tectonics, seafloor spreading's dynamics was jettisoned. The tight tie between rising convection currents and the formation of ridges was severed. Plate tectonics was a kinematic theory, it said nothing about mechanism. The reality of continental drift was established and plate tectonics was accepted before its mechanism was revealed.

1.2 Wilson, the man

John Tuzo Wilson (1908–93), known professionally as Tuzo, was born in Ottawa, Canada.[1] His father, J. A. Wilson, was apprenticed as an engineer, and worked in India and western Canada before taking a job with the Canadian government in Ottawa in 1910. By 1918, he was in charge of developing civil aviation in Canada. Wilson's mother, Henrietta Tuzo, in an expedition in 1906 with the Alpine Club of Canada to the Valley of the Ten Peaks in the Rocky Mountains, made the first ascent of the seventh peak, later named Mount Tuzo in her honor. At the camp she met her future husband. They married three years later. Wilson's early life prepared him for hard work and the outdoors. He and his younger sister were taught the primary school curriculum at home by a governess. Well taught, he was ahead of his class when he went to school in 1915. Forced to play with older and bigger boys, he (1990: 267) developed "an independent cast of mind," and "a poor opinion of team sports." Beginning in 1924 at age fifteen, he spent fourteen summers working for the Geological Survey of Canada (GSC). In summer 1925, he had "the good fortune to become a field assistant to Noel Odell," "a fine geologist who had been the hero of the 1924 Mount Everest expedition" (Wilson, 1982: 4; 1990: 268).

In 1926, Wilson enrolled in an honors program in mathematics and physics at the University of Toronto. Compared to fieldwork in geology and inspired by Odell, he found the laboratory work boring (Wilson, 1982: 4). He told his professors that he wanted to study geology. Except for Lachlan Gilchrist, they were surprised and disappointed. Gilchrist designed a program for him to combine physics and geology. He was struck by the contrast between the global approach of physicists and the regional approach of geologists. He wanted geologists to develop a more global approach and underpin their theories with physics. Gilchrist also found work for Wilson as a geophysical assistant during summers.

Graduating in 1930 as the first Canadian student in physics and geology, Wilson, supported by a Massey Fellowship (1930–2), enrolled at the University of Cambridge for a second B.A. degree, where he (1990: 269) had "the good fortune to take lectures from Sir Harold Jeffreys . . . my grasp of mathematics was limited; I understood only a fraction of what he said." Wilson was scheduled to learn exploratory geophysics from Bullard, but worked with James Wordie, because Bullard was delayed in Africa doing a gravity survey of the Great Rift Valley (III, §4.11). They met when Bullard returned, becoming lifelong friends. His supervisor was John Cockcroft, later Sir John and Nobel Laureate.

He obtained his degree, and returned to Ottawa. After a year working with W. H. Collins, Director of the GSC, he was ready to do a Ph.D. in geophysics. Wilson chose Princeton University because he learned that Field, who had encouraged Hess (III, §5.2), Bullard (III, §4.11), and Ewing (§2.2) to work in marine geology, hoped to start offering geophysics at Princeton. But it did not happen. Field at least arranged for Wilson and Woollard, a fellow graduate student, to work with Ewing on his seismic study of New Jersey's coastal plain. He spent a few weekends with Ewing. He also got to know Hess. Wilson was assigned to Professor T. Thom, a structural geologist, and expert on the Beartooth Mountains. Thom gave Wilson $200 to buy a used car, drive to Montana, and survey a section of the Beartooth Mountains. Wilson spent the summers of 1934 and 1935 on the project. He was the first to ascend Mount Hague (12 300 feet). What he found "strengthened" his "belief that geology was rife for change" (Wilson, 1990: 270).

When I reached the top, I was astonished to find the summit flat, about the size of a football field, and only gently tilted . . . Pondering on the climb later, I realized that, to preserve its flat top, the mountain must have been recently uplifted vertically and not squeezed up like toothpaste some sixty million years ago, which was then the conventional view. This strengthened my belief that geology was rife for change.

(Wilson, 1990: 270)

Wilson did not think that "rife for change" included continental drift.

With his Ph.D. in hand, he rejoined the GSC in 1936. He worked in Nova Scotia, Quebec, and the Northwest Territories, and developed a longstanding interest in the GSC. Under Collins, the survey had begun using aerial photographs and Wilson used them to trace large structures, bedding of stratified rocks, and glacial deposits, but their use was resisted by older geologists who regarded it "as a form of cheating" (Wilson, 1982: 8). He argued that the Canadian Shield was not a monolithic structure but a mosaic of increasingly younger provinces surrounding very old nuclei. He came to appreciate that aerial photographs allowed one to see large-scale structural features from afar and helped geologists see regional and not just local patterns.

In 1939 he joined the Canadian Army as a member of the Royal Canadian Engineers. Commissioned a lieutenant, he spent four years overseas in Britain. He continued working for the Royal Canadian Engineers until 1946 when he became professor of geophysics in the Department of Physics at the University of Toronto. He did not become a university professor until age 38. When invited to accept the position, Wilson sought the advice of C. J. Mackenzie, President of the National Research Council (NRC) of Canada.

In 1946 I had to choose whether to remain in the army engineers, where I had reached the rank of colonel, return to the Survey where I was promised that I could soon be Director, enter industry, or succeed Gilchrist as Professor of Geophysics at Toronto. I sought the advice of C. J. Mackenzie, then the wise president of the National Research Council of Canada ... He advised me to go back to the university and to take no administrative job for twenty years for he predicted that I would be successful in research. I accepted his advice and he rewarded me with ample opportunities to travel and organize projects.

(Wilson, 1982: 8–9)

Wilson wisely accepted Mackenzie's prophetic advice.

Mackenize's patronage helped Wilson become an influential science advisor in "high" places, especially the NRC. In 1945, Wilson was appointed chair of an Associate Committee of NRC to advise it on geophysical matters. In 1946, this committee joined with Canada's International Union of Geodesy and Geophysics (IUGG) committee to form the Associate Committee of Geodesy and Geophysics of the NRC. He also became a prominent member of IUGG, serving as president (1957–60). The 1957 IUGG meeting was held in Toronto mainly because of his efforts. Wilson was appointed a member of the NRC 1958–64. He used his membership of IUGG and NRC committees well, attending many meetings, giving many lectures, seeing many rocks, and speaking with many geologists worldwide. He eventually took nine trips around the world, continually educating himself about world geology.

In 1950, Wilson embarked on his first world-tour, with extensive stop-offs in Australia and Africa. He was "guided through every state in Australia by local geologists, who explained the regional geology" (Wilson, 1990: 276). They were then, like Wilson, staunchly fixist. King and Plumstead, two avid mobilists, showed him much of South Africa's geology; they failed to convince him that mobilism was correct. Despite his close-mindedness about mobilism, Wilson (1990: 278) later claimed that the trip reinforced his conviction that geologists should take a global approach, studying Earth "as a whole." So he was becoming a globalist in 1950 even though he did not become a mobilist until the 1960s.

Wilson continued his high-paced research efforts until about 1967, when he (1982: 12) thought it "an appropriate time to move on … the twenty years of scientific research which C. J. Mackenzie had suggested and which I had enjoyed was up." He accepted an offer to become Principal of Erindale College, a new suburban college of the University of Toronto. Both Tuzo and his wife, Isabel, found the experience rewarding, and Erindale College had become a thriving institution in 1974 when Wilson reached the mandatory retirement age of sixty-five for holding leadership roles in administration. Before he had a chance to retire, however, the Premier of Ontario asked him to be the Director General of the new Ontario Science Centre. Wilson remained an enthusiastic and successful Director General until he retired in 1985.

Wilson was greatly honored for his research and service to his country. In 1946 he became Officer of the Order of the British Empire and Officer of the Legion of Merit, United States. He was elected a fellow of the Royal Society of Canada in 1948, and served as its President in 1972–3. He was elected to the Royal Society of London in 1968. In 1970 he was named Officer of the Order of Canada, and four years later promoted to Companion, Canada's highest civilian honor. He was given a score of medals and awards including: Willet G. Miller Medal, Royal Society of Canada (1958); Civic Award of Merit and Gold Medal, City of Toronto (1960); Logan Medal, Geological Association of Canada (1968); Bucher Medal, American Geophysical Union (AGU) (1968); Penrose Medal, Geological Society of America (GSA) (1968); J. J. Carty Medal, US National Academy of Sciences (1974); Gold Medal, Royal Canadian Geographical Society (1978), Wollaston Medal, Geological Society of London (1978); Vetlesen Prize, Columbia University (1978); Ewing Medal, AGU (1980); Maurice Ewing Medal, Society of Exploration Geophysics (1980); and A. Wegener Medal, European Union of Geosciences (1989). In 1978 the Canadian Geophysical Union established the J. Tuzo Wilson Medal, and he was the first recipient.

1.3 Wilson champions contractionism and continental accretion, 1949–1954

Wilson loved to speculate. He wanted to find general solutions, and was willing to try, even if many, geologists especially, thought his efforts were misguided. Consider, for example, remarks about Wilson by Philip B. King, a prominent, highly respected field geologist, who spent most of his career with the United States Geological Survey (USGS). King constructed a marvelous map of North America (1944) based partly on his own fieldwork. Like Kay, he championed the geosynclinal theory with the evolution of island arcs into mountain belts, and growth of continents around an ancient central core (1959). Wilson made good use of both Kay's and King's maps. In King's 1959 *The Evolution of North America*, he recommended several fixist publications, among them Kay's (1951) *North American Geosynclines* (I, §7.3), Umbgrove's *The Pulse of the Earth* (1947; I, §8.14), and Wilson's (1954) "Development and structure of the crust," his most extensive account of contractionism and continental accretion. King (1959: ix) said of Wilson's work, "It sets forth some bold syntheses of classification and interpretation. Interesting reading, even in those parts which do not altogether inspire belief."

In his defense of contractionism and continental accretion from the late 1940s and first half of the 1950s Wilson makes plain his desire to satisfy physics and geology.

A comparison is made between the different approaches to the study of the Earth by the field geologist and by the physicist interested in broad terrestrial problems. On one hand, the rapid increase in geological knowledge is giving an increasingly clear picture of the progress of geological history and the nature of the processes which have been involved. On the other hand, the application of new and rapid methods of obtaining measurements of physical properties of the Earth are serving to limit and direct physical speculation about the nature of the Earth's mechanisms. These different approaches must lead to the same conclusion. When reached, it will be a history of the Earth and an explanation of its processes satisfying both to the geologist and to the physicist.

(Wilson, 1952b: 444)

He was preoccupied with the Canadian Shield and thus with Precambrian geology. Emphasizing uniformitarianism, he argued that mountain belts had formed throughout the Precambrian and in much the same way as they had during the Phanerozoic. Unlike Dana, for example, he rejected the idea that the Canadian Shield formed as a single unit. This led him, as it did King, to argue that the Canadian Shield had itself formed around original old nuclei as ever younger, newly formed, marginal mountain belts were added.

Wilson was enough of a field geologist to want to apply Jeffreys' idealized theory to presently active mountain belts and island arcs, and to adjacent continents. He also wanted to utilize developments in the physics of continuous matter to help explain how Earth's outer layers fail as they contract. Realizing that he needed someone with a stronger theoretical background, he approached Adrian Scheidegger, an applied mathematician at the University of Toronto, who spent 1950 and 1951 on the task. Together, and separately, they argued that contractionism offered the best available solution to the origin of island arcs and mountain belts, and the origin and evolution of continents (Scheidegger and Wilson, 1950; Wilson, 1949a, 1949b, 1949c, 1950, 1951a, 1951b, 1952a, 1952b, 1953, and 1954; Scheidegger, 1953a, 1953b).

Jeffreys maintained that below a depth of 700 km Earth has not appreciably contracted, that it is "contracting, and becoming thinner from a depth of 700 km to about 100 km," and that its outermost layer is no longer contracting, and is separated from the contracting layer by a level of no strain. Scheidegger determined that the outermost layer is under compression, and "fails by thrust faulting in conical zones" along dip angles of less than 45°, while the layer underlying the level of no strain is under tension, and fails by normal faulting at dip angles of more than 45°. Wilson summarized Scheidegger's findings.

Jeffreys (1929, pp. 278–279) suggested that cooling has not affected the interior of the earth and that, as a consequence, from the center of the earth to within about 700 km of the surface no appreciable change in volume has yet taken place. He proposed that the layers were cooling, contracting, and becoming thinner from a depth of 700 km to about 100 km and that, because the interior was not altering, these layers were stretched out horizontally. He suggested that there was a level of no strain at about 100 km depth, above which the layers had already largely cooled and were therefore under a horizontal crushing stress ... The shell between 70 and 700 km is considered to be contracting because of cooling, as Jeffreys visualized; and the earthquakes in it are thought to be due to normal faulting, that is, sliding fracture along conical fault zones, as shown in Figure 13. Since these are due to relief of horizontal pressure, these deep cones dip at angles of rather more than 45° ... On the other hand, the shell above 70 km is in compression due to the contraction below it and fails by thrust-faulting in conical zones, which usually lie immediately above those just mentioned. Since the upper faults are due to compression, the cones dip at less than 45°. The direction of motion in shallow and deep earthquakes is opposed, one set being due to normal faulting, the other to thrust faulting.

(Wilson, 1954: 170–172; Figure 13 is reproduced below as Figure 1.1)

Scheidegger applied Mohr's (1928) classical theory of fracture to spherical shells, and obtained an excellent match with the geometrical features of island arcs, mountain belts, the trend of earthquakes, and with types of faults. He determined that the outermost shell fails conically around a point of weakness. Assuming that weak regions are located along continental margins, Scheidegger and Wilson argued that contraction explained how island arcs and mountain belts formed (RS1); the fracture's conical shape determined the arcuate shape of island arcs and young mountain belts; the fracture's incline corresponds to the Wadati–Benioff zone, the zone of earthquakes that descends beneath island arcs and active mountain belts. Wilson claimed:

No system of forces acting upon a uniform spherical shell has been discovered which can explain a system of arcuate failures such as occur on the earth. On the other hand, a non-uniform spherical shell with points of weakness or zones of weakness along continental margins may be capable of failing so that there might be produced a series of arcuate failures near those margins. It seemed to Scheidegger and Wilson most likely that such a system of arcuate failures would rise as a result of sliding fractures due to cooling and tension occurring below the crust in the upper part of the mantle.

(Wilson, 1950: 141)

They argued that Jeffrey's theory explained the inclined descent of the Wadati–Benioff zone with thrust faults near the surface caused by compression, normal faults below the level of no strain caused by tension, and the absence of earthquakes at the level of no strain (RS1) (Figure 1.2).

Wilson applied this explanation of the origin of present day island arcs and mountain belts globally, grouping them into common types, and explaining their type-differences mainly in terms of their distance from continental margins (RS1). As expected, he emphasized geometry. He emphasized, as others had before, that

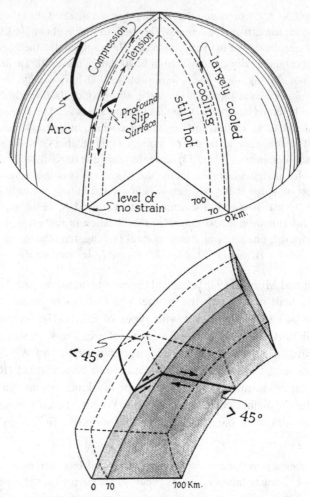

Figure 1.1 Wilson's Figure 13 (1954: 172). His caption reads: "Scheidegger's development of the contraction theory, with detailed sketch of part of one arc." The outermost shell is under compression, earthquakes are due to thrust faulting with shallow cone angles being < 45°, the next shell is contracting because of cooling. It is under tension, earthquakes are due to normal faulting with deep cone angles being > 45°. The two shells are separated by a zone of no strain.

both mega-belts of Mesozoic–Cenozoic mountain and island arcs approximate great circles. One borders (Figure 1.3) the Pacific (East Asian–Cordilleran belt); the other extends from the Alps through the Himalayas to Oceania (South Eurasian–Melanesian belt). Together they form a T, meeting orthogonally at the Banda Sea. They consist of primary and secondary mobile belts. Primary belts have deep-seated connections; secondary ones are more superficial. Both are arcuate.

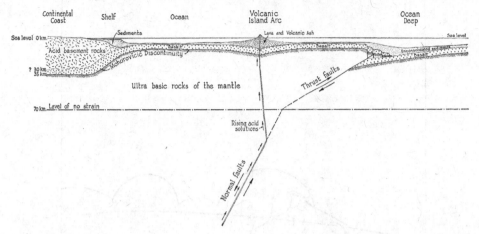

Figure 1.2 Wilson's Figure 4 (1954: 156) illustrates the descending zone of earthquakes beneath single island arcs. Shallow and deep earthquakes are respectively due to thrust and normal faults. The reported scarcity of earthquake foci between 70 and 90 km deep coincided with Jeffreys' zone of no strain.

Wilson identified six diagnostic characteristics of primary arcs:

(1) they are underlain by all the world's deep earthquakes and most of its major shallow ones, (2) they have associated with them most of the world's active volcanoes which give acid [continental-like] lavas and ... young intrusive rocks or batholiths of similar composition; (3) they are followed by strips of large negative gravity anomalies; (4) they are accompanied by the world's greatest oceanic trenches; (5) they rest upon ... visible basement of older gneissic [continental] rocks; (6) they contain peculiar sediments, called *ophiolites*.

(Wilson, 1954: 153–154; my bracketed additions)

He (1954: 155) defined ophiolites, again following Kay and King, as consisting "of lava, greywacke, and other sediments derived from lava and cherts, containing oceanic rather than coastal fossils and cut by intrusives of ultrabasic composition."

He divided primary mobile belts into five types: single and double island arcs, single and double mountain arcs, and fractured arcs. All but fractured arcs are circular with their convex side facing the ocean. Single island arcs are volcanic (Aleutian, East Asian, South Sandwich Islands); they are associated with ocean trenches, large negative gravity anomalies, and Wadati–Benioff zones. Double arcs (Kodiak, Timor, and West Indian arcs) have an inner volcanic arc and an outer sedimentary (ophiolitic) arc instead of an ocean trench. Single arcs that are located near continents become double island arcs as their trenches fill with sediments, which are squeezed and elevated, transforming them into outer ophiolitic arcs, which are associated with negative gravity anomalies. Single mountain arcs "are similar to single island arcs except that they form part of the continent;" they are volcanic, have deep offshore trenches, and are associated with Wadati–Benioff earthquake zones; the western ranges of the Central Andes served as the defining example

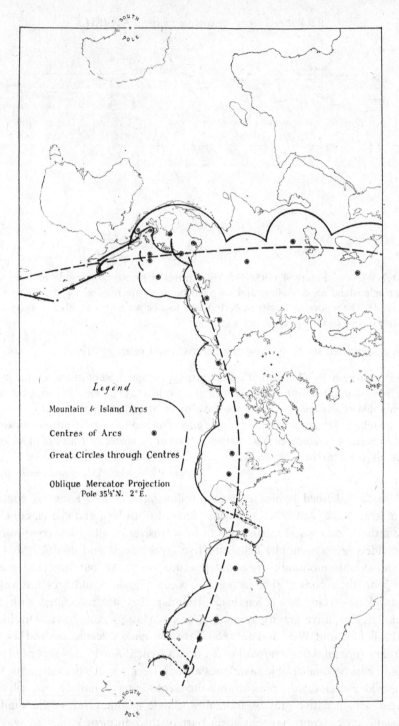

Figure 1.3 Wilson's Figure 3 (1954: 153) illustrates the two worldwide active belts of island arcs and mountains. The circum-Pacific mega-belt is shown in its entirety. Only part of the South Eurasian–Melanesian mega-belt is shown; the South Eurasian arm extends toward Spain; the Melanesian arm extends to New Zealand. The belts meet orthogonally at the Banda Sea, where they form a T. All island arcs and mountain ranges are arcuate except for the straight arcs along the Melanesian arm.

(Wilson, 1954: 158). Double mountain arcs have two circular arcuate arcs, the inner is volcanic and the outer ophiolitic; Wilson selected the Coast–Cascade and Coast–Sierra Nevada ranges along the western coast of Canada and the United States as his defining example. Fractured arcs are straight; either they were straight originally or, originally arcuate, were later torn apart by transcurrent faulting and straightened; he selected the Melanesian arcs from the Philippines to New Zealand as his defining examples (Wilson, 1954: 159).

Speculating further about the origin of mobile belts, Wilson proposed that continental shelves form "by deltaic accretion," zones of weakness arise, Scheidegger's conical fracture zones form, volcanism ensues, and island arcs are created. Wilson thought that in this way passive continental margins, over time, are converted into mobile belts.

First a continental shelf forms by deltaic accretion. The Gulf and Atlantic Coast shelves of the United States are examples ... Next a conical zone of fractures forms (see Fig. 4 [my Figure 1.1]) ... The existence and location of the fracture zone ... is marked by earthquake foci. Volcanism, which gives rise to the volcanic arc, starts as a consequence of fracturing. The Aleutian arc is an excellent example of this stage. Most of the arc is far off shore in deep water. In the northern and eastern parts of the Bering Sea is a large shelf, which was, no doubt, fed in the past as at present, by the Yukon and Anadyr rivers.

(Wilson, 1954: 159; my bracketed addition)

Single arcs sometimes evolve into double arcs. Wilson, following Umbgrove, suggested that double arcs

occur where there is a source of sediment to fill the trench and where later movements have squeezed these sediments and elevated them into a string of islands. The Chirikof and Kodiak Islands and Kenai Peninsula provided a splendid example of this, formed where the Aleutian arc approaches and joins North America.

(Wilson, 1954: 159)

Furthermore, island arcs evolve into mountain belts.

The next step in this evolution is that the rivers of the continents and the volcanoes of the arc both contribute material to fill in the sea behind the arc and often to fill the trough in front of the arc. Examples of arcs in which the inland sea is now shallow are the Japanese and Ryukyu arcs and ... the North American Coast Range arcs at the beginning of the Mesozoic.

(Wilson, 1954: 159–160)

Mountain belts are accreted onto continental shields. Borrowing from Kay, King, and other proponents of geosynclinal theory (I, §7.3), he made use of the notion of eugeosynclines.

With further deposition, compression, and uplift, the inland seas become land, and the arcs become joined to the continent as double or single mountain arcs. The great thicknesses of volcanic rocks and sediments derived from them which emerge are sometimes called *geosynclines.*

(Wilson, 1954: 161)

Turning to secondary arcs, which lack deep-seated features, he, again borrowing from geosynclinal theory, described them as miogeosynclines.

These [secondary] arcs all lie on the concave side of the primary arcs at their junctions. They are almost all curved in the opposite direction to primary arcs but are more irregular. They have been formed by the accumulation and subsequent folding and thrust-faulting of thick deposits of sedimentary rocks. Each was laid down upon a basement of gneissic rocks forming a pre-existing part of the continent. These deposits are largely nonvolcanic, include no ophiolites, and are what might be called "normal" sediments. They have been called *miogeosynclines*. Where the sedimentary cover is thin, great blocks of the basement may be faulted up to form ranges in these arcs as happened in the Rocky Mountains of the United States. They are without any deep-seated seismic activity. Examples are the Rocky Mountains of Canada and the United States, the Pyrenees, and the Carpathians.

(Wilson, 1954: 165–166)

Scheidegger (1953b: 1148) also showed that, given his theory of contraction and Mohr's theory of fracture, secondary mountain belts would be expected to form where primary arcs join at an obtuse angle. Wilson then identified examples in both the East Asian–Cordilleran and South Eurasian–Melanesian mobile belts (RS1).

Turning to fractured arcs and noting that they are restricted to the Melanesian arm of the South Eurasian–Melanesian mobile belt, Wilson explained their origin in this way (RS1).

It is not likely that this distribution is due to chance. The proposed explanation is that the Melanesian arm, and only that arm, has had to undergo two sets of movements of one belt. The interaction of these forces has caused the complexity ... This can be seen by reference to Figure 16 [my Figure 1.4] in which the two belts are shown diagrammatically. Along two of the three limbs of the T, contraction below the level of no strain and compression above it can take place without horizontal shearing, but along the third side shearing must accompany the shrinkage or compression. This can be demonstrated by arranging three books on a table separated by 1-inch gaps to form a T-shaped pattern and then moving them together or apart.

(Wilson, 1954: 179; my bracketed addition)

This shows again his preoccupation with the geometry of features. His use of simple, broad-brushed heuristic models foreshadows his later use of them to contrast transform and transcurrent faulting (§8.9).

Wilson now extended Scheidegger's and his explanation of current orogenies to old orogenic belts, doing so on a strictly uniformitarian basis arguing that old inactive belts had formed in the same way as currently active ones (RS1) (1954: 187). Because old mobile belts are not seismically active, have been metamorphosed and eroded over time, their component parts are very difficult to identify.

To gain an idea of what an ancient range might be expected to look like, consider the probable appearance of the Cordillera a few hundred million years hence. There would be no volcanism, only shallow and minor earthquakes. The height of mountains would be reduced, more batholiths would be unroofed, and much of the present coast might be covered by a coastal

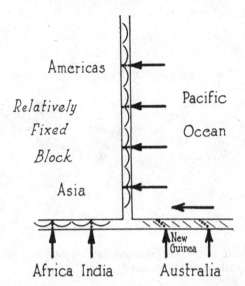

Figure 1.4 Wilson's Figure 16 (1954: 179) shows "direction of motion along belts to explain why all the Melanesian arcs near Australia have been sheared into fractured arcs." Only the Melanesian arm undergoes two sets of movements, which are roughly orthogonal to each other. Their interaction produces straight fractured arcs instead of the arcuate roughly circular arcs.

plain and shelf of sediments eroded from the mountains. This shelf might extend in places for scores of kilometers out to sea. Weathered down to roots, without volcanic or seismic activity and partly hidden, the primary arcs might be scarcely recognizable. Further inland the median-land, which in some other mountain systems is wholly lacking, would be a complex area of metamorphic rocks whose interpretation would prove puzzling. It might be difficult to distinguish from the primary arcs. Inland from these irregular igneous belts and separated from them by a fault zone, one would expect to find the curved outlines of eroded secondary mountains such as the Alberta Rocky Mountains, forming a series of large arcuate basins concave toward the coast and composed of sedimentary deposits without batholithic intrusions. They would be less likely to be covered than the primary arcs. Relatively mild metamorphism would not have destroyed their folding or their fossils or confused their stratigraphy. The present basins are so deep that much erosion would not have destroyed them. One would expect them to stand out as the most prominent features and the only ones with a clear structure.

(Wilson, 1954: 187)

What did this portray? "The Appalachian Mountains of today . . ." Wilson claimed, as recently described by Kay and P. B. King. He first identified structures as secondary and then "presumed" they were originally secondary arcs, all, as predicted by Scheidegger (1953b), "at the cusps, where . . . primary arcs join one another" at obtuse angles.

It is suggested that this description suits the Appalachian Mountains of today, which have recently been very clearly described by King (1951) and are illustrated in Figure 14 [reproduced as Figure 1.5]. It also corresponds, in general, to their history as suggested by Kay (1951). On the geological or tectonic maps of the United States and Canada a series of sedimentary

Figure 1.5 Wilson's Figure 14 (1954: 176) shows the "primary and secondary arcs and fault zones of the Inactive Appalachian Mountains."

basins, curved and concave toward the ocean, strike one immediately as secondary arcs. Conspicuous are the arcs of the Valley and Ridge Province in Tennessee and in Pennsylvania – New York and the Ouachita Mountains in Oklahoma. According to King, the Marathon uplift in Texas, the hills east of Lake Champlain, the extreme north coast of Gaspé and the Long Range of northwestern Newfoundland are all narrower but similar basins of altered sediments. They can be presumed to be seven secondary arcs and must be supposed to lie at the cusps, where a series of six primary arcs join one another.

(Wilson, 1954: 188; my bracketed addition)

He then identified what he believed to be the five primary arcs, and indicated their locations.

King's description of a belt of metamorphosed sediments in central Newfoundland intruded by many large batholiths corresponds precisely with that of part of an inner primary arc. In the Avalon Peninsula of southeastern Newfoundland the oldest rocks are a vast thickness of slates, conglomerates, and greywackes resting upon volcanics and meeting the requirements for an outer ophiolitic primary arc. There is no medianland. This primary arc, which may be called the "Maritime arc," strikes southwest across Newfoundland under the Gulf of St. Lawrence and presumably turns farther west across northern Nova Scotia to New Brunswick, in so doing giving rise to the large offset of the Newfoundland succession relative to that south of the St. Lawrence. Southern Nova Scotia is underlain by the Maguma series, which is the counterpart of the ophiolitic Avalon series of Newfoundland. In New Brunswick, opposite the Gaspé secondary arc, the inner primary belt of metamorphic rocks changes direction abruptly to form an arc across Maine … Two more primary arcs lie to the south, forming the metamorphic rocks of New England-Pennsylvania and of the Piedmont. The next primary arc must be assumed to turn sharply west in Alabama under the cover of young rocks in the Mississippi Embayment to reach the Ouachita cusp … South of Nova Scotia the outer ophiolitic parts of the primary arcs are hidden. The southwestern extension of the Appalachian arcs is much less certain. It is, however, clear that the present arcs form belts which are continuous around the world, and it might be expected that former belts had also been continuous. Kay's illustration (1951, Pl. 1) shows the Cordilleran and Appalachian systems as having both been in similar states in early Paleozoic time. He separates them in Mexico but gives no clear evidence to show that a Mexican land existed at that time. A connection between the Cordilleran and Appalachian arcs in Paleozoic time may therefore be speculated upon. Central America may have grown subsequently.

(Wilson, 1954: 188–189; see Volume I, Figure 7.1 for Kay's figure)

He did not detail the worldwide array of Paleozoic primary and secondary arcs, but referred readers to Umbgrove (1947) and others, and provided little more than a sketch of the locations of the principal Phanerozoic orogenic belts elsewhere in the world (1954: 180).

Wilson then extended his account of continental evolution to the Precambrian, continuing to do so in terms of Scheidegger–Jeffreys' contractionism, and uniformitarianism (RS1). Stretching the notion of continental accretion to its limit, he speculated that orogenesis and the accretion that he thought resulted from it are the fundamental processes that shaped Earth's crust. Rejecting the notion that continental shields were part of the primeval Earth's crust, he argued that they had formed throughout the Precambrian by progressive accretion of island arcs around ancient nuclei.

In the first place, the idea that continent blocks are part of an original unaltered crust is quite untenable. The outer parts of most continents are young and active mountain ranges like the Cordillera or older, quiescent mountains like the Appalachians. The central parts are Precambrian shields, covered to a greater or less extent, as in the prairies or central United States, by a relatively thick veneer of young and flat rocks. The shields themselves have often been referred to as the root of ancient mountains and they contain abundant traces of altered and folded sedimentary material. The pattern of these ancient ranges does not look at all like that of an original part of the crust.

(Wilson, 1951a: 4)

His extension of continental accretion to the Precambrian was in keeping with the principle of uniformitarianism.

It is considered that the continents are entirely built up from the roots of former primary orogenies. The idea that they have grown thus since Precambrian time is already widely accepted ... The principle of uniformitarianism suggests that the same processes went on in Precambrian time ...

(Wilson, 1951b: 107)

From firsthand knowledge, Wilson applied Scheidegger's version of contractionism to the growth of the Canadian Shield in the Precambrian. "Interestingly enough there are few places better situated to study the Precambrian rocks than Toronto, 60 miles north of which lies a well exposed and vast area of ancient rocks" (Wilson, 1951a: 7). He was well placed to do so, arguing (1953: 66) that extensive mapping of the Canadian Shield and radiometric age determinations of different provinces of the

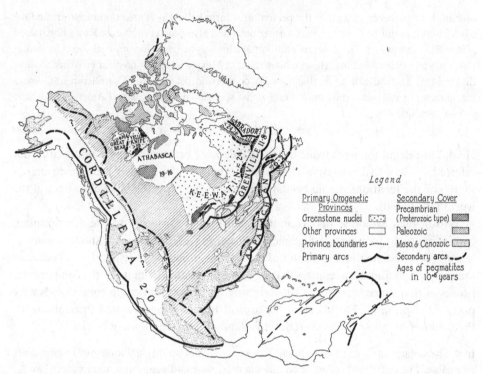

Figure 1.6 Wilson's Figure 20 (1954: 196). Known orogenic provinces of North America (others may be assumed to exist under the interior plains and in the Arctic).

shield had led "to a clear understanding of Precambrian history." Radiometric dating revealed that the duration of the Precambrian was much longer than previously thought. Citing Holmes as his authority, Wilson noted:

The vast span of Pre-Cambrian time from 28 to 5×10^8 years ago and the patterns revealed by age determinations have led Holmes (1948) to write: "Obviously the time has come to liberate Precambrian geology from the tyranny of a telescoped classification in which the Archean ... is regarded as a single era of worldwide distribution." Five "Archean" cycles are already known and there has probably been time enough in the geological past for double that number.

(Wilson, 1954: 191)

Radiometric dating of Canadian provinces revealed that two provinces, the Keewatin (Superior) and Yellowknife (Slave), were the oldest nuclei, and that the age of other provinces decreases with distance from the older province; he believed them to be original continental nuclei, and claimed that they became part of the same landmass when the gap between them was bridged by the Athabasca (Churchill) province (Figure 1.6). Geological mapping provided evidence that provinces other than the Athabasca formed by accretion of mountain belts that had evolved from island arcs around these continental nuclei.

All these conclusions agree with one another and suggest that the Pre-Cambrian provinces represent "a long series of orogenic cycles, each of which, though having distinctive peculiarities of its own, is essentially of the same kind as the later Caledonian, Hercynian and Alpine cycles" (Holmes, 1948: op. 254). This restatement of the great "principle of uniformitarianism" suggests that the ideas which were used to guide the classification of the present and Appalachian belts may also be applied to the shield area.

(Wilson, 1954: 192)

The two oldest nuclei differed from other provinces: both possess linear belts of greenstone rocks which were then the oldest known rocks. Greenstones, originally lavas, are low in silica, are interbedded with cherts, and often show pillow structure, indicative of eruption beneath the sea. They also lack limestones. Their low silica content (basic or mafic) indicates that they began as single volcanic arcs as opposed to typical continental rock that is high in silica content (acidic or felsic). The products of erosion of these nuclei accumulated around their margins and the surrounding mobile arcs formed, which became "joined to the nuclei as marginal ranges."

Erosion of the nuclei produced sediments of normal kinds, including the first sandstones and limestones which accumulated in great shelves around the margins. The shelves acted as centers controlling the location of new systems of arcuate fractures offshore … These arcs, in turn, grew and became joined to the nuclei as marginal ranges. This process has been repeated in changing patterns to expand the continents ever since and is still proceeding today.

(Wilson, 1954: 206)

Wilson made what was very much a token attempt to show that other continents had formed by continental accretion around ancient nuclei (RS1). But the Canadian Shield provided the basis of his ideas and the center of his concerns; discoveries from other continents did not change his ideas about geological history. He (1952a: 199–200) noted that greenstone regions had been found in Western Australia (Kalgoorlie Series) and Africa (Transvaal). He had even seen them, if only briefly. He speculated that Europe and Asia began as separate continents and grew into each other "along the Urals and later along the Alpine-Himalayan." He fitted them to his North American model. Although his approach to current processes was global, his approach to processes in deep time (Phanerozoic and Precambrian) was regional, not global. So far as these latter "deep-time" questions were concerned, questions that were central to the mobilism debate, Wilson, at this stage of his career, was a regionalist not a globalist, preoccupied with matters North American which were the extent of his practical experiences.

The writer has seen four of these greenstone regions and suggests that they are the nuclei of older rocks about which the continents have grown. Other nuclei undoubtedly exist, some of which may be younger, but the main continents may be supposed to have started to form in about the same manner at about the same time. Most of them are probably composed like North America with its two suggested nuclei, or like Eurasia which may have been united along the Urals and later along the Alpine-Himalayan.

(Wilson, 1951a: 8)

Wilson (1951b: 107–108; 1952b: 446–447) developed another argument for Precambrian continental growth based on the extrapolation of estimated rates of erosion for the Mesozoic and Cenozoic to the Paleozoic and Precambrian (RS1). If the rate of erosion has been constant throughout geological time, then there should be abundant continental shelves surrounding ancient continents, but there are not, so erosion must have been much less before the Mesozoic. But, this conclusion clashed with the principle of uniformitarianism, which for Wilson was sacrosanct. How can this dilemma be avoided? He argued, by supposing that such shelves once did exist but had subsequently been transformed beyond recognition into mountain belts.

1.4 Wilson and Scheidegger raise difficulties with continental drift and mantle convection as the cause of island and mountain arcs, 1949–1954

In 1949, Wilson attacked continental drift as a solution to mountain building (RS2). He characterized drift as one of those theories that requires "removal of a part" of an original sialic crust, and piggybacked his objections to it onto his dislike of theories involving partition of the Moon early in Earth's history. His attack shows again his desire to see geological theories grounded on a sound physical basis, his commitment to the principle of uniformitarianism, and his high regard for Jeffreys.

Theories depending upon removal of part of a sialic crust. The first group of theories we sufficiently discussed in the previous section to indicate that, even if the separation of the moon from the earth during its early history is physically reasonable, which is now open to doubt (Jeffreys, 1948), difficulties are involved in using that disruption to form the continents. These difficulties would be sufficiently great if the continental fragments had taken up their present positions immediately after the time of the moon's birth. They are made in many ways more difficult in Wegener's theory of continental drift. The physical forces that could cause a single continent to break up and separate in relatively recent geological time are quite unknown. Why such an event should happen only once in geological history is unexplained. Even if the theory provided a satisfactory explanation of the world's recent mountain ranges, some other explanation would have to be advanced to explain the older ranges including those whose roots are not considered to make up the shields (Jeffreys, 1929, p. 304).

(Wilson, 1949c: 173)

Wilson did not mention Holmes' theory of mantle convection, though he surely knew about it because he cited Holmes' *Principles of Physical Geology* (Holmes, 1944).

He argued that contractionism offered a better explanation of the formation of continents and mountains than mantle convection currents (RS2, RS3). Admitting that the "suggestions put forward have many attractive features," he added, "they do not explain the structures of the Canadian Shield" and "leave at least two major geological difficulties unexplained" (1949c: 176). The first difficulty, which again shows his preoccupation with geometrical matters, was the failure of proponents of mantle convection to work out "the distribution of these currents ... in plan."

Griggs (1939), and Vening Meinesz (1948) have discussed the formation of cross-sections of mountain ranges formed in this manner in a most convincing fashion, but it is very difficult to visualize a system of world-wide currents which could explain in plan the arcuate mountain belts of the world. Where arcs like the Aleutians and Kuriles join the currents have to intersect one another. Until an explanation is offered the theory is seriously incomplete.

(Wilson, 1949c: 176)

A second difficulty was that mantle convection does not explain why parts of the world once active become passive, and others once passive become active. He drew support from his regional knowledge of the Canadian Shield.

Another difficulty with currents is to understand why some parts of the world should be so active and yet leave other parts formerly active now so stable. For example, the greater part of the Canadian shield is a relatively low plain rising less than 2000 feet. At many places near the borders and around Hudson Bay there are outliers of Ordovician and Silurian age ... These infaulted patches are sufficiently numerous and widespread to suggest that the Shield was extensively covered during parts of Ordovician, Silurian and perhaps Devonian time, but the absence of inliers of all other ages ... suggests that the Shield was not covered by the sea at other periods or some record would have been left ... This suggests, if it does not prove, that the Shield has been very stable, neither depressed enough to allow the sea to enter often, nor raised sufficiently for erosion to have removed all trace of those few principal inundations.

(Wilson, 1949c: 176)

Three years later, Wilson again attacked continental drift, raising the standard mechanism difficulty. He (1952b: 446) first noted that there "is now considerable doubt as to whether any hypothesis involving continental drift is a satisfactory explanation," implying that the case against drift was strengthening; he emphasized that any adequate account "must be shown to be physically valid," and added, "When that can be done, geology will have been placed upon a sound physical basis."

Two years later, he renewed his attack on convection raising six difficulties (RS2).

Although it is agreed that convection currents may exist within the earth's mantle, several difficulties have been pointed out which prevent their acceptance as a complete explanation for mountain- and continent-building. These difficulties include (a) the great length and continuity of orogenetic belts; (b) the sharp deflections in the belts; (c) the absence of any evidence of stretching in the crust which should accompany compression if convection currents were operating; (d) the apparently layered nature of the mantle; (e) the evidence provided by deep-focus earthquakes of great strength and stress differences in the mantle which flow should tend to reduce; and (f) the absence of any direct evidence for the existence of currents. This last has led to speculations involving a variety of models of currents and disagreement as to whether convection currents rise or sink under continents and mountains and whether they are continuous or intermittent. Particularly unsatisfactory from the point of view of a résumé such as this is the fact that no paper has yet described the details of the present mountain system in terms of currents and that no attempt at all has been made to use them to explain the former mountains of Pre-Cambrian times.

(Wilson, 1954: 168–169)

Convection provided no detailed account either "of the present mountain system" or of the "former mountains of Pre-Cambrian times" because he believed that Scheidegger/Wilson contractionism had both (RS2). He (1954: 169) acknowledged "some of these difficulties can probably be overcome" but added that contraction "appears to provide a more complete explanation" (RS3). One gets the sense that he thought drift theory was on its last legs.

Scheidegger (1953a) also attacked continental drift (RS2). He began by noting that "because of serious difficulties it is now almost abandoned." Like Wilson, Scheidegger took as "a most serious difficulty" that drift at best explained only one mountain-building episode.

A most serious difficulty is that all drift theories can explain only *one* orogenetic diastrophism. If two continents are drifting towards the equator for example, they may proceed until they collide. This can explain the folding of, say, the Alpide mountains, but once the continents have arrived at the equator, they would simply stay there and nothing more would happen, nor could anything have happened before they arrived at the equator. The diastrophisms of the Paleozoic era and of the Precambrian thus remain totally unaccounted for.

(Scheidegger, 1953a: 143)

Also, like Wilson, Scheidegger apparently was not familiar with Argand's proto-Atlantic (I, §8.7). He was, however, familiar with Holmes' appeal to mantle convection as the cause of continental drift and mountain-building, and raised the same difficulty against it (RS2).

Large continental shifts have been postulated before, always with the understanding that the "collisions" gave rise to the mountains. The search for the forces led to the postulate of subcrustal convection currents (Holmes, 1929) whose drag would have enough power to throw the continents together in such a manner that mountains would be folded up. However, the old objection that only single orogenetic diastrophisms would be explained in this way still holds.

(Scheidegger, 1953a: 143)

He also discussed mantle convection without continental drift as the cause of island and mountain arc formation, claiming that it was preferable to continental drift.

We have reviewed the theories of orogenesis. In doing so, we tried to be as impartial as possible, discussing the merits and faults of each of the hypotheses from the viewpoint of a physicist. The principal competitors of the true theory of orogenesis are the contraction hypothesis (with sliding fracture along arcs) and the hypothesis of thermal convection. Both theories are based upon the assumption that the earth is losing heat. (On the other hand, the theory of continental drift would hold for any model of the earth which is sufficiently fluid independently from thermal considerations, but it now is almost abandoned.)

(Scheidegger, 1953a: 148)

He thought that contractionism but not mantle convection "gives a reasonable explanation for the shape of the island arcs and mountain belts, for the earthquakes, and for the fact that orogenesis occurs in single diastrophisms" (RS2, RS3). He admitted that contractionism fails to explain high heat flow measurements while mantle convection succeeds (RS2, RS3), which led him to suggest that both may occur.

Both the two principal theories, thus, seem able to account for part of the observed phenomena. Heretofore, it has always been assumed that the two theories preclude each other. This is, however, not necessarily the case. The mechanism that produces island arcs, mountain belts, and earthquakes may be different from the mechanism causing the heat flow and movements on a continental scale. The two phenomena – thermal convection currents and sliding fracture – could exist side by side if the characteristic time for the convection is long compared with the characteristic time of the fractures – in other words, if the movements around the island arcs are measured at least in centimeters per year, as compared with the few millimeters per year necessary to account for the heat flow by convection.

(Scheidegger, 1953a: 149–150)

Did Scheidegger include continental drift among "movements on a continental scale?" I do not think so. In discussing convection he mentioned the antipodal positions of continents and oceans and their initial formation, but he did not mention continental drift in his discussion of "movements on a continental scale." Scheidegger concluded on a cautionary note; giving the impression that he was less committed than Wilson to their version of contractionism.

Thus, a combination of the two theories might in the end provide a satisfactory explanation for the observed patterns. However, it should be kept in mind that all the theories of orogenesis remain purely speculative until more factual data about our planet are known. The question as to whether the earth is heating or cooling is most fundamental and has not been decided yet. The directions of motion of the deep focus earthquakes are known only for one or two cases which are hardly conclusive. Finally, it might be hoped that some day one will be able to get really substantial information about the rheological state of the earth's interior. As long as we are so much in the dark about these fundamental facts, we can hardly arrive at more than rather vague conclusions as to the cause of orogenesis.

(Scheidegger, 1953a: 149)

In retrospect, it is unfortunate that other geophysicists such as Jeffreys (I, §4.3; III, §4.10), Birch (I, §5.10) and MacDonald (III, §1.6, §2.7) were not as cautious as Scheidegger regarding the possibility of long-term movements within Earth's mantle; insisting as he did that geological hypothesizing be put on a firm physical foundation, and yet realizing how little was known about the physical properties of Earth's mantle and the behavior over time of material under high pressure and high temperature.

1.5 Wilson continues to support contractionism and reject mobilism, 1959

Warren Hamilton (I, §7.6–§7.9) recalled that Wilson defended his account of the evolution of island arcs into mountain arcs and their accretion onto continents when they sat together on two legs of a flight from San Francisco to Christchurch, New Zealand, in November 1958.

In 1958, I flew from San Francisco to Christchurch on a Navy prop plane, which took 3 interminable hops to do it. My seatmate on 2 of the legs was Tuzo Wilson. I was going as a young field geologist to work in the Dry Valley region, whereas he was on a VIP tour as President of the International Geophysical Year (IGY) (which provided the impetus for the big push into Antarctic work that year). Tuzo gave me a manuscript to read. North America is sort of ringed – if you ignore all the details – by outward-younging orogenic belts. Tuzo's grand scheme was that each belt represented a marginal pile of sediments that accumulated until it became unstable and the sides pushed in and, voila, an orogenic belt, no mobilism required except for the local collapse. I protested that the Urals and all the Tethyan systems did not fit this scheme. He dismissed this: he had the explanation, and anything purported to the contrary represented merely misinterpretation and bad data. He was outspokenly arrogant. And, of course, a few years later Tuzo was a major contributor to plate tectonics ... I got home from my first Antarctic trip about the first of February of 1959. I had been gone 3 months.

(Hamilton, August 24, 2002 email to author)

Wilson probably had given him the manuscript to his forthcoming paper in *American Scientist*, the published (March 1959) version of his 1957–8 Sigma Xi National Lecture. Much of what he had to say was unchanged. He (1959) reemphasized the need to underpin geological theories with physics and stressed the importance of uniformitarianism. As he had in the early 1950s, he continued to champion contractionism, offering the same solutions to the origin of island and mountain arcs, secondary mountain arcs, Mesozoic and Paleozoic arcs, and continents. He even used identical or almost identical figures to explain island arc formation on a contracting Earth, the Wadati–Benioff earthquake zone beneath single island arcs, and North America's active and inactive mobile belts (my Figures 1.1, 1.2, 1.5, and 1.6) (Wilson, 1959: 16, 6, 10, and 22).

Not everything, however, remained the same. He removed a new difficulty facing his explanation of fault mechanisms beneath island arcs. He (1959: 16) noted, without giving any specific references, that Hodgson had recently shown that "most earthquakes are due to neither normal nor thrust faults" as predicted by contractionism "but to horizontally-moving shears called transcurrent faults." He argued that even if Hodgson was correct, initial faulting still may be thrust or normal, subsequently changing to transcurrent.

This does not disprove Scheidegger's explanation because Mohr's theory and Scheidegger's only apply to movement at the moment of initial fracturing. Once the factures have been formed, other forces come into play and different movements are channeled along the existing fractures.

(Wilson, 1959: 16)

Wilson was also forced to confront two new issues: increasing evidence that Earth has not cooled sufficiently to explain the long history of orogenic activity, and the recently discovered world-encircling mid-ocean ridge system with its central rift valley. He needed to find a new contraction process and to show that the presence of the mid-ocean ridge system was or was not consistent with it.

Admitting that cooling was insufficient, he proposed that "emission of the crust" by early volcanism shrunk what became the layer immediately underlying Earth's newly formed crust. He no longer had to maintain that Earth's radius had decreased. Earth's interior could even be expanding. He simply wanted an Earth whose outermost layers were contracting relative to the interior, regardless of whether or not the interior contracted more slowly than the outer layers, remained the same size, or expanded.

If this [initial volcanism covered Earth's original surface], it follows that the Mohorovičić discontinuity represents the original surface of the Earth. Since this original surface is now overlain by an average thickness of 15 km of crust, it must have shrunk or been reduced in radius by that amount. The emission of the crust would therefore have produced about 100 km shortening in the circumference of the original surface which would be available to cause mountain building. If, in addition, the Earth is cooling, the estimates ... are that, at most, there might have been a further 50 km of shortening in any circumference, but this is not necessary. One can in fact visualize an Earth whose central parts are warming and expanding, whose outer parts may either be cooling or warming, but whose surface is in any case contracting as the result of the escape of volcanic matter. This indeed is the closest estimate which we can make at the present time of the probable behavior of the Earth.

(Wilson, 1959: 15)

Wilson offered no explanation of the original volcanism, which caused contraction in the first place. Once contraction began, it generated "two vast systems of fractures." One became the system of mid-ocean ridges; the other became island arcs and mountain belts.

Thus, it is suggested that the Earth, whether it is cooling or not, is contracting as the result of steady volcanism at about the present rate, and this contraction has produced two vast systems of fractures which form a network over its surface. One of the systems is feeble, shallow and hidden by the oceans; the other is active, deep and clearly the more important. No very striking characteristics have been observed in the mid-ocean system, but the continental system has one outstanding peculiarity: It is formed of a scalloped series of conical fractures. The basic problem of mountain building may thus be reduced in its simplest terms to one question in theoretical physics. How can a conical fracture be formed on the surface of a shrinking sphere? To this, A. E. Scheidegger has given an answer.

(Wilson, 1959: 15)

But his and Scheidegger's account of island arcs and mountains was too dear to abandon. Based on physics, or so he believed, it explained the origin and shape of island arcs, their evolution into mountain belts, the position of secondary arcs, the formation of fractured arcs, and continental growth. He even found a way through this new construction to explain the origin of the system of mid-ocean ridges.

   Wilson was quite taken with the vast extent of mid-ocean ridges, remarking that discovery of such vast new features "is an interesting commentary upon the state of our knowledge about some parts of the Earth." Accepting Heezen and Ewing at their

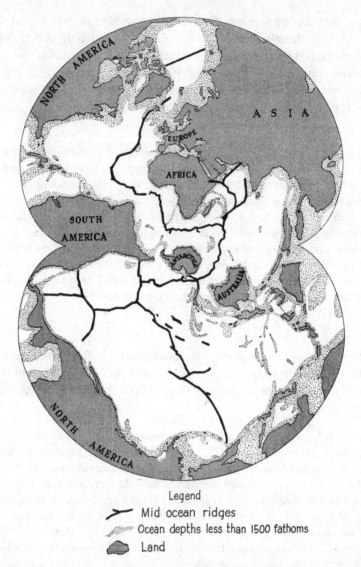

Legend

⤚ Mid ocean ridges

⬚ Ocean depths less than 1500 fathoms

◆ Land

Figure 1.7 Wilson's Figure 2 (1959: Frontispiece, opposite p. 1). His caption reads: "It seems probable that a continuous ridge a few hundred miles wide, from 10 000 to 33 000 feet high and perhaps 40 000 miles long, winds its way across the central ocean floors. The discovery, only two years ago, of the continuity of this, by far the greatest mountain range on Earth, is an interesting commentary upon the state of our knowledge about some parts of the Earth."

word, he agreed that the system of ridges is "by far the greatest mountain range on Earth." He illustrated the ridge in his Figure 2 (Figure 1.7), placing Antarctica at the center, and showing that it is almost entirely surrounded by ridges, a fact that he would later invoke as a difficulty facing mantle convection as the driving force of seafloor spreading (§1.9).

Wilson fleshed out his explanation of the mid-ocean ridge system. After noting its basaltic composition, apparent central rift valley, and presence of only shallow earthquakes, he suggested:

The ridge has been formed by the escape of basalt lava along a fracture zone and that basalt has been formed by partial melting of pockets in the ultra-basic rock of the upper mantle at a depth of not more than 70 km.

(Wilson, 1959: 5)

Once the mid-ocean fracture system formed, basalt had a natural escape route provided by Earth's initial contraction. Contrary to Hess (§1.14, §1.15) and Menard (§1.7), he thought ridges are very old.

The lack of abandoned ridges and the slow rate of the ridge's volcanism suggest that it has been in its present position for a very long time, perhaps most of the Earth's history.

(Wilson, 1959: 5)

Was Wilson uneasy with this explanation of mid-ocean ridges? I do not know; however, it certainly lacked the geometrical precision that Scheidegger and his explanation of the origin of orogenic regions had brought to the discussion. Moreover, Hess (1954, 1955a) (III, §3.8–§3.10), Menard (1958a) (III, §5.7), and Heezen (1957) (III, §6.9) were beginning to offer more robust solutions.

Wilson did not claim that his theory warranted outright acceptance; although "tentative," he believed it was the best available.

Earth due to its emission of lava and volcanic gases provides a tentative theory for the building of mountains and continents which is capable of explaining more of the details of these features than any other theory yet proposed.

(Wilson, 1959: 23)

Once again he objected to continental drift because it was without a mechanism and contravened uniformitarianism, to which he was rigidly committed as he was to contractionism. Still under Jeffreys' influence and unfamiliar with Argand's work, he had not yet even begun to consider the possibility that continental drift could have occurred more than once.

It has been shown that contraction could lead to fracturing of a kind which might show many of the principal features observed in existing and past mountains. A vast amount remains to be done, but no other theory can explain so much. Continental drift is without a cause or a physical theory. It has never been applied to any but the last part of geological time. Convection currents may exist in the mantle, but this is a pure hypothesis and it has never been shown how they form mountains with many of the properties of existing mountains.

(Wilson, 1959: 23)

Wilson (1959: 23) also mentioned polar wandering. "Polar wandering (that is, movement of the Earth as a whole about its axis) can have occurred without its any way affecting the arguments here put forward." Polar wandering, unlike continental drift, was, he thought, not physically impossible.

I suspect that Wilson mentioned polar wandering because he was following the development of paleomagnetism's growing support for mobilism. I say this because he definitely knew about it, and wanted geologists to make and to take seriously measurements of Earth's physical properties. Irving received the Canadian Geophysical Union's J. Tuzo Wilson Medal for 1984, and in his acceptance speech he recalled his "first encounter with J. T. Wilson."

My first encounter with J. T. Wilson was through Canada Post. In the 1950's Wilson was a fixist: he believed in fixed continents, permanent oceans. There was no better advocate for fixism than J. T. Wilson. At the time I was in Australia, composing my very first paper, in which I was able to show, as Runcorn was also showing, that on a scale of hundreds of millions of years the motions of the geographical pole were different when viewed from Europe, North America, Australia and India. They could be reconciled if the continents had moved as Wegener said. My paper was published in 1956, and I [wanting to test the waters] sent Tuzo the paper. J. T. Wilson wrote me about it. The letter was lost long ago, but I remember part of it very well. It went something like this, "Dear Irving – I have read your paper – For a long time I have tried to persuade geologists to take physical measurements, and now look what you people have turned up." Evidently he was surprised, even a little indignant. He did not exactly welcome the news, but he (unlike so many fixists of the day) did not ignore it either: he had bothered to write to me, a very junior scientist.

He was evidently concerned about it, because when I actually met J. T. Wilson in the following year 1957 he (as president of IUGG) was already publically advocating a full review of the problem of continental drift. And, over the next decade, that is just what he did, in the process of abandoning his fixist views and making, in the mid-1960's, his profound contributions to mobilist theory. I believe that this story is interesting, Mr. President, because through the example of J. T. Wilson it defines good science: there is the need to have firmly-held closely-argued opinions and the need also to have an open mind. J. T. Wilson dared to be wrong, and he (along with almost everyone else) was dead wrong for much of his career, but he never closed his mind. And this enabled him, when the time came, to rethink his world-view and ultimately to become gloriously right.

> *(Irving, 1984: 33–34; bracketed expression added by Irving on*
> *October 30, 2008; see II, §3.13 for discussion of Irving's paper)*

Irving's rhetoric is too charitable. After all he was receiving the Wilson Medal. Wilson in the early 1950s had essentially closed his mind to the possibility of continental drift. However, by 1957 he had definitely begun wondering about drift and perhaps its paleomagnetic support. But he was not ready to switch. Three years later, he (§1.7) still maintained that continental drift "is supported to a small degree but not to the extent desired by ... some students of palaeomagnetism" (Wilson, 1960: 881). Wilson would take four more years before he became a mobilist (§1.8). Moreover, as I shall show presently, he vehemently attacked paleomagnetism's support of mobilism in his co-authored 1959 textbook, *Physics and Geology*. I suspect that he still found his and Scheidegger's explanation of island arcs, mountain belts, and continental growth too appealing to toss aside, and found continental drift unappealing because it was "without a cause or a physical theory" and until he began to see pre-Wegenerian drift as a possibility, thought that drift clashed with uniformitarianism. What Irving and other paleomagnetists "turned up" was not what Wilson wanted to hear. Paleomagnetists had not provided a physically acceptable mechanism, but they had provided physical evidence that continental drift had occurred, and Wilson must have felt uneasy as he hinted to in his comment to Irving, "I have tried to persuade geologists to take physical measurements, and now look what you people have turned up." He already had found a way to reconcile with his contractionism, the system of mid-ocean ridges, and his doubts about a cooling Earth. He could even acknowledge polar wandering as a possibility and its paleomagnetic support, but he could not find a place for continental drift.

Wilson co-authored a textbook, *Physics and Geology*, with J. A. Jacobs and R. D. Russell (Jacobs *et al.*, 1959). Russell, a former student of Wilson's who worked on radiometric dating of the Canadian Shield, and Jacobs were at the University of Toronto when they began the book, but moved to the University of British Columbia before its publication. They (1959: 361) cited Wilson's Sigma Xi paper, so the textbook was completed after the lecture was published. They addressed both continental drift and its paleomagnetic support and found both wanting. I do not know if Wilson was the lead author of either discussion. All three physicists, Wilson brought "the additional training of a geologist, Jacobs of a mathematician, and Russell of a chemist" (Jacobs *et al.*, 1959: viii). They (1959: viii) all had a hand in "extensively" rewriting chapters so that "no chapter can be said to be the sole responsibility of any one author." I suspect that Wilson was primarily responsible for the chapter on orogenesis, which championed his and Scheidegger's explanation of mountain building, and which included a section on continental drift. Either Wilson or Jacobs was primarily responsible for the section on paleomagnetism's support of polar wandering and continental drift. If Jacobs was primarily responsible, Wilson still shared responsibility.

Their treatment of mobilism's paleomagnetic support was slipshod. They ignored significant new findings that strengthened that support, raised difficulties that already had been answered, got some of their references wrong, and did not mention Irving's results or his arguments for mobilism found in the paper he sent Wilson. Wilson still seemed "surprised, even a little indignant" about paleomagnetism's support of mobilism. Wilson (1990) later remarked of himself that until he began to favor mobilism he "never examined it dispassionately." His review of its paleomagnetic support substantiates this confession.

As physicists, Wilson, Jacobs, and Russell definitely understood the paleomagnetic method; they cannot, like many geologists, be partly excused for their shoddy review because they could not understand the geocentric axial dipole (GAD) hypothesis and the geometry of calculating poles. Indeed, they welcomed the paleomagnetic evidence for polar wandering; constructing a figure after Runcorn's 1956 figure in which he drew one apparent polar wander (APW) path for both North America and Europe (see Volume II, Figure 3.7) and introducing T. Gold's (1955) explanation of polar wandering (II, §3.9), they announced their strong support.

The paleomagnetic and paleoclimatic evidence leaves little doubt that the position of the pole relative to the land masses has changed through geologic time. The preceding discussion also indicates the extreme probability of polar wandering.

(Jacobs et al., 1959: 141)

Thus they thought the paleomagnetic method sound.

Turning to paleomagnetism's support for mobilism, they (1959: 141) correctly noted that Clegg *et al.* (1954) had "concluded that England must have rotated clockwise some 34° since the rocks were magnetized some 150 million yr. ago and that England was probably much nearer the equator than it is today," and added, "it would be nice to know whether the rest of Europe has shared in this view." Engaging in selective reporting and giving no reference, they noted, "Work on rocks in France and Spain has so far been inconclusive." Clegg and some of his co-workers had done the work. What Wilson and his co-authors failed to mention is that Clegg and company had tentatively concluded that Argand's and Carey's claim that Spain had rotated 35° counterclockwise relative to France was correct (II, §3.13). The physicists summarized the support in favor of a separation of North America and Europe, and concluded:

The two positions lie approximately 20° apart, and this suggests that the North American continent has drifted about 300 km westward relative to Europe. The drift seems to have occurred since the late Triassic period and to have been completed before the end of the Miocene. It must be emphasized, however, that the dispersion of some of the results and the uncertainty of the dating must cast doubt on the significance of the discrepancy in the pole positions.

(Jacobs et al., 1959: 141)

However, paleomagnetists already had constructed APW paths for India and Australia. Jacobs *et al.* reported:

Measurements of the basaltic lavas of the Eocene Deccan traps of India (J. A. Clegg and E. R. Deutsch, 1955) indicate a position for the North Pole close to Florida, which would imply that at the time when the lavas were laid down, some 70 to 100 million yr ago, the Indian subcontinent must have been situated considerably to the south of the equator. This result is supported by the paleoclimatic evidence of heavy glaciation (the ice flowing northward) during Permo-Carboniferous times, but on the other hand there is no evidence for cold conditions in Florida during the Eocene. Furthermore, the pole position corresponding to these results does not agree with that found by other workers for British and American rocks of the same period. Clegg and Deutsch conclude that a movement of India relative to North America and England has taken place at some time during the past 70 million yr.

(Jacobs et al., 1959: 141–142)

Their summary is astonishingly bad. Clegg and Deutsch's paper appeared in 1956, not 1955, and was co-authored with Griffiths. Clegg and company (1956: 429) claimed that India's Eocene-Cretaceous pole was at 28°N, 85° W, which is approximately 5° east of

Florida's *current* position, not that Florida was near the North Pole during the Eocene when in fact it was not (II, §3.13). They raised no difficulty about "cold conditions in Florida during the Eocene." They also neglected to inform the reader that Clegg and company's Indian results agreed with Irving's. They did not mention Irving's results, even though he had reported them in the paper he had sent to Wilson, and Clegg and company cited them. They failed to report Deutsch and company's new results that were in good agreement with the previous ones (Deutsch *et al.*, 1958). The three fixists (1959: 142) noted, "preliminary results from Tasmania appear to indicate a relative movement between Tasmania and India some time during the past 100 million years." What they neglected to report is that Irving and Green (1958) already had constructed an APW path for Australia very different from those of India, North America, and Europe (II, §5.3). There were as yet no APW paths from South America or Africa, but there were individual results from the Serra Geral Basalts of Brazil and Uruguay (Creer, 1958) (II, §5.6), and from the Karroo dolerites (Nairn, 1957) and dolerites of South Africa (Graham and Hales, 1957) (II, §5.4). This was a significant omission because Australia's APW path greatly diverged from the others. Moreover, Wilson probably knew about it. Irving most probably told Wilson of their work when he and Sheila, his fiancée, met him in Toronto in 1957.

At the IUGG meeting in Toronto Sept 1957 I talked with Tuzo and I am sure paleomagnetism came up. I KNOW I talked with him because it was at a reception in the Royal Ontario Museum and Sheila was with me. I introduced her to him (both being Torontonians) explaining that we were about to marry. There was a man in full regalia playing bagpipes at the Museum and Tuzo's response was that we should have a piper at our wedding. Tuzo then promptly on the spot arranged that with the man! Sheila remembers all this so yes he and I did definitely talk. I would presumably tell him what I was telling everyone who would listen at the time, that I understood there was resistance to separation of APW paths from Europe and North America, but now we had an Australian path that was widely different, without doubt significantly different. I gave a talk on this at an UGGI session that Bullard chaired. I don't know if Tuzo was there, I suspect not. Hess, who was certainly at the meeting, was I think at my talk but I am not sure.
(Irving, October 3, 2008 email to author; see II, §8.12
for a discussion of Irving's presentation)

They (1959: 142) also claimed, "Since we do not yet fully understand the behavior of the earth's magnetic field, extreme caution must be used in accepting these results." Apparently, they (1959: 141) believed enough was known about the geomagnetic field to claim that the paleomagnetic and paleoclimatic results "leaves little doubt that the position of the pole relative to the land masses has changed through geologic time." Hardly "extreme caution"! They also thought enough was known to support field reversals.

There is no *a priori* reason why the earth's field should have a particular polarity, and in the light of modern views on its origin there is no fundamental reason why its polarity should not change.
(Jacobs et al., 1959: 135)

Their discussion of the paleomagnetic case for drift was inconsistent and incomplete, and seriously so in crucial ways for establishing mobilism. Wilson and his co-authors were just not ready to judge fairly the strength of paleomagnetism's support for mobilism. Wanting to delay such a judgment for future years, they stalled.

> There is no doubt, however, of the importance of rock-magnetism studies. In future years they may become one of the most valuable tools of geophysical research, but to date results are sparse and often inconsistent.
>
> *(Jacobs et al., 1959: 132)*

The results were not sparse. The future had already arrived; their account of mobilism's paleomagnetic support was obsolete before it was published, and Wilson surely knew it.

Their review of mobilism was by and large standard fixist fare. They listed the problems contractionism purportedly solved, and claimed that mobilism, which they banished to Earth's remote past, gave solutions that were based on questionable data or assumptions (RS2), and that fixism had alternative solutions.

> Some of the points in support of the theory, in addition to paleomagnetic evidence, are the great shortening observed across the Alpine and other mountain ranges, the approximate parallelism of the coasts of the Atlantic Ocean, the need to provide offshore lands as sources of sediments, the apparent matching of formations, structures, and glacial striae across the Atlantic and on opposite shores of the several southern continents, and the necessity for furnishing migration routes of land animals and plants.
>
> It is believed that all these arguments can be answered by other explanations. The great apparent shortening of the Alps may be superficial and not represent buckling of the crust to the extent indicated by surface folding. An alternative explanation for the parallelism of at least some of the Atlantic coasts may be the original fracture pattern of the earth ... The source of offshore sediments has been indicated to be old arcs, not vanished continents, and in any case continents are poor sources of sediments compared with young arcs or ranges, which are eroded far more quickly. The idea that formations, structures, and glacial patterns match is not agreed to by all stratigraphers, and the argument for matching is much weakened by the several different ways in which it has been suggested that the southern continents once joined. Alternative migration routes can be found for fossil creatures.
>
> *(Jacobs et al., 1959: 348)*

Closer to home, Wilson and colleagues claimed that mobilism could not explain both the stability of the Canadian Shield and its peripheral Phanerozoic mountain belts (RS2).

> It is not apparent how the Canadian shield or any other shield could have been moved horizontally for thousands of kilometers during Mesozoic time without disturbing the lower Paleozoic rocks that overlap the shield when the same movement is supposed to have led to the building of whole Cordilleran system.
>
> *(Jacobs et al., 1959: 348)*

They concluded their attack on mobilism by arguing that the few identified lengthy transcurrent faults, the San Andreas in California, Great Glen in Scotland, and Alpine in New Zealand, constituted "a further argument against the drift of continents."

If faults with these movements are so conspicuous, the much greater movements of continents proposed by Wegener would have left clear evidence, but none has been found.

(Jacobs et al., 1959: 349)

This objection against mobilism was anything but standard; most viewed the horizontal movements along transcurrent faults as mildly supportive of mobilism. Indeed, once Wilson became a mobilist, he (1962c: 136) invoked transcurrent motions "of more than 1400 km. on the Pacific Ocean floor" as "evidence supporting drift." In addition, given mobilism, landmasses typically separated or collided with each other as opposed to sliding past each other. So evidence of such movements was rare. Wegmann actually tried to think of some and proposed that the De Geer line marked the movement of Greenland relative to Europe (I, §8.9). Wilson and his co-authors said nothing about Wegmann. Wilson (1963d) would later propose a left-lateral transcurrent fault, which he called the Wegener Fault, trending east–west between northwest Greenland and Ellesmere Island to account for Greenland's supposed eastward movement relative to Ellesmere Island.

1.6 Scheidegger acknowledges paleomagnetic support for mobilism, 1958, 1963

Once he became aware of paleomagnetic support for mobilism, Scheidegger, who was not, like Wilson, blinded by a longstanding distaste of continental drift, wondered if there was something to it. In the first edition of his *Principles of Geodynamics* (1958), he briefly discussed that support. Written a year earlier than Jacobs, Russell, and Wilson's book, Scheidegger's is less strident. He was mildly sympathetic to drift, and did not raise frivolous difficulties. However, it is just as incomplete. For example, he omitted Irving and Green's Australian results, widely different from those for Europe or North America that appeared in 1956 and 1957 (II, §5.3). After displaying, like Jacobs, Russell, and Wilson, Runcorn's 1956 figure showing his single APW path for Europe and North America (II, Figure 3.7), and discussing mobilism's paleoclimatic support, he (1958: 61) turned to Clegg, Deutsch, and Griffiths' (1956) initial results from India that indicated drift relative to Europe and North America (II, §5.2), and noted that the northward drift was "in the same direction as that suggested by paleoclimatic evidence."

The paleoclimatic evidence for polar wandering suggests, as was outlined earlier, not only that shift of the poles took place, but also that large continental drifts occurred, particularly in view of the simultaneous glaciation of several southern continents which are now widely separated. It is interesting to observe that similar conclusions are suggested by paleomagnetic work. Clegg *et al.* [1956] made a very careful study of this question, using rock samples collected in India. They arrived at a suggested position of the North Pole for the Eocene at 28°N, 85°W. This is widely different from the position of 75°N, 120°E as show in Fig. 27 [the figure illustrating Runcorn's single APW path for North America and Europe]. The discrepancy can be resolved if a drift of India is assumed with regard to Europe and North America. The postulated drift of India would be northward[7], which is the same direction as that suggested by paleoclimatic evidence. However, because of the relative newness of paleomagnetic studies, the results quoted in this section should perhaps still be regarded with some caution.

(Scheidegger, 1958: 60; my bracketed additions. Scheidegger stated in his note 7 that India had moved 600 km towards North America since the Eocene.)

He did not read the results in a straightforward way as evidence for drift, but because of their "newness" took a wait-and-see attitude, implying that "newness" diminished their reliability. It is "old" results that generally are considered less reliable. Results in a new subject cannot help being new. I take his comment to mean that he regarded them "with some caution" because he had not yet digested them.

In the second edition of his *Principles of Geodynamics* (1963), Scheidegger's discussion of mobilism's paleomagnetic support was hopelessly out of date (1963: 100–101); it was no more than a mild edit of his first edition with references to recent papers. However, when he discussed orogenesis, he spoke favorably of continental drift powered by mantle convection.

To sum up, it is almost inevitable to admit that the physiographic evidence regarding features on the Earth's surface is such that it must be assumed that some parts of the Earth's crust have been subject of compression (cf. crustal shortening), others have been subject to tension (mid-ocean rifts) and yet others to shearing (fracture zones). Most theories have been proposed for the purpose of explaining only one of these features and are therefore inadequate with regard to the others. A theory of orogenesis, to be acceptable, must be flexible enough to allow for compression, tension *and* shear of sufficient magnitude to be produced in the Earth's crust. Setting aside those hypotheses which are inadequate for lack of self-consistency, the only acceptable concepts at present remaining are the continental drift hypothesis and the convection current hypothesis. The convection hypothesis has its own unsatisfactory features if it is used alone, but if the continental drift hypothesis is adopted, particularly with the assumption that the drift is (to a degree at least) random, a somewhat more satisfactory explanation for the various oceanic features is obtained. However, the reality of continental drift has by no means been established beyond reasonable doubt and, furthermore, the forces that could cause such drift are still somewhat of a mystery, although convection currents might be advocated as a possibility.

(Scheidegger, 1963: 291)

He still thought that neither the paleomagnetic nor any other evidence had answered all the difficulties – "the reality of continental drift has by no means been established beyond reasonable doubt." He certainly favored mobilism but stopped short of outright acceptance. He seemed uncertain whether mobilism's lack of mechanism was an unsolved problem or a serious difficulty. As I shall soon describe, through an interesting turn of events, Wilson, whose opposition to mobilism had been far stronger than Scheidegger's, began in 1961 arguing in favor of mobilism, leaving behind Scheidegger and his tepid support of it.

1.7 Wilson combines slow Earth expansion and his contractionist account of orogenic belts, 1960

Wilson, more than ever impressed with the discovery of the world-encircling system of mid-ocean ridges and still retaining Scheidegger and his explanation of orogenic belts and continental growth, jettisoned contractionism, replacing it with slow Earth expansion. In March 1960, he argued that it was expansion that opened rifts, which were the central feature of mid-ocean ridges. Expansion also buckled Earth's crust "at some distance from" the rifts "where the continents with their mountain system are found."

Opening of rifts along the mid-ocean ridges would tend to make the rest of the Earth's surface buckle at a distance from them. That is where the continents with their mountain systems are found. There the increase in radius of curvature would produce local compression above a level of no strain (or neutral surface) along with tension below it. A similar general situation was postulated by Jeffreys as results of contraction in the Earth, and it was used by Scheidegger and Wilson to explain the details of mountains, island arcs, grabens and other features. Their explanations are still applicable in principle to the new hypothesis. At the same time, since the brittle shells on the Earth's surface are spherical not cylindrical, the buckling which is considered to produce mountain building would shift in direction as well as in position from time to time.

> *(Wilson, 1960: 881; Wilson referenced Jeffreys (1929),*
> *Scheidegger and Wilson (1950), and Scheidegger (1953b))*

Still anxious to underpin geological theories with physics, he invoked Dirac's idea (1938) that the gravitational constant G is decreasing with time. Dicke (1957, 1959), who recalculated G's decrease based on recent age determinations of the universe, estimated that it would cause an increase of Earth's circumference of about 4.5 percent or 1700 km at an average rate of about 0.5 mm per year. Wilson (1960: 880), following Heezen's estimate that mid-ocean ridges occupy about 2.4×10^7 square miles, calculated that their formation "would require an increase in the Earth's circumference of about 6 percent" (2360 km) over "all geological time" at an annual rate of 0.66 mm. He (1960: 881) then noted that this estimate "is close to Dicke's estimate" of 0.5 mm per year.

Wilson expressed sympathy for Egyed's slow Earth expansion (1963; II, §6.2). Egyed's estimate of an expansion of 0.38 mm per year, based on geological considerations, was in keeping with Dicke's estimate. However, he (1960: 880) rejected Heezen's rapid earth expansion (III, §6.9, §6.11) (RS2); it was, he argued, impossible, given current views in physics and astronomy, requiring "nearly doubling the Earth's radius during geological time." He did not mention Carey's rapid Earth expansion (II, §6.13).

Wilson argued that this hybrid theory of very slow expansion, coupled with his and Scheidegger's account of orogeny and continental accretion, was preferable to his earlier contractionism, which, with its growing continents and shrinking ocean basins, faced the prospect of flooding continents permanently (RS2, RS3). He alleged that his new theory explained the origin of huge transcurrent faults (which he saw as an advantage of mantle convection (RS2, RS3)), and also "the horizontal first motions observed in circum-Pacific earthquakes," which his previous contractionist theory had difficulty accounting for.

One of the chief arguments used in support of convection currents has been that they might explain the rapid horizontal motions observed along large faults such as the San Andreas in California, and the horizontal first motion observed in circum-Pacific earthquakes, although the manner in which they could do this has not been explained. But such motions are also a consequence of the present hypothesis. An active mid-ocean ridge extends across the South Pacific Ocean from the vicinity of New Zealand to South America, where it terminates in two or three branches against the Andes. If this ridge is widening at the rate of about 0.5 mm. a year and if along the Andes and Cordillera there is an existing line of fracture due to buckling, then a northward shearing motion of the Pacific Ocean relative to the Americas could take place along the mountains north of the ridge. Such a northward motion of the Pacific has long been postulated, and the rate is perhaps compatible with the rather uncertain geological evidence. That this shear motion would not occur south of Chile may perhaps explain the absence of any equivalent to the Isthmus of Panama across Drake Passage. Similar motions in the past could account for other shifts, such as those along the Great Glen fault or the Murray scarp.

(Wilson, 1960: 881)

Taking a global approach and stressing the geometry of the situation were, along with his flights of fancy, becoming second nature to Wilson.

It is perhaps worth noting that to a first approximation the mid-ocean ridges lie in the shape of a reversed J folded about the Earth, the stem being in the Atlantic and hook being in the Southern Ocean. Two branches extend northward into the Indian and Pacific Oceans [see Figure 1.6]. Also to a first approximation the continents lie in a similar but complementary pattern of a J inverted and folded about the Earth with the stem through Antarctica and the Americas and the hook through Eurasia. Two branches extend southwards into Africa and Australia.

(Wilson, 1960: 881; my bracketed addition)

He also attacked continental drift, and what he judged to be the over enthusiasm of paleomagnetists.

The view that continents have drifted apart is supported to a small degree but not to the extent desired by some advocates of continental drift and some students of palaeomagnetism.

(Wilson, 1960: 881)

Wilson greatly underestimated both the degree of such support and the number of paleomagnetists in its favor. At the time, however, there were few paleomagnetists in North America, and, by my count, none favored drift. Wilson had, in the ordinary course of events, little opportunity to have the paleomagnetic case for drift explained to him sympathetically. But he had begun reading papers on paleomagnetism, and I wonder what reasons he had for claiming that paleomagnetists had overestimated the degree of their support for mobilism. He could hardly have been influenced by Cox and Doell's 1960 review because his paper appeared three months before theirs. Perhaps his admitted inability to review evidence for continental drift dispassionately stood in the way of his acknowledging the strength of its paleomagnetic support. Nonetheless, Wilson was at least looking at paleomagnetism well before most other fixists, and within a few years would (in the form of geomagnetic field reversals) become one of its most fervent proponents.

1.8 Bernal and Dietz discuss seafloor spreading, October 1961

The first published exchange about seafloor spreading, that between the British crystallographer and polymath J. D. Bernal and Dietz, is placed here because it served as the occasion of Wilson's first favorable assessment of mobilism. Bernal, one of Britain's best known scientists because of his brilliance and communist sympathies, wrote a note to *Nature* assessing Dietz's version of seafloor spreading. Dietz responded, Bernal answered, and the exchange was published in *Nature* along with a note by Wilson, which first announced his support of mobilism. Bernal (1961a: 123) initially said that Dietz successfully combined discoveries about mid-ocean ridges with older ideas of mountain formation "set out specially by Tuzo Wilson." I suspect that Wilson refereed Bernal's original note, which prompted him to author a response of his own. Wilson definitely saw Bernal's first communication before it was published because he (1961: 125) noted that "J. D. Bernal has welcomed these views (first communication) and commented on particular points and I should like to do the same."

Bernal (1961a: 123) characterized Dietz's contribution as "a marked contribution to what might be called the oceanographic revolution in geotectonics," and added:

One gets the impression that with it a stage has been reached like the last one in fitting together a jigsaw puzzle, with the various parts of the puzzle having been laboriously assembled, piece by piece, at last seeming to look as if they are all part of one single picture.

After noting that Dietz's view "removes many difficulties and makes it possible to envisage a mantle which can expand under some parts of the crust and contract under others at the same time," Bernal (1961a: 123–124) stated that "two major difficulties, however, are still left" (RS2). The first was that Dietz could not explain the occurrence of deep earthquakes beneath trenches and island arcs, if earthquakes can occur only within rigid layers; Dietz maintained that the rigid lithosphere, which is carried atop convection currents in the underlying plastic asthenosphere, is only 70 km thick, and thus there can be no rigid layer below 70 km. Bernal's second difficulty was that seafloor spreading did not provide a way for continents to thicken from below even though it allowed for peripheral continental growth by welding on of island arcs and ocean floor sediments.

Dietz (1961e: 124), "pleased to learn that he [Bernal] finds the concept of seafloor spreading ... at least plausible," attempted to answer Bernal's difficulties (RS1).

He disposed of the first by arguing that deep-focus earthquakes can be caused by the release of short-term stresses, which apparently can build up in the asthenosphere.

Regarding the question of deep-focus earthquakes, I am inclined to favour the idea of a lithosphere extending to about 70 km. both under the continents and ocean basins. This is, of course, derived from isostatic evidence. Below this level, in the asthenosphere, permanent stresses seem not to accumulate. Deep-focus earthquakes down to 800 km. apparently indicated that stresses do accumulate for short periods of at least a few years, but this does not prove accumulation for thousands or millions of years. Hence deep-focus earthquakes do not disprove the existence of an asthenosphere at moderate depths.

(Dietz, 1961e: 124)

Dietz proposed that juvenile sialic material is welded onto the bottom of continents as it slides underneath them on descending convective currents.

Regarding the sialization of the continents, I visualize that the sima, rising from the deep mantle contains some juvenile sialic material. By spreading this "sial-sima" is eventually slid under the continents. On subsidence of the sima beneath the continent the sialic fraction gets largely squeezed out (gravitationally differentiated) and is plastered to the underside of the continent.

(Dietz, 1961e: 124)

Bernal (1961b: 125) found Dietz's explanation of the sialization of continents "most illuminating," but his explanation of deep-focus earthquakes was unacceptable.

These [deep-focus earthquakes] cannot, in my opinion, represent merely temporary accumulation of stresses in a long-term easy creep material. The location of the conical 45° failure zones must be at least as old as the system of deep sea trenches, island arcs and continental mountain systems they underlie.

(Bernal, 1961b: 125; my bracketed addition)

Resolving both these difficulties was seven years in the future; it would require recognition that the slabs of lithosphere bend downward and descend deep within the asthenosphere along Wadati–Benioff zones, and that deep earthquakes are caused by the release of stresses within these descending slabs of lithosphere (Isacks, Oliver, and Sykes, 1968; Isacks and Molnar, 1969).

Bernal proposed his own model of mantle convection as the cause of seafloor spreading. Dismissing the idea of convection in the upper mantle, he (1961b: 125) imagined lower mantle convection with downwards displacement "of successive blocks of the upper mantle into the presumably secularly softer material of the lower mantle." This scenario with its rigid upper mantle would provide a deep rigid layer allowing for deep earthquakes in certain places.

1.9 Wilson becomes a mobilist, 1961

I was too stupid to accept, until I was fifty, the explanation which Frank Taylor and Alfred Wegener advanced in the year I was born. In my youth scarcely anyone mentioned Wegener's ideas of a mobile earth and moving continents. Only in middle age was I converted and understood why it was impossible to theorize successfully about the earth as a static body if in reality it was a mobile one. The great impediment was that geologists only studied that one quarter of the earth's surface not covered by ice or water; at that time no one had any means for exploring the great interior or the ocean floors.

(Wilson, 1982: 6)

Whatever the exact date of his [Tuzo's] conversion, it was the most important event in global tectonics at the time.

(Menard, 1986: 174)

Wilson was a bit sloppy with his dates. He was born in October 1908; Wegener first advanced continental drift in 1912. Although Wilson may have been momentarily uneasy about his highly negative attitude toward mobilism when he wrote to Irving in 1956 (§1.5) and began monitoring paleomagnetism's growing support for mobilism, he still opposed it in March 1960, and still thought that paleomagnetism supported mobilism only "to a small degree." So he was at least fifty-one when he became favorably inclined toward mobilism.

Wilson's first announcement that he had become a mobilist, as just explained, appeared in *Nature* on October 14, 1961, ten days before his fifty-third birthday, immediately following the Bernal–Dietz exchange. Like Bernal, Wilson (1961: 125) welcomed seafloor spreading. Unlike Bernal, who did not know about Hess's seafloor spreading, he welcomed Hess's version. Wilson had heard Hess introduce seafloor spreading during a twenty-five minute talk in the Great Hall of the National Academy of Sciences on April 26, 1961 (Hess, March 25, 1968 letter to Wilson). He also had a preprint of Hess's 1962 paper, which he cited as forthcoming in *The Sea* where it was originally supposed to appear (§2.10). Wilson also supported mobilism in a paper published in the proceeding of a symposium held in Quebec City on June 6 and 7, 1961; he made several suggestions of his own, making "them [Hess and Dietz's ideas about seafloor spreading] more readily acceptable" (Wilson, 1962a: 178; my bracketed addition). Again he cited Hess's paper as appearing in 1961 in *The Sea*. However, I doubt that Wilson assessed seafloor spreading during his original Quebec talk because he gave it just three or four days after Dietz's paper was published. I suspect it was revised soon after he submitted his note to *Nature*, which he did not cite in his Quebec paper. So Wilson appears to have changed his mind after he submitted his March 1960 paper to *Nature* but before he submitted his October 1961 note to *Nature*.

Thirty years later Wilson described his reaction to a visit he made to South Africa in 1950 where King and Plumstead, among others, introduced him to the geology of South Africa, and tried to convince him of mobilism; he recalled that back then he still "remained inflexible for another nine years." This sounds plausible. Perhaps he began to question his views about mobilism as early as 1959, and took two more years before becoming inclined to believe it. His recollection also underscores his earlier strong resistance to mobilism.

Because of this (introduction of air travel), in 1950 I took four and a half days to reach Johannesburg by flying boat from London ... In South Africa, Mrs. Plumstead, [H. J.] Nel, and King ... showed me the geology of that Country.

(Wilson, 1982: 9; see I, §6.3 for Plumstead and I, §6.10 for King)

In South Africa, all the geologists [I met in 1950] were disciples of Alfred Wegener and A. L. du Toit, and they were anxious to correct my failure to accept continental drift, but I remained inflexible for another nine years. At all the three universities which I had attended [University of Toronto, Princeton University, and the University of Cambridge], this concept had been so strongly condemned and so little discussed that I had never examined it dispassionately. It is indeed true that science can be tested in impartial ways, but it is also true that scientists habitually avoid any dispassionate examinations which might cause them to change their minds.

(Wilson, 1990: 277; my bracketed additions)

So why did Wilson now finally welcome mobilism? I believe that paleomagnetism's growing support eroded his fixism, finally forcing him to question his close-minded attitude toward mobilism, and that he decided to examine that support seriously. His slipshod reviews of paleomagnetism would be things of the past. Indeed, he asked Irving to lecture extensively on paleomagnetism at the University of Toronto in fall, 1961.

I gave a series of lectures at the University of Toronto in the Fall of 1961. We were on sabbatical leave from the ANU [Australian National University]. I was writing my book staying with Sheila's [Irving's wife] relatives. I had made no arrangements with the University but went in one day to say hello and Tuzo soon had me working.

(Irving, October 30, 2008 email to author; my bracketed additions)

Irving recalled what happened. Given his remarks, it is clear that Wilson had begun thinking seriously about mobilism.

In Toronto I visited Tuzo Wilson at the University. He was having second thoughts about mobilism but it was my recollection that he was not yet a committed mobilist. As President of IUGG he had traveled widely and seen a lot of global geology. He had an assistant collecting information on oceanic islands, and another doing translations of South American geology. He asked me to give about a dozen lectures in his course on tectonics, and the stipend was very helpful. Much of them were from the beginnings of my [1964] book [*Paleomagnetism and Its Application to Geological and Geophysical Problems*]. There was a class of about 20, graduate and final year undergraduates. Tuzo attended and took notes energetically. He also persuaded me to get two MSc graduate students started on rock magnetism. They had to work on something physical because geophysics was in the Physics Department. I protested that I was a geologist but Tuzo said that did not matter. Henry Gross, I got started on looking experimentally at the effect of using different frequencies for AF demagnetization of rocks. He did a neat thesis. Jo Hodych, I got started experimentally on effects of stress. He eventually did a PhD on this. Seemed a strange way of getting students started, by a transient like me. I think all this was characteristic of the way Tuzo operated, very opportunistic. His regular attendance at my lectures and his energetic note taking makes clear that he wanted himself to hear an extended account of paleomagnetic work.

(Irving, December, 2008 email to author; my bracketed additions)

Nonetheless, paleomagnetism alone did not convince Wilson that continental drift had occurred. Irving's retrospective remarks make it clear that Wilson wanted to "hear an extended account of paleomagnetic work" but had already begun investigating possible consequences of seafloor spreading. I believe that Wilson seriously began to wonder about mobilism once he found it could accommodate his long-standing attachment to his and Scheidegger's explanation of mountain building (orogenesis) and continental growth, his emphasis on the principle of uniformitarianism, his emphasis on the need to underpin geological theories with physics, and his long-term interest in Precambrian geology and the Canadian Shield. He did this (1) by recognizing continental drift as a continuous process, a process that occurred throughout geological time, (2) by finding a way to combine at least the remnants of his and Scheidegger's account of orogenesis with mobilism, (3) by realizing that adjacent structural provinces in the Canadian Shield not only formed at different times but at places not necessarily close together, and (4) by viewing mantle convection as a possible mechanism for seafloor spreading and thus continental drift. (1) enabled him to continue emphasizing uniformitarianism; it was no longer a difficulty facing mobilism. (1) and (2) allowed him to keep the idea that island arcs evolve into mountain belts, and that mobilism offers an explanation of orogenesis throughout geological time – that mobilism could at best explain only one orogenic episode was no longer a difficulty. Even if he did not realize (2) on his own, Bernal's comments told him that it was so. (3) was crucially important in allowing him to retain his view that continental shields formed during the Precambrian either through the accretion of island arcs or collision of drifting landmasses. (4) allowed him to maintain that mobilism was not incompatible with physics; it is not that he was obliged to accept mantle convection, only to see it as somewhat reasonable and not in conflict with physics. Once he realized that he did not have to give up some of his cherished attachments and beliefs, he quickly accepted mobilism, and wasted no time drawing out new consequences of seafloor spreading or fleshing out ones recognized by Hess and Dietz.

I now turn to the two papers in which he announced his change of mind. I first consider his, October 14 1961, note to *Nature*. It was exploratory. He mentioned some of seafloor spreading's solutions, and raised several difficulties. He occasionally mentioned an advantage it had over contractionism. He worked out the implications that the new view had for his own longstanding ideas on the formation of island arcs and mountain belts, and continental growth. He first gave his blessing to both Hess's and Dietz's versions of seafloor spreading, announcing that they

have made similar proposals in which, guided by discoveries made about palaeomagnetism and the ocean floors have combined features of several older theories into one which appears to fit many observations.

(Wilson, 1961: 126)

In sketching out some of the consequences of their views, I suspect he did just what he said could be done. He took a globe and constructed "the approximate direction of motion" of the convective currents he thought responsible for seafloor spreading.

These currents are held to rise under the active mid-ocean ridges and sink under the belts of active mountains and island arcs which lie along some continental margins. The locations of these are well known and so it is possible to sketch on a globe the approximate directions of motion of the upper surfaces of these currents flowing from the mid-ocean ridges towards the active continental mountain systems.

(Wilson, 1961: 126)

Island arcs and the formation of mountains came to the forefront as did convection cell interactions. He first noted that seafloor spreading explained the formation of the Atlantic Basin and the central position of the Mid-Atlantic Ridge. Turning to island arcs and mountain ranges, he (1961: 125–126) described how seafloor spreading explained them. The Andes, located above where "the current flowing east from the East Pacific Rise meets that from the Atlantic," formed "as arcs off the coast of South America" were "overrun and swept up by the westward motion of the continent." The Alpine-Himalayas "are due to compression over downward flowing limbs of huge convection cells." The East and Southeast Asian island arcs, and the shearing observed in New Zealand and the Philippines, which he had previously explained in terms of contractionism (§1.3), he now claimed were formed by convection currents "in much the same manner postulated by Scheidegger."

The current flowing north-west from the East Pacific rise presumably meets the current flowing south-east from the ridges in the North Atlantic Ocean and Arctic Sea. The resulting downflow and compression can be held to have produced the east Asian arcs in much the same manner postulated by Scheidegger; indeed, no other explanation has yet been offered why failure should take the form of circular margins. Thus some features of the contraction theory are preserved. This movement of the Pacific basin north-west past the end of the Indonesian-Melanesian belt of mountains and island arcs fits the shearing observed between the Philippines and New Zealand.

(Wilson, 1961: 126)

Wilson argued that seafloor spreading "admits of continental growth."

Besides admitting of some features of compression in mountains, the new theory admits of continental growth, both by sweeping up sediments on the ocean floor and by the emission of andesitic lavas. This growth along some continental margins implies that coasts separated by rifting need not necessarily fit if later brought together again.

(Wilson, 1961: 127)

He envisioned continental drift as a continuous process, not a one-time event. Although, seemingly still unaware of Argand's proto-Atlantic, he most likely had come to accept paleomagnetism's accumulating evidence that continental drift had occurred before the breakup of Pangea.

Not only does this marginal growth make it difficult to find the coasts which formerly fitted together, but also old theories of continental drift included the notion that all continents had formerly been united into one or two ... The blocks of continents should rather be thought of as sometimes moving together in one place and sometimes splitting and moving apart at another place, always producing scattered patterns.

(Wilson, 1961: 127)

Mobilism explained how mobile belts were formed throughout geological time; it no longer clashed with uniformitarianism (RS2).

Wilson also claimed that mobilism's solutions to the origin and evolution of island arcs into mountain belts avoided two difficulties faced by competing solutions (RS3).

This theory offers an explanation of two formerly puzzling features of older arcs. The first is that it was never apparent why island arcs, so symmetrical in youth as for example the Kurils, should lose this with age, for example in the Andes. It now appears likely that the Andes formed as arcs off the coast of South America and that these arcs have been overrun and swept up by the westward motion of the continent. The contrast between the position of deep-focus earthquakes under the heart of the South American continents and off the margin of east Asia is another indication of this. To be thus caught up and overridden appears to [be] the fate of arcs.

(Wilson, 1961: 126; my bracketed addition)

Thus he highlighted mobilism's improved relationship with uniformitarianism.

Another feature, which was hard to explain by earlier theories and which appeared to be contrary to uniformitarianism, is that, whereas the active mountains form one continuous t-shaped pattern, older mountain systems only form short lengths. It is difficult to fit the Hercynian Appalachian mountains together into a single system and quite impossible to do it for Precambrian orogenies.

(Wilson, 1961: 126–127)

Wilson raised two difficulties with seafloor spreading. The first, much like Bernal's second difficulty, concerned the dire prospect of continents drowning. Given that continents and oceans, at least in the long run, are in isostatic equilibrium, and that continued erosion of continents will wear them down to sea level, there must be a way to thicken continents. Wilson acknowledged that both Hess and Dietz proposed ways to increase continental thickness under mountains such as the folding and overthrusting of mountain belts, but they did not have a way to increase thicknesses beneath interior plains. Wilson (1961: 128), still thinking favorably of very slow Earth expansion, suggested it as a way out of the dilemma; expansion at a "rate of a few mm/yr." would increase the size of ocean basins and "would suffice to maintain continents above sea-level."

Wilson's second difficulty pertained to Antarctica being surrounded by ridges. This was serious. Ewing and Heezen raised the same difficulty within the context of continental drift without seafloor spreading in 1957 when announcing the discovery of the rift valley (§4.8). Heezen (1959b) restated it in 1959 at the International Oceanographic Congress in New York (§4.9), and in his 1960 *Scientific American* paper (Heezen, 1960b).

The extension of this concept to the ridge through the Southern Ocean poses a difficulty. From the Atlantic the ridge passes south of Africa and Australia to reach the Americas, but a current spreading south from this part of the ridge would flow towards Antarctica from nearly every direction and have no sink into which to descend. The only compressional mountains in Antarctica lie along the Antarctic peninsula (Grahamland) and these are scarcely adequate.

(Wilson, 1961: 126)

To avoid rapid expansion, he then proposed the novel idea of migrating ridges.

To escape the dilemma one must suppose that, instead of a current flowing south from the ridge, the ridge itself is migrating northwards and that the ring which it forms around most of Antarctica is expanding in radius. If so, the current flowing northwards on the other side of the ridge is moving at twice the normal rate (taken as of the order of 1cm./yr. by both Dietz and Hess).

(Wilson, 1961: 126)

His response shows that he had begun reading paleomagnetists and listening to Irving.

This [northwards migration of the ridge] would cause a relatively rapid northward movement of all the southern continents except South America, a fact which is borne out by palaeomagnetic observations. Blackett, Clegg and Stubbs have recently shown that for the past 2×10^8 years Africa, India, and Australia have all been moving north at rates of 2–8 cm./yr. It will be observed that, since the mid-ocean ridge meets the west coast of South America and does not pass south of it, there is not the same reason for South America to have moved north. Nevertheless, its fit with Africa suggests such a motion. It would be useful to have palaeomagnetic data for South America. (Note added in proof. E. Irving has directed my attention to palaeomagnetic work by K. M. Creer on the Jurassic (Serra Geral) lava of Uruguay and southern Brazil which suggests that the Brazilian Shield has not changed appreciably in latitude since the Jurassic – a result consistent with the view outlined here.)

(Wilson, 1961: 126; my bracketed addition; see II, §5.19
for Blackett et al. and II, §5.5 for Creer)

Wilson had become a committed mobilist and, in typical Wilsonian fashion, was using his fertile imagination to tease out new consequences of seafloor spreading.

A notable example was his presentation at a symposium on the tectonics of the Canadian Shield in June 1961, which was revised for publication the next year. He listed seven problems that any adequate tectonic theory should solve: development of continents, lack of pre-Cretaceous sediments on the seafloor, conversion of sedimentary rock into metamorphic, formation of island arcs, the origin of mid-ocean ridges, large horizontal movements along lengthy faults, and shortness of past mountain belts in contrast to current ones. Beginning with the development of continents, he turned to home ground, the Canadian Shield, and, as always, emphasized uniformitarianism.

A major problem in tectonics, particularly suitable for investigation in the Canadian Shield, is to explain how the continents developed and what part the unstable zones played in its development. In discussing this problem it is usual to assume the truth of uniformitarianism … The continents are zoned. Young mountains surround shields. Age determinations and differences in geology show shields to be divided into areas, each having a uniform history, different from its neighbors.

(Wilson, 1962a: 177)

Wilson did not explicitly propose a complete solution. Some continental growth was achieved by the welding of island arcs onto continents as new mountain belts. He also now recognized that continental drift has occurred throughout Earth's history.

The present random arrangement of the continents can be expected to be normal. That is to say, continental blocks can join and rift at random. There is no need to imagine that all the continents were ever joined in one or two Pangeas or Laurasias although age determinations on granitic intrusions in shield areas of different continents suggest that the shields may once have been continuous. India and Asia have just joined as Europe and Asia did at an earlier time, while along the Red Sea and Rift valleys Africa is starting to break up as did the Atlantic earlier. Such events have occurred sporadically throughout geological time.

(Wilson, 1962a: 179)

Ironically noting that his interesting comments on the tectonics of the Canadian Shield are "rather speculative," he (1962a: 179) suggested that "the possible former union of Canada with Greenland and Europe should be more carefully investigated," and he identified the two key assumptions that he had formerly made about the Shield, the rejection of which had freed him to consider mobilism.

The fact that two provinces of the Canadian Shield have been together during post-Cambrian time does not necessarily mean that they were formed close together or that the sediments lying on one province were divided from the province now beside it.

(Wilson, 1962a: 179–180)

Wilson (1959: 348) had previously claimed that the drifting of continents through the seafloor for thousands of kilometers would have disturbed the now flat lying "lower Paleozoic rocks that overlap the shield." Perhaps he thought that seafloor spreading removed this difficulty because continents do not plow their way through the seafloor but, as he (1962a: 178) emphasized, "are carried like rafts by currents in the mantle."

Wilson now turned to the conversion over time of sedimentary into metamorphic rocks, the lack of pre-Cretaceous sediments on the seafloor, and formation of island arcs. He noted (1962a: 176) that in "successively older periods fewer sedimentary rocks and more metamorphic rocks are exposed on the continents" and added that some "process appears to be converting sedimentary rocks into metamorphic ones." He noted, "No fossils older than Cretaceous have ever been found in the deep oceans" and, catching up with what others had been wondering about for some time, rhetorically asked, "Can some process be removing all old sediment from the ocean floor?" He then went on to discuss the formation of island arcs and their evolution into mountain belts, arguing that Scheidegger/Wilson contractionism got high marks in explaining their origin but not their subsequent distortion.

The contraction theory provides a more detailed explanation than any other yet advanced of the shape and properties of island arcs at the moment of their creation, but it does not predict the fault-motions observed and with age island arcs seem to be steadily distorted and to lose their pristine regularity. For example, no former island arcs have been satisfactorily identified in the Canadian Shield.

(Wilson, 1962a: 176)

According to Wilson (1962a: 179), seafloor spreading provided solutions to all three problems. Ocean basins are young; sediment is removed from seafloor where sinking mantle convection currents "form trenches and arcs," and "continental growth occurs along such coasts, the trench providing the mill to metamorphose sediments." It also explained the steady distortion in shape of island arcs: "Where the sinking current has been overridden by the continent the island arcs have been, as it were, pushed ashore and distorted" (1962a: 179).[2]

Moving to the fifth problem, the origin of mid-ocean ridges, the cornerstone of seafloor spreading, Wilson now acknowledged:

The contraction theory has never been extended to explain the mid-ocean ridges which seem much more likely to be due to expansion. A rift across Iceland which lies on the ridge is steadily expanding ...

(Wilson, 1962a: 178)

He next considered known large horizontal movements.

Any tectonic theory must explain the large horizontal movements now known or suspected on several dozen faults ... Such motion is apparently confined to limited periods of time and to limited lengths along strike. The Great Glen fault, for example, does not extend into Norway and the Great Slave fault does not cross the Rockies. Both are now inactive. The greatest horizontal displacement along the Murray escarpment appears to be 800 miles and yet this fault terminates abruptly at the California Coast.

(Wilson, 1962a: 176–177)

What an about-face: just two years before (§1.5) Wilson had argued that if continental drift had occurred, more indicators of past horizontal movement should be evident. Needless to say, these movements, especially the displacement along the Murray escarpment, he now touted as support for mobilism because they showed that large horizontal movements had definitely happened. In 1965 Wilson would, through his concept of transform faults, go on to reform understanding of active strike slip faults, such as the San Andreas and Alpine (New Zealand), through his concept of transform faults. However, at this point, he was content to repeat the explanation he had offered in his first mobilism paper.

The current which is supposed to rise up along the East Pacific Rise must flow northwest across nearly the whole Pacific Ocean before sinking in East Asia. Such a motion must shear the island chains from New Zealand to the Philippines and also the coast of California, and could thus explain the great faults and their directions of motion. The fact that the earthquake evidence suggests that the ridge has three branches reaching the coasts of Chile, Ecuador, and Mexico suggests that three rising currents may lead the east side of the Pacific to move northwards faster than the west side producing the counter-clockwise motion of the Pacific Ocean basin.

(Wilson, 1962a: 179)

Moreover, he was already investigating whether or not the Great Glen fault of Scotland was originally much longer and had a counterpart in North America. He may well already have identified what he considered its mate as the Cabot Fault in Canada, because he argued as such in a forthcoming paper published on July 14, 1962 (Wilson, 1962c). Indeed, after introducing this problem, he went on to consider the shortness of past mountain belts compared to current ones. He wondered whether they were tectonic disjuncts, with a passing reference only to E. B. Bailey.

Today the young mountain systems, both continental and mid-ocean, form continuous zones about the earth for great distances. Backwards in time it becomes progressively harder to discern more than short sections of mountain systems. It is particularly hard to establish connections between continents. Were ancient mountain systems formed in a different way or have the chains been disrupted?

(Wilson, 1962a: 177)

He did not, however, have an explanation for the horizontal movement along the great fracture zones such as the Murray and Mendocino. Wilson had previously objected to mobilism's account of mountain formation because he believed it clashed with uniformitarianism. Having accepted mobilism's explanation of modern mountain belts, he now asked rhetorically if former mountain belts had been split and moved apart, as Wegener and du Toit had done. Perhaps he had begun to read or reread and better understand the works of Bailey (I, §8.13), Wegmann (I, §8.10), Holtedahl (I, §8.11) and F. E. Suess (I, §8.12), viewing them with an open mind, and now recognizing the Caledonides, as they did, as one continuous orogenic belt extending from Spitsbergen through Scandinavia and Greenland to Scotland and beyond to eastern North America.

Wilson now turned to the paleomagnetic and paleoclimatic data, and finally was willing to accept their support for mobilism. He had been sitting in on Irving's fall 1961 lectures on paleomagnetism, and now was receptive.

As palaeomagnetic results accumulate, the pattern which they show steadily increases in consistency. They suggest that for the past 200 million years Europe, North America, India, and Australia have all been moving steadily northward with average velocities of from 2 to 8 cm per year, but that they have rotated relative to one another – Europe by 50° clockwise relative to North America in the past 300 million years (Blackett *et al.*, 1960). One cannot measure changes in longitude but similar rates of movement east or west would have served to open the Atlantic Ocean.[3]

(Wilson, 1962a: 177; see II, §5.19 for discussion of Blackett et al.)

Wilson, it seems, was either unaware of or rejected Irving's idea that changes in longitude can be fixed for continents if they were once part of a moving supercontinent that subsequently split apart; the fragments record their earlier but now separate APW paths (II, §5.16).

So long unpersuaded, Wilson, writing now like an old-time mobilist, recognized the potency of the paleoclimatic case for continental drift.

It is curious that so little sense is made of palaeoclimatic data. Although it is understandable that there might be arguments about conclusions based upon the habitats and migration of fossil species, surely it is probable that extensive tillites were formed in polar or near-polar regions and that evaporites were formed in the tropics. If polar wandering or horizontal displacements of the crust have occurred and if they are not taken into account, the failure to make sense of palaeoclimatic data is explicable. It seems impossible to explain evidence for continental glaciation in the present tropics or think of coal seams in the present polar regions in any other way.

(Wilson, 1962a: 177)

What a difference five years makes. This, and its consilience with the paleomagnetic data, had been Irving's argument in the paper he had sent Wilson in 1956.

Returning to mantle convection as a cause for mobilism and having already raised the difficulty of Antarctica being surrounded by ridges, Wilson now added Africa, which had ridges on three sides. Quite pleased with his suggestion of migrating ridges, he characterized the difficulty as "at first sight ... insuperable."

The currents rising in the southern and India oceans appear at first sight to offer an insuperable difficulty in applying any scheme of convection currents. Antarctica and Africa are each nearly surrounded by rising currents which have no place into which to sink. If, however, the diameters of the rings of rising current are steadily increasing there is no need for a sinking current. The alternatives can be thought of in this way. If the mid-ocean ridge that surrounds Antarctica from the South Atlantic through the Indian and Pacific oceans to the west coast of Chile is thought of as being fixed in location and if it is also the locus of rising current then there must be a sink, but if the rising current has existed only for a limited period of time and if its radius is expanding then no sink is necessary. A similar type of expanding eddy is frequently seen in fast rivers. In this fashion the broad features of the currents around Africa and Antarctica can be accounted for.

(Wilson, 1962a: 178)

He thought that convective cells must expand but did not know how or why they could do so. Worried perhaps that mechanism difficulties would be raised against seafloor spreading on this account, he sought a far-fetched analogy with expanding eddies in fast rivers. He was not going to let the absence of a real answer as to why convective cells should expand stand in the way of seafloor spreading and mantle convection. Wilson had become a fully fledged mobilist.

1.10 Wilson matches the Cabot and Great Glen faults in support of mobilism, 1962

A year or so ago I noticed that there is a striking resemblance between the Great Glen Fault in Scotland and a system of faults which I found I could trace in the literature from northern Newfoundland to Boston. This aroused my interest in continental drift.[4]

(Wilson, March, 1963 letter to H.W. Menard)

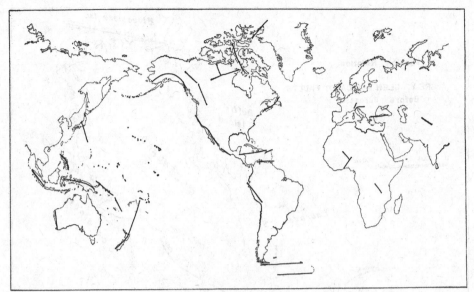

FIG. 2-13. Some of the world's major transcurrent faults.

Figure 1.8 Wilson's Figure 2-13 (Jacobs *et al.*, 1959: 40) of large transcurrent faults showing the later-named Cabot Fault extending from Newfoundland across the Bay of Fundy to Rhode Island.

Wilson discussed this "striking resemblance" between the Great Glen Fault in Scotland and what he named the Cabot Fault in eastern North America in two papers (Wilson, 1962b, c) and a letter (Wilson, 1963a) in which he responded to an objection. He (1962b: 135) matched the well-known Great Glen fault in Scotland with the Cabot Fault, which ran from "northern Newfoundland across the Bay of Fundy to Massachusetts and Rhode Island." He knew of the fault before he became a mobilist; it was figured in the 1959 textbook (Figure 2-13 of Jacobs *et al.*, 1959; reproduced here as Figure 1.8) but was neither named nor discussed there.

Wilson recognized it from aerial photographs. He (1962c: 31) saw that the individual faults, studied on the ground by others, "might be parts of a single fault zone." He estimated that the Cabot Fault extended for 1100 miles, was probably a left-lateral transcurrent fault, active during the Late Paleozoic. Comparing it with the San Andreas Fault, he argued that the Cabot and Great Glen Faults were once "two ends of the same fault."

Most reconstructions of the continents based on the assumption of continental drift bring Newfoundland into juxtaposition with the British Isles. On any such map it will immediately be seen that the Cabot and Great Glen faults are opposite to one another and have the same orientation. Since both are the same age (uppermost Devonian and Lower Carboniferous), are transcurrent, are large, are perhaps sinistral and cut rocks well known to be similar, it is reasonable to speculate that, if drift has occurred, they are two ends of the same fault.

(Wilson, 1962b: 136)

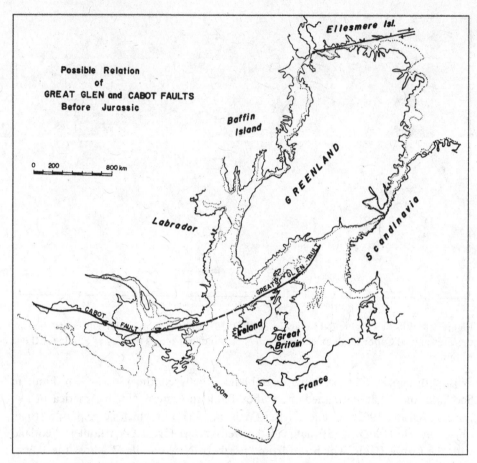

Figure 1.9 Wilson's Figure 2 (1962c: 34). His caption reads: "Sketch map showing the position of the Great Glen and Cabot faults in the usual pre-Jurassic reconstruction of continents." Scanned image courtesy of Sally Croston.

In his second paper, he offered a Jurassic reconstruction of the continents, showing the relative positions of northeastern North America, the British Isles, Greenland, and part of northwest Europe. Wilson, sensitive as he was throughout his career to shearing motions between blocks, was pleased to learn that a fault between Ellesmere Island and Greenland had been "recently mapped" (Figure 1.9).

According to this reconstruction the separation of Europe from Greenland and North America along the line of the mid-Atlantic ridge would have been by motion of the opposing lands directly away from one another, but the separation of Greenland from North America would have been by a shearing motion. It is therefore of interest to find that V. K. Prest has recently mapped the fault between northwestern Greenland and Ellesmere Island which this reconstruction would require.

                                                                        (Wilson, 1962c: 33)

A year later Wilson named the fault after Wegener in a paper which he wrote for *Scientific American* (Wilson, 1963d: 100) (§3.2).

Sounding once more like an old-time mobilist not a new convert, he recalled the standard arguments against mobilism, arguments he now regarded as bogus (§1.4, §1.5).

When the theory of continental drift was first proposed it was supported by the parallelism of some adjacent coasts on opposite sides of oceans and by the faunal, structural and stratigraphic coincidences which would result if these coasts had once been in contact. Opponents pointed out that to have fitted more than a few coasts together would have required unacceptable distortions of the continents. They felt that, in fauna, coincidences could be explained by migration, in structure by mountain chains crossing oceans, and in stratigraphy, by the same reasons plus climatic similarities. Most physicists considered that the crust of the Earth was too strong and the interior too viscous for continents to have been moved by the proposed dynamic forces.

(Wilson, 1962b: 136)

He (1962b: 136) now cited favorably mobilism's paleomagnetic support and the discovery of "transcurrent motions of more than 1400 km. on the Pacific Ocean floor ..." He also included, as a tectonic disjunct, his newly proposed Cabot–Great Glen mega-fault, and added another possible connection, the continuation of the Moine Thrust into western Newfoundland. F. E. Suess, Bailey, and Collet had made a similar suggestion (I, §8.12, §8.13, §8.9). He referenced Bailey.

Wilson also gave impetus to convection theory, claiming (1962b: 136) that "the discovery of a low velocity layer at a depth of 100–200 km in the upper mantle has provided a possible lubricating channel on which the crust may move." Thinking further about the possibility of migrating ridges, and the differing patterns of mantle convection that pre-Wegenerian drift would have entailed, he appealed in an ill-considered way to the work of Raymond Hide (1953d) on convection of liquids in rotating bodies (II, §1.15). Giving no references and misspelling Hide's name, Wilson noted:

While considering convection currents, R. Hyde demonstrated by a model that the currents between two rotating cylinders tended to follow one stable mode or pattern until the thermal conditions were changed sufficiently to cause the currents to change abruptly to some other mode.

(Wilson, 1962b: 137)

Hide's work may relate to convection in the liquid core or atmosphere, but not the mantle. Apparently, Wilson did not think it mattered. Three years later Holmes would also appeal in similar far-fetched manner to Hide's work while discussing mantle convection in the second edition of his *Principles of Physical Geology*.

Wilson (1962a: 136), I suspect, was unaware of Argand's proto-Atlantic because he continued to complain about the "great harm" mobilists did to their cause by holding to "their too rigid concepts of the proto-continents of Pangea or Laurasia." But, he was reading work by other mobilists. He referred to Carey's work, and, quite fittingly, had read Hamilton's mobilist account of the opening of the Gulf of California (I, §7.8). Recall that Hamilton had been lectured not long before by Wilson as they flew toward Antarctica at the beginning of IGY, on the merits of contractionism and accretion of fixed continents (§1.5).

With his proposal of the Cabot–Great Glen disjunct, Wilson was just warming up. During the next few years, he would continue finding further support for mobilism, culminating in his 1965 idea of transform faults. He was fast becoming seafloor spreading's most active proponent.

1.11 Menard and Hess correspond about seafloor spreading, 1961, 1962

On May 25, 1961 Menard wrote Hess questioning his version of seafloor spreading. Menard already knew about Dietz's version, which was published eight days after he wrote Hess, but said nothing to Hess about that. Menard noted, "I have read your Evolution, Ocean Basins with greatest interest, but it will hardly surprise you that I would like to take issue with a few points." He raised eight. I consider seven. The first (1) pertained to the Mesozoic Mid-Pacific Ridge that Hess had postulated to explain the formation of guyots and atolls in the western Pacific; they formed from volcanic peaks that arise at ridge crests, and have their tops flattened by erosion as they move off the ridge down its flanks. Hess proposed that the guyots formed on the northern flank of the ridge closer to the Mesozoic North Pole while atolls formed on the south flank in lower latitude and warmer water (§1.15). Menard raised a difficulty: there is no clear north–south separation of atolls and guyots with atolls on the south and guyots on the north; they are intermixed (RS2).

I agree with your former location of a Mid-Pacific Ridge in a general way and I have been inclined to think that I over emphasized the exact median position of the Tuamotu-line Mid-Pacific Mountains. However I can see no justification for excluding the Emperor seamounts from the subsiding region. Your point on climatic control of distribution of atolls and guyots skips rather lightly over the intermixing of the two – not only in the northern Marshalls but also in the Tuamotus and Tubuais. I wonder if random minor fluctions [fluctuations?] of sea level connected with a general subsidence of the region might not cause some guyots to escape from the zone of active coral growth while others remained in it. In this connection it is noteworthy that the region west and southwest of the former ridge is full of drowned atolls mixed in with normal atolls. Is the regional subsidence continuing?

(Menard, May 25, 1961 letter to Hess; my bracketed addition)

Menard agreed that there was a former Mesozoic Mid-Pacific Ridge. He discussed it in his forthcoming paper in *Scientific American* (Menard, 1961), and he (1964: 138) later named it the Darwin Rise after Charles Darwin because it was he who "recognized from the prevalence of atolls that a vast area of the southwestern Pacific had subsided." Hess and Menard agreed about the relative ages of the Mid-Pacific,

Mid-Atlantic, and East Pacific Rise. Both thought the Mid-Pacific the oldest, and the East Pacific Rise the youngest. They also agreed that it trended northwest–southeast roughly in the middle of the Pacific Basin. Nor did they disagree significantly about its dimensions; Hess (1962: 604) estimated a width of 3000 km and length of 14 000 km, and Menard (1964: 138) that it was 4000 km by 10 000 km. Menard's second disagreement, (2), concerned the width of *present-day* active ridges. Hess (1962: 604) thought they "have an average width of 1300 km," while Menard thought the East Pacific Rise much wider, 2000–4000 km.

You make some point of the difference in width between the former Mid-Pacific ridge and present ones. I believe that this is not well taken. My own thinking is that a ridge or rise centered in an ocean basin is already rather old, whereas a youthful one could be randomly distributed with regard to contents and ocean basins. There are, of course, other criteria which can be used to separate different stages of development. Consequently, the East Pacific Rise seems to be the only clearly youthful one. It is 2000–4000 km wide.

(Menard, May 25, 1961 letter to Hess)

Menard later remarked when considering this point of disagreement:

Like Wilson, Hess had some misconception that mid-ocean ridges are of different widths that are unrelated to the total width of an ocean basin. He did not realize that contour maps can be very misleading in this regard.

(Menard, 1986: 176)

Hess, as far as I can determine, did not think that the width of the Mid-Atlantic and East Pacific Rise greatly differed. Moreover, Hess's seafloor spreading could just as easily explain the origin of a 1300 km wide ridge as a 2000 km wide one.

So, why was Menard making such a fuss about the width of the East Pacific Rise? His next point (3) hints at why.

You mention the seismic results of Lamont and Scripps but in fact you completely ignore the crustal thickness on the rise measured by Raitt and Shor. The crust is about 1 km thinner on the crest of the rise.

(Menard, May 25, 1961 letter to Hess)

The decrease in thickness of crust (layer 3) beneath the East Pacific Rise was central to Menard's notion of seafloor thinning. He attributed the thinning to crustal stretching caused by mantle convection which formed the rise. He (1961) even claimed that he had a "semiquantitative test" of his notion. Given the 1 km decrease in crustal thickness over the ridge crest and the 2800 km average width of the ridge, he determined that the crust had stretched approximately 600 km, which was of the same order of magnitude as the lengths of offsets that crossed the East Pacific Rise. But Hess had undercut Menard's explanation by underestimating the width of the East Pacific Rise and by ignoring Raitt and Shor's key determination that the crust was "about 1 km thinner on the crest of the rise."

Menard's next two points, (4) and (5), concerned the fracture zones in the Pacific. He rightly criticized Hess for finding no place for them in his account of seafloor evolution (RS2).

> Granting some connection between fracture zones and the East Pacific Rise, the rough correlation between crustal thinning on the crest of the rise and displacement on the fracture zones should be explained by your hypothesis.
>
> *(Menard, May 25, 1961 letter to Hess)*

Naturally Menard emphasized fracture zones because he had extensively researched them and because they were central to his own explanation of ridges. He reminded Hess (5) that in this regard his own solution was superior to Hess's (RS3).

Menard's sixth point (6) was, as he (1986: 176) later noted, actually directed at Heezen and Ewing, as is evidenced by his reference to the "great publicity" that they had received when they announced the discovery of the continuous "world-girdling submarine mountain system" with its central rift valley (§4.8).

> I think that more attention might be given to the fact that oceanic rises have ends. The great publicity about a world-girdling submarine mountain system has obscured the fact that the various parts may be of quite different age. My attitude is that the bulges in the mantle are so big that if three or more exist at once they must inevitably overlap and give a continuous topographic high. I think that the East Pacific Rise ends at about 45°S where the crest is offset. We explored the Pacific-Antarctic Ridge south of New Zealand on [expedition] Monsoon. It has normal (about 1.9) heat flow at three places on the crest, it has pronounced ridge and trough topography, and it is centered in the basin. In short, it is nothing like the East Pacific Rise. If rises have ends it is possible to account for the Western Pacific trenches more easily than if the rise continued under Asia. Moreover, if rises have ends, some longitudinal extension might be expected which would facilitate wrench fault displacements.
>
> *(Menard, May 25, 1961 letter to Hess; my bracketed addition)*

Hess and Menard thought that different parts of the "world-girdling submarine mountain system" were of different ages. Hess did not distinguish the East Pacific Rise from the Pacific–Antarctic Ridge, but his failure to do as such did not affect the viability of seafloor spreading.

The last of Menard's points was not critical of Hess's seafloor spreading, but describes his growing doubts about continental paleomagnetic data as evidence for mobilism.

> The question of the location of ancient rises is left dangling and I wish that we could find another one. I suppose that if one could believe the paleomagnetic evidence in detail, the position of a rise might be deduced from the movement of continents. Have you tried this – to save me the trouble?
>
> *(Menard, May 25, 1961 letter to Hess)*

When he was applying his theory of mantle convection to the development of the Mid-Atlantic Ridge, Menard (1960) wrote that continental drift had been "revitalized by the paleomagnetic evidence for polar shifts and possible drift" (III, §5.12). However, in his *Scientific American* paper, which appeared about seven months after writing this letter to Hess, he did not mention paleomagnetism. Nor did he mention paleomagnetism in his *Marine Geology of the Pacific* (III, §5.12), his last work in favor of mobilism before he reverted to fixism at the Royal Society's March, 1964 symposium on continental drift (§3.6). Menard was impressed by Munk and MacDonald's 1960 attack against paleomagnetism (II, §7.11). Because Munk was a major figure, an eloquent advocate, and a colleague at Scripps Institution of Oceanography (hereafter, Scripps), it was perhaps under his close influence that Menard began to doubt and later dismiss mobilism's paleomagnetic support.

Hess responded on June 12, 1961. His praise of Menard was high.

I appreciate your comments on Evolution of Ocean Basins and place more weight on what you have to say about it than I would on anyone else in the world. Several of your comments are brief but it is easy to see that there are complex arguments behind·them which would take several hours of conversation to weigh and understand. I do not have any strong objections to your comments. Some of them I would adopt outright.

(Hess, June 12, 1961 letter to Menard)[5]

Hess, however, took exception with Menard's account of the great fracture zones, but admitted that he "had no solution" of his own.

The one bone of contention is the relationship of the great fracture zones to everything else. I am not quite prepared to accept your views but do not have a solution of my own.

(Hess, June 12, 1961 letter to Menard)

This private discussion between Menard and Hess appears to be Menard's only response to seafloor spreading until March 1962, when he raised what he took to be a significant objection (§3.7). He said nothing about it in his 1964 *Marine Geology of the Pacific* or in any of his other publications until accepting mobilism in late 1966 (§6.12). Nor at this time did he say anything about Dietz's version of seafloor spreading, even though he retrospectively remarked, "It is a curious fact that none of my objections to Hess's version of seafloor spreading applied to Dietz's version" (Menard, 1986: 176). He did not recognize it as a "curious fact" until later. The lack, between 1962 and 1966, of any published discussion by him of seafloor spreading is puzzling given its triumphal acceptance in 1966. Much later Menard (1986: 176) remarked that during the first half of the 1960s seafloor spreading was, "just one more hypothesis, like expanding earth, or mantle convection."

1.12 Menard defends seafloor thinning and attacks Earth expansion, 1962

Menard (1964: x) wrote *Marine Geology of the Pacific* "in the stimulating surroundings of the Department of Geodesy and Geophysics, Cambridge University," where he spent part of the 1962–3 academic year. He left for Cambridge in April 1962,

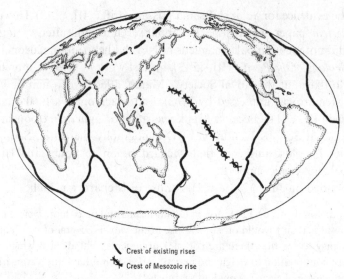

Figure 1.10 Menard's Figure 6.1 (1964: 119). His caption reads: "Existing and ancient rise system (Atlantic and Indian Oceans. After Heezen and Ewing, 1961.) Doubtful rise under Asia discussed." The proposed Melanesian Rise, Pacific–Antarctic Ridge, and Mid-Indian Ridge meet south of New Zealand. Menard speculated that the three largest continental plateaus, in eastern Africa, western North America, and central Asia might overlay ridges. The Mesozoic rise is the Darwin Rise.

arriving with an outline of the book, which he completed approximately a year later (Menard, 1986: 207). Although the book was published in1964, it represents his views in late 1962 and early 1963; there are no references to work dated 1964, and only a few in 1963: his own co-authored papers and a Ph.D. thesis by R. F. Dill at Scripps.

As already noted, Menard, like Hess, believed in the existence of a Mesozoic, mid-Pacific rise, which he named the Darwin Rise (Figure 1.10), which he invoked to explain the origin of the many volcanic islands, atolls, and guyots in the central Pacific. He also suggested that many of the East Asian island arcs formed where westward-moving convection currents from the Darwin Rise met eastward-moving convective currents from beneath Asia.

He (1964: 148) identified five active oceanic rises (Figure 1.10). The East Pacific Rise is the youngest; Menard envisioned it extending through the Gulf of California, underneath the western United States and reappearing off the coast of Oregon; it is seismically active with high heat flow along its crest, and except for its northern extension which he thought might be the oldest part of the rise, it lacks a rift valley. The next youngest ridge is "the problematical Melanesian Rise." Like the East Pacific Rise, it is not located in the center of an ocean basin. It is seismically active in some but not all places, and is topped with guyots. The remaining and older three, the Mid-Atlantic, Mid-Indian, and Pacific–Antarctic ridges are highly faulted, seismically active, and centered in ocean basins. The Mid-Atlantic Ridge was characterized by

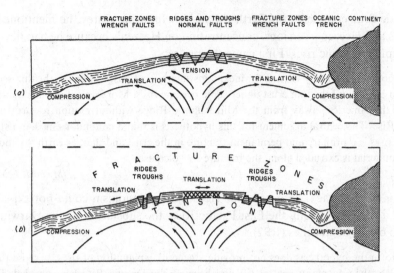

Figure 1.11 Menard's Figure 6.20 (1964: 150). His caption reads: "Convection-current hypothesis of origin of topography and structure of oceanic rises. (*a*) Rifting at crest, translation of crustal blocks between fracture zones on flanks. (*b*) Rifting on flanks, translation outward on flanks beyond rifted regions, translation in either direction on crest but less than on flanks."

high heat flow. Menard (1964: 148) speculated that the northern part of the Mid-Indian Ridge, which "trends under Africa, may be as young as the East Pacific Rise."

He thought that ridges form as overheated material from the mantle rises and bulges the crust outward (Figure 1.11). Upward-moving convecting material divides, the crust splits longitudinally apart, normal faulting occurs, differential spreading away from the ridge crest occurs as seafloor stretches (this is not seafloor spreading) and latitudinal wrench (transcurrent) faults (fracture zone) form. Extensive volcanic islands form. Trenches arise where blocks of translated seafloor are pushed against continents. After about 100 million years, convection ceases, ridges subside, and many volcanic islands become guyots or atolls. Menard allows for the possibility that the crust does not always split longitudinally apart, with ensuing normal faulting. When it does, a rift valley forms; when it does not, there is no rift valley.

Menard's enthusiasm for mobilism was waning. In his discussion of the formation and evolution of the Atlantic Ocean and its ridge, as already noted, there was no mention of paleomagnetism, no mention of the now well-established systematic difference between the APW paths of Europe and North America as evidence of its former closure. His support of drift that remained was tepid.

If the [convective] cells [that cause formation of a ridge or rise] form under a continent they may tear the crust asunder and float the continent toward the flanks of the rise. This may have occurred in the Atlantic; the margins of the continents on the two sides match, and there is no normal oceanic crustal layer – merely a thick layer of volcanic extrusives.

(Menard, 1964: 150; my bracketed additions)

He made four comments on Earth expansion. Ignoring Carey, he mentioned only Egyed's and Heezen's versions, concentrating on Heezen's because he had developed it to explain oceanic rises. First came the praise:

This hypothesis has many virtues, for it explains such varied observations as the thin sediment of ocean basins and the thin crust under some oceanic rises. Moreover, it permits continental drift of the Americas away from the Mid-Atlantic Ridge without motion toward the East Pacific Rise. The central argument for this hypothesis is that a continuous median rift in the crest of rises is a primary and permanent fracture in the crust and that, as earth expands, new crustal material is extruded along the fracture by volcanism.

(Menard, 1964: 149)

Turning to the attack, he noted that Heezen's expansionism could not explain the lack of a rift valley along the East Pacific Rise, the subsidence of the Darwin Rise, and the origin of trenches (RS2).

In this form the hypothesis does not provide a general explanation of rises, because the East Pacific Rise has no major central rift and because the Darwin Rise has subsided. Even if modified, it probably could not account for the downbowing of trenches on the flanks of the rises, because for this compression is essential.

(Menard, 1964: 149)

He (1964: 149) also noted that other explanations of the oceanic ridges, in particular his own, explained the origin of rift valleys as well as "all the other diverse facts." His version of mantle convection, seafloor thinning, solved all the problems solved by Heezen's Earth expansion, and many that were not (RS3). Menard closed his attack with what he took to be the most serious obstacle facing Heezen's rapid Earth expansion.

Nevertheless, the great weakness of this hypothesis is the improbability of a large expansion of the earth. Adducing the paleomagnetic evidence of continental drift, Heezen (1960) estimated that the earth had doubled its diameter. This means that the initial density was 44 gm/cc – which is impossible in the present state of the universe for a body with the mass of the earth. Considering that most of this expansion is supposed to have occurred during the last 5 percent of geological time, when the universe cannot have been very different from what it is now, the hypothesis is not acceptable.

(Menard, 1964: 149; reference to Heezen is same as my 1960b)

Menard did not mention Cox and Doell's paleomagnetic assault on rapid Earth expansion (II, §8.14). Perhaps he thought he did not need it; perhaps, he did not believe it, for to do so depended on believing "the paleomagnetic evidence in detail" which, given his comment in his May 25, 1961 letter to Hess, he could no longer do.

1.13 Heezen attacks seafloor spreading and seafloor thinning, 1962

Heezen raised difficulties with seafloor spreading and seafloor thinning in the defense of Earth expansion he gave in Runcorn's 1962 anthology *Continental Drift* (III, §1.6) (RS2). He attacked Hess's forthcoming paper (Hess, 1962) (III, §3.15),

Dietz's *Nature* paper (Dietz, 1961a) (III, §4.9), and Menard's articles in *Science* and *Scientific American* (Menard, 1960, 1961) (III, §5.12). He also objected to Bernal's views on mantle convection (Bernal, 1961a) (§1.8).

Heezen began by raising his favorite difficulty, that of Antarctica being surrounded by active ridges (RS2).

If one takes the view that the Mid-Atlantic Ridge represents a rift line from which the new and old hemispheres drifted apart, the pattern looks favourable until one considers the relationship of the Mid-Indian and Easter Island segments of the ridge in relation to this drift. If Africa has moved east relative to the Mid-Atlantic Ridge, it must be running into the Mid-Indian Ridge. If South America has moved west relative to the Mid-Atlantic Ridge, it must be colliding with the Easter Island Ridge; and if one considers the drift of Antarctica relative to the Mid-Atlantic, Mid-Indian and Easter Island portions of the Ridge, one must only conclude that Antarctica has shrunk, for the pattern of the Ridge would indicate that Antarctica must have drifted towards its geographical centre.

(Heezen, 1962: 278–279)

Wilson had already proposed that ridges migrate (§1.9). Heezen either had not read Wilson's answer to this difficulty, thought little of it, or chose to ignore it. As Menard retrospectively declared, it is:

A pity that Heezen and those who followed did not absorb Tuzo Wilson's paper on sea-floor spreading in *Nature* in 1961. Heezen read the adjacent paper by Bernal because he cites it. If he had read and remembered Wilson he would have realized that within the developing theory, no problem existed. The spreading rift merely drifted away from Antarctica. This may not have been easy to accept because, in the model of Dietz and Hess, the giant convection cells in the mantle would have to drift with the rift.

(Menard, 1986: 183–184)

I suspect that Heezen thought that migrating mantle convective cells and "drift with the rift" were unlikely. However, to have said so would have called attention to Earth expansion's lack of an acceptable cause. Heezen then raised three other difficulties with the patterns of convection cells postulated by Hess, Dietz, and Menard (RS2).

When such a pattern of convection cells is envisaged in a single cross-section of one ocean, it is relatively easy to reconstruct the geometry. But when one examines the continuous 35 000 mile-long mid-oceanic ridge on the surface of the earth, it becomes apparent that the convection cells must be in the unlikely form of long tubes of varying diameter which everywhere rise beneath the mid-ocean ridge and sink somewhere beneath the continent. Although convection cells rising beneath the ridges and sinking somewhere beneath the continents supply a plausible explanation of the origin of any one ocean basin, it becomes very difficult to suggest a pattern which adequately explains the entire mid-oceanic ridge.

(Heezen, 1962: 280–281)

Two obstacles pertaining to the location of descending currents remained. Heezen directed the first at Menard, Dietz, and Hess.

Convection cells in the upper mantle have often been proposed as a genetic explanation of mountains, trenches, and geosynclines. Recently Menard, Dietz and Hess have each advanced similar hypotheses of mantle convection in an attempt to explain the formation of the mid-oceanic ridge. The evidence of high heat-flow near the crest of the ridge suggests that the ridge generally lies over the rising limbs of a pair of convection cells. More difficult is the question of where the convection cells are sinking. By what criteria do we identify descending arms of the current? Menard, Dietz and Hess have each suggested that the marginal trench of Central and South America marks a line of compression where the Atlantic and Pacific cells converge and sink. It seems a bit difficult to understand the geometry of these cells, particularly in Mexico and the Caribbean.

(Heezen, 1962: 281)

His final difficulty was aimed especially at Menard.

Still more difficult is the problem of the western descending limb of the east Pacific convection cell. Menard has proposed that a zone of compression exists in the deep-sea floor parallel to the axis of the mid-oceanic ridge. However, no physiographic or structural evidence for this zone has been found. Indeed, if a really significant amount of compression occurred in the oceanic crust comparable in magnitude to the extension of the ridge inferred from its physiography, isostatic considerations would demand a topographic ridge marking the zone of crustal thickening. Also, it would seem likely that if the convection cell is rising under the crest of the ridge with sufficient energy to cause the relatively high seismicity of the mid-oceanic ridge, a line of seismic activity should mark the descending limb of the cell as well. No such line of earthquakes has been observed in the deep Pacific.

(Heezen, 1962: 281–282)

Menard's proposal was that currents descend along the western margin of the East Pacific Rise, approximately 2000 km from its crest, where there should be some seismic, physiographic, and structural evidence of them: Heezen claimed there was none. By contrast Hess and Dietz posited that the currents do not descend until reaching the trenches and island arcs of eastern Asia and Aleutian Islands, thus avoiding Heezen's last difficulty. Menard (1986: 184) retrospectively remarked that Heezen's criticisms "were quite reasonable."

Finally, Heezen (1962: 282) criticized Bernal's idea of deep mantle convection as discussed in his favorable review (1961a) of Dietz's ideas; even if it explained the occurrence of deep foci earthquakes beneath trenches and island arcs, the origin of shallow earthquakes along ridge axes remained a mystery.

1.14 Irving cautions Hess, 1961

Hess gave Irving a preprint of his "History of ocean basins" when Irving passed through Princeton in early summer 1961. Later Irving wrote Hess a letter. He was glad "to see large scale relative continental movement back in vogue," and favored "convection in the mantle." It was hardly "in vogue" and Irving worried that the controversy over continental drift would again get bogged down in fruitless arguments about whether or not mantle convection occurred. Continental paleomagnetism had shifted the fixism/mobilism debate by providing measurements of continental displacements. Even so most, unlike Hess, were unwilling to accept continental drift because of its paleomagnetic support. Irving wanted Earth scientists to concentrate on finding other ways of measuring displacements.[6]

I read your preprint on "The Evolution of Ocean Basins" some time ago and I've been meaning to write for some time. It's nice to see large scale relative continental movement back in vogue ... I would be very grateful if you will send me any further work that you do on this topic. Please don't feel that I am not in favour of convection in the mantle; I am ... I do however harbour a slight fear (which I remember expressing to you in Princeton) and this is that the reality or otherwise of continental drift might come to be associated in the minds of geologists with the question of whether or not there is or is not convection in the mantle. Evolution was quite wrongly dismissed to limbo for 50 years after the demolition of Lamarck's premature theory as to its cause; it was not dismissed on the basis of facts but on the failure of a theory which purported quite erroneously to explain these facts. A similar thing happened in the 1930's when Jeffreys and others dismissed Wegener's premature theory of the mechanism for drift and convinced geologists who had temporarily lost confidence in their skills as well as themselves that the earth obeyed a set of differential equations. The case of the drift theory was then attacked *piecemeal* by geologists anxious for a share of the spoils. The reality or otherwise of continental drift depends on the interpretation of data. My fears are probably quite groundless, however, and my thinking may be too strongly affected by the historical analogue. Please give my regards to your wife.

(Irving, August 7, 1961 letter to Hess)

Looking backwards, Irving was certainly correct about the futility of endless debates about the cause of continental drift. Wegener's proposed forces were severely inadequate, and Jeffreys and others demonstrated their inadequacy, and as I have documented, the mechanism difficulty became the most serious impediment to mobilism's acceptance. Moreover, even though Holmes' theory of mantle convection provided mobilists with a mechanism that Jeffreys acknowledged as not impossible but highly improbable (I, §5.5), it had no significant effect on the debate. Mobilists and almost all fixists ignored Holmes' contribution. Ewing, Heezen, and Wilson, for example, all criticized, even mocked, the idea that continents plow their way through the seafloor, but did not consider Holmes' far more plausible alternative. Hess, who knew about Holmes' work, did not arrive at his own theory of mantle convection from reading Holmes. Moreover, as this chapter has made clear, the reactions by Dietz, Ewing, Hess, Heezen, Menard, and Wilson reveal that the idea of mantle convection as the driving force of continental drift once again was fast becoming widely debated. Furthermore, MacDonald, Jeffreys, and other geophysicists would continue to reject mobilism because they rejected mantle convection. Irving's concern was well-founded.

Irving did not, however, see the fecundity of Hess's seafloor spreading. He would have been a lonely prophet had he done so. Seafloor spreading spawned two then unknown corollaries, the Vine–Matthews hypothesis (1963) and Wilson's idea of ridge-ridge transform faults (1965b) (§2.13, §4.9). Their confirmation, respectively in 1966 and 1967, established the mobility of oceanic crust. The Vine–Matthews hypothesis together with the development of the timescale of magnetic reversals provided accurate measures of displacement of ocean crust. At this point, the vast majority of Earth scientists who were then actively participating in the mobilist/fixist controversy and who had not already accepted mobilism because of the paleomagnetic or geological support, now accepted mobilism. Moreover, many who now accepted seafloor spreading believed that it reflected mantle convection, which was, they thought, the long awaited mechanism for continental drift. Even though "the reality ... of continental drift" became "associated in the minds of geologists with the question of whether there is or is not convection in the mantle" it was still accepted, albeit because it was crustal displacements, not mantle convection, that had been demonstrated.

But Irving was also right about the fruitlessness of the earlier quest for mobilism's mechanism; when Hess proposed seafloor spreading, it still was not a tractable problem, and it remained intractable after the Vine–Matthews hypothesis and the idea of transform faults were confirmed. Seafloor spreading did not, through mantle convection, provide mobilism's mechanism. The two corollaries of seafloor spreading were kinematic. This is not to belittle Hess's tremendous contribution, let me be clear about that. Seafloor spreading was crucial. It did establish that continents do not plow their way through the seafloor but ride passively atop new mantle material which is created at ridge crests and is destroyed beneath the leading edges of drifting continents. But at risk of repeating myself, establishing seafloor spreading did not establish mantle convection as the driving force. Perceptive workers, including Wilson, realized that ridges migrate, because Antarctica entirely and Africa largely are surrounded by spreading ridges, and this was incompatible with them being the surface expression of rising mantle convection currents, which could not "know" that they had to move away from each other at just the right rate to stay beneath the ridges (§4.9). Indeed, this problem was not sorted out fully until over twenty years after confirmation of the Vine–Matthews hypothesis (McKenzie and Bickle, 1988).

One of the great ironies of mobilism's triumph is that even though great efforts were made unsuccessfully over four decades to find an acceptable mechanism, and even though this failure was one of the major reasons why continental drift was overwhelmingly rejected in the 1920s through the 1950s, and it (mobilism) was, by the late 1960s, overwhelmingly accepted as a result of the confirmation of the two key corollaries of seafloor spreading, the mechanism was still unknown. A further irony is that many who accepted seafloor spreading at the time continued, mistakenly, to think that it did provide a mechanism and many may still do; the tight tie between rising convection currents and the formation of ridges was not easily severed.

But as we shall see, the tie was severed when seafloor spreading was transformed into plate tectonics. Plate tectonics is a kinematic theory that says nothing about mechanism; it is the geometrification of geology and blind to mechanism. With the acceptance of plate tectonics and the establishment of the reality of continental drift, the absence of a mechanism question was no longer a stumbling block; it simply became unequivocally identified as an *unsolved problem* instead of a seemingly overwhelming *difficulty*. The reality of continental drift was established and plate tectonics was accepted even though the mechanism had not been revealed.

Notes

1 This synopsis is based primarily on Garland's biography of Wilson (Garland, 1995). I have also drawn from Wilson (1982, 1985, and 1990).
2 Wilson said nothing more about fault motions observed beneath island arcs. Hodgson's analysis that such motions are transcurrent instead of tensional (normal) or compressional (thrust) was difficult to explain for both contractionism and seafloor spreading. Wilson rejected his 1959 contractionist attempt to avoid the difficulty (§1.5), but he did not propose a mobilist explanation. Hess, taking Hodgson's work seriously and disagreeing with Fisher, dismissed Benioff's idea of thrust faults along diagonally descending fault planes that descended underneath trenches and continued underneath continental margins, and posited a downward vertical descent of convection currents (III, §3.16). The difficulty would later disappear once it was shown that Hodgson's analysis was based on unreliable data (§6.9).
3 Apparently Wilson was not entirely receptive to Irving's and K. W. T. Graham and colleagues' views about analyzing paleomagnetic data because Irving argued that changes in longitude of several landmasses can be tracked by paleomagnetism if the landmasses had been fragments of a supercontinent. The fragments will have a record of the earlier common APW path, and one will be able to fix the positions of continents within supercontinent assemblies (II, §5.16).
4 Menard (1986: 174) quoted from this letter in his *The Ocean of Truth*.
5 I want to thank Scripps archivists Carolyn Rainey and Deborah Day for sending me this letter, which is in Menard's papers at Scripps. I thank Scripps for letting me quote from it.
6 Hess kept Irving's letter. I suspect that he was surprised at Irving's response because he knew Irving was a vehement defender of continental drift. Indeed, Irving's work helped convince Hess of the reality of continental drift.

2

The origin of marine magnetic anomalies, 1958–1963

2.1 Introduction

One of the most important events in the mobilist debate, an event crucial to the development of plate tectonics, was the publication of the eponymous Vine–Matthews hypothesis. Frederick Vine proposed the idea while a first-year graduate student under the supervision of Drummond Matthews in the Department of Geodesy and Geophysics at Madingley Rise, the University of Cambridge. Vine's hypothesis was based on new results of a survey over the Carlsberg Ridge, which was then unequalled in detail, and which was carried out in 1962 by the Cambridge marine geophysics group headed by Matthews. His hypothesis explained the origin of marine magnetic anomalies, and provided mobilism with its second difficulty-free solution. Unlike mobilism's first difficulty-free solution (through the discovery of divergent APW paths (III, §2.17), the Vine–Matthews hypothesis (hereafter, V-M), when it was confirmed three years later (§6.8), was quickly recognized as difficulty-free by most active researchers engaged in the mobilist controversy.

The hypothesis is essentially a corollary of seafloor spreading and polarity reversals of the Earth's magnetic field and it provided the first unified solution to two topical problems: the origin of the pronounced magnetic anomaly over the axis of mid-ocean ridges, and the origin of the remarkable striped pattern of magnetic-field highs and lows observed in the northeastern Pacific off the North American coast from California to British Columbia. Given seafloor spreading, newly created seafloor acquires a thermo-remanent magnetization in the direction of the geomagnetic field as it cools through the Curie point. As seafloor spreading proceeds and the field reverses polarity, new seafloor becomes permanently magnetized in a direction opposite to previously created seafloor. With continued seafloor spreading and further field reversals, alternating normal and reversely magnetized strips of crust running roughly parallel to the crest of the ridge are produced.

Lawrence Morley of the GSC independently proposed a hypothesis similar to V-M. They are not twins, only cousins. Contrary to what many, including myself, have mistakenly claimed (Frankel, 1982: 1; 1987: 228; Glen, 1982: 297; Lear, 1967: 48: Oreskes, 1999: 271; 2001: 22) the Morley and Vine–Matthews hypotheses are not identical and one of the purposes of this chapter is to draw a distinction between them. Some of us suggested that V-M be called the Vine–Matthews–Morley hypothesis (Allègre, 1988; Frankel, 1982; Glen, 1982; Wyllie, 1976). We all saw things in a similar way: Morley came up with the same idea independently of Vine and Matthews, but his paper was twice rejected, first by *Nature* and then by the *Journal of Geophysical Research (JGR)*. Wanting to set the record straight, we claimed that he deserved credit for coming up with the same hypothesis as Vine and Matthews.

Morley did invoke reversals of Earth's magnetic field and seafloor spreading and he did claim that newly created seafloor acquires a permanent thermo-remanent magnetization as it cools at the ridge crest. However, unlike Vine and Matthews, who proposed that the magnetization of the crust was due predominately to remanent magnetization, the effect of induction by the present field being small, Morley attributed approximately half of seafloor's magnetization to induction. V-M, with its simplifying proposal of dominant remanent magnetization, leads naturally to a straightforward explanation of the observed 50/50 distribution of positive and negative anomalies of the northeast Pacific as a direct reflection of the even distribution of normal and reversed polarity of the geomagnetic field during the later Cenozoic. Morley's presentations contained no critical new results, his hypothesis was not physically the same as, and was far less complete than, V-M.

There also were at least two other occasions when a hypothesis that was either similar or identical to V-M was proposed. Neither was submitted as a paper. Gough, McElhinny, and Opdyke came up with it while working together in Salisbury (Harare) on an APW path for Africa (II, §5.4). Geoffrey Dickson, a Ph.D. student at Lamont Geological Observatory (hereafter, Lamont), not knowing of Vine and Matthews' paper, came up with the hypothesis or one much like it after their paper was published (Glen, 1982: 297). Dickson had worked at Irving's laboratory at ANU before arriving at Lamont and knew what reversed geomagnetic field anomalies looked like, and a great deal about the magnetization of mafic igneous rocks.

I devote most of this chapter to V-M. I review earlier attempts to explain marine magnetic anomalies, explain Matthews' earlier work on the magnetization of oceanic igneous rocks and his very important, very detailed survey of the Indian Ocean, Carlsbad Ridge, and explain why he was such an excellent supervisor for Vine. I then describe how Vine developed his hypothesis and how Morley developed his. I also comment on the limited early reception of V-M.

2.2 Explaining marine magnetic anomalies

There were many variables to take account of in interpreting marine magnetic anomalies and computer programs were just coming into use to deal with them – the magnetism of the ship, geomagnetic storms, daily variations in field strength, secular variation – and regional trends have to be removed from the raw data before magnetic anomaly or bathymetric contour maps can be produced. Once the maps (or profiles if single lines are being run) are compiled, many questions remain. What are the shapes, sizes, and depths of the causative magnetic bodies? How was magnetization acquired, is it induced or remanent magnetization, or a combination? If it is a combination, what percentage is remanent? If it is remanent, is its polarity normal or reversed? If it is reversed, is it caused by a field reversal or intrinsic self-reversal? There are also questions about the body's origin and composition.

Bathymetric and magnetic anomaly data have to be compared. Are magnetic anomalies correlated with gross topography – with ridge crests, seamounts, trenches and such like? The more detailed the surveys, the more accurately the relation, or absence of relation, between topography and magnetic anomalies can be established. For example, Matthews foresaw the dividends to be reaped from detailed surveys and planned and completed what was probably the most detailed marine magnetic survey carried out at that time. He restricted the buoy-controlled 1962 *Owen* survey of the Carlsberg Ridge in the Indian Ocean to 40 × 50 nautical miles, and tracked the ship at intervals of one nautical mile, five times closer than surveys by Scripps and by the US Coast and Geodetic Survey (USCGS) in the northeastern Pacific (Vine and Matthews, 1963; Matthews, Vine, and Cann, 1965). But comparing magnetic and bathymetric maps alone does not yield the depth of the causative body because it may not extend to the surface. Researchers typically assumed that the sharpness of the magnetic anomaly is a function of the body's depth. They attempted to present interpretations that were consistent with received models of the oceanic crust. Questions about the spatial dimensions of the magnetic body are tied to questions about the intensity of its magnetism. The more magnetic the body, the less thick it needs to be. Although most workers believed that ocean-bottom magnetic bodies were basalt, at first, little was known about the magnetization of oceanic basalts. Most dredged basalts, such as those analyzed by Matthews for his Ph.D., were altered, which made them less magnetic than fresh basalts (Matthews, 1961a, 1961b). Until about 1960, most workers assumed the magnetized bodies were magnetized in the direction of the current geomagnetic field implying that induced magnetization dominated.

Researchers developed interpretative models and compared them with observed anomalies to estimate the shape, depth, and magnetization of the causative magnetic body. Conversely, they assumed these parameters and determined the resultant anomaly. Vacquier constructed standard models (Vacquier and Affleck, 1949; Vacquier *et al.*, 1951) when he worked at Gulf Oil before moving to Scripps; Press and Ewing (1952), at Lamont, did so too. Bullard (1961) developed one called *Attractions (1)*, which he programmed for Cambridge's EDSAC 2 (electronic delay storage automatic calculator) computer. These early interpretative models could not deal separately with magnetization caused by remanence and by susceptibility (induced). Researchers assumed that the direction of remanent magnetization is that of the present-day field; they determined k_{ap}, the apparent susceptibility, from the equation $k_{ap} = (I + kH)(H^{-1})$, where I is the intensity of remanent magnetization, k is the true susceptibility, and H is the intensity of the field. If the remanence of the magnetized body is directed along the present field, the effects of induced and remanent magnetization add enhancing apparent susceptibility; if it is reversely magnetized, remanence diminishes its apparent susceptibility. Looking ahead, before Vine began working on the problem, only two groups invoked reversely magnetized magnetic bodies as the cause of anomalies at sea (Laughton, Hill, and Allan, 1960; Girdler and Peter, 1960). Vine learned of such an interpretative model in 1962 when he undertook an exhaustive literature search; K. Kunaratnam, a Ph.D. student of Mason's at Imperial College, developed one and allowed Vine to use it, and Vine adapted it for use on EDSAC 2. Kunaratnam had also developed two- and three-dimensional programs. The three-dimensional program was just what Vine needed to analyze the magnetic anomalies from Matthews' detailed survey over the Carlsberg Ridge; it allowed him to experiment with varying strengths of remanent and induced magnetization, and direction of remanent magnetization.

2.3 Interpretation of northeast Pacific marine magnetic anomalies prior to Vine's proposal, 1958–1962

Before Vine and Matthews began working on the problem, the best known marine magnetic anomalies were in the northeastern Pacific. Matching magnetic anomalies across the Mendocino, Pioneer, and Murray fracture zones, Mason, Raff, Vacquier, and Menard argued that seafloor moves as rigid blocks (III, §5.9). They speculated about the source of the anomalies. In 1958, Mason, with help from Steenland and Nettleton, proposed that the source of positive anomalies was within, or mainly within, the 2 km thick, supposedly volcanic Layer 2 of oceanic crust. This required a very high susceptibility k of 0.0109 c.g.s. units, which was improbably high; Steenland and Nettleton concluded, "It is our opinion that such a high value must be indicative of appreciable remanent magnetization of the slabs obtained during cooling in the presence of Earth's magnetic field" (Mason, 1958: 323). Induced magnetization alone in a 2 km thick magnetic layer would be insufficient. He could lower the susceptibility to a more reasonable 0.005 c.g.s. units by tripling the thickness of the magnetic body, extending it to the Moho discontinuity, approximately 11 km below the ocean bottom. This was unreasonable; from the seismic evidence, Layer 2, the supposed volcanic layer and the most likely candidate, is only 2 km thick. This led him to claim that the total magnetization (apparent susceptibility) is a combination of induced and remanent magnetization.

This strongly suggests that the rocks involved here are basic igneous rocks, which are characterized not only by a high susceptibility but also by a high intensity of remanent magnetization, commonly several times greater than the intensity of induced magnetization in the Earth's field. Thus in the present case, where the Earth's field is $50\,000\gamma$, an apparent susceptibility of 0.015 could be accounted for by a true susceptibility of 0.005 together with an intensity of remanent magnetization of 0.005 G in the direction of the Earth's field, both acceptable values for a basic igneous rock.

(Mason, 1958: 327)

Having accepted the idea that the magnetic bodies responsible for the positive anomalies had a substantial component of remanent magnetization, he thought that it would be directed along the Earth's field at the time the bodies were emplaced. Paleomagnetic work on land had shown that the orientation and polarity of Earth's magnetic field had changed and he wondered about using the direction of the remanent component to date Layer 2 beneath the positive anomalies.

The interesting question arises as to whether, since the direction of remanent magnetization will correspond to the direction of the Earth's field at the time when the rocks were formed, in all probability different from the present direction, anything can be learned about the direction of magnetization that may throw light on the age of the formations … Palaeomagnetic measurements (see Runcorn 1956 for a summary of relevant information) suggest that since Jurassic times the direction of the Earth's magnetic field in North America has centred around that of a dipole field, with normal or reversed polarity, but that during earlier periods it was quite different.

(Mason, 1958: 327; Runcorn reference is same as mine)

Mason then determined what effect changes in the direction of Earth's magnetic field might have. Keeping the susceptibility at 0.005, the intensity of remanent magnetization at 0.005 G, and the dip (inclination) unchanged, he found that the effect of varying the horizontal direction (declination) of remanent magnetization was negligible. Still, if I am correct, referring to positive anomalies he remarked:

A more thorough investigation is beyond the scope of this paper. However, it seems clear that, since any remanent magnetization is required to augment rather than to oppose the present field, the vertical component must be substantially downwards and this makes any appreciable degree of reversed magnetization unlikely if the magnetic poles were anywhere near their present positions.

(Mason, 1958: 328)

Turning to the negative anomalies (the magnetic lows), Mason imagined that the seafloor beneath them was not magnetized or very weakly magnetized. He apparently did not entertain the possibility that the negative magnetic anomalies were caused by magnetic slabs of reversed polarity that had formed when the geomagnetic field was reversed. If he had, he would surely have attributed the steep gradients between the high and low magnetic anomalies to reversals of the geomagnetic field. Although he believed that Earth's magnetic field reversed repeatedly (Mason, 2001: 41), he never picked up the stick from that end; he never began with the assumption that Earth's magnetic field has undergone repeated reversal and that these reversals would have been imprinted in volcanic rocks of Layer 2 as they cooled, and then used it to explain the zebra pattern of positive and negative magnetic anomalies in the northeastern Pacific. It seems unimaginable that he would not have done so had he had the benefit (as Vine had) of the geological idea of seafloor spreading, and this points to the crucial importance of Hess's proposal by means of which, as we shall see, Vine linked geomagnetic reversals and volcanic processes at the ridge.

Returning to magnetic slabs underlying the positive anomalies, Mason (1958: 328) claimed that they probably are at least 1 km thick and comprise basic igneous rocks that are more highly magnetic than adjacent crust. He also noted that there was no correlation between anomalies and topography, and this became a significant difficulty for all explanations that depended on supposing that the seafloor is made up of stripes of material of alternating high and low intensity of magnetization.

Despite its thickness it shows little or no correlation with topography, as far as can be judged from the best available topographic maps, suggesting that it is either buried in consolidated sediments of relatively high seismic velocity or emplaced in less magnetic igneous rocks.

(Mason, 1958: 328)

Mason went on to propose that these highly magnetized bodies were huge lava flows that had filled pre-existing troughs. He compared them to the Columbia River basalts.

The most obvious and acceptable geological explanation is that the excess material represents lava flows which have spread out over the floor of the ocean possibly filling pre-existing troughs, or depressions produced by the weight of the accumulating flows. This picture is consistent with the [shape of the derived slabs] which thin out at the edges in a manner characteristic of lava flows. Within the limits of the survey the major anomalies cover an area of about 50 000 sq km. In extent therefore, and in thickness, the flows would rank with the largest continental lava fields, such as the Columbia River basalts which cover an area of 120 000 sq km and in places reach thicknesses approaching 2 km.

(Mason, 1958: 328; my bracketed addition)

He did not say how the troughs had formed and had no explanation why they were so evenly aligned.

In 1961, Mason and Raff considered the origin of the magnetic anomalies after extending their survey and contouring the magnetic anomalies off the west coast of North America from 32° N to 42° N latitude (see III, Figure 5.5). Their paper was received on December 16, 1960, before Hess's unpublished version of seafloor spreading was distributed and before Dietz's paper appeared. In it they suggested three possibilities, essentially Mason's two previous suggestions (repeated as (1) and (2) below) and a new one.

The geological possibilities fall into three categories: (1) isolated bodies of magnetically anomalous material, for example basic lava flows, within the second layer; (2) elevated folds or fault blocks of the main crustal layer; and (3) zones of intrusion of highly magnetic material from the mantle, extending from top to bottom of the crust. Category (1) would fit in well with current views about the somewhat enigmatic second layer. Such lavas would constitute 10 per cent of the local crustal volume. For (2), a parallel might be drawn with the fault-block topography of the neighboring Basin and Range province. For (3), the anomalies might arise from magnetic contrast within both the second and the main crustal layers.

(Raff and Mason, 1961: 1263)

They admitted that their previous explanations were not satisfactory and faced difficulties (RS2).

There is as yet no satisfactory explanation of what sort of material bodies or physical configurations exist to give the very long magnetic anomalies. It has been suggested that they are troughs and ridges buried under sediment, platelike lava flows that filled older troughs, unique groupings of vertical dikes, high iron-content derivatives of weathering that filled older troughs, or a thermal pattern that has highs and lows about the Curie point of ocean rocks. Two of the principal difficulties in imagining a body that could cause these north-south features are that the measured magnetic anomalies and computations about them require that the upper surface of the body be within 1 km of the sedimentary layer and that the body have what is considered an unusually strong magnetic moment. Such features were not revealed on the basis of topography nor have they been detected by seismic methods (R.W. Raitt, personal communication).

(Raff and Mason, 1961: 1269)

In an attempt to explain the lack of seismic detection (RS2), they (1961: 1269) appealed to unpublished[1] measurement of the magnetization of rocks collected in the Scripps dredge hauls indicating that igneous oceanic rocks have a remanent magnetization and susceptibility several times greater than the median value of continental igneous rocks, and raised the possibility that the causative magnetic body may be too thin to be detected seismically.

In an attempt to remove a second difficulty (RS2), they claimed that there was at least a weak correlation between the magnetic anomalies and topography in the surveyed region north of the Mendocino fracture zone.

If there originally had been a trough and ridge system, these troughs and ridges should show at least slightly in the topography of the heavily sedimented ocean floor. If the magnetic bodies are the original ridges, the magnetic highs should correspond with the slight topographic ridges; but if the magnetic bodies are accumulated material in the valleys, then the magnetic highs should be over the slight topographic valleys. A statistical correlation analysis made of the magnetic and topographic relief for some of this area showed that the topographic ridges and magnetic highs correspond in about 65 per cent of the cases as against 35 per cent of negative correspondence. This is certainly not conclusive, but does suggest that the magnetic bodies are the original igneous ridges or elevated areas that survive in a modified, perhaps flattened, form.

(Raff and Mason, 1961: 1269)

Raff and Mason were still working on the assumption that the magnetic lows were underlain by weakly magnetized rocks. The idea that reversely and strongly magnetized rocks could cause magnetic lows apparently had not occurred to them, although they were aware of these puzzling results from seamounts.

In the general region ... there are many seamounts, some of which cause strong magnetic anomalies, while others do not. It is not at all clear why volcanic seamounts in the same area should give magnetic anomalies of such different magnitudes.

(Raff and Mason, 1961: 1269)

Despite this tiny hint, they again did not pick up the stick from the other end, beginning with the idea of geomagnetic field reversals and working out its

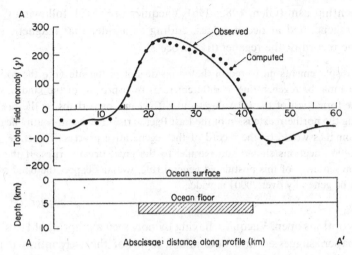

Figure 2.1 Vacquier's Figure 3 (1962: 138) showing his interpretation of the magnetic profile A–A′ north of the Mendocino Fracture Zone. The solid line represents the observed profile with its pronounced positive anomaly. The dots represent the anomaly produced by an infinite north–south block of susceptibility 0.0109, computed for a field of 50 000λ, dip 59°, declination 17°. The hatched rectangular block represents the magnetized slab whose computed profile models well the observed profile.

implications for the origin of marine magnetic anomalies. They continued to believe that weakly magnetized bodies cause the anomaly lows even after Laughton, Hill, and Allan (1960) had interpreted the magnetic anomalies over a seamount in the Atlantic as caused by a reversely magnetized body (§2.4, §2.9)

Vacquier directed his attention primarily to measuring the huge horizontal displacements of adjacent blocks of seafloor by matching magnetic anomaly patterns (III, §1.6, §5.9); however, he also speculated about the origin of the anomalies in Runcorn's 1962 edited collection *Continental Drift*. He began with what he described as the customary assumption that the causative bodies were magnetized in the direction of the current field. Using his interpretative model, he showed that a magnetic high in a typical profile just north of the Murray Fracture Zone could be accurately explained in terms of a magnetic slab of 1.5 km basalt on the ocean floor (Figure 2.1). He speculated the contrast between the magnetic highs and lows arose from thermal and petrological differences.

Very broad magnetic anomalies may be caused by both local differences in geothermal gradient and by differences in the Curie temperatures due to minor mineralogical impurities such as titanium.

(*Vacquier, 1962: 137*)

Favorably disposed to mobilism because of the huge displacements of adjacent blocks of seafloor along the great northeast Pacific fracture zones, and influenced by Carey's enthusiasm (Glen, 1982: 296), Vacquier proposed, following Carey, that seafloor is regenerated at ocean ridges, adding as a rider that magnetic anomalies would be the record of the regenerative process.

Because the displacements measured in the ocean do not propagate onto the continent, the oceanic crust must be regenerating at different rates by a process of the kind postulated by Carey for the formation of the rift valleys and rift oceans, or perhaps by the rise of mantle material along the northern extension of the East Pacific rise. The north-south lineation of the magnetic anomalies would be the record of the regenerative process. Therefore, a band of strong magnetic lineations should run parallel to the great oceanic rises. If the north-east Pacific is representative of this postulated general rule, we may expect the band of magnetic anomalies to be generally over 2000 km wide.

(Vacquier, 1962: 144–145)

Keeping his options open, Vacquier, having by now seen a preprint of Hess's seafloor spreading paper, suggested in a note added in proof that serpentinized peridotite causes the magnetic anomalies.

The magnetic anomalies occurring in bands parallel to the oceanic rises can very well be caused by the hydration of the olivine of the peridotitic mantle [14] into serpentine and magnetite by juvenal water.

(Vacquier, 1962: 144; reference [14] refers to Hess's forthcoming paper)

Just how seriously Vacquier took this idea at the time is not clear. When asked seventeen years after proposing it, Vacquier, as reported by Glen, did not remember doing so. Nor did he appeal to reversed magnetization of newly created seafloor or reversals of the Earth's magnetic field. His idea is not a foreshadowing of V-M, which he rejected two years later at the Royal Society's (London) symposium on continental drift.

2.4 Interpretation of magnetic anomalies over ridges and seamounts prior to Vine's proposal, 1953–1962

During the 1950s workers from Lamont and Cambridge crossed the Mid-Atlantic Ridge towing magnetometers. Heezen, at Ewing's request, had made the first crossing while chief scientist of *Atlantis* cruise 153 (November, 1948) (Heezen, Ewing, and Miller, 1953). Heezen recalled:

Now, magnetics – the magnetic story of the world is because of him [Ewing] too. At his insistence I, working as his assistant, went down to the Geological Survey and got the first magnetometer and towed it first behind a ship in 1948. It wasn't my idea to do it. I hardly knew what magnetics was. I did it for him. It was the first successful profile and I goofed and I missed the magnetic positive over the rift valley, the first time I took it – correct it out of the record. But four years later I noticed it on the record north, further north. This positive anomaly that's over the rift valley ...

(Lear, BCH 08B: 11; my bracketed addition; see III, Chapter 6, note 8 for Lear)

Heezen and his co-authors noted, "while the anomaly curve is generally the same across the Atlantic there is a slight but noticeable difference over the centre of the Mid-Atlantic Ridge" (Heezen, Ewing and Miller, 1953: 29). They (1953: 31) did, however, notice several large positive and negative anomalies associated with the Cape Verde Islands, and suggested, applying Press and Ewing's interpretative model, that the anomalies are the result of changes in the magnetic properties of the crust. They made no mention of remanence, and talked only of susceptibility (induced magnetization).

Three years later, Ewing, Heezen, and Hirshman (1957) found a substantial positive anomaly which they claimed coincides with the central rift valley and which "has proven to be an excellent indicator of the location of the crest of the Mid-Atlantic Ridge." They suggested in their brief abstract that it can be accounted for in terms of the variation of magnetic susceptibility within the crust. Two years later, Heezen, Tharp, and Ewing (1959: 100) remained silent about its cause but noted that it was surrounded by negative anomalies. At the first International Oceanographic Congress in late 1959, Ewing, Hirshman, and Heezen (1959) discussed the strong, positive anomaly observed in many of their profiles of the Mid-Atlantic Ridge. They (1959: 24) again attributed the anomaly to a subsurface body of high magnetic susceptibility and did not mention remanence. They also claimed that they had observed positive anomalies in the Gulf of Aden and Red Sea.

In the early 1960s, Lamont workers, Manik Talwani, Heezen, and Worzel (1961) discussed the cause of the central magnetic anomaly. Combining gravity, heat flow, and magnetic analyses, they proposed that a low-density mixture of crust and mantle material resided beneath ridge crests, and that if its temperature rose above the Curie point and then cooled it would become strongly magnetized. The degree to which this mixture became magnetized depended on the depth of the Curie point isotherm.

Doubt was expressed about the rift valley of the Mid-Atlantic Ridge being always characterized by a pronounced positive anomaly. In 1956, Hill at Cambridge surveyed an area of the Mid-Atlantic Ridge at approximately 47° N. Hill (1960: 201) found no noticeable magnetic anomaly above the rift valley, adding that his results conflicted with Heezen's findings.

A notable advance was made by Girdler's work on the magnetic anomaly in the middle of the Red Sea. Ronald Girdler (1930–2001) earned a B.Sc. degree at Reading University; he read mathematics, chemistry, geology, and structural geology. Awarded a Dutch Shell Studentship in 1955, he began working on his Ph.D. in the Department of Geodesy and Geophysics at Cambridge. He had to change research topics because of the 1956 Suez crisis. "I started in oceanography interpreting the gravity over the Red Sea. I had to change to paleomagnetism because of the Suez crisis! The war meant no ships there" (Girdler, 1985 letter to author). He wrote (1958a) a two-part Ph.D. dissertation, "Interpretation of gravity anomalies in the Red Sea area & the measurement of anisotropy of rocks and a study of the magnetic properties of some Jurassic rocks."

Given his undergraduate work in geology and his experience in two areas of geophysics, Girdler was adept at offering geological interpretations of geophysical data. As Hess later said in recommending Girdler for an academic position at the University of Bristol:

I would say that Dr. Girdler's chief asset is his ability to apply the reasoning of a geologist to the data of geophysics, a very rare attribute today. This was perhaps best illustrated by his paper on the Red Sea written several years ago. It was a very intelligent and imaginative analysis. I don't happen to agree with all of his conclusions but that does not lessen my respect for him.

(Hess, March 6, 1967 letter to H.C. Butterfield)

Girdler accepted mobilism after reading Blackett's 1956 *Lectures on Rock Magnetism* (Girdler, 1985 letter to author) (II, §3.15), and Hospers' work convinced him of reversals of the geomagnetic field. His own work on Jurassic aged rocks from the West of England also led him (1959a: 541) to conclude that there may have been a reversal of the Earth's magnetic field in the Lower Jurassic period. Furthermore, following Hospers (1954a), he estimated that the reversal lasted for approximately 6×10^5 years, which, he added, compares favorably with Hospers' estimates for reversals in the Late Cenozoic (1959a: 541).

Girdler (1958b) presented a geological interpretation of the pronounced negative gravity anomalies associated with the rift valley running the length of the Red Sea. After writing most of the paper, he learned of an aeromagnetic survey over the Gulf of Aden, which also revealed a pronounced negative magnetic anomaly over the rift valley.[2] Using Press and Ewing's interpretative model, he argued that the magnetic anomaly in both the Red Sea and the Gulf of Aden was caused by intrusion of a basaltic material through tension cracks within the rift valley, material that was magnetized in the direction of the present Earth's field.

At Ewing's invitation, Girdler spent much of 1959 at Lamont. There he examined the magnetic profile obtained on the eighteenth cruise of RV *Vema* in 1958 across a ridge segment in the Gulf of Aden (Figure 2.2). He saw the expected pronounced negative anomaly over the ridge's rift valley, but there was also a strong positive anomaly to the north of the rift valley. He and G. Peter, his co-author, described the anomaly (Figure 2.2).

Over the rift valley there is a strong negative anomaly (1) which has an amplitude of about 1000 gammas in this locality. The profile shows a sharp positive anomaly (2) of 900 gammas to the north of the valley and it is likely that this is associated with the topography feature (x). On either side of the positive anomaly are smaller negative anomalies (3) and (4) of amplitude 200 gammas.

(Girdler and Peter, 1960: 476–477)

They (1960: 478) emphasized that the strong, positive flanking anomaly ((2) of Figure 2. 2) was surprising, especially because of the small, positive inclination and northern declination of the geomagnetic field at this magnetic latitude. Using an

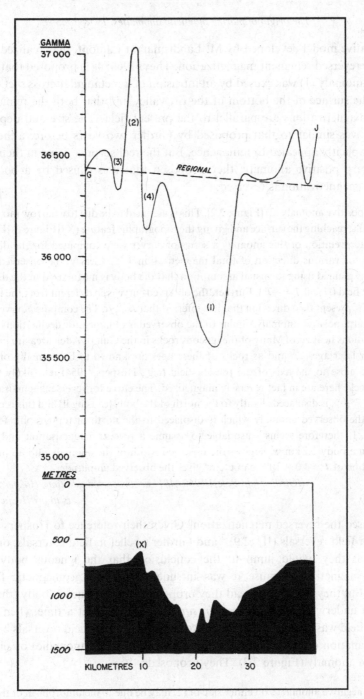

Figure 2.2 Girdler and Peter's Figure 2 (1960: 477) showing the magnetic profile GJ and bathymetry beneath. (1) is the pronounced negative anomaly over the rift valley, (2) is the sharp positive anomaly, and (3) and (4) are small negative anomalies. (X) is the topographic feature purportedly responsible for (2). North is to the viewer's left.

interpretative model developed by M. Landisman at Lamont, they realized the need to invoke reversed remanent magnetization. They (1960: 481) proposed that the large negative anomaly (1) was caused by an intrusion of a rectangular cross section which reaches the surface of the bottom of the rift valley and that both the remanent and induced magnetizations are parallel to the present field. The size and depth of the intrusion was similar to that proposed by Girdler two years before, although this time he explicitly appealed to remanence. But the real surprise came in their analysis of the sharp positive anomaly; they proposed that it is caused by a body whose remanent magnetization is reversed.

The sharp positive anomaly (2) [Figure 2.2]. This is assumed to be due to a narrow intrusion with vertical walls, reaching the surface and giving the topographic feature (X) (Figure 2) [Figure 2.2]. For the interpretation of this anomaly a series of curves were computed for the direction of profile GJ for various direction of total magnetisation J ... Clearly, the observed anomaly cannot be explained using the usual assumption that the body is magnetized in the direction of the present field ($D = 0, I = -7°$). Further, the observed curve is so different from the theoretical curve for the present field direction that it is inferred that $Jr > Ji$. The computed curves ... show that the sharp, positive anomaly similar to the observed is obtained for declinations near 180° and inclinations near zero. Many of the igneous rocks in the Gulf of Aden area are probably of Tertiary or later ages ... and as rocks of these ages are known to be normally or reversely magnetized close to the axis of the present field (e.g., Hospers, 1954) it is likely that some intrusive rocks here are in fact reversely magnetized. The curve for reverse magnetisation ($D = 180°, I = -7°$) ... is displaced slightly to the north of the body [causing it] and this is exactly the case with the observed anomaly which is displaced to the north of topographic feature (X) (Figure 2). It therefore seems reasonable to assume a reverse magnetisation and with this assumption a body 1.2 km wide gives the observed width of the anomaly and an intensity of magnetization of $J = 4.0 \times 10^{-3}$ e.m.u/cm^3 gives the observed amplitude.

(Girdler and Peter, 1960: 481; my bracketed addition; Hospers reference
is to my Hospers, 1954a)

What caused the reversed magnetization? Given their reference to Hospers who had argued for field reversals (II, §2.9),[3] and Girdler's belief in field reversals, one would expect that they would jump to the conclusion that the igneous body became reversely magnetized because it was intruded while the geomagnetic field was reversed. But they did not. Instead they proposed that the igneous body (the narrow intrusion) underwent a self-reversal *in situ* after intrusion at a time when the geomagnetic field was normal. Why? They were forced to reject field reversals because of their explanation of features (3) and (4), the small negative anomalies on either side of positive anomaly (Figure 2.3). They proposed:

The small, negative anomalies on either side of (2) may be due to metamorphism of the country rock by the narrow intrusion. As these baked rocks are normally magnetized, it is unlikely that the Earth's magnetic field was reversed at the time of cooling of the volcanic rocks and the data favour a chemical mechanism for the reverse magnetization.

(Girdler and Peter, 1960: 482)

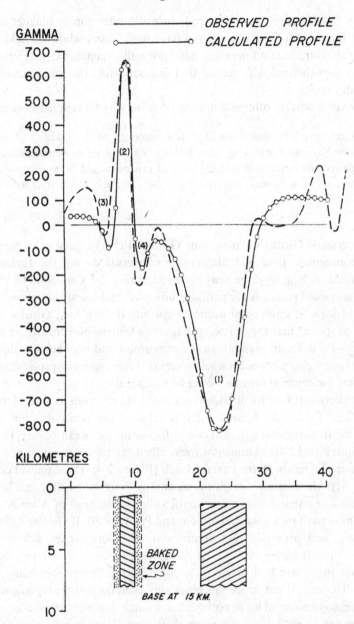

Figure 2.3 Girdler and Peter's Figure 5 (1960: 482) showing their interpretation of magnetic profile GJ with the hypothesized intrusions and baked country rock beneath. (1) is the pronounced negative anomaly over the rift valley, (2) is the sharp positive anomaly, and (3) and (4) are small negative anomalies. The remanent magnetization of igneous intrusion beneath (1) is normal. The narrow igneous intrusion beneath (2) sandwiched by the baked zones of country rock is reversely magnetized having undergone by chemical means a self-reversal after intrusion; when intruded it became normally magnetized as did the country rock it baked. North is to the viewer's left.

Because the baked country rock surrounding the intrusion is magnetized in the direction of the current field, it could not have been baked while the field had been reversed. So the intruded rock was originally normally magnetized, and subsequently underwent a non-thermal self-reversal that did not affect the magnetization of the baked country rock.

Girdler retrospectively offered a different reason why he rejected field reversals.

I was keen on reversals and considered self reversals to be oddities. The enclosed photocopy from Nature illustrates my 1959 thinking. You will see we were searching for some kind of regularity and never suspected the reversal process would be so irregular. This was a remarkable asset as we could never have got the correlation with time as Vine did so brilliantly.

(Girdler, April 19, 1985 letter to author)

I do not understand Girdler's explanation. Girdler (1959a) argued in his *Nature* paper that the geomagnetic field had undergone a reversal during the Jurassic which probably lasted as long as those that occurred during the Cenozoic. Girdler could have still maintained reversals are periodic, however, and simply maintained that the igneous intrusions he envisioned occurred episodically. In fact, Girdler and Peter (1960: 482) proposed that the normally magnetized intrusion beneath the rift valley "is probably of a different mineralogical composition and may be of a different age from the intrusion that underwent a self-reversal." Vine, when he proposed V-M, did not think that the reversal process would be so irregular.

I think it likely that Girdler ultimately rejected field reversals as an explanation for anomaly (2) of Figure 2.3 because he was reluctant to reexamine his own 1958 hypothesis for the origin of rift valleys and formation of ocean basins. He claimed that the country rock, the continental rock, albeit stretched, thinned, and cracked, remains within the newly formed axial trough (Figure 2.4). He remained committed to his 1958 hypothesis even though his co-author Peter and colleagues had shown through seismic refraction work that the rift valley in the Gulf of Aden is devoid of country (continental) rock (Nafe, Hennion, and Peter, 1959). If Girdler had amended his hypothesis and proposed that continental rock was carried sideways to the periphery as the rift valley evolved, he could have invoked field reversals because he would not then have had to worry about baked sediments becoming reversely magnetized. He seemed not to see this possibility even though the seismic work gave him ample reason to amend his hypothesis accordingly before Peter and he submitted their 1960 paper arguing for a self-reversal. Perhaps I am being unfair to Girdler; perhaps there was no way to remove continental rock from young rift valleys. But he did have another way, Earth expansion (Girdler, 1960), which provided a way of removing sialic rock from a developing rift valley. Two years later, after Girdler took a position in the Geology Department at the University of Durham and was befriended by Runcorn, he switched to Runcorn's version of mantle convection as the cause of rift valleys and continental drift (III, §1.5) and the means whereby

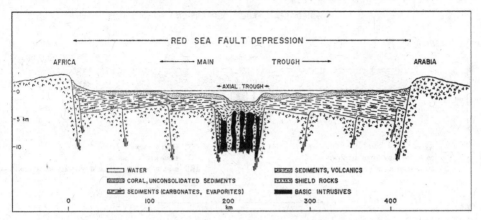

Figure 2.4 Drake and Girdler's Figure 14 (1964: 489) and Girdler's Figure 8 (1967: 1267) representing the Red Sea.

continental crust could be removed from newly formed rift valleys. Girdler did not amend his hypothesis and appeal to field reversals

Even after V-M appeared, Girdler did not amend his hypothesis and appeal to field reversals. Drake and Girdler (1964) and Girdler (1967) used Figure 2.4 to illustrate the structure of the Red Sea, which they thought similar to the rift valley associated with ridge segments in the Gulf of Aden. Girdler later told Vine that Peter stopped him from appealing to field reversals.

Ron Girdler ... published a paper with George Peter ... I think that is the first paper ... they ... invoked reversed magnetization, and they attributed it to baked contacts and self-reversals ... And he insists, and I sympathize with him, I'm sure it could be true, he wanted to go further, he wanted to invoke field reversals, but George Peter held him back. Now that could be true.

(Vine, August 1979 interview with author)

Perhaps Girdler sincerely came to believe that George Peter held him back. Did Drake also hold him back? I think Girdler held himself back.[4]

Looking ahead, Girdler did make the needed changes four years later when he incorporated V-M, and redrew his and Drake's original figure (Figure 2.5).

This was also, I believe, only the second time that Girdler referred in print to Vine's interpretation. In a short review of Hill's organized 1966 edited volume, *A Discussion Concerning the Floor of the Northwest Indian Ocean,* he wrote:

J. R. Cann and F. J. Vine give details of the petrology of rocks dredged from an area on the crest of the Carlsberg Ridge and also describe a magnetic survey of the same area. They support the spreading sea floor hypothesis, finding the magnetic anomalies parallel to the trend of the ridge regardless of bathymetry. The rocks seem to become progressively older on either side of the crest and are normally and reversely magnetized depending on their age.

(Girdler, 1967: 225)

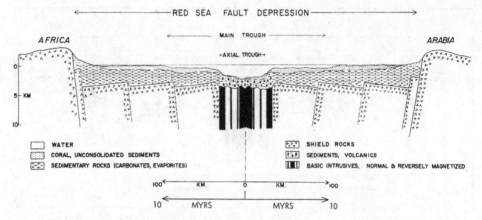

Figure 2.5 Girdler's Figure 2 (1968: 1103) summarizing the seismic results. His caption reads: "Structural section of the southern part of the Red Sea modified after Drake and Girdler (1964) and incorporating Vine's interpretation of the axial magnetic anomalies. The igneous rocks have been intruded progressively over the past 3 to 4 million years."

Despite the fact that Girdler and Peter did not appeal to field reversals, they still were one of only two groups (the second now to be described) to appeal to reversed magnetization to explain marine magnetic anomalies before Vine began working on the problem.

In summer 1956, A. S. Laughton participated in the first cruise by the Cambridge group using Hill's proton precession magnetometer. Results from the cruise led Laughton, Hill, and Allan (1960) to argue that the magnetic anomalies were associated with a seamount situated north of Madeira in the east Atlantic, composed of reversely magnetized rock formed while the geomagnetic field was reversed. Laughton described the cruise, and explained his own role in this way.

The 1956 cruise of RRS *Discovery II* was led by Maurice Hill, but we also had on board Sir Edward Bullard with his heat probe, Russ Raitt and Ronald Mason. It was the first cruise by Cambridge using a self built proton magnetometer and Ron came along to compare the results with his fluxgate magnetometer. They agreed!! But the objective of the cruise was to examine and map part of the mid Atlantic Ridge and some seamounts that lay away from it. My personal role was involved in the mapping and in photographing the seabed using a camera I had just designed and built. So a large part of my contribution to the 1960 paper was in the charts and the photos. Maurice took a great interest in the magnetics having helped build the proton magnetometer.

(Laughton, March 28, 2009 email to author)

Laughton recorded the discovery of the seamount that Allan, Hill, and he later argued was reversely magnetized. He also explained how Allan got involved.

Reading from my diary of the cruise, on the 16th August we went to investigate a large magnetic anomaly that has, so far, no apparent related topography. Overnight we surveyed and found no seamount that we could associate with the anomaly. For three days we studied a nearby seamount and found what appeared to be an anomaly that could possibly be associated with reversed magnetisation.

All the magnetic data from this cruise was given to a new student, Tom Allan, who had just joined the department, to work up under the supervision of Maurice Hill. Hence Tom's name on the paper.

(Laughton, March 28, 2009 email to author)

I first introduce Laughton and Allan before describing their appeal to reversed magnetization. Laughton, later Sir Anthony, was an undergraduate at King's College, Cambridge. He read natural sciences, doing physics, chemistry, and maths (whole subjects) and mineralogy (half subject) in his first two years. In his final year he read physics and, as he noted, did not read geology (Laughton, March 28, 2009 email to author). He originally planned to work in nuclear physics. Later, Maurice Hill suggested that he work in marine geology. Laughton agreed, and he started graduate work in the Department of Geodesy and Geophysics at Cambridge in 1951.

My tutor was Edward Shire, who encouraged me to do post graduate work at Cambridge and obtained for me a place in the Cavendish laboratory working in nuclear physics under Prof Otto Frisch. After a few months of this I had blood tests to check on my liability to suffer from radiation. It turned out that I had too few white corpuscles, and the doctors advised me not to start a career in nuclear physics. I was friendly with Maurice Hill in College, who suggested, that as I had been in the Navy and had read physics, I should join him in the Department of Geodesy and Geophysics and research with him. This was a turning point in my career. Maurice was hereafter to have a considerable influence on me.

(Laughton, March 28, 2009 email to author)

Laughton obtained his Ph.D. in 1956, working under Hill on compaction of ocean sediments. He simulated the compaction of ocean sediments in the laboratory and simultaneously measured the velocity of sound through them (Laughton, 1954). Laughton left Cambridge in 1954, spent a year at Lamont, and then took a position as a research scientist at the National Institute of Oceanography (NIO) in Wormley, England. While at NIO, he continued to work closely with Hill and the marine geophysics group at Cambridge. In 1955 he developed a deep-sea underwater camera, and used it on many of the cruises he took with NIO and Cambridge scientists.

Without a background in geology, Laughton had no prejudices for or against continental drift. He got to know fellow graduate students Hospers, Irving, Creer, and Hide. Because of what he heard about their work, he thought continental drift had occurred. He also thought Hospers' Icelandic work established that the Earth's magnetic field reverses its polarity.

Not having read geology I had no views on continental drift while an undergraduate ... I knew Jan Hospers and heard about his work on magnetic reversals in the Department. His work in Iceland was very convincing ... I also knew of Ted and Ken's work with Runcorn although

I was not directly involved with it ... Runcorn never tried to recruit me to his group. My awareness of what they were doing came from department seminars and coffee table talk. Reversals and polar wandering were widely discussed and I believed in what I heard.

(Laughton, March 28, 2009 email to author)

Thus Laughton, already favoring field reversals, was quite comfortable proposing that the seamount had formed while the geomagnetic field had been reversed, once he and his co-workers realized that the magnetic anomaly associated with the seamount was best explained by invoking reversed magnetization.

Tom Allan also had a background in physics.[5] Allan grew up in Perth, Scotland. His parents could not afford to send him to Perth Academy, the best school in Perth, but Allan was admitted as a bursary boy in 1943, and remained at the Academy until 1949 when he was offered a place at the University of St. Andrews to study natural philosophy (physics), maths, and chemistry. After five years at St. Andrews, he earned an honors degree in physics, and in fall 1955 he began working on a two-year M.Sc. degree in applied geophysics at the Royal School of Mines, Imperial College, London, where he attended lectures by Bruckshaw and Mason. The first year was devoted entirely to class work; the second year, to applying what one had learned during the first year. Allan enjoyed class work, and decided to pursue a Ph.D. in geophysics at Cambridge. Receiving a Shell Studentship to pursue his Ph.D. in the Department of Geodesy and Geophysics at Cambridge, he joined Maurice Hill's marine magnetic group in October 1956 after receiving his Diploma of Imperial College (DIC) in applied geophysics.

He had spent the summer assisting in a detailed magnetic survey in northern Norway, but this is not why he ended up working in marine magnetics. He happened to arrive at Madingley Rise when the RRS *Discovery II* team returned with the magnetic data obtained with Hill's new precession proton magnetometer. His assignment was to reduce the data.

When, thanks to a Shell Studentship, I cycled up the drive to Madingley Rise in October 1956. RRS *Discovery II* had recently returned from a geophysical cruise to the Atlantic. One of the instruments Hill's group had tried out for the first time was an ingenious new device to measure variation in the Earth's magnetic field. The proton magnetometer consisted of little more than a cylindrical bottle of water around which was wound a coil of copper wire ... Soon I was handed a pile of blue-covered log books from the recently completed Atlantic Cruise where the frequency of or count from the magnetometer had been written in by the watchkeeper at 5-minute intervals. I was given the job of converting these counts to magnetic field measurements via a simple conversion table. These then had to be corrected for the normal diurnal variations in the Earth's field and corrected again for the residual effect of the ship's own magnetism according to the course steered.

(Allan, 2008: 58)

Allan (2008: 59) soon began trying to match the observed anomalies to models of rock formations of various shapes, sizes, and orientations. He went to sea, and used Hill's magnetometer. He also worked with Bullard compiling programs to read and eventually interpret data on EDSAC 2.

I used it subsequently on 4 cruises from 1957–1959 on *Sarsia*, *Discovery* and HMS *Dalrymple* ...
Teddy [Bullard] added the punched paper tape and he and I compiled programmes to read and
eventually interpret the data on the University's first (immense) computer called EDSAC.

(Allan, June 6, 2009 email to author)

These pre-Kunaratnam programs, as already noted, did not allow for separating
remanence and induced magnetization in modeling marine magnetic anomalies and
he ended up using Press and Ewing's 1952 interpretative model.

Allan had no background in geology, and he had, like Laughton, no prejudices for
or against continental drift. He read Holmes' *Principles of Physical Geology* and
learned about continental drift (Allan, June 6, 2009 email to author). He also read
about field reversals, including Hospers' work. He believed that they had occurred.
"I also was aware of field reversals and had read Hospers' work on Icelandic lavas"
(Allan, June 6, 2009 email to author). He learned of Girdler and Peter's 1960 work on
the Gulf of Aden and gradually realized that the seamount was reversely polarized.

Yes I did think at times that interpretation would be easier by assuming reversed rather than
normal polarity. This conversion was a gradual process however built up over a few years.
I don't recall any Eureka moment when I leapt from my desk. We knew reversals had been
found on land so the idea of reversals at sea fell on fertile ground in my case. This is to take
nothing away from Fred Vine and Drum Matthews who clicked the whole thing into place
from the jigsaw pieces we had provided.

(Allan, June 6, 2009 email to author)

In their paper they displayed the results of their magnetic survey over the seamount
as shown in Figure 2.6.

Using Press and Ewing's 1952 interpretative model, Allan tried to match the
magnetic observed anomaly along Section AB, and found that they could not get a
decent match unless he supposed that part of the seamount was reversely magnetized.

Measuring the magnetization of their dredged basalts, he found that their remanent
magnetization averaged out to 5.0×10^{-3} and their magnetic susceptibility an order of
magnitude less. Importantly this allowed him, while calculating the best fit of the anomaly,
to attribute the seamount's magnetization entirely to remanence. Allan found that if he
assumed that the seamount is normally magnetized and varied the declination, he could
not get a good fit. He reversed the remanent magnetization and the fit became very good.

Comparison of Figs. 15(a) and 15(b) show that none of the computed curves could be made to
fit the observed anomaly. In fact, the dissimilarity was so striking that a curve was computed
assuming reversed polarity in the structure, that is, with the assumption that during the
formation of the rock, the magnetic north pole was in the position of the present magnetic
south pole. The results are plotted in Fig. 15(c).

(Laughton et al., 1960: 136. Figure 15 is reproduced as my Figure 2.7)

He made one correction. He raised the "zero" of the observed curve by 140γ, which he
(1960: 136) thought justifiable because the measured value of the geomagnetic field

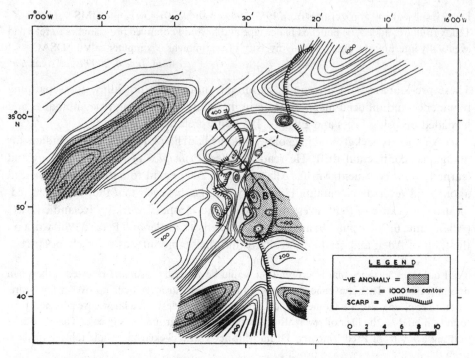

Figure 2.6 Laughton, Hill, and Allan's Figure 14 (1960: 134). Their caption reads: "The magnetic contour map of the area showing the anomalies in total field, after removal of regional gradient (contour interval 50γ). The largest magnetic anomaly is over the northeast part of the seamount, just southwest of A of section AB. There is a negative magnetic anomaly on the southeast side of the seamount at B."

may have been subject to large inaccuracies over the oceans. Drawing on the strong evidence that field reversals have often occurred, the Cambridge workers argued that the seamount had formed when the field was reversed and noted that they would not be able to date the seamount because reversals had occurred as recently as the Pleistocene:

Dredging has shown that the exposed rocks are principally volcanic and this is confirmed by the magnetic anomalies associated with it and by the seismic velocities. The possibility that the magnetization of the rocks was acquired at a time when the earth's field was reversed is plausible in the light of recent work on palaeomagnetism where there is strong evidence that reversals in the earth's dipole field have occurred often throughout geological time (Irving, 1959). However, reversals have been observed as recently as the Pleistocene and so it is not possible to use this evidence as a means of dating the seamount.

(Laughton et al., *1960: 136; Irving reference is my 1959)*

Dating oceanic objects by polarity came much later when long anomaly sequences were assembled.

His dissertation on Atlantic marine magnetic anomalies almost finished, Allan was invited by Hill to participate in a cruise to the Red Sea, promising to find funds

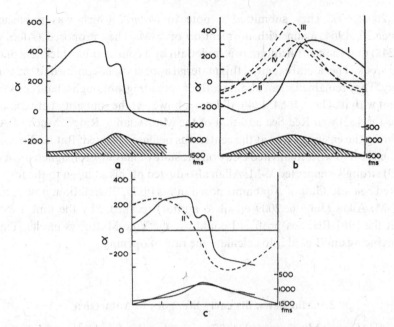

Figure 2.7 Laughton, Hill, and Allan's Figure 15 (1960: 135). Their caption reads: "Fig. 15. (a) The observed magnetic anomaly over the seamount. (b) The calculated curves for the trapezoid for I = 52° but with the horizontal component of remanent magnetisation in four different azimuths (i) 13½° W of N (in direction of the present earth's field) (ii) 13½° N of E (iii) 13½° N of S (iv) 13½° N of W. (c) (i) The observed anomaly with the zero increased by 140 γ. (ii) The calculated curve for the trapezoid with reversed polarity. The seamount and trapezoid are shown superimposed."

to support him, his Shell Studentship being close to expiry. Allan agreed; he obtained twenty-three magnetic transverse profiles from the Gulf of Suez to Aden, and argued (1960: 156) that the Red Sea formed as the Arabian block rotated relative to the African block with a fulcrum somewhere in Palestine. He argued that the pronounced positive magnetic anomaly was caused by a normally magnetized block of basalt that arose from beneath Earth's crust through the tension crack running along the axis of the rift valley. He claimed (1960) that the areas beneath the slightly negative anomalies surrounding the central positive anomaly lack such a basaltic block; the slightly positive anomalies on the landward side of the negative anomalies were caused by thin blocks of basalt that broke off from the main block beneath the central positive anomaly.

Allan passed his orals with Bullard as internal and Laughton as external examiners, and took a position as a researcher at the newly opened NATO Scalant anti-submarine-warfare (ASW) Center in La Spezia, Italy. While there, he wrote up his Red Sea paper making no mention of V-M (Allan, 1964). While at the center, he and two of his fellow scientists undertook a gravimetric and magnetic survey of the Red Sea between October and December 1961, crossing the sea fifty-eight times

(Allan, 2008: 97). They submitted a note to *Nature*, which was published on December 26, 1964, again with no mention of V-M. They proposed (Allan *et al.*, 1964: 1248) that the central anomaly is underlain by a considerable thickness of dense, magnetic rock. The lateral extent of this material appears to be confined to the width of the valley. They remained silent about seafloor spreading, although what they said was consistent with it. They (1964: 1246) did note, however, the similarity between the rift valley of the southern Red Sea and that of the Mid-Atlantic Ridge. Much of Allan's work during the ensuing years at the center was defense-oriented. But he did receive a brief sabbatical to write an invited essay on the state of marine geomagnetism in which he (1969) strongly supported V-M. Allan also dusted off his solution to the formation of the Red Sea and Gulf of Aqaba proposed in his Ph.D. dissertation, now incorporating V-M. Allan (June 6, 2009 email to author) recalled, "by the time I came to interpret the 1961 Red Sea profiles I could use the Vine-Matthews model. This was very convincing and I used it to calculate the rate of opening."

2.5 Matthews, his early life; goes to Antarctica[6]

Drummond (Drum) Matthews (1931–97) grew up at his family's horse-riding school in Porlock on the Somerset coast, where he also learned to sail small boats. His father, a fighter pilot in World War I, was badly wounded while flying his Sopwith Camel and suffered a nervous breakdown. With the outbreak of World War II, he insisted on reenlisting. Stationed in an office in Plymouth that was heavily bombed, the experience led to his father's suicide when Drum was eleven years old.

The next year he went to Bryanston School, where he became interested in science. He also sang, acted, played rugby and rowed, and rejuvenated the Sea Cadets. He became Head Boy, and was noted for his willingness to help others, his integrity, and his modesty, characteristics that later made him an excellent supervisor and leader of the Cambridge Marine Group (1966–80). He won a State Scholarship. For his National Service, he served in the Royal Navy, and rose to RNVR Sub-Lieutenant. Thinking of continuing full-time in the Navy, his mother insisted that he take up his State Scholarship. He enrolled at King's College, Cambridge, in 1951 and read natural sciences. He earned a First in Part I of the exams, and was elected a College Exhibitioner in June 1954. During his undergraduate years he suffered from depression, which was not helped by his characteristic self-deprecation (White, 1999: 278). He also suffered from severe migraines. He was befriended by Maurice Hill and his wife Philippa who helped him with his depression, and Hill encouraged Matthews to pursue a Ph.D. under him. Matthews, however, did poorly on Part II of his exams; specializing in geology and petrology, he graduated in 1955 with a lower second class degree. Matthews summed up his undergraduate career in this way, and noted that there was one occasion when he should not have listened to Hill.

I was an undergraduate at Cambridge. I've always been at Cambridge, which is not a clever thing. I was an undergraduate from '51 to '55. I was a little bit rebellious; I was not a clever undergraduate. I got a two-two finally. I had a first on the way. I got a two-two because Maurice Hill, who is my boss here, persuaded me to do two sorts of Part II at once, which was a bit much for me. Anyway, I got a two-two. So nowadays I would be totally eliminated from science. There is no way I could go on in science in this country. In those days I was offered a studentship here [in the Department of Geodesy and Geophysics to do graduate work].

(Matthews, August 1979 interview with author; my bracketed addition)

Severely disheartened by his poor performance, he decided not to go to graduate school, and took a job with the Falkland Islands Dependencies Survey (FIDS). Assigned to make a geological map of Coronation Island, the largest of the South Orkney Islands, he spent two summers and one winter in the Antarctic, returning to the UK in June 1957. Matthews later said that his time on Coronation Island helped him develop perseverance, self-reliance, and his leadership skills. The isolation, beauty, and severity of Coronation Island encouraged self-reflection. Recording his thoughts in a diary, he wrote the following entry on Christmas Eve, 1955.

This is a fabulous life. Mountains, perfect snow, skis and crampons. To be paid for such things is good enough, but in this place with rocks, snow, cliffs, glaciers, sea and sunshine ... to say nothing of the garden of Eden animals. The seals with their big wild eyes and sleepy flippers, ridiculous penguins and friendly bouncing huskies. Birds and animals, none yet recognize man for an enemy. I pray it may remain so ... But one fears the hard brutal sealers & whalers – may they wither away. Christmas Eve – I wish I knew what to think of it all. The most beautiful story with the world, round it have gathered the most beautiful lives & one must love them highest. Yet it makes so little sense, God omnipotent in the amoral world. It can only be so in a special spiritual sense, poetical truth & emotional knowledge. Closely akin to nonsense.

(Matthews, diary entry on Christmas Eve, 1955)

January 15, 1956 was a particularly good day. His companions finally agreed with him that they would be better served by moving their depot and camp. He and Norman, the surveyor who sometimes accompanied him, also found a way to amuse themselves.

Feel tired but more satisfied than any day since we got here. Why, today a small suggestion of mine, to move the depot as well as the camp, wasn't turned down ... It seems so nice & calm, with only occasional snow that we've tied the tunnel of the tent to an upended ration box so that we can keep the door open all night. N [Norman] calls it the verandah & we amused ourselves with the usual game of composing an advertisement – delightful country residence 75 feet above the sea. Plenty of wild life & game. Excellent skiing. Infrequent mail service. Only seven thousand miles from London.

Matthews read widely while there, from *Winnie the Pooh*, a lifelong favorite, to the Bible, from Descartes to the metaphysical poets – his favorites. Thinking of Marvel's "The Garden" and perhaps already missing England's green and pleasant land, he mused:

Read from poems Donne to Dryden. Herrick (Gather ye rosebuds while ye may) Marvel [in The Garden] annihilating all that made to a green thought in a green shade. Despite these superb stanzas – I think almost my favorite unless it be to His Coy Mistress – the poem is strangely unfinished. The last stanza about the sundial seems to be quite out of place. I wondered when I packed it how the metaphysicals would wash here but they're better than ever.

(Matthews' diary, January 15, 1956; my bracketed addition)

With "To his Coy Mistress" in mind, he reflected about losing his youth and what to do with his life.

Here thousands of miles from any one and even more isolated because of my own aura of scorn the mind ought to blossom into primal self knowledge. But it doesn't, or, if it does it grows like the expanding cauliflower and not the unfolding rose. Perhaps it would be better after all to tear ones pleasures with rough strife through the iron gates of life. How sad it is to be prisoned inside this canvas pyramid while the unpausing stream bears youthful days away. How glad I shall be to escape. But what to do next? One thing is clear – snobbish as it sounds it must be a gentleman's occupation in which my companions are possible friends.

(Matthews' diary, February 24, 1956)

Still smarting from his poor performance on his exams, he wondered if he should become a schoolmaster.

I don't feel ready yet to go back to Cambridge after the disappointment of June. I could never work well in the Min[eralogy] & Pet[rology] Dept again thou' I still regret the geophysics. But I was thinking last night of chucking all this that is so hard for me, going off to schoolmaster somewhere, perhaps having a little garden ... & a boat somewhere and maybe ... for a wife if she'd have me. The garden should be very small with high walls, some long grass & fruit trees. Just big enough to sit & read in. Roses, a few shrubs, pinks maybe and lavender.

(Matthews' diary, January 15, 1956; my bracketed additions)

A month later he, with a good dose of self-irony, wrote again about his lost youth and faith.

Then after tea I went for a walk and sat on a rock watching the come and go of the tireless sea & the light gold tipping the further peaks. I thought so many great thoughts that I felt like a lavatory calendar and may as well tear off a few to record here. After all being 25 it is time to emerge from the dog end of boyhood which had lost its security with its religion & degenerated into a breathless chase after ephemeral targets of commissions and blues and firsts. The peace of the Antarctic evening calmed me. But neither in the swift flash of the enquiring mind, nor in the beauty of the remote mountains, nor in the security of friends, but only in the depths of the love of God will the questioning soul find peace and cessation of loneliness. But I cannot believe in God – it is too monstrous to have created this amoral universe. Why do we select so stubbornly the light of reason only? For reason stultifies, freezes warm life, makes the dry prig, is sterile and life destroying. Reason defies art. Is she not a mere tool, an accessory for living, making man more efficient at cracking nuts than are his fellow apes? It is because she is cold steel blue light like glacier dawns, because she is the sharp point of the halliard thrusting in the gloom to procreation of knowledge. And in knowledge we touch immortality. As it grows fragment by fragment painfully illuminating each stick and stone it makes a little clearing in the darkness. It forces back fear and doubt and faith. At each feeble stroke the encircling eyes withdraw a little further into the darkness, their age long confidence a little weakened ... But can't it be combined with life? The young Jesus astounded them because of his understanding, but the man Jesus did more. Reason there was plenty cleaving through curtains of enshrined hypocrisy. (It is written My house shall be a house of praise, but ye have made it a den of thieves.) But there was so much more. That is the third age of man, after the enquiry of late boyhood, a blending of all one was as a boy ... So I mused on my rock and it did me good, even if it's crammed with purple passages. Even if we are twisted by a lazy boar there's much contemplation to be done & much stealing back to boyhood's wonder. On the calm of confident summer, courage on a Gallop. Oh to believe again & feel the night vibrating with heaven again. Or is it just to be young again? I wonder and not to have eaten of the fruit of the tree. Not to understand that most people love one more for a pretty body than a serene mind. There are gulls crying. I will sleep and be out easily.

(Matthews' diary, February 15, 1956)

By the next Christmas, Matthews seems to have recovered his faith.

I climbed up the soft snow and sat down in the warm sunshine ... It was still and warm there, at first anyway, and I pondered the immutability of desert places. How when Christ was born the same fickle sunshine shone on the same barren black rocks of the point ... I doubt if the busy vacant breeding penguins would have cared if a predatory leopard [seal] paused to hear the angels singing. I read the account of the nativity in Luke's gospel and thought of King's and home. A pagan Christmas is an unsatisfactory feast, a mere pointless extravagant gorge, an advertiser's stunt. At Christmas everyone wants to be on the side of the angels.

(Matthews' diary, Christmas, 1956; my bracketed addition[7])

Spirituality remained important to him. He retired when only fifty-nine, partly, as he himself said, to discover more about God and to develop his spiritual life (White, 1999: 292). In his late forties, he and Derek Blundell took the initiative in organizing the deep seismic profiling of continental crust around Britain (BIRPS). He led the program until retirement.

2.6 Matthews visits the Falkland Islands and favors continental drift

As an undergraduate, Matthews did not believe in continental drift. He recalled a particular meeting:

Oh continental drift was totally out while I was an undergraduate [1951–5]. I remember they ran a rather high powered one day meeting in order to persuade finally a class of undergraduates that continental drift was totally inconceivable. It really was quite high powered. It was Arkell, who was a Jurassic paleontologist actually working in Cambridge, Gold, an astronomer, Jeffreys, and I don't know who else. I think there was an American paleobotanist, paleobiologist of some sort ... I remember Tommy Gold talking about building – it was the days of the cold war – skyscrapers in Moscow and skyscrapers on the equator with the Astronomer Royal being the most important observatory in the world saying we've lost another second of an arc today trying to move the other continent toward the pole. So, no, I didn't believe in continental drift.

(Matthews, August 1979 interview with author; my bracketed addition)

I think that the meeting Matthews remembered occurred between January and March 1955 and was sponsored by Cambridge's Department of Geodesy and Geophysics. Arkell, Gold, Runcorn, and the American paleobotanist Durham spoke against continental drift, and when a vote on continental drift was called among the audience, only one person supported it. But Jeffreys did not speak at the meeting (see II, §3.10 for discussion of it). Regardless, he clearly did not believe in continental drift during his undergraduate years.

   Matthews, however, became sympathetic toward continental drift on his return to the UK from Coronation Island. He stopped at the Falkland Islands, and saw firsthand, albeit with help from Adie and du Toit, what he took to be strong support for mobilism. Adie, a mobilist, was employed by FIDS, and later remained with the organization after it was renamed the British Antarctic Survey in 1962 (see I, §6.11 and §7.5 for discussion of Adie's support of mobilism).

On the way back we spent time in the Falkland Islands. The survey was based on the Falkland Islands. And, I asked . . . Adie what one should read about the Falkland Islands. (It was so nice to walk around without a rope.) He is South African. He said, Oh, well, you've got du Toit . . . if you don't believe in continental drift just take out a tape measure and measure the Devonian sections in the Falkland Islands. I did, and they were very much impressively the same as the description [given by du Toit for South Africa] . . . inch for inch they measured up. So, I came back quite enthusiastic.

(Matthews, August 1979 interview with author)

Once Matthews returned, he, Adie, and Maling, who had mapped the South Orkney Islands with FIDS before Matthews, planned to submit two papers. Matthews wrote up his two parts, but only one paper was published and not until ten years later (Matthews, 1967). He also wrote another paper on the geology of the Scotia Arc, which includes the South Orkney Islands (Matthews, 1959). After arguing as others had before that the islands of the Scotia Arc are not oceanic but continental, he (1959: 436) suggested that the arc might represent fragments of a disrupted continental mass brought about by continental drift or part of a submerged continental mass occupying much of the Scotia Sea area. Proposing continental drift as the cause of the Scotia Arc was not new; Wegener (1929: 94–95), du Toit (1937: 194–197), and Barth and Holmsen (1939) had already done as such.[8] Matthews did not accept either hypothesis, raising a difficulty against each (RS2).

If they [i.e. the islands of the Scotia Arc] are fragments of a disrupted mass, then breaks are to be expected in the submarine ridge connecting the islands . . . The bathymetric chart gives little support to this hypothesis . . . [there is] but one such break . . . On the other hand, if there is a submerged continent under the Scotia Sea, it must be associated with a large gravity anomaly. Inspection of the bathymetric chart shows that the average depth of the Scotia Sea is [only] about 3 ½ km ... However, the forces required to depress and hold down a continental block of a size comparable with the Scotia Sea to a depth of even 3 ½ km. would be impossibly great unless the mass of the block was radically altered by intrusion or spreading.

(Matthews, 1959: 436–437; my bracketed additions)

Even though he was unwilling to argue that mobilism explained the Scotia Arc, he had begun thinking about it, even citing Adie and adding that continental drift is an attractive hypothesis to apply to the Falkland Islands.

2.7 Matthews, his graduate work in the Department of Geodesy and Geophysics at Cambridge, 1958–1961

Matthews was awarded a Fellowship at King's College, Cambridge, and began graduate work in January 1958. He was only the third geology graduate to do so in the Department of Geodesy and Geophysics; Hospers was the first, Irving the second (II, §1.20). Although I have no idea if Matthews had thought much about the geomagnetic field and whether it undergoes polarity reversals, a key assumption of V-M, he soon found that for most in the department there was no doubt about it (II, §1.12; §2.9).

At that time [1958] we already knew that the Earth's magnetic field reversed ... We were put off by the Japanese discovery of lavas that reversed by themselves, but Jan Hospers ... had done a Ph.D. in Iceland very soon after the end of World War II in which he came back with quite definite reversely magnetized lavas. Japan was a long way away; we thought there must be some other explanation [for the self-reversing rock they had found]. Indeed, Néel had an explanation for self-reversals, and it was basically believed. There were a lot of people working here on paleomagnetism in 1958. The magnetometers were very much part of the lab. Runcorn had left but there were still people working. There was really no doubt [here] that the Earth's magnetic field reversed.

(Matthews, August 1979 interview with author; my bracketed additions)

Thus Matthews was already sympathetic if not favorably disposed toward mobilism and field reversals before he met Vine.

Matthews joined the marine geophysics group, intending to work with Hill on the velocity of sound in rocks.

I came back from the Antarctic and joined this lab [at the Department of Geodesy and Geophysics] in January of '58 and worked under Maurice Hill ... It was suggested that I should work on the velocity of sound in rocks ... That was the idea, and I spent the first term here, couple of terms here, going through the literature and concluded from preprint information that the people at Gulf in Houston were just about to mop the whole subject up, which they did do ... So I was a bit at a loose end.

(Matthews, August 1979 interview with author; my bracketed addition)

In the preface to his Ph.D. thesis, "Rocks from the Eastern North Atlantic," he explained how he came to work on ocean rocks dredged from an abyssal hill called Swallow Bank, located 300 miles northwest of Lisbon in the Atlantic Ocean.

On May 30th 1958, when a large heap of black stones, mud smeared and evil smelling ... was poured from the dredge bag on to the poop deck of *R.R.S. Discovery II*, we were faced with a choice: whether to seek for a petrologist to work on the dredged rocks, or for the writer to abandon the study of the velocity of sound in rocks under pressure to which he was already apprenticed, and take on the dredge collection himself. Several considerations seemed to urge the second course. Even to a renegade petrologist the collection looked exciting, since it was unique – basalts had been reported dredged from greater depths on the flanks of deep trenches but no substantial collection had ever before been obtained from a depth as great as 5 km in an apparently typical area of ocean basin floor. Besides, those who go on expeditions ought to work up the results themselves if they can, and it did not seem likely that a better qualified petrologist would be found with time to spare for all the diverse questions posed by the unprepossessing altered rocks from Swallow Bank.

(Matthews, 1961b: i)

Through his study of the dredged rocks Matthews reached two conclusions, one being that Layer 2 of oceanic crust was basalt.

There was absolutely no doubt around the coffee table here ... We knew the sea floor was made of basalts, we knew the mid-ocean ridge was made of basalt because I had done a lot of reading ... I looked at all the literature of every rock that had ever been dredged at that point. It is all in the first chapter of my thesis. There weren't very many in those days, but there were several thousand, and it was perfectly clear that, what I did basically was to rediscover what was known at the end of the 19th century by the people laying telephone cables. We knew a great deal about the geology of the mid-ocean ridge which was subsequently forgotten. They knew that whenever they went over the mid-ocean ridge and got bits of rock back, they were glassy basalts. So they knew that the mid-ocean ridge was volcanic.

(Matthews, August 1979 interview with author)

There was absolutely no doubt around the coffee table at Madingley Rise that Layer 2 of the seafloor was made of basalt. Hill's seismic refraction work on the Mid-Atlantic Ridge led him to conclude that a layer of basalt underlay its thin veneer of sediments (Hill, 1960: 202). In addition, Laughton, Hill, and Allan strongly believed that the seafloor was basaltic given their analysis of the dredged rocks from the seamount 150 miles north of Madeira Island (Laughton *et al.*, 1960).

His second finding, which Matthews also thought was perfectly available in the literature, was that ocean basalts are remanently magnetized.

One chapter [Chapter 6] is about the magnetics, which I did on the astatic, measuring susceptibility and remanence and sorting them out. And from that and the literature thereof – and there were a great deal of measurements already in the lab –I got a lot of ideas ... The primary idea about magnetization was that basalts were remanently magnetized. It was perfectly available in the literature but there is a gap between what is available in the literature and what people apply to other things. Everybody else knew that, I think. I suppose that Ron Mason must have known it ... I talked about remanent magnetism because of my Swallow Bank basalts.

(Matthews, August 1979 interview with author; my bracketed addition)

He measured the remanent magnetization of his basalts on the departmental astatic magnetometer rebuilt by Flavill and Belshé from the remains of Creer's magnetometer at Cambridge, and previously used by Girdler to measure his Jurassic rocks from the West of England. Belshé showed Matthews how to make the measurements, and M. Fuller assisted (Matthews, 1961b: iii). Like Laughton and company, Matthews found that the magnetization of the basalts was primarily attributable to remanence, the induced magnetization being an order of magnitude less. He measured forty-four cylinders of altered lavas cut from twenty-five dredged stones (Matthews, 1961b: 125). He found that they

are slightly less magnetic than normal basalts although their susceptibility (median value 5×10^{-4} e.m.u./oersteds/cm^3) and remanent magnetization (median 5×10^{-3} e.m.u./oersteds/cm^3) lie within the known range of variation for basalts.

(Matthews, 1961a: 158)

Because the basalts were altered there was the possibility that the intensity of their magnetization and ratio of remanent to induced magnetization might not be representative of unaltered oceanic basalts. He claimed that their remanent magnetization was probably slightly less than fresh basalts.

The available evidence indicates that the remanence and susceptibility of the altered marine lavas falls within the range of variation observed in sub-aerial lavas, but that the values are lower than average. Thus, whilst it seems possible that the alteration which the lavas have suffered has resulted in some slight reduction in their thermo-remanent and induced magnetisation, it is certain that the effect is small: the susceptibility has not been reduced by more than a single order of magnitude, and the remanence has not been reduced by as much as that.

(Matthews, 1961b: 138)

He was pleased with the results, and that Vine used the same value for remanent magnetization when proposing V-M.

Thus Matthews' beliefs about continental drift, reversals of the geomagnetic field, basaltic nature of the seafloor, and dominance of remanence in the magnetization of oceanic basalt matched Vine's before he arrived at their hypothesis. Unlike Vine, however, he was not enamored with Hess's seafloor spreading. It is not that Matthews was opposed, he just did not have a strong feeling that seafloor spreading was correct. Matthews also recognized that detailed magnetic surveys should be coordinated with similarly detailed bathymetric surveys. Indeed, his detailed survey of a small section of the Carlsberg Ridge provided Vine with the data whose interpretation led to Vine's version of V-M.

2.8 Vine's early interest in continental drift and undergraduate years at Cambridge, 1959–1962[9]

I was very favorably disposed towards drift much to the chagrin of some of my supervisors at Cambridge as an undergraduate – not all, for there was just one notable exception [Brian Harland] ... I think it derives from – it is a very stupid story – I was in the fifth form at that time, i.e. the year at the end of which one took O (ordinary) level exams (by which time I was 16). It was during the Easter holiday in 1955 and I was staying with my great uncle and aunt in Dartford, Kent. The incident occurred on a cycle ride in the countryside to the south of Dartford. I had a new textbook in geography. In general, one didn't look at textbooks, I suppose, but I happened to open it to the first page during the Easter vacation . . . for want of something better to do. For some unearthly reason it talked about the fit of South America and Africa, and I never got beyond the first page. I looked at this picture and I thought it absolutely fascinating. Yet, it went on to say that geologists don't really know whether this is significant. I thought, well, come on, wait a minute. Surely, it is rather important to know whether these two were ever alongside. It seemed rather basic to me, rather fundamental. From that point on, I decided it had to be true; it was too simple and elegant. So I was always very favorably disposed. I guess, I was always looking in studying geology for evidence that confirmed my prejudice. So, in doing things, in fact, to quite a high level as an undergraduate, I interpreted things in terms of it. I remember Normal Hughes, a palynologist. He was my former supervisor. He set essays on things like: write a critique of the [Henri] Termier and [Genevieve] Termier paleogeographic maps. He would get an essay back from me which tore them apart because they didn't start with continental drift reconstructions. Then he would crawl up the wall because he didn't believe in continental drift. This sort of thing went on. But, the notable exception was Brian Harland, a structural geologist. I think he was rather like Arthur Holmes.[10]

(Vine, August 1979 interview with author with additions from Vine, 2008
email to author; also see Vine, 2001: 49)

Thus Vine, like Irving who played a large role in developing mobilism's paleomagnetic support during the 1950s and 1960s (II, §1.14), became favorably disposed toward mobilism at school.

Vine attended a primary school in Chiswick in West London. He then obtained a scholarship to attend Latymer Upper School, which was then a direct grant school.

Latymer is now an independent or public school (i.e. private and fee paying). However this was not the case when I was there in the 1950's. It was then a direct grant school. It had an independent board of governors, 14% fee paying pupils and 86% scholarship boys (the scholarships paid for from public funds). I was incredibly lucky in being the only boy from my primary school in Chiswick to obtain a scholarship at Latymer that year (1950).

(Vine, 2008 email to author)

Vine's interests lay in the sciences, and while at Latymer, he read everything he could find, for example, on IGY, which ran from July 1957 to December 1958 (Vine, 2001: 49). He does not recall, however, reading anything more about continental drift or discussing it with his teachers (Vine, 2008 email to author). After doing very well on his O (ordinary) level exams in mathematics, physics, and geography, he was persuaded to take his A (advanced) level exams in pure and applied mathematics, physics, and chemistry, even though he had wanted to take them in mathematics, physics, and geography (Vine, 2001: 49). After A-levels and the Cambridge entrance exam he spent the next eighteen months teaching mathematics at Latymer. He not only survived the teaching but enjoyed it, and planned to become a school teacher (Vine, 2001: 50). He did sufficiently well in his A-levels to obtain a state scholarship.

Vine entered St John's College, Cambridge, in fall 1959. He planned to study mathematics and physics as an undergraduate but learned that he needed an additional subject. He recalled what led him to choose geology when he talked to his advisor.

He ran down the list and alighted on geology. He said, How about geology? There was a slight pause, and he looked up. Do you like the open air? Would you enjoy the fieldwork? Why, yes, I said, of course. Well, that's done then, geology it is.

(Vine, 2001: 50)

He obtained a B.A., reading geology, mineralogy-petrology, mathematics, and physics. During his final year, he concentrated on geology with emphasis on petrology-mineralogy.

From his reading, Vine continued to favor mobilism; however, except for his supervisions with Hughes, he heard little about it from undergraduate teachers at Cambridge. Even Harland, as far as Vine remembers, said nothing about it during his lectures on structural geology. But Vine did learn about paleomagnetism, and presumably its support of mobilism from Harland, who with his student Bidgood had recently found interesting results from Norway (III, §1.11).

I distinctly remember that on the last day of our first-year field course on the Isle of Arran, at Easter 1960, Brian Harland demonstrated taking paleomagnetic samples using a sun compass. Presumably he went on to mention some of the results. I think this may well have been my first introduction to paleomagnetism because it made such an impression on me. I also had lectures from Brian Harland on structural geology but I don't recall that these included anything on continental drift. I don't think I ever discussed continental drift with Brian but I guess I did realize by 1962 that he was an advocate, possibly the only one on the faculty of the geology department. In general there was not much discussion of continental drift although it did crop up in tutorials with Norman Hughes and I think one could say that we agreed to differ about it.

(Vine, 2008 email to author)

Vine read works by Holmes, Wegener, du Toit, and Carey, four of continental drift's major supporters. He also read Jeffreys on mobilism, and thought him wrong. However, except for Holmes, he cannot recall whether he read the others as an undergraduate or as a first-year graduate student.

I certainly read Holmes as an undergraduate. I also read some of du Toit, Wegener and Carey but I am not sure whether this was as an undergraduate or as a first year graduate student. I did read Jeffreys and thought he was wrong.

(Vine, 2008 email to author)

Although none of Vine's undergraduate teachers did anything to boost his interest in continental drift, he was encouraged by talks by Hess and O. T. Jones that he heard while a third (and final) year undergraduate at the 10th Inter-University Geological Congress in January 1962, where the topic was the evolution of the North Atlantic. Except for Hess, all speakers were from Earth science departments at Cambridge. Jones and Hughes spoke on Marginal Stratigraphy; W. A. Deer, on Igneous Provinces; Bullard, on Structure of the Ocean Floor; M. Black, on Oceanic Sedimentation; Belshé, on Comparison of Palaeomagnetic Data; and Harland, on Comparison of Marginal Structures, disjuncts that is. The tenor was pro-drift (Vine, 2001: 51). Hess, Jones, Belshé, and, presumably, Harland spoke in favor of mobilism, and I suspect that Bullard expressed sympathy but without being fully committed (III, §2.14). Belshé, Vine recalled (1979 interview with author), summarized some paleomagnetic evidence for drift around the North Atlantic area, but his talk was fairly standard. Jones reviewed trans-Atlantic disjuncts and Vine was struck by his enthusiastic defense of mobilism.

I do not remember O. T. Jones mentioning the paleomagnetic data, only that he made a very convincing case, by comparing the lithologies and the stratigraphy on either side, for the closure of the North Atlantic (despite the fact that his co-author was N. F. Hughes!). The other aspect that struck one was O. T.'s enthusiasm for the idea.[11]

(Vine, 2008 email to author)

Vine, already familiar with Hess's work in petrology, was most impressed by his talk on seafloor spreading, which was for him the highlight of the meeting.

I was very excited by [Hess's talk]. I think it was the first time I'd heard Hess's hypothesis in depth. I cannot imagine how I would have come across it any earlier, and basically I was very excited by it ... I think it was a very successful meeting, a very stimulating meeting, primarily because of Hess's contribution.

(Vine, 1979 interview with author; my bracketed addition)

Hearing Hess reminded Vine of Holmes' appeal to mantle convection. He also thinks he began and continued to talk about Hess's ideas.

Hess's talk did indeed remind me of Holmes' mechanism for continental drift in terms of convection in the mantle ... I am sure that I did talk to others about Hess's idea because I was very excited by it but, again, I cannot remember specifics.

(Vine, 2008 email to author)

Elected President of the Sedgwick Club, Cambridge's geology club, he devoted his Presidential Address, which he gave at its 870th meeting on May 15, 1962, to HypotHESSes. Vine wrote a summary for the minutes. After discussing the development of Hess's ideas from his early career through to his January talk, he turned to his new hypothesis. He noted that it provided a mechanism for continental drift (more below). He also showed that he had been thinking about the geomagnetic field.

Hess favours convection and impermanence, and this theory certainly has its attractions. Convection could help to explain the apparent dearth of sediment on the ocean floors. The distinctive and probably quite ubiquitous wrinkles on the ocean floor may well be due to

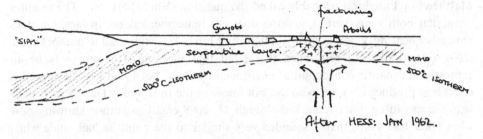

Figure 2.8 Vine (minutes of Sedgwick Club's May 10, 1962 meeting), "After Hess: Jan' 1962." The figure shows the formation of layer three of oceanic crust above the raised 500° isotherm at ridge crests with serpentinization of olivine, the fractured crust at the ridge crest, and rising convection currents. Hess's solutions to the formation of guyots and atolls are also illustrated.

convection. It may well explain the embarrassingly large heat flow beneath the oceans, especially over mid-ocean ridges. Coupled with this it may well provide a mechanism for the removal of heat from the fluid core. Gymnastic convulsions of this almost certainly give rise to secular variations in the magnetic field, e.g., the appearance and movement of isoporic foci. It also provides a mechanism for continental drift.

(Vine, Minutes of meeting of the Sedgwick Club, May 10, 1962, Vine's personal files)

Continuing his summary of Hess's new hypothesis, he mentioned other problems it solved.

In January, as ever Professor Hess favoured convection, impermanence, and a mineralogical transition at the Moho rather than a basalt-eclogite phase change. Petrologists and geophysicists alike are expressing doubts as to whether the latter can explain the Moho beneath the oceans and this may well lead to its downfall as regards the transition beneath the continents. His current theory accounts for the following observations: (i) the abnormally high heat flow over the mid-ocean ridges, (ii) the concentration of earth-quake epicenters over the ridge, (iii) the anomalous seismic velocities obtained over the ridges, (iv) the formation and distribution of guyots and atolls, (v) the deficiency of oceanic sediment and (vi) the paucity of seamounts in view of their present frequency of occurrence. Hess considers that such convection cells have a life-span ~ 300 m. yrs. And that the surface movement is ~ 1 cm per yr.

(Vine, Minutes of meeting of the Sedgwick Club, May 10, 1962, Vine's personal files)

He concluded with a figure illustrating Hess's view of seafloor spreading (Figure 2.8).

I am pretty sure that I drew the diagram on a blackboard. It was derived from the notes I took during Harry's talk. The "+'s" signify fracturing of the newly formed crust to explain, together with the higher temperatures, the anomalously low seismic velocities obtained at ridge crests.

(Vine, 2008 email to author)

Vine's Presidential Address confirms his enthusiasm for Hess's seafloor spreading and continental drift.

After his address he was asked if Hess's seafloor spreading could explain the zebra pattern of magnetic anomalies in the northeast Pacific. He thinks that either

Matthews or Laughton probably asked the question (Vine, 2001: 53). The minutes show that both were there. Laughton does not remember asking, in fact, but does remember hearing Vine's address (Laughton, March 27, 2009 email to author). Vine (2001: 53) recalled, "I responded that I felt that they must in some way be an expression of mantle convection as envisioned by Hess, but I had no idea how this effect was produced." Vine, who did not know at the time about Dietz's version of seafloor spreading, later remarked (March 11, 2009 email to author knowing now what Dietz had said), that it sounded very similar to the point he had made when answering Drum's or Tony's question – Dietz claimed that the magnetic anomalies were normal to the movement of the seafloor, and suggested that convection was somehow involved (§2.9).

Although Vine had already been introduced to paleomagnetism's support of mobilism by Harland and Belshé, he did hear a likely out-of-date but very lively review by Blackett who, in March 1962, gave a talk at the Natural Sciences Club at Cambridge.

Blackett came and gave a talk, and that was rather late in a way; one felt it was almost out of date. He just talked in terms of latitudinal changes of the continents throughout the Phanerozoic, and I remember Australia going up and down, and I guess it was generally pro-drift.

(Vine, 1979 interview with author)

Vine also read Jeffreys' attack on paleomagnetism (II, §7.13) and again thought it wrong.

As regards the paleomagnetic evidence, I thought that it should be taken seriously and Jeffreys' dismissive view of paleomagnetic studies ignored. At some stage I did read Cox and Doell's 1960 paper but again I cannot remember whether this was in my final year as an undergraduate or my first year as a graduate student.

(Vine, 2008 email to author)

Reading the fence-sitting GSA review did not deter him from believing that the paleomagnetic support for mobilism should be taken seriously.

He was familiar with Hess's seafloor spreading; he knew about paleomagnetism and thought it gave substantial support to mobilism. He was a geologist, had studied mineralogy and petrology, and was good at mathematics – training that set him apart from physicists Vacquier, Mason, and Raff, helping him realize the implausibility of their explanations of marine magnetic anomalies. His mathematical abilities helped him with computer analysis which, at the time, set him apart from most geologists. Thus, when he began to read about marine magnetic anomalies, he was well positioned to reinterpret them.

2.9 Vine begins research, 1962

Bullard gave a short introductory course to recruit undergraduates at Cambridge to the Department of Geodesy and Geophysics. Vine liked what he heard. Hill gave a course to Cambridge undergraduates on geophysical exploration. Vine again liked what he heard. In April 1962 he was awarded a three-year Shell Studentship to pursue a Ph.D. in the Department of Geodesy and Geophysics. Notwithstanding, he still thought he might become a teacher, and enquired at Cambridge University's Department of Education about later obtaining a Certificate of Education (Vine, 2001: 50).

Soon after he was awarded the Shell Studentship he spoke with Hill and decided on marine geology. He recognized that despite the many discoveries of the 1950s, there was still so much to learn about ocean basins. Moreover, Vine thought that there must be a record of drifting continents on the seafloor; there must be something peculiar about the ocean basins that would give some clues.

I remember in my last year as an undergraduate I went to Madingley Rise and spoke to Maurice Hill. Well, I distinctly remember talking to Maurice Hill, and Maurice Hill just told me some of the things they were doing as a marine geophysics group, and the great impression I came away with from that interview ... from that discussion was: well, whatever you did at that time, you turned up something new. There were all these new instruments there ... a few years before they had measured heat flow for the first time, and eureka, got a surprising result. And then he talked about someone putting down magnetometers in the Western Approaches to the English Channel, and voilà, they got surprising results and interesting results. And I thought, well, I cannot go wrong here ... virgin ground, you couldn't go wrong there. I mean this was true particularly in the fifties and even in the sixties when I began. I went to that department because I felt it was an exciting area, being at the forefront of geology and geophysics. I particularly wanted to work in geophysics. But it was quite a difficult field to get into as a geologist. I was primarily a geologist by training because geophysicists were ... even to this day, are more commonly physicists or mathematicians by training, and only very recently have you had undergraduates in geophysics ... So one could only approach through one route or the other, and it was more common to approach it through mathematical physics or physics. I also wanted to work on the marine side because I just felt that this was a closed book; you just had to open it somewhere and there must be something new. You know two-thirds of the Earth's surface was deep ocean crust, and we really didn't know much about that and it had been only fifteen years before that we discovered how thick it was ... And yet people had all sorts of nonsense ideas about how permanent it was and how uninteresting it was. It seemed to me that if drift had occurred at all, then there should be some record of it in the ocean basins; there must be something peculiar about the ocean basins that would give some clues. I think it was obvious, so obvious that most people didn't do it or point themselves in that direction.

(Vine, 1979 interview with author)

Hill told Matthews about Vine, and Matthews noted in his diary that he had discussed programmes for him.

Discussed a programme for F. J. Vine – a geologist from John's [St. John's College] who wishes to join us. If he wants to do petrology, works with Joe [J. R. Cann]. Palaeontology, Brian [Funnell]. Other geology work with me viz: Sed-petrology of cores, especially clay mineralogy and area of a turbidity current. Radioactivity of hard rocks. Geophysics with me viz: Model studies on interpretation of magnetic and gravity anomalies.

(Matthews, diary, May 7, 1962; my bracketed additions; I thank Vine for
identifying Cann and Funnell)

When Vine later saw Matthews' entry, he (2009 email to author) remarked, "I am not sure why Drum spelt out so many options because I am sure that I wanted to do geophysics."

Matthews was the best person in the department to be his supervisor. Like him Matthews was trained in geology. Matthews was the third geology graduate to be admitted to the Department of Geodesy and Geophysics; Hospers, Irving, Matthews, and now Vine; geology served Cambridge geophysics well. There was Bullard, who later influenced and helped Vine enormously, but Bullard's students were physicists and theoreticians, which Vine was not. Then there were the instrument makers, which Vine was not. Vine later spoke about the department, about his being only the third, not counting Hospers, undergraduate geology major to be admitted for graduate work, and, looking ahead, about how he managed to obtain his Ph.D. without building an instrument, which he thought had been an unwritten requirement for obtaining a Ph.D. at Madingley Rise, and which he thought had been why Irving had not been awarded his Ph.D.

[The Department of Geodesy and Geophysics] was isolated from the other Earth science departments. Within that department, within that physical setup, so many of the people in geodesy and geophysics were theoreticians or physicists. In fact, Geodesy and Geophysics ... was an entirely graduate department. There was a little bit of undergraduate teaching, which was just a sales talk by Teddy Bullard to bring graduates like myself in. To some extent there was a geodesy aspect ... and gravity measurements were an important aspect ... with Ben Browne ... and sea floor gravity. But, in general, people were either theoreticians, typically working under Teddy Bullard, or they were physicists who were building instruments. And, of course, that was very appropriate to the forties and the fifties because one needed to build or adapt geophysical instruments for use at sea, and this is typically what they did. But really, as many of us foresaw, that sort of philosophy was drying up round about 1960. You were running out of new things, new ideas to build. I mean the heat flow equipment had been built, bottom magnetometers had been built, deep-tow magnetometers had been built, surface magnetometers had been built; they were pretty well running out of things to do. So I think the first time they took on a geologist was Drummond Matthews ... I think he must have been the first geologist to do a Ph.D. in that department. And then I as his student essentially was the second geologist. No! Well actually there was an earlier geologist, and this was all part of the incredible story; there was an earlier geologist, and that was Ted Irving[12]... To put the story in a nutshell. The story or the legend always is, and I think it is essentially true, that Ted Irving did not get his Ph.D. because he did not build an instrument.[13] With the personalities involved, it was said that to get a Ph.D. in the department you had to build an instrument. I don't know if Drum did build an instrument; yes, he may have built a device just to satisfy the requirement for measuring sound velocity or seismic velocity in sediments. I'm not sure if he did that or Tony Laughton did that or both of them had a go at it.[14] I think he might have done something to satisfy this requirement ... In a way I was the next test case after Ted Irving because I didn't build equipment; I used equipment, I maintained equipment at sea .... And for that reason, although the significance is lost on almost everybody, the introduction to my thesis starts, "The apparatus used in this work was computer programs because what I did do was build or write computer programs."[15] This was the new thing obviously at the time with the advent of digital computers and that was the breakthrough, certainly in relation to interpreting magnetics and gravity was that now you could generate potential fields over rather complicated bodies using computers, which opened up a whole new ballpark.

(Vine, 1979 interview with author; my bracketed addition and notes)

In October 1962 Vine joined the department, which had now been consolidated at Madingley Rise close to the Observatory. Matthews was away in the northwest Indian Ocean, completing detailed bathymetric and magnetic surveys aboard the HMS *Owen* over the Carlsberg Ridge. This was part of the International Indian Ocean Expedition and it was Matthews' results that led Vine to their eponymous hypothesis. Matthews would have taken Vine with him; however, Vine, not knowing about the opportunity, had already made plans to work in Spitsbergen.

When I went to Madingley Rise . . . in October 1962, Drum was away. He was at sea. He was at sea an awful lot in the Indian Ocean. He was annoyed with me because I had – I didn't know and had arranged quite innocently to go to Spitsbergen on a geological expedition – well no, actually I did geophysics on Spitsbergen. Spitsbergen is Brian Harland's stamping ground. He has run expeditions to Spitsbergen for as long as anybody can remember, including he himself. He went this year, yet again. I went to Spitsbergen in 1962 after graduating. This had to be arranged up to a year beforehand. I thought it would be rather fun, a completely new experience, and, in fact, they proposed for the first time to do some gravity and magnetics . . . When Drum heard about this – when I had eventually arranged to go to Madingley Rise about six months later after arranging to go to Spitsbergen, he said, "What the hell are you wasting your time going to Spitsbergen, you come to sea with us." Well I didn't know I could come to sea with him. I did not know what their plans were. So I didn't go. Had I been able to start earlier, I probably would have gone to sea. Anyway, he was away.

(Vine, 1979 interview with author)

With Matthews at sea, Hill served as Vine's supervisor during his first term. Hill was too busy to give him much direction.

Maurice Hill was my supervisor for the first term, and he was never available . . . He was up to his ears in administration. I always had to catch him when he was dashing out the door . . . I'd go into his office on the rare occasion when I could get an interview, and he would be on the phone all the time.

(Vine, 1979 interview with author)

Once Matthews returned, he and Bullard became Vine's biggest supporters. Matthews encouraged Vine to work on marine magnetic anomalies, strongly believing that remanence was important, and that the geomagnetic field reversed from time to time. And, perhaps most importantly, Matthews' decision to make a highly detailed magnetic survey over a small section of the Carlsberg Ridge provided Vine with the data that led to his coming up with their hypothesis. Matthews also favored continental drift, but Vine needed no prompting about drift. Bullard offered general encouragement, and helped Vine with his work on Cambridge's computer. Bullard also favored field reversals.

Discussing their influence on him, Vine emphasized Bullard's willingness to work with graduate students, the help he gave him with Cambridge's computer, and how he talked to him about the hypothesis.

So the people who influenced me were very few. I think Bullard and Matthews most of all. Hill, as I've indicated was primarily an instrument man and a physicist by training. (Maurice was a tremendous person and it was a tragedy when he died and the way he died.) He looked a bit askance at me as a geologist. I don't know if he felt he didn't understand what I was doing or didn't understand me but he seemed to be unimpressed. When I had the idea, Drum Matthews was away – he wasn't at sea this time, he was on his honeymoon – there was hardly anyone to explain it to, and I remember explaining it to Teddy Bullard. He was tremendous, particularly with graduate students. He liked graduate students very much, and if you had something you wanted to see him about, whether it was a computer program or an idea, he would drop everything; he would have you in his room, and talk about it. There was nothing he liked more; he was absolutely fantastic. I think Teddy could see straight away the potential significance of the idea. Although he was a geophysicist he had done some geology. However there was always a certain amount of antagonism between him and geologists, particularly in this country. But he had a feel for geology, which few geophysicists had. So Bullard recognized the significance. He was very encouraging.

(Vine, 1979 interview with author)

Bullard and Matthews believed that the geomagnetic field undergoes reversal. Vine raised this point while discussing Matthews' role in helping him come up with the hypothesis. Matthews linked Vine with Hospers' work on reversals.

Drum Matthews' role was very significant. Obviously, I spoke to him most when he was there. And, I think, in rather subtle ways in which I've only come to realize subsequently, he did influence the way I thought. His own ideas on reversals ... He had been at Cambridge when Hospers was there and worked on the magnetics of Iceland, in which he documented reversals.[16] There were reversals, and they most certainly were field reversals. Now I met Hospers subsequently very briefly, but I didn't know Hospers at Cambridge, we didn't overlap. So I had not been exposed to him or his work directly, but Drum sort of passed that on. He passed on rather subconsciously the idea that field reversals were probably very likely. I was never adverse to the idea of reversals. I think probably because of Bullard and Matthews. They both had an open mind and thought it was quite possible, really no strong reason for disbelieving.

(Vine, 1979 interview with author)

Vine also recalled what Bullard used to say about Uyeda's discovery of the self-reversing Haruna dacite (II, §1.10, §2.8).

The hang-up was self-reversal, and there is this incredible business – it was one of Bullard's rules. The Japanese had studied some rocks in the laboratory for self-reversal by heating them up and cooling them down, and sure enough the second rock they looked at did this phenomenon of self-reversal. My God, obviously it is very common, the second rock you look at ... Well, it turns out that this was the Haruna dacite. Well, a dacite is not a very common rock-type ... And they went through about 2000 rocks which did not exhibit a self-reversal. One of Bullard's rules: the exception always turns up when you've done the thing two or three times;

the statistics look incredible. When you've done it 2000 times, it looks completely different. But it is amazing how the exception comes up in the first two or three times.

(Vine, 1979 interview with author)

Matthews also insisted that remanence was important.

This is another almost hidden influence of Drum Matthews. Drum had worked on dredged basalts, primarily on their alteration, in his thesis. He was very familiar with magnetic properties of rocks, partly because of what Jim Hall was doing on basalts but also because there was the other work of Girdler, and Irving, and Belshé, and Runcorn, and some of it earlier work at Cambridge before my time. So Drum probably instilled in me at quite an earlier stage that remanence was or could be important.

(Vine, 1979 interview with author; Jim Hall later went by Ade-Hall)

It was Matthew's influence that decided Vine to work on marine magnetics. "I was certainly directed toward [marine magnetics] by Drum Matthews" (Vine, 1979 interview with author). The choice was an excellent one. So was Matthews' approach: if it turned out that progress could not be made, Vine could forget about it, there were other things to do.

And the philosophy at Cambridge [Vine recalled], certainly Drum's [Drummond Matthews] philosophy, which I think was a very good one in a way was that: Well, look, this is a mine of data that we have got, we quite obviously don't know really what to do with it, it's all very well we're accumulating it but it's a bit difficult to interpret and nobody seems to make any sense out of it. But anyway have a go at it for six to eight months or a year and if it doesn't look like it is getting anywhere we will forget about it and try something else.

(Vine, August 1979 interview with author; my bracketed additions)

Matthews also told him to do a literature review, which he did from October to December 1962. Before describing that, I return to Mason's and Raff's continuing work on marine magnetic anomalies; it was through researching that review that Vine learned of their vitally important work. He wrote the review before writing the first draft of his and Matthews' 1963 paper in which they introduced their hypothesis.

2.10 Mason and Raff on magnetic anomalies, 1962–1963

I was familiar with ideas about sea floor spreading and reversals of the earth's magnetic field, and I could have kicked myself for not thinking of the idea, particularly because, had I looked more carefully at our map, I would have realized that some of the seamounts might be reversely magnetized, and this might have headed my thoughts in the right direction.

(Mason, 2001: 41)

Both Mason and Raff continued working on marine magnetic anomalies after completing their survey of the northeastern Pacific. Neither changed his position even though both were offered further hints that reversely magnetized bodies might be responsible for some magnetic anomalies. I begin with Mason. He and Bullard agreed to co-author a review of marine magnetic anomalies for *The Sea* (III, §3.20). Their manuscript was received by the editors in October, 1960. However, Mason continued working on the manuscript, which was not published until 1963, and discussed some forthcoming work by J. H. Ade-Hall, a student in the Geology Department at Imperial College with interests in the petrology of the iron minerals (Ade-Hall, 1964). Mason also discussed his unpublished work on the magnetization of a seamount with an associated magnetic low, which he thought could be interpreted as caused, not by reversed magnetization (as would appear at first sight), but by normal magnetization; the magnetic low arose, he argued, from a deficiency of magnetic material at the roots of the seamounts as compared to normal crust.

In some cases the effect of the visible seamount may be overshadowed by the effect of its roots on the normal crustal structure. Fig. 23 [his figure] shows a plot of the anomaly over a row of four seamounts in the north-east Pacific. It is dominated by the magnetic lows to the south of the topographic peaks and the highs to the north. At first sight it might appear that the seamounts are negatively magnetized, but, when the observed anomaly is compared with the theoretical anomaly computed for normal magnetization of the topography, it is seen that anomalies of the correct sign do in fact occur where they should be, but are masked by larger negative anomalies of greater wave-length. It is evident that the latter anomalies represent a deficiency of magnetic material in the roots of the seamounts as compared to the normal crust.

(Bullard and Mason, 1963: 208–209)

He also rejected Laughton and company's 1960 appeal to reversed magnetization (§2.4) to explain the magnetic anomaly associated with a seamount 350 miles north of Madeira Island, correctly noting that they had not conclusively shown that the seamount was reversely magnetized.

There is no seamount for which the magnetization has been conclusively shown to be reversed, though Laughton *et al.* (1960) have suspected that the one shown in Figure 20 may be.

(Bullard and Mason, 1963: 209)

Given Mason's retrospective comment that he was sympathetic to field reversals, and his 1958 remark that he was not confident about his explanation of the magnetic highs and lows as representing an excess and deficiency of magnetization (§2.3), it is surprising that he unreservedly rejected the idea that some seamounts are reversely magnetized, claiming, "It is evident that the latter anomalies represent a deficiency of magnetic material in the roots of the seamounts as compared to the normal crust" (Bullard and Mason, 1963: 209).

Raff was presented with an even stronger hint that oceanic crust contained reversely magnetized material. And he recognized it. Raff and D. C. Krause, a Menard student, were asked to select a drilling site during the experimental drilling phase of the Mohole project in the northeastern Pacific (Raff, 1963). They chose a site, designated as EM7, beneath a positive magnetic anomaly. When core was recovered, Cox and Doell (1962) found that the oriented sample of cored basalt

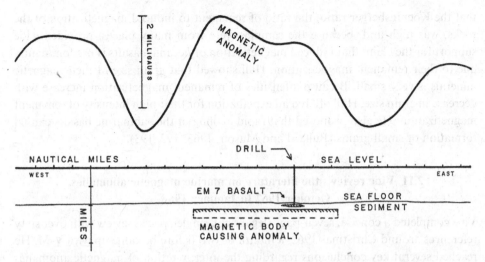

Figure 2.9 Raff's Figure 2 (1963: 956). The magnetic body causing the positive magnetic anomaly is magnetized in the direction of Earth's present field. The reversely magnetized basalt, the EM7 basalt, is situated above the principal rock mass that causes the positive magnetic anomaly as suggested with Raff's first solution.

from drill hole EM7 is reversely magnetized, not normal as expected. Raff suggested three possible explanations.

Thus, the measured magnetic anomaly and the measured remanent magnetization of the EM7 samples allow for only three model solutions. (1) The massive rock causing the magnetic anomaly lies under the positive portion of the anomaly and is magnetized with the present field, and the basalt samples of EM7 do not represent the main rock mass of the area; (2) the main rock mass of the area has a gap where the positive magnetic anomaly is shown and is reversely magnetized like the EM7 samples, and the EM7 samples are from a small isolated body or from a thinly tapered edge of the principal rock mass; or (3) EM7 was drilled into the main body of rock associated with the positive magnetic anomaly over the drill site (an unlikely possibility), and the accompanying magnetic anomaly is positive because this rock mass is bounded on each side by an equally large mass of rock that is also reversely magnetized and approximately twice as strongly magnetized.

(Raff, 1963: 956)

Raff clearly preferred (1) or (2) over (3) as an explanation of the positive anomaly. With either (1) or (2) the main mass of rock beneath the positive anomaly is magnetized in the direction of the Earth's current field. Raff probably preferred (1) for he illustrated only it (Figure 2.9).

Mason also discussed his unpublished findings and those of Hall on dredged basalts from the Mendocino Fracture Zone. The recovered basalts, unlike those available to Matthews, were fresh and Mason found that their intensity of remanent magnetization was exceptionally high, higher than both that of continental basalts and that of Matthews' altered basalts from Swallow Bank. Their findings established that the Koenigsberger ratio, the ratio of remanent to induced magnetization of the rocks, was high and, because the samples were from many places, provided wide support for the claim that induced magnetization of oceanic basalts is predominantly due to their remanent magnetization. Hall showed that grain size of their magnetic minerals is very small. Because intensities of remanent magnetization increase with decrease in grain size, Hall offered an explanation for their high intensity of remanent magnetization. He also proposed that rapid cooling of the submarine basalts caused formation of small grains (Bullard and Mason, 1963: 192–193).

2.11 Vine reviews the literature on marine magnetic anomalies, October 1962 to January 1963

Vine completed a concise, seven-page (legal size), single-spaced review with over sixty references around Christmas 1962, a month or two before he came up with V-M. He reached several key conclusions regarding the interpretation of magnetic anomalies over deep ocean: it is better to use computer simulations rather than physical models, to invoke remanent not induced magnetization, and to appeal to reversed magnetization brought about by field reversals not self-reversals. He came close to coming up with his hypothesis even before he saw the results of Matthews' detailed survey over the Carlsberg Ridge.

Interpreting magnetic anomalies was difficult because there were many parameters whose values were unknown, for instance the shape, intensity, and direction of magnetization and depth of magnetic bodies.

With the invention of the total-intensity fluxgate and proton precession magnetometers it has become a comparatively easy matter to record the earth's magnetic field over the oceans with an accuracy approaching one part in 10^5.... Clearly the technique itself is approaching perfection and, dependent on the success of the interpretation of the results, this could be a very powerful method. However, as is generally known, interpretation of magnetic surveys is particularly difficult because of the many parameters involved, essentially those of shape, intensity of magnetisation and depth of burial of the magnetic body.

(Vine, 1963a: 1)

Simulating anomalies by computers allowed the many variable parameters to be manipulated much more readily than by using physical models.

The purpose of this essay is to review published magnetic surveys and traverses at sea, methods used in interpreting them and current lines of approach. The use of computers in more recent years to simulate anomalies, and conversely, magnetic bodies from anomalies, may herald a breakthrough in interpretation methods. Computer techniques are probably more potent than model studies and easier to handle, judging by the dearth of model studies in the literature.

(Vine, 1963a: 1)

The unknowns were many and any additional information that could reduce their number making them "more realistic" was "highly desirable."

As stressed earlier, the number of unknowns, and hence assumptions, that have to be made in magnetic interpretations are so numerous that any additional information which reduces these, or at least makes them more realistic, is highly desirable.

(Vine, 1963a: 3)

He saw it as especially important to know more about the magnetic properties of oceanic crust, to know more about the relative contributions of induced and reman-ent magnetization, and, if the latter dominates, is it of normal or reversed polarity.

Vine was sure that the anomalies could not be explained solely in terms of induced magnetization and the role of remanent magnetization was crucial. As already noted, Matthews had stressed the importance of remanence to Vine. In his literature review, he mentioned Matthews' (1961a) and Laughton and company's (1960) work on dredged basalts, and he agreed with Mason (1958) that appeal to remanence was needed to keep susceptibility contrasts within reasonable limits. But he also thought that more work had to be done on ocean basalts to get a more reliable estimate of their remanent and induced magnetization. Vine also referred to Mason's and Hall's unpublished work on dredged basalts from the Mendocino Fracture Zone, which Mason was to describe in his upcoming paper with Bullard, cautioning that it had not been confirmed.

All too little is known about the magnetic properties of oceanic basalts. Work on dredged samples suggests that they are not essentially different from exposed basalts but would indicate that they invariably have very strong remanent components, such that the remanence is very much greater than the induced intensity (18), (21) ... Values of susceptibility contrasts assumed in models simulating magnetic anomalies often, necessarily, have to be high (commonly $\sim 10^{-2}$ (19), (26)), higher than is reasonable in the light of existing measurements on basalt samples. Bullard and Mason (7), however, report the dredging of basalt from the Mendocino scarp, with an exceptionally high intensity of magnetisation due to the presence of maghemite (γFe_2O_3) in the sample.

*(Vine, 1963a: 3: (18) refers to Laughton et al. (1960); (21) to Matthews (1961b); (19) to Mason
(1958); (26) to Raitt, Fisher, and Mason (1955); (7) to Bullard and Mason (1963))*

Vine later recalled that he is pretty sure he briefly met Hall at Imperial College before finishing the review (Vine, March 11, 2009 email to author).

Vine's insistence that remanence is important was not new. He, however, was also convinced that reversed magnetization should be invoked, and favored geomagnetic field reversals over self-reversals. Vine, as already noted, had been strongly influ-enced in these matters by Matthews and Bullard, both of whom favored reversals, and were familiar with Hospers' work at Cambridge a decade earlier. But Vine also had his own reasons: seamounts. He claimed:

Seamounts and volcanic islands give rise to large and obvious anomalies but these can rarely be explained by models assuming uniform magnetization throughout and directed parallel to the present earth's field ... The discrepancy [between computed and observed anomalies] also suggests that there may be a large thermo-remanent component of magnetisation, probably often reversed relative to the present earth's field. Several anomalies may well have been misinterpreted as a result of neglecting this possibility.

(Vine, 1963a: 2; my bracketed addition)

He discussed Girdler and Peter's appeal to a self-reversal to explain the pronounced positive magnetic anomaly just north of the rift valley in the Gulf of Aden (§2.4). He agreed that the bodies were likely reversely magnetized, but rejected their appeal to a self-reversal.

Girdler and Peter consider it essential to assume reversed magnetisation in order to interpret a linear anomaly in the Gulf of Aden and support this by convincing computations (12). They postulate that the reversely magnetized intrusive body has normally magnetized baked margins and therefore favour a chemical mechanism for the reversed magnetisation. This does not strike one as being a necessary corollary.

(Vine, 1963a: 2; (12) refers to Girdler and Peter (1960))

Vine followed up his criticism of Girdler and Peter by edging closer to what would become V-M.

If current theories, regarding the impermanence of the ocean floor, are correct (27), palaeomagnetic evidence would suggest that the thermoremanent component of oceanic basalts should, in most cases, be approximately normal or reversed.

(Vine 1963a: 2; (27) refers to Runcorn's edited 1962 Continental Drift*)*

Vine's reference was to the whole volume, not to any specific paper within it. The volume included a paper by Dietz with his version of seafloor spreading (III, §4.9). Vine, however, did not have seafloor spreading in mind specifically, only the ideas of continental drift and youthfulness of ocean floor. Explaining why he referenced the whole volume and what he had in mind by the foregoing passage, Vine recalled:

I think I quoted Runcorn's edited volume because the whole volume concerns Continental Drift and hence implies impermanence of the ocean floor which in turn implies that the ocean floors are very young, geologically speaking. The paleomagnetic results I am referring to are all those for geologically young material, i.e. formed millions or tens of millions of years ago for which the remanent magnetization direction is not very different to that at the present day (normal or reversed) ...

(Vine, March 11, 2009 email to author)

So he had not yet combined seafloor spreading and reversals of the Earth's magnetic field. As Vine (March 11, 2009 email to author) later put it, "Clearly I too was a bit slow in putting together seafloor spreading and reversals of the Earth's magnetic field, although it must have dawned on me fairly soon after writing the review."

Continuing his discussion of reversed magnetization, Vine stressed the need to obtain oriented samples of oceanic basalt. The magnetization induced by the present earth's field and the intensity of remanence had been measured on a small number of unoriented dredged rocks, but the *direction* of remanence *in situ* (specifically whether it is normal or reversed) can only be determined in an oriented sample and oriented samples had not yet been retrieved from the ocean floor (Vine, 1963a: 2). He also suggested how this might be done.

More detailed knowledge of the magnetic properties of the crystalline basement rocks would be of great value in postulating models to account for magnetic anomalies at sea. Results from the future Mohole cores will clearly be of considerable value. It is also suggested that the proposed Mini-Mohole Rig, from which 12 to 18 inch solid rock cores now seem feasible, if the rig could be alighted on a suitable surface, would also be invaluable in this connection. As well as providing material for the petrological and dating laboratories, the core could probably be taken as an orientated specimen. The orientation of the rig could be telemetered, and the core inscribed in some way, for example by a line down its length on one side.[17]

(Vine, 1963a: 3)

Vine did not know that a Mohole core had already been obtained, and that Cox and Doell (1962) had determined that it was reversely magnetized (§2.10).

Vine discussed interpretation methods by identifying two obstacles, which he believed explain the lack of success in interpreting marine magnetic anomalies: the interpretative mathematical models were tedious to apply because of the many assumptions, and the assumptions had little geological validity.

It is noticeable from the literature that there has been a considerable gap between the theory of interpretation and that which is applied: possibly because most oceanographers are not mathematicians but more probably because the methods have often been tedious to apply and of little geological validity. The numerous assumptions that must be made and the infinity of possible solutions have obviously deterred many.

(Vine, 1963a: 3)

Vine had a background in geology and mathematics. He also knew that K. Kunaratnam, a student of Bruckshaw and Mason's at Imperial College had the needed program. He no longer remembers just how he found out about Kunaratnam's program, but thinks that it was not through Bruckshaw or Mason, but through Vacquier or Grossling.

I am not sure how I found out about Kunaratnam's program. It may well have been via Vic Vacquier or Bernardo Grossling in the States. I do not recall meeting or having any contact with either Bruckshaw or Mason at this time. When I went to Imperial (presumably before I wrote the review) I spent most of the time with Kunaratnam, and I think I also met Jim Hall, briefly, as I mentioned earlier.

(Vine, March 12, 2009 email to author)

Vine visited Kunaratnam at Imperial College, and was impressed with his programs, one of which allowed possible configurations of the magnetic body to be computed from observed anomaly profiles:

Kunaratnam at Imperial College, London has devised a program ... to compute possible basement rock configurations from observed anomaly profiles, (this should be available within a month or so). The program was evolved in connection with the East-West traverses obtained off the west coast of America to suggest possible geological models for the crust in these areas. Here, since there is an upper limit to the susceptibility contrast it is reasonable to assume, slabs of igneous material limited to the crustal layer 3 do not give sharp enough anomalies and slabs in layer 2 are suggested (20). This type of program strikes one as being particularly useful.

(Vine, 1963a: 4; (20) refers to Mason and Raff (1961))

Vine added that Kunaratnam had also devised a program which allowed one to assess the magnitude and direction of magnetization of a three-dimensional body of known dimensions if the magnetic anomaly developed over it was well known.

Recently Sprague (for Vacquier) at Scripps and Kunaratnam (for Mason) at London have devised programs to assess the magnitude and direction of the intensity of magnetisation of a magnetic body whose boundaries are well known and using a magnetic survey carried out over it. Clearly, this type of program is of limited applicability but as suggested by Vacquier, and carried out by Kunaratnam, may, for example, be applied to accurately surveyed seamounts. If a value of susceptibility is assumed, an idea of the remanent vector may be obtained. This program has been successfully run by Kunaratnam for seven seamounts occurring within the area surveyed off the west coast of North America [and is now available at Cambridge].

(Vine, 1963a: 5; Vine added the bracketed addition after he had adapted Kunaratnam's original program for the EDSAC digital computer at Cambridge)

In an appendix he explained the program in more detail:

Anomalies due to horizontal sheets defined by the topographic contours are first calculated for unit magnetisation along each of three mutually perpendicular directions. A simple numerical integration is then performed with respect to depth to obtain the anomaly due to the body. The contours themselves are approximated by rectangular sections, the rectangular axes being chosen in such a way that number of sections required to represent the contours is kept to a minimum. The actual components of magnetisation are then obtained from a best fit with the observed anomaly.

(Vine, 1963a: 7)

Vine closed his review by again accenting the importance of using computers. But he added a caveat, wondering if he was up to the task of working with such programs.

In conclusion it seems highly desirable that any interpretation technique should be able to take account of remanent magnetic intensity, even if unknown. Certain computer programs would appear to be capable of doing this. Possibly model studies could also, but there is no evidence for this. Model studies would also appear to necessitate very elaborate, cumbersome and, presumably, costly apparatus.

As a geologist it seems more appropriate to attempt model studies but a consensus of opinion would appear to favour computer techniques. As to whether a geologist could satisfactorily grapple with these remains an open question.

(Vine, 1963a: 6)

He was up to the task. Indeed, the first thing he did with Bullard's encouragement was adapt Kunaratnam's seamount program for the Cambridge computer.

2.12 Matthews' meticulous survey over the Carlsberg Ridge, November 1962

Vine first described Matthews' survey in a draft of his and Matthews' paper in which they introduced their eponymous hypothesis.

In November, 1962, *H.M.S. Owen* as part of the INTERNATIONAL INDIAN OCEAN EXPEDITION carried out a detailed magnetic survey over a part of the Carlsberg Ridge. The survey, centred on 5° 21′N, 61° 45′E, covers a rectangular area of 52 × 37 nautical miles, its longer side being orientated N40°E, that is, approximately at right angles to the supposed trend of the Ridge in this area. The average track spacing is one nautical mile, and this, together with cross-tracks, represents nearly 2000 miles of continuous profiling within the survey area. Complete bathymetric records and virtually complete total magnetic field profiles were obtained throughout.

Eight anchored beacons were put down in the area to facilitate fixing by radar ranges and visual/radar bearings. In places however dead reckoning had to be relied on. Owing to the movement of the beacons the probable accuracy of tracks relative to each other is considered to be ± ½ mile. Star sights were taken at various points to determine the geographical position of the tracks, again the probable error in this is ½ mile.

(Vine, 1963b: 1)

Vine (1963b: 2) added, "The magnetic anomaly contour map arising out of this computer reduction is probably the most detailed obtained to date over the deep ocean."

A month after Vine began studying at Madingley Rise, Matthews was over the Carlsberg Ridge aboard HMS *Owen* conducting a magnetic survey of the seafloor. The survey differed strikingly in both areal extent and detail from magnetic profiling at Lamont and surveying at Scripps. At Lamont, Ewing preferred recording profiles routinely as his vessels crisscrossed the oceans. Workers at Scripps, like those at Cambridge, undertook magnetic surveys, devoting most of their attention to the superb multi-staged survey of the northeastern Pacific. Their survey covered a much larger area than that by Cambridge workers, but was less detailed. Both Lamont and Scripps had more funding to support their work in marine geology than those at Cambridge. The US Office of Naval Research provided abundant support to Lamont and Scripps, and even helped, albeit to a lesser extent, Cambridge marine geologists. With only modest support, marine geologists at Cambridge wisely decided to conduct small but highly detailed surveys.

Thus Matthews, following the maxim to its extreme, undertook the most detailed magnetic survey yet in deep water aided by a crew schooled in seamanship. Track spacing was five times closer than the most detailed magnetic surveys in the northeastern Pacific undertaken by Scripps and by the USCGS. The survey, partly supported by the US Office of Naval Research, took ten days. Magnetic anomalies were measured with Hill's proton precession magnetometer, completing a cycle at one minute intervals. Matthews combined the magnetic survey with bathymetric and gravimetric surveys, which allowed Vine to determine whether there was any meaningful correlation between magnetics and bathymetry.

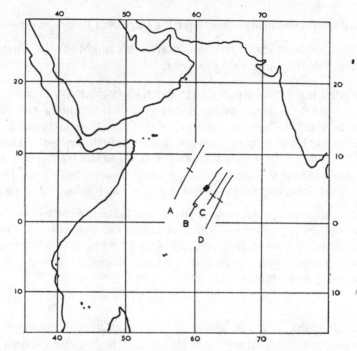

Figure 2.10 Matthews, Vine, and Cann's Figure 1 (1965: 676). Lines A through D show positions of magnetic/bathymetric profiles taken across the Carlsberg Ridge. Median rift valley is marked by ticks across lines showing profiles. Black rectangular area on line B marks the area of Matthews' detailed survey. Profile D was obtained by HMS *Owen* in 1961. Profiles B and C were obtained by HMS *Owen* in 1962, several weeks after the detailed survey was completed.

Matthews wanted to make sure that the survey centered over the crest of the Carlsberg Ridge. Guided by a sparse record of shallow earthquakes over the Carlsberg Ridge, and two previous profiles that showed a well-developed rift valley, he chose an area which he thought was directly over a section of the ridge approximately midway between where two previous profiles had crossed the ridge's rift valley (see Figures 2.10 and 2.11). At first he thought they had missed the ridge. The bathymetric data as they came aboard did not reveal a definite median valley, and if indeed there was a ridge, it certainly was not continuous. Vine later recalled Matthews' worry about being in the wrong place.

When Drum was doing the survey he thought he was in the wrong place because the median valley wasn't very well developed, in fact it's blocked by volcanism ... and the other thing is that he had a transform fault [transform faults had not yet been recognized] in it. It didn't have a major offset on it but it was enough to complicate both the bathymetry and the magnetics over almost half the survey area.

(Vine, 1997 interview with Robert White; slightly altered by Vine,
July 23, 2009; my bracketed addition)

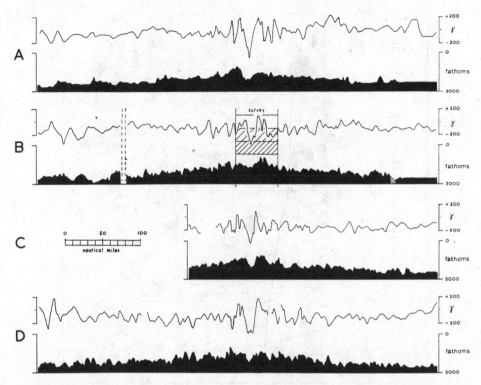

Figure 2.11 Matthews, Vine, and Cann's Figure 2 (1965: 676). Profiles A through D showing magnetic anomalies and bathymetry. Positions of profiles are shown in Figure 2.10. All profiles exhibit large negative magnetic anomalies over ridge crest. Median rift valley is prominent in Profiles A, C, and D, but partially blocked by seamount in B. Profile B crosses over the seamount just north of ridge crest and is associated with large positive magnetic anomaly. Profile D was obtained by HMS *Owen* in 1961, and was one of the profiles used by Matthews to guide where to survey. Profiles B and C were obtained by HMS *Owen* in 1962, several weeks after the detailed survey was completed. Vine had access to all profiles while writing draft to Vine and Matthews (1963).

Perhaps because of his concern, HMS *Owen* also obtained additional profiles a few weeks after completion of the survey, which were available to Vine (Figures 2.10 and 2.11, 2.12, and 2.13). But all turned out well; in fact, better than expected. HMS *Owen* had not only found the ridge's crest but, as luck would have it, also discovered two seamounts, which played a crucial role. As for the left-lateral transform fault, which they initially interpreted as a right-lateral transcurrent fault, it offset the ridge by approximately ten nautical miles (Figure 2.12).

The pattern of magnetic anomalies obtained in the survey was striking (Figure 2.13). Visual inspection revealed a pronounced negative anomaly flanked by positive anomalies. The anomalies appeared to be about ten nautical miles wide,

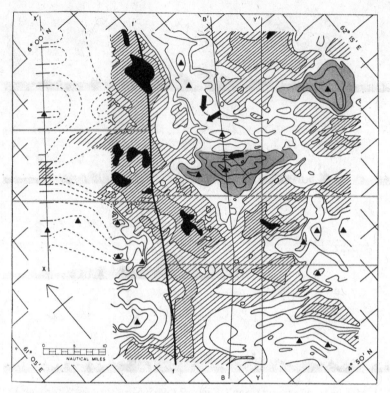

Figure 2.12 Redrawn from Matthews, Vine, and Cann's Figure 3 (1965: 677). Figure shows bottom topography from Matthews' detailed survey over the Carlsberg Ridge. Figure has been redrawn to show approximate position of seamounts (cross-hatched) and position of the ridge crest/median valley (between two sets of parallel lines). Solid triangles mark peaks shallower than 1300 fathoms; widely spaced lines show depth 1700–1900 fathoms; closely spaced lines show depth 1900–2100 fathoms; solid black shows depth greater than 2100 fathoms. Vine returned to the area aboard the RRS *Discovery* in September 1963, collecting dredged samples, additional profiles, and extending the survey along the northeast edge of the original survey [see §4.3]. The new results are shown to the left of the original survey. The three solid arrows indicate, from north to south, *Discovery* dredge stations 5106, 5123, and 5111. The line ff′ marks the position of the fault; line BB′ is the Profile B in Figure 2.11. It crosses over the reversely magnetized seamount in the center of survey. Lines XX′ and YY′ mark new profiles obtained by RRS *Discovery*. Profile marked by XX′ crossed the rift valley of the ridge north of the fault; profile marked by YY′ crossed the rift valley and central seamount. None of the results obtained from the RRS *Discovery* voyage were available to Vine while developing the hypothesis.

and ran parallel to the ridge. The pronounced negative anomaly gave a better fix on the position of the ridge than did the bathymetry (Figure 2.13). As Vine later recalled:

But what is so striking, once you get away from the transform fault area is that the central negative magnetic anomaly and the flanking positives, and particularly the gradients between them, sort of go marching through irrespective of the topography and the valley is not

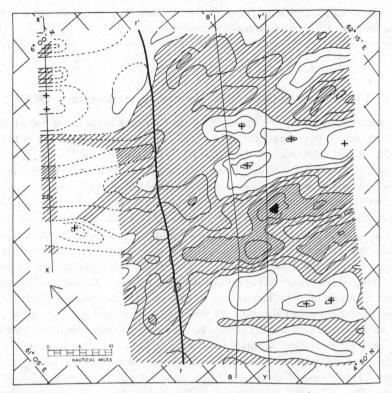

Figure 2.13 Redrawn from Matthews, Vine, and Cann's Figure 4 (1965: 678). Vine constructed this figure after coming up with his hypothesis. Their caption reads: "Chart showing magnetic anomalies in Carlsberg Ridge surveyed area and lines of the fault and sections shown in Figure 3 [Figure 2.12]. The contour interval is 100 gammas ($1\gamma = 10^{-5}$oersted). Anomalies greater than -350γ are shown by solid black; -350 to -150γ, by the closely spaced lines; -150γ to $+50\gamma$, by the widely spaced lines. Areas of positive anomaly are not lined. Field maxima greater than $+250\gamma$ are indicated by crosses."

well developed, it's partly blocked by volcanism. So the topography is really quite irregular, but the magnetics are marvelous.

(Vine, 1997 interview with Robert White)

Indeed the magnetics were so striking that Vine could see that you got great positive anomalies of topographic features which must be reversed. He could see this directly but he put it through the computer to make it generally acceptable; this was a time when it was becoming increasingly necessary do so to be believed.

And you can see that the topography is to some extent influencing the magnetics and you can see that in some cases you got positive anomalies over topographic features which must be reversely magnetized. Then I just put it through a computer so that people might believe it a bit more ... If you did something on a computer, particularly then, perhaps more so than now, they believed you – Oh, that's impressive, sort of thing ...

(Vine, 1997 interview with Robert White; slightly altered by Vine, July 23, 2009)

2.13 Vine develops the Vine–Matthews hypothesis, early 1963

I came back with this data [from the survey of the Carlsberg Ridge in the Fall of 1962] ... I said, "Well, here is this magnificent survey ... Cannot we remove the effects of the topography which is very substantial for the magnetics ... in order to look deeper?" ... Fred very quickly showed what I had already guessed, I think, that if you had a normally magnetized block it would produce a negative anomaly in this latitude. So that was settled very soon – in days. Then Fred said, "What are you going to do about the magnetization of the seamounts? How do we know what direction they're magnetized in?" We had no idea what age they were. I said, "I don't know." And I suspect that it was all him. I really don't know. We talked for hours and hours. There was a great deal of conversation ... My major contribution was always that we had to understand the mid-ocean ridge in terms of fissure eruptions ... We knew people at Imperial [R. Mason was at Imperial College, London]. Anyway, he [Mason] had a Ceylonese, Kunaratnam, who had written programmes for determining magnetization of arbitrarily shaped bodies. Fred came back with this program and got it going on EDSAC 2 ... We knew that the central valley is underlain by a normally magnetized body. We wanted to see if we could determine the direction of magnetization of ... the seamounts ... Well, we did that, and we came out ... with normally magnetized vectors on either side of the seamount ... He determined the magnetization, came out with the story, and then it was as obvious as falling off a log ... Other people in the lab had already made use of normal and reversed magnetized blocks. Fred had the Harry Hess story in his mind which he was always talking about. I was quite certain that it was fissure eruptions that we were dealing with. We knew it was remanence; I was quite certain it was remanence, and so was Fred. So, if you think of it, the hypothesis wasn't really there ... Well, I guess it was a hypothesis. It was as good a guess as any for explaining the magnetization in this little corner of the Carlsberg Ridge. And then we said, "Well, OK, the only survey we have shows these stripes [the survey from the northeastern Pacific]." And, of course, we'll explain this. So could we compute what the magnetization would look like in the northeast Pacific ...? Well, that was the story.

(Matthews, 1979 interview with author; my bracketed additions)

He came back in ... I think it must have been December. The first term Maurice [Hill] supervised me nominally, although in practice Teddy [Bullard] did. I spent the first term as one does looking into the problem and doing a review and deciding how on earth we were going to approach it ... Anyway, I reviewed the problem and, not surprisingly at the time,'62, decided the way forward was by computers, which seems obvious now but wasn't then because digital computers used for scientific purposes were pretty new, we were on EDSAC 2 ... So I wrote up this review and this conclusion by Christmas '62 and Drum took one look at this and said well if you're going to use computers, forget it, he said, I mean I cannot help you at all. I think Drum was really rather crestfallen, I saw his face drop, because I think Drum envisaged sort of playing around with iron filings and putty or something and making some analogue model of the ocean.

(Vine, 1997 interview with Robert White; my bracketed additions)

Vine is unable to recall precisely when he came up with V-M. He thinks it was in February or March 1963. He believes that it was from visual inspection of the survey results before he actually ran them through the computer. This is not surprising given his favorable attitude toward seafloor spreading and reversals of the geomagnetic field. The survey was all he needed: a pronounced negative anomaly over the ridge crest flanked by positive anomalies and anomalies associated with two seamounts. The seamount in the center of the surveyed area just north of the ridge crest appeared to be associated with a positive anomaly; the other seamount, located in the upper right corner of the survey, appeared to be associated with a negative one. Given the latitude of the surveyed region, the negative anomalies, Vine argued, are caused by normally magnetized bodies, and the positive flanking ones, by reversely magnetized bodies. So given seafloor spreading, new seafloor at the ridge crest is normally magnetized, the older material flanking the ridge crest, including the centrally located seamount, is reversely magnetized, and the oldest material, including the seamount in the upper right of the survey, is normally magnetized. Perhaps it was the combination of a ridge, a spreading center, of normally magnetized rock, flanked on both sides by oppositely magnetized rock, and the oppositely magnetized seamount that led Vine to finally relate together seafloor spreading and field reversals. Regardless of precisely what happened, Vine conceived the hypothesis before ·doing the computer analysis. Despite his emphasis on computer modeling, he himself did not need it to have the idea. As he retrospectively remarked, he needed computer modeling to convince others.

Vine had begun reprogramming Kunaratnam's program in autocode to use on EDSAC 2 before finishing his review of the literature. He fondly recalled his nights working on EDSAC 2 at the old Mathematics Laboratory, the help and encouragement he got from Bullard, and the importance of Kunaratnam's program.

So what I was really getting round to is that it became clear that after a few months what I was really most interested in was writing computer programmes or getting hold of them, and the only person that had actually done certainly two-dimensional programmes that were relevant was Teddy Bullard ... He was fantastic, particularly as far as students were concerned. If you went as a student, a first-year graduate student as I would have been, he dropped everything if he possibly could and dealt with it; he just loved dealing with students and helping students – very memorable occasions in those days. There was this EDSAC 2 at the old Mathematics Laboratory. There was batch processing, but particularly for development, you could go down pretty well on the hour for five minutes, you could run test programmes. It was great fun, you had this very primitive paneling and there were all these toggle switches on it and things, and you'd operate the computer yourself ... Tape·readers, you know, no cards or anything, all the old magnetic tape decks which barely worked, so I would have to keep going down there to run these tests – a hell of a business trying to get programmes running. But one evening a week, I think it was Thursday, we had the thing to ourselves – well not to ourselves, I think we shared it with Radio Astronomy, and they said you can come down and as long as the thing runs – it was crashing every 5 minutes of course – you can use it. If it pegs out at 10 o'clock that's too bad, if it goes on till 2 or 3 you're lucky ... So several evenings I'd be down there and Teddy would be there and he'd be a tremendous help ... But it was such fun because you were actually controlling the machine yourself in effect. Teddy was more useful because I got the mathematics for a 3-dimensional programme from Kunaratnam ... that was crucial; it was a very elegant programme which was what you needed in those days if you didn't have the speed or capacity. It was a very elegant programme and I wrote that in. When I started we had to learn machine code and all Teddy's programmes were in machine code; they were a nightmare. But the autocode had just come in because it was literally *code exec* autocode so I actually wrote in autocode and I translated these things, both two and three dimensional programmes.[18]

(Vine, 1997 interview with Robert White)

After rewriting Kunaratnam's program in autocode for EDSAC 2, Vine confirmed what he already had surmised from visual inspection of the survey's results: that the positive anomaly over the seamount adjacent to the crest of the ridge was definitely best explained by supposing that it was reversely magnetized while the other seamount was most likely normally magnetized. He discussed his findings in the first draft of what became Vine and Matthews (1963).[19] Vine wrote the draft in May 1963 while Matthews was on his honeymoon. After summarizing the nuts and bolts of the survey, he turned to the anomaly over the crest of the ridge, and argued that it is magnetized in the present direction of the Earth's field. He also noted that the intensity of remanent magnetization was particularly high, and appealed to the work of Hall and Mason on oceanic basalts by referring to Bullard and Mason's forthcoming 1963 paper in *The Sea* (§2.10 and §2.11).

For this low magnetic latitude (inclination −6°) a body magnetized in the present direction of the earth's field will produce a pronounced negative anomaly with a small positive to the north of it. Here, over the centre of the Ridge, it is generally accepted that the bottom topography reflects the relief of basic extrusives such as volcanoes and fissure eruptions, and that there is little sediment fill. The bathymetry, therefore, defines the extent of magnetic material having a considerable intensity of magnetisation potentially as high as any known igneous rock types[1], and probably higher, because it is extrusive, than the main crustal layer beneath. That the features are capable of producing anomalies is immediately apparent on comparing the bathymetric and anomaly charts [see Figures 2.13 and 2.14]. Several features have well defined anomalies associated with them.

(Vine, 1963b: 3; footnote 1 is to Bullard and Mason (1963); my bracketed addition)

Having established that the ridge crest is normally magnetized, he turned to the seamounts and his computer analysis.

Two comparatively isolated volcano-like bodies were singled out and considered in detail. One had an associated negative anomaly as one would expect for normal magnetisation, the other, completely the reverse anomaly pattern, that is, a pronounced positive anomaly suggesting reversed magnetisation. The topography of each feature and its associated anomaly were fed into ~~EDSAC 2~~ [computer] and an intensity and direction of magnetisation for each obtained.[2] The directions of the resulting vectors are shown in fig. 1 on a stereographic projection. Having computed the magnetic vector by a best fit process the computer re-calculates the anomaly

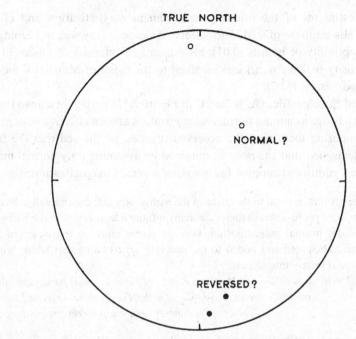

TRUE NORTH

NORMAL ?

REVERSED ?

Figure 2.14 Vine's Figure 1 (1963b). The figure is identical to Figure 2 in Vine and Matthews (1963: 948). There is no caption for the figure in Vine (1963b). The caption for the figure in Vine and Matthews is: "Directions of the magnetic vectors obtained by computer programme plotted on a stereographic projection, together with the present field vector and its reverse. Bearings and inclinations: present field vector 356°; −6° (up); computed vectors 038°; −40° (up); 166°30'; +13° (down)." The figure was reproduced with slight changes as Figure 6 in Cann and Vine (1966: 211).

over the body, assuming this vector, thus giving an indication of the accuracy of the fit. For the case of reversed magnetisation this was quite good but for that of normal magnetisation notably bad. This discrepancy is hardly surprising in that one has ignored all other topography, and the interference of all other anomalies, in the vicinity. The example of normal magnetisation is also near a corner of the area where the control at all stages of reduction, and hence ~~at all stages of reduction, and hence~~ on contouring, is less precise. In both cases the intensity of magnetisation deduced was of the order of 0.005 e.m.u. (the intensity of the normal vector being the greater); this is equivalent to an effective susceptibility of ± 0.0133.

(Vine, 1963b: 3; Vine crossed out the struck-through parts; bracketed word replaced first cross-out; footnote 2 thanks Mason and Kunaratnam for permission to rewrite Kunaratnam's program for EDSAC 2; Figure 1 reproduced as Figure 2.14; see Vine and Matthews (1963: 947–974) for equivalent paragraph.)

Thus Vine not only computed the reversed magnetization of the centrally located seamount and normal magnetization of the seamount in the upper right-hand corner of the survey (see Figure 2.12), but he also explained why the fit for the normally magnetized seamount was poor, and in doing so he did not have to resort to unreasonable estimates of the intensity of remanent magnetization and effective susceptibility. His estimate of 0.0133 for effective susceptibility was in keeping with the mean susceptibility of basalts (0.01); his estimate of intensity of remanent magnetization intensity (0.005 e.m.u.) was identical to the mean of Matthews' measurement of dredged basalts (§2.7).

He computed three profiles (A, B, and C in Figure 2.15) perpendicular to the crest of the Carlsberg Ridge assuming normal magnetization and an effective susceptibility of 0.0133. Comparing them with the observed profiles, he showed that the fit was poor (RS2). He argued that the poor fit obtained by assuming only normal magnetization provided additional support for invoking reversed magnetization.

In addition, three profiles, normal to the strike of the bathymetry and the anomalies, have been considered. Computed profiles across these, assuming infinite lateral extent of the bathymetric profile, and uniform normal magnetisation, bore no resemblance to the observed profile (Fig. 2). This was anticipated and added to the necessity of assuming that whole blocks of the survey area are reversely magnetized.

> *(Vine, 1963b: 3; see Vine and Matthews (1963: 948) for a similar paragraph, although the last sentence was replaced with "These results suggested that whole blocks of the survey area might be reversely magnetized.")*

He then recomputed profile B assuming normal and reversed magnetization and the fit was strikingly improved (see Figure 2.15).

With this analysis of the magnetic anomalies from the survey, Vine presented his first version of V-M. He claimed (Vine, 1963b: 3) that 50% of the oceanic crust might

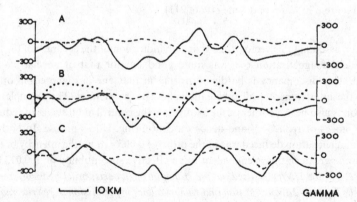

Figure 2.15 Vine's Figure 2 (1963b). The figure is identical to Figure 3 in Vine and Matthews (1963: 948). There is no caption for the figure in Vine (1963b). The caption for the figure in Vine and Matthews (1963) is: "Observed and computed profiles across the crest of the Carlsberg Ridge. Solid lines, observed anomaly; broken lines, computed profile assuming uniform normal magnetization and an effective susceptibility of 0.0133; dotted line, assuming reversals ... The computed profiles were obtained assuming infinite lateral extent of the bathymetric profiles."

well be reversely magnetized, which in turn has suggested a new theory to account for certain magnetic anomalies over the oceans. He then introduced his new idea.

The theory is entirely consistent with, in fact virtually a corollary of, current ideas on ocean floor spreadings[3] and periodic reversals in the earth's magnetic field.[4] As the oceanic crust forms over a convective up-current in the mantle, and at the centre of an oceanic ridge, it is magnetized in the prevailing direction of the earth's field; assuming impermanence of the ocean floor, the whole of the oceanic crust is comparatively young, probably not older than 150 million years, and its thermo-remanent component is therefore either essentially normal, or reversed, with respect to the present earth's field. This component almost certainly predominates for the majority of basic rocks, although it is not essential for the purpose of the theory that it should for all. Thus as ocean floor spreading occurs, blocks of alternately normal and reversely magnetized material drift away from the centre of the ridge and parallel the locus of it.

> *(Vine, 1963b: 3–4; footnote 3 references Dietz (1961a); he left footnote 4 blank; the passage is almost exactly the same as what appears in Vine and Matthews (1963: 948), except the penultimate sentence was deleted.)*

Vine left the reference for periodic reversals of the geomagnetic field blank; he needed a trip to the library. As already noted, he knew about Hospers' work, and he, like Matthews and Bullard, believed in field reversals (§2.9). In the final version, Vine and Matthews, somewhat oddly, referred to Hospers' short paper (Hospers, 1954b) instead of his definitive one (Hospers, 1954a) in which he showed that his reversely magnetized Icelandic rocks had not undergone self-reversal. Vine also referred only to Dietz's work. This was not a slight of Hess as he did not know that Hess's 1962 paper had already been published in the volume honoring Buddington, a peculiar and obscure place for it to appear (Hess, 1962).

Vine next applied the hypothesis, which he continued to refer to as a theory, to the magnetic anomalies in the northeastern Pacific. He now had a definite answer to the question that either Matthews or Laughton had asked him a year before during his Presidential Address to the Sedgwick Club.

This configuration of magnetic material could explain the lineation or grain of magnetic anomalies observed over the Eastern Pacific to the west of North America.[5] Here north-south highs and lows of varying width, usually of the order of ten miles, are bounded by steep gradients. Whereas the amplitude and form of these anomalies is difficult to simulate assuming normal magnetisation alone,[6] they are readily reproduced assuming the above model as shown in Fig 3 [Figure 2.16]

> *(Vine, 1963b: 4; footnote 5 refers to Mason (1958); 6, to Mason and Raff (1961); an expanded version of the paragraph appears in Vine and Matthews (1963: 948–949))*

Normally magnetized bodies produce positive anomalies in profile 1 from the northeast Pacific because the inclination is positive (inclination = 60°).

Vine then returned to what he supposed happened at ridge crests and attempted to explain why the anomaly associated with ridge crests is always indicative of normally magnetized crust and why it is much more pronounced than adjacent anomalies.

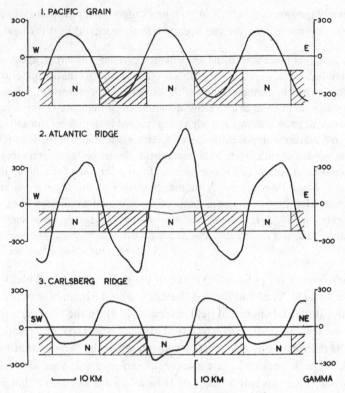

Figure 2.16 Vine and Matthews' Figure 4 (1963: 948). Figure 3 in Vine (1963b) is essentially Profile 1 of this figure. Figure 4 in Vine (1963b) is essentially Profile 2 and Profile 3. The figures have no caption in Vine (1963b). The caption for Figure 4 in Vine and Matthews (1963) is: "Magnetic profiles computed for various crustal models. Crustal blocks marked N, normally magnetized: diagonally shaded blocks, reversely magnetized. Effective susceptibility of blocks, 0.0027, except for the block under the medium valley in profiles 2 and 3, 0.0053. (1) Pacific grain, (2) Mid-Atlantic Ridge, (3) Carlsberg Ridge."

The crust at the crest of the ridge is uniformly magnetized because being young it is uncontaminated by subsequent addition of oppositely magnetized rock, whereas the sides, being subject to continuing volcanism, were contaminated, thus reducing the mean intensity of magnetization of the crust. He used the Carlsberg and North Atlantic ridges as examples (see profiles 2 and 3, Figure 2.16).

It has been recognized for some time that the pronounced anomalies recorded centrally over oceanic ridges are explicable in terms of a normally magnetized block filling the median valley and producing a positive susceptibility contrast with the adjacent crust. On the above theory this block, being most recent, is the only one which has a uniformly directed magnetic vector (cf. the area of later Quaternary basics in central Iceland). Adjacent, and all other, older blocks have doubtless been subjected to subsequent volcanism in the form of volcanoes, fissure eruptions and lava flows, often of reverse sign and hence reducing the effective susceptibility of the block, whether essentially normal or reverse. This is illustrated for the North Atlantic and Carlsberg Ridges where pronounced positive and negative anomalies respectively have been observed (Fig. 4).

(Vine, 1963b: 5)

Because the east–west trending profile of the Mid-Atlantic Ridge is at an inclination of 65°, the positive magnetic anomalies are due to a normally magnetized central block; because the southwest–northeast trending profile of the Carlsberg Ridge is at −6° inclination, the negative magnetic anomalies are due to a normally magnetized central block. The effective susceptibility of the block on both ridges is greater than that of the other blocks.

Vine also noted that the effective susceptibility values he used to compute anomalies (0.0053 and 0.0027) were two to five times lower than the value he used to compute the anomalies produced by the seamounts (0.0133). But, this was consistent with Hall's recent work showing that fine-grained oceanic basalts are more highly magnetized than the coarse-grained intrusive material at depth. Vine again referenced Bullard and Mason's forthcoming paper in *The Sea*.

Vine added that the particular anomaly pattern produced is also a function of magnetic latitude and orientation and he could generate these patterns by computer.

It can be shown that this type of anomaly pattern will be produced for virtually all orientations and magnetic latitudes, the amplitude decreasing as the locus of the ridge approaches magnetic north-south or the profile approaches the magnetic equator.

(Vine, 1963b: 5; Vine and Matthews, 1963: 948)

This helped make his hypothesis testable because he could generate the anomaly pattern that should be produced given any orientation and magnetic latitude and compare it with results from surveys.

In his final paragraph, Vine noted that he raised the susceptibility of oceanic blocks only when he had reason, namely, when blocks (central blocks of ridge crest) are uncontaminated or when they (seamounts) are made of a greater proportion of fine-grained extrusive rock to non-fine-grained intrusive rock than typical oceanic crust. This maneuver allowed him to make clear that he, unlike those espousing alternative hypotheses, did not have to invoke unreasonable susceptibility contrasts to explain the anomalies (RS3). Moreover, his hypothesis, unlike other much less specific ones, was clearly testable.

This point is important in that the steep gradients and amplitudes of the anomalies simulated above result from the vertical boundaries and net susceptibility contrasts between adjacent crustal blocks. It is appreciated that other theories based on variations in the intensity of the earth's magnetic field or periodic changes in the composition of the material forming the oceanic crust can also produce these conditions, but we venture to suggest that in the light of present knowledge, the above theory is more plausible. Above all it has the merit of suggesting further experiments which may enhance or refute it.

(Vine, 1963b: 7)

He could have expressed what he had in mind more clearly, and he did so in the corresponding paragraph in the revised version of the paper.

In order to explain the steep gradients and large amplitudes of magnetic anomalies observed over oceanic ridges all authors have been compelled to assume vertical boundaries and high-susceptibility contrasts between adjacent crustal blocks. It is appreciated that magnetic contrasts within the oceanic crust can be explained without postulating reversals of the Earth's magnetic field; for example, the crust might contain blocks of very strongly magnetized material adjacent to blocks of weakly magnetized material in the same direction. However, the model suggested in this article seems to be more plausible because high-susceptibility contrasts between adjacent blocks can be explained without recourse to major inhomogeneities of rock type within the main crustal layer or to unusually strongly magnetized rocks.

(Vine and Matthews, 1963: 949)

Vine and Matthews, unlike those who did not appeal to geomagnetic reversals, did not need huge susceptibility contrasts that implied "major inhomogeneities of rock type within the main crustal layer." As heirs to Hospers' work on field reversals done a decade earlier in their department, they were confident that the geomagnetic field did reverse. Apparently they no longer thought it worth mentioning that their hypothesis has the merit of suggesting being eminently testable. V-M might have seemed fantastic but, as Vine later recalled, he certainly thought it was better than the competition.

You have to go to real contortions to try to explain … [the magnetic anomalies] … any other way. As I say, this is one of the reasons that maintained my faith in the idea. The competitors were really very implausible. However fantastic my idea might seem, in terms of its capability in explaining the anomalies and the lateral magnetization contrast you had to invoke it was a much more powerful idea, much more effective an idea than anything that had been suggested. The only thing we had then were those great linear anomalies in the northeast Pacific. As a geologist I couldn't see that any petrologic or structural contrast would be that regular. It just seemed totally unreasonable.

(Vine 1979 interview with author; my bracketed addition)

Reactions within the department differed. Vine is not sure what Matthews first thought. Bullard was enthusiastic. He even asked Bullard if he would be a co-author, but he declined.

I remember writing the first draft when he [Matthews] was on his honeymoon. Well, there are several aspects, some people were disgusted by the whole business, they wondered who had been allowed into the research group, a raving lunatic who believed these stupid ideas … I don't know quite what Drum thought. The interesting reaction was, well not surprisingly really, Teddy's, because Teddy thought it was great and I think, being Teddy, he could see the implications if it was right – and it was really rather exciting if it was right … And again I started talking to him and walking back from the Maths Lab at some unearthly hour on a Thursday; but no, Teddy was tremendous. Something I have admitted to several people is that when I wrote the first draft, I was quite keen to have Teddy's name on it but he wouldn't. He declined, which is odd in a way because, as I say, he had contributed ... he had given me quite a lot of help with the programming.

(Vine, 1997 interview with Robert White; my bracketed addition)

Vine made changes and added new material to the final paper.

[The first draft] differed from the published paper in that it did not include the first two paragraphs and the penultimate paragraph (excluding acknowledgements). It did, however, include more details of the acquisition and reduction of the ... survey. It was reviewed internally by Maurice Hill, Teddy Bullard, and ultimately Drum. I cannot be certain, but I suspect that Hill was very unhappy with it, that Teddy was quite excited about it (recognizing the tremendous implications if it turned out to be correct), and that Drum, caught in the middle, did not know what to think except perhaps that having a research student was something of mixed blessing. All agreed that the full details of the acquisition and reduction of the ... survey [data] were inappropriate to a letter to *Nature* and so this section was removed. The problem then was that it became rather long on interpretation and speculation and short on original data.[20]

(Vine, 2001: 57–58; my bracketed addition)

Hill contributed an unpublished profile from the Mid-Atlantic Ridge and Matthews added one from the Carlsberg Ridge (Figure 2.17). The title of the paper was changed from "Magnetic anomalies over the oceans" to "Magnetic anomalies over oceanic ridges." Vine also distinguished three types of magnetic anomalies.

Typical profiles showing bathymetry and the associated total magnetic field anomaly observed on crossing the North Atlantic and North-West Indian Oceans are shown in Fig. 1 [Figure 2.17]. They illustrate the essential features of magnetic anomalies over the oceanic ridges: (1) long-period anomalies over the exposed or buried foothills of the ridge; (2) shorter-period anomalies over the rugged flanks of the ridge; (3) a pronounced central anomaly associated with the median valley. This pattern has now been observed in the North Atlantic, the Antarctic, and the Indian Oceans. In this article we describe an attempt to account for it.

(Vine and Matthews, 1963: 947; my bracketed addition)

Vine's last addition to the paper was an appeal to the preliminary Mohole drilling efforts of Raff, Cox, and Doell, who had discovered (§2.10) that some oceanic basalt was reversely magnetized. The situation was complicated however because that would not be expected from the sea-surface anomaly, an apparent contradiction that Vine and Matthews deftly rationalized as not unexpected on account of the complexity of the observed profiles.

In Fig 4 [Figure 2.16] no attempt has been made to reproduce observed profiles in detail, the computations simply show that the essential form of the anomalies is readily achieved. The whole of the magnetic material of the oceanic crust is probably of basic igneous composition; however, variation in its intensity of magnetization and in the topography and direction of magnetization of surface extrusives could account for the complexity of the observed profiles. The results from the preliminary Mohole drilling[13, 14] are considered to substantiate this conception. The drill penetrated 40 ft. into a basalt lava flow at the bottom of the hole, and

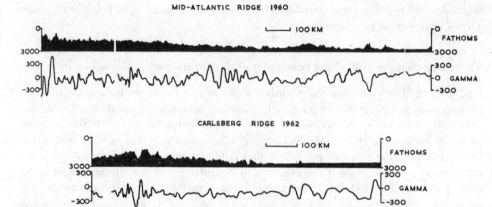

Figure 2.17 Vine and Matthews' Figure 1 (1963: 948). Profiles showing bathymetry and magnetic anomalies over the Mid-Atlantic Ridge and Carlsberg Ridge. Hill supplied the profile from the Atlantic; Matthews supplied the one from the Indian Ocean. The Atlantic profile is from 45° 17′ N, 28° 27′ W to 45° 19′ N, 11° 29′ W; the Indian Ocean one is from 3° 5′ N, 61° 57′ E to 10° 10′ N 66° 27′ E. The most prominent positive magnetic anomaly in the upper profile coincides with the rift valley of the Mid-Atlantic Ridge; the most prominent negative magnetic anomaly in the lower profile coincides with the rift valley of the Carlsberg Ridge. The lower profile also appears as Profile C, Figure 2 in Matthews, Vine, and Cann (1965). It was obtained by HMS *Owen* in 1962, several weeks after the detailed survey was completed (Figure 2.11).

this proved to be reversely magnetized[13]. Since the only reasonable explanation of the magnetic anomalies mapped near the site of the drilling is that the area is underlain by a block of normally magnetized crustal material[14], it appears that the drill penetrated a layer of reversely magnetized lava overlying a normally magnetized block.

　　(Vine and Matthews, 1963: 949; references are to Cox and Doell (1962) and Raff (1963))

The revised paper was submitted to *Nature* in late June or early July. Vine cannot remember submitting it; he thinks that either Matthews or Hill sent it off (Vine, 2001: 59). The paper was published in *Nature* on September 7, 1963, by which time Vine and Matthews were at sea aboard the RRS *Discovery* in the northwest Indian Ocean, taking additional profiles over the Carlsberg Ridge, dredging samples, and extending the detailed magnetic/bathymetric survey along the northeast edge of the original survey to determine the position of the ridge west of the fault and to see if the pronounced negative magnetic anomaly remained above it (Figure 2.12).

2.14 Morley's education and early work in paleomagnetism and aeromagnetic surveying

Lawrence Morley proposed his hypothesis in February 1963.[21] Although Morley had not worked on marine magnetic anomalies, he had the requisite background knowledge and training to come up with the hypothesis, for he was an expert in interpreting continental aeromagnetic maps, had worked in paleomagnetism, and believed in geopolarity reversals and continental drift. Although Morley's hypothesis differed from the V-M hypothesis in that he attributed about half of the magnetization of oceanic crust to induced magnetization, he, like them, made the crucial moves of invoking remanence and field reversals to explain the steep gradients between the adjacent magnetic highs and lows that Mason and Raff had discovered off the west coast of North America.

In 1938, Morley entered the University of Toronto, and enrolled in the same interdisciplinary physics and geology program that Gilchrist had designed for Tuzo Wilson a decade earlier. The outbreak of World War II interrupted his studies, and he became a radar officer in the Royal Canadian Navy, and learned enough electronics to later build his own spinner magnetometer. After the war, he obtained his undergraduate degree, and spent the summer working in the bush taking magnetic readings with a field magnetometer. Wanting to continue working in magnetism but hating ground-based fieldwork, he learned that Gulf Research and Development Company had developed an airborne magnetometer. Although Gulf did not hire him, he got a job in 1946 with Fairchild Aerial Surveys, a subsidiary of Gulf, to undertake an aeromagnetic survey in South America.

I took geophysics at the University of Toronto before the War. I took the interdisciplinary course which was physics and geology. Tuzo Wilson had been the first ... ten years earlier. There hadn't been any students for a few years, and I came in 1938. I was attracted to the interdisciplinary course, so I took it for 2½ years. Then the War came and the Royal Navy was looking for graduates of math and physics, and electrical engineering to get into radar ... I was only in my third year, but they took me anyway. I spent most of the War as a radar officer in the Royal Navy. That is where I got my electronics ... I was responsible for maintaining the radar and training the operators and explaining the whole thing to the officers on the ship. So it was very exciting; it was a new technology, and everybody thought it was magic.

After the war I figured I might as well finish my geophysics course. The fellow that started it was Lachlan Gilchrist. He had done his Ph.D. at the University of Chicago working on the speed of light, and he had just missed the Nobel Prize. He became very interested in magnetic prospecting. He had introduced geophysics at the University of Toronto. He saw the point of getting interdisciplinary courses between physics and engineering. So he set up this liaison with the engineering faculty, and he started engineering physics. He also started the interdisciplinary courses of physics and chemistry, and physics and geology. I was interested in physics and geology. Perhaps the first teacher I had was Arthur Brant. He was a Canadian who took his geophysics in Germany; he was a hockey player, and got a hockey scholarship in Berlin before the War. After the War, he taught us magnetometry and electrical surveying – exploration geophysics as opposed to Earth-geophysics.

The only magnetometers in existence then were the old tripod jobs with needle and a vertical magnetometer. My first summer job when I graduated was with [Norman] Keevil – Keevil and Brant were the two geophysics professors in 1946, the year after the War. Keevil had started a radioactive age-dating, and they were both consulting for mining companies. He gave me a job doing magnetometer-work ... just south of Hudson Bay in northern Ontario. I hated the magnetometer work ... I spent months by myself in the bush totally isolated by myself ... swatting mosquitoes. I was practically a raving lunatic when I returned. I heard Arthur Brant talk about the Gulf airborne magnetometer. I said, "I am going to go down there and get a job. I'd much rather fly instead of taking five minutes to get an individual point-reading." So [in 1946] I went down. I had a lot of trouble getting through the border because they wouldn't let me through looking for a job. I couldn't get a job unless I had a visa, and I couldn't get a visa unless I had a job. I actually sneaked across the border. I went down to Gulf Research and Development Corporation ... They had this airborne stabilized magnetometer which had been invented by Vacquier before the War but was given to the US Navy for submarine detection. After the War, Gulf wanted it back to use for exploration. [Leo Peters] interviewed me. He said, "No you don't have your Ph.D.; you're in competition with these fellows. We'll give you a job in seismic." I'd had a taste in seismic, and there was no way I wanted that. So I said, "No." I was going out the door, and he said, "Oh just a minute. There is a guy here from Fairchild Aerial Surveys. They have a contract with us in South America, and he would like to talk with you." The guy came in – I was twenty-six at the time – interviewed me for five minutes and said, "You're hired." So I became the Party Chief of this airborne magnetometer crew ... to do a large part of the Llanos in Venezuela and Columbia ... We had planned it to the nth degree before we went down. The first day everything went like clockwork, and a lot of techniques that were used are still used to this day, actually evolved on that survey. In 1948, we finished down there.

(Morley, 1987 interview with author; my bracketed additions)

Morley returned to Canada. He was anxious to undertake aeromagnetic surveys in his own country because of their application to mineral exploration. Exploration geophysics was and remained his chief interest. He worked for Dominion Gulf Company, a new subsidiary of Gulf Research and Development Company. Realizing his need to further his education he returned to the University of Toronto in 1949, and Wilson became his supervisor. He got interested in paleomagnetism from reading John Graham. Learning from Graham that paleomagnetism could be used to test continental drift, he decided to give it a try. He knew little about continental drift, but he liked the idea that he could test it through paleomagnetism.

I had had a speeded up degree in geophysics because of the war, and by this time some of the graduates who were coming out knew a lot more geophysics than I did, so I thought I'd better go back and study. I went back in '49 to do my Masters. This is where my interest in paleomagnetism came in. Tuzo Wilson was my supervisor. He said, "Well what do you want to do?" I said, "Something in magnetism. I would like to do rock magnetic studies to be able to interpret all of this aeromagnetic data." [We] agreed. I had to pick a thesis topic. I started to read. The first paper of significance I ran across was John Graham's [paper] published in 1949. I was impressed. At the very end he said that his method can be used to test the theory of continental drift. Well, that really grabbed me. So I said that's what I want to do ... I never had any introduction to continental drift as an undergraduate, but I was taking a graduate course from Tuzo Wilson in '49, and he had mentioned it, although at that time he didn't believe in it. But I was very much taken by ... continental drift ... I never had any deep knowledge of it; I thought it was an intriguing idea and, I guess the thing that intrigued me most about it, which I could see from reading Graham's paper, was that if you could plot pole positions, you could prove it ... Test it. By using this technology, I could prove or disprove it.

(Morley, 1987 interview with author; my bracketed additions)

He constructed a spinner magnetometer patterned after one used by Graham. He decided to start with the Precambrian. If he got different pole positions, he would work toward the present. In summer 1950, he began collecting mostly gabbros in the Bancroft area of the Precambrian. It was a disaster. He cut his samples into ten to twelve cubes, measured them, and to his horror, he could not get coherent results. He could not even get repeatable results. He'd take a sample, bang it on the table, measure it and it would change up to 90°. But he also found that the volcanics he collected were stable. He read Bozorth's 1951 *Ferromagnetism*, learned that there are soft (unstable) and hard (stable) components of magnetization, and suspected that his gabbros possessed a huge amount of soft magnetization which swamped their hard magnetization. He also learned from Bozorth that soft magnetization can be removed by exposing rocks to an alternating current (AC). He did not, however, have access to AC. So he improvised.

I didn't have an AC. But I had this spinner magnetometer, and I had a big permanent radar magnet. So I thought, "Well, if I put that radar magnet in while the sample is spinning, it will be the same as an AC demagnetizer, and I'll slowly draw the sample away." I had some success. I could down the unstable component by a factor of maybe 50 or 60, but I could not get a constant direction. I couldn't clean them enough to get a consistent direction.

(Morley, 1987 interview with author)

So Morley was unable to test mobilism through paleomagnetism. He completed his Ph.D. thesis, "Correlation of the susceptibility and remanent magnetism with the petrology of rocks from some pre-Cambrian areas in Ontario," and took a position in May 1962 with the GSC, and got swallowed up supervising the aeromagnetic surveys.

2.15 Morley accepts reversals of the geomagnetic field and continental drift

Morley attended the spring 1953 AGU meeting, and heard Runcorn talk about work in geomagnetism at Cambridge, about Hide's work on core convection, about Hospers' work on Icelandic lavas, and without naming Irving, his work on the Torridonian (II, §2.14, §2.9, §2.10, §2.5, §2.6). He was impressed with Runcorn's confidence, and probably wished he had more of it himself.

I was very shy and retiring, and scared out of my skull. There were a lot of high-powered scientists there. That was the symposium at which Runcorn was the guest speaker. He was younger than I was. He was an absolutely marvelous speaker, super confident and so forth.

Wilson, who was sitting next to Morley, prodded him to say something about his own work.

Then there was a question period, and I was sitting beside Tuzo. He jabbed me – I don't think he remembers this – and said, "Get up and say something!" So I did, and I talked about the unstable rocks, and that they couldn't be relied on ... I think I said that volcanics were more stable. Somebody heard me; it went in the record, and as a result of that I was invited the following year to this Idyllwild Conference at UCLA.

(Morley, 1987 interview with author)

Here is what the record shows Morley said.

For the purpose of inferring the direction of the geomagnetic field at various times during geological history, the choice of rock formation selected for study is very important. It would seem that Runcorn's group made a happy choice in the selection of Icelandic lava flows. Their rocks, because of their rapid cooling and consequent fineness of grain, possess a high coercive force and are thus able to retain their original thermoremanental directions with great tenacity ... As I now know, I made a particularly bad choice of rock for my thesis study at the University of Toronto. The Grenville gabbros which I investigated had such a low coercive force that the mere dropping of a cube of this rock on the floor would in some cases change the polarization direction. Most samples, however, evidently withstood the effect of cutting with the diamond wheel because in each case a number of contiguous cubes cut from the same hand specimen gave the same polarization directions. Samples taken from within a foot or two of each other generally varied by as much as 30° or 40° while at separations of six to seven feet, remanence directions would often be at variance by as much as 180°.

(Morley, 1954)

Morley next discussed his Ph.D. work at the Idyllwild Conference. Runcorn told Morley to use sediments, if he wanted to get coherent results.

I more or less presented my Ph.D. thesis. I remember Runcorn dominated that meeting. When I talked about these unstable rocks, he said, "Ah yes, you have to have a green thumb." I remember this. You pick the rocks that are stable. Sediments. Stay away from igneous rocks. Even at that time his experience had been in the red sandstones. They were quite stable. Stick with sediments. Unfortunately, the remanent magnetometer I had was not sensitive enough, it couldn't touch sediments.

He also recalled that most were concerned with whether or not the geomagnetic field undergoes reversals.

They also had a ferromagnetic physicist there. The reason for meeting was [to discuss the question] Is the Earth field reversing or is there something self-reversing in the rocks itself? There were two camps. Buddington and Balsley said that it is the self-reversing in the rocks, and Runcorn ... said, "No. It must be reversing because we've done various kinds of tests."

(Morley, 1987 interview with author; my bracketed addition)

Morley favored field reversals. He had been convinced by Roche's work on baked sediments and by Runcorn's talk. In 1953, Roche found that clay, baked by overlying reversely magnetized lava flows was invariably reversely magnetized in the same direction. He argued that the underlying clay was magnetized by the molten lava in the reversed magnetic field at the time of extrusion, and took this as strong support of geopolarity reversals (II, §2.8). Morley agreed.

You know what convinced me was that somebody [Roche] had found a dyke rock that was reversely polarized and the country rock next to the dyke rock was also reversely polarized, where the dyke had come in and heated the rock. Now to me that was the clincher ... I was [also] sold by Runcorn on this.

> *(Morley, 1987 interview with author; my bracketed addition; Morley later remembered Roche's name)*

Morley was also becoming more interested in the implications of his aeromagnetic surveys. Of course, there were the economic implications, but he also wondered about the negative anomalies and what could cause them.

But I had become very interested in aeromagnetic surveying ... A new hunk of a survey would be done; we would get the maps out: we would look at these things, and compare them with the geology. I was interested in these negative anomalies from my experience in paleomagnetism.

> *(Morley, 1987 interview with author)*

He began wondering if some negative anomalies were caused by reversely magnetized rocks. In 1957, he was particularly intrigued by negative anomalies over some volcanic plugs in the Yamaska Mountains south of Montreal, and he suggested to Andre Larochelle, who had joined GSC in 1958, that he should obtain his Ph.D. and work on the plugs. Larochelle returned to McGill University, measured rocks from one of the plugs, and confirmed Morley's educated guess that the negative anomaly was caused by reversely magnetized rocks.

Larochelle worked on negatively polarized rocks around the Yamaska. It was a plug just south of Montreal. There was a whole series of them, and we had looked at them on the aeromagnetic maps. By this time I had been able to recognize negative anomalies as remanence from aeromagnetic surveys. Certain anomalies, from the look of them, I could tell they were inversely polarized. I looked at this Yamaska anomaly, and there were no higher positive anomalies around. It just went sort of straight down. So I said, "That has got to be a negatively polarized neck or dike."

> *(Morley, 1987 interview with author)*

Morley also became a mobilist. Although he had lost track of the development of paleomagnetism's growing support of mobilism after the Idyllwild meeting, missing, for example, 1956 papers by Irving and Runcorn in support of mobilism (II, §3.12, §4.3), he did see a preprint of a paper by Du Bois (1957) favoring mobilism when reviewing his application for a position at GSC.

I had wanted to believe it was continental drift. As a matter of fact the first paper I remember which actually showed a polar wander curve for Europe and one for North America was a preprint handed to me by du Bois. As a student of Runcorn's he had published this. And the two curves were 45° apart, but with the same shape. And, it looked very convincing – 45°. That was the reason I hired du Bois. I was so impressed by that paper.

> *(Morley, 1987 interview with author)*

Thus Morley, like Vine, had come to favor geomagnetic field reversals and continental drift before working on marine magnetic anomalies.

2.16 Morley's hypothesis

Morley first learned of Mason and Raff's magnetic surveying of the northeastern Pacific by reading their 1961 papers (Mason and Raff, 1961; and Raff and Mason, 1961). They appeared back to back, along with Vacquier *et al.* (1961) in August. Given his administrative responsibilities and consuming interest in exploration geophysics, he missed Mason's 1958 paper, and shorter papers by Menard and Vacquier on the magnetic anomalies (Mason, 1958; Menard and Vacquier, 1958; Vacquier, 1959). Morley found Mason and Raff's explanations wanting; he immediately concluded that the negative anomalies were caused by reversely magnetized rock. What perplexed him was why anomalies were so linear.

And then in 1961 Raff and Mason came out with those anomalies on the west coast. It drove me bananas. What the hell is that? I'd looked at all these confused anomalies in the Canadian Shield; I saw these linear anomalies, and said, "My God that is something weird." So I couldn't leave that alone … And then it came out in *Scientific American*, in GSA [Bulletin]; Raff and Mason had no explanation. They tried. They said, maybe it is due to topography. But it did not correlate with sea bottom topography. They said, maybe there are some dykes that have intruded there. But there was no evidence of that. I knew it was negatively polarized because I had been looking at the negatively polarized stuff in Canada, in the Canadian Shield. I said that has got to be negative polarization, positive and negative polarization. But why in the hell is it so linear?

(Morley, 1987 interview with author; my bracketed addition)

He also thought that Mason and Raff simply assumed that if remanence were involved, it would be in the direction of the current field. Morley traced their assumption back to Vacquier and company's *Memoir 47*. Given, however, his and Larochelle's work on the Canadian Shield, he was convinced that the magnetization was remanence and that it was reversed.

So I recognized this polarization and they had never mentioned that because it was thanks to people like Vic Vacquier … He had come out with *Memoir 47* of the GSA on the interpretation of aeromagnetic data —he and Isi[dore] Zietz, etc. – which was a classic for aeromagnetic interpretation. And, the assumption was that any remanence in the rock was in the same direction as the Earth's field; they couldn't separate it. They said, therefore, forget about remanence. So the whole theory of aeromagnetic interpretation ignored remanence. They said there is remanence, but it is in the same direction, and for practical purposes, for mathematical purposes, we have to forget about it. So everybody forgot about it. But I had been working in my thesis work on the Shield, and I knew it [reversely magnetized rock] was there, and the work with Larochelle. It was my favorite thing to look at aeromagnetic surveys and say, OK, there is a negative polarization. When it came out, I knew damn well that those positive and negative things were remanence. Hell, they had no explanation for it.

(Morley, 1987 interview with author; my bracketed addition)

Morley might be right about Vacquier's influence. In *Memoir 47*, he explicitly stated that his interpretative method assumed that permanently magnetized rocks were magnetized in the same direction as the present direction of the Earth's magnetic field.

The last physical assumption on which this work is based is that the magnetic polarization of the basement rocks is in the same direction as the present direction of the earth's magnetic field. Samples of igneous rocks possessing permanent polarization of the same magnitude as the induced polarization are mentioned in the literature ... and there is no reason to expect that this permanent magnetization was acquired when the earth's magnetic field had the same direction as it has today. The vector sum of the permanent and the induced polarizations can therefore differ from the present orientation of the earth's field. There is no real evidence that the polarization of large igneous rock bodies is not aligned with the earth's field. However, if evidence to the contrary is found, it must be recognized as an additional limitation. Vacquier (1938) suggested that, by carefully repeating a survey across the span of a decade in regions where the secular variation is pronounced, it might be possible to determine in situ the ratio of permanent to induced polarization of contrasts giving rise to the large magnetic anomalies.

(Vacquier et al., 1951: 7–8)

As already explained, Mason did not entertain the possibility that the negative anomalies were caused by reversely magnetized bodies even though he believed that the geomagnetic field undergoes reversals (§2.3). He instead presupposed that the seafloor beneath the negative magnetic anomalies was either not magnetized or insignificantly magnetized. He worked from what he described as the customary assumption that bodies causing the magnetic anomalies are magnetized in the direction of the current field. This was the customary assumption because there was no information about whether or not oceanic rocks are remanently magnetized. Thus, I believe that Vacquier did not think about appealing to reverse magnetization, even after he was favorably disposed to the generation of seafloor at ridges. Moreover, Vacquier rejected V-M the year after it was proposed (§2.3, §3.8).

Morley, confident as he was about reversed magnetization, had no idea why the anomalies were so linear. He had to figure out how the reversely and normally magnetized stripes of seafloor are generated. He started to read papers in marine geology. Once he came across Dietz on seafloor spreading, he had the answer.

Then I thought, well, I'll start looking ... Most of my research at that time – I snatched between administrative jobs – was in geophysics. So I started going through the geological literature to find out all I could about ocean basins. My God, I ran across Dietz in 1961 on ocean floor spreading. Bingo. You know, as the rock rose it had to be magnetized. I knew about the mid-ocean ridges. I had believed in polarity reversals ... going right back to Idyllwild. So I said, "That has got to be it." So I was so excited, I thought, "God, I've got to get this in." So I wrote that short paper, thinking that everybody would read it and say, "Yeah that it is it!" Because with my background, I was so God damn sure.

(Morley, 1987 interview with author)

Retrospectively, Morley said (1986: 666) that he knew immediately while reading Dietz that seafloor spreading explained the linearity of the anomalies, and that his eureka moment occurred in December, 1962. However he is not certain when, and in what order, he read Dietz's papers. In his retrospective comment to me, he said that he ran across Dietz's work in 1961. He told Glen (1982: 301) that he read the paper that Dietz (1962d) presented at the 10th Pacific Science Congress (August 12 through September 6, 1961); he did not hear Dietz's presentation, but likely read the published version eight months later. However, he did not reference it in either the *Nature* version or the *JGR* version. (I shall explain these versions anon.) In the former, he referenced the paper Dietz published in the *20th Anniversary Volume* of the *Journal of the Oceanographical Society of Japan* (Dietz, 1962c). In the *JGR* version he made no reference to Dietz, only to a paper by Hess, which, I suspect, he may have learned about from seeing it referred to in Dietz (1962c). Morley also claimed that he saw Mason and Raff's survey maps nearly two years before he encountered Dietz's work (Morley, 2001: 80). Mason and Raff's two joint 1961 papers appeared in August 1961. Although Morley told Glen that he read Dietz's Pacific Science Conference paper, he did not reference it in either the *Nature* or the *JGR* version. If Morley's comment to Glen is correct, then he did not read Dietz until well into 1963, which is too late because he submitted his paper to *Nature* in February 1963. Assuming that he submitted it soon after he came up with his hypothesis, it is likely that he first read Dietz at the end of 1962 or very beginning of 1963.

Morley submitted his paper to *Nature* in February 1963. It was rejected. After making several changes, Morley submitted the paper to the *JGR*. It was again rejected. There are three published versions, all different, of the paper that Morley is said to have submitted to *Nature*: John Lear reproduced one in the *Saturday Review* (Lear, 1967), Emiliani another in Appendix I of the seventh volume of *The Sea* (Morley, 1981), and a third in Morley's own retrospective in *Plate Tectonics* (Morley, in Oreskes, 2001). The *Saturday Review* and *The Sea* versions are very similar. The wordings of the last paragraph differ slightly, and *The Sea* has a reproduction of Figure 1 from Mason and Raff (1961). The *Saturday Review* version contains no references, but both the figure and references may simply have been eliminated because of lack of space in the *Saturday Review*. I suspect that the version in *The Sea* is an accurate reproduction of the paper Morley submitted to *Nature*. The version in *Plate Tectonics* includes a reference to a paper by J. Tuzo Wilson that did not appear until April 1963. Because Morley submitted his paper to *Nature* in February 1963, the *Plate Tectonics* version cannot be a reproduction of the paper Morley sent to *Nature*. I suspect that the version in *Plate Tectonics* is a reproduction of the paper Morley sent to *JGR*. I shall refer to the *Saturday Review* version as the *Nature* version and the *Plate Tectonics* version as the *JGR* version.

There are several minor, and one major, differences between the two versions. Unlike the *Nature* version, the *JGR* version contains no figures. And, as already noted, the references are different. The *Nature* version references Dietz (1962c), R. L. Wilson (1962), Einarsson (1957), and Mason and Raff (1961); the *JGR* version references Mason and Raff (1961), Vacquier (1962), J. T. Wilson (1963d), R. L. Wilson (1962), Einarsson (1957), and Hess (1946). The wording of corresponding paragraphs in both versions often differs slightly. The major difference is that Morley added three paragraphs to the *JGR* version in which he raised a difficulty with his explanation of the marine magnetic anomalies (RS2), and then found a way to eliminate it (RS1).

Morley began both versions by noting that several investigators and authors had related mantle convection and continental drift with the linear pattern of magnetic anomalies in the Pacific (Morley, 1981: 1717; Morley, 2001: 80). He then introduced his explanation of the zebra-pattern of anomalies, both their linearity and alternating sign.

If one accepts, in principle, the concept of mantle convection currents rising under the ocean ridges, traveling horizontally under the ocean floor and sinking at ocean troughs,[1] one cannot escape the argument that the upwelling rock under the ocean ridge, as it rises above the Curie Point geotherm, must become magnetized in the direction of the earth's field prevailing at the time. If this portion of the rock moves upward and then horizontally to make room for new upwelling material and if, in the meantime, the earth's field has reversed, and the same process continues, it stands to reason that a linear magnetic anomaly pattern of the type observed would result.

    (Morley, Nature *version, 1981: 1717; Morley,* JGR *version, 2001: 80; two versions differ only in that the* Nature *version contains footnote 1, which references Dietz, 1962c)*

He then claimed that his explanation was preferable to the alternatives (RS3). Unlike them, it did not appeal to dubious assumptions for which there was no evidence (RS2). He claimed that the assumptions embedded within his explanation, field reversals and large convection cells, were more reasonable.

This explanation has the advantage, over many others put forward, that it does not require a petrologically, structurally, thermally or strain-banded oceanic crust. It requires a convection cell whose axis of rotation is at least as long as the linear magnetic anomalies and whose horizontal distance-of-travel stretches from ocean rise to ocean trough. In addition to this, it requires a large number of reversals of the earth's magnetic field from at least the Cretaceous period (which ended 68 000 to 72 000 years ago) to the present (since no older rocks than the Cretaceous have been found in the ocean basins).

    (Morley, Nature *version, 1981: 1717; Morley,* JGR *version, 2001: 80, with slight changes)*

Although he offered no independent support for mantle convection, he appealed to T. Einarsson's work on Icelandic lavas and a report by Mrs. J. Cox indicating that the geomagnetic field had been normally and reversely magnetized almost the same number of times from the Carboniferous. Cox's results were published in a paper by R. L. Wilson (1962) who defended field reversals. Einarsson, who extended the work that he had done with Sigurgeirsson and that of Hospers (II, §2.9), claimed, as had Hospers, that reversals lasted on average approximately 500 000 years. Morley extrapolated, and estimated that there had been as many as 180 reversals since the Lower Cretaceous.

R. L. Wilson reported that Mrs. J. Cox, in a recent search of the palaeomagnetic literature, was able to find 136 normally polarized cases and 141 reversely polarized from the Carboniferous to the present. Since there is no evidence to suggest that the Earth's field should have been normally polarized for any more periods or for longer periods than it has been reversely polarized, it is entirely possible that there may have been as many as 180 reversals since the Lower Cretaceous. This would be one reversal about every half million years on the average, a figure which T. Einarsson[3] gives from his investigation of Icelandic lavas. He also suggests that the time taken for a reversal of the field is geologically very short – a few centuries to 10 000 years.

(Morley, Nature *version, 1981: 1717–1718; Morley,* JGR *version, 2001: 81, with only stylistic changes; footnote 3 refers to Einarsson (1957))*

Morley then estimated that seafloor spreading in the northeastern Pacific had occurred at a rate of 3.5 cm per year, given that the average width of a pair of positive and negative anomalies equaled 35 km, and that it took 1 000 000 years to complete a reversal cycle. He admitted, however, that there was a better way to determine the rate of seafloor spreading and duration of a reversal period: measure the ages of rocks at widely spaced locations in the Pacific and count the number of reversals occurring between these points (*Nature* version, 1981: 1719; *JGR* version, 2001: 81). Thus Morley recognized the geochronological consequences of the zebra pattern of magnetic anomalies. If his explanation for the anomalies was correct, they represented a continuous record of field reversals since at least the Cretaceous, and therefore could be used to construct a reversal timescale. Indeed, Morley took this aspect of his explanation to be one of its important merits. Vine said nothing about it.

The purpose of this letter is to point up the possibility of calibrating the frequency and duration of reversals of the earth's magnetic field in geologic history from a study of magnetic surveys of ocean basins; and the idea presented is considered to add support to the theory of convection in the Earth's mantle.

(Morley, Nature *version, 1981: 1719; Morley,* JGR *version, 2001: 82)*

Morley also extended his explanation of the magnetic anomalies to account for the magnetization of seamounts, in particular guyots, and it is here that Morley invoked induced magnetization of the seafloor. He explained why some of the guyots in Mason and Raff's survey seemed magnetized. Aware of Dietz's explanation of the origin of guyots, Morley argued that they become magnetized in the same way as the seafloor. Seamounts that appear to be non-magnetic actually are reversely magnetized to an intensity which nearly equals their magnetization induced by the present Earth's field. Seamounts which appear to be strongly magnetized formed when Earth's magnetic field was not reversed, and therefore their remanent and induced magnetization is in the same direction.

R. G. Mason and A. D. Raff (who made the magnetic survey for Scripps) report that some of the many guyots (flat-topped undersea islands) that were detected on the echo sounder produced magnetic anomalies, while others apparently had little or no effect.[4] It seems unlikely that these guyots would be divided into two classes – those containing magnetite and those containing little or none. A more likely explanation would be that the ones which gave little or no effect are negatively polarized to an intensity which nearly equalizes their magnetization induced by the present Earth's field. If the non-magnetic guyots always occur in the negative anomaly bands and the magnetic ones in the positive bands, this would be evidence that they cooled below the Curie point at approximately the same time as the rock surrounding them because they were magnetized in the same direction.

(Morley, Nature version, 1981: 1719; Morley, JGR version, 2001: 81)

Morley's appeal to induced magnetization enabled him to explain the difference between what appeared to be strongly magnetized and weakly or non-magnetized seamounts, but it left unexplained the source of the induced magnetization: it also differentiated his hypothesis from Vine's. Morley had to identify a rock type whose seismic velocity was consistent with seismic models of oceanic crust, and whose induced magnetization approximately equaled in intensity that of the remanent magnetism possessed by the seafloor. Morley, like Vine, believed that the remanent magnetization of the seafloor was possessed by the 2 km thick slab of oceanic basalt (Layer 2). Vine, however, claimed that the basaltic layer had enough remanent magnetization to produce the anomalies. Morley did not. Vine knew about Hall's unpublished work on ocean basalts, and therefore knew that the Koenigsberger ratio of ocean basalts was much higher than that of continental basalts. Morley did not know about Hall's work. Morley knew about very old continental basalts from the Canadian Shield. Their Koenigsberger ratio is low. So Morley had either to claim that the basaltic layer was three times thicker than indicated by the seismological evidence or find a rock type in the main layer of oceanic crust that could possess enough remanent or induced magnetization to produce the anomalies.

I do not know if Morley had found his solution or even if he had recognized the difficulty before submitting the *Nature* version of his paper; however, he definitely had a solution by the time he wrote the *JGR* version. He began by raising the difficulty within the context of Mason and Raff's three interpretations of a magnetic profile from their survey (§2.3).

There are a few difficulties. The seismic results postulating 3 layers about the Moho must be incorporated into the theory. Mason and Raff (1) offer three models to marry the seismic and magnetic results.

(1) A 2 km-thick slab of intensely magnetized lava of $K = 0.15$ units underlain by a relatively non-magnetic crustal layer.
(2) A topographical plateau 2 km high composed of material $K = 0.15$ underlain by a main crustal layer of the same magnetic susceptibility.
(3) A 6 km-thick slab extending from the bottom of the unconsolidated sediment to the Moho composed of 2 seismic layers, but all of the same magnetic susceptibility $K = .005$.

From measurements of several thousand of basaltic lavas from the Canadian Shield (6), none have been shown to possess a magnetic susceptibility of as great as .015 c.g.s. This would mitigate against accepting models (1) and (2). On the other hand, many lavas have a susceptibility as high as .005.

(Morley, JGR version, 2001: 81–82; reference (1) is my Mason and Raff, 1961; reference (6) is to a personal communication with A. S. MacLaren)

Morley found his solution by reading Dietz (1962c). He learned from Dietz that Hess had proposed that the main layer of oceanic crust was serpentinized peridotite, which would satisfy both the magnetic and seismic requirements.

If this layer [between the 2 km basaltic layer and Moho] were unaltered ultramafic rock, it would not be sufficiently magnetic to cause the observed anomalies, nor would it have a significant seismic velocity contrast with the mantle material. Hess (7) has suggested that the main crustal layer beneath the oceanic basalt could be serpentinized ultramafic rock. This would satisfy both the magnetic and seismic requirements, since the serpentinization process both increases the magnetic susceptibility of the ultramafic rock and lowers the seismic velocity.

(Morley, JGR version, 2001: 82; reference to Hess is Hess (1946);[22] my bracketed addition)

Morley then re-stated his explanation of the magnetic anomalies, putting it in terms of Mason and Raff's third alternative.

Thus Mason and Raff's (1) model number 3 is favored, with the modification that adjacent prisms would be magnetized oppositely. The prism producing the positive anomaly would be normally polarized with a total magnetization (remanent plus induced) equal to >.005 c.g.s. The prism producing the adjacent negative anomaly would be inversely polarized, so that the remanent magnetization would approximately cancel the induced.

(Morley, JGR version, 2001: 82; reference (1) is my Mason and Raff (1961))

Vine and Matthews' and Morley's hypotheses are not identical. Morley's introduction of induced magnetization as producing approximately half of the magnetization of the seafloor distinguishes it from what Vine and Matthews proposed.

2.17 Morley's paper is twice rejected[23]

Morley submitted his paper to *Nature* in February 1963. He learned, probably in April, two months later, that his paper had been rejected. I say probably in April, because Morley has also said that the paper was rejected in May and June. In both his retrospective papers he recalled that the paper had been rejected two months after he submitted it, which would have been in April (Morley, 1986: 666; Morley, 2001: 83). He also told Glen that the paper had been rejected about two months later (Glen, 1982: 298). However, he wrote me that he learned in late June that the paper had been rejected.

I wrote to *Nature* in February 1963 ... It was in late June '63 that I received the reject notice from *Nature* saying that they did not have room to print my letter. The letter would have occupied three-quarters of one page ... After I got the rejection notice from *Nature*, I sent the letter to the *Journal of Geophysical Research*.

(Morley, October, 1976 letter to author)

He apparently told Emiliani (1981: 1717) that the paper was rejected three months later, which would have meant that he received the rejection notice in May. Morley has given two reasons for the rejection. He told me (Morley, 1979 letter to author) and others, except, perhaps, Emiliani, that the paper was rejected because *Nature* did not have enough room. Emiliani (1981) reported that Morley submitted his paper to *Nature* in February 1963, but the paper was rejected three months later because "it did not include original experimental data."

He then submitted his paper, slightly revised, after reading Wilson's April 1963 paper in *Scientific American*, to *JGR*. He had also been invited to give a talk with Larochelle at the 1963 Annual Meeting of the Royal Society of Canada, which was going to be held in Quebec City on June 4, and he decided to present his explanation of marine magnetic anomalies at the end of their talk.[24] To Morley's surprise, he got a negative response. Even Larochelle was not supportive. Morley found that he had no allies in the Geological Survey.

When I got that negative reply, I couldn't believe it. And then I got more and more nervous as time went on. This was February when I wrote the thing. I was invited to the Royal Society [of Canada], and I gave the paper with Larochelle. Larochelle incidentally was more a physicist and mathematician; he was not really that interested in geology. He didn't really care or understand about ocean floor spreading. He didn't really believe in this, what I had come up with. So I had essentially no allies in the Geological Survey in this whole business. There was nobody there. I presented it at the Royal Society, and I got a negative response.

(Morley, 1987 interview with author; my bracketed addition)

He finally heard back from *JGR* in late August or September 1963.[25] He was told, "This is the sort of thing you would talk about at a cocktail party, but you would not write a letter on it."

Here again there was a long delay before the reader got around to assessing the paper and then followed his famous remark, This is the sort of thing you would talk about at a cocktail party, but you would not write a letter on it. He obviously did not understand the full significance of what was being divulged.[26]

(Morley, October, 1976 letter to author)

Morley (1986: 66) also said that the anonymous referee's report was sent along with the rejection notice. The name was cut off, but he learned that the referee had been on fieldwork in Hawaii all summer and had not received the paper until he returned to his laboratory.

While Morley was waiting to hear from *JGR*, he began to wonder what was taking so long, and started to think about where else he could publish his paper. Once he learned that *JGR* had rejected his paper, he decided to submit his paper to a Canadian journal, but Vine and Matthews' paper appeared before he sent it off.

And the summer wore on, and I thought, "Where else can I publish?" And then I was still waiting for a response from *JGR*. It came in and I was so God damn mad. I thought, "Well they won't publish it in Britain; they won't publish it in the States, I've got to publish it somewhere in Canada." I had just begun to think about that when Vine and Matthews appeared. So it was a big disappointment to me. Then at this time, that was in '63,

I had missed out on it and I was occupied with other things and I thought, "Well that is the luck of the thing."

(Morley, 1987 interview with author)

Morley appended a truncated description of his explanation of the anomalies onto Larochelle's and his joint paper (Morley and Larochelle, 1964). This was the only publication of his version of V-M until it was published three years later in the *Saturday Review* (Lear, 1967). He restricted his discussion to the geochronological implications of his explanation, which was appropriate because their talk was about using paleomagnetism as a means of dating geological events.

However, Morley suggests that a nearly-unbroken record of these reversals may exist in the permanent magnetization of the rocks on the floor of the ocean basins. Mason and Raff (1961) have outlined a wide sequence of long north-south striking magnetic anomalies whose origin is puzzling. If one refers to recent papers of several authors. (J. T. Wilson 1963; Vacquier 1962; Dietz 1961) in which the concept of mantle convection currents rising under ocean floors and sinking at ocean troughs is presented, it stands to reason that the rising rock, as it reaches the Curie point geotherm, will become permanently magnetized in the direction of the Earth's field prevailing at the time. As this portion of rock moves upward and then horizontally to make room for the following rising material and if, in the meantime, the Earth's field has reversed and the same process is repeated many times throughout geological history, a linear magnetic anomaly pattern of the type observed would result. The present rate of travel of this mantle convection is believed to be of the order of a few centimeters per year. If this figure could be determined accurately for the various geological periods, it would then be possible, using the data from a magnetometer survey of the ocean basin floors, to reconstruct not only the history of reversals of the Earth's field but also the direction and rate of the drift of continents. This time-reversal information could then be applied to problems of geochronology on the continents. Although such thinking is still very speculative, the study of paleomagnetism and magnetometer surveys in the oceanic islands and ocean basins will no doubt prove or disprove it in the near future.

(Morley and Larochelle, 1964: 50; references are my Wilson (1963d),
Vacquier (1962), and Dietz (1961a)

In an attempt to get the record straight, he noted in a footnote that he had submitted his original rejected paper to *Nature* in February 1963.

This paper was written after the Letter that I wrote to *Nature* in February 1963 and on page 50 at the end of the paper I made reference to the concept of magnetic imprinting of the ocean floor. This was made verbally before the Vine-Matthews paper appeared which was in September '63. I did not want to repeat the whole letter that I wrote for *Nature* because at that time it was still pending acceptance.

(Morley and Larochelle, 1964: 50)

Morley also attended the 13th IUGG meeting in San Francisco that was held in August 1963, from the 19th to the 31st, where he talked about his hypothesis with Hess and Runcorn.

Another little comment which is very interesting from a human point of view; at the meeting of the International Union of Geodesy and Geophysics held in the Berkeley campus in August 1963, I sat at tea with Runcorn and Hess. Since my paper was already submitted to the *Journal of Geophysical Research*, I figured it was safe to divulge my ideas to both of these men to see what they thought of them. Hess became extremely interested in the whole concept and kept asking me more and more questions while Runcorn, who certainly knew what we were talking about, and must have had many ideas on the subject himself, merely looked off into space and did not participate in the conversations. The whole thing somewhat embittered me and perhaps had something to do with my decision to leave the pure side of the field and become more embroiled in applied science. This was possible since I was not in the academic field but in the government service and had the choice of doing this.

(Morley, October 5, 1976 letter to author)

In his second retrospective paper, Morley added:

Runcorn was either bored or distracted, because he was obviously not listening, but Hess, who had been pushing his ocean floor spreading theory, was very interested and expressed the desire to meet with me again. Unfortunately, he died before we had the opportunity to do so.

(Morley, 2001: 82)

Hess died six years later. Morley was likely right about Runcorn. He was not listening. Later, when asked if he remembered the conversation, Runcorn said no, but admitted that he might have forgotten.

No. I've got a very good memory, but I suppose we all forget embarrassing moments, Maybe later on I realized that I had not paid attention to something very important. But that is about it. I don't know where this was supposed to be.

(Runcorn, August, 28, 1984 interview with author)

2.18 Why Morley's paper was rejected and Vine and Matthews' paper was accepted

There has been substantial discussion of this question: who were the referees, what were the grounds for the editors' verdicts? Peer review is not a perfect process, editors and their choice of referees contributing variable human factors. Some mention of these has already been made (Frankel, 1982: 1, 1987: 228; Glen, 1982: 297; Lear, 1967: 48; Oreskes, 1999: 271: Oreskes, 2001: 22). *Nature's* records are not available and the identity of the reviewers is unknown, so there is little to be gained by extending these discussions. Instead I shall concentrate on two central questions: were there good scientific reasons for *Nature* and *JGR* to reject Morley's paper and were there good reasons for *Nature* to accept Vine and Matthews' paper?

Both hypotheses were highly controversial. Several commentators were surprised that Vine and Matthews' paper was accepted and Morley's twice rejected. Vine certainly thought that the chances of theirs being accepted were slim.

One of the rather delightful conversations I remember at coffee at Madingley Rise – it must have been in June or July '63. I remember Jim Beaumont was there, who was an American working on gravity – I think he's director of IBM or something. He has some incredible history. It was Saturday morning. I'm sure it was a Saturday morning for there weren't many of us there. We were sitting around at coffee, and not an awful lot was being said. Somebody said, "Do you know if *Nature* gets their articles reviewed, or do they publish almost anything?" I said, "Well, we're just about to find out because, you know, I just put my paper in, and if they publish that they'll publish anything."

(Vine, 1979 interview with author)

John Sclater, then a graduate student in the Department of Geodesy and Geophysics working under Hill in heat flow studies, believes most in the department were surprised that *Nature* published the paper.

After Fred and Drum came up with their hypothesis of sea floor spreading it was often a topic of conversation at coffee (at 11 am) or tea (at 4 pm) at the Department of Geodesy and Geophysics at Madingley Rise. Most of the comments about the theory were humorous. Most of us still believed that the continents were fixed and we were surprised that *Nature* published what we considered idle speculation. We gave both Drum and Fred a lot of friendly grief over their success at conning *Nature*.

(Sclater, September 3, 1992 letter to author; Sclater's response to author's questions are accessible through Scripps at www.escholarship.org/uc/item/4xj8c69c)

Both hypotheses are corollaries of seafloor spreading and reversals of the geomagnetic field. Suppose for a moment that the referees were unfavorably inclined toward seafloor spreading and field reversals, then they probably would not be inclined to favor publication, unless there were other reasons to do so. Were there positive reasons to recommend their work for publication? I suggest that there were with Vine and Matthews and there were not with Morley. Morley's *Nature* version had the added difficulty of finding a plausible source for the induced magnetization that his hypothesis needed to explain the intensity of the magnetic anomalies. Vine and Matthews' hypothesis required that ocean basalts have a much higher Koenigsberger ratio than that generally found in continental basalt; in support they appealed to Matthews' work on Swallow Bank (§2.7) and, as background, they likely knew of rock magnetism work at Imperial College (§2.9) then unpublished.

Most importantly, Vine and Matthews had new data – Matthews' detailed survey over the Carlsberg Ridge and new magnetic profiles over the Carlsberg and Mid-Atlantic; Morley had none. Even if the referees thought the hypothesis very speculative, these new, detailed, directly relevant results obtained by the Cambridge group gave them very strong reason to recommend publication. After all, Mason and Raff did not have a good explanation of the magnetic anomalies, yet their results certainly were worth publishing. Moreover, Morley told Emiliani that *Nature* rejected the paper because it did not include original data (§2.17). Thus, I think it is possible that *Nature* rejected Morley's paper mainly because it lacked new, directly pertinent data, while Vine and Matthew's paper was accepted because it included new and potentially critical data.

It was Vine's very good fortune that, serendipitously, he had a suitable computer program that he had obtained from Kunaratnam. Adapting this, he skillfully separated the effects of remanent and induced magnetization; he was able to generate anomaly patterns and showed how they vary depending on their orientation and magnetic latitude – essential ingredients for the analysis of marine magnetic anomalies. He modeled Matthews' data, applied it to two seamounts, to some results from Mason and Raff's survey, and to the new Carlsberg and Mid-Atlantic Ridge profiles, thereby demonstrating the generality of the Cambridge hypothesis. Computer modeling was then in its infancy, and the referee may have thought that Vine's novel application was another reason to recommend publication, recognizing that he was blazing a new trail in studies of ocean crust.

Vine's presentation had other advantages. He attempted to explain why the central anomaly over active ridge crests was more pronounced than other anomalies; Morley's offered none: explanations of magnetic anomalies over active ridges were (implied) embedded within his hypothesis, but his analysis did not reveal them. Finally, Morley, unlike Vine, invoked induced magnetization, and left its origin unexplained in the *Nature* version of his paper.

These differences, especially the first, could, I think, have led to the rejection of Morley's and acceptance of Vine and Matthews' paper. Matthews provided a splendid set of critical observations, Vine an elegant and original analysis; they embellished their hypothesis, addressing and providing solutions (not always correct but reasonable and testable) to many outstanding problems, thus giving referees good reasons to recommend publication even if they thought their hypothesis highly speculative. Morley's paper had little more to recommend it other than a skeletal hypothesis, and I think a referee would have good reasons for rejecting it. I also think that despite (with the telling exception of Bullard) misgivings back home in Cambridge, a referee would have had good scientific reasons to accept Vine and Matthews' paper.

2.19 Two other Vine–Matthews-like hypotheses

The idea that marine magnetic anomalies are explicable in terms of field reversals and seafloor spreading was proposed on at least two other occasions; however, neither was published. Gough, McElhinny, and Opdyke came up with the idea in 1962. At the time, they were working out the APW path for Africa (III, §1.17). Opdyke briefly recalled the incident.

The tea-time reversal story was interesting but too late. Ian [Gough] had traveled to the US and visited Scripps where he became aware of the lineated magnetic anomalies. He brought this back to Africa and we were discussing this data and someone (I don't know who) suggested reversals of the field.

(*Opdyke, January 25, 2009 email to author; my bracketed addition*)

Prompted by Opdyke's recollection, Gough further described what happened.

I remember the conversation in Salisbury, Rhodesia (now Harare, Zimbabwe) during one of our tea breaks. As Neil has told you, I visited Scripps at the end of 1962 – actually Walter Munk's IGPP [Cecil H. and Ida M. Green Institute of Geophysics and Planetary Physics] where I gave a seminar on the Rhodesian group's paleomagnetic work. During this brief visit I saw a draft of the Raff, Mason & Vacquier map of the linear magnetic anomalies over the ocean floor of the eastern Pacific. Like its authors, I realized that this map had something very significant to say about the planet, without quite seeing what that something was, and I told my colleagues about it when I returned to Salisbury. I can't quite recall how the meaning came to us, but my impression is that Neil Opdyke and/or Mike McElhinny thought of what is now known as the Vine–Matthews hypothesis first. Once suggested, it captured us all, not only Neil, Mike and me, but also our younger colleagues, Dai Jones and Andrew Brock. We talked about writing to *Nature*, but we were very busy that year writing the six papers that appeared in *JGR* in 1964 on the paleomagnetic work of the preceding decade on the southern African shield, and Neil with his great journey to Tanzania, with Margery [Opdyke] and their eldest son, that added a Lower Permian pole position (Opdyke, JGR, 69, 2495–7, 1964) to our Phanerozoic APW path (Gough, Opdyke & McElhinny, JGR, 69, 2509–19, 1964). The same paper gives a Precambrian APW path relative to Africa from 2.51 Ga [10^9 years] to the Pilansberg pole near the horn of Africa at 1.31 Ga. In South Africa the Pilansberg dykes give a reversed vertical anomaly [in the present geomagnetic field], but a normal horizontal anomaly. Gelletich could have found this out, as his Askania magnetometer had a sensor for horizontal fields, but he seems never to have used this. Their reversed vertical-field anomaly distinguished the Pilansberg dykes from the normally-magnetized dykes of Karroo age, and so served his purposes for mapping.

> (Gough, January 27, 2009 email to author; my bracketed additions; see III,
> §1.17 for their Phanerozoic APW path for Africa)

McElhinny, who spent a sabbatical after they came up with the idea in Princeton where he set up a paleomagnetic laboratory and had access to Hess's preprint on seafloor spreading, believes that Gough's account, as far as he can recall, is correct. He also is quite sure that he did not connect the dots and thinks that it probably was Ian.

I do recall this discussion over tea with Neil and Ian. I am sure it was in 1962 before I went on sabbatical to Princeton. I left in December 1962 if I recall correctly so it would have been before that. More than that, I cannot recall unfortunately. I think it was probably Ian who connected the dots so to speak. It certainly wasn't me! While I was in Princeton I immediately had access to Hess's preprint on the History of Ocean Basins. I really enjoyed my time there, Hess was such an inspiring person and I got to know him and his wife quite well.

> (McElhinny, May 19, 2009 email to author)

Gough, McElhinny, and Opdyke combined seafloor spreading and field reversals and came up with the core idea that the zebra-patterns of marine magnetic anomalies are caused by alternating stripes of normally and reversely magnetized rocks. They were in an excellent position to do so; they worked in paleomagnetism and were familiar with reversely magnetized rocks. Gough had studied the Pilansberg dykes of South Africa, which because of the anomalies in the geomagnetic field associated with them were thought by Gelletich to be reversely magnetized. Gough (1956) sampled them and found strongly oblique, steep downward magnetizations, confirming they were reversely magnetized. The obliqueness he attributed to polar wandering, continental drift, or their combination (II, §5.4). They all believed continental drift and reversals of the geomagnetic field, and had read Dietz's paper on seafloor spreading. Thus they were well primed to put two and two together once Gough returned to Salisbury with the news of the Raff–Mason linear magnetic anomalies of the northeast Pacific. The September 1963 issue of *Nature* arrived with Vine and Matthews' note before they had a chance to write up their idea.

Would their paper have been accepted if they had written up their idea? Gough is not so sure because they did not have any relevant data of their own.

I'm not sure that we were well placed to anticipate the Vine–Matthews hypothesis in print, as we had no ocean-floor magnetic maps of our own: our work was in the African shield. We knew of Dietz's version of seafloor spreading, but had no knowledge of the great Hess papers. While we were writing our papers on the African evidence for continental displacements, the Vine–Matthews paper appeared.

(Gough, January 27, 2009 email to author)

As far as the data were concerned, they were in the same position as Morley. Unlike Morley, however, McElhinny (June 5, 2006) did not recall ever discussing this with Harry Hess while he was in Princeton: "It was a while before I got to know him on a personal basis."

Geoffrey Dickson also came up with a Vine–Matthews-like hypothesis before their paper was published. Dickson, then a graduate student at Lamont, told Jim Heirtzler, who headed the group working on marine magnetic anomalies.

It has been nearly 50 years since Geoff Dickson strolled into my office at Lamont to speak about the origin of marine magnetic anomalies. It is impossible to tell you exactly what was said by either of us or the exact date of our discussion. I can only give you my impression of what was said, filtered by that great span of time and events.

Basically he suggested that the linear marine magnetic anomalies could be caused by past reversals of the geomagnetic field and could thus mark the age of the ocean floor. I do not think he had prior knowledge of the Vine–Matthews paper. He may have had knowledge of the paper by George Peter and Ron Girdler, which showed that a reversely magnetized body caused an anomaly in the Gulf of Aden. Geoff Dickson was a bright, pleasant young man who I hoped would stay on at Lamont after graduation but didn't. I don't think he was terribly concerned whether the linear marine anomalies were induced or remanent. There may have been others at Lamont with similar ideas but they never mentioned them to me or put such ideas in a paper. Looking back do I feel guilty in not encouraging him more? Sure but at that time there were many ideas about and we were concentrating on those that could be firmly nailed down with data. So far as I know I am the only one that Geoff spoke to about his ideas.

(Heirtzler, June 4, 2009 email to author)

Like Gough, McElhinny, and Opdyke, Dickson was also in an excellent position to come up with the idea that marine magnetic anomalies can be explained in terms of field reversals. Dickson arrived at Lamont with an M.Sc. he obtained in 1962 from the University of Sydney under Alan Day, who was familiar with the paleomagnetic support of mobilism (II, §3.10) and inclined toward mantle convection (III, §2.12). But, more importantly, Dickson did most of the work for his M.Sc. at ANU, where he studied paleomagnetism and rock magnetism with Frank Stacey.

Geoff Dickson worked extensively at ANU (I remember him there) with Stacey. Stacey had a lab for physical properties in the main building at ANU. Dickson used my lab (away from campus in a sheep paddock) for his remanence measurements. Dickson did an M.Sc. on TRM magnetization of igneous rocks and wrote 4 short papers [Dickson, 1962a, 1962b, 1963, and Dickson *et al.*, 1966].

(*Irving, January 9, 2009 email to author; my bracketed addition*)

All but Dickson (1962b) were theoretical papers. In the first paper, Irving recalled that Dickson showed that he and his co-workers had been wrong.

Irving *et al.* 1961 had suggested that the random magnetizations remaining after demagnetization were intrinsic. Dickson (who was then working at ANU with Stacey who strongly influenced him) showed we were incorrect.

(*Irving, January 9, 2009 email to author*)

Dickson's 1962b paper reported the thermo-magnetic properties of some reversely magnetized rocks (Irving, January 9, 2009 email to author). He knew and understood mobilism's paleomagnetic support, and was familiar with the APW path for Australia, which greatly diverged from APW paths from other continents. Thus, by the time Dickson arrived at Lamont in late 1962 or early 1963, he was sympathetic toward mobilism, had worked with reversely magnetized samples, and knew a great deal about the TRM properties of mafic igneous rocks. Once he became familiar with the zebra pattern of magnetic anomalies in the northeastern Pacific and some of Lamont's magnetic profiles over ridges, and devoted his full attention to figuring out their origin, he was amply prepared to make the connection between marine anomalies and reversals. As I shall later show, Dickson helped break down the resistance of some in marine magnetics at Lamont against V-M (§6.6).

Notes

1 The work in question, I suspect, was that of Ade-Hall (1964). Ade-Hall was a Ph.D. student of Mason and Bruckshaw at Imperial College, London. Mason discussed his still unpublished work in a review paper that Bullard and he wrote on marine magnetic anomalies (Bullard and Mason, 1963: 192).

2 In 1967, this ridge segment which extended from the Carlsberg Ridge to the East African Rift System was named the Sheba Ridge by Laughton (Laughton, April 2009 email to author; see Matthews, Williams, and Laughton, 1967).

3 See also Irving (2008) for his account of Hospers' essentially difficulty-free solution to the origin of rocks with a reversed natural remanent magnetization.

4 The year of publication of Girdler (1967) is misleading. This paper was probably submitted to the editorial office of the *Dictionary of Geophysics* before July 1964. There are no references after 1964, Girdler did not provide the pages for his forthcoming paper with Drake which appeared in July 1964, and he referred to a paper of his that appeared in 1964 as appearing in 1963. Girdler had come to accept seafloor spreading and V-M by 1967.

5 This brief biography of Allan is drawn from his privately published *Memories From a Life* (2008) and several emails to the author. I thank Allan for giving me a copy of his engaging *Memories*.

6 This short biography of Matthews is almost entirely drawn from R. S. White's sensitive biography of Matthews (White, 1999). I have also had the benefit of reading Matthews' diary. I would like to thank Sarah Finney, Conservator, Sedgwick Museum of Earth Sciences, and David Norman, Director of the Sedgwick Museum of Earth Sciences, for their patience and kindness in letting me examine and quote from Matthews' diary.
I thank Matthews' daughter and Bob White of Bullard Laboratories, Department of Earth Sciences, University of Cambridge, for giving me permission to quote from his diaries.

7 White also reproduces this passage.

8 The Norwegians Tom F. W. Barth (1899–1971) and Per Holmsen agreed to analyze a collection of rocks that Olaf Holtedahl had collected in 1928 while briefly visiting some Antarctic and Sub-Antarctic Islands. Holtedahl was an avid mobilist (I, §8.11). Barth, who became a celebrated mineralogist, winning the Roebling Medal of the American Mineralogical Society, was then Director of Norway's Geological Institute (Dickson and Krauskopf, 1973). Barth earned his Ph.D. under V. Goldschmidt. An international scientist, he worked at the Technische Hochschule in Berlin and the University of Leipzig, before spending a year at Harvard where he met R. A. Daly, another avid mobilist, and then working (1929–36) at the Geophysical Laboratory of the Carnegie Institution of Washington (I, §4.6, §4.7). Barth returned to Norway shortly before the German occupation, and eventually was confined for a short time in a concentration camp before World War II ended. Barth and Holmsen's analysis of the rocks led them to argue that the islands of the South Antillean Arc, originally sialic (continental), were becoming simatic (oceanic). Using a metaphor appropriate to the times, they concluded that the arc formed with the westward drift of Antarctica and South America.

Let us now make the assumption that one such fragment was represented by South America and Antarctis united by a sialic land-bridge so as to form a twin continent tied together with a narrow band. If now this twin continent was drifting westward, then the connecting landbridge, due to its lesser moment of inertia, would lag behind, retarded by the friction against the viscous sub-stratum. This result would be an arcuate ridge pulled out to a large loop that eventually would break up to pieces. We can think of this process as a battle between continent and ocean-basin. The continental mass that used to connect South America and Antarctis has been conquered by the ocean-basin. The Southern Antilles represent patches of the mother-continent torn off and left behind in a foreign environment; they are spoils of war which sooner or later will become integral parts of the ocean-basin.
(Barth and Holmsen, 1939: 62)

9 This account of Vine's development of his version of the hypothesis is based on various documents that Vine had kindly sent to me, a 1979 interview with Vine, his answers to questions I have asked over the last half-dozen years, and his own retrospective paper with its entertaining title, "Reversals of fortune" (Vine, 2001).

10 Neither Vine nor his former geography teachers can remember the book. Vine also thinks that the picture showed a fit of the coastlines not the margins.

Neither I nor my geography teachers at that time can trace this book. I think it must have been a library book rather than a textbook. I do however distinctly remember the incident but not whether it was a fit of the coastlines or margins – I suspect it was the former.
(Vine, 2008 email to author)

11 Jones' strong support of mobilism represented a change in attitude toward mobilism. Jones, who supervised Irving and Creer while Runcorn was away from Cambridge, and served on Creer's Ph.D. committee, had remained a fixist through 1956, supporting only polar wandering (II, §3.6, §3.10).

12 Vine, I suspect, did not realize that Hospers received his undergraduate degree in geology. Hospers had left the Department of Geodesy and Geophysics almost a decade before Vine arrived, and immediately went into industry.

13 Irving and Runcorn suggest other reasons why Irving was not awarded a Ph.D. for his work (II, §3.4).

14 Matthews actually built only a pressure pot, stating in the preface to his Ph.D. thesis, "With the exception of one new pressure pot, no new apparatus has been built by me for use in the work recorded in this dissertation." However, perhaps to be on the safe side, he added, "Experience gained during the work has indicated the need for new techniques, and machinery for collection of submarine rocks is under construction, but it does not figure here" (Matthews, 1961b: iv).

15 Vine's actual first sentence of his Ph.D. dissertation was "The 'apparatus' used in this work has been a small library of computer programs" (Vine, 1965).

16 Vine is mistaken here. Hospers left the department in1953. Matthews arrived in 1958. Matthews would know of Hospers' work from reading about the magnetic properties of basalt and presumably from reading Hospers' thesis in the departmental library. See also Note 12.

17 Vine explained what he had in mind by the Mini-Mohole rig. It had nothing to do with the device that had been used to obtain the sample whose remanent magnetism Cox and Doell measured.

The Mini-Mohole rig was to be a very small rig that could be lowered to the seafloor from a research ship and remotely drill a short core. At the time it only existed as a set of technical drawings at Cambridge and it was never built at Cambridge. Some years later, however, it was built at the Bedford Institute, Dartmouth, Nova Scotia, and much later by the BGS [British Geological Survey] in the U.K. In both cases, I think, it was successfully deployed.
(Vine, March 11, 2009 email to author; my bracketed addition)

18 EDSAC 2 was designed by a team led by Maurice Wilkes. It first became operational in early 1958, replacing EDSAC 1, and was closed down in 1965. It was the first computer to have a microprogrammmed control unit. Autocode, a high-level programming language was introduced to EDSAC 2 by D. F. Hartley, University of Cambridge Mathematical Laboratory, in 1961. See Wilkes (1992) and Wheeler (1992).

19 I shall reproduce sections mostly from Vine's draft instead of Vine and Matthews' actual paper because the draft better shows Vine's views soon after he came up with the hypothesis and before Matthews, Hill, and perhaps, Bullard suggested changes and additions. The draft also gives more information about Vine's analysis of the survey results on EDSAC 2 than is included in the published version.

20 Vine's feeling about Hill's attitude toward the Vine–Matthews hypothesis is correct (see John Sclater's remarks about Hill's negative attitude toward the hypothesis in §4.3).

21 This account of Morley's development of his hypothesis is primarily based on brief correspondence I had with Morley in 1976 and an extensive interview in 1987. In the interview he explained matters in his background and training relevant to his coming up with his version of the hypothesis. Morley (1986, 2001) has written two retrospective accounts. Glen (1982) also provided a detailed account of Morley's development of the hypothesis. There is some disagreement among accounts, but the accounts agree on key points.

22 Morley mistakenly referenced Hess (1946). This was the paper in which Hess discussed his discovery of guyots and offered his first solution to their origin, which he later rejected (III, §3.5, §3.12). Morley got the reference from reading Dietz (1962c) because Dietz referenced Hess (1946) on page 296 while discussing the origin of guyots. But, in 1946,

Hess did not maintain that the main crustal layer beneath the oceanic basalt could be serpentinized. Nor did Dietz claim that Hess did. Dietz also referenced Hess's 1960 note "Scientific objectives of Mohole and predicted section." Dietz referenced Hess because he had come to agree with his analysis of oceanic crust (III, §4.9). Here is what Dietz wrote.

The mantle itself must be ultramafic peridotite. The transition at the M-discontinuity must be, in agreement with Hess's (1960) proposal, a transformation mainly by hydration of peridotite into serpentine forming the virtually ocean-wide thick Oceanic Layer.

(Dietz, 1962c: 6; Hess (1960) is my Hess (1960b))

Morley referenced the wrong paper by Hess.

23 This account of the rejection of Morley's paper is based on his 1979 letter to me, his 1987 interview with me, his own retrospective papers (1986, 2001), his comments to Emiliani (1981) and Glen's account (1982). There are some discrepancies. Unfortunately there is no way to compare the various retrospective comments with the original correspondence from the journals because it was destroyed in a fire in 1978 (Glen, 1982: 299).

24 Glen mistakenly claimed that the meeting in question was held on June 4, 1964.

25 Morley told Emiliani that he received the rejection notice from *JGR* in September (Emiliani, 1981: 1717). He told Glen that it was late September (Glen, 1982: 298). He told Emiliani that it was rejected in September. Morley himself stated that he received the rejection notice in late August (Morley, 1986: 666; Morley, 2001: 82). Because Morley was still thinking about submitting his paper to another journal after he learned of the rejection by *JGR* but before he found out about Vine and Matthews' paper, the late August date is probably correct unless Morley initially missed Vine and Matthews' paper which appeared on September 7, 1963. Morley also attended the 1963 IUGG meeting in San Francisco, which ran from August 19th to the 31st. It is unclear, however, if he stayed for the whole meeting. It is also possible that the rejection notice was waiting for him once he returned from the meeting.

26 There is some variation as to the wording of the rejection notice from *JGR*. However, the point is always the same: this is an interesting but highly speculative idea better fit for discussion over drinks instead of in a respectable scientific journal. Here is what Morley told Emiliani (1981: 1717):

The paper however was rejected in September 1963, for the reason that (as the editor communicated to Morley) it was the sort of thing one would talk about at a cocktail party, but one would not try to publish such thoughts under serious scientific aegis.

3

Disagreements over continental drift, ocean floor evolution, and mantle convection continue, 1963–1965

3.1 Introduction

The aim of this chapter is to trace the continuing controversies over continental drift, seafloor spreading, and mantle convection during 1963 and 1964, reserving, for the most part, those matters relating to the origin of marine magnetic anomalies until the next chapter. I shall describe attempts to expand seafloor spreading's problem-solving effectiveness. I shall then review the Royal Society's (London) 1964 symposium on continental drift. Symposiasts argued about the merits of mobilism and fixism sometimes with results and arguments we have heard before, sometimes with new ones. I shall also devote a lot of attention to the computer fit of the continents across the Atlantic. It was by far the most influential presentation at the symposium and the reconstruction of peri-Atlantic continents it provided became an icon of plate tectonics. I shall explain how, as evidence in favor of mobilism increased, more Earth scientists began to appreciate the impressive support offered by paleomagnetism, even if most did not recognize its difficulty-free status. As a result, the importance of the mechanism difficulty that had plagued mobilism since Wegener's and Taylor's day began to diminish. It wasn't just old-time mobilists who began to argue that continental drift and seafloor spreading are worthy of acceptance even if their ultimate cause remained a mystery. But it was not all plain sailing; one notable worker even when faced with the new evidence saw fit to return to fixism after a brief time as a mobilist. I shall then discuss Holmes' treatment of mantle convection at this time, and close the chapter with a section on the mounting evidence against rapid Earth expansion, especially the introduction of global methods to use paleomagnetic data to test it.

3.2 Wilson continues seeking further support for mobilism[1]

Tuzo Wilson was off in a new direction. Many of us were collecting or had easy access to new observations of the age and nature of the oceanic crust. He had no ships, but it occurred to him that oceanographers were not taking advantage of the existence of islands that might be samples of the sea floor. Not that we did not sample islands; no geological oceanographer ever passes one by if he can help it. Moreover, nineteenth-century geologists had repeatedly visited almost every island in the deep-ocean basins. With the expansion of oceanography, however, the islands had been relatively neglected, and many had not been studied in the twentieth century. Wilson obtained a research grant contract from the U.S. Air Force and set out to compile geological information about the hundreds of oceanic islands. Volume I, *The Atlantic and Indian Oceans*, was mimeographed and bound in January 1963, and the Pacific volume followed a month later. Tuzo kindly sent me copies, and it became almost impossible to talk to him without studying these books. Geologists are expected to know vast amounts of geography, but until Tuzo no one I knew had ever been so casual about references to Kerguelen or Tristan da Cunha or Fernando Po. Perhaps, come to think of it, I was that way, too, regarding the Pacific Islands.[2]

(Menard, 1986: 192)

Wilson, now a fully committed mobilist and a vocal champion of seafloor spreading, expanded its problem-solving effectiveness by seeking solutions to problems not previously identified. Without ships, he looked to the oceans and to the vast litera-ture of oceanic islands, reasoning that from them there is much to learn about past movements of the seafloor upon which they sit. In the first half of 1963 he had published five mobilistic papers. His "Evidence from islands on the spreading of ocean floors" and "Pattern of uplifted islands in the main ocean basins" appeared in February, "A possible origin of the Hawaiian Islands" and "Continental drift," a popular piece in *Scientific American*, appeared in March, and "Hypothesis of Earth's behavior," a summary paper with additional ideas and his own reconstruction of the continents, appeared in June (Wilson, 1963a, 1963b, 1963c, 1963d, 1963e).

As he read about ocean islands and ocean floors he had one central question in mind: are there any unnoticed patterns that are explicable, if seafloor spreading and continental drift have occurred? He found that there were, in effect solving previously unrecognized problems in terms of seafloor spreading (RS1), and arguing that it was either the only solution or a better one than fixism (RS3). In each case it is not always clear whether Wilson reasoned from seafloor spreading to the data, or in reverse from data to seafloor spreading; which end of the stick he picked up. Did he think about possible consequences of seafloor spreading and see if they showed up in the data, or did he move from the data to seafloor spreading, scouring the data looking for patterns and then asking if they could be explained by seafloor spreading. He first asked if there is a correlation between the age of oceanic islands and their distance from oceanic ridges. He argued (1963a: 536) that if seafloor spreading has occurred then "the distance of islands from the ridge increases with the age of the islands." I suspect that in this case he picked up the stick at the seafloor spreading end, and then narrowed his focus on the ages of oceanic islands, seeing if the predicted correlation held. I say this because Hess had already proposed that guyots are formed at ridge crests and eventually drown as they are carried along with the seafloor on the backs of diverging convection currents.

With good reason, he (1963a: 536) restricted his attention to islands "within main ocean basins" thus excluding "such islands as New Zealand, Fiji and the West Indies."

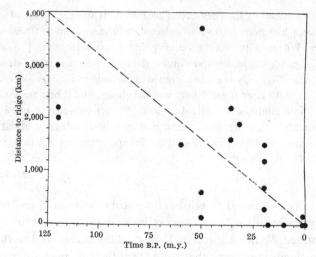

Figure 3.1 After Wilson's Figure 2 (1963a: 537). His caption reads: "Graph of oldest ages reported for some islands in the Atlantic and Indian Oceans against distance of the islands from the nearest point on a mid-ocean ridge." He identified the islands in a separate table.

He (1963a: 536) also, with equally good reason, excluded islands taken to be continental fragments, such as Madagascar and the Seychelles of the Indian Ocean, and the Falklands of the South Atlantic. Turning to the Atlantic Ocean, where he also excluded island arcs such as the West Indies and South Sandwich Islands, he argued that the correlation, although "not perfect," held.

Although the correlation is not perfect it seems adequate to suggest as one possible explanation that many ocean islands have originated over the mid-ocean ridges and that the older ones have been steadily carried away from the ridge. In the Atlantic Ocean it is noticeable that six active or very recently active volcanic islands lie along the ridge. They are Jan Mayen, Iceland, the Azores, Ascension, Tristan da Cunha and Bouvet. On the other hand, islands with Cretaceous rocks, the Bahamas, Fernando de Noronha, and Cape Verde, lie close to continents, while islands of Miocene to Eocene age, like the Faeroes, Bermuda, Saint Helena and Madeira, lie at intermediate positions. All of these are inactive.

(Wilson, 1963a: 536)

Wilson listed the age and distance to the relevant ridge for seven islands in the Indian Ocean (Kerguelen, Heard, St. Paul, Madagascar, Seychelles, Providence, and Christmas) and fifteen islands in the Atlantic Ocean (Jan Mayen, Iceland, Faeroes, Azores, Madeira, Canary, Cape Verde, Fernando de Noronha, Bermuda, Bahamas, Ascension, St. Helena, Tristan da Cunha, Bouvet, and the Falklands). Discounting the Falklands, Madagascar, and the Seychelles, he graphed the remaining nineteen islands in both oceans (Figure 3.1).

Wilson admitted that Iceland and the Faeroes are too close to the Mid-Atlantic Ridge given their age (RS2), and in an attempt to remove the difficulty, proposed

(1963a: 536), "this may be due to the fact that, if the Atlantic Ocean spread, it must have done so by the least amount in the north. Another contributing factor was perhaps the opening of Baffin Bay." Christmas Island presented another anomaly; it was too far away from the ridge given its age. Wilson (1963a: 536), citing a source, suggested that its real age might be much greater than the age of its exposed rocks.

He then turned his attention to aseismic ridges that branch off active ridges. Using the Walvis and Rio Grande Ridges, which respectively extend from South America and Africa and meet at the Mid-Atlantic Ridge by Tristan da Cunha, he explained their formation in terms of seafloor spreading and a non-migrating continuously active volcanic source. He noted that his explanation is testable.

Convection also introduces horizontal motion. This would slowly move volcanoes away from their sources on the mid-ocean ridges, so that those volcanoes would cease to be active. The sources would remain fixed over the vertical currents and would produce fresh volcanoes. Thus, in time, each source on the mid-ocean ridge would produce a chain of progressively older extinct volcanoes, or perhaps two such chains, one on either side of the mid-ocean ridge ... Here is a possible explanation for ridges like the Walvis and Rio Grande. This proposal could be tested because the oldest seamounts on the branches should lie farthest from the mid-ocean ridge.

(Wilson, 1963a: 537)

Wilson noted that if this explanation is correct then these aseismic ridges tracked the movement of continents as they moved away from the Mid-Atlantic Ridge, and indicated the flow of underlying convection currents. He (1963a: 537) also claimed, without identifying them in this paper, that the Walvis and Rio Grande Ridges should parallel other aseismic ridges shooting off from the Mid-Atlantic Ridge. "All such branches might be flow lines. This hypothesis could explain the general parallelism of all the rises on either side of the Atlantic." He (1963d: 88) sketched the position of the other aseismic ridges in a crude figure of the Atlantic and surrounding continents in his *Scientific American* paper. One extended from Bouvet Island on the Mid-Atlantic Ridge to Africa's Cape Peninsula; another, from Ascension Island on the Mid-Atlantic Ridge through the islands of Annobón, São Tomé, de Principe, and Fernando Po (now Bioko) to the African coast, and another, from the east coast of Greenland across Iceland where it intersected with the Mid-Atlantic Ridge and extended to the Faeroe Islands. Unable to contain himself, he (1963a: 537) further proposed that the herringbone pattern of the aseismic ridges served as another directional marker of past movements. They indicated a northward movement to the drifting apart of North America and Africa, which, he added, had been independently confirmed by paleomagnetic studies of Africa's past movement (RS1).

The well-known herringbone pattern of the ridges could be an expression of the fact that Africa and North America did not move directly apart, but that each had a northward component of motion as well. For Africa at least palaeomagnetic evidence has already suggested this.[49]

(Wilson, 1963a: 537; endnote 49 refers to Blackett et al., 1960; see II, §5.19)

Wilson now turned his speculative eye to the Pacific, looking for further stream lines. He found them in island chains, and previewed his hypothesis, which he already had written in detail, that these island chains form from lava sources in the mantle as they intermittently produced lava while remaining relatively fixed as new seafloor moves over them.

If this view is adopted it seems likely that the Austral, Society, Tuamotu, and Hawaiian Islands chains also lie along stream lines that the progression in age of the islands along each chain is due to this cause. The Hawaiian Islands and some of the other chains are exceptional in that their source does not lie on the mid-ocean ridge. A fuller discussion of this suggests[48] that chains of islands could still be found if the source of lava lay within the stable cores of convection cells ... The recent discovery[52] of Miocene rocks on Oahu supports, but does not prove, the view that the Hawaiian Islands have been formed successively over a common source beneath the present position of the Island of Hawaii, and that they have been successively borne away toward the west. The dates given by Menard *et al.* indicate that the rate of movement is not greater than 10 cm a year.

> *(Wilson, 1963a: 538; endnote 48 refers to Wilson (1963c), which was*
> *in press, and endnote 52 refers to Menard* et al. *(1962))*

Menard recalled how he discovered Miocene rocks on Oahu, and how Wilson knew what do with his discovery.

Tuzo Wilson ... took a new approach. He rejected the idea that the islands had all been active at the same time; that had been tenable only when none were dated, and I had dated one. I had conducted a one-day cruise off Honolulu in 1961 to show foreign oceanographers at the Pacific Science Congress how we used our ships. I had the luck to dredge fossil reef corals at the edge of the insular shelf at 500 m. My paleontological colleagues thought the fossils were probably of Late Miocene age. Thus, reasoned Wilson, the age progression was caused by volcanism at the tip of a propagating crack. The crack was not needed once he proposed the correct origin, namely, that the archipelagoes are produced by drifting the ocean crust over a relatively fixed source of magma in the mantle. He began by explaining how a source could be fixed when Hess's model required a convecting mantle to cause the drift. The sources were within immobile cores above which moved relatively thin horizontal surface flow and below which moved a horizontal return flow.

> *(Menard, 1986: 195–196)*

Wilson devoted an entire paper to the origin of the Hawaiian Islands, and other island chains he claimed formed from a relatively fixed source within the mantle (a blow torch as he called it) as seafloor drifted above. He first submitted the paper to *JGR*, but it was rejected (Glen, 1982: 376). He then submitted it to the *Canadian Journal of Physics* where it appeared in March. The paper is particularly interesting because Wilson hypothesized the existence of what later became known as hotspots, which Jason Morgan, acknowledging his debt to Wilson, reintroduced eight years later (Morgan, 1971).

Wilson identified two groups of such islands, both in the Pacific basin on opposite sides of the East Pacific Rise (Figure 3.2). The larger group with seven island chains

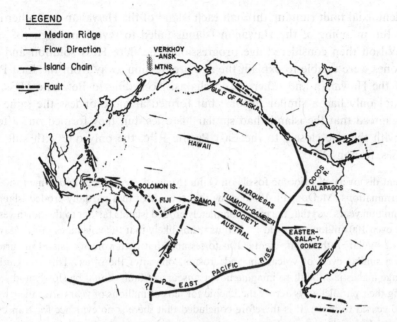

Figure 3.2 Wilson's Figure 1 (1963c: 865). Nine linear chains displayed with heavy arrows. Age of islands in each chain increases in direction of arrow. Single-headed arrows indicate direction of motion along transcurrent faults; small arrows illustrate proposed direction of flow from median ridges.

is west of the rise; the other, east of it. He emphasized that chains in each group were parallel with each other, that the seven west of the rise trended northwest–southeast, and that:

All except Samoa, which is particularly situated, have active or recent volcanoes on the most easterly island and appear to get progressively older towards the northwest in the direction away from the East Pacific Rise.

(Wilson, 1963c: 865)

Wilson, citing Menard, added that Samoa appears to be situated above diverging convection currents. Wilson (1963a: 865) claimed that the two island chains in the other group "have mirror-image properties to the other seven so that these chains increase in age towards the east away from the East Pacific Rise."

Wilson argued that his solution was superior to the still-prevailing view, one proposed in 1942 by Betz and Hess (III, §3.4). They had argued that the Hawaiian Islands and other similar island chains had formed by extrusion of lava through large faults. Wilson raised three difficulties with this (RS2). The first, a longstanding one that Betz and Hess themselves had recognized and Dietz and Menard (1953) later raised (III, §4.5), was that the elusive fault had not been found despite more surveys with more sophisticated equipment. Wilson (1963c: 866) next argued that if the Betz–Hess solution is correct, the original transcurrent fault should reveal itself as an ancient axial fault running through each island of the Hawaiian and other island chains. But mapping of the Hawaiian Islands failed to reveal any signs of such a fault. Wilson then considered age progression. The Betz–Hess solution and more recent ones were unable to explain the age progression away from the East Pacific Rise of the Hawaiian and other island chains. According to Betz and Hess, each island not only had a similar history, but formed at more or less the same time. Wilson agreed that the islands had similar histories, but they formed one after the other, with the one closest to the East Pacific Rise, the only one with still active volcanoes, forming last.

The recent discovery of Miocene fossils on Oahu (Menard *et al.*, 1962) and some radioactive age determinations (McDougall, 1963) show that the age of that partially eroded island is at least 20 million years and that the more completely eroded islands farther to the northwest may be 50 or even 100 million years old ... It thus seems likely that the volcanoes of the Hawaiian chain had similar, rather than identical histories and that each volcanic island in turn went through a similar cycle of volcanism and erosion, one after the other. This recognition of greater age makes it difficult to imagine how transcurrent faults could be slowly and steadily extending their parallel ways across the Pacific for tens of millions of years after their western ends had ceased to move. It is therefore concluded that there is no evidence for transcurrent faults beneath the islands and that some other explanation must be found.

(Wilson, 1963c: 867; references are the same as mine)

Wilson introduced his solution, placing it within the context of mantle convection and seafloor spreading. He claimed that recent estimates of the mantle's viscosity permit mantle convection, introduced Hess's and Dietz's versions of seafloor spreading, mentioned Menard's view that convection currents rise under the East Pacific Rise and flow away from the rise in both directions, and then (1963c: 867) introduced the key issue of his paper, "how to explain the Hawaiian Islands in terms of such currents."

The difficulty Wilson faced was how to keep the source of the upwelling lava from moving along with the convection current. Wilson adopted what he referred to as "jet-stream" convective cells as opposed to the more prevalent "uniformly rotating" ones. He referred to Hide's work, as he had done in 1962, although this time spelling Hide's name correctly (§1.10). Jet-stream convective cells, unlike uniformly rotating ones, have a stable core. Wilson proposed that the source of the upwelling lava that formed the Hawaiian Islands is within the stable core.

It is difficult to imagine how a source not over a mid-ocean ridge can remain relatively fixed ... [If] the source of the lava is within the relatively stagnant center of a jet-stream type of cell and if the surface layer is moving past the source, then a chain of volcanoes could result, as shown in figure 5 [Figure 3.3]. It is not necessary for the source to be immobile. It need only move more slowly than the near-surface current.[3]

(Wilson, 1963c: 869; my bracketed additions)

Wilson's solution explained why the active or recent volcanoes are located on the island closest to the East Pacific Rise, why the islands are progressively older the

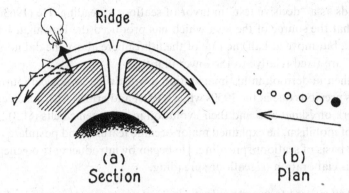

Figure 3.3 Wilson's Figure 5 (1963c: 869). The cross-hatched area represents the stable core. Intermittent rising basalt creates a line of islands as the jet-stream convective currents carry the seafloor away from the ridge.

further they are from the rise, why the island chains are linear and parallel to each other, why they are not associated with huge transcurrent faults, and why axial fissures had not been found on individual islands.

He (1963b) presented another hypothesis to account for a line of uplifted ocean islands in the central Pacific that were on a line extending from the channel between Samoa and Fiji to Easter Island. The line also appeared to extend through a great fracture zone just north of Easter Island to San Felix Island off the coast of Chile (Fisher, 1958). Wilson tentatively proposed that the islands are uplifted because they lie on an uplifted fault, itself uplifted in response to the down-warping of Pacific seafloor as it descended into the mantle. This particular explanation caught the attention of few.

Wilson's two summary papers, a general one in *Scientific American* (1963d) and a more technical one in *Nature* (1963e), both show him as committed a mobilist as he once was a fixist. In the *Nature* paper, he cited the work of Wegener, du Toit, van der Gracht, Carey, and King. In both papers, he summarized Hess's and Dietz's versions of seafloor spreading. He introduced Daly's distinction between the lithosphere and asthenosphere, and proposed that convection currents travel through the astheno-sphere, moving away from ridges dragging the overlying lithosphere and seafloor. In the *Scientific American* paper he (1963d: 90–93) talked about the recent discovery of *Glossopteris* flora in Antarctica (I, §7.4), and his own proposed connection of the Cabot and Great Glen faults (§1.10). In both papers he gave figures of the reassembled continents.

He summarized the key, new support of mobilism, first mentioning the horizontal displacement of up to 750 km along the northeast Pacific fracture zones. Turning to mobilism's paleomagnetic support from the continents, he (1963d: 95) claimed, "Continental drift offers the only explanation of these findings that has withstood analysis." That is, the paleomagnetists' solution to the mobilism question was diffi-culty-free. He (1963d: 95) offered his solutions to the origin of aseismic ridges and ocean islands as a "decisive test" in favor of seafloor spreading. He (1963d: 92) also proposed that the source of the lava which has produced the Hawaiian Islands may not be fixed, but move at half the rate of the lithosphere above; he did not insist that "hot spots" are fixed relative to the lower mantle.

Wilson then undertook in his imagination a worldwide geological journey reminiscent of his previous one in the 1950s while a fixist and armed with Scheidegger's and his own work on island arcs and their evolution into orogenic belts (§1.3). Now, as a champion of mobilism, he explained major oceanic features and postulated unknown ones on the basis of seafloor spreading. He began by introducing two generalizations based on his elaboration of seafloor spreading.

These observations lead to two generalizations. First where a mid-ocean ridge lies half-way between two continents, they were once in contact. Secondly, the ends of lateral ridges and the fit of shore lines may be used to re-assemble them in the positions in which they once lay.

(Wilson, 1963e: 926; similar passage in Wilson, 1963d: 96)

Starting in the North Atlantic, he used his first generalization in reverse, picked up the stick at the seafloor spreading end, and postulated that there was a ridge extending up Baffin Bay between Baffin Island and Greenland, which had drifted apart. His evidential support amounted to no more than "six large and unexplained earthquakes."

The reverse case can also be argued. The rifting of the Atlantic Ocean not only involved separating South America from Africa and Europe, but also the separation of Greenland from North America. Therefore a longitudinal, median ridge should extend up Baffin Bay. Six large and otherwise unexplained earthquakes have been recorded along this line; but the bathymetry is as yet inadequately charged to prove that such a ridge exists. This can be checked.

(Wilson, 1963e: 927)

He reintroduced the Wegener Fault, which he had proposed the previous year (§1.10), identifying it as a sinistral transcurrent fault that marked Greenland's eastward movement relative to Canada. He further speculated about other happenings in the Arctic coeval with the opening of the Atlantic.

Shortly before the start of the Cretaceous period, about 120 million years ago, the continent [Pangea] developed a rift that opened up to form the Atlantic Ocean. The rift spread more widely in the south, with the result that the continents must have rotated slightly about the fulcrum near the New Siberian Islands ... Soviet geologists have found that the compression and uplift that raised the Verkhoyansk Mountains across eastern Siberia began at about that time ... It seems reasonable to suggest, particularly from the geology of the Verkhoyansk Mountains and of Iceland, that at the start of Tertiary time, about 60 million years ago, this convection system became less active and that rifting started elsewhere.

(Wilson, 1963d: 99; my bracketed addition)

Thus Wilson maintained that seafloor spreading rates are not constant and that seafloor spreading may stop in one area and begin elsewhere. It may also restart in

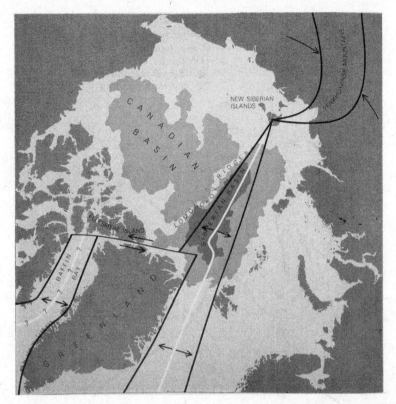

Figure 3.4 Wilson's figure "Rifting of Supercontinent" in the Arctic (1963d: 100). The Wegener Fault, a transcurrent fault, is just north of Greenland's northern coast. The opening of the Atlantic Basin around the pivot point is balanced by the compressing wedge that forms the Verkhoyansk Mountains. Opposing arrows indicate seafloor spreading with formation of Baffin Bay, Siberian Basin, Greenland and Norwegian seas.

the same place, which is what he (1963e: 927) claimed had happened in the North Atlantic, arguing that the "Pliocene to Recent rocks in the central graben of Iceland indicate a second period of rifting."

Wilson (1963e: 927) no longer viewed the worldwide system of mid-ocean ridges as continuous. He noted, presciently, "Mid-ocean ridges appear to end either at large transcurrent faults or by wedging out." He pointed to clear examples in the Arctic. The Wegener Fault connects the "Baffin Bay" Ridge with the "Siberian Basin" Ridge, which itself terminates by wedging out with transference of motion to a compressing wedge that caused formation of the Verkhoyansk Mountains (Figure 3.4).[4] Wilson went on to speculate about what would be discovered in the forthcoming "International Indian Ocean Expeditions." Believing that India and Australia were once juxtaposed, he proposed that a median ridge would be found between them.

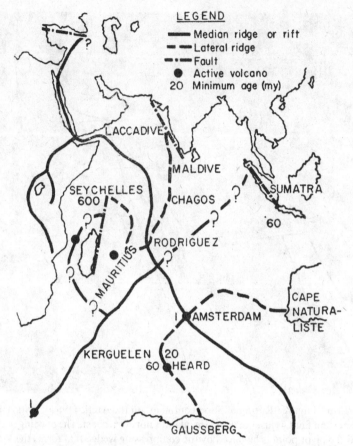

Figure 3.5 Wilson's Figure 2 (1963e: 925) with the following caption: "Ridges of the Indian Ocean showing two postulated median ridges, between Africa and Madagascar and Rodriguez Island and Sumatra respectively. Four aseismic lateral ridges connect points believed to have been separated by continental motion."

The continents around the Indian Ocean have been re-assembled by various authors to form Gondwanaland. The patterns have been very varied. [Wilson here referenced Wegener, 1924; du Toit, 1937, and van der Gracht, 1928.] Applying my new generalizations one can see that to create the Indian Ocean by rifting required the separation of four continents, Africa, India, Australia and Antarctica, and that in addition to the three known median ridges there should be a fourth extending north-easterly between India and Australia (Fig 2) [Figure 3.5]. Several earthquakes and some shallow soundings shown by Gutenberg and Richter support the proposal that such a ridge may exist. Menard has included what may be an indication of it in his sketch of mid-ocean ridges. If it does exist it will no doubt be discovered during the International Indian Ocean Expeditions.

(Wilson, 1963e: 927; my bracketed additions)

Using as guides his idea of lateral ridges and his thesis that the distance of islands from ridges increases with age, he argued that Antarctica and Australia had once been joined.

The opposing coasts of Australia and Antarctica lack large indentations by which they may be matched, but the Amsterdam-Cape Naturaliste ridge and the Amsterdam-Kerguelen-Gaussberg ridge can be seen to form a pair of lateral ridges on either side of the south-eastern median ridge. One is roughly a mirror image of the other. Our premise implies that Cape Naturaliste was once in contact with the Gaussberg coast and this gives a precise method of re-assembling Australia with Antarctica ... The known ages of Amsterdam, Kerguelen and Heard Islands get progressively greater. The Gaussberg volcanics and those at Bunbury near Cape Naturaliste should be older still and related in composition. Many such predictions enable this hypothesis to be checked.

(Wilson, 1963e: 927)

He proposed that "southern India was once adjacent to western Australia based on the positions of the Chagos–Maldive–Laccadive and Amsterdam-Cape Naturaliste lateral ridges, and gave an additional reason for the existence of a previously postulated but unconfirmed fault along India's west coast.

The Chagos-Maldive-Laccadive ridge forms another lateral ridge paired with the Amsterdam-Cap Naturaliste ridge across the postulated north-east median ridge. This suggests that southern India was once adjacent to western Australia and hence close to Gaussberg. If, when India was carried north and collided with the rest of Asia, the floor of the Arabian Gulf had continued to move northward the fault long postulated along the west coast of India would be explained.

(Wilson, 1963e: 927)

Clearly enjoying himself and undeterred by the possibility that his daring proposals of undiscovered spreading ridges might be wrong, he postulated yet another between Madagascar and Africa.

Symmetry suggests that there should be a fourth lateral ridge. It must obviously be the Rodriguez-Mauritius-Seychelles ridge, and again a large fault, known for fifty years, would have allowed the floor of the Arabian Sea to move northwards past Madagascar. If this fault be extended along a line of atolls, one of which, Providence, has a volcanic base, it would pass the Seychelles and may have carried this group away from Madagascar, thus explaining the unusual continental nature of those Islands. A median ridge through the active volcano on Comores Island may exist and have separated Madagascar from Africa, and bathymetry suggests that such a ridge may exist and extend southwards.

(Wilson, 1963e: 927)

He also tied the termination of two median ridges to transcurrent faults.

Two median ridges apparently terminate in the Indian Ocean area. It may be more than a coincidence that there is a large transcurrent fault in Sumatra where one ridge should end. The other ridge passes into the African and Jordan rift valleys. Again there is a large transcurrent fault in Turkey, but it seems to lie some distance from the end of the Jordan rift.

(Wilson, 1963e: 927)

Wilson renewed his discussion of the Pacific in terms of the basic tenets of seafloor spreading and his explanation of the origin of Hawaiian and Society Islands.

The East Pacific Rise trends north-easterly. Thus the greatest contraction and flow is in a generally west-north-westerly direction as indicated by the orientation of the Hawaiian and Society Islands. The ocean floor must therefore be disappearing in trenches lying at right angles to the direction. Such trenches are those of the East Asian Arcs, the Tonga-Kermadec Islands and the Andes. These are the most active seismic regions in the world with most of the deep earthquakes.

(Wilson, 1963e: 928)

He imagined the seafloor spreading away from the East Pacific Rise northwestward eventually passing under East Asia, and, less accurately, the "expansion in a north-ward direction due to the median ridge through the southern seas" to account for compression in the Alpine–Himalayas–Indonesian belts.

The large shears in Melanesia have been supposed to be left-handed in the Philippines, New Guinea, and between the New Hebrides and Tonga. Thus all the directions of shearing in California, off British Columbia and in Melanesia, and all the directions of dip of the planes on which deep earthquakes lie in the Pacific, may possibly be explained by supposing that the floor of the Pacific Ocean is moving away from the East Pacific Rise, northwesterly to pass under East Asia and New Zealand and south-easterly beneath the Andes. Likewise, the lesser expansion in a northward direction due to the median ridge through the southern seas can explain the compression and activity of the Alpine-Himalayan-Indonesian region.

(Wilson, 1963e: 928)

Introducing another generalization, he (1963e: 928; my bracketed additions) claimed, "Where two horizontal [convection] currents meet at about 180° angles [head-on], arcs form, but where currents meet at about 90°, shears form." Reproducing as Figure 3 his Figure 1 from Wilson (1963c) (Figure 3.2 above), he applied his new generalization to the Pacific, and even extended the Alpine Fault southward to the East Pacific Rise.

The right-hand shear through New Zealand, like those in Melanesia and the Western Americas, results from two directions of flow meeting at right angles. Because Macquarie Island is the only island in the deep oceans to be well folded, prophyllitized and to have views of sulphides, it seems to be partly continental in characters. (So are the several groups of islands in shallow water south-east of New Zealand.) It is suggested that the Alpine Fault of New Zealand extends south past Macquarie Island and offsets the mid-ocean ridge near the Balleny Islands.

(Wilson, 1963e: 928)

Recalling Scheidegger and their earlier success explaining the geometry of island arcs during his previous imaginative worldwide tour, he attempted to explain varying differences in the angle of descent of earthquake planes beneath island arcs.

Attempts to derive a simple interpretation of the results [of the work by Gutenberg and Richter, by Benioff, and by Hodgson] have been inconclusive. It is suggested that whereas

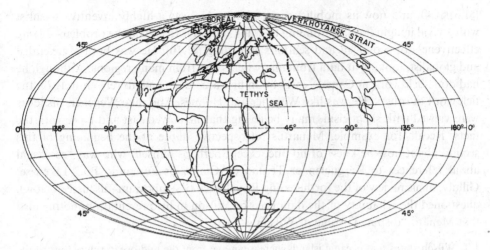

Figure 3.6 Wilson's Figure 6 (1963e: 928). "The dashed lines represent some earlier mountains formed by unions of older continental blocks."

Scheidegger's explanation of the circular shape of island arcs by fracture is tenable for a brittle and uniform ocean floor in the shallow lithosphere, for deeper earthquakes an element of flow intervenes and that no consistent pattern exists. It is suggested that the reason why the planes marking the foci of deep earthquakes dip west under the East Asian area and the Tonga Islands, east under the Andes and nearly vertically under the Solomons and Bonin Islands is not decided by any special angle of fracture, but by plastic failure in places decided by the direction of flow.

(*Wilson, 1963e: 928–929; my bracketed addition*)

He concluded his discussion of the Pacific by reintroducing his explanation of Hawaiian Islands.

Wilson ended his worldwide survey by presenting a reconstruction of the continents in mid-Mesozoic time based on his analysis of the Atlantic, Indian, and Pacific Oceans. He introduced his reconstruction with the following:

This use of median and lateral ridges and the recognition of the Verhoyansk Mountains as the proper hinge between Eurasia and North America instead of Alaska enables a unique and new reconstruction of continents to be made for Lower Jurassic and probably also Triassic time (Fig. 6 [Figure 3.6]).

(*Wilson, 1963e: 928; my bracketed addition*)

Wilson (1963e: 928) thought it "especially important" that his reconstruction was "assembled according to rules, and not merely by arbitrary matching of coasts." He also claimed that he had obtained approximate positions of the continents from paleomagnetic data, although he gave no details, and acknowledged that the positions "could be improved," although in the quote he had described it as "unique."

Tuzo Wilson never seemed at a loss for new ideas. He was a master at developing auxiliary hypotheses both as a fixist/contractionist in the late 1940s and early 1950s

(§1.3, §1.4), and now as mobilist/convectionist. He was a highly inventive scientist with a vivid imagination, continuing to crank out ways to extend the problem-solving effectiveness of mobilism and seafloor spreading (§1.10). He was both a generalist and globalist, looking at data from many fields and from many regions. Although he had done fieldwork in Canada early on in his career, he came to rely entirely on the fieldwork of others. He was like Wegener and Holmes. And, like Wegener, his ideas were viewed with skepticism, many believing that he played fast and loose with the facts. Recall, for example, Menard's retrospective quote at the beginning of this section about how he knew of no one, except perhaps himself, who was "so casual about references to Kerguelen or Tristan da Cunha or Fernando Po." Likewise, Gilluly, in his review of the forthcoming Royal Society symposium (discussed below), questioned the reliability of the data used by Wilson in his presentation, noting also that Menard "disputed" some of Wilson's "facts."

J. T. Wilson states that oceanic islands increase in age from the mid-ocean ridges toward the continents and the andesite line, with a maximum age of Jurassic. Several pairs of aseismic ridges diverge from the Mid-Atlantic Ridge to join the continents at points that would match on Wegener's scheme. Holmes has shown that each such pair of ridges tends to have distinctive chemical characteristics. Possibly these lateral ridges are streamlines of the convection cells that carry the continents apart. (In a later paragraph Wilson suggests a wholly different theory: Ridges such as the Walvis and Rio Grande may diverge symmetrically from the mid-ocean ridge because they were formed by eruptions from a central volcanic source that migrated along the ridge.) The radiometric and chemical evidence presented in support of these contentions seems to be rather tenuous. Wilson is skillful and fertile in hypotheses, but the "precise geological observations" that he offers in support seem far from convincing to a geologist, however appealing they may be to a geophysicist. Several of Wilson's "facts" were disputed by Menard during the later discussions, more recently, the discovery of Miocene rocks on the Mid-Atlantic Ridge weakens his first argument.

(Gilluly, 1966: 948)

3.3 The Royal Society's 1964 symposium on continental drift

In the next four sections I shall review the papers and discussions presented at this meeting. I pay special attention to the computer-generated reconstruction of the peri-Atlantic continents, which, with its use of Euler's fixed point theorem, was the harbinger of plate tectonics and by far the most influential paper of the meeting.

Runcorn had just organized the huge NATO meeting at Newcastle in January 1963 (III, §1.9) that drew together over seventy speakers, paleomagnetists, paleontologists, paleoclimatologists, and geologists. Nonetheless, he decided that another meeting was needed "to advance the campaign." He envisaged one with a different slant, one devoted to the new evidence for mobilism, especially from paleomagnetism and marine geology, a meeting that reflected a progressive outlook. He wanted to have discussions of seafloor spreading and mantle convection. As it turned out, he got a bonus – what has become the iconic continental reconstruction just referred to.

Runcorn recounts that he spoke with Blackett about having the Royal Society of London sponsor such a meeting. He agreed, approached the Society, and recommended that Bullard be an organizer. They were asked to proceed.

I said to Blackett, "We really ought to try and have a meeting in England to properly lay out all the different arguments that are now beginning to be considered." And I suggested the Royal Society. Of course, I wasn't then a fellow, but I thought that was the body which would carry weight, and would perhaps advance the campaign. Blackett, of course, was very enthusiastic. So he suggested it to the Royal Society ... When Blackett kind of asked the Royal Society to do it, he suggested that Bullard and he, being fellows, and I should be the organizers.

(Runcorn, 1984 interview with author)

The symposium was held in London on March 19 and 20, 1964. Papers and comments were submitted by November and the proceedings published in 1965. Blackett and Bullard "delegated the task of organizing the meeting to Runcorn." Runcorn, anxious to present a balanced program, made sure to invite MacDonald and Jeffreys, two of mobilism's most vociferous critics. MacDonald accepted; Jeffreys refused but submitted a brief statement.

Bullard and Blackett let me organize the meeting ... and I said to them, "Well if you want me to do this, I'm going to do it myself. I don't like working in committees. But as the programme takes shape, I'll send you continually around the latest state of play, and you can add any other people you want." You see that is how it was done. Of course Bullard and Blackett were quite happy with me to go ahead and organize it, and write all of the letters, etc. But I didn't want to make it appear to be just a recital of the evidence for continental drift and that is why I invited Gordon MacDonald who was a very vigorous opponent. But I was obviously a little anxious. It was a new venture for the Royal Society. I was rather anxious to make it quite fair, and invite MacDonald and other critics. So it was an attempt to strike a bit of a balance, and that is why I invited Gordon MacDonald. And we asked Harold Jeffreys. I don't think he came but he did contribute to it.

(Runcorn, 1984 interview with author)

In this section I summarize the symposium itself. Several papers of durable interest are dealt with in more detail elsewhere (§3.4, §3.6). There were four sessions which I deal with in turn. I also report comments by discussants. Most participants definitely favored mobilism, which questions just how successful Runcorn was in achieving a balanced program. There was, however, certainly a slant in the direction he intended, away from the climatological and paleontological aspects that were so extensively covered in the NATO meeting two years earlier. Notably many participants singled out paleomagnetism's strong support of mobilism, indicating that it was still in their eyes the strongest evidence for drift. Earth expansion was put forth, but few paid attention.

The first session was entitled "Continental reconstructions." There were four presentations, all favoring mobilism. Runcorn (1965a) and Creer (1965), then a member of Runcorn's Physics Department at the University of Newcastle, gave talks on paleomagnetism summarizing its support for mobilism. No doubt wanting a review of the classical case for mobilism, Runcorn invited Westoll of the Geology Department, University of Newcastle, and a long time advocate of continental drift, to speak. Westoll (1965: 25), acknowledged that the geological "evidence taken piece by piece cannot prove or disprove drift," but taken "collectively form a linked network that argues powerfully in its favour." He welcomed paleomagnetism's support, and praised new geophysical studies that have "gone far to justify Arthur Holmes's use of convection in the mantle." Bullard closed the session with an account of the Cambridge assembly of continents around the Atlantic. This was published under the co-authorship of Bullard, Everett, and Smith (1965).

In the second session, entitled "Horizontal displacements in the Earth's crust" there were five presentations. Wanting to showcase large transcurrent motions, Runcorn asked V. Vacquier (III, §1.6, §5.9) to cover transcurrent faults at sea, and C. R. Allen (Caltech) to cover continental transcurrent faults on the continents. To cover horizontal displacements, I know that he invited Heezen, Menard, and Hess who declined. I don't know if he invited Ewing and Dietz. He may have asked Ewing, who may have suggested Worzel, then at sabbatical at Madingley Rise, as replacement. Worzel was a discussant. Runcorn also invited Girdler, from his own department.

Vacquier (1965) gave an updated version of the paper he presented in Runcorn's edited volume, *Continental Drift* (III, §1.6); he discussed the huge displacements along the great fracture zones of the northeastern Pacific as revealed by matching the zebra patterns of magnetic anomalies. Allen marshaled much evidence for large terrestrial transcurrent faults in regions surrounding the Pacific, identifying them in Chile, California, New Zealand, Mexico, Taiwan, and Alaska (Allen, 1965). He thought their presence and that of the huge transcurrent faults in the northeastern Pacific enhanced support of continental drift, even though it was not currently possible to reconcile the east–west trend of the oceanic transcurrent faults with the northeast–southwest and northwest–southeast trend of the continental ones.

Thus continental drift – in a limited sense – is taking place before our eyes, and at very finite rates. Such observed horizontal movements of continental crustal blocks, taken together with evidence for very large displacements in the past, certainly add to the attractiveness of the continental drift hypothesis, for they destroy the argument that continental drift of any type is mechanically absurd. On the other hand, the directions of displacement along major transcurrent faults of the circum-Pacific rim are largely parallel to the oceanic borders and thus fail to fit neatly with most of the specific hypotheses of convection and drift that have been presented.

                     (Allen, 1965: 88)

In their presentation, Heezen and Tharp continued to support Earth expansion as an explanation of continental drift. Citing Dietz (1961a) and Wilson (1963e), they claimed that the "tectonic pattern of the Indian Ocean floor seems particularly difficult to explain in terms of oceanic spreading." As for Earth expansion, Heezen admitted that it faced difficulties but still preferred it to mantle convection.

Although there are many unsolved problems and some evidence which apparently is adverse to this concept, the writer believes that a general expansion of the earth better explains the sea floor tectonic fabric than the recently popular convection current hypothesis.

(Heezen and Tharp, 1965: 105)

Menard spoke next and it was in this presentation (discussed in detail later, §3.7) that he ended his brief flirtation with mobilism. He still argued for mantle convection, but placed it within a new framework which did not permit continental drift (Menard, 1965a). Girdler, fully supportive of continental drift, presented his own version of seafloor generation, again applying it to the Red Sea (Girdler, 1965). Thus three symposiasts (Vacquier, Allen, and Girdler) preferred continental drift; one (Menard) announced he had become a fixist.

There were six main speakers in the third session, "Convection currents and continental drift." G. D. Nicholls, a petrologist from the University of Manchester, offered petrological support for seafloor spreading based on analyses of different kinds of oceanic basalts (Nicholls, 1965). G. P. L. Walker (Imperial College, London) discussed the formation of seafloor along the Mid-Atlantic Ridge as it crossed Iceland, in terms of dyke injection. Although he estimated that the rate at which dykes are carried off the ridge axis as they make way for newer ones is appreciably less than is required to separate America from Europe, he claimed that his analysis showed how seafloor spreading can be achieved through dyke injection (Walker, 1965: 199). Geochemist J. A. Miller (Department of Geodesy and Geophysics, University of Cambridge) showed that the general distribution of age zones around the North Atlantic based on radiometric studies matched well if North America, Greenland, and Europe were juxtaposed, as proposed in the Bullard, Everett, and Smith (1965) reconstruction (Miller, 1965a); in an appendix to Miller's paper, F. J. Fitch (1965) (Birkbeck College, London) added further age dates in support. J. Goguel, Professor of the School of Mines in Paris and Director of the Geological Map Service of France, was the only main speaker not from the UK, Canada, or the United States. Wilson (1965a) argued that the ages of oceanic islands support mobilism. Goguel (1965) maintained that the tectonic evidence in favor of mobilism had actually decreased since Argand and Wegener's day because they had probably overestimated the amount of lateral movement needed to explain formation of the Alps. However, he also claimed that the evidence of horizontal displacements based on huge transcurrent faults was solid. Although at the symposium, he said nothing about the new paleo-magnetic support of mobilism, as he had done so not long before in an updated version of the 1962 English translation of his 1952 *Traité de Tectonique*, noting however that much more work was needed (Goguel, 1962: 359). He also favored mantle convection as the cause of continental drift, but nonetheless, did not commit himself to mobilism. I think he was still sitting on the fence, but leaning toward mobilism. Thus, four of the six speakers (Wilson, Nicholls, Miller, and Fitch) clearly favored mobilism. Walker, I believe, was more inclined toward mobilism than Goguel. For counting purposes, I count Walker as inclined toward mobilism and Goguel as neutral.

The fourth session, "Physics of convection currents in the Earth's mantle" opened with MacDonald continuing his attack on mobilism and mantle convection (III, §2.7) and concluding (1965: 225), "Both the deep structure of continents and the mechanical properties of the mantle make untenable any hypothesis linking large scale convective currents with continental drift." He dismissed mobilism's paleomagnetic support, essentially because the origin of the geomagnetic field was not fully understood. MacDonald apparently believed that, at least when it came to paleomagnetism, nothing can be known until everything is known. Runcorn (1965b) presented his account of changing patterns of mantle convection as the core grew in size (III, §1.5). D. C. Tozer of the Physics Department at Newcastle argued that radiogenically fueled upper mantle convection, perhaps augmented by lateral chemical and physical inhomogeneity, may cause continental drift. Disapprovingly, he (1965: 254) argued that Runcorn based his theory "primarily on geometrical considerations" and did not take "into account the variation of physical parameters, in particular viscosity." According to Tozer, the core was created entirely early in Earth's history and has not grown since. E. Orowan, of the Department of Mechanical Engineering, Massachusetts Institute of Technology (MIT), agreed with MacDonald and Jeffreys that mantle convection is highly unlikely, if the mantle is treated as a Newtonian fluid, a fluid in which the relation between shear stress and strain rate is linear and thus whose flow properties are determined by its viscosity (Orowan, 1965). But, he argued that the mantle, being substantially crystalline, should be treated as non-Newtonian, as a material whose shear stress and strain rate are non-linear, and as one that creeps rather than flows. Orowan then argued that given reasonable temperate estimates within the mantle, convection can take place and cause seafloor spreading and continental drift. J. Verhoogen, from the University of California, Berkeley, considered the question as to whether or not phase changes in the mantle hinder or enhance convection (Verhoogen, 1965). He argued that it depends on the circumstances, but remained silent about continental drift. Of the speakers in this last session, only MacDonald opposed mobilism. The others either supported mobilism (Runcorn, Orowan, and, perhaps, Tozer), or remained neutral (Verhoogen and, perhaps, Tozer).

Thus, as far as the main presentations go, Runcorn did arrange for "a bit of a balance," but not a very even one, perhaps a true reflection of opinion among active workers in the field. Excluding presentations by Blackett, Bullard, and Runcorn from the count, there were sixteen major presentations. Eleven (Allen, Creer, Fitch, Girdler, Miller, Nicholls, Orowan, Vacquier, Walker, Westoll, and Wilson) favored mobilism; Heezen supported continental drift and Earth expansion as its cause; MacDonald and Menard supported fixism; Goguel, Tozer, and Verhoogen were neutral.

What about those who contributed to the discussion, or declined an invitation to speak but submitted brief comments? Six discussants, J. H. Taylor (King's College London), Rutten (University of Utrecht), Hales (Southwest Center for Advanced Studies, Dallas), R. M. Shackleton (University of Leeds), Nairn (University of

Newcastle), and Harland (University of Cambridge), all favored continental drift. Of course, it should be no surprise that Rutten, Hales, Nairn, and Harland should do so because their work was in paleomagnetism (Rutten, 1965a; Hales, 1965; Nairn, 1965; Harland, 1965a, b). Shackleton was impressed with Bullard's fit of the reassembly of the continents surrounding the Atlantic (Shackleton, 1965). He did not think most continents would fit as well given crustal shortening in orogenic belts. J. H. Taylor also singled out paleomagnetism, claiming that the paleomagnetic support was the prime mover in turning the tide in favor of mobilism.

Small wonder that, thirty years ago, these uncertainties produced a position of stalemate. Paleomagnetism more than anything else has been responsible for the changed outlook of today. If one can accept the proposition that it is justifiable to extrapolate the known behavior of the Earth's field in geologically recent times back into the more remote past – and this comes relatively easily to geologists brought up in the school of uniformitarianism – then one is faced with a formidable weight of evidence for drift. The fact that the pole positions calculated for ancient rocks of the same age in different continents don't agree makes it impossible any longer to consider polar wandering by itself as adequate. The evidence presented this morning by Professor Runcorn and Dr. Creer is impressive even though there are still uncertainties to be resolved and discrepancies to be explained.

(Taylor, 1965: 52)

During the second session there were seven discussants. J. Sutton (Imperial College) supported mobilism and singled out the importance of paleomagnetism (Sutton, 1965). Gaskell (British Petroleum Company) and R. L. Fisher (Scripps), who was on sabbatical at Madingley Rise, had already signaled their support of mobilism. Gaskell had co-authored a paper with Irving in 1962 (III, §1.18); contributed to Runcorn's 1962 edited volume, *Continental Drift* (III, §1.6); had spoken favorably of mobilism's paleomagnetic support at a 1959 meeting of the Royal Astronomical Society (RAS), and now suggested that the time had come to "compile and analyze information on coastal belts to link the relatively well known geology of the continental areas with the intensive research currently in progress on the ocean floors" (Gaskell, 1965: 141). R. L. Fisher had co-authored a paper with Hess in 1963 in which they discussed trenches in terms of seafloor spreading (III, §3.20). He reprimanded those who theorized about the origin of the Indian Ocean, saying they should not overlook what was already known about it from recent studies, no doubt wanting, with good reason, to place some factual restrictions on Heezen's and Wilson's speculations.

My primary intention in entering the discussion is to point out that the median ridge and topographic highs that occur so well developed in the Indian Ocean are by no means the unknown features that several of the speakers thus far obviously assumed ... Before we spend too much time philosophizing and glibly talking about "micro-continents" for example, we should review what is known about some of the crustal thickness and velocities. To get some micro-insights on these "micro-continents," I shall draw my information mostly from the results of Russell Raitt, George Shor and colleagues of Scripps Institution ...

(Fisher, 1965: 139)

David Davies (University of Cambridge) added to Fisher's remarks by reporting on some of the recent British work in the western Indian Ocean on RRS *Discovery* and HMS *Owen* (Davies, 1965). He did not say whether he preferred mobilism or fixism. The geologist S. E. Hollingworth (University College London) asked several specific questions; however, he (1965) expressed no opinion about mobilism or fixism; earlier he had positively reviewed Holmes' 1944 *Principles of Physical Geology* (I, §5.9) and had more recently claimed that Bullard, in his address to the London Geological Society, underestimated the support by geologists for mobilism (III, §2.15). E. R. Oxburgh (University of Oxford), who had been mildly supportive of mobilism the year before at Bullard's address but was not impressed with mobilism's paleomagnetic support (III, §2.15), argued in favor of Hess's seafloor spreading, claiming that it explains the origin of basaltic magma at mid-ocean ridges. The other discussant, Worzel, Ewing's longtime associate at Lamont, vehemently opposed continental drift and attacked Bullard and company's reassembly of the continents bordering the Atlantic. Thus, four discussants in the second session (Sutton, Gaskell, Fisher, and Oxburgh) favored mobilism; two (Hollingworth and Davies) expressed no opinion, and one (Worzel) was strongly opposed.

There were four discussants at the third session. The structural geologist, L. U. De Sitter (University of Leiden), was mildly supportive of mobilism (De Sitter, 1965). He characterized as "excellent" arguments favoring mobilism based on marine geology, paleomagnetism, and fit of the continents. He was unimpressed, however, with arguments for continental drift based on transform faults: the continental examples did not show enough movement to support mobilism, and the oceanic ones did not continue onto the continents. Neither M. J. Graindor (1965) of the Collège de France nor P. Henderson (1965) of the University of Oxford stated a preference for mobilism or fixism.

Five discussants spoke at the fourth session. Two favored mobilism, and two expressed no opinion. M. H. P. Bott (University of Durham) strongly favored mobilism and was impressed with its paleomagnetic support.

It has become clear in recent years that there is a formidable case for continental drift. The most convincing single piece of evidence comes from palaeomagnetism, but this is amply supported by a wealth of geological evidence.

(Bott, 1965: 212)

Bott commented on his own model of mantle convection, shortly to appear in *Nature*.

J. A. O'Keefe, from NASA's Goddard Space Flight Center, Greenbelt, Maryland, expressed his doubts about mantle convection as the cause of continental drift, arguing that if mantle convection caused continental drift, then the continents should be stacked up either at the poles or at the equator (1965). Bernal, the final discussant, supported mobilism, as he had three years before when discussing Dietz's version of seafloor spreading (III, §4.8) (Bernal, 1965).

Runcorn also received brief remarks from Vening Meinesz and Jeffreys. Vening Meinesz had become a mobilist in 1962 when he accepted its paleomagnetic support

(III, §2.6). He addressed only issues concerned with mantle convection, and said nothing about continental drift (Vening Meinesz, 1965). Jeffreys remained adamantly opposed to mobilism (Jeffreys, 1965). Runcorn also read a letter that the Soviet paleomagnetist P. N. Kropotkin had sent to Bernal. Kropotkin favored mobilism, and briefly described recent paleomagnetic work supportive of mobilism by Khramov and Shmaleov, by Kalashnikov, and by himself (see II, §5.9 for discussion of Khramov's early paleomagnetic work) (Kropotkin, 1965).

Counting Blackett, Bullard, Runcorn, and co-authors, almost five times more supported mobilism than opposed it: thirty-two supported mobilism; seven opposed it; six were neutral. Had Runcorn really tried to get a balance of speakers, then either most anti-mobilists he asked declined to speak, or most who had been working on the topics he thought worth covering were inclined toward mobilism. Interestingly, Bullard later claimed that most workers in the field in England had become mobilists by the time of the meeting (Bullard, 1975a: 19).

In March 1964 ... the Royal Society held a discussion of continental drift, the proceedings of which were later published ... A number of people from the USA attended and there were some complaints that the meeting had been packed with believers and that the opponents had not been given a proportionate representation. This was not a subtle plot; the fact was that almost everyone working in the field in England had become convinced a year or two before a comparable near unanimity was reached in the USA.

By my reckoning, twenty-five out of the thirty symposiasts who were working in the UK supported mobilism; Jeffreys was the only contributor working in the UK who was definitely opposed; Davies, Henderson, Hollingsworth, and Tozer expressed no opinion; perhaps they were sitting on the fence. If, however, Runcorn had sought the opinion of paleontologists, structural geologists, and paleoclimatologists, he could have found many workers in the UK who were still opposed to mobilism. However, it was not his purpose to gather opinions of workers in such fields because he had just got them together with paleomagnetists the previous year at the huge NATO meeting in Newcastle.

It is also evident from this summary that paleomagnetism influenced the views of approximately 40 percent of the participants, and 56 percent of those who favored continental drift. Eight had worked in paleomagnetism (Blackett, Creer, Girdler, Harland, Kropotkin, Nairn, Runcorn, and Rutten) and ten had either become mobilists because of paleomagnetism's support or thought its support was vital (Bott, Bullard, De Sitter, Gaskell, Goguel, Sutton, Taylor, Vening Meinesz, Weston, and Wilson).

This summary also shows Earth expansion had almost no support at the symposium. Almost everyone ignored it. Only Heezen (and Tharp) spoke in its favor, and his once great enthusiasm continued to wane (Heezen, 1962; see III, §6.15). Westoll and Tozer argued against rapid Earth expansion. Westoll argued that Bullard, Everett, and Smith's excellent fit showed that Earth expansion was not needed.

Carey, in particular, has constructed a remarkable synthesis (1958) from which he concludes that not only the Atlantic and Indian oceans, but also the Pacific Ocean, have actually expanded ... Indeed, he considers that the diameter of the Earth in the Late Palaeozoic was only about three-quarters of the present diameter, its surface area about half the present, and its density more than twice the present density (Carey, 1958, p. 347). This seems quite opposed by the excellence of fit on the present globe; any expansion (or contraction) since drift would be very small indeed. On an expanding earth, a pre-expansion fissure will tend to "gape," and so would the sundered margins of continents if re-fitted on the expanded globe. It should be possible with careful "best-fitting" to estimate the maximum expansion possible since drift.

(Westoll, 1965; Carey reference is same as mine)

Tozer said what many had already concluded.

Theories of continental drift may be divided into two classes depending on whether they do or do not involve convection in the earth. At the present time, theories in the former class are the most favoured since they do not conflict strongly with any notions that may be reasonably held about the physical properties and state of the Earth's interior. On the other hand, theories that attribute continental drift to a general expansion of the Earth have an *ad hoc* nature since they require either rather fantastic physical properties for the major constituents of the planet, or a premature meddling with the foundations of physics. They should therefore be dismissed until their assumptions can be independently justified or the inadequacy of convection theories demonstrated.

(Tozer, 1965: 253)

Tozer might not have been confident about whether mantle convection could cause continental drift, but he certainly was not ready to entertain Earth expansion.

3.4 The Everett, Bullard, and Smith fit of the continents surrounding the Atlantic

Of all the presentations at the symposium, Everett, Bullard, and Smith's fit of the continents (hereafter, EBS fit) around the Atlantic was the most influential. Appealing to the fit of the continents, especially to Africa and South America, had been the first and best known argument in favor of mobilism.[5] Everett, Bullard, and Smith were not solving a new problem, but they were certainly doing something new; they were presenting a quantified fit, based on a new methodology that had little chance of being affected by the beliefs of those doing the matching. They treated the problem as a mathematical one. Instead of sliding cutouts of the continents around a globe, they used a computer, digitizing the data, and using an application of Euler's theorem to rotate the continents and match their continental margins.

Carey (1955b, II, §6.11) had already proposed an excellent fit of Africa and South America which he again presented in his first paper favoring Earth expansion (1958, II, §6.14). He had hoped to lay to rest Jeffreys' objection that the fit between Africa and South America was poor. Jeffreys, however, announced in 1959 that he had no intention of ever reading any of Carey's papers (III, §2.10). Carey's fit also did not carry great weight among other fixists and Bullard, Everett, and Smith suggested why this was. Although they made it seem as if Jeffreys remained skeptical because of Carey's method, I suspect that Bullard was not very optimistic about changing Jeffreys' mind but hoped to change the minds of less steadfast fixists.

Carey (1958) was the first to show that the fit of Africa and South America is much closer at the continental edges than it is at the coastline. In spite of this, Jeffreys has expressed a total disbelief in the reality of the fit; he says (1964): "I simply deny there is an agreement." The reason for this scepticism is not clear; perhaps it is connected with doubts about the accuracy of Carey's fits carried out on a globe provided with moveable transparent caps.

(Bullard et al., 1965: 41)

Bullard later offered two other reasons why Carey's fit failed to attract much attention.

The first man to make careful fits was Carey (1955b, 1958) who worked with transparent caps which could be moved over the surface of a globe. The resulting fits around the Atlantic were much better than anyone expected: the impact of his 1958 paper was less than it might have been because the maps were published only on a small scale in a duplicated conference report which contained much other material of a more or less controversial character with Carey's theoretical views on the expansion of the earth.

(Bullard, 1975a: 20)

I believe that Bullard is correct, especially about "Carey's theoretical views on the expansion of the earth." The case for rapid Earth expansion had never been strong and it had few advocates, and fixists likely did not disassociate the great value of Carey's fit from his version of Earth expansion. Carey was so adamant about his correctness, so prone to self-promotion, that I suspect the correctness of his fits became obscured in many people's minds by the wild improbabilities of his notions on expansion: they came to doubt his judgment.

Once Bullard decided to test Carey's fit, he asked Jim Everett to work on the problem, and Everett then took the lead. Jim Everett had been an undergraduate at Clare College, University of Cambridge, from 1958 through 1961. He planned to read physics, chemistry, and maths, but found out that maths did not count as one of the three required science subjects. He needed a third science. He chose geology.

I was an undergraduate at Clare College, Cambridge 1958–61. I went up to read physics, maths and chemistry, but we had to do a third science (maths didn't count) so I chose geology because it had field trips on Sunday afternoons, ending up at the pub.

(Everett, August 3, 2009 email to author)

Everett liked the field trips. He decided to pursue a Ph.D. in geophysics, and Bullard agreed to be his supervisor. However, he won a fellowship to study in the United States. After a summer stopover in La Spezia, Italy, where he worked with Tom Allan, he obtained an M.Sc. at MIT working under Ted Madden.

In December 1960 John Belshé (then at Madingley Rise) organised a geophysical field excursion to Charnwood Forest [Leicestershire], introducing us to a range of geophysical techniques. Following that I made enquiries about doing a PhD in the department, and Teddy Bullard offered me a place. However, I won a Harkness Fellowship to study in the US for one or two years. Through the department, I spent the summer 1961 working at the NATO Saclant base in La Spezia, Italy, doing oceanographic work. Then I went to MIT and completed my MSc, with a magneto-telluric thesis under Ted Madden.

(Everett, August 3, 2009 email to author; my bracketed addition)

Everett had a favorable attitude toward continental drift before he began working on the circum-Atlantic fit. "I knew of Wegener's work and thought the hypothesis reasonable. Carey's work also seemed to support it. There were other theories around, such as an expanding earth" (Everett, April 13, 2010 email to author). Continental drift never came up for discussion while he was at MIT, but after he returned to Cambridge in September 1962, he became acquainted with some of its paleomagnetic support.

I first became aware of their work on my return to Cambridge, in late 1962. I have trouble distinguishing between what I read and what I learned of their work in the coffee room discussions. They and/or their students were frequent visitors, for short and long stays. I must admit that I was not directly familiar with Hospers' work, and was more familiar with that of Irving, Runcorn and Blackett.

(Everett, April 13, 2010 email to author)

So Everett was not a strong proponent of continental drift, someone who had found support for it through his own work; he thought drift a good idea, and was familiar with some of its support, including its paleomagnetic support.

When Everett returned to the Department of Geodesy and Geophysics, Bullard had two projects for him. Everett accepted both, never thinking for a moment of working with anyone else because Bullard "was the god professor of the department."

I returned to Cambridge (England) in September 1962, to start the PhD scholarship that had been held over. Teddy then showed me Carey's qualitative Atlantic fit and suggested that I work on developing a quantitative method. He also asked me to work with John Osemeikhian, a Nigerian student, developing a three-dimensional proton magnetometer for magneto-telluric studies. My PhD covered both topics. Other PhD students were starting on projects under Hill, but the thought of any alternative to Teddy Bullard never crossed my mind – he was the god professor of the department.

(Everett, August 3, 2009 email to author)

Bullard told Everett that he was "impressed by Carey's fit." But Jeffreys, whom Bullard greatly respected, had dismissed Carey's fit. It was "worth following up."

He was certainly impressed by Carey's fit and believed it to be a hypothesis well worth following up. It must be remembered that Jeffreys was his "god professor," and highly regarded by all. I remember Jeffreys coming to give a seminar at the department, early in 1963. He appeared very old (though he was only a couple of years older than I am now!), his delivery was not very intelligible, but we all respected him immensely.

(Everett, August 3, 2009 email to author)

Bullard wanted a procedure that would allow for keeping continents rigid, one that would not be "carried out on a globe provided with moveable transparent caps." He knew just what to do: use EDSAC 2. He had already used EDSAC 2 to reduce magnetic data, and had encouraged and helped Vine to use EDSAC 2. Bullard also discussed the project with Sclater, but decided that Everett was better qualified. Sclater recalled what happened.

I was a close friend of Jim Everett's in Cambridge. He worked on building a magnetometer for use on land, and with Teddy on the paper that became Bullard, Everett, and Smith. Teddy discussed this paper with Jim and me before asking Jim to work on it with him. Jim was a better theoretician than I was, and Teddy was very impressed that Jim had spent a year at MIT before coming back to Cambridge as a graduate student. Teddy was impressed by the work of Sam Carey on reconstructing Africa and South America on a globe. He wanted to test this idea using a computer technique ... Had Jim Everett not been given the opportunity to work with Teddy Bullard on the computer fit of the continents, Teddy would have asked me. Teddy correctly believed Jim Everett was a better applied mathematician than I, and therefore asked Jim to work with him on the problem. Teddy discussed it openly one morning over coffee with the two of us.

(Sclater, October, 1992 answers to author's questions.
His answers are on line at eScholarship, University of California)

Everett's work at MIT and his mathematical abilities made him a perfect fit for the project.

He got to work digitizing the data, and programming EDSAC 2. Like Vine, he had to work at night. Bullard did not involve himself with the programming or "developing the optimization method."

I had had some programming experience from my year at MIT. At the start of the project, October 1962, a research assistant Ivor (I forget his surname) and I digitised the coastline and contour data manually ... Then (November 1962 to early 1963) I set about programming EDSAC II in autocode (which I learned by doing) to carry out the spherical geometry rotations and iterative optimisation, as described in my thesis and the recent Earth Sciences paper. Yes, research students had to book half-hour night time slots on EDSAC, and I spent many post-midnight hours there. Autocode was a sort of assembler language. Coding and data were on punched paper tape which fed into a basket. Program errors and bugs (and paper tape tangles) meant the end of one's run, to resume another night after correction. Bullard did not involve himself in the programming nor in developing the optimisation method. There were many useful coffee room discussions: we PhD students all used to discuss each others' research topics, and many ideas bounced around.

(Everett, August 3, 2009 email to author)

Euler's Point Theorem, which describes the motion of rigid bodies on a sphere, is central to Everett's method. But Everett did not know of Euler's theorem; however, he developed an equivalent method. Indeed, use of the method seemed "fairly self-evident" to him, much easier to figure out than how to "home in upon the optimum pole."

I worked the method out without knowing Euler's Point (or displacement) theorem, and was not aware that the method was equivalent until later. The theorem, that any displacement of a curved line on the surface of the sphere is a rotation about a pole, actually seems fairly self-evident, as was using a least-squares measure of fit. What was harder was working out the successive iteration to home in upon the optimum pole, and establishing that this was not just a local optimum (although I acknowledge that similar iteration methods had been used many times before: Newton's approximation is not that different). I remember discussing the method with Teddy, and his approval of it, but do not recall his ever mentioning Euler. However, my working out of the method was certainly helped by coffee room discussion with other PhD students, though again I have no recollection of Euler being mentioned. Throughout my career, I have been better at working things out than with keeping abreast of the literature. This has been a mixed blessing, but on balance I am glad of it, since it has helped me to publish across a variety of fields.

(Everett, August 3, 2009 email to author)

Bullard did remember Euler's Point Theorem (1975a: 21).

Influenced by Carey's results, and stung by Jeffreys' oft repeated disbelief in the closeness of the fits, I began to consider how one could best make fits that were, in some defined sense, best fits using the data from maps, which are much more reliable and detailed than any globe. In this I was joined by Everett and Smith. We realized that the movement of a continent could be defined by three parameters and that the problem was to choose a criterion of goodness of fit and use it to determine the parameters. By analogy with least squares fitting it is convenient to express the criterion as the finding of the values of the three parameters that minimize a symmetric function of the coordinates of points on the two continental edges after one of them has been displaced. A vague memory of undergraduate mechanics led me to Routh's *Rigid Dynamics* and to Euler's (1776) theorem that any motion of a sphere over itself can be regarded as a rotation about some pole of rotation. The problem was then to find the position of this pole and the amount of the rotation (many other choices of parameters are possible, for example the "Euler Angles," but our choice seems to have established itself). The results for the fit around the Atlantic were spectacular, the root mean square gaps and overlaps being only about 50 km.

(Bullard, 1975a: 21)

So Everett came up with what amounted to Euler's Point Theorem on his own, and Bullard independently realized that Euler's Point Theorem provided a description of past movements of a rigid continent on a spherical surface.[6] Everett learned of Euler's Point Theorem by the time he wrote his Ph.D. thesis, for he states it at the beginning of his thesis.

By the fixed point theorem (Euler's theorem), any displacement of a line restricted to the surface of the earth may be represented as a rigid rotation about a vertical axis through some point on the surface of the earth. We shall call this point the "centre of rotation."

Everett recalled that Bullard may have told him about Euler's theorem, but again emphasized that he did not know about it "when doing the original work in 1962/3."

So by the time I came to write up the thesis in early 1965 I must have realised that the rotation was equivalent to Euler's theorem, though I did not know that when doing the original work in 1962/3. Teddy may have pointed it out when reviewing my draft.

(Everett, August 3, 2009 email to author)

Their use of Euler's Point Theorem was not the first but the second time it was used to describe past movements of continents relative to each other, for Creer, Irving, Nairn, and Runcorn had done so seven years earlier (Creer *et al.*, 1958; II, §5.16) but they did not develop it into a useful tool. However, Bullard, Everett, and Smith were the first to state that they were using Euler's Point Theorem. Like Creer *et al.*, however, Bullard, Everett, and Smith did not view Euler's theorem as central to understanding past movements of continents relative to each other. McKenzie put it this way in his biographical memoir of Bullard:

Teddy [Bullard] needed a convenient description of the movement of a rigid continent on the surface of the Earth. He ... used Euler's theorem for the purpose, and obtained the pole of rotation angle required by minimizing the misfit. Though he regarded this description simply as a convenient way of describing the motions, this theorem later became the cornerstone of plate tectonics, and was not widely known to geophysicists before his work.

(McKenzie, 1987: 86; my bracketed addition)

Everett first used his program to find the best fit between the contour lines of Africa and South America. He believes that he "digitized data from the Times Atlas, in 1962/3" (Everett, August 3, 2009 email to author). His preliminary results showed that Carey was right. Encouraged by the results, he turned to the North Atlantic, by which time Bullard had asked Smith to join the project.

The results for the South Atlantic were extremely good (with residual misfits such as the Niger delta being readily explainable), but less so for the North Atlantic. Alan Smith came to the department in late 1963 (I think) and Teddy asked him to work with me on the project. His geological knowledge provided much improvement for the North Atlantic, including justification for rotating the Iberian peninsula.

(Everett, August 3, 2009 email to author)

Smith also did his undergraduate work at Cambridge. In the late 1950s he attended St. John's College, where he read physics and geology. He graduated in 1959 with a B.A. in physics. While an undergraduate he was most influenced by Harland and Bullard. Bullard had not then become a mobilist (III, §2.13), while Harland had been a longstanding drifter and was adding to the paleomagnetic support of mobilism. Smith had been exposed to continental drift, but learned very little about it. He did not, for example, have any extended conversations with Harland about it and Harland did not tell him about his own work in paleomagnetism (III, §1.11). Moreover, Smith read none of his work with Bidgood in paleomagnetism, which was "generally considered graduate level material." His attitude toward continental drift was shaped more by Jeffreys than whatever Harland had to say. Smith recalled, "Jeffreys thought the [paleomagnetic] evidence [for mobilism] poor" and added, "I was influenced by him" (April 6, 2010 email to author; my bracketed additions). Smith also thought Jeffreys' mechanism objections were correct. He did read Holmes' account of mantle convection as presented in his *Principles of Physical Geology*. He thought "it was an interesting idea but it lacked predictability."

Smith considered working on a Ph.D. in geophysics in the Department of Geodesy and Geophysics, but decided against it.

I actually had a discussion with Maurice Hill about a Ph.D. at Madingley Rise in marine geophysics, but despite all the attractions of another 4 years in Cambridge, I wanted something different. It was one of the best decisions I ever made.

(Smith, April 6, 2010 email to author)

He ended up at Princeton University. A personal note from Hess did the trick.

I looked at the Poldervaart volume published by GSA to celebrate an anniversary at Columbia, selected half a dozen departments or so that seemed interesting and applied to all of them (UBC [University of British Columbia]; Scripps; Columbia and Princeton among them). I chose Princeton because I had an understanding, handwritten letter from Harry Hess, rather than a set of forms to fill in.

(Smith, April 6, 2010 email to author; my bracketed addition)

Circumstances led him to choose geology over geophysics.

I found it difficult to choose between geophysics and geology. In fact I spent my first summer working for a seismic project headed by Bob Meyer and John Steinhart at Wisconsin, for which Princeton was also providing some support. We shot the Moho in Wisconsin, Montana and Wyoming. It became clear that Steinhart would have first crack at the data, and that, together with other reasons made me look for another project. When John Maxwell offered a mapping project in the Montana Rockies I was delighted because it was interesting large-scale crustal geology and I had seen what Montana was like.

(Smith, April 6, 2010 email to author)

Maxwell became his supervisor, and in 1963 Smith obtained his Ph.D., "The structure and stratigraphy of the Whitefish Range, Montana," which was based on his work with Maxwell.

Smith was introduced to Hess's seafloor spreading while at Princeton. He was fascinated but remained unconvinced; Jeffreys' shadow extended across the Atlantic.

I really thought Jeffreys was right because he was a mathematical physicist. As a physics graduate I went along with his views. All of us at Princeton were fascinated by Harry Hess's views but for myself I took them on board as yet another thing that was interesting but lacked proof. It was in the realm of a possibility and one could discuss its implications, but it was not part of one's thinking: there was no physical evidence to support it.

(Smith, April 6, 2010 email to author)

Smith (April 6, 2010 email to author) was also influenced by MacDonald, who deeply respected Jeffreys, and had begun to assume Jeffreys' mantle as mobilism's sternest critic (III, §1.6, §2.7, §2.8).

He originally planned to go into industry; he did not plan on becoming an academic; however, concerns about renewing his visa led him back to Cambridge and academe. He was hired in 1963 as a research assistant in the Department of Geodesy and Geophysics. He was supposed to help Miller radiometrically date rock on opposing margins of South America and Africa, but ended up working with Everett.

I thought I could not stay in the US because of my visa, but probably could have since my wife, whom I met at Princeton, is American. Anyway, when I saw my time ending I wrote to my former supervisors in Cambridge and had the offer of a research assistant at Madingley Rise. I had had no plans to become an academic. In fact, had I stayed in the States I would most probably have landed up in the oil industry, which was doing some very interesting work at the time. Anyway, in Cambridge I started with Jim's map of the S America-Africa fit; plotted on the geology as it was then known and I wrote to all the surveys in the countries along the margins asking for samples, together with some oil companies, which many duly provided. Jack Miller was my supervisor, but I never actually got round to running a date.

(Smith, April 6, 2010 email to author)

With this work in tectonics and stratigraphy, Smith had become well versed in geology. He also had the foresight to take a class at Princeton in computer programming. Both put him in an excellent position to help Everett determine the optimum fit of the continents around the Atlantic. In addition, he did not believe in continental drift. Thus, he had no vested interest in making sure that the results favored continental drift.

He summarized his contributions to the project.

As I recollect, I fitted the N Atlantic continents together (I still have the paper tapes to do this, but the language in which Jim's program was written is now extinct). My approach was partly empirical and partly intuitive. I was most concerned to get the best fit I could, so if something got in the way such as Iceland or the Davis Strait I excised it, but other things that seemed to have a space for themselves, such as Rockall, were retained. In all cases I found that I could justify any omission or inclusion on reasonable geological grounds. After this I fitted the N and S Atlantic continents together, which required a rotation of Iberia to get a good fit. (Jim may also have done some of this independently or jointly, but I cannot remember.) I cannot remember if Carey's closure of the Bay of Biscay guided me to doing that or whether it was geometrically so obvious that it was inevitable that one would close the Bay of Biscay. I think others may have suggested it much earlier (Argand??).

(Smith, April 6, 2010 email to author; Argand did in fact make such a suggestion (I, §8.7))

Smith was also "not unduly worried" about leaving Central America out of the fit. "We were concerned mostly with the Atlantic, and had little to offer to say about Central America" (Smith, April 6, 2010 email to author.) Everett presented their results in the first chapter entitled "The fit of the continents around the Atlantic," of his Ph.D. dissertation "Magnetic variations." He devoted the remaining seven chapters to the development of a three-dimensional proton magnetometer for magneto-telluric studies. He finished his thesis in April 1965, and took his orals the following month. His two examiners, Hill and Runcorn, spent more time arguing with each other; Everett had it easy.

My internal and external examiners were Maurice Hill and Keith Runcorn. Runcorn had got his PhD from Cambridge some years before, and was then Professor at Newcastle. The examination, in May 1965, was an all day viva, but Hill and Runcorn quickly lapsed into arguing with each other, and I did not have to do much defending. We adjourned for lunch at Maurice Hill's. During lunch Maurice asked me if I would tutor his daughter in A-level maths. I said I would be happy to if I did not have to rewrite. It was all very jolly. I was very grateful that neither of them gave me a hard time.

(Everett, August 3, 2009 email to author)

I now turn to their results as presented in their joint paper, the one Bullard delivered at the Royal Society meeting. Bullard wrote the paper. Everett does not remember when Bullard told him that he wanted to present their fit at the Royal Society Meeting, but he does remember that Bullard told him he had to be the first author because only he was a Fellow of the Royal Society. "I do not recall how far in advance he told me. He did tell me that the first author had to be an FRS (i.e. himself)" (Everett, August 3, 2009 email to author). I believe that Bullard was mistaken about having to be the first author. For example, M. N. Hill, FRS, organizer of the FRS sponsored 1964 symposium "A discussion concerning the floor of the Northwest Indian Ocean," was not the first author of the paper he co-authored with T. J. G. Francis and D. Davies, neither of whom were at the time fellows of the Royal Society (Francis *et al.*, 1966).

Once Smith joined the project, they refitted South America with Africa. The results again "fully" confirmed Carey's results. They also fitted together North America and Greenland, and then fitted together the whole Atlantic.

Africa and South America were first fitted, then a second block was assembled from North America, Greenland and Europe; the closeness of these fits exceeded our expectations and fully confirms the work of Carey. An attempt was then made to fit the two blocks together; here the fit was less good.

(Bullard et al., 1965: 42)

Using data supplied by the US Hydrographic Office, which gave contours of the continental margins at 100, 500, 1000, and 2000 fathoms, they achieved their best overall fit at the 500-fathom contour; the 1000-fathom fit was second best. Carey had fitted together South America and Africa at the 2000-meter (approximately 1100 fathoms) contour. They thought the data were accurate within $0.5°$, and nobody seriously questioned the reliability of their data. Defining continental contour lines by the latitudes and longitudes of a set of points along them, they applied Euler's

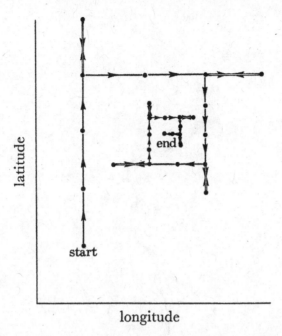

Figure 3.7 Bullard *et al.*'s Figure 2 (1965: 43). "Approach to the best fit."

point theorem. Using EDSAC 2, they determined the pivot point that gave the least misfit by homing in on the point where they obtained the best fit. They also repeated the process several times, and got the same result.

The misfit with this rotation will depend on the position of the centre of rotation. The relation between the position and the misfit is complicated and the most convenient method of finding the minimum is to start from an estimated position and search systematically about it. This was done with computer EDSAC 2 by the process illustrated in Figure 2 ... The process starts for a point in the neighborhood of a minimum and "homes in" on it by changing the latitude and longitude of the centre of rotation alternately, as shown in Figure 2 [my Figure 3.7].

(Bullard et al., *1965: 43)*

They found that the misfit at the 500-fathom depth contour of South America and Africa could be matched with an average misfit of only 88 km if they removed overlaps at the Niger Delta and at the ends of the Walvis Ridge (Figure 3.8), places where Carey also had found overlaps. Bullard *et al.* (1965: 45) described them as "recent excrescences on the edge of the African continent."

They broke down their fit in the North Atlantic to fitting Greenland to northern Europe and Greenland to North America. They omitted Iceland and the Faeroes Ridge in the fit of northern Europe and Greenland, being of Cenozoic origin. At the time the age of the Rockall Bank was unknown but they kept it because it accurately fills a gap.

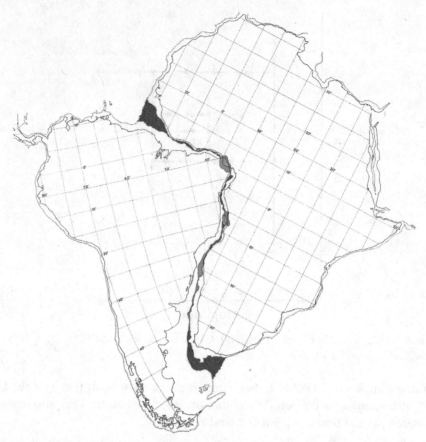

Figure 3.8 Bullard *et al.*'s Figure 5 (1965: foldout, facing p. 48). "The fit of Africa and South America at the 500 fm. contour." The original diagram showed the overlaps in red and gaps in blue. Overlaps in this diagram are in grey, gaps in black.

The fitting of the lands around the North Atlantic requires the bringing together of three major continental masses, North America, Greenland and Europe. A fit of Greenland to northern Europe on the 500 fm. contour was first attempted. In this fit Iceland was ignored altogether, as were also the ridges joining it to Greenland and to the Faeroes (these ridges are shallower than 500 fm.) Iceland is composed of Tertiary and Recent igneous rocks and its omission is clearly justified. There is no direct evidence as to the age of the ridges, but they, like the Walvis ridge, are typical of the "transverse ridges" associated with the mid-ocean ridges and there is little doubt that they are Tertiary. The Rockall Bank is a more doubtful case, Rockall itself is Tertiary (Miller 1965a), but it may well be intrusive into older rocks as are many of the Tertiary igneous rocks of western Scotland. We have retained the Rockall Bank in the fit, largely because it fills what would otherwise be a gap; further study of the bank is very desirable to determine whether it does, in fact, contain older rocks.

(Bullard et al., 1965: 47–48)

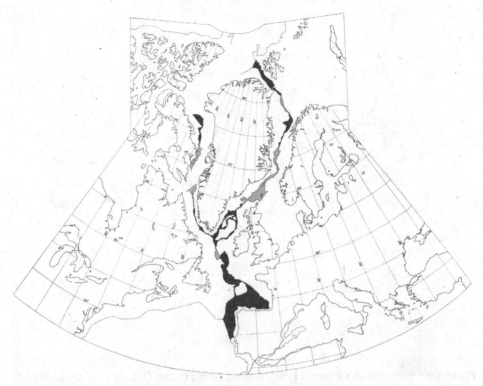

Figure 3.9 Bullard *et al.*'s Figure 7 (1965: foldout, facing Figure 8). "The fit of North America, Greenland and Europe at the 500 fm. contour." The original diagram showed the overlaps in red and gaps in blue. Overlaps in this diagram are in grey, gaps in black.

Their best result gave a misfit of only 43 km. They next combined their fit of Greenland and Europe with a fit of North America and Greenland (Figure 3.9). They omitted sections of the Davis Strait shallower than 500 fm, ending up with a misfit of 34 km.

They next fitted the entire Atlantic margin, and immediately saw that there was a huge overlap between southern Spain and Africa. They eliminated this by a counter-clockwise rotation of Iberia, which led to formation of the Pyrenees and opening of the Bay of Biscay. This rotation had been first proposed by Argand in 1924 for tectonic reasons (I, §8.7), seconded by du Toit (I, §6.7, Note 6), amplified by Carey in 1955 (II, §6.7), and had been confirmed to a limited degree by Clegg and company in 1957 (II, §3.14). Bullard, Everett, and Smith (1965: 48–49) referenced Carey and du Toit, and the newer paleomagnetic work through Irving's recently published mono-graph *Paleomagnetism and Its Application to Geological and Geophysical Problems* (1964) (§3.5). This rotation of the Iberian Peninsula removed the overlap between southern Spain and Africa (1965: 49), but there remained a 100 km gap between the west coast of Iberia and the Grand Banks of Newfoundland which was outside their statistical errors; they removed the gap by moving Iberia westward without worsening

Figure 3.10 Bullard *et al.*'s Figure 8 (1965: foldout, facing Figure 7). "The fit of the continents surrounding the Atlantic at the 500 fm. contour." The original diagram showed the overlaps in red and gaps in blue. Overlaps in this diagram are in grey, gaps in black.

its fit with the French coast. After making a few minor adjustments, they fitted all the continents around the Atlantic at the 500-fathom contour (Figure 3.10).

They did not consider Central America, which borders not the Atlantic but the Caribbean Sea. Moreover, the area is complicated by the currently active Puerto Rico Trench and Lesser Antilles Island Arc. Smith (April 6, 2010 email to author) recalled he "was not unduly worried. We were concerned mostly with Atlantic, and had little to offer about Central America." This was the difficulty that Worzel referred to above.

Turning to the overall significance of their fit, they acknowledged that they did not have a statistical criterion to decide if the closeness of their fit arose from chance or because the continents, once united, had separated. But, was one really needed? They (1965: 49) claimed that separately the fits across the southern and northern Atlantic are "striking," and admitted that the combined fit "is somewhat less convincing – perhaps because there has been distortion represented by the Tertiary folding of southern Spain and North Africa." Their reconstruction also had large gaps in the Caribbean and the Mediterranean, which they considered not unexpected because of younger deformation (RS2).

It is noteworthy that the reconstruction shows large gaps in the Caribbean and the Mediterranean which is just where they would be expected in view of the considerable Mesozoic and Tertiary deformation in these regions.

(Bullard et al., *1965: 49–50)*

Like old-time mobilists, they proposed that their fit could be further substantiated by seeing if geological disjunctions are consistent with it (RS1); consilience would make "negligible" the "probability that the fits are due to chance."

It is perhaps more profitable to approach the problem of significance by considering other aspects of the fits. If the continents were once joined, then not only the shapes but the ages, structures and petrology of the rocks must match across the joins; if they do, the probability that the fits are due to chance is negligible. The importance of the geometrical fits is that they position the continental blocks with an accuracy of the order of a degree and leave little room for adjustments to fit other evidence.

(Bullard et al., *1965: 50)*

They (1965: 50) mentioned some geological disjuncts between Africa and South America discussed by du Toit (1937) and updated by Martin (1961). They noted Miller's matching of "the ages around the north Atlantic in relation to" their fit. They also mentioned the well-known matching of pre-Jurassic orogenic belts on both sides of the north Atlantic, which do not continue on the seafloor, and contrasted them with "some recent Tertiary features" that "seem to have extensions on the seafloor."

By contrast some Tertiary features seem to have extensions on the deep sea floor. For example, the folding in southern Spain appears to continue as a not very well marked ridge to the Azores, and the volcanoes of the Cameroons form part of a chain crossing the continental edge and extending to St Helena.

(Bullard et al., *1965: 50)*

They also found recently acquired and independent evidence that the continental margins facing the Atlantic had changed little through erosion or sedimentation.

If the present shapes of the continents do really give an indication of how they once fitted together, then those shapes cannot have been greatly affected by erosion or sedimentation since the separation took place. That, in most places, the continental shelves are not being built outwards is well known (Heezen, Tharp & Ewing, 1959) as a result of dredging and of the study of their stratigraphy by sparkers and similar devices; that erosion is also very slow is not so obvious from direct observation.

(Bullard et al., *1965: 50)*

Although trans-Atlantic matching of continental margins and geological disjunctions was hardly new, they were adding a new methodology and utilizing new findings based on radiometric dating and geophysical investigations of the seafloor. Indeed, they closed with an offer to anyone who wanted to test their fit by undertaking detailed regional studies of geological disjuncts.

Clearly, a great deal of work needs to be done before we can fully accept the hypothesis that the Atlantic Ocean was formed by continental fragmentation and we have merely sketched a few geological implications of this theory. Some of the most important data bearing on it will probably come from detailed comparative geological studies of geometrically matching areas that have structures truncated at the continental margin, particularly where this is narrow. For such studies the authors can supply dyeline prints of figures 5, 7 and 8 on a larger scale.

(Bullard et al., *1965: 50)*

Their offer was not ignored. P. M. Hurley (MIT) and colleagues combined forces with F. F. M. de Almeida and colleagues at the University of São Paulo, and tested their reconstruction against new radiometric dating of Precambrian basement ages in West Africa, Brazil, the Guianas and Venezuela (Hurley and Almeida *et al.*, 1967). Bullard, Everett, and Smith planned to match other coastlines around the Indian Ocean and between Australia and Antarctica. Their fits were "ill defined in one direction" because the continental margins approximate arcs of circles.

We hope to see if the continents around the Indian Ocean can be assembled with fits as good as those around the Atlantic. A good fit of Australia and Antarctica has already been obtained, but since the fitted parts are approximately arcs of circles the solution is somewhat ill defined in one direction. Similar difficulties are to be expected from the straight coasts of eastern Madagascar and western India.

By contrast, their fit around the Atlantic was well defined in both directions.

Everett became convinced of continental drift once he completed the fit of the South Atlantic. In his case, as he noted, he "never thought it to be unlikely."

When my calculations mapped out to such a good fit for the South Atlantic, I was convinced. To be fair, I had never thought it to be unlikely. I was surprised at Jeffreys' adamant opposition to it, but did not give him much credence, and did not then realise that his doubts were so widely shared.

(Everett, April 14, 2010 email to author)

Everett was definitely well primed to accept V-M when he first heard about it, probably in early summer 1963.

I vividly remember the day they descended from the hay loft to the coffee room and announced their discovery. It was so beautiful and so elegant that I fully accepted it. I cannot remember if it was discussed earlier.

(Everett, April 14, 2010 email to author)

Surprisingly, Smith, who came into the project a fixist, remained one, although his resistance weakened "a little." He did not doubt the quality of the fit or even that the continents surrounding the Atlantic had once been united, but found a way to still remain a fixist and not deviate from Jeffreys' assessment of mantle convection.

Although it was largely an entertaining geometrical exercise, it did raise the question of what had happened and the only way I could reconcile it with Jeffreys' view was to put the fit and its subsequent break up back into some remote period when the Earth could still convect, but not to place it in the Mesozoic and Cenozoic.

(Smith, April 6, 2010 email to author)

Indeed, Jeffreys did allow for mantle convection very early in Earth's history (I, §5.5).

At the symposium, the EBS fit was favorably received by most who discussed it. Some simply offered praise, others provided geological support by identifying geological disjuncts consistent with the fit or recommended that such studies be undertaken. Heezen and Tharp (1965: 91) remarked that the Carey and EBS fits show that the match of South America and Africa "is good indeed." Shackleton (1965: 59) said it "was of great geological significance." Taylor, who characterized the paleomagnetic evidence as impressive, and thought that continental drift "has to be retained as a working hypothesis until either its existence can be confirmed or makes way for a more acceptable theory," welcomed the EBS fit, but did not think it by itself proved a former connection; its validity had to be tested by finding well-established geological disjuncts consistent with the fit. He recommended studying Precambrian fold belts, especially if studied through "structural analysis allied to radiometric age determinations – surely one of the most exciting combinations in geological research today," adding that it had already produced results in "the northern group of continents" in a "comparison of Labrador with western Greenland." He wanted to know if there was any radiometric comparison of the Brazilian shield with west and equatorial Africa, and concluded:

It would be helpful to see the latest structural and radiometric data plotted on the various units of Gondwanaland for it is only by a synthesis of this kind that we can assess the validity of our continental reconstructions.

(Taylor, 1965: 53)

Harland (1965a: 63) characterized the EBS fit of the north Atlantic as "excellent" and noted its consistency with Bidgood and his own paleomagnetic results.

The EBS fit also received sharp criticism from Worzel who with Jeffreys and MacDonald were the severest critics of mobilism at the symposium. Worzel raised a serious difficulty (RS2).

We have seen the continents authoritatively reconstructed by a computer. This seems most convincing except of course it seems necessary to discard Central America, Mexico, the Gulf of Mexico, the Caribbean Sea and the West Indies, along with their pre-Mesozoic rock!

(Worzel, 1965a: 137)

The consilience with many disjuncts was remarkable but many puzzles still had to be worked out. As for Jeffreys, he did not mention the fit in his brief written contribution. Perhaps he saw no reason to read their paper before submitting his contribution.

James Gilluly (1966) discussed the fit in his review of the symposium. He (1963) had by then switched to mobilism. He was now impressed with the fit, but emphasized the omissions and rotations that had been made to improve the fit.

One of the most striking papers of the conference is that by Sir Edward Bullard, J. E. Everett, and A. Gilbert Smith on the fit of the continents around the Atlantic. If the continents have drifted as rigid bodies, their pre-drift relations can be restored by rotations about a radius through some point on the surface (Euler's theorem). Matching fairly closely spaced points on the 500-fathom depth contours of Africa and South America by successive approximations by computer, the continents can be matched ... This fit neglects the Niger delta (a post-drift addition) and the junction of the Walvis ridge with Africa (a young tectonic feature). If we omit Iceland and the ridges joining it to Greenland and the Faroes, and rotate Spain north-ward to fill the Bay of Biscay, North America, Greenland and Europe may be grouped to yield a misfit of 52 km ... If we omit Iceland, Mexico, Central America, and the West Indies and rotate Spain, the boundaries of both North and South Atlantic can be matched ... The authors concede there is no statistical criterion, but they consider these fits too close to be accidental. Certainly they are impressive on the Mercator projections that display them ... Despite the impressive fits of the subsea contours, it will still be a highly subjective matter to join the respective structural trends of the continents; South African land is still nearly 800 km from the nearest land in South America and the Falklands.

(Gilluly, 1966: 974)

Looking ahead, this difficulty of the elimination of Central America from the fit of continents remained an unsolved problem until after the acceptance of mobilism and the development of plate tectonics. It was resolved by G. W. White in 1980. It was made difficult by the lack of clear marine magnetic anomalies in the area and because of the abundance of post-Jurassic deformations and rotations of the small continental blocks that eventually became southern Mexico and Central America, making interpretation of paleomagnetic data tricky (White, 1980: 825). Working with the constraints of a reassembly much like that of Everett, Bullard, and Smith, and fitting together the margins of the pre-Central America blocks (Yucatan and Chortis) in terms of their margin morphology and geological continuity, he positioned the blocks within the area now occupied by the Gulf of Mexico. A subsequent clockwise rotation of the blocks and continued drifting of North and South America led to the formation of Central America. (See Kearney, Klepeis, and Vine (2009: 57) for a summary of White's work and its relation to the EBS fit.)

3.5 Paleomagnetism, other new evidence for continental drift, and mobilism's mechanism difficulty

In closing this discussion of the Royal Society's symposium on continental drift, I want to discuss the status of the mobilism debate; in particular, three interlocking issues: the strength of paleomagnetism's support for mobilism, the strength of other new evidence for mobilism, and the decreasing importance of the oft-repeated diffi-culty that there is no viable mechanism for continental drift or seafloor spreading. I argued in Volume III (§2.17) that mobilism's solution to diverging APW paths had attained difficulty-free status by 1959. A few inside paleomagnetism, most notably

Doell and Cox, mistakenly did not recognize its difficulty-free status (III, §2.14, §2.18). As expected, mobilists welcomed the paleomagnetic support (III, §2.18), and some, like Holmes (III, §2.16) and Gutenberg (III, §2.4), were particularly impressed by it. However, most fixists were not impressed and remained fixists, but there were some very notable exceptions: Hess (III, §3.15), Wilson (§3.9), Vening Meinesz (III, §2.6), and Dietz (III, §4.6). Bullard, perhaps best characterized originally as sitting on the fence, became a mobilist persuaded by its paleomagnetic support (III, §2.14).

Mobilism's solution to divergent APW paths also improved throughout the early 1960s. The use of thermal and AF demagnetization became standard practice. Gough, Opdyke, McElhinny, and others at University College of Rhodesia and Nyasaland (UCRN) constructed an APW path for Africa based on magnetically cleaned rocks often from radiometrically dated formations, which agreed with the paleoclimatic picture of Africa (III, §1.17). Blackett (III, §1.7), Irving, Briden, and Brown (III, §1.16) strengthened the consilience between paleomagnetic, paleoclimatic, and paleobiogeographic results. Paleomagnetists also continued to concern themselves with the GAD hypothesis. Creer, Irving, and Ward, and paleomagnetists at UCRN increased support for the GAD hypothesis by observing the dispersion associated with secular variations, which showed that statistically it varied as a function of latitude in accordance with models of the current field. Finding agreement, they argued in favor of the GAD hypothesis (II, §5.17). Blackett, taking a phenomenological approach, argued in favor of continental drift without, he claimed, having to appeal to the GAD hypothesis (II, §5.19). Irving, on the other hand, influenced by Popper, argued that those who objected to the GAD hypothesis, replacing it with some unspecified non-dipole hypothesis, were proposing non-falsifiable and thus non-scientific hypotheses as a desperate attempt to avoid continental drift (Irving, 1964: vii).

Even though paleomagnetism provided mobilism with its best support, and fixists had no explanation of the diverging APW paths from different continents, it was a hard sell, especially in North America and Australia where resistance to mobilism was the strongest. Many did not understand the GAD hypothesis, believed that it was not correct, and questioned the reliability of paleomagnetic data. Geologists had difficulty understanding the method. As Bullard later remarked:

The clarity which was finally achieved in the interpretation of paleomagnetism should not obscure the complexity and difficulty of the route by which it was attained. To establish the facts of continental movement and field reversal in the face of doubts raised by the existence of many unstable rocks, the existence of self-reversing rocks and the complexity of the relations between movements of the continents and the pole is a major achievement ... All this necessarily took some years to elucidate; the surprising thing is that by 1960 the case was substantially complete.

(Bullard, 1975a: 14)

Irving's *Paleomagnetism and Its Application to Geological and Geophysical Problems*, paleomagnetism's first full-length monograph, was published in 1964 after the Royal

Society's symposium in April but before the November deadline for submission of its proceeding's papers. Irving designed it to help geologists and physicists with little knowledge of each other's field understand paleomagnetism, its method, and its defense of mobilism.

In this book I have tried to summarize our present knowledge of paleomagnetism and to outline its relation to some other studies in the Earth sciences. Paleomagnetism is a composite subject with both geological and physical aspects, and I have written this book in a manner which may be followed by a geologist equipped with an elementary text on magnetism or by a physicist who has at hand a dictionary of geological terms.

(Irving, 1964: vii)

He provided a clear and thorough account of mobilism's paleomagnetic support, which he described as independent of any previous evidence of mobilism, as the first physical or quantitative evidence of mobilism, and as the first to make rigorous predictions which, if not corroborated, would have led to mobilism's certain rejection.

The study of paleomagnetism has revealed phenomena which have not been previously suspected (for example, reversals of the geomagnetic field), but perhaps its greatest contribution so far has been that it provides numerical tests of hypotheses which have been formulated by workers in other subjects. As an illustration I will say a little about the hypothesis of continental drift which dates back in the geological literature for about a century. This hypothesis supposes that the continents have changed their relative positions in the past few hundred million years. Prior to the rapid developments in paleomagnetism in the past dozen years there was, so far as I am aware, no method by which the hypothesis could be firmly rejected – it made no predictions which could be rigorously tested by methods then available. Evidence (usually of a type related to fit and misfit arguments) was brought forward by the proponents of drift, and while this evidence was considered consistent with the hypothesis, it was similar to what had first been the basis for the formulation of the hypothesis; this "new" evidence was, by and large, only a proliferation of the type already available and which the critic has found insufficient.

Now the paleomagnetic method provides a test which is independent of any previous evidence. The test is that if the continents had *not* changed their relative positions, then the equivalent paleomagnetic poles obtained from rocks of the same age from all parts of the Earth should agree. If this agreement had been found from the observations of numerous workers on rocks of many ages in all continents, there seems little doubt that the hypothesis would have been set aside. So far as I know this was the first time the hypothesis had been exposed to a physical test which, if successful, would have disproved it. However, when the test was made on rocks from all continents, large divergences between equivalent poles were found, and these divergences occurred in a way which was roughly consistent with the hypothesis of drift. Of course this does not prove that drift occurred – it merely means that the drift hypothesis survives and merits further critical discussion. The need now is for rigorous tests of special aspects of the drift hypothesis, and in this book I have tried to show that such tests are now well in hand.

(Irving, 1964: vii–viii)

Rutten agreed with Irving. He put the importance of paleomagnetism's support of mobilism in proper perspective at the end of the symposium, emphasizing both its difficulty-free status and its superiority over mobilism's other support.

In view of Blackett's introductory remarks on the violently opposed continental drift discussions during the 1920's, subsequent papers of this symposium have shown that, apart from paleomagnetic data, nothing much has been changed. The geological – and this includes geophysical and geochemical – arguments pro and contra drift are still based each on their own particular sets of data. These either "prove" or "refute" drift, but cannot be readily compared with those used in the opposite camp.

To give two examples only, the papers by Wilson and Bullard will be cited. Wilson, to obtain a good fit between continents, does away with the Caribbean and most of Central America. This is done, "because the area is young, the sea having reached Cuba only during the Jurassic." Of course, this does not exclude the area being formed by older continental masses, which have since foundered, such as is presumed by [Jacques] Butterlin. This alternative hypothesis could incidentally well explain the rock salt layers supposed to exist underneath the present floor of the Caribbean by workers of Lamont.

Bullard, to obtain a nice fit of continents, does away with Iceland, "because it is young," but keeps the Rockall bank, "because it fits nicely." The older part of Iceland belongs to the early Tertiary, as dated on paleobotanical evidence from [Oswald] Heer. This is just as old as the 60 My absolute dates for the Rockall bank supplied by Miller. Moreover, the eastern part of Iceland, as studied by Walker, is truly continental in its volcanics.

The inheritance from the twenties is clear. It still depends on which part of the geological data one finds most strongly heuristic, if one is a "drifter" or a "fixist." It is only the measurements of paleomagnetism which have introduced a really new set of values. The only way to remain fixist now, is to disbelieve paleomagnetism, a position which becomes more and more awkward as its methods tend to become better substantiated. In future we shall have to base all of our geological theoremata on the data supplied by paleomagnetism.

(Rutten, 1965c: 321; my bracketed additions)

Rutten spoke from experience. He had traveled far since arguing with Irving at the 1957 IUGG meeting in Toronto. Then a fixist, he was displeased with Irving's APW path of Australia (II, §8.12). It diverged so greatly from the APW paths of Europe and North America that it was not easy to reconcile with fixism. He later became a mobilist because of its paleomagnetic support, and worked with Van Bemmelen in helping Veldkamp and Zijderveld develop paleomagnetics in Holland (II, §5.5). Of course, Rutten did not foresee that confirmation of V-M and of Wilson's not yet conceived idea of transform faults would lead to the acceptance of mobilism by most who remained unconvinced by its paleomagnetic support. He also did not foresee the development of plate tectonics. He was not, however, so far off base about ultimately assessing "geological theoremata on data supplied by paleomagnetism," if the theoremata involve or are related to ancient latitudes, because, as I shall later explain, plate tectonics is blind to paleolatitudes.

Blackett agreed with Rutten about the paleomagnetic evidence, but he also thought that "the evidence of the great transcurrent faults" and from the study of marine geology was excellent.

Round about 1950, as pointed out by Bullard, a certain weariness set in, of repeating the old seemingly inconclusive arguments, both observational and theoretical. Perhaps the last pitched battle between the "drifters" and the "anti-drifters" took place in 1948 at the meeting of the British Association in Birmingham: the proceedings make entertaining reading. Since then advances in two virtually new subjects, the study of the magnetism of rocks and the study of the floors of the oceans, have thrown exciting new light on the subject of drift, and have done something to overcome the former widespread objections to it ... Then there was the evidence of the great transcurrent faults. Kennedy's demonstration of the 100 km shift along the Great Glen in Scotland, together with the probable shifts of 500 km or so at the San Andreas Fault in California and at the New Zealand Fault – not forgetting the several hundred kilometers of compression necessary to produce the great mountain ranges such as the Himalayas – all this suggested that the question was not the qualitative one "Have or have not the continents drifted?" but the quantitative one "How much have they drifted and when?" ... So the new evidence acquired during the last decade from oceanography and rock magnetism will, I think appear in the history of the subject as supporting and making more quantitative an already rather strong qualitative case.[7]

(Blackett, 1965: 5–6)

Blackett also discussed the search for "the possible mechanisms for drift." Here is what he said about the previous work.

Another and very important aspect of the controversy between the "drifters" and the "anti-drifters" lay in the possible mechanisms for drift. It must be admitted that some proponents of drift weakened their case for it, derived from very impressive observational material, by rather weak theorizing about how it might have occurred. When their opponents demolished the theoretical models of the drifters, they tended to ignore the weight of observational facts which had to be explained. However, the "anti-drifters" weakened their own case by producing theoretical arguments why drift cannot have occurred, which, though perhaps more sophisticated, were probably as fallacious as those theoretical arguments put forward by the "drifters" that it must have occurred.

(Blackett, 1965: 46–47)

He expressed (1965: 46) caution about the future search for continental drift's "ultimate explanation." The work will be even more complex than before, it is good to do because it suggests new observations, "but simplified models which prove that some supposed phenomenon cannot have occurred, must be treated with caution, for they may discourage new observations."

Jeffreys took issue with the quality of mobilism's new support, and certainly took the "impossibility" arguments against mobilism and mantle convection as serious impediments to their acceptance. A Bayesian probability theorist, he pronounced that all evidence in favor of continental drift is "considerably lower than is usual for a new phenomenon," while it should meet a higher standard because continental drift and mantle convection are forbidden.

I have been asked to make a written contribution to the discussion, but have nothing new to say. My views will be found in the 1961 version of *The Earth* and in my lecture to the Royal Astronomical Society last October, which will be published in their *Quarterly Journal* [see III, §2.10].

My main points are that the only type of imperfection of elasticity considered in convection and drift theories is the elastico-viscous law, which has been found to lead to numerous contradictions when confronted with actual evidence. Different phenomena led to values of the effective viscosity differing by factors of millions. On the other hand, a modified law, chosen to fit two quantitative data and applied far beyond the range of the periods related to those data, has steered its way nicely among the other evidence for some sort of imperfection of elasticity, without giving any contradiction. But it does forbid convection and continental drift.

I should be disposed to agree that inability to explain an alleged phenomenon is not necessary a disproof of that phenomenon; but it does require a higher standard of scrutiny of the evidence for that phenomenon. The standard actually applied to evidence for continental drift seems to be considerably lower than is usual for a new phenomenon, and is not associated with any alternative explanations of things that can be explained.

(Jeffreys, 1965: 314; my bracketed addition)

The new evidence was not good enough; even if better than the old evidence, it still had to be much better if it was to overcome the evidence stacked against it. Nothing more need be said.

Bullard (1965) took on Jeffreys in his closing, two-page summary of the Royal Society's symposium, echoing Blackett, Irving, and Rutten. Bullard viewed the evidence in favor of mobilism as impressive, recognized serious difficulties with mantle convection, the only viable explanation of continental drift, but because of "the scanty and hypothetical nature of our knowledge of the Earth's interior," concluded "that it seems best not to be too much influenced by the theoretical difficulties in the interpretation of the facts of observation." He began with his favorable assessment of the new evidence in favor of mobilism and its concordance with the old evidence.

There have been two threads running through this Symposium; the interpretation of observations and the discussion of mechanisms. Nearly all the speakers concerned with the evidence derived from the comparison of the continents and from palaeomagnetism have interpreted their results in terms of movement of the continents. It is difficult not to be impressed by this agreement of many lines of study leading to compatible conclusions, though there have been some dissenting views.

(Bullard, 1965: 322)

Bullard, like Irving, was impressed by the consilience of the paleomagnetic support of mobilism and other independently based lines of evidence. Before the rise of paleomagnetism, old-time mobilists had also been impressed with the agreement of independent lines of evidence, but fixists had raised serious unresolved difficulties, and relentlessly offered fixist solutions. But the paleomagnetic support of mobilism could not be so easily dismissed. Even if most still failed to recognize its difficulty-free status by the time of the Royal Society's symposium, many had come to appreciate its strength and the failure of fixists to offer a reasoned alternative explanation of the diverging APW paths of different continents.

Bullard then turned to the other thread of the symposium, "discussion of possible mechanisms."

The second theme of the conference has been the discussion of possible mechanisms of movement. Here there is no agreement as to whether movement is possible and great difficulty in providing a convincing discussion that does not involve arbitrary assumptions about the interior of the Earth. Such discussions are different in kind from discussions of the evidence for and against the movement of continents. There are phenomena, such as ice ages and thunderstorms, for whose occurrence there is incontrovertible evidence, but for which there is no theory that is not open to substantial objections. Difficulties in accounting for a phenomenon do not provide a proof of its non-existence, though they may give a strong indication that the evidence is being misinterpreted.

(Bullard, 1965: 322)

This argument was not new. Old-time mobilists such as du Toit, van der Gracht, King, Rastall, and Wooldridge and Morgan had presented the same core argument (I, §5.13). But the argument did not have much bite before the development and defense of mobilism's paleomagnetic support. The existence of thunderstorms and ice ages is unassailable; the occurrence of continental drift was subject to reasonable doubt before maturation of its paleomagnetic support. With the maturity of mobilism's impressive paleomagnetic support, mobilism had, for many, become worthy of acceptance. But there was more. Evidence against mobilism based on the unlikelihood of mantle convection had not increased. Little was still known definitely about Earth's interior, and therefore, Bullard correctly argued that even though there are serious difficulties against mantle convection in light of "the scanty and hypothetical nature of our knowledge of Earth's interior, it seems best not to be too much influenced" by them.

There are very real difficulties in explaining continental movement, but it must be remembered that the explanations and the difficulties depend on the composition, properties and temperature of materials within the Earth of which our knowledge is very indirect; also the history and processes in the Earth are doubtless more complicated than the theories. In view of the scanty and hypothetical nature of our knowledge of the Earth's interior, it seems best not to be too much influenced by the theoretical difficulties in the interpretation of the fact of observation. If the facts are correctly observed there must be some means of explaining and co-ordinating them and many precedents suggest the un-wisdom of being too sure of conclusions based on the supposed properties of imperfectly understood material in accessible regions of the Earth.

(Bullard, 1965: 323)

If continental drift and seafloor spreading were to be accepted, finding their ultimate cause would become an unsolved problem rather than an impediment to acceptance.

Bullard was recommending that the time had come to make the change regardless of what Jeffreys and MacDonald proclaimed impossible.

In his review of the symposium, Gilluly came very close to agreeing with Rutten, Bullard, and Blackett. Gilluly (1966: 949) agreed with Rutten that paleomagnetism offered the only "new argument" in favor of mobilism, and added that paleomagnetism's support "seems to be better established owing to improved methods of investigation." Turning to Blackett's opening comments and Bullard's closing ones, Gilluly wrote:

Bullard closes the conference, pointing out that there is no agreement about whether drift is possible, and that much depends on arbitrary assumptions about the interior of the earth. There are other phenomena such as ice ages and thunderstorms whose existence is incontrovertible but for which no theory advance is free from serious objections. Difficulties encountered in accounting for a phenomenon do not disprove its existence, although they do indicate that some evidence is being misinterpreted ... There are thus real difficulties in the drift hypothesis, but it must be remembered that they depend on properties of material about which we have only indirect knowledge. The true processes are doubtless far more complex than our theories. It thus seems that Blackett's opening remarks are well justified. The only new evidence bearing on the problem is from paleomagnetism and oceanography. Paleomagnetism seems to carry more conviction to physicists than to geologists, but despite the special pleading obvious in many paleomagnetic papers, there remains a rather persuasive residuum. The oceanographers are divided about whether the midocean ridges arise from localized convection or from expansion of the earth. The geophysicists differ by five orders of magnitude in their estimates of mantle viscosity.

(Gilluly, 1966: 949)

The once mighty mechanism difficulty was beginning to lose its force. Yes, there was no adequate model of mantle convection that explained continental drift and seafloor spreading before their acceptance, and there would be no adequate model to explain plate tectonics before its acceptance. They were accepted because the evidence that continents had and were moving relative to each other had become too impressive to dismiss or ignore.

3.6 Menard ends his flirtation with mobilism

Gordon MacDonald and I flew home to California by the polar route, and somewhere over the arctic wastes, as we admired the glacially rebounded shorelines, I remarked that it was odd that he and I, people from a state being torn in half by the San Andreas fault, should be less inclined toward drift than the residents of the earthquakeless England. Geology, once locally oriented, had indeed become global.

(Menard, 1986: 237)

It was at the Royal Society's symposium on continental drift that Menard ended his four-year flirtation with mobilism, where he re-embraced fixism but retained mantle convection. At the time, Menard was preparing a lengthy paper for *Physics and*

Chemistry of the Earth, but split it in two unequal parts, giving the much shorter one at the Royal Society symposium, and saving for the journal the lengthier one. It was in the latter that he raised difficulties with seafloor spreading and expanded his critique of Wilson's explanation of the origin of oceanic islands in terms of seafloor spreading (May 18, 1966 letter to J. Gilluly). In 1958, Menard proposed that mid-ocean ridges truly are mid-oceanic. He plotted their positions and found that most bisected or nearly bisected oceanic basins, and pointed out that all except the Pacific have mid-ocean ridges (§1.7). In late 1963 or early 1964, he noticed another pattern.

Any major novelty in this paper lies in the suggestion that most oceanic rises are distributed in circles of relatively uniform radius around ancient continental shields. Geotectonicists commonly "discover" major patterns of this sort, and the speed with which the patterns descend into oblivion should properly cause skepticism when a new one is offered. We may justifiably feel such skepticism concerning the circular pattern of rises because the location of rises in some places is only conjectural. Nevertheless, on present evidence the pattern seems real and a brief discussion of the implications appears warranted.

(Menard, 1965c: 354)

He prepared polar projections to display the new pattern. He described the system of ridges surrounding Africa, and the two Americas.

Only about half the length of rises is centered relative to ocean basins. A much better correlation is obtained if rises are visualized as encircling continental shields. The configuration is difficult to see on a Mercator Map. Consequently a series of polar projections maps have been prepared by measuring distances and bearings on a 24 in. diameter globe on which locations of rises have been plotted ... The centers of projection were selected to give an approximate circle corresponding to some sections of the rise system. It was noticed later that the centers correspond to several continental nuclei. Long sections of rises plot along arcs of circles. Qualitatively the fit is striking around South Africa from the equatorial Atlantic to the Arabian Sea and to a lesser extent in the Red Sea and eastern Mediterranean [see Figure 3.11]. The qualitative agreement around South America is even better [see Figure 3.12]. All segments of the proposed Galapagos-Chile Rise lie near an arc and so does the Scotia Ridge. Moreover the southern and equatorial parts of the Mid-Atlantic Ridge lie along arcs of the same circle. The Brazilian shield would be completely circled by rises were it not for an apparent gap in the eastern Caribbean Sea. If the southern Canadian shield is taken as a center, both the northern East Pacific Rise and the northern Mid-Atlantic Ridge are found to circle it with connections across the Canadian Arctic, where earthquakes occur, and the Caribbean [see Figure 3.13].

(Menard, 1965c: 347; my bracketed additions)

He calculated (1965a: 111) that 50–56 percent of the total length of all rises were located in the centers of ocean basins, while 85–87 percent "followed arcs of circles centered on continental shields." This new pattern arose from the recent discovery of more rises. Indeed, Menard, T. E. Chase and S. M. Smith's 1963 discovery of the Galapagos Rise, which he linked to the previously discovered Chile Rise, may have been what prompted Menard to look for new patterns in ocean ridge distribution and

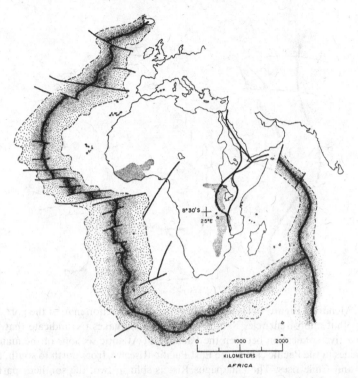

Figure 3.11 Menard's Figure 15 (1965c: 348). "A polar projection map of the part of the rise-ridge system that almost encircles Africa." Continental nuclei shown in grey in Africa. Crests of ocean ridges are shown as bold lines; fracture zones, as thin lines; edges of rises, as dotted lines.

seek a new explanation for their origin. Here is what they wrote about their discovery:

> The implications of the existence of the Galapagos Rise may be quite important with regard to the origin and history of oceanic rises and ridges. Two rises separated crest to crest by only 1400–1600 km and roughly parallel for several thousand kilometers exist in the southeastern Pacific. One is obvious, but, because of sounding distribution, the other has only just been discovered even though it is as broad as the Mid-Atlantic Ridge in the North Atlantic. It may be anticipated that other obscure oceanic rises and fragments of rises exist and that the distribution of active rises has varied markedly in geological time. Moreover, adjacent parallel rises imply complications in the origin of oceanic rises which have not previously required explanation.
>
> *(Menard* et al., *1963)*

Menard claimed that the orientation and position of the convection currents in the upper mantle which he thought responsible for the formation of oceanic ridges are controlled by the position of the continents. Building on an idea of R. Ramberg (1963), that upper mantle convection could be brought about by vertical inhomogeneities, Menard proposed that the formation of low-density continents released high-density

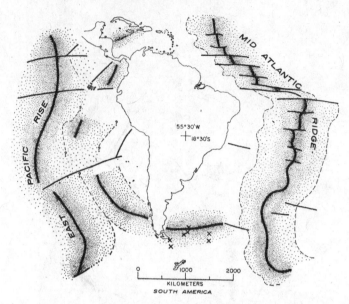

Figure 3.12 Menard's Figure 16 (1965c: 349). "A polar projection map of the part of the rise-ridge system that almost encircles South America. Earthquakes (X) indicate that the Scotia Ridge is an active connection between the Pacific and Atlantic sections of the main system." The three ridges in the Pacific east of the East Pacific Rise are, from north to south, the Cocos, Galapagos, and Chile rises. The Galapagos Rise is split in two; the southern part joins the Chile Rise.

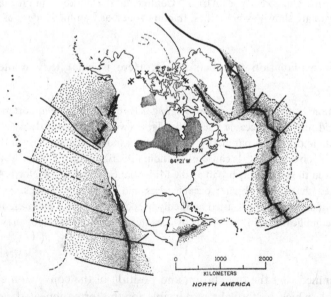

Figure 3.13 Menard's Figure 17 (1965c: 349). "A polar projection map of the rise-ridge system around North America. Continental nuclei shown in grey. Note earthquakes (X) which seem to give a connection between the Pacific and Atlantic parts of the system." Note also the clear positioning of the Gorda, Juan de Fuca, and connecting fracture zones.

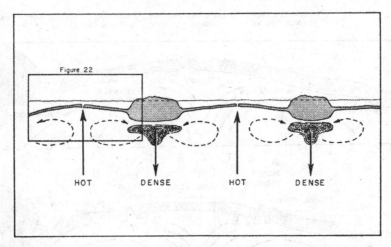

Figure 3.14 Menard's Figure 21 (1965c: 355). Menard's proposed model of upper mantle convection.

material into the upper mantle beneath, material which subsequently sank initiating the formation of convection cells. Here is how he put it in his Royal Society paper:

If mantle convection forms the rise-ridge system, many of its characteristics can be deduced. The chief conclusion to be derived from the distribution of rises is that they roughly circle continental nuclei which suggests that some phenomenon related to continents exerts a primary control over their locations (Menard, in the press). Continuing differentiation of continents from the underlying mantle might be such a phenomenon. Separation of a light component of the mantle which rises to form a continent may leave a dense residue which sinks. Ramberg (1963) has shown experimentally that a heavy body sinking through a viscous material produces flow around it and that an overlying competent crust is stretched in a ring around the sinking body. If this occurred under Africa and South America the rings of stretching would meet at the Mid-Atlantic Ridge in the middle of the Atlantic basin. The flow around the sinking bodies is shallow convection of a sort and it has the right motion to produce the observed faulting and stretching of the Mid-Atlantic Ridge, and the high heat flow on the crest.

(Menard, 1965a: 120–121)

So Menard basically reused his former solution to the origin of oceanic ridges (III, 5.12), placed it within a new convective pattern, and dropped continental drift. In fact, when illustrating his new hypothesis in his more extensive paper, he used the diagrams lifted from his *Marine Geology of the Pacific* (Figures 3.14 and 3.15).

Menard also postulated secondary convection, especially in the Pacific.

The transverse symmetry, high heat flow, and exceptional volcanism of oceanic rises, especially in the Pacific, seem to require some secondary convection. We may further suppose that in the circular zones of tension around continents, magma is generated in profusion in the upper mantle. As this magma rises and spreads under the crust, it produces some of the displacements

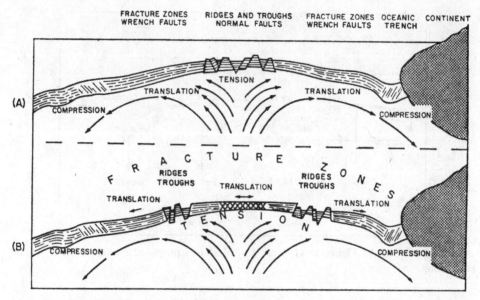

Figure 3.15 Menard's Figure 22 (1965c: 355). His caption reads: "Topographic and tectonic effects resulting from rising and lateral spreading of hot material under the rise-ridge system (from Menard, 1964). Both types of topography are observed."

observed on fracture zones. Where it breaks through cracks in the crust it forms volcanoes and a thickened second seismic layer.

(Menard, 1965c: 355)

Where did this secondary convection in the Pacific originate? Menard speculated that it may have been caused by the subsidence of the Darwin Rise. He plotted the Darwin Rise on a polar projection, and conjectured that it "may also be a center partially circled by other rises." This led him to propose that its subsidence "generated surrounding rises."

The Darwin Rise has subsided. We may consider the possibility that this subsidence generated surrounding rises. The Melanesian Rise and the East Pacific Rise south of Easter Island appear to encircle the ancient rise in the way that would be expected by this hypothesis.

(Menard, 1965c: 351)

3.7 Menard attacks seafloor spreading and Wilson's work on oceanic islands

It was at the 1964 Royal Society symposium that Menard (1965b) first attacked Wilson's work. He recalled that it was during the discussion following Wilson's talk.

Tuzo Wilson spoke in the following session. He recapitulated his papers about islands published in 1963. He had perfected his explanation of the uplifted islands beside trenches. The distribution suggested "that they may be on the crest of a standing wave in front of the trenches and that the crust is rigid." He thought, however, that the islands in the eastern Pacific must therefore form on the crest of such a wave and are uplifted. This was simply wrong. In the discussion I questioned his observation that island age increases with distance from a ridge crest. It was "doubtful in itself and, if correct, does not require migration of volcanoes from a hearth at the crest of the ridge." Island-groups at the edge of the Atlantic are still active. Moreover, there was no evidence of deep submergence – such as guyots – of the Canary and Cape Verde islands as would occur if they had formed at the ridge crest. My voice was full of tension, I recall. Perhaps I had had enough of this hypothesis.

(Menard, 1986: 236)

Menard also questioned Wilson's claim that the Hawaiian Islands and other island chains in the Pacific mark the direction of convective flow away from the East Pacific Rise. Menard cited two island groups that did not follow the pattern of its islands and accompanying guyots increasing in age toward the east away from the rise (RS2).

In the Pacific, he has suggested an ingenious explanation for the common occurrence of atolls at one end of an archipelago and active volcanoes at the other. The volcanoes are thought to be carried away, by convection, from a hearth in the mantle and to become inactive and subside as they move. A critical test of this hypothesis is given by the Austral or Tubuai Islands. They display the usual atoll to recent volcano pattern about sea level. However, below sea level and between the islands are several large guyots at a uniform and great depth. They must be older than the existing islands and therefore volcanism and subsidence have not proceeded in any simple manner from one end of the chain to the other.

(Menard, 1965b: 206)

Menard also expanded his attack on Wilson's claim that the age of islands in the Atlantic varies with distance from the ridge. He raised three difficulties (RS2). First there was the difficulty that he had privately mentioned to Hess in 1962 about the distribution of guyots atop the subsided Darwin Rise (§1.11). If Hess is correct, guyots which are truncated while moving off the ridge crest should all be approximately the same height. But they are not. Their height increases systematically the further they are from the former crest of the rise, a distribution that probably arose because guyots that formed on the rise's flanks had to grow higher before reaching sea level than those born atop the crest. Second he reproduced Wilson's figure (Figure 3.1) omitting Indian Ocean islands and adding seamounts and core samples of sediments from the Atlantic (Figure 3.16).

Menard claimed that there was no correlation.

If all the dated points in the Atlantic are plotted together, it is difficult to see any relation between age and distance from center (Fig. 24 [Figure 3.16]). The main basis for correlation is that three of the oldest samples are among the most distant from the crest. This may be an artifact related to the method of dating in which fossils identified as Eocene or Lower Cretaceous are assumed to have ages of 50 or 120 million years respectively (Wilson, 1963a, b). This is reasonable but the ages in years may be much closer together and the points on the figure may be even more clustered. It is by no means clear that the correlation of the whole Atlantic suggested by Wilson on the basis of a similar figure is real.

(Menard, 1965c: 361)

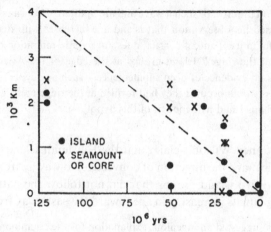

Figure 3.16 Menard's Figure 24 (1965c: 360). He added nine dated core samples from the Atlantic to Wilson's Figure 2 (1963a: 537).

Finally he considered islands, seamounts, and guyots in the region between the Azores and Africa.

A second test region lies between the Azores and Africa. The oldest fossils from the Azores, Madeira, and Amphere Seamount are all Miocene, from the Canary Islands they are Oligocene, from the Great Meteor Guyot they are Neogene. Thus the earliest datable material on the crest and flanks of the Mid-Atlantic Ridge and the whole width of the adjacent Atlantic Basin is about the same age. Migration of oceanic volcanoes in this region is not evident.

(Menard, 1965c: 361)

Menard (1965c: 361) also objected to Wilson's "supposition that volcanoes tend to form in an area of high heat flow on the crest of a rise and that their presence elsewhere in an ocean basin suggests migration from the crest."

On this basis most of the active volcanic islands in the Atlantic are outside the belt of high heat flow associated with the crest of the ridge. Even the contention that all active volcanoes in the Atlantic are on the Mid-Atlantic Ridge is mistaken. Active volcanoes occur even at the edge of the basin in the Cape Verde and Canary Islands.

(Menard, 1965c: 362)

Perhaps Menard had had enough of Wilson's speculations, believing them insecurely based on facts.

Menard next raised difficulties with both Hess's and Dietz's versions of seafloor spreading. Neither could successfully explain the origin of Layer 3 of oceanic crust. Hess maintained that the 5 km deep layer was created by the serpentinization of peridotite above the 500 °C isotherm which, according to Hess, cannot rise above 5 km below the seafloor at ridge crests. However, recent heat measurements show that it can rise to 3 km.

There remains the possibility that the oceanic crust is produced at a serpentinization boundary in some special place, where heat flow is high, and is then transported elsewhere to form a uniform cover of the ocean basins. The crests of oceanic ridges and rises have just such high heat flow. If the crust is created at the crest and moves laterally from it as suggested by the Hess hypothesis, the present uniform crustal thickness might be produced. However, this seems unlikely on two counts. First, heat flow on rises reaches 8×10^{-6} cal/cm^{2}sec^{-1} on the East Pacific Rise and this gives 770 °C at the base of the oceanic crust (von Herzen, 1960). The 500° isotherm is at a depth of only about 3 km below the sea floor. Thus the suggestion of Hess (1962) that this isotherm cannot rise above 5 km below the sea floor does not appear correct.

(Menard, 1965c: 338)

Menard admitted, however, that high heat of 8×10^{-6} cal/cm^{2}sec^{-1} on the East Pacific Rise was unusual, and that "it is possible that a heat flow of about 4.4×10^{-6} cal/cm^{2}sec^{-1} required to form the present oceanic crust by serpentinization mechanism is a common maximum of the crest of rises." This led to the second difficulty he had with seafloor spreading.

Even so, we come to the second objection to this hypothesis in this form, namely the support for the hypothesis derived from the uniformity of crustal thickness is destroyed. The concept that the crust is uniform because it is limited by an isotherm determined by normal heat flow and parallel to the sea floor is very plausible. The concept that it is uniform because it was limited by such an isotherm in a thermally unstable environment on the crest of a rise and has since been transported for several thousand kilometers without change in thickness is much less plausible.

(Menard, 1965c: 338)

Menard also objected to Dietz's proposal of the eclogite–basalt phase change. Noting that laboratory experiments have shown that the phase change "requires a correlation between the thickness of the crust and temperature and pressure," he (1965c: 357) rejected the phase change because "oceanic crust has a remarkably consistent thickness at locations at which the depth differs by a factor of two and the temperature gradient by a factor of forty."

As I see it there is one glaring omission from Menard's symposium paper: it made no mention of mobilism's paleomagnetic support. When they flew over the North Pole, he and MacDonald shared more in common than working in California and rejecting mobilism; neither let paleomagnetism's support of mobilism stand in their way of favoring fixism. Nor did Menard seem too bothered by the lack of pre-Cretaceous sediment on the ocean floor. He took comfort in E. L. Hamilton's suggestion that Layer 2 might contain consolidated sediments. As he (1964: 211) put it in his *Marine Geology of the Pacific*, and retrospectively quoted (1986: 169), "In any event there is no compelling reason to believe that the Pacific Basin has not been receiving sediment from the continents for most of geological time." Moreover, Menard could still maintain that Layer 2 could contain some volcanics.

Nor is it necessary to have a wholly sedimentary second layer in this ocean to achieve the balance. Other lines of evidence suggest that this layer is composed of volcanic extrusives and shallow intrusives interbedded with consolidated sediment ...

(Menard, 1964: 169)

As for his new proposal about the origin of mantle convection, it received scant attention; there was not much time anyway because little more than a year elapsed between the publication of the symposium proceedings and the confirmation of seafloor spreading. Menard, however, did receive an encouraging letter from David Griggs, a longtime proponent of mantle convection (I, §5.6).

I have read with great interest your two articles on Phys. and Chem. of the Earth vol 6 and in the Roy. Soc. Symposium on Continental drift ... Both Bill Ruby and I are intrigued with your idea of the sinking of sub-continental mantle residue as the cause of "convection" and its relation to mid-ocean ridges. I fumbled around the fringes of this idea some twelve years, and published some aspects of it that may not have occurred to you in the attached discussion.

(Griggs, March 24, 1966 letter to Menard)

Menard retrospectively noted that his paper in *Physics and Chemistry of the Earth* "was my last grasp before converting to seafloor spreading," and added that Runcorn, then editor of the journal, "certainly would have been happy if I had withdrawn it."

3.8 Early responses to the Vine–Matthews hypothesis, 1964

Vine and Matthews' paper appeared in September 1963. It did not go entirely unnoticed; it created a few ripples but no waves. In 1964, only Hess and George Backus discussed it in print, and at the 1964 Royal Society's symposium only Vacquier discussed it. Manik Talwani at Lamont mentioned but did not discuss the hypothesis in a paper he submitted to *Marine Geology* in May, 1964. A harbinger of Lamont's negative response to the hypothesis, Talwani devoted three sentences to it, and characterized it as "startling."

Perhaps the only attempt to explain these anomalies in any detail has been made by Vine and Matthews (1963) who describe magnetic anomalies over the Carlsberg Ridge in the Indian Ocean. They ascribe ridge magnetic anomalies to a horizontal succession of alternate zones of normally and reversely magnetized rocks. It should, however, be pointed out that less startling explanations are also possible for these anomalies.

(Talwani, 1964: 73)

These were the other responses in chronological order.

Backus received all his degrees at the University of Chicago, starting with a B.Phil. (1947), followed by a B.Sci. (1948), M.Sci. in mathematics (1950), M.Sci. in physics (1954), and Ph.D. in physics (1956). Backus recalled his Ph.D. work under Chandrasekhar. He worked on the origin of the geomagnetic field. Although Elsasser and Bullard developed the idea of a self-exciting dynamo (II, §1.4), despite Cowling's theorem that a symmetric self-exciting dynamo is an impossibility, there was "the suspicion that a self-excited fluid dynamo might not explain the existence of geomagnetic field." Backus (1956a, 1956b, 1957a) showed that it can by considering a time-dependent conducting material within Earth's core.

My PhD thesis at Chicago was under Chandrasekhar, and gave about 40 000 years as a rigorous upper limit on the lifetime of an axisymmetric magnetic field maintainable by an axisymmetric fluid flow in the earth's core. This was an extension of the Cowling theorem that the lifetime could not be infinite. His theorem had led to the suspicion that a self-excited fluid dynamo might not explain the existence of a geomagnetic field. Pursuing my thesis, I looked for, and at the same time as A. Herzenberg, found that relaxing axisymmetry permitted fluid dynamos to exist.

(Backus, July 1, 2009 email to author)

Backus had no coursework in geology or geophysics at the University of Chicago. He worked on his Ph.D. at The University of Chicago's Yerkes Observatory in Williams Bay, Wisconsin, where Chandrasekhar lived, and it was there that he got his introduction to continental drift, polar wandering, their developing paleomagnetic support, and reversals of the geomagnetic field. Runcorn gave a seminar, and Backus was impressed.

I first heard about continental drift and geomagnetic reversals in a seminar by Runcorn at Yerkes. I think this was about 1955. I was more impressed by Runcorn's data than were many who knew the geological and physical uncertainties involved in interpreting them ... Runcorn's 1955 seminar at Yerkes was my introduction to reversals. I thought it was a fascinating but mysterious idea. Runcorn showed Hospers' data and some of his own, and it looked convincing to me, but I did not have the trained observer's skepticism.

(Backus, July 1, 2009 email to author)

Runcorn could not have hoped for a better listener than Backus. Runcorn had himself worked on the origin of the geomagnetic field. Backus, as Runcorn might have put it, had not been persuaded by geologists to dismiss extensive polar wandering or continental drift, and, unlike so many who heard Runcorn talk about mobilism's support for polar wandering or continental drift, and reversals, Backus thoroughly understood the nature of the geomagnetic field and support for the GAD hypothesis. Moreover, there was another reason why Backus was receptive to Runcorn's ideas.

Chandra had been interested in convection in viscous fluid spheres and his interest grew because he thought Runcorn might be onto something. While I was his PhD student, I did some related mathematical work which he managed to have published. I had great respect for Chandra, and his good opinion added weight to Runcorn's position.

(Backus, July 1, 2009 email to author)

In the early 1950s, Chandrasekhar had begun working on the more theoretical aspects of thermal convection, examining the way in which it is affected by rotational and magnetic forces (Chandrasekhar, 1952, 1953a, 1953b, 1954, 1956). (Hide did a post-doc at Yerkes with Chandra 1953–4.) I suspect that Chandrasekhar was also interested in Hide's work on core convection, for he thanked Hide for pointing out to him the relevance of a particular theorem in dealing with slow motions in a rotating fluid which has no viscosity (Chandrasekhar, 1957: 329). Moreover, Backus also had begun working on thermal convection, writing a paper (Backus, 1957b) in which he tightened up some of the mathematics in Chandrasekhar's work on convection in spheres (Backus, July 1, 2009 email to author).

After spending two years at Princeton University, Backus joined the mathematics department at MIT. He was warned by Gordon MacDonald that he should stay away from working on mantle convection. Backus ignored him.

When I was in the Math dept at MIT, MacDonald warned me (1958) that interest in mantle convection would mark me as a non-serious student of the earth. I ignored him but, in retrospect, not for any good reason.

(Backus, July 1, 2009 email to author)

Backus accepted a position at Scripps and the IGPP at the University of California, San Diego (UCSD), in 1960. He believes that Munk wanted to hire him because of his "dynamo work"; Chandrasekhar's recommendation certainly did not hurt.

Roger Revelle had just given Walter Munk the appointment as first director of IGPP in La Jolla, in order to counter Harvard's offer to Walter. Walter thought that geomagnetism was a promising field, and hired me (I think) on the basis of my dynamo work and Chandrasekhar's recommendation to Revelle and Carl Eckart (then the éminence grise of theoretical fluid dynamics at Scripps).

(Backus, July 1, 2009 email to author)

Backus continued his theoretical work on geomagnetism and mantle convection at Scripps.

Like others at Scripps, he took advantage of Bullard's ties, and took a sabbatical in the Department of Geodesy and Geophysics at Cambridge. He arrived at Madingley Rise in September 1963 and left in January 1964 (Backus, July 1, 2009 email to author). He first learned of V-M at tea. Their presentation convinced him of continental drift and seafloor spreading.

Soon after hearing Vine and Matthews announce their hypothesis at Madingley Rise, Backus attended a seminar given by Everett on his ongoing work on the fit of South America and Africa.

I was on sabbatical at Madingley Rise (the geophysics department of the U of Cambridge) when Vine and Matthews came in to afternoon tea with the announcement that they had an explanation for the magnetic stripes ... That tea may have been the first group to hear them. Shortly thereafter, Everett gave a seminar at Madingley Rise on the fit of the African and South American continental shelves.

(Backus, July 1, 2009 email to author)

Backus was convinced by Vine and Matthews that seafloor spreading was probably correct. He was not in a position to decide between Hess's or Dietz's versions of seafloor spreading, or what should be excised from either version, but he thought the basic idea was correct. He was surprised about the skepticism at IGPP after returning in 1964.

I didn't know enough about Hess's or Dietz's versions to dismiss anything. I became a convert at the tea at Madingley Rise when Vine and Matthews first described their work. It seemed to be a good argument for both sea-floor spreading and geomagnetic reversal. When I returned to IGPP in 1964 I was surprised at the skepticism still present there.

(Backus, July 1, 2009 email to author)

Backus also recalled that Jeffreys, who heard Everett's talk, declined Everett's request to look at "computer processed maps." Backus then decided to stop listening to what Jeffreys had to say about the fit of the continents.

As I walked out with the audience, I heard Everett invite Jeffreys upstairs to look at the computer-processed maps. Jeffreys said he was too busy. After that I lost interest in Jeffreys' opinions on the subject.

(Backus, July 1, 2009 email to author)

So Jeffreys once again decided not to look at an improved fit of North America and Africa. He would not look at what Carey had done, and would not look at what Everett was doing, even though Everett was one of Bullard's students, and Bullard himself was even involved (III, §2.10). Jeffreys, I admit, was seventy-two or seventy-three at the time, but he was still quite active, as evidenced by his text of the first Sir Harold Jeffreys Lecture of the RAS, which he gave either the same year or a year later.

Backus put Everett and Bullard's fit of South America and Africa together with V-M, and proposed a test of both. He did not view himself as starting a research program on continental drift or marine magnetics. Still interested in geomagnetism, he was interested in the implications of V-M for "paleomagnetic history," for the same reason that Blackett and Runcorn were first interested in paleomagnetism.

I was really an outsider looking in and making comments that seemed interesting. I never pursued a research program about continental drift, and was interested in it mainly as the source of a "tape-recording" of paleomagnetic history.

(Backus, July 1, 2009 email to author)

Nonetheless, his lack of fundamental interest in continental drift did not deter him from proposing an inventive test. Moreover, he was the first to apply Euler's theorem to analyze the width of marine magnetic anomalies and determine that their width varies as a function of the sine of their angular distance from the pole of rotation – one of the cornerstones of plate tectonics.

Referencing Vine and Matthews (1963), Dietz (1961a), Hospers (1954b), Cox, Doell and Dalrymple (1963a), and Irving (1958b), Backus began with a description of the generation of magnetic anomalies in accordance with V-M especially suited for his application of Euler's theorem. He also assumed that the rate of seafloor spreading was the same on both sides of the ridge, that the zebra pattern of magnetic anomalies exhibited bilateral symmetry with the ridge axis running down the center.

Vine and Matthews suggest that the mantle material cools as it rises convectively under a ridge and then spreads horizontally outward. As the material cools through its Curie point it is magnetized parallel to the contemporary local geomagnetic field. Because this field reverses quasi-periodically with a period $2T$, T being of the order of 0.5–1.0 million years, stripes of alternate permanent magnetization are produced the width of which is vT, v being the local horizontal velocity with which material at the surface of the mantle spreads away from the center of the ridge. The stripes are observed to have widths of the order of 20 km. If T is 0.5 million years, v is 4 cm/yr.

(Backus, 1964: 591)

With this general description of seafloor spreading and V-M, he narrowed his discussion to the South Atlantic, introduced the separation of South America and Africa, and explained the purpose of his note.

The purpose of this communication is to suggest some measurements in the South Atlantic which would test simultaneously Vine and Matthews's hypothesis and the hypothesis that South America and Africa have drifted apart as flotsam on a convection current in the mantle which rises under the Mid-Atlantic Ridge. If these two hypotheses are correct, and if geomagnetic field reversals occurred regularly during the period in which South America and Africa were drifting apart, there should be magnetic stripes on the flanks of the south Mid-Atlantic Ridge, running parallel to that ridge. Furthermore, these stripes should be wider at more southerly latitudes, because the South Atlantic is wider there. According to the hypotheses of continental drift and renewal of the ocean bottom at ridges, this increased oceanic width results from a southward increase of the horizontal velocity where the ocean floor diverges from the south Mid-Atlantic Ridge.

(Backus, 1964: 591)

If that prospect wasn't exciting enough, he came up with an even more exciting aim. He wanted to test the two hypotheses by making a quantitative prediction. He could explain the rate of increase in the width of the magnetic stripes with angular distance from the northern pole of rotation to the equator of rotation. True, the "magnetic stripes should be wider at more southerly latitudes, because the South Atlantic is wider . . ." But there was a more interesting explanation. Given Euler's Point Theorem, the width of the magnetic stripes should increase until the angular distance from the rotational poles is 90°. Working partly from Carey's fit of South America and Africa, but primarily from a pre-publication version of what became Bullard, Everett, and Smith (1965), and what Bullard would soon present at the forthcoming Royal Society symposium on continental drift, Backus estimated the latitude of the northern pole of the rotation of South America relative to Africa to be at 43.3° N. Thus, he reasoned that the magnetic stripes increase their width at more southerly latitudes and, although he did not explicitly state it, should reach their maximum width at 46.7° S.

In fact, it is possible to make a quantitative prediction of how the width vT of the magnetic stripes should vary with south latitude. Carey found, by sliding model continents on a large world-globe, that South America could be fitted to Africa by rotating it toward Africa about a point near the Azores. Following a suggestion of Sir Edward Bullard, Everett has used Edsac to calculate a least-squares fit of the two coast lines. Everett finds that the 1000-fathom line of the east coast of South America can be made to fit the 1000-fathom line of the west coast of Africa from the Ivory Coast to Cape Town with a root mean square error about 100 km. To obtain the fit, South America must be rotated through 56.4° toward Africa about a pole P at 43.3°N, 30.25°W.

(Backus, 1964: 591)

Backus now introduced, albeit without naming it, Euler's Point Theorem. He also made three simplifying assumptions, two explicit, one implicit. He moved the northern rotational pole to a position "on the south Mid-Atlantic Ridge's meridian of longitude," and assumed that the seafloor spreading is normal to the axis of the Mid-Atlantic Ridge. These assumptions allowed him to move from rotational to geographical latitudes with ease. He also assumed that the rate of seafloor spreading was constant.

Of course, any rigid displacement of a continent on a spherical globe is a rotation about some point, but the simplest hypothesis is that convection current rising between South America and Africa was steady and produced a rotation of South America away from Africa with constant angular velocity Ω about the pole P near the Azores. The south Mid-Atlantic Ridge runs almost north and south along the meridian of 15°W longitude. The pole P is so far north of the equator that only a small error is committed by placing it on the south Mid-Atlantic Ridge's meridian of longitude, 15°W, at 43.3° north latitude. Then the velocity with which the ocean floor spread away from the south Mid-Atlantic Ridge should vary with south latitude l along the ridge as follows:

$$\nu = a\Omega \sin(l + 43.3°) \qquad (1)$$

a being the radius of the Earth.

Backus (1964: 592) next, assuming that "reversals of the Earth's field were exactly periodic with period $2T$," derived the equation for determining the width of magnetic stripes at any latitude between rotational poles:

$$\nu T = a\Omega T \sin(l + 43.3°) \qquad (2)$$

and noted that if "l varies from 5° to 30°, νT should increase by 28 percent," which "may be large enough to measure." At its greatest width, 46.7° S, the width should increase an additional 6 percent.

Backus next suggested what had to be done to test his prediction. The ideal situation would be a detailed magnetic survey such as the one Mason and Raff had done in the northeastern Pacific. This would definitely determine if the magnetic stripes were there as predicted by Vine and Matthews.

The measurements suggested by the foregoing argument are as follows: First, it would be interesting to try to see whether indeed there are magnetic stripes on the flanks of the south Mid-Atlantic Ridge. Ideally, such an investigation would require a detailed magnetic survey of the sort carried out by Mason off the Pacific coast of North America.

(Backus, 1964: 592)

But if this became cost prohibitive, Backus suggested Fourier analysis as a less expensive alternative for a preliminary study.

A much less expensive preliminary test would be to measure the local magnetic anomaly along a line perhaps 1000 km long running roughly parallel to the ridge and within 100 km of its centre. If the power spectrum of this record showed much less energy at short wave-lengths (say less than 30 km) than the power spectra of magnetic anomalies on lines normal to the ridge, such a difference in spectra would be an indication of the existence of magnetic stripes running parallel to the ridge and having widths less than 30 km.

(Backus, 1964: 592)

If the magnetic stripes were there, there was the issue of whether or not their width increases along a sine curve toward the rotational equator. With an extensive survey, this would be easy to check. But, if such a survey were too expensive, comparison of power spectra of magnetic anomaly profiles normal to the ridge "at latitudes ranging from 5 to 30° S" would (might) suffice.

To measure whether the stripes increase in width as in equation (2), it is necessary to analyse the local magnetic anomalies obtained on several lines 200–400 km long, crossing the ridge normally on parallels of latitude ranging from 5° to 30° S. It is not necessary to be able to detect correlations between these records visually, as in Vacquier's discussion of the lateral displacement along the Mendocino escarpment. If the period T between reversals really is roughly constant, the power spectra of the magnetic anomaly records obtained on normal crossings of the ridge at various latitudes would show peaks at a fundamental wave number k, and its harmonics, nk, and k^{-1} would vary with latitude as does vT in equation (2).

(Backus, 1964: 592)

Thus far, Backus had assumed that the reversals are periodic. If "the period between geomagnetic field reversals" is "highly variable," Backus proposed a "subtler data analysis" than he had recommended for periodic reversals.

Menard (1986: 190) described Backus' paper as visionary. He was right. Backus recognized that the movement of continents relative to each other is best described in accordance with Euler's Point Theorem. But this already had been suggested by Creer, Irving, Nairn, and Runcorn (1958) (II, §5.16), and, of course, as Backus knew, by Bullard, Everett, and Smith. Backus, however, did more. Exploiting Euler's Point Theorem further, he realized that the linear velocity of one block to another will increase along a sine curve toward the rotational equator, and thus predicted that the widths of the magnetic stripes would increase accordingly. Then, working from the north pole of rotation, as determined by Everett and Bullard, and estimates of the rates of seafloor spreading and geomagnetic reversals, he calculated the width of the magnetic stripes at different latitudes. If Backus had known the widths of the magnetic stripes, he could have reversed the process, calculating the position of the rotational poles from the widths of the magnetic stripes, which is what Jason Morgan did three years later when proposing plate tectonics (Morgan, 1968).

Backus applied to the National Science Foundation (NSF) to obtain the needed data. NSF rejected his proposal because, as Menard (1986: 222) later reported, it was judged as "too speculative." Backus can no longer remember the details of his proposal, but suggested another reason for its rejection, and added that he was actually relieved because "then Gilbert [Freeman] and I could get on with our seismic inversion work."

I don't remember what I proposed to NSF and have not kept any records. As I mentioned above, my interest in the subject was rather dilettantish, and I had no experience of running a marine survey, so I was not particularly surprised at NSF's rejection. In fact I do remember being rather relieved, because then Gilbert [Freeman] and I could get on with our seismic inversion work.

(Backus, July 1, 2009 email to author; my bracketed addition)

Backus may be right. Menard noted that Scripps certainly could have done the surveying, but nonetheless, Backus "had no experience running a marine survey." Another factor that might have led some reviewers to reject Backus' proposal is that he might not have been insistent about avoiding low latitudes. At low latitudes where the inclination of the geomagnetic field is 0 or close to 0, magnetic anomalies surrounding a north–south trending ridge are very flat. Backus, at least in his paper, spoke of taking profiles from 5° to 30° S. Now it may be the case that some spectral difference might show itself, but Backus' purpose would have been better served if he had proposed taking profiles at higher latitudes, say from 30° S through the rotational equator at 46.7° S to 55° S where the anomalies are much more pronounced. As Backus (July 1, 2009 email to author) later noted, "I did want to avoid low latitudes, but I don't remember calculating how low. Maybe this was one reason for the NSF turndown."

Vacquier was the next to discuss V–M. Vacquier and R.P. Von Herzen first mentioned the hypothesis in a paper they submitted to the *JGR* in late October 1963, not even two months after Vine and Matthews' paper appeared. They took fourteen magnetic profiles across the South Atlantic Ridge from approximately 6° S to 30° S, almost exactly the area later singled out by Backus. They (1964: 96) noted, "the magnetic anomaly associated with the ridge crest may be a general property" as already shown in the North Atlantic "as well as in the Indian Ocean [Vine and Matthews, 1963]," but said nothing about whether or not the profiles revealed the predicted zebra pattern of anomalies.

Vacquier had much more to say at the Royal Society's symposium on continental drift. The only symposiast to mention the hypothesis, he raised two difficulties (RS2), both pertained to the lack of a clearly identified zebra pattern of magnetic anomalies near ocean ridges in lower latitudes.

Unfortunately this [Vine–Matthews hypothesis] mechanism is probably not adequate to account for all the facts of observation. As previously mentioned, in the eastern Pacific the lineated magnetic anomaly pattern appears to the eye to end south of 25° N latitude. Where the East Pacific Rise can actually be seen, a lineated magnetic pattern has so far not been detected by visual inspection of the magnetic records. The same is true of the Mid-Atlantic Ridge between 6 and 30° S latitude where 13 profiles failed to reveal visual evidence of magnetic and topographic features parallel to the ridge which could be correlated from one profile to the next, except for a single prominent broad magnetic anomaly which seems to form the only clearly continuous feature (Vacquier & von Herzen 1964).

(Vacquier, 1965: 80)

Vine was quite surprised by Vacquier's criticism. He thought it was "bad science."

In particular he [Vacquier] pointed to a survey of the Mid-Atlantic Ridge in the equatorial Atlantic which did not reveal any linear anomalies. This I thought was quite extraordinary because it was bad science – the linear anomalies will not be developed over a north-south trending ridge at the equator, where the Earth's magnetic field is directed horizontally.

(Vine, January 2009 email to author)

It was bad science, and surprising that Vacquier said it, because Vacquier's profiles, as discussed above, were from low latitudes, where the inclination is near zero; if it is near zero and the ridge trends north–south, then anomalies at the sea surface are flat. Thus Vacquier should not have expected to observe the predicted pattern of magnetic anomalies surrounding the Mid-Atlantic Ridge at latitudes he surveyed. Vine's model predicted that the anomalies would be difficult to observe (see Fowler, 2000: 49–50 and 52–53, and McKenzie and Sclater, 1971). Vacquier's second difficulty about the lack of well-defined magnetic anomalies in the eastern Pacific south of 25° N latitude was subject to the same rebuttal. The anomalies along the East Pacific Rise, which trends north–south, are poorly developed in low latitudes. There are recognizable anomalies just south of 25° N, but marine magnetic data from the region were scarce in the early 1960s. It is no wonder that Vacquier put the cutoff point at 25° N. In contrast the zebra pattern of magnetic anomalies is very pronounced along the Galapagos Ridge, which trends east–west along the equator (see Fowler, 2000: 48 and 52–53). Vacquier's difficulties were pseudo-difficulties.

Vacquier, aware of Backus' work, added that a spectral analysis "might conceivably reveal a pattern where visual inspection does not."

It has been pointed out (Backus, 1964) that spectral analysis of these records might conceivably reveal a pattern where visual inspection does not. Such an analysis has not yet been carried out ... In the Atlantic the magnetic anomaly on the crest of the ridge appears to be the only reliable marker unless extensive numerical data processing should reveal others.

(Vacquier, 1965: 80)

Backus did not do the spectral analysis because he saw no need for it. As he (July 9, 2009 email to author) later recalled, "I didn't do any work on Vacquier's data. I thought the correspondence across the fault was evident to the eye."

Hess, quite fittingly, also discussed V-M, which he (1964a: 639) said was "very fruitful." Hess was quite proud of his pun (Vine, 2001: 59). Hess found what he interpreted as new evidence for mantle convection. He linked results of laboratory work on the seismic velocities of olivine, which he thought made up the upper mantle, and recent evidence obtained by Raitt and Shor of seismic anisotropy in the oceanic

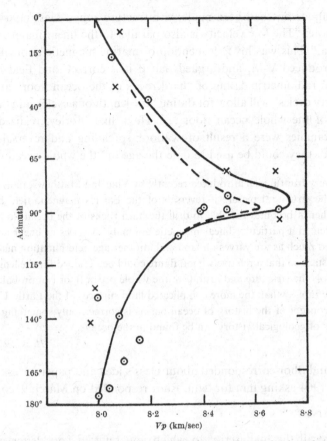

Figure 3.17 Hess's Figure 3 (1964a: 630). His caption reads: "Seismic velocity, *Vp* versus azimuth. Circles, Mendocino area; crosses, Maui area; dashed curve for former, solid curve both areas combined."

crust. Laboratory studies in the 1940s had shown that olivine is seismically anisotropic, and Hess confirmed the results, working with deformed peridotites from St. Paul's Rock (Hess, 1959b). Seismic waves travel faster through olivine's glide plane than the two other orthogonal planes: 9.87 km/s as compared to 8.65 km/s and 7.73 km/s. Because the glide plane "tends to line up parallel to the major shear zone," Hess (1964a: 629–630) reasoned that "low seismic velocities would be perpendicular to it and the higher velocities parallel to it." He found that the results of Raitt and of Shor were consistent with his hypothesis. Raitt had found that the seismic velocity of material of the upper mantle was faster when shots were taken parallel to the Mendocino fracture than when taken almost orthogonally to it. Hess noted that Shor, in a study just north of the Molokai Fracture Zone near Maui, found faster velocities were obtained from shots parallel to the fracture zone. Hess combined the results (Figure 3.17), and argued (1964a: 630) that they show "a striking agreement between the hypothesis and the observations."

So far comparisons had been made with fracture zones. Hess (1964a: 630) now went on to note, "The low velocity is also parallel to the linear magnetic anomalies off California." This was his first mention of marine magnetic anomalies. He then explicitly introduced V-M, and argued that if it is correct, and field reversals are periodic, then radiometric dating of "hard rocks of the ocean floor" at intervals of "perhaps thirty cycles" will allow for dating "the length of a cycle" and "the structure and history of the whole ocean floor ..." Hess, like Morley, realized that if the magnetic anomalies were a result of seafloor spreading and reversals of Earth's magnetic field, they could be used to date the age of "the whole ocean floor."

Applying the very fruitful idea introduced recently by Vine and Matthews, rates of flow could be added to the structural picture if reversals of the Earth's magnetic field had a regular periodicity. When it becomes possible to drill the hard rocks of the ocean floor an absolute age can be given to a particular linear magnetic anomaly. A series of linear anomalies may then be counted much as are varves on land and another age determination made at perhaps thirty cycles distant so that a relatively good figure could be obtained for the length of a cycle. Using these tools the structure and history of the whole ocean floor can probably be worked out much more rapidly than the more complicated land surfaces of the Earth. Unfortunately, if my general concept of the history of ocean basins is correct, only something less than the last 10 per cent of geological history can be found in the oceans.

(Hess, 1964a: 630–631)

Hess, Raitt, and Shor corresponded about Hess's idea and paper.[8] Hess wrote Raitt on March 5, 1964 asking him for data. Raitt responded on March 9, copying Shor.

> Dear Harry,
>
> I don't recall the manuscript to which you refer in your letter of 5 March. As I didn't go to the Woods Hole meeting last, but George Shor did, perhaps you are thinking of the anisotropy in the upper mantle velocity which George observed on crossed profiles recorded north of Maui on the Mohole site selection survey. We have some unpublished profiles taken off California showing only a small difference, probably not significant, between velocities recorded north to south and east to west. I have no objection to your referring to this, unpublished or not, and I will pass your letter on to George.

Hess then sent the following note to Shor, dated March 20, 1964. He copied Raitt and also included a draft of his paper on anisotropy, which he changed only slightly before submitting it.

> Dear George:
>
> I am enclosing a note which I would like to publish on seismic anisotropy based on your work north of Maui, Russ Raitt's off California and my own observations on the St. Pauls rock material. If you have no objection I'll send it off but if you would prefer that I wait until you have published your data, I am quite agreeable.

I would appreciate any comments you and Russ might care to make on the manuscript.

Best regards,

H. H. Hess

Shor replied to Hess on March 31. Shor was not sure whether the directional differences were real, or whether they were "just observational errors." He also suggested that Hess add "a sentence urging the proper field testing be done."

Dear Harry,

I am not as convinced about the validity of the data on seismic anisotropy as you are, possibly because I am closer to the original data. I think the theory is an interesting one, however, and should be submitted to a proper test – which would mean executing a number of profiles at assorted azimuths through the same point, with very detailed shooting of large charges in the range of mantle velocity arrivals, and proper care to ensure that true reversals on the mantle arrivals were obtained. The data you have cited thus far merely indicated that your hypothesis is a possible one that such a test should be made.

If the differences in velocity are truly due to variation in condition of the mantle, and are not just observational errors, one should obtain the same depth to mantle in both directions using the different velocities. This is not always the case, and I am not sure whether the error is in the velocity or the depth. Anyhow, the theory is interesting and the data give it some support; I would suggest that you enclose a sentence urging the proper field testing be done.

My paper will be in print in two weeks, so you are perfectly welcome to quote it; it's due for the April issue of JGR.

Sincerely yours,

George G. Shor. Jr.

Following Shor's suggestion, Hess added this to his manuscript, but no cautionary remarks that the seismic anisotropy might be due to experimental error.

Whereas tests for seismic anisotropy should be made in other parts of the oceans, it seems likely that the oceans everywhere have a fabric such that directions in which the material has flowed in the solid state can be determined. The low velocity is perpendicular to the shear direction and high velocity parallel to it. The low velocity is also parallel to the linear magnetic anomalies off California.

(Hess, 1964a: 630)

Looking ahead, Raitt and Shor tested Hess's hypothesis, and obtained results consistent with it (Raitt *et al.*, 1969, 1971). But they had a question about the design of their test, and asked Backus for help. Backus obliged and ended up writing a paper in which he (1965: 3429) explained "how an arbitrary small anisotropy in the upper mantle will affect the azimuth dependence of measured speeds of P and S head-waves just below the M discontinuity." At the end of the paper, he (1965: 3439) acknowledged that the paper "is essentially the answer to a question Raitt asked me in connection with the design of further experiments to test whether the upper mantle is isotropic." Hess let Backus know how much he appreciated his efforts, writing him on August 5, 1965, "I am enormously indebted to you for defining the problem so nicely. I am also happy that Raitt and Shor are going to test your analysis and my hypothesis at sea." Later work has continued to demonstrate seismic anisotropy (Davis, 2003).

Hess previously had paid little attention to Layer 2 of the oceanic crust and thought it was probably compacted sediment overlying Layer 3, the main layer of oceanic crust, which he believed was serpentinized peridotite. He now changed his mind about Layer 2 in light of V-M and recent findings from the AMSOC (American Miscellaneous Society) 1000-foot bore-core obtained near Mayaguez, Puerto Rico. Hess had chaired AMSOC's Site Selection Committee. The project, a predecessor of the Mohole Project, was undertaken by Brown and Root, Inc. It was completed in November 1962; one month before Brown and Root was acquired by Halliburton Energy Services. Hess noted that Cox, Doell, and Thompson (1964) had found that the magnetic susceptibility and intensity of the remanent magnetization of the serpentinized ultramafic rocks from the core were insufficient to account for "the magnitude of linear magnetic anomalies found in many parts of the oceans thus far investigated." Hess concluded that Layer 2 was not consolidated sediment, but mainly basalt, whose remanent magnetization is of sufficient intensity to cause the magnetic anomalies.

It is now necessary to modify in detail the general picture for the generation of oceanic crust on mid-ocean ridges as suggested by the writer (1962). The magnetic properties of the Mayaguez serpentinites suggest that a more magnetic material must be present on the ocean floor to account for the magnetic intensity of the magnetic anomalies. Therefore it is proposed that "layer 2" of the oceanic crust is basalt, at least in part, but that "layer 3" is serpentinites as before. Basalt, as well as water to produce serpentinization, is released along the axis of the mid-ocean ridges. The result might be 1 km of basalt overlying 4 ½ km of serpentinized mantle material. In this case part of the thermal anomaly over mid-oceanic ridges could be due to the rapid vertical transfer of heat and basaltic magma, generated by partial melting at perhaps 100 km below the sea floor. The remainder of the thermal anomaly may be attributable to convective overturn in the upper 600 km of the mantle. This arrangement would be in better accord with the relatively narrow width of the thermal anomaly.

                                                                    (Hess, 1964b: 173–174)

Hess also claimed that there was independent evidence that the second layer was basaltic.

One must conclude the "layer 2" consists at least in part of basalt with much higher susceptibility and intensity of remanent magnetization. The basalt samples dredged by Woods Hole investigators from the north wall of Puerto Rico Trench and also from near the top of

Barracuda fault scarp are consistent with this conclusion, as, of course, would be the basalt found by drilling at the Mohole Guadalupe site three years ago.

(Hess, 1964b: 173–174)

Thus there were three strongly positive and one negative response in 1964 to V-M. But, there were no positive responses from Mason or Girdler. Perhaps they had no new data. Nonetheless, if Girdler had been held back by Peter from invoking field reversals, I would think that he might have written a note suggesting that their results could also be explained in terms of field reversals rather than self-reversal. Similarly, if Mason believed in field reversals and seafloor spreading, he might have reexamined his explanation of the marine magnetic anomalies in the northeastern Pacific in terms of V-M.

In closing this section on early responses to V-M, I want to consider why Irving did not mention it in his 1964 monograph *Paleomagnetism* (II, §8.13). As he explains, he could have done so in late 1963 when working on the galley-proofs. This surprised me because I would have thought that Irving should have favored their hypothesis. He strongly supported continental drift, and believed in field reversals. Irving retrospectively discussed why he decided not to add a section.

I considered adding a new section devoted to V & M in the galleys, which I could easily have done, but decided against it. The V & M paper is really spare, on the rock magnetism side. The thing I found difficult was that they proposed flooring the oceans with a rock whose properties differed from basalts I knew about; magnetically a very different rock. The Koenigsberger ratio had to be generally 5 or more whereas many of us had spent a lot of time showing that this ratio was generally around one or less for continental basalts. Many sequences had made extensive studies of the K ratio in Tertiary basalts because it was a rough guide to magnetic stability. I had built and calibrated the magnetometer I used myself. I knew it worked. At the time I just didn't believe V & M likely because they were proposing that a large part of Earth was covered with a rock whose properties were very different from those we had spent a lot of time and effort measuring – Australian continental tholeiitic basalts.

Drum or Fred may have known more than they were letting on in their paper, and I suspect they did. Drum had dredge samples from his expeditions with Maurice Hill, and he probably measured them on our old magnetometer in the gravel pit at Madingley Rise; I have no recollection of having read their paper in 1963. Also, Jim Ade-Hall at Imperial College London had begun a global study of ocean bottom rocks especially basalts, and showed in his 1964 paper (which of course I knew nothing about in 1963) that their K ratios were systematically much higher, very different from continental rocks. His work was supported at the time by the formidable Professor Tilley (both Fred and Drum had done petrology in Tilley's Department, as I had but only for Part I) and according to his 1964 paper he had obtained early results as early as 1961 [Muir, Tilley, Scoon, 1964]. Perhaps Drum or Fred had got wind of these? I suspect they had. Jim must have given a seminar on his data in London. [Vine believes that he briefly talked with Jim at Imperial College.] Anyway Jim's 1964 paper makes it plain that V & M is reasonable from a rock magnetic property standpoint, and that my 1963 doubts were groundless. But of course I wasn't to know that in 1963. So I didn't include V & M in my book "Paleomagnetism." Silly me, or was I silly?

Of course, had I really thought about quenching of hot, vesicular, wet, basalt, I would have had less difficulty in following V & M, but I was at the time very skeptical about theorizing to predict magnetic properties having read so much patent nonsense on this matter. In fact I was a confirmed empiricist in general not believing all that nonsense Jeffreys put about.

(Irving, March 2001, note to author, amended August 8, 2011)

So the rub for Irving was whether or not oceanic basalts consistently had enough remanence to explain the magnitude of the anomalies over vast areas of ocean. Given his and Green's Australian experience with Cenozoic continental basalts (Green and Irving, 1958), Irving did not think the Konigsberg ratio of basalts was sufficiently high to produce the anomalies. Ade-Hall had shown otherwise for a range of oceanic basalts, but Irving did not know about his work when he was correcting the galleys to his book. Given these doubts he decided to say nothing about V-M.

3.9 Holmes on mantle convection, seafloor spreading, and Earth expansion

Holmes argued in his new and fully revised edition of *Principles of Physical Geology* (1965), which he finished in October 1964, that paleomagnetism's support of mobilism had reached difficulty-free status, although he did not say it quite like that (III, §2.16). He also continued to support mantle convection and showed sympathy toward Egyed's slow Earth expansion, which is in keeping with the encouragement and help he gave Egyed during the late 1950s in developing his ideas (II, §6.3).

Holmes welcomed the new evidence from marine geology indicative of mantle convection. He (1965: 998–1002) appealed to studies indicating high heat flow values on the East Pacific Rise and Mid-Atlantic Ridges, and low values in the Japanese Trench and Middle America Trench, and he argued that convection currents rose at oceanic ridges and sank at trenches. He (1965: 1022) appealed to mantle convection to account for the belt of intermediate and deep earthquakes associated with trenches, the buildup of andesitic lavas, and the concentration of volcanoes above the area of intermediate earthquakes. He (1965: 1009–1021) was particularly impressed with Wilson's work on oceanic islands: that island chains in the Pacific such as the Hawaiian Islands mark the movement of convection currents from the East Pacific Rise, and that the age of islands in the Atlantic tends to increase the more distant they are from the Mid-Atlantic Ridge. He favorably reviewed (1965: 1017) Wilson's explanation of the formation of the Hawaiian Islands and other such island chains with the upwelling of lava from a hotspot within the stable core of a "jet-stream" convective cell. He (1965: 1017) also favored Wilson's explanation of the formation of branch ridges such as the Walvis and Rio Grande ridges in the Atlantic by continuous upwelling of lava from a source on or near the Mid-Atlantic Ridge that remains fixed relative to the ridge.

Holmes favored seafloor spreading. He (1965: 1001) reproduced the 1928/1931 diagram he used to illustrate his earlier appeal to seafloor thinning and mantle convection as the mechanism of continental drift and added that he had proposed it "when it was thought that the oceanic crust was a thick continuation of the continental basaltic layer" (I, §5.4, Figure 5.1). He did not mention Hess's version of seafloor spreading or reference his 1962 benchmark paper. Perhaps he had not seen it. He twice (1965: 959, 993) referenced the paper in which Dietz (1961) presented his version of seafloor spreading but mentioned seafloor spreading only twice by name, noting (1965: 1021) that Dietz was "exploring the possible effects of subcrustal currents" with his "spreading of the seafloor" hypothesis, and relating (1965: 1015–1016) the fact that all volcanoes "that stand on the Mid-Atlantic Ridge and its flanks" are active . . . "to the suspected spreading of the ocean floors." Nonetheless, I believe that he definitely supported seafloor spreading. Here, for example, is what he wrote about the opening of the Atlantic.

If the continents on each side of the Atlantic have been moved apart by a convective mechanism . . . it is to be expected that the ocean floor has been largely constructed of material that was formerly part of the Ridge: material brought up from the mantle by hot, ascending currents and spread out laterally where the currents turned and flowed horizontally.

(Holmes, 1965: 1016)

Holmes' appeal to Wilson's application of seafloor spreading to the origin of oceanic islands and branch ridges also strongly suggests that he favored seafloor spreading.

Holmes (1965: 942) also discussed the magnetic anomalies in the northeastern Pacific, accepting, as others had proposed, that they showed horizontal movement of great slabs of oceanic crust relative to each other along the great fracture zones, and further claimed that the Pacific crust is gliding under the North American crust. He did not mention V-M. Perhaps he had not seen their paper. He did try to explain the origin of the magnetic anomalies, offering a "Scripps-like" explanation with no appeal to remanence or reversely magnetized seafloor.

The source of the magnetic anomalies is, as yet, only a matter of inference by analogy. The anomalies themselves are sharply bounded, showing that whatever is responsible for them comes up to the ocean floor or nearly so. They can also be continuously traced for long distances. If they had not been interrupted by faults, some of the strips of high positive anomaly would reach a length of 1500 km. These considerations strongly suggest that the anomalies are caused by the presence of dyke-like intrusions composed of iron-rich basic rocks. If so, the great swarms of such dykes that are indicated by the magnetic pattern must have added enormously to the area occupied by this part of the Pacific crust before the suspected intrusions were emplaced.

(Holmes, 1965: 942)

Holmes (1965: 1212) knew about reversals of the geomagnetic field, and claimed that there was good evidence in their favor. So I suspect that he did not know about V-M.

After presenting the geological and geophysical evidence favoring mantle convection, Holmes considered whether or not mantle convection is even possible. He began by briefly recalling debates over his own appeal to mantle convection. Although he mentioned neither Jeffreys nor himself by name, he clearly had his longtime adversary and friend in mind.

In 1912 and for years afterwards, while Wegener was looking for natural processes that might be equal to the task of displacing continents he overlooked the most effective one – sub-crustal currents ... However, it is only fair to remember that at the time convection in the mantle was widely supposed to have ceased at an early stage in the earth's history. Moreover, it was argued by geophysicists that there was no more possibility of continents overcoming the resistance of the ocean floor than there was of icebergs ploughing through layers of boulder clay thicker and stronger than themselves. Convection in the "solid" earth seemed to be as impossible as continental drift. And even if sub-crustal convection were to be admitted as a physical possibility and not merely regarded as an *ad hoc* hypothesis invented to bolster up the continental drift hypothesis, it could not explain displacements of the continents unless the ocean-floor obstruction could be overcome, e.g., by transformation of basaltic rock into eclogite, which was heavy enough to sink out of the way by joining the down-turning currents.

(Holmes, 1965: 1030)

But Holmes had more than reminiscing in mind. He wanted to draw a contrast between then and now over the evidence for mobilism and the eclogite-basalt transformation, which he had supposed in his own account of seafloor thinning, and Dietz had proposed in his version of seafloor spreading. Both, once questionable, are now, Holmes claimed, beyond reasonable doubt.

Today the position is very different. A wealth of geological and experimental evidence has demonstrated the reality of the basalt-eclogite transformation. Continental drift has ceased to be hypothesis, since it is happening before our eyes. The recognition of continuing movements along great transcurrent or wrench faults such as those described on pp. 226–30, leaves no doubt that crustal displacements already amounting to hundreds of miles are still actively in progress.

(Holmes, 1965: 1030)

Holmes also claimed that the recent estimates of the viscosity of the mantle, even given Jeffreys' standards, allowed for mantle convection.

The sub-crustal currents that are interpreted as the outer horizontal parts of convection cells have speeds that are all the same order, however they are estimated. In 1928 Harold Jeffreys showed that the viscosity of the mantle would have to be 10^{26} or 10^{27} (c.g.s. units) to prevent convectional movements. From rates of isostatic recovery and certain astronomical indications it is known that the actual viscosity is of the order 10^{23}. Convection is therefore physically possible, given slight differences of density such as might be due to differences of temperature in the same material, or differences of material at the same temperature. Since 1928 various theoretical studies based on a wide range of geophysical data have all indicated that speeds of a few centimeters a year would suffice to exert an adequate drag on the undersurface of the crust, and that such speeds are easily reached in all the earth-models investigated.

(Holmes, 1965: 1031)

So he believed that continental drift is a reality and mantle convection is physically possible.

Holmes then looked at different models of mantle convection. Like Wilson, he was intrigued by Hide's jet-stream model. Hide, who had worked on Runcorn's mine experiments at the University of Manchester, and went to Cambridge in 1950 to work on his Ph.D. with Runcorn, designed an elegant set of experiments to examine convective flow in the outer core, their original motivation being Elsasser's and Bullard's work on the self-exciting dynamo (II, §1.4). Hide (2006: 261) recalled the encouragement he received from Blackett and Runcorn to study at Cambridge, and how he and Leslie Flavill, who later helped Creer build Cambridge's first astatic magnetometer (II, §2.12), pieced together the needed laboratory apparatus.

In 1948, as an impecunious undergraduate studying physics at the University of Manchester needing part-time paid employment, I joined the "mine experiment" team as an assistant. The experience of working with Runcorn and his team stimulated my interest in geomagnetism and introduced me to the literature of the subject. Encouraged by Blackett and Runcorn, on graduating in 1950 I enrolled as a PhD student in the small Department of Geodesy and Geophysics at the University of Cambridge, where research in geodesy and seismology was already well established and new (and highly fruitful) initiatives were being taken in other areas – in marine geophysics by M. N. Hill and in palaeomagnetism by J. Hospers and Runcorn (who had moved from Manchester to Cambridge).

With some experience in experimental physics (but none in fluid dynamics), on reaching Cambridge I started some laboratory experiments on thermal convection in a cylindrical annulus of liquid (water) spinning about a vertical axis and subjected to an impressed axisymmetric horizontal temperature gradient. The necessary apparatus was quickly designed and constructed using equipment and other resources available in the department, including a war-surplus synchronous electric motor, a steel turntable used previously for grinding rocks, a supply of brass and glass tubing up to about 10 cm. in diameter and a recording camera incorporating a set of galvanometers which was no longer needed for field work in seismology. The resources also included, crucially, the facilities of a small workshop where research students could design and construct apparatus under the guidance of an experienced technician, L. Flavill.

(Hide, 2006: 261)

Hide's work, however, turned out to be of much greater importance to the behavior of a fluid of much less viscosity than the outer core or mantle, namely, the atmosphere, and it was none other than Jeffreys who told Hide that this might be so. Hide recalled what happened,

His [Jeffreys'] brilliant reputation as a writer was never quite matched by his lecturing and conversational skills, but as students we were advised that "his grunts and murmurs should always be taken seriously, because they were likely to contain pearls of wisdom." It was a remark consisting of no more than four words uttered by Jeffreys in 1951 that prompted my own interest and subsequent work in dynamical meteorology, a subject in which Jeffreys had made seminal contributions a quarter of a century earlier. The remark was made as he passed through the large hut where several graduate students in the Department of Geodesy and Geophysics at Cambridge University were engaged in various unrelated laboratory studies. When I showed him some of the flow patterns produced in an apparatus I had designed for investigating thermal convection in a rotating fluid, he muttered "looks like the atmosphere" and wandered off into the field outside the building, leaving me to ponder the implications of what he had said.

(Hide, 1997: 295)

Hide obtained his Ph.D. in 1953 when he left Cambridge University, completed his compulsory National Service, and spent a year working with Chandrasekhar at Yerkes Observatory before taking a position at MIT. His work showed the importance of undertaking laboratory studies to understanding non-linear fluid dynamics.

Holmes wrote Hide in April 1963 requesting permission to reproduce a photograph of "jet stream" convection from Hide's paper "Fluid motion in the Earth's core and some experiments on thermal convection in a rotating liquid." Hide had written the paper for a 1953 symposium on the use of models in geophysical fluid dynamics (Hide, 1953b). Holmes particularly wanted to show "the reader that there are many possibilities" of mantle convection "besides those conventionally drawn by myself (since 1929), Vening Meinesz, Runcorn and others."

Some years ago I bought a copy of "Fluid Models in Geophysics," hoping it would provide some clues to the behaviour of "our puzzling earth." I was delighted to find in it your contribution of "Fluid Motion in the Earth's core ..." which at once cured me of the naïve notion of convection cells in the mantle (of course) as necessarily being like smoke rings.

Since retiring from the Chair of Geology at Edinburgh I have been re-writing my "Principles of Physical Geology" and am now busy on the last few chapters, in which I must revise my treatment of convection currents. I should very much like to illustrate the new edition with one of your admirable pictures of the patterns produced by rotation. Probably No. 6 (Fig 1) on page 109 would be most effective in making clear to the reader that there are many possibilities besides those conventionally drawn by myself (since 1929), Vening Meinesz, Runcorn and others.

(Holmes, April 25, 1963 letter to Hide)

Holmes, suspecting that Hide may have written something new on convection, added that if Hide had something better to send, he would be "most happy to leave the choice" to him.

Hide did have something better, and proposed that Holmes use photographs from a more recent paper (Hide, 1958). But he also pointed out to Holmes that the convective patterns produced in his experiment would not be produced in the mantle, unless the mantle's viscosity were very much smaller than the representative value of 10^{21} cm^2 sec^{-1}. Although Hide's experiment was designed to test the effect of Coriolis force on convection in the outer core, whose viscosity is considerably less than that of the mantle, his apparatus produced, as Jeffreys had observed, a pattern like that in the "atmosphere." Moreover, Hide had shown that the rotational effect of Coriolis force on mantle convection is insignificant.

You comment on the need to make clear that possibilities of convection patterns other than those conventionally drawn by you, and by the more recent workers who have now revived general interest in the whole subject of convection in the mantle. The flow patterns of my experiments are characteristic of a highly rotating fluid, namely one for which $\Omega L^2/v$ is very much larger than unity, where Ω is the angular velocity of rotation, L is a characteristic linear dimension of the system, and v is the kinematic viscosity (actual viscosity divided by density). For the Earth's mantle, taking L equal to its depth, $\Omega L^2/v$ is equal to $6 \times 10^{12}/v$. If we accept the usually quoted value for v, namely $10^{21} \text{cm}^2 \text{ sec}^{-1}$, then $\Omega L^2/v = 6 \times 10^{-9}$, so that rotation would be utterly negligible. There is, of course, the possibility that $10^{21} \text{cm}^2 \text{ sec}^{-1}$ is not a representative value for v, but very much smaller values would be required for rotational effects to be significant.

(Hide, May 12, 1963 letter to Holmes)

Hide also raised a difficulty with conventional models of mantle convection such as those proposed by Runcorn and Vening Meinesz (RS2).

One strong objection I have always had to the conventional model, especially as elaborated in a quantitative way by Runcorn, using Chandrasekhar's theoretical calculations, is that this model neglects the effects of strong gradients of v and also of mean density in the mantle. In the absence of such effects the horizontal scale of the flow depends essentially on the depth of the mantle, and it is the gradual change of this depth over geological time that gives rise to the essential features of Runcorn's mechanism. Theoretical difficulties are such that no-one has yet extended Chandrasekhar's work to the case where v varies strongly with depth. Moreover, there are general reasons for expecting the effects of such a variation to be important. In particular, it might reduce the horizontal dimensions of the convection quite significantly (although this is not certain).

He expressed his hope for "progress in the near future," and ended his letter "on a more personal note."

I suppose that until theoretical refinements of the kind required to render the model more realistic have been made, one has to proceed using the conventional, over idealized model. Many good theoretical fluid dynamicists are now showing an interest in "geophysical fluid dynamics" and we can be hopeful of further progress in the near future.

If I might end on a more personal note, I have always treasured my copy of "Principles of Physical Geology" and am delighted, therefore, at the prospect of a new edition.

(Hide, May 12, 1963 letter to Holmes)

Holmes responded on June 3. He told Hide, "I don't know how I overlooked [your 1959 paper and] its importance ... since I [have been] actively interested in your work ..." Clearly disappointed to learn that Hide's experiment did not apply to mantle, Holmes pressed Hide on the point.

... I have now read it [your paper] with very great interest, though the mathematics is rather beyond my limited range. However, I was impressed with your remark that your experiments are characteristic of a highly rotating fluid, whereas in the mantle the effect of rotation would be utterly negligible. Am I wrong in supposing that if your experiments are regarded as dynamically scaled-down models of the mantle in, say, equatorial section, the viscosity in the model would have to be reduced to a value not much above that of water. If 2900 km is represented by 2.9 cm (ratio 10^8) and 1 day by 1 second (ratio about 10^5), then the viscosity must come down by at least 10^{13}. Time also comes in in another way, I believe – the effect of geological time being to make the mantle act in accordance with the principles of fluid mechanics. I'm not sure how to deal with this without learning more rheology and hydro-dynamics than I have ever known!

(Holmes, June 3, 1963 letter to Hide; my bracketed addition)

Holmes requested use of several photographs, including a series of three "to illustrate the effect of reducing the radius of the core." Why? Holmes was also supporting Egyed's slow Earth expansion. Egyed's idea of phase changes would provide the needed energy "to keep convection going effectively."

I am following Egyed in referring the source of energy (required to keep convention going effectively) to phase changes at various levels in the mantle. This means the core decreases in radius during geological time, instead of increasing, as Urey and Runcorn suppose. All this is highly speculative, of course, but the very possibility should prove stimulating, and a few pictures of actual experiments will be worth pages of description: although a few words of warning that the earth's real behavior depends on the interaction of a great many other factors will be necessary.

(Holmes, June 3, 1963 letter to Hide)

Hide said nothing about Egyed's slow Earth expansion. However, he again explained to Holmes that "the effects of rotation on mantle convection will be negligible."

As regards the question of how seriously rotation affects mantle convection, the straightfor-ward way to work this out is to compare the magnitude of Coriolis force/ unit volume, $2\rho\Omega \times u$ with viscous forces, $\eta\nabla^2 u$. Since $\rho \sim 4\,\mathrm{gm\,cm^{-3}}$, $2\Omega = 10^{-4}$ rad/sec, $\nabla^2 \cong$ (1/typical length scale)$^2 \sim 10^{-16}\,\mathrm{cm^{-1}}$, irrespective of the magnitude of u, the flow velocity, the ratio (R, say) of $2\rho\Omega \times u$ to $\eta\nabla^2 u$ is $4 \times 10^{12}/(\eta)$ for the mantle, where (η/ρ) is the kinematical viscosity, dimensions $\mathrm{cm^2\,sec^{-1}}$. Thus unless ρ is much less than $4 \times 10^{12}\,\mathrm{gm\,cm^{-1}\,sec^{-1}}$, effects of rotation on mantle convection will be negligible. In contrast, typical values of R in my experiments range from 10^2 upwards, and then, as in the case of the Earth's atmosphere, rotational effects dominate and we have so called "geostrophic motion."

(Hide, July 1, 1963 letter to Holmes)

Holmes definitely got the point; he agreed that Hide had shown that the effects of rotation on mantle convection will be negligible. However, he had decided that he was not going to tell his readers. He maintained with "armchair" assurance that if the mantle is partly composed of eclogite, and the eclogite melts and becomes basalt, the basaltic magma would have "a very low viscosity even at great depth." With a good dose of self-irony, Holmes told Hide what he planned to do.

Your demonstration that the effects of rotation on mantle convection will be negligible seems to be conclusive, but I do not propose to say so, because it may well be that parts of the mantle composed of eclogite become fused to make basaltic magma which may have a very low viscosity even at great depth. At times in the earth's history when the amounts of such magma have been considerable (e.g. to account for the great flood basalts, such as the Deccan and the Brito-Arctic province [i.e. the North Atlantic Tertiary Volcanic Province]) they may have played a leading part in controlling the convective pattern. *However, this is pure speculation at present and in my book I am simply using your spectacular results to show that it is impossible to sit in an armchair and think out what would happen in a rotating sphere!*

(*Holmes, July 20, 1963 letter to Hide; Holmes' emphasis; my bracketed addition*)

Holmes was true to his word. He highlighted Hide's experiment on thermal convection in a rotating liquid, reproduced four of Hide's photographs, one showing "jet stream" convection, and three showing changing flow patterns which "may simulate those produced in the earth's mantle if the radius of the core at successive stages in the earth's history has been diminishing" (Holmes, 1965: 1038). As for whether or not Hide's experiments captured what occurs in the mantle, Holmes left it up in the air.

It is not possible to say to what extent the top-surface patterns produced in Hide's experiments correspond to those developed in the mantle: say, for example, through an equatorial section of the earth. But at least they serve to suggest possibilities differing widely from the simple cellular patterns of non-rotating layers

(*Holmes, 1965: 1038–1039*)

The idea of "jet stream" mantle convection was too much for Holmes to resist. He agreed with Wilson that it provided just the sort of convection needed to explain the origin of island chains of the Pacific (Holmes, 1965: 1039).

One obvious complication arises. If Africa moved away from the Mid-Atlantic Ridge along the route indicated by the Walvis Ridge, it cannot at the same time have moved away from the Mid-Indian-Ocean Ridge ... Three ways out of the difficulty suggest themselves: expansion of the earth; or outward migration of the convection cell responsible for the growth of the Mid-Atlantic Ridge and its continuation into the Indian Ocean; or rejection of the hypothesis tentatively adopted for want of a better. The first of these is unacceptable because of the abnormal amount of global expansion that would be necessary within a geologically short period of time. The second may be physically possible, being analogous to the increase in diameter of a smoke ring as it drifts away from its source, though it poses problems of relative motion in three dimensions that are at present intractable. The third is a hazard to which all hypotheses are subject; it must not be forgotten.

(*Holmes, 1965: 1018*)

Holmes had already come to view continental drift as a reality given its paleomagnetic support. He also had come to view seafloor spreading, or at least the idea that the ocean floor is made up of material that spreads out horizontally from ridge axes where it is created, as well established, but realized that its purported cause, mantle convection, which may now be "tentatively adopted for want of a better" may turn out to be incorrect. As far as Holmes was concerned, finding the mechanism for continental drift and seafloor spreading was not a difficulty for either view but simply an unsolved problem.

Holmes, as he indicated in his June 3, 1963 letter to Hide, had adopted Egyed's hypothesis in referring to the source of energy (required to keep convection going effectively) to phase changes at various levels in the mantle. After introducing several hypotheses of Earth expansion, including those of Egyed, Carey, and Heezen, Holmes emphasized that expansion and convection are compatible, even complementary. "Convection does not exclude global expansion. Global expansion does not exclude convection. And the combination is stronger than either separately" (Holmes, 1965: 967). Perhaps Holmes felt it necessary to accent the complementarity of the two hypotheses because, although he had encouraged Egyed and expressed sympathy with his Earth expansion when Egyed had asked for his help in 1957, he still rejected Earth expansion in favor of mantle convection (II, §6.3). This time, however, he adopted Egyed's hypothesis, in part, because it provides a source of the energy needed to produce mantle convection. According to Egyed, energy is produced by continuous transformation of the core's unstable, ultra-high-pressure-phase matter to a stable lower-density phase, with consequent shrinkage of the core and expansion of Earth.

Holmes (1965: 967–970) also adopted Egyed's determination of the rate of expansion based on Strahow's and the Termiers' paleogeographical maps, even though he had objected to it when previously corresponding with Egyed. They claimed that the continental areas covered by water had decreased over time. Egyed, assuming a 4 percent increase in the hydrosphere's volume since the Early Cambrian, calculated an expansion rate of about 0.5 mm/year since then. Holmes, again following Egyed, estimated the rate of expansion to be about 0.58 mm/year based on the assumption that the present area of the sialic continents equaled the surface area of the primordial Earth. However, Holmes, noted that this is

a maximum estimate, because if there had been any ocean basins 4500 m.y. ago, r [former radius of Earth] would be correspondingly greater than the value just calculated, and the annual increase to R [current radius] would be less.

(Holmes, 1965, 970–971; my bracketed addition)

Egyed also calculated the rate of expansion given the general belief that the length of the day has been increasing. Egyed, ignoring the appeal to tidal friction to explain the increasing length of the day, estimated a radial increase of 0.38 mm/year. In 1957, Holmes did not agree with Egyed's dismissal of tidal friction as an explanation of the day's increase in duration (II, §6.3). Holmes, however, had changed his mind.

Slowing down of the earth's rotation has generally been ascribed to tidal friction, especially in shallow seas such as the North Sea, where the tides act like a brake on the rotating earth. However, at best, this accounts for only part of the slowing down. Moreover, the braking effect is more or less compensated by the tidal effect exerted on the atmosphere by the Sun. It was first noticed by Kelvin that the atmosphere tide caused by solar attraction speeds up the earth's rotation. Another explanation has therefore to be found for the fact that, nevertheless, the earth's rotation is slowing down and the day is growing longer. Expansion of the earth at a rate which is consistent with other evidence appears to meet the requirements.

(Holmes, 1965: 972)

Holmes developed a new argument in support of slow Earth expansion. It rested on the work by John W. Wells of Cornell University on rugose corals. They provided independent support for the radiometrically derived Phanerozoic timescale and for the estimate of the deceleration of Earth's rate of rotation due to tidal forces. Wells, particularly pleased that his study established that paleontology helped determine absolute ages, gleefully discussed how nice it would be if paleontology could make such a contribution.

Absolute age determinations of points on the geological time-scale, based on radioactivity, are generally accepted as the best approximations now known, even though they rest on a series of assumptions, any one of which may be upset at any time. At present there is no means of confirming or denying the accuracy of these determinations by independent methods. Can palaeontology provide anything by way of verifying the pronouncements now emanating, at very considerable expense, from the black boxes? Can palaeontology give any support to the shaky chronometric creation of the geophysicists and astronomers?

Except for providing a relative geochronology; palaeontology has thus far offered no data on absolute geochronometry. Several geochronologists have noted that there is no fossil evidence concerning variations in the length of the day[1], the rotation of the Earth, or diurnal variations, in geological past – variations that are linked to the passage of time ... How nice it would be if instead of paying a large sum for an isotopic analysis we could examine a fossil and estimate directly, with luck, not only its relative but also its absolute age – every palaeontologist a geochronometrist; every fossil a geochronometer!

(Wells, 1963: 948; his reference 1 is to Munk and MacDonald, 1960)

Wells then grafted the "simple relation between the geological time-scale and the number of days per year" and explained why it was important to find an independent "means of determining the number of days per year" for geological periods.

The two chief approaches to geochronometry are based on radioactive isotopes and on astronomical data. The most recent estimates[2] of geological time based on the rates of radioactive decay give the Cenozoic a length of 65 million years, the Mesozoic 165 million years, and Palaeozoic 370 million years, and so on ... To accept these figures is an act of faith that few would have the temerity to refuse to make. The other approach, radically different, involves the astronomical record. Astronomers seem to be generally agreed that while the period of the Earth's revolution around the Sun has been constant, its period of rotation on its polar axis, at present 24 h, has not been constant throughout Earth's history, and that there has been a deceleration attributable to the dissipation of rotational energy by tidal forces on the surface and in the interior, a slow-down of about 2 sec per 1 00 000 years according to the most recent estimates. It thus appears that the length of the day has been increasing throughout geological time and that the number of days in the years has been decreasing. At the beginning of the Cambrian the length of the day would have been 21 h (ref. 1). After the first, a second act of faith is easy, and we can develop a simpler relation between the geological time-scale and the number of days per year (Fig. 1 [Figure 3.18]). Accepting this relation, in the absence of evidence to the contrary, it now follows that if we can find some means of determining the number of days per year for the different geological periods, we have a ligation between the results of geophysical and astronomical deductions.

(Wells, 1963: 948–949; his reference is to Kulp, 1961; my bracketed addition)

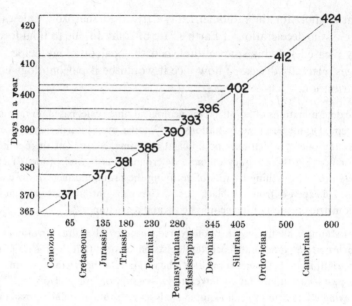

Figure 3.18 Wells' Figure 1 (1963: 949): Relation between days in each year and geological time.

Wells, following a suggestion of T. Y. H. Ma, the paleontologist who argued in favor of mobilism because of his work on fossil corals (II, §3.6), attempted to determine the number of days per year from studying fossil corals. He found by counting the growth rings of Middle Devonian and Pennsylvanian-aged fossil specimens of rugose corals that the Middle Devonian year had approximately 400 days while the Upper Carboniferous year had between 385 and 390 days. He (1963: 950) also found that the number of ridges on the epitheca of living coral (*Manicina areolata*) "hovers about 360 in a year's growth." He concluded that his results

> imply that the number of days a year has decreased with the passage of time since the Devonian, as postulated by astronomers, and hence that values of the isotropic dates of the geophysicists agree well with the astronomical estimates of the age of the Earth. It is not claimed that coral growth proves that either is right; but is suggested that paleontology may well be able to supply a third stabilizing, and much cheaper, clue to the problem of geochronometry ...
>
> *(Wells, 1963: 950)*

Paleontology, much less costly than radiometric dating, was still an important dating tool.

Holmes attributed much of this decrease in the rate of Earth's rotation based on Well's study to Earth expansion rather than tidal friction. The Devonian average gave an average radial increase of 0.66 mm/year; the Carboniferous results, an increase of 0.6 mm/year. Holmes acknowledged that the rates are maxima, because tidal friction also lowered Earth's rotational rate.

With the establishment of a maximum expansion of 0.6mm/year, Holmes rejected the rapid Earth expansion of Heezen and Carey. He also argued that earth expansion is auxiliary to mantle convection in moving continents relative to each other. Only a small proportion of newly created seafloor can be attributed to expansion.

This can only mean that circulation processes operating in the mantle, such as convection currents, have been of greater importance than global expansion in "engineering" the latest separation of the continents.

(Holmes, 1965: 975)

Indeed, Holmes settled on a radial expansion of only 0.25 mm/year, which would produce only a 50 km increase of Earth's radius over the past 200 million years and a surface increase of 8×10^6 km^2, less than one-twentieth the area of the Atlantic, Arctic, and Indian oceans. In Holmes' scheme, the major role of Earth expansion was to provide the energy for mantle convection, not to move the continents.

Holmes closed his discussion of Earth expansion by speculating about its cause.

It was stated above that energy would be released and the earth would expand "*if there should be a relief of pressure.*" We have now to consider how such a relief of pressure could be brought about. There seems to be only one possibility in the light of present knowledge: that the terrestrial force of gravity, which can be represented by g, has systematically decreased as the earth has grown older. The universal constant of gravitation, G, may have decreased with time, as inferred by P. A. M. Dirac in 1938; or matter may have steadily vanished from every part of the earth (and from all other material throughout the universe), as proposed by R. O. Kapp in 1960.

(Holmes, 1965: 983)

Holmes did not pretend that either Dirac's or Kapp's ideas were well founded; neither, he (1965: 983) admitted, "yet deserves the honoured title of 'theory'."

3.10 Rapid Earth expansion under attack, 1963–1964

Rapid Earth expansion was attacked on several fronts. Westoll and Tozer, digressing from the main focus of their presentations at the Royal Society's symposium on continental drift, both raised difficulties with Earth expansion. They were not, however, the only ones to object to rapid Earth expansion; Martin A. Ward and Runcorn launched full-blown attacks. Ward (1963) generalized Egyed's paleomagnetic test of Earth expansion and argued that paleomagnetic results eliminated rapid expansion but still allowed for Egyed's slow expansion (II, §6.4). Ward started working part-time at the beginning of 1959 for Irving at the ANU. He stayed for three years, leaving at the end of 1961 or the beginning of 1962 after he completed his undergraduate mathematics degree at Canberra University College. After graduation, he did a Ph.D. on group theory and became a lecturer and professor at the College, which later was incorporated into ANU.

Ward was originally hired to do routine measurements. Irving soon recognized Ward's abilities, and Jaeger arranged for him to take a programming course so he could use ANU's newly acquired IBM 620. He wrote programs for Fisher's statistics and pole calculations. Irving had a task in mind for Ward to employ his newly acquired computing skills.

It was plain that it would be a waste of his talents for me to keep Martin doing routine measurements all the time. In any case he had soon produced so many that I was having difficulty keeping up my thinking about them. He did many calculations for my book. I discussed with Jaeger putting him on some project of his own. Jaeger agreed. We could monitor progress and then he would consider taking Martin on as a student himself in mathematical geophysics. Earth expansion was being discussed and Egyed had given a 2-dimensional paleomagnetic method of calculating the radius of the ancient earth. I asked Martin to think about putting this in its proper spherical environment using Fisher's stats and paleomagnetic poles – into three dimensions that is, and so employing data globally. Martin came up with the idea of monitoring pole dispersion as one changed earth's radius keeping continent-size fixed. This became known as Ward's method. Martin found minimum dispersion for radius equal to the present radius for several periods. That is, Earth's radius had within statistical error, remained unchanged since the Late Paleozoic, in strong disagreement with Carey's and Heezen's proposal. Recalculation today would yield the same result.

(Irving, July 24, 2009 email to author)

Ward developed a procedure that, like Egyed's, required comparison from stations on the same undeformed landmass, but, unlike his, did not require that they all be located close to the same paleomeridian. Ward's "minimum dispersion method" was as follows: first, poles are calculated for each station for a given period on an Earth of present radius and their dispersion estimated by Fisher statistics (II, §1.13); then the radius is decreased and increased recalculating the pole positions and their dispersion at each incremental change of radius. Ward spoke in terms of Fisher's precision or concentration parameter and its relation to dispersion.

Thus for a chosen hypothetical Earth's radius ρ, we may calculate the set of paleomagnetic pole positions from the rock-data, and then calculate the precision k of this set; hence k appears as a function of ρ, and the most probable ancient earth's radius P is that ρ which k is maximum. P is then, in fact, the ancient Earth's radius of maximum likelihood on the assumption of a Fisher distribution of paleomagnetic pole positions.

(Ward, 1963: 217)

The centrality of Fisher's statistics for dispersion on a sphere to Ward's method had the advantage of making it "tailor-made to the spherical environment of the data." It had other advantages.

This analysis has the advantages, firstly that all the rock-data can be treated together, reducing the amount of calculation necessary, secondly that the statistics used … are tailor-made to the spherical environment of the data, a fact which is reflected in the greater degree of confidence

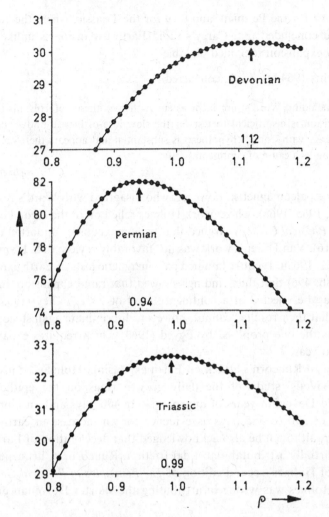

Figure 3.19 Ward's Figure 3 (1963: 221). His caption reads: "Pole precision (k) as a function of the hypothetical Earth's radius (ρ), where $\rho = 1$ for the present radius. The arrows indicate the point on each graph at which k is maximum."

in the results ... and thirdly that no previous weighting of the rock data is necessary, since this is automatically done by the statistics used.

(Ward, 1963: 218)

Like Bullard and Vine, Ward became skilled in the use of the new computer technology, doing many, long, tedious calculations on the newly installed IBM 1620 at ANU.

Ward applied his procedure to Devonian, Permian, and Triassic data from Europe and Siberia. He (1963: 217) also tried to work with Triassic and Carboniferous data from North America, but at the time there were too few "to yield a meaningful result." He found that the most probable values of Earth's radius are 1.12 for the

Devonian, 0.94 for the Permian, and 0.99 for the Triassic, where the current radius $\rho = 1$. Ward concluded that Carey's rapid Earth expansion is unlikely, although Egyed's slow expansion remained possible.

Irving, in his 1964 textbook, concurred.

The results [including Ward's] are inconsistent with hypotheses of Hilgenberg and Carey, but their precision is insufficient to test for the slow rate of increase advocated by Egyed. The inconsistency with Carey's hypothesis is substantial and appears therefore to invalidate Earth expansion as a cause of continental drift.

(Irving, 1964: 292)

There was one paleomagnetist, however, who disagreed with Ward's results. D. van Hilten (1963, 1965, 1968), whose work Holmes relied on in the second edition of his *Principles of Physical Geology*, argued that paleomagnetism supported rapid expansion (III, §2.16). Van Hilten's work was not favorably reviewed. S. I. van Andel and Hospers (1967, 1968a, 1968b) examined paleomagnetic tests of Earth expansion; they argued (1968b: 496) that their findings showed that rapid expansion, including van Hilten's, "may be rejected at a confidence level of 95%." They (1968b: 496) also maintained that their results "cannot, however, discriminate against slow expansion rates, such as the one proposed by Egyed (1963) which requires expansion of the order of 1mm/year."[9]

I now turn to Runcorn's attack on Earth expansion. Holmes, as just explained, appealed to Wells' study of the daily growth rings on the epithea of corals that indicated Devonian years of approximately 400 days and an Upper Carboniferous year of 390 to 385 days as evidence for an increase in earth's radius of 0.66 mm/year, although he also acknowledged that deceleration of Earth's angular velocity is partially attributable to tidal friction. Runcorn, in his generally favorable review of Holmes' second edition of *The Principles of Physical Geology*, flatly stated that Holmes was wrong in attributing the shorter Devonian day to Earth expansion.

The second edition is up to date in its subject matter. Holmes was remarkably appreciative of new developments and enthusiastically discusses Wells' suggestion that coral growth rings show that the day was two hours shorter in middle Devonian Times. He is wrong in supposing that this slowing down of the Earth's rotation, approximately at the same rate as astronomers find at the present day, could be due to Earth expansion. The effect of changes of moment of inertia, such as should be expected from the latter, is separable from the effect of lunar tidal friction and Wells' result is largely explained by the latter.

(Runcorn, 1968)

What made Runcorn so sure? Earth's rate of rotation is equal to its angular momentum/its moment of inertia. Its moment of inertia is a function of its mass times its radius cubed. Because its mass has basically remained constant, any change in its moment of inertia must be attributable to a change in its radius. So if the lengthening of the day is attributable to an increase in the moment of inertia, Earth is expanding. But there is another explanation for deceleration of Earth's angular velocity, the tidal interaction of the Moon on Earth. The torque produced in the opposite sense to Earth's rotation by the tidal lag acts as a brake on Earth's rate of rotation.

Holmes, of course, knew this, but, appealing to Wells' work on Devonian and Carboniferous rugose corals, attributed part of the lengthening day to tidal interaction. Runcorn, however, had new information that allowed him to attribute the decrease in Earth's angular velocity almost entirely to tidal interaction. Interestingly, the new information was based on a new study of the same corals that Wells had studied. Colin T. Scrutton of Oxford University Museum confirmed Wells' finding: the daily growth ridges were there. But he (Scrutton, 1964) noticed another periodicity: the diurnal growth ridges were grouped together in sets of 30.59. He interpreted the newly discovered groups as lunar months. Agreeing with Wells that there were approximately 400 days to the Devonian year, he reasoned that there were thirteen months in the Devonian year. He also noted that work on contemporary corals had shown an analogous periodicity of twenty-eight days, the length of the present-day lunar month (Scrutton, 1964: 555). Moreover, he also coordinated a monthly growth cycle with skeleton building of the corals.

Scrutton gave Runcorn all he needed. He argued that Scrutton's work

is very important geophysically because it opens up the possibility of separating changes in the Earth's rotation due to changes in its moment of inertia from those due to lunar tidal friction, because the latter changes the length of the month as well as the day.

(Runcorn, 1964: 824)

Given estimates of the lunar month, Runcorn (1964: 824) calculated the ratio of the present value of the orbital angular momentum of the Moon to that at any time in the past. Given the above ratio, and that lunar "tidal friction results in angular momentum being transferred from the Earth's rotation about its axis to the Moon's orbit," he calculated that "the average lunar tidal deceleration torque on the Earth over 370 m.y. is 3.9×10^{23} dyne cm (since the Devonian)" which "agrees with that given for measurements of the longitudes of the Sun and Moon over the past 2–3 centuries (3.9×10^{23} dyne cm)." Using the ratio of the Moon's orbital angular momentum to the Earth's spin angular momentum, Runcorn calculated that Earth's moment of inertia during the Devonian was 99.9 ± 0.003 to 99.4 ± 0.003 percent of its current value. By contrast, Carey's and Heezen's rapid Earth expansion requires that Earth's Devonian moment of inertia be about half its current value. Runcorn, noting that his calculations agreed with those of M. A. Ward (RS1), concluded:

These results dispose convincingly of Carey's and Heezen's extreme hypotheses. This is in agreement with the palaeomagnetic evidence, which has been used by Ward to determine the radius of the Earth in the Permian by relating within one continent the angles of magnetic inclination and the distances between the sites at which they were determined.

(Runcorn, 1964: 824)

Egyed's slow expansion required that Earth's moment of inertia be 94 to 89 percent of its current value and these results did not favor it. However, Runcorn acknowledged that they did not exclude Egyed's slow expansion, unless, as Dirac had suggested, *G* does vary inversely over time. Given that Holmes rejected rapid Earth expansion and insisted that *G* must so vary for Egyed's slow expansion to work, I suspect that he would have not objected to Runcorn's conclusion.

Notes

1 Menard (1986: 192–197) discussed the development of Wilson's ideas during this period. I have benefited greatly from Menard's account.
2 Wilson also received funding from the NRC and the GSC (Wilson, 1963c: 870).
3 Given continued controversy over whether hot spots are fixed relative to Earth's mantle or simply move more slowly than the lithosphere above, it is of interest that Wilson did not insist that they are fixed relative to the mantle.
4 Menard (1986: 196) suggested that Wilson's proposal that ridges are connected by transcurrent faults may have helped him later propose the idea of transform faults.
5 Everett and Smith (2008) wrote their own retrospective account of the genesis and development of their computer fit of the circum-Atlantic continents. Everett and Smith also answered questions that I asked them about their work on the computer fit. My account of what transpired could not have been written without Everett and Smith's paper and their subsequent answers. I thank them for their answers. Given my retrospective account, I propose that what has been called the Bullard, Everett, Smith fit, be renamed the Everett, Bullard, Smith fit.
6 This account of Everett's and Bullard's contributions to the methodology differs from what Bullard himself retrospectively claimed. In wanting to give Everett and Smith the credit they deserved, Bullard (1975a: 21) said that he "only provided the methodology and wrote the paper." Bullard certainly did write the paper that appeared in the Royal Society's symposium. However, Everett provided the methodology: he came up with an equivalent to Euler's Point Theorem before Bullard told him about the theorem; he worked out the iteration method to obtain the optimum pole and did the programming. I agree that Bullard also told Everett about Euler's Point Theorem, which is probably the basis for Bullard's remark that he provided the methodology, and in that sense they independently provided the methodology.
7 Blackett was mistaken about the date of the meeting. The British Association for the Advancement of Science did have a session on continental drift at its meeting in Birmingham, but the meeting was in 1950, not 1948 (I, §5.7, §5.13).
8 These letters are in Hess's papers at the Firestone Library at Princeton University. I thank Firestone Library and George Hess for allowing me to reproduce them.
9 Since Ward's work, others have carried out estimates of Earth's paleoradius based on global paleomagnetic data from stable continental blocks. Their work has been reviewed most recently by McElhinny and McFadden (2000: 331–332). McElhinny and McFadden analyzed the data available to them and, using Ward's minimum dispersion method they "found that for the past 400 Myr the average paleoradius was 1.020 +/− 0,028 times the present radius" not different from the present radius and confirming Ward's result.

4

Further work on the Vine–Matthews hypothesis and development of the idea of transform faults, 1964–1965

4.1 Introduction

My two major aims of this chapter are to continue tracing the reception of V-M through summer 1965, and to introduce the idea of transform faults, the second key corollary of seafloor spreading. After outlining the difficulties facing V-M when originally proposed, I consider favorable work on the hypothesis by Vine, Matthews, and Cann at Cambridge. I then consider mild support of the hypothesis by two new groups of marine geologists from New Zealand and Canada, and mild opposition by George Peter who, with Ron Girdler, had already appealed to self-reversal to explain an anomaly in the Gulf of Aden (§2.4). I then turn to work in the Indian Ocean by Carl O. Bowin, who had obtained his Ph.D. at Princeton in 1960 and taken a position at Woods Hole Oceanographic Institution (WHOI), and Peter Vogt, who was at the Geophysical and Polar Research Center, Department of Geology at the University of Wisconsin. They were unwilling to support V-M. I next discuss the first of a series of three papers by Lamont researchers strongly opposed to V-M. Heirtzler and Le Pichon wrote the first paper, and argued that their work on the Mid-Atlantic Ridge raised difficulties for V-M.

Before resuming discussion of Lamont's resistance, I turn to new work by Vine, Wilson, and Hess. Wilson and Hess spent the first half of 1965 at the Department of Geodesy and Geophysics at Cambridge where they worked primarily with Vine. Hess tweaked his ideas on seafloor spreading. More importantly, Wilson developed the idea of transform faults, which was of tremendous importance to the advancement of seafloor spreading, and constitutes Wilson's greatest conceptual contribution to the Earth sciences. Wilson reasoned that if seafloor was both created at ridges and destroyed at trenches then there must be another type of fault, which he named a transform fault. Vertical motion is transformed to horizontal at ridge axes as new seafloor is created and moves horizontally away from ridge axes; horizontal motion is transformed to vertical as old seafloor descends back into the mantle at trenches. He claimed that what earth scientists such as Menard and Vacquier had viewed as transcurrent faults in the Northeast Pacific were actually transform faults.

As it turns out, Wilson was not the only person to envisage the idea of transform faults. The idea was proposed by at least three others. Vine actually thought of the idea before Wilson, and mentioned it to him. In addition, Alan Coode came up with the idea while a graduate student working on his Ph.D. under Runcorn at the University of Newcastle-upon-Tyne. Coode also told Wilson of the idea. I shall describe Vine and Coode's work, however, after discussing Wilson's because Vine did not publish his idea, and because Coode's paper, rejected by *Nature* and eventually published in *Canadian Journal of the Earth Sciences*, then a backwater and fledging journal, was completely ignored by Earth scientists.

4.2 Initial difficulties facing the Vine–Matthews hypothesis

When proposed, V-M depended on three questionable assumptions or "premises" as Vine later called them at a conference held in New York City at the Goddard Institute for Space Studies in November 1966, and as a consequence faced three corresponding major difficulties (Vine, 1968). The three assumptions were that from time to time the geomagnetic field reversed polarity, that the remanent magnetization of oceanic basalts greatly exceeded that induced by the present field, and that seafloor spreading occurred; the last was the most dubious. Although when V-M was proposed, field reversals were almost universally accepted among paleomagnetists as firmly established, most other Earth scientists, at least in North America, had merely heard of the possibility of both field reversals and self-reversals and generally doubted the existence of field reversals. Quite simply, most Earth scientists outside of paleomagnetism had no idea of the substantial work that had been done on reversals; they may not have heard about, let alone read, the relevant literature. In addition, when Vine and Matthews presented their hypothesis, little was known about the details of the reversal timescale even for the last several million years, details that were essential for correctly dating the anomalies. These details became known in early 1966 from work by Cox, Doell, and Dalrymple at the USGS and MacDougall, Tarling, and others at ANU on dating continental lava flows, and by Opdyke and co-workers at Lamont based on their study of deep-sea sediment cores (§6.3, §6.4). The assumption about the dominance of remanence of ocean floor basalts was actually settled by Ade-Hall (published 1964) just before Vine came up with V-M; as already related, Vine knew of Ade-Hall's results, but most, including those like Irving, who knew about continental basalts, did not know of his work on ocean floor basalts until after it was published (§3.8). Further work by Engel and Engel (1964), Muir, Tilley, and Scoon (1964), Engel, Engel, and Havens (1965), Engel, Fisher, and Engel (1965), Cann and Vine (1966), and Opdyke and Hekinian (1967) unquestionably established that oceanic basalts possess sufficient remanence to cause the steep gradients of marine magnetic anomalies. The third premise, seafloor spreading, was still viewed by most as "geopoetry," as Hess himself had called it.

Vine put it this way in his 1966 blockbuster paper in which he presented the new evidence in favor of the hypothesis.

Of the three basic assumptions of the Vine and Matthews hypothesis, field reversals (7) and the importance of remanence (13) have recently become more firmly established and widely held; thus in demonstrating the effectiveness of the idea one might provide virtual proof of the third assumption: ocean-floor spreading, and its various implications.

> *(Vine, 1966: 1406; (7) referenced Cox, Doell, and Dalrymple (1964)*
> *and Doell and Dalrymple (1966); (13) referenced Ade-Hall (1964))*

Although Wilson had extended seafloor spreading's problem-solving effectiveness, Menard and others had raised difficulties with his extensions (§3.7). The tie between mantle convection and seafloor spreading was not nearly as simple as Hess had originally envisioned because the complete encirclement of Antarctica and partial encirclement of Africa required migration of ridges. Some decoupling of mantle convection and ridge formation was needed. Moreover, most Earth scientists had no more than a passing acquaintance with the rapid developments in the paleomagnetic case for continental drift; they did not follow it and hence did not understand it, or they erroneously thought Cox and Doell's criticism (II, §8.4) or even Jeffreys' (II, §7.13; III, §2.10), and MacDonald's attacks (II, §7.11; III, §1.6, §2.8) had shown that it still faced serious difficulties; these Earth scientists generally opposed continental drift and seafloor spreading. The huge improvement in the problem-solving effectiveness of seafloor spreading that shortly was to occur did so because of the eventual success of V-M.

Vine also delineated three key empirical difficulties that the hypothesis faced when originally proposed. Beginning a section entitled "Difficulties," he described them as follows:

At the time this concept was proposed there was very little concrete evidence to support it, and in some ways it posed more problems than it solved. There were, for example, at least three rather awkward points that it did not explain:

1) Many workers felt and feel that the northeast Pacific anomalies do not parallel any existing or preexisting oceanic ridge (14).
2) Whereas one can visually correlate anomalies on widely spaced profiles in the northeastern Pacific, one cannot do this over ridge crests, except for the central anomaly. Vacquier (12) maintained, therefore, that there are no linear anomalies paralleling the central anomaly over the crests of ridges.
3) The idea did not, very obviously, explain the fact that the low-amplitude, short-wavelength anomalies observed on either side of the axis of a ridge give way to higher-amplitude, long-wavelength anomalies over the more distant flanks – an observation made by Vine and Matthews (6) and emphasized by Heirtzler and Le Pichon (15). With the increase in depth of the magnetic material as one moves from the ridge crest to the flanks, one would expect disappearance of shorter wavelengths but not an increase in amplitude.

> *(Vine, 1966: 1406; (14) refers to Peter and Stewart (1965); (12), Vacquier (1965); (6),*
> *Vine and Matthews (1963); (15), Heirtzler and Le Pichon (1965))*

All three difficulties were not removed until 1966 (§6.6). The first two were turned into exemplary confirmations of V-M by Vine (1966) and Pitman and Heirtzler (1966), and Vine (1966) removed the third difficulty.

4.3 Vine, Matthews, and Cann defend and further develop the Vine–Matthews hypothesis, June 1964 to May 1965

Vine and Matthews teamed up with J. R. Cann, a petrologist from the Department of Mineralogy and Petrology, University of Cambridge, and they co-authored two papers discussing Matthews' detailed 1962 (§2.12) survey over the Carlsberg Ridge. They submitted their paper in June 1964, and argued for V-M (Matthews, Vine, and Cann, 1965). Cann and Vine gave an extended account of the petrology of dredged rocks, and offered a slight revision of V-M, presenting them in London at the Royal Society's November 1964 meeting on the floor of the northwest Indian Ocean. Their written paper was received by the society in May 1965, but did not appear until April 1966 (Cann and Vine, 1966).

Both these papers included new paleomagnetic results from a new survey of the Carlsberg Ridge area by RRS *Discovery* (see Figures 2.11–2.13). This cruise took place between August 23 and December 4 and was part of the International Indian Ocean Expedition. Further extension of Matthews' original magnetic survey over the Carlsberg Ridge was not a priority. John Sclater, then a new graduate student in the Department of Geodesy and Geophysics at Cambridge who was working under Hill on heat flow, and Vine were cabin mates (Sclater, September 3, 1992 letter to author; see www.escholarship.org/uc/item/4xj8c69c for his answers to my questions). Matthews was also onboard. Sclater recalled that Vine almost did not get the ship time he thought was needed to extend the original Matthews survey. Sclater's remembrances also confirm Vine's feeling that Hill was not in favor of V-M, and show that Hill also opposed continental drift and Hess's seafloor spreading. Sclater also acknowledged that he did not at the time favor mobilism or V-M.

The expedition on which I participated was led by Maurice Hill and concentrated on two major seismological experiments. The objectives were two fold; one to determine the crustal thickness between the African mainland and the Seychelles involving Tim Francis, and the other to determine the crustal thinness of the Seychelles Plateau. [See Francis and Shor *et al.*, 1966.] There was little time for heat flow work and even less for the magnetic anomaly profiles and seamount surveys that Fred Vine wanted to run. I certainly felt on this cruise that Fred did not get enough ship time. I had not bothered to study his paper with Drummond Matthews on sea floor spreading that had been submitted to *Nature*, but felt he should at least get a chance to do some seamount surveys. My arguments were based on fairness, not science. I was not popular for challenging my seniors such as Hill and Laughton, but Fred got his ship time. Actually the captain of *RRS Discovery* who was a decorated war hero supported us and Maurice had a lot of respect for his opinion. In retrospect Hill, Laughton and Matthews treated me as an equal and though it was hard for a 23-year-old student to take their criticism, they respected those who stood up and argued with them. Also, they were sufficiently open minded that if they felt we had a good case, they would change their cruise plan.

In retrospect, the surprise on this whole cruise was how little effort was made to check out the paper by Drum and Fred and how little Drum and Fred fought to do this. If my memory is correct, I was the only one who argued for more time for magnetics and I think even Drum [Matthews] initially argued against me! Fred was disappointed that we all didn't take his paper more seriously, but was too much of a gentleman ever to let us know this in public. I only picked it up because I was rooming with him.

Maurice Hill, who was perhaps the major critic, did not believe it at all. He discussed Vine's thesis along with me one Saturday morning in 1965 just after he read it. He felt the whole theory from Hess' idea through Vine and Matthews was unsubstantiated speculation.

Maurice Hill's comments about Fred Vine's thesis had an effect upon me. They made me reconsider the whole hypothesis and certainly made it harder for me to accept the concept. At the time I was critical of all theoretical ideas. All I wanted to do was to obtain reliable heat flow observations from the ocean floor that I could write up for my thesis and to understand the variability of the values I had obtained. My horizons were very limited at this time.

(Sclater, September 3, 1992 letter to author)

I suspect that many, even in the UK where more were inclined toward mobilism than in North America because of its paleomagnetic support, shared Sclater's attitude toward V-M.

The new magnetic survey showed that the Carlsberg Ridge was offset about ten nautical miles along what Matthews, Vine, and Cann now unambiguously identified as a right-lateral transcurrent or shear fault (Matthews *et al.*, 1965: 679). They also noticed that the zebra pattern of magnetic anomalies "is expunged near the fault." This, they noted, was in keeping with what had been observed along the fracture zones in the northeast Pacific, and with what Matthews (1966a: 182) had noticed along small fracture zones in the Gulf of Aden and also along the Owen Fracture Zone, which he had just discovered offsetting the Carlsberg Ridge near Socotra (the large island off the tip of the Horn of Africa).

The width of this zone is surprising; apparently, the underlying blocks of magnetized rocks are not merely truncated by the fault, but the pattern itself is expunged near the fault. Magnetic surveys recently completed over large transcurrent faults which displace the axis of the mid-ocean ridge near Socotra and in the Gulf of Aden suggest the existence of similar wide zones devoid of magnetic relief near the faults (Matthews, in press). Zones of subdued magnetic anomalies can be seen in places along the lines of the well-known transcurrent faults off the coast of California ... but often these faults have elongate anomalies parallel to them (Mason and Raff, 1961 p. 1263).

(Matthews et al., *1965; the Matthews, in press, refers to Matthews (1966a);*
the Mason and Raff reference is the same as mine)

Cann undertook a petrological investigation of the dredged rocks. Those from the more northerly stations 5106 and 5123, both within ten nautical miles of the fracture zone, showed extensive brecciation and hydrothermal alteration, which, they argued, explained the expurgation of the linear pattern of magnetic anomalies (see Figure 2.12).

Some degree of brecciation is common to all the rocks from these two hauls, many of them are severely brecciated and have been considerably modified by the action of hydrothermal solutions. The occurrence of such definitely metamorphosed rocks on the ocean floor seems to be previously unrecorded. There can be little doubt that the metamorphism is associated with the transcurrent faulting which displaces the median valley and truncates the pattern of magnetic anomalies. Apparently unusually wide zones of brecciation, accompanied by hydro-thermal activity, may be developed where relatively minor faults transect the thick oceanic crust. If the permanently magnetized-rock fragments have been rotated relative to one another within the breccia, the result might be to reduce the gross effect of the remanent magnetization of the brecciated rocks in the fault zone.

(Matthews et al., 1965: 680)

Cann and Vine (1966) defended V-M. They explained the higher amplitude of the central anomaly by proposing, as Vine had done, that only the crust beneath it is uncontaminated by younger material of opposite polarity.

On the basis of this model, only the central "block" will be composed exclusively of young material which is magnetized normally (i.e., approximately parallel to the present direction of the axial dipole field) except for the minor probability of self-reversals. All other blocks will be contaminated with younger material, often of reverse polarity to that of the initial block, and hence lowering or modifying its resultant magnetic effect.

(Cann and Vine, 1966: 214–215)

Vine and Matthews (1963) had originally proposed that the layer responsible for the anomalies is 1.6 to 6 miles (2.6–9.7 km) thick; they now narrowed it to 2.2 miles (3.5 km). Either version would work, it just depended on whether the magnetization was entirely or primarily restricted to Layer 2 or included part of Layer 3 (Cann and Vine, 1966: 215).

They compared their hypothesis with the competition. They began at the beginning.

The basic tenet of the idea is the combination of ocean-floor spreading and periodic reversals, which provides a plausible and effective mechanism to produce a considerable magnetic contrast across an approximately vertical boundary within the oceanic crust, without implying any change laterally in the petrology of the crustal material. Such boundaries would appear to be essential if one is to simulate the steep magnetic gradients so often observed over the oceans, and notably at the centre of ridges. Having accepted this basic principle, there is no difficulty in explaining the anomalies, but only in deciding on the distribution of magnetization within layers 2 and 3 of the oceanic crust.

(Cann and Vine, 1966: 215)

They noted that Hess had described the hypothesis as "fruitful," while Talwani found it "improbable and startling" (1966: 215). They also noted that even though competing views can easily explain the central anomaly in and of itself, it is difficult to explain why the central anomaly "should be so different from the remaining oceanic crust in producing such a high magnetic contrast with it" unless reversed magnetization is involved (RS2). Talwani, Heezen, and Worzel's alternative was a case in point (§6.4).

It has been suggested that the central anomaly might be due to a magnetic body at depth, the extent of which is controlled by variations in the depth to the Curie point isotherm (Talwani, Heezen & Worzel 1961). It seems inconceivable that this, or any other effect at depth (for example, local depressions or elevations of the Curie point isotherm), could account for the steep magnetic gradients and short wavelengths of the observed anomalies. As mentioned previously such suggestions are unsubstantiated by detailed models or computations.

(Cann and Matthews, 1966: 215–216)

Having focused on the ability of V-M to explain the axial anomaly of ridges, they added (1966: 216) that it "might well account for the enigmatic, but possibly ubiquitous lineation of 'grain' of the magnetic anomalies ... in the eastern Pacific" (RS1). They also discussed the rates of seafloor spreading calculated from the anomalies as interpreted by V-M and the recent radiometrically determined reversal timescale of Cox, Doell, and Dalrymple (RS1).

If one were to assume that the average width of these anomalies (i.e., half-wavelength is 20 km, approximately 10 mi), and that a major reversal of the earth's magnetic field occurs once every million years (Cox, Doell & Dalrymple 1964), then the model implies a rate of spreading of 2 cm/y for each limb of the spreading system or in the case of the Atlantic, say, a rate of opening of 4 cm/y. If the opening-up has occurred within the last 150 My then this rate of spreading implies a width of 6000 km. Clearly such rough calculations show that the model is consistent with other heretical ideas on ocean-floor spreading and continental drift.

(Cann and Vine, 1966: 216)

Claiming this as a virtue of their hypothesis, they, nonetheless, let readers know that fixists would hold it against V-M in part because it was consistent with the "heretical ideas on ocean-floor spreading and continental drift." Fixists were unlikely to accept V-M and seafloor spreading unless they were forced to do so, unless V-M or some other corollary of seafloor spreading became difficulty-free and fixists could no longer hide their heads in the sand, as many were doing to avoid mobilism's paleomagnetic support.

Matthews addressed V-M in a review article he wrote for the *International Dictionary of Geophysics* the week before. Arguing in its favor, he appealed to the mechanism proposed by Bodvarsson and Walker (1964) for the formation by injection of parallel dykes (sheeted dykes) of new crust in Iceland, which is situated astride the crest of the Mid-Atlantic Ridge. Matthews believed this neatly explained the linearity of the magnetic anomalies running parallel to the ridge axes as envisioned by V-M.

Iceland, the largest of the world's volcanic islands, stands astride the crest of the Mid-Atlantic Ridge where the ridge passes between the continental blocks of Greenland and Scandinavia, and its structure should throw light on the structure of mid-ocean ridges in the deep oceans. Bodvarsson and Walker (1964) have written as follows – "Iceland has an area of about 10^5 km^2 and is composed almost exclusively of Tertiary and Quaternary volcanic rocks, predominantly basalt lavas ..., with Quaternary volcanic rocks in a belt crossing the centre of the island, and the Tertiary occupying large areas on either side and generally inclined at a few degrees towards the younger rocks." The volcanic rocks above and below sea-level must attain a thickness of several kilometers and the authors believe that the width of the base of the lava pile has been extended by the injection of the feeder dykes associate with the lavas ... They conclude that "The oldest rocks in the extreme east of Iceland and rocks of corresponding age (possibly Eocene) in the west may originally have been erupted in an active belt a few tens of kilometers wide, but since then have been carried apart by some 400 km or more, by crustal drift."

(Matthews, 1967: 985)

Matthews went on to say:

These results have suggested a mechanism capable of explaining the pattern of elongated magnetic anomalies found on the ridge (Vine and Matthews 1963). The mechanism envisaged is the one postulated by Bodvarsson and Walker (1964) for Iceland ... at the present time the crust under the crest of the ridge is being extended by the intrusion of a dyke swarm which reaches its maximum concentration under the median valley. In this way the older crustal rocks are shouldered aside while continued intrusion of feeder dykes results in the formation of volcanoes upon them. The youngest crustal block, situated at the crest of the ridge under the median valley, consists of recent dykes and lava flows like the central graben of Iceland, and is homogeneously magnetized in the present direction of the Earth's field. The older blocks adjacent to the valley consist of reversely magnetized dyke rocks which were emplaced at the crest of the ridge during a period when the field was reversed but which have been subsequently intruded by recent, normally magnetized, dykes and overlayed by normally magnetized volcanics. The magnetic effect of such an inhomogeneously magnetized body will be weaker than that of the young central block. In this way a pattern of alternately normal and reversely magnetized blocks of basalt and dolerite might be built up in the crust.

(Matthews, 1967: 989)

In this way, Matthews, who had heard Walker speak at the Royal Society's symposium in March, 1964 (§6.3), favorably summarized Bodvarsson and Walker's explanation of crustal extension by dyke injection. Although Hess did not speak of dyke injection, Matthews viewed Bodvarsson and Walker's account as a refinement of Hess's seafloor spreading (Matthews, 1979 interview with author).

In his review, Matthews included twenty calculated magnetic profiles perpendicular to mid-ocean ridges, for which he thanked Mr. and Mrs. F. J. Vine for computing on EDSAC II. He also noted (1967: 989), perhaps in criticism of Vacquier's objection to V-M (§7.8), that "the form of calculated anomalies depends upon the latitude and bearing of the axis of the ridge." He (1967: 989) also claimed that V-M "seems consistent" with the seismic, gravity, magnetic data, and heat flow measurements over ocean ridges.

Matthews favored mantle convection. Citing Hess (1962), Menard (1964), Wilson (1963e), and noting Fisher and Hess's (1963) extension of Hess's seafloor spreading to trenches, he (1967: 990) claimed that even though mantle convection faced difficulties, it "provides the best available guide to future research."

4.4 Mild support and criticism of the Vine–Matthews hypothesis during the first half of 1965

As we have seen, Hess and Backus supported V-M in 1964 (§3.8). Vacquier argued against it at the Royal Society's 1964 symposium on continental drift (§3.8), Talwani dismissed it as improbable and startling (§3.8), and Vine, Matthews, and Cann offered further support for the hypothesis when revisiting magnetic anomalies over the Carlsberg Ridge and extending Matthews' original survey (§4.3). Matthews defended the hypothesis and appealed to Bodvarsson and Walker's account of crustal extension through multiple dyke injection along ridge crests. However, in 1964 and 1965, Menard, Dietz, Holmes, Ewing, and Carey did not discuss it. Menard had written about marine magnetic anomalies, and Dietz, Holmes, and Carey, nonetheless, could, with profit, have evaluated V-M in terms of their own account of the evolution of ocean basins. Most surprisingly, Mason was not sufficiently motivated to reinterpret his analysis in terms of V-M despite the fact that it provided him with a way to avoid appealing to huge susceptibility contrasts laterally across alternating stripes of oceanic crust and the existence of highly contrasting rock types that this implied, something for which there was no independent evidence. Raff wrote a paper in 1966 in which he discussed several new magnetic profiles in the northeast Pacific and suggested a left-lateral slip along the Molokai Fracture Zone, but said nothing about V-M (Raff, 1966). Neither Allan nor Girdler, the two former Ph.D. students in the Department of Geodesy and Geophysics at Cambridge who had worked on ocean magnetic anomalies (§2.4), mentioned V-M at this time; Allan did not in his two 1964 papers on the magnetics of the Red Sea, and Girdler did not mention V-M in his 1964, 1965, and 1967 papers on the Red Sea and formation of oceanic crust (Allan, 1964; Allan *et al.*, 1964; Drake and Girdler, 1964; Girdler, 1964, 1967). Thus, V-M clearly failed to motivate many who had worked on marine magnetic anomalies or the evolution of ocean basins; they were not prepared to reevaluate their earlier analyses of the anomalies or to incorporate V-M into their current views about the evolution of ocean basins.

V-M, however, received support from studies in New Zealand and Canada. Both claimed that they had identified an oceanic ridge with linear anomalies paralleling the central anomaly over its crest. I first consider the New Zealanders.

D. A. Christoffel and D. I. Ross (1965) from the Physics Department, Victoria University of Wellington, appealed to V-M to explain sharp negative anomalies that they correlated across four north–south trending profiles taken with a proton precession magnetometer aboard HMNZS *Endeavour* during the outward and homeward runs of its supply cruises of 1958–9 and 1963 between New Zealand and McMurdo Sound. The outward run crossed the Campbell Plateau, a deep basin, and the Pacific-Antarctic Ridge, before entering the Ross Sea. They submitted their paper in February 1965, revised it a month later, and it was published in June 1965. I believe that they were the first, other than Vine and Matthews, to appeal to the hypothesis as an explanation of a group of marine magnetic anomalies that they themselves had discovered. Christoffel first worked with R. D. Adams, who was at the Seismological Observatory, Department of Scientific and Industrial Research, Wellington, New Zealand (Adams and Christoffel, 1962). They (1962) analyzed the two profiles from the first supply runs. Coordinating results from the profiles revealed an east–west trending zebra pattern of alternating positive and negative magnetic anomalies over the deep basin. The stripes also paralleled the east–west section of the Pacific–Antarctic Ridge just south of the basin. They found no correlation between the anomalies and bathymetry. Pointing to the similarity between what they had discovered and the anomalies found in the northeastern Pacific, they (1962: 811) noted that if, as Mason (1958) proposed, "the anomalies are due to a thin sheet of volcanic rocks, the susceptibility would have to be as high as 10×10^{-3} emu." Over the ridge itself, they were unable to correlate individual anomalies from the two profiles.

On the second supply run between New Zealand and McMurdo Sound, Christoffel and his Ph.D. student D. I. Ross obtained two more magnetic profiles. They matched magnetic anomalies on all four profiles across the deep ocean basin and over the crest of the ridge, but they were unable to match them over the ridge's foothills. This led to two options: dipping dykes that became tilted after acquiring their remanent magnetization or V-M. The dipping-dyke model was unable to explain the sharpness of the observed negative anomalies; "a complete reversal of the magnetization vector" was needed, which is what Vine and Matthews proposed.

To obtain the sharp negative anomalies actually observed requires a complete reversal of the magnetization vector. Vine and Matthews [1963] have postulated a mechanism for the explanation of the anomaly pattern associated with ocean ridge systems which includes such a reversed direction of magnetization. Their basic assumptions are, first a periodic reversal of the earth's magnetic field and, second, an impermanent ocean floor as postulated by Dietz [1961]. Such a mechanism could be used to explain the anomaly pattern observed here, particularly over the basin region.

> (Christoffel and Ross, 1965: 2861; the Vine and Matthews reference
> is identical to mine; the Dietz reference is to my Dietz, 1961a)

V-M also received support from a two-stage aeromagnetic reconnaissance of the Labrador Sea. In all, four researchers were involved: E. A. Godby and R. C. Baker of the National Aeronautical Establishment, Ottawa, Ontario, Canada, and M. E. Bower and P. J. Hood of the GSC. They were prompted to do the surveys because Wilson (1963) had suggested there ought to be an oceanic ridge extending from the Mid-Atlantic Ridge northwestward through the Labrador Sea and Baffin Bay (§3.2). Wilson's evidence for the ridge was no more than a half-dozen unexplained earthquakes. However, from seismic profiling, Drake *et al.* (1963) at Lamont found evidence in the Labrador Sea of a buried ridge that appeared to be aseismic and no longer active; Menard (1986: 205) retrospectively characterized this as "a reasonable confirmation of Wilson's prediction."

Hood and Godby (1964) first obtained two aeromagnetic profiles between Labrador and Greenland during a Royal Canadian Air Force reconnaissance flight in February 1963. They found a good correlation between the two profiles, especially near the coast of Greenland; they eventually obtained ten more profiles. They presented their results at the Annual Meeting of the AGU in Washington, DC, in April 1965, and submitted a paper three months later, which was published in January 1966 (Godby *et al.*, 1966). They correlated the magnetic anomalies across many of the profiles and argued (1966: 516) that there "are two anomalous magnetic zones which extend for hundreds of miles in a northwest-southeast direction." They (1966: 514) noted that the revealed trends of magnetic anomalies in the Labrador Sea "are reminiscent of a similar pattern" of magnetic anomalies found in the northwest Pacific, and they (1966: 516) proposed that their finding "agrees quite well with the observations of Vine and Matthews [1963]."

Many individual magnetic anomalies in these zones appear to be continuous from line to line, and determinations of depths of the anomalies indicated that the causative geologic bodies are close to the ocean floor in the southern end of the zone and plunge to the north toward Davis Strait. A typical width for the causative bodies is apparently about 5 miles (9 km) and the distance between their centers is approximately 10 to 15 miles (19 to 28 km), which agrees quite well with the observations of Vine and Matthews [1963].

(Godby et al., 1966: 516)

Unlike Drake, they claimed that Wilson's predicted ridge is active. They did not mention seafloor spreading by name, but spoke in terms of continental drift.

The line of epicenters in the central part of the Labrador Sea bounded on either side by zones of parallel magnetic anomalies and additional seismic refraction and profiler evidence appear to support the hypothesis that there is an active buried median ridge in the Labrador Sea which is further evidence of a continuing continental drift between Greenland and North America.

(Godby et al., 1966: 516)

Both New Zealand and Canadian workers found zebra patterns of magnetic anomalies. Both were reminded of anomalies in the northeastern Pacific but were unaware of their relationship to ridges. More and better examples of them flanking ridges would be needed before opponents of seafloor spreading and continental drift would consider changing their minds. Unbeknownst to both groups Vine, Wilson, and Hess had, as we shall see, already identified the Juan de Fuca and Gorda ridges, two extensions of the East Pacific Rise off the Oregon–Washington–Vancouver Island coast, and had associated them with the magnetic anomalies of the northeastern Pacific (§4.7). Talwani, Le Pichon, and Heirtzler also recognized the two ridges and their telltale zebra pattern of magnetic anomalies, but remained opposed to V-M (§5.3).

Peter, who with Girdler had proposed self-reversal to explain a magnetic profile over a ridge segment in the Gulf of Aden (§2.4), and his co-author, H. B. Stewart, Jr., added more fuel to the second difficulty facing V-M. They found linear magnetic anomalies in the northeast Pacific that did not appear to be associated with a mid-ocean ridge. They (1965) favored a Vacquier-type explanation claiming that they could "reproduce faithfully the magnetic anomalies" by supposing that "normally magnetized blocks" are "separated by non-magnetized blocks." They saw no need for V-M. They also recommended that single-track profiling of magnetic anomalies should be replaced by "carefully planned systematic areal surveys of the deep sea."

At the time Vine and Matthews[7] suggested that linear magnetic anomaly trends probably exist parallel to mid-oceanic ridges, only the 1955–1956 *Pioneer* and later Scripps areal survey results had been published. According to Vine and Matthews, the anomaly trends represent areas of normal and reversed magnetization that are related to the outward spreading of the ocean floor, and to the periodic reversals of the Earth's magnetic field. Backus[8] suggested an areal survey of the mid-Atlantic ridge to test these and other hypotheses at the same time, but the necessary survey has not yet materialized. The significance of the anomaly pattern shown in Fig. 1 is that it does not parallel a mid-oceanic ridge; it is nearly parallel to the continental margin of North America (900 nautical miles away), and on the west, to the Emperor Seamount chain (1500 nautical miles away). Calculations have shown[3] that in this area normally magnetized blocks, separated by non-magnetized blocks, can reproduce faithfully the observed anomalies. These suggest that linear anomaly patterns may exist in areas far away from mid-oceanic ridges and also from the continental margins. To locate, examine, and understand the meaning of these trends, the establishment of more areal surveys is imperative. If it is found that these are characteristic throughout all oceans, and reveal other large displacements of the ocean floor as well, new ideas may evolve in regard to the structure of the crust or upper mantle under the oceans.

No one questions the need for further oceanic investigations, but on the approach, only a few agree. We believe that to resolve existing controversial hypotheses single track-line exploratory expeditions now must be supplemented by carefully planned systematic areal surveys of the deep sea.

> *(Peter and Stewart, 1965: 1018; reference 7 is to Vine and Matthews (1963); reference 8, to Backus (1964); reference 3 is to Elvers et al., (1964))*

Regardless of whether or not they agreed with Vine and Matthews, they realized that more detailed areal surveys like those undertaken by Matthews, Scripps, and the USCGS were needed.

Carl O. Bowin and Peter Vogt reported new bathymetric and magnetic findings from the Indian Ocean nearby to where Matthews conducted his detailed magnetic survey. Their paper (Bowin and Vogt, 1966) was received on November 13, 1965. They obtained forty-eight profiles taken along a course between the Carlsberg Ridge and Seychelles Bank during a May 1964 cruise of WHOI's RV *Chain*. They found the typical pattern of alternating high and low magnetic anomalies, coordinated them with profiles from cruises of HMS *Owen* in 1962 and 1963 and RRS *Discovery* in 1963 by Cambridge and NIO workers, and claimed (1966: 2627) that "a magnetic lineation trending between N55° W and N65° W exists between the Carlsberg Ridge and the Seychelles Bank over an area of at least 6 00 000 km^2." They noted that the Carlsberg

Ridge just north of the area and the Seychelles and Saya de Malha banks, which they categorized as ridges, trends close to N45° W. Thus the trends of the ridges and magnetic anomalies, although similar, were not the same. Bowin and Vogt accented the "distinct difference" between the trends and mentioned, without supporting it, V-M (1966: 2629). If the trends had been the same, or at least if the trend of the Carlsberg Ridge and magnetic anomalies had been the same, they might have favored the V-M hypothesis:

There appears to be a distinct difference between the regional trends of the topography of the two bordering ridges and that of the magnetic anomalies, although the trends are similar. The significance of this difference in trend, if real, is unknown, as is the nature and origin of the source of the magnetic anomalies. There is a possibility that, if the magnetization (either induced or remanent) of the source is parallel to the current geomagnetic vector and of appropriate dimensions, the source might trend near N45° W but the isoanomaly lines would trend closer to N55° to 65° W ... *Vine and Matthews* [1963] postulate that the strong magnetic anomalies result from alternating normally and reversely magnetized strips of the crust (usually about 20 km wide). Such strips would have large apparent susceptibility contrasts and would explain the anomalies without the necessity of assuming unusual magnetic properties of the crustal rocks.

(Bowin and Vogt, 1966: 2628)

Although not supporting V-M, they understood its key advantage, namely, explaining the "large apparent susceptibility contrasts ... without the necessity of assuming unusual magnetic properties of the crustal rocks."

4.5 Heirtzler, Le Pichon, and Talwani at Lamont Geological Observatory

Events now began to occur in quick succession. In 1965, James Heirtzler, Xavier Le Pichon, and Manik Talwani co-authored three papers on marine magnetic anomalies (Heirtzler and Le Pichon, 1965; Talwani, Le Pichon, and Heirtzler, 1965; Heirtzler, Le Pichon, and Baron, 1966).

They proposed a fixist explanation of the anomalies, and rejected seafloor spreading and its corollary, V-M. The first paper was by Heirtzler and Le Pichon (1965) and was concerned with magnetic anomalies over the Mid-Atlantic Ridge; it was submitted to *JGR* in April 1965, revised in May, and published on August 15, 1965. The second, by Talwani, Le Pichon, and Heirtzler, (1965) was concerned with "the magnetic pattern and the fracture zones," of the East Pacific Rise; it was published in *Science* at the end of November 1965 without reception date, and no indication of revisions. They mentioned Wilson's first paper on transform faults (Wilson, 1965b) which was published on July 24, so they probably submitted or resubmitted their paper between late July and mid September (Wilson, 1965b). Wilson and Vine also wrote two papers on the magnetics of the same area of the northeast Pacific, and their papers were submitted on June 17, at least a month before Talwani and company submitted theirs (Wilson, 1965c; Vine and

Wilson, 1965). Wilson's and Vine and Wilson's papers were published on October 22. The third Lamont paper, by Heirtzler, Le Pichon, and Baron (1966), concentrated on the Reykjanes Ridge south of Iceland; it was received by *Deep-Sea Research* in November 1965 and published in early 1966; however, Heirtzler and company had at least some of their data from the Reykjanes Ridge while writing the first two papers. Talwani also discussed their work on the northeastern Pacific and Reykjanes Ridge at the Ottawa meeting on September 5 sponsored by the International Upper Mantle Committee (Talwani *et al.*, 1966) (§5.7). To make clear the relative timing of Wilson and Vine's work and that of the Lamont group, I describe events in what I believe to be their chronological order. I begin with Lamont's work on the Mid-Atlantic Ridge, then discuss Wilson and Vine's work on the northeast Pacific, and end with work by the Lamont group on the northeast Pacific and the Reykjanes Ridge.

Manik Talwani was born in India in 1933. He obtained his B.S. and M.S. degrees in physics from the University of Delhi in 1951 and 1953. He obtained his Ph.D. in 1959 at Columbia University, where he worked with Ewing, Worzel, and others at Lamont, concentrating primarily on gravity studies. Ewing had wanted Talwani to work in paleomagnetism, analyzing Lamont's huge collection of deep-sea cores. Neil Opdyke learned this after he had been hired to work on the cores. Opdyke learned that Talwani worked with John Graham, but returned to Lamont discouraged, and "essentially refused to do paleomagnetism" (Opdyke, March, 1997 interview with R. E. Doel, Session 1, p. 113). In 1972, Talwani was later selected as Director of Lamont–Doherty Geological Observatory after Ewing's departure. He left Lamont–Doherty in 1981, and worked in the petroleum industry before accepting a position in 1985 at Rice University, Texas. He received the first Krishnan Medal (1964) from the Indian Geophysical Union, the James B. Macelwane Medal (1967) and Maurice Ewing Medal (1981) from the AGU, the George P. Woollard Award (1984) from the GSA, and the Alfred Wegener Medal from the European Union of Geosciences (1993). He is a fellow of the AGU (1971), Fellow of the Norwegian Academy of Arts and Sciences (1987), Foreign Member, Russian Academy of Natural Sciences (1992).

James Heirtzler was born in 1925 in Baton Rouge, Louisiana. He earned B.S. and M.S. degrees in physics from Louisiana State University in 1947 and 1948, and his Ph.D. in physics in 1953 from New York University. After spending three years teaching physics at the American University of Beirut, he worked at General Dynamics Corporation until 1959. Hired at Lamont in 1960, he headed the magnetics laboratory through 1967 when he became Director of Hudson Laboratories of Columbia University. Before leaving Lamont, Heirtzler mined Lamont's data on marine magnetics and developed the first long-term geomagnetic timescale (Heirtzler *et al.*, 1968). He remained as Director of Hudson Laboratories until they closed and then took a position at WHOI, where he remained until retirement. Heirtzler was elected a fellow of the AGU in 1960, and served as President of its Geomagnetism and Paleomagnetism Section from 1982 through 1984.

Xavier Le Pichon was born in 1937 in Qui Nhon, Vietnam, and remained there until age nine, when his mother took him with her upon her return to Cherbourg, France. He attended a private Catholic school. Initially undecided as to whether he should concentrate on literature or science, he rejected both in favor of becoming an officer in the French Navy because his father and other relatives had been officers in the armed forces. Poor eyesight, however, led to a change in plans. He decided to pursue a career in the sciences, attended the University of Caen, and obtained a B.S. degree in physics in 1959. He earned a degree in geophysical engineering a year later at the University of Strasbourg, where he had the good fortune to study with the seismologist P.J. Rothé, whose work during the early 1950s had strengthened the association between shallow earthquakes and oceanic ridges (III, §6.7). Rothé, who had met Ewing in 1954 in London at the Royal Society's discussion about the floor of the Atlantic, helped Le Pichon secure a Fulbright Fellowship in 1959 to study at Columbia University. Le Pichon originally planned on taking courses, but once Columbia professors realized that he was sufficiently prepared to do research, Ewing immediately arranged for him to go to sea. Except for a hiatus in the French Navy (1961–2), he remained at Lamont until 1968. After briefly getting his feet wet in physical oceanography, he worked with Talwani on gravity studies, with John Ewing and Drake on seismic profiling, and with Heirtzler on marine magnetics. Concentrating on the Mid-Atlantic Ridge, he received his Ph.D. from the University of Strasburg in 1966. His dissertation was based on his work at Lamont on the Mid-Atlantic Ridge (see, for example, Heirtzler and Le Pichon, 1965; Le Pichon and Talwani, 1965; Talwani, Le Pichon, and Ewing, 1965; Talwani, Le Pichon, and Heirtzler, 1965; Heirtzler, Le Pichon, and Baron, 1966). Le Pichon was a diehard fixist until his conversion to mobilism in 1966. One of the few to understand Jason Morgan's first talk on plate tectonics at the April 1967 AGU meeting, Le Pichon exploited Lamont's database, and presented the first rigorous test of Morgan's ideas (Le Pichon, 1968a). During the next sixteen years, he held positions at France's National Center for Exploitation of Oceans, the Middle Solid Oceanographic Center of Brittany, University of Paris, University Pierre and Marie Curie, the College of France, École Normale Supérieure, Lamont–Doherty Geological Observatory, and the IGPP, University of California. He was Director of the Geological Laboratory at École Normale Supérieure from 1984 through 2000, and became Chair of Geodynamics at the College of France in 1986. Le Pichon is a member of the French Académie des Sciences (1985), associate member of the US National Academy of Sciences (1995) and Commandeur de l'Ordre National du Mérite (2001). He also is a fellow of various professional societies including the RAS (London) (1971), the AGU (1975), and the Geological Society of London (1982). He has been showered with prizes and medals. They include the Hirn Prize of the French Académie des Sciences (1969), Belgium's Paul Fourmarier Medal of the Royal Academy of Science of Belgium (1970), the Silver Medal of the Centre National (France) de la Recherche Scientifique (1973), the French Oceanography Medal (1975), the Fondation de France Scientific Prize (1981), the Ewing Medal of the AGU (1984), the A. G. Huntsman Award for Excellence in Marine Science of the Bedford Institute of Oceanography (1987), the Japan Prize (1990), the Wollaston Medal of the Geological Society of London (1991), the Balzan Prize for Geology (2002), and the Wegener Medal of the European Geosciences Union (2003).

4.6 Lamont's view of mid-ocean ridges and rejection of the Vine–Matthews hypothesis: work on the Mid-Atlantic Ridge

In examining the work by Heirtzler, Le Pichon, and Talwani on marine magnetic anomalies, I shall concentrate on their alternative to V-M, their objections to it, and how much of what they discovered also supported V-M.

In the first Lamont paper "Magnetic anomalies over the Mid-Atlantic Ridge," Heirtzler and Le Pichon reviewed fifty-eight profiles across the Mid-Atlantic Ridge from 60° N to 42° S. Twelve were from Lamont's unpublished files, twelve from previously published Lamont papers, fourteen from surveys by Vacquier and Von Herzen (1964) in the South Atlantic, and the remainder from previously published work by non-Lamont workers including Hill (1960) and Vine and Matthews (1963). They found, as earlier workers had, positive and negative linear anomalies running parallel to the ridge axis. They distinguished between *axial anomalies* within the roughly 1000 km wide axial zone, and beyond that, *flank anomalies* typically of greater wavelength and amplitude. They distinguished two patterns of axial anomalies. In the North Atlantic, they comprise a large amplitude central anomaly astride the ridge crest and anomalies of considerably smaller amplitudes on either side. In the South Atlantic, axial anomalies typically comprise several large anomalies surrounded by anomalies whose amplitude decreases with distance from the ridge axis. They also noted that the linearity of the axial anomalies is particularly evident over the Reykjanes Ridge.

> In the vicinity of … the Reykjanes ridge south of Iceland, a detailed aeromagnetic survey was flown in 1963 as a joint Lamont-U.S. Naval Oceanographic Office endeavor. The area covered was approximately 400 by 400 km, with a flight line spacing of about 10 km. That work will be reported in another paper, but it is important to mention here that the survey revealed that not only the axial anomaly but most details of the magnetic profiles are continuous, linear features.
>
> *(Heirtzler and Le Pichon, 1965: 4016)*

So Heirtzler and Le Pichon had established throughout much of its length that the Mid-Atlantic Ridge is flanked by alternately high and low linear magnetic anomalies which also extend out beyond the axial zone. This is precisely what was needed to lessen the severity of the difficulty that faced V-M at its inception, regarding the lack of an unmistakably recognized mid-ocean ridge flanked by arrays of linear magnetic anomalies. They, like Vine and Matthews (1963) and unlike Vacquier (§3.8), realized that the pattern of magnetic anomalies should be less conspicuous in equatorial regions where the inclination of the geomagnetic field is low.

Heirtzler and Le Pichon raised two new difficulties with V-M (RS2). It could not explain the diminution in amplitude of the axial zone anomalies with distance away from the ridge axis in the South Atlantic.

Vine and Matthews [1963] have hypothesized that the whole ocean crust is made of strips of material parallel to the ridge axis, having alternatively reversed and normal magnetization. They see their hypothesis as a corollary of the "spreading floor" hypothesis of *Dietz* [1961]. Reversely magnetized material is probably present in the ridge, and if incorporated in our models, somewhat smaller apparent susceptibilities could have been used. However, the pattern of the anomalies is such that we had no basis for including reversed magnetization in our models. The recent study shows that the effect of the change in water depth is insufficient to explain the decrease in amplitude of the axial anomaly in the South Atlantic (see Figure 9) and that the lower boundaries of the source bodies must be shallower than the probable depth of the Curie-point isotherm. It is clear from this study that most of the profiles do not follow the pattern assumed by Vine and Matthews (for example see Figure 5).

(*Heirtzler and Le Pichon, 1965: 4028; Vine and Matthews reference same as mine;*
Dietz reference is my Dietz (1961a); Figures 9 and 5 are reproduced below)

It also failed to explain, they claimed, the origin of the flank anomalies, which had larger wavelengths and amplitudes than the axial anomalies. This they argued meant that the axial and lower flank anomalies had different origins.

Model studies allow a wide range of bodies to explain the flank anomalies, in contrast to the sharp limitations they put on the axial anomaly. However, the large increase in wavelengths and amplitudes over the lower flanks of the ridge suggests an important difference in the magnetic sources which may be deeper and have a larger extent.

(*Heirtzler and Le Pichon, 1965: 4028*)

They concluded that Vine and Matthews were mistaken in proposing that the axial and lower flank anomalies had a common origin.

Heirtzler and Le Pichon (1965: 4027) then laid the groundwork for their own solution by noting the correlation between the axial magnetic anomalies, high heat flow, shallow earthquakes, and seismically anomalous mantle zone, characterized by abnormally low compressional seismic wave velocities. They proposed that the central axial anomaly over the Mid-Atlantic Ridge in the North Atlantic is caused by a shallow magnetized volcanic body which possesses higher magnetic suscepti-bility than surrounding material. The body formed as the volcanic material filled a fracture in the oceanic crust beneath the rift valley.

Consequently, in the North Atlantic, the origin of the axial anomaly is the presence in the thick volcanic basement layer of a body, 10 to 15 km wide, having a high essentially constant apparent susceptibility contrast of about 0.01 cgs. This body is just under the floor of the rift valley, where the valley is clearly defined, and has a longitudinal extent of thousands of kilometers ... The correlation of this body with the axis of the high heat flow belt and the

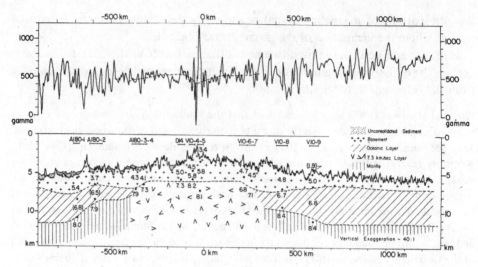

Figure 4.1 Heirtzler and Le Pichon's Figure 5 (1965: 4022). Their caption in part reads: "V-17 magnetic profile at 32° N as would be recorded at a constant height of 3 km over the ocean bottom. The composite crustal section is from Talwani *et al.* [1965]." The pattern of magnetic anomalies is characteristic of those found in the North Atlantic with a pronounced central anomaly surrounded by small anomalies. The axial region extends approximately 300 km to the east (right), and 500 km to the west (left) from the ridge axis, and coincides with the anomalous mantle zone. The transition from the axial to flank anomalies with the increase in amplitude and wavelength is evident. The dotted line represents what Heirtzler and Le Pichon identified as the depressed regional anomaly associated with the axial zone of the Mid-Atlantic Ridge and noticeable at higher latitudes.

earthquake epicenter belt has been indicated earlier. The high apparent susceptibility contrast one has to assume is an indication of important remanent magnetization, and the existence of zones having low magnetic anomalies on each side demonstrates the uniqueness of this feature. All these facts suggest that the origin of this body is the filling of a fracture in the crust by volcanic material which has been highly magnetized, either because of its physical properties or because of its mode of implacement in the crust.

(Heirtzler and Le Pichon, 1965: 4027)

The fracturing was caused by tensional forces.

Making a minor adjustment for the South Atlantic (Figure 4.2), they (1965: 4028) proposed multiple fractures within the axial zone and a "swarm of bodies which increase in depth toward the axis."

They (1965: 4028) extended their hypothesis to other ridges, claiming that one or the other of the two patterns found on the Mid-Atlantic Ridge is found on other ridges.

We have demonstrated that the axial anomaly along the whole mid-Atlantic Ridge falls into only two main patterns, which probably correspond to two different types of fracturing of the ridge crust under the effect of tensional forces. These facts are even more significant because

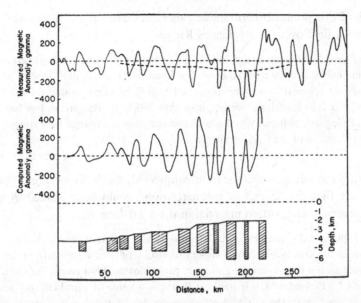

Figure 4.2 Heirtzler and Le Pichon's Figure 9 (1965: 4027). The uppermost profile is the measured profile of axial zone for the Z-2 ridge crossing at 39° S. The profile is characteristic of the South Atlantic with several pronounced axial anomalies instead of one as found in the North Atlantic. The amplitude of the anomalies also decreases with distance from the ridge axis. The dotted line represents what Heirtzler and Le Pichon identified as the depressed regional anomaly associated with the axial zone of the Mid-Atlantic Ridge and noticeable at higher latitudes. The middle profile is their calculated profile. They assumed an apparent susceptibility of 0.013 c.g.s. The cross-hatched areas in the bottom portion of the figure represent remanently magnetized bodies causing the positive anomalies. They are the "swarm of bodies which increase in depth toward the axis."

these two patterns have now been recognized on many crossings of the mid-oceanic ridge in other oceans. In particular the "south axial pattern" is characteristic of the Pacific-Antarctic Ridge near 60° S [Heirtzler, 1961] and of the extension of the mid-ocean ridge under the continental rise off Vancouver Island (unpublished data). The absence of any recognizable pattern across the East Pacific rise near 10° S [Talwani *et al.*, 1965] is due to the low inclination of the earth's field there. The "north axial pattern" is seen, for example over the Carlsberg ridge ... and over the Pacific-Antarctic ridge at 175° W.

> (*Heirtzler and Le Pichon, 1965: 4028; Heirtzler (1961) is the same as mine;*
> *Talwani* et al. *(1965) refers to my Talwani, Le Pichon, and Ewing (1965)*)

Heirtzler and Le Pichon (1965: 4015) also proposed that the regional geomagnetic field over the Mid-Atlantic Ridge "is a region of relatively low amplitude anomalies." They defined the regional field as "the magnetic field as it would be without the disturbances due to the shallow magnetic sources situated over the Curie-point isotherm"; the regional field is essentially the field originating in the Earth's core. They attributed the decreased strength of the regional field to the abnormally shallow depth of the Curie point isotherm under ridge axes. They found a similar depression of the regional field over the Reykjanes Ridge.

It has been noticed that the regional value of the magnetic field strength, when averaged over the axial zone anomalies, may be reduced by several hundred gammas (10^{-5} oersteds = 1 gamma) from the adjacent regional value. This depressed region, localized around the axial anomaly, corresponds to a region of high average heat flow where a comparison has been possible. The deficit of magnetic material which causes this lowering of regional field strength could be caused by an elevation of the Curie isotherm.

(Heirtzler et al., 1966: 429)

Although they did not, in this context, mention V-M, they acknowledged but rejected the possibility that the "deficit of magnetization" could be explained by an abundance of "reversely magnetized material in the axial zone ..."

The cause of this broad minimum has to be a progressive deficit of magnetization, reaching a maximum under the axis. While a much larger proportion of reversed material in the axial zone would produce such a low, a simpler and more probable cause is a rise in the Curie isotherm (Heirtzler and Le Pichon, 1965). If the heat flow has a value of 1 μcal/cm^2 per sec over the flank, and reaches 5 under the axis, with a thermal conductivity of 5×10^{-3} cal/°C. cm. sec, the Curie isotherm (500° C) would rise from a depth of 27 km under the flanks to a depth of 6 km under the axis. Such a configuration of the Curie isotherm would produce the regional anomaly observed with a large minimum over the axis and a smaller maximum on the west side. It is interesting to note that this rise in the Curie isotherm has to be continuous from about 100 km on each side toward the axis in order to match the shape of this regional anomaly.

It can then be concluded that the bodies causing the magnetic anomalies in the axial zone are probably entirely within the 2–4 km thick upper volcanic layer except for the body causing the large axial anomaly which is about twice as wide as the other bodies and may be several times thicker, its lower end corresponding to the Curie isotherm. This whole 2-km wide axial zone seems to have a distribution of temperature such that the Curie isotherm progressively rises to a minimum depth under the axis.

(Heirtzler et al., 1966: 439–440)

They determined the value of the regional field by taking the mean of the anomaly values, and it was unclear whether their estimate was an artifact of the analysis or actually revealed low regional field. Vine retrospectively remarked:

There were real problems then in determining the "regional field" (i.e. the field originating in the Earth's core) and for this reason I did not attach any significance to such discrepancies. (It is probably better now that we have so much satellite data.) If they were correct then the explanation in terms of elevation of the Curie point isotherm is probably a good one but I did not consider it relevant to the point I was trying to make.

(Vine, October, 2009 email to author)

Vine (October, 2009 email to author) also noted that the "elevation of the Curie point isotherm is very likely beneath the Reykjanes Ridge because of its proximity to the Iceland hot spot."

Heirtzler and Le Pichon appealed to the petrological work of the wife and husband team of Celeste and Al Engel. Al was at Scripps; Celeste, who was employed by the USGS, had her own laboratory at Scripps. The Engels, busy analyzing igneous rocks from the Mid-Atlantic Ridge, East Pacific Rise, and Mid-Indian Ocean Ridge (Engel and Engel, 1964; Engel, Engel, and Havens, 1965; Engel, Fisher, and Engel, 1965), found that they were typically tholeiitic basalts. Engel and Engel (1964: 1330) found this somewhat surprising because, except for the Hawaiian Islands, most volcanic oceanic islands, including those "scattered along" the crest and flanks of the Mid-Atlantic Ridge, are predominately alkali basalts. Appealing to their work in the Pacific and Indian oceans, they (1964: 1333) proposed that mid-ocean ridges are composed primarily of tholeiitic basalts, and tholeiites "appear to be the dominant igneous rock throughout at least the upper parts of the entire oceanic crust." But they also found that, along the Mid-Atlantic Ridge, alkali basalts "clearly dominate all volcanic cones" that had been "built from 2500 to 5500 m above the ocean floor." This led them to propose that "elevated alkali basalt cones" formed "from a parent tholeiitic magma by gravity differentiation" (1964: 1333). They also noted that alkali basalts have on average appreciably higher concentrations (by weight) of titanium dioxide and higher ratios of Fe^{3+} to Fe^{2+} than tholeiitic basalts. Although the Engels said nothing about the magnetic properties of either type of basalt, Heirtzler and Le Pichon conjectured that higher concentrations of TiO_2 and higher ratios of Fe^{3+} to Fe^2 of alkali basalts gives them a higher susceptibility and hence greater induced magnetization, and suggested that the "possibility exists that the volcanic material filling" the axial fracture zone "is an alkali basalt, which is known to be more magnetic than the tholeiitic basalt that probably forms most of the ocean floor" (Heirtzler and Le Pichon, 1964: 4028).

Throughout, Heirtzler and Le Pichon assumed that the anomalies arose from a strongly contrasting induced component. They invoked an "apparent susceptibility contrast of about 0.01 cgs." I close this discussion of their work by considering whether or not they thought remanence played a role. They first wrote that the magnetization is "only by induction."

A digital computer program [Talwani and Heirtzler, 1964] for the calculation of anomalies over a polygonal body, infinite in one direction, was used. The body was assumed to be 10 km in width and 10 km in depth, to have its upper surface at the ocean floor, to have a strike of geographic north-south, to have a magnetic susceptibility of 0.01 cgs, and to be magnetized only by induction.

> *(Heirtzler and Le Pichon, 1965: 4018; Talwani and*
> *Heirtzler, 1964 reference is the same as mine)*

But they also maintained, as quoted above, that the "high apparent susceptibility contrast one has to assume is an indication of important remanent magnetization." However, they (1965: 4027) did not lower the apparently susceptibility contrast, but kept it at "about 0.01 cgs." Thus, even though they thought remanence might be important, they did not actually include it in their analysis. Heirtzler recalled why they did not appeal to remanence.

Our (then new) computer program showed that all along the axis of the Mid-Atlantic Ridge there was a body about 10 km wide with a magnetization of about 0.01 cgs. When Le Pichon and I showed this it would have been pure speculation as to whether this was remanent or induced. In fact all remanent magnetism starts out as induced magnetism. It just keeps that magnetization after the ambient inducing field changes. Of course, if one assumed that the sea floor moved over geologic time, and the geomagnetic field is changing with geologic time we can speak about remanent and induced as two separate things. However before the Vine–Matthews hypothesis we didn't think the seafloor was moving ... ergo it didn't make much sense to argue whether the central anomaly was induced or remanent. This was the essence of our fuzzy wording. For the off-axis anomalies, after the acceptance of the V-M theory, we can discuss the two types of magnetization.

(Heirtzler, October 18, 2009 email to author)

Heirtzler's comment, "it didn't make much sense to argue whether the central anomaly was induced or remanent" before V-M appeared, is reasonable if one did not believe in seafloor spreading; however, Vine and Matthews' original paper had already been in the literature for approximately twenty months before Heirtzler and Le Pichon submitted their paper. Opdyke, on staff at Lamont since December 1963, was asked to be an internal examiner. Originally a fixist student from Columbia, he had travelled the world seeing a lot of global geology, and was now a confirmed mobilist because of his work with Runcorn on aeolian sandstones (II, §5.11–§5.15), with Irving in paleomagnetism at ANU, and with Gough and McElhinny in Rhodesia, where he filled in significant gaps in the APW path for Africa (III, §1.17). He talked to Le Pichon and Heirtzler about Ade-Hall's work on remanence in which he established the predominance of remanence in a wide range of oceanic basalts. They did not change the manuscript. Opdyke's response was to work on remanence of ocean basalts with R. Hekinian, also at Lamont. Opdyke recalled:

Le Pichon and Heirtzler were busy with a paper describing the magnetic anomalies of the Mid-Atlantic Ridge [Heirtzler and Le Pichon, 1966]. The paper was eventually internally reviewed by Hans Wensink and myself. I recall that a meeting between Le Pichon, Heirtzler, Wensink, and myself to discuss our review was held in Heirtzler's office. In the paper, Le Pichon and Heirtzler had dismissed reversal of the field and permanent remanence as a causative agent for the observed anomalies. Wensink and I contended that it could not be so easily dismissed since Ade-Hall [1964] ... had shown that some dredged basalts, admittedly not many, had high remanent intensities and were stably magnetized. Our arguments, however, were not accepted, and I left the meeting and immediately set up a program with Roger Hekinian to measure the magnetic properties of basalts available in the Lamont dredge collection.

(Opdyke, 1985: 1177; references are same as mine; for similar comments see Opdyke, 2001: 103–104, and Lamont oral history project, Opdyke, March 17, 1997 interview with R. Doel)

Opdyke's remarks were more pointed when he again recalled what had happened.

[Xavier Le Pichon is a] very strong-minded Frenchman, you know, and he'll stonewall you. And I also used to socialize with him and his wife, but at the time, now he's a big gun in France. Everybody crawls to him. In fact he was – he makes believe he invented the whole business, but in fact he didn't. But he was very reticent to do it. I think it was just emotional and personal reasons. When he published his thesis; when he finished it, I was one of the readers on the papers, and he had done the magnetic anomalies as I said on the Mid-Atlantic Ridge . . . I had to be the internal reviewer for the paper, so I read the paper, and he had made some statements there that I didn't believe, and so I went in to talk to him and Jim Heirtzler about it. In particular he said the magnetic intensity of the rocks on the bottom of the ocean floor was not high enough to give rise to the magnetic anomalies for seafloor spreading. I said, "That's not true!" because Ade-Hall had published a few results . . . and there was plenty of intensity in magnetization . . . And so we got in this big argument. So at the time I left and went out. I began to study the Lamont dredge collection. The proper way to solve a scientific problem is to get more data.

(Opdyke, December 8, 2009 email to author; my bracketed addition)

Opdyke and Hekinian (1967) found a high ratio (over 10 in 92% of the samples) of magnetic intensity to susceptibility, which was in line with Ade-Hall's findings, and therefore further substantiated the importance of remanence in the magnetization of oceanic basalts.

The study of the magnetic properties of the basalts presented in this paper substantiates the conclusion that the remanence value of volcanic basalts is very important and that in the interpretation of marine magnetic anomalies it may be of primary importance.

(Opdyke and Hekinian, 1967: 2258)

Ironically, Le Pichon and Heirtzler's refusal to change their manuscript eventually helped add to the support for V-M.

4.7 Vine, Wilson, and Hess at Madingley Rise, late 1964 to middle 1965

Wilson and Hess spent several months together in 1965 in the Department of Geodesy and Geophysics at Madingley Rise where they worked especially with Vine. Their visit and interaction with Vine was a key moment in the entire mobilism debate and I shall deal with it at some length. Once Vine learned that Hess planned to spend the early part of 1965 at Madingley Rise, he wrote to Hess on January 18, 1965 asking if there might be a position for him at Princeton "for a year or two as of October next." He wanted to warn Hess ahead of time of his "audacity." He also told Hess of his own "vested interest (!*) in ocean-floor spreading, and hence, continental drift." Hess was well aware of Vine's vested interest, and had already referred to V-M as "very fruitful" (Hess, 1964a) (§7.8).

Dear Professor Hess,

I am now in my third post-graduate year and about to write up my thesis in order to submit, I hope, by September next. I graduated here in Cambridge as a petrologist and have since been working in the Department of Geophysics on the interpretations of magnetic surveys at sea. My research supervisors have been Dr. Drummond Matthews and Dr. Maurice Hill.

I have one paper in print on magnetic anomalies over ocean ridges* and several others in preparation. Two of these concern an area of approximately 50 to 40 nautical miles on the crest of the Carlsberg Ridge and surveyed in considerable detail by H.M.S. OWEN during the early stages of the International Indian Ocean Expedition. One paper, written in conjunction with Dr Matthews and Dr J.R. Cann, concerns the structure and petrology of the area and has been accepted for publication in the Bull. Geol. Soc. Amer. The second was delivered at a discussion meeting of the Royal Society here in London in November and relates my efforts in interpretation of the detailed magnetic survey of the area using three-dimensional computations. Two other papers in preparation give a qualitative interpretation of a preliminary magnetic survey of the Western Approaches to the English Channel, and a quantitative analysis of surveys over an isolated "magnetic" seamount to the south-west of the Seychelles. A large part of my thesis, in fact, will concern "vector fits" on isolated seamounts assuming (initially) that they are uniformly magnetised and that the anomaly associated with them is caused by the visible topography alone.

Having now a vested interest (!*) in ocean-floor spreading and hence, continental drift, I am particularly anxious to continue studying the ocean floor (not necessarily "by magnetic alone") believing that it holds the answers to some of the biggest and most fascinating problems of petrology and structural geology. Since your visit in January '62 I have been particularly interested in your own work and ideas and in the same delivered a Presidential Address about your work to the Sedgwick Club, which organised the Congress.

Clearly my interests are very close to your own and I wondered if there might be any possibility of joining your department for a year or two as from October next. I hope that you will not think it strange that I should write to you when you are about to visit us but I thought you might like warning of such audacity.

Yours very sincerely,

F. J. Vine

*Vine, F.J. and D.H. Matthews, 1963 Magnetic Anomalies over the oceanic Ridges Nature, 199 947–949.[1]

Vine airmailed the letter the next day, and Hess received it before leaving Princeton. Vine recalled what happened when he first met Hess.

I did not receive a written reply, which was unsurprising, but it meant that I was on tenterhooks when his arrival was imminent. Much to my delight, on his arrival the first thing he said to me was that he thought that our idea was great, and the second thing he said was that he thought he would be able to find a position for me at Princeton.

(Vine, 2001: 59)

Because Hess was still at Princeton when he received Vine's letter, Hess probably did not arrive before the end of the first week in February. Precisely when he arrived is unclear, but he was definitely at Cambridge by February 20. A. Hallam wrote to Hess on February 15, 1965 to confirm Hess's earlier offer to give a talk before the Edinburgh Geological Society. Hess agreed to give a talk on May 5, and noted on Hallam's letter that he answered it on February 20, which fell on the last day of the third week.[2] However, he probably arrived before the 20th. Hess gave a paper in Bristol at the Seventeenth Symposium of the Colston Research Society on April 8, 1965. Hess wrote Menard recalling that the Colston paper "was written during the first two weeks I was in England to meet a March 1 deadline" (September 6, 1966 letter to Menard). W. F. Whittard, Chaning Wills Professor of Geology, University of Bristol, and symposium director, wrote to Hess on February 23, telling him that he was "delighted to know that you will definitely be joining us" and extended the deadline for receipt of Hess's paper until March 8.[3] Just when Hess actually submitted his manuscript is unclear; however, Whittard acknowledged its receipt in a letter of March 15.

Hess went on a lecture tour beginning April 28 at Imperial College and ending May 5 at Edinburgh as he promised Hallam before leaving for England. He also lectured at the Universities of Leeds, Durham, and Newcastle.[4] Of all his talks, the most poignant was that to the Edinburgh Geological Society, not because of the lecture itself, but because Holmes was in the audience, and "was quite delighted to learn of Hess's work" (Hallam, October 4, 2004 email to author). Hallam arranged a meeting between Hess and Holmes, but Doris Reynolds, Holmes' wife, concerned about Holmes' health, cut the meeting short.

I engineered a short meeting with Holmes after the lecture. I forget the details but neither Dietz nor V & M were referred to, and Holmes never mentioned them with me on any other occasion. By the time Hess came Holmes was getting rather frail and tired quickly in the evenings. Consequently his ever-protective wife whisked him away back home fairly soon. Even at his fittest, he was a shy and retiring man who abhorred the public stage. Thus when he was awarded the Penrose Medal by the GSA he was horrified at the prospect of giving an acceptance speech at the ceremony and found someone to substitute for him.

(Hallam, October 4, 2004 email to author)

It is sad that Hess and Holmes did not meet before he became so frail. Hess remained in England at least through May 12, the day he gave a colloquium at Madingley Rise on "Tectonics of the ocean floor."

Tuzo Wilson arrived in the UK just before Christmas 1964 so that his family could spend the holiday in London.[5] He also had planned to lecture at Newcastle during his stay, as evidenced by a brief summer correspondence with Runcorn, who wrote Wilson on July 2, 1964, inviting him to lecture at Newcastle.

As you know, we would like you to spend a short time here next year and give us some lectures. Perhaps in due course you will suggest to us what the best time would be. In addition, the University is wanting to invite you to give a college general lecture possibly about our visit to

China, and I wonder whether you would still be in England on March 18th which seems to be the most convenient date for this lecture.

(Runcorn, July 2, 1964 letter to Wilson[6])

Wilson replied about two weeks later, letting Runcorn know when he planned to be in England, and tentatively agreeing to give a general lecture on March 18.

Thanks also for the invitation to lecture at Newcastle. I hope to be in England from about December 22nd to mid-June except for a month on the Continent during the Easter vacation. I am sure that March 18th would be a convenient date to lecture in Newcastle.

(Wilson, July 17, 1964 letter to Runcorn[7])

Wilson also planned to attend the Twenty-Second Session of the International Geological Congress in New Delhi. His name appears on the list of provisional registrants. The Congress began on December 14 and ended on December 22. Presumably, he attended the Congress, left before it was over, and went straight to England, arriving as planned before Christmas. Wilson and his family also attended Winston Churchill's State Funeral on January 30, 1965. Wilson managed to obtain a ticket intended for the Lady Mayoress of Athens, who was to attend the service at St. Paul's Cathedral. Patricia Proctor (June 8, 2010 email to author) remembers that Wilson said that "he rather enjoyed where he was seated – not the front row or anywhere near it but high up – not a wide view but the choir's singing and sometimes the congregation too was wonderful from where he was." Wilson settled into the "visitors" office at Madingley Rise before Hess. Vine is certain because he remembers what happened when Hess first entered the office set aside for Wilson and himself. "They were supposed to share an office, but by the time Harry arrived Tuzo had filled it completely. Harry took one look at it and never set foot in it again" (Vine, October 12, 2009 email to author).

As planned, Wilson spent the Easter vacation on the continent with his family. They left London on March 12, spent a day in Vienna, and continued by train to Istanbul, where they arrived on the 19th. They toured Turkey visiting many historical places and, most notably for this narrative, the southern resort town of Antanya. Overnight, they took a boat from Izmir to Athens. Wilson then flew to Munich to attend a meeting of the International Council of Scientific Unions from April 4th to the 7th. They then set sail for approximately a week, touring the Cyclades and disembarked at Piraeus on April 15. Next they took a boat to Italy, eventually arriving in Rome on the night of the 17th, so that they could, as planned, spend Easter in Rome. They "went to St. Peter's catching a glimpse of 'il Papa'" (Proctor, June 8, 2010 email to author). They returned to England by train, arriving at Cambridge on Wednesday, April 21.[8] Like Hess, Wilson also gave lectures in England, at Leicester University for example on February 25 (Susan Wilson, June 6, 2010 email to author). Alan Coode is quite sure that Wilson lectured at Newcastle at the physics department in early February (§4.11). He must have been able to be in Newcastle on March 18, as he and Runcorn had originally planned being then en route to Istanbul.

Wilson's scientific work at Cambridge during the first half of 1965 was as energetic as his family trip, for he not only developed the idea of transform faults, but with Vine and help from Hess, found strong support for V-M as they examined the Ridge and Trough Province of the northeast Pacific (§5.2).

4.8 Hess fine tunes and extends seafloor spreading, 1965

Hess furthered the development and widened the application of seafloor spreading in light of V-M, which he (1964a) had characterized as "very fruitful." He had proposed that determination by Scripps of seismic anisotropy in the eastern Pacific indicated that the upper mantle was moving in a direction parallel to the great fracture zones (§3.8). And, in order to accommodate V-M and provide a source for the magnetic anomalies, he now agreed (1964b) that Layer 2 of oceanic crust was mainly basalt, not consolidated sediment, as he originally thought (§3.8).

But Hess had not finished tinkering with seafloor spreading, for he amended and extended it at the Colston Research Society's Bristol meeting in early April. Invitees included marine geologists and geophysicists from several generations. Pioneers such as Kuenen, Shepard, and Wiseman, established figures such as Bullard, Heezen, Hersey, Hill, Menard, and Worzel, and up-and-coming researchers such as Bott, Laughton, and Loncarevic attended.

Hess made several important changes to his account of mantle convection, explicitly restricting convection to the upper mantle. He likened Earth to a tennis ball, its seams representing the descending currents of the world-wide convective system, while the major axes of the two dumbbell shapes outlined by the seams represent ascending currents. He rejected his own previous proposal that seafloor plunges vertically downwards into the mantle (Fisher and Hess, 1963), and adopted Fisher's 1953 field-based view that seafloor descends diagonally downward beneath trenches, the angle of descent varying from 30° to 45°. Reminiscent of Carey (II, §6.10), he began to think of convection as solid creep instead of liquid flow. He rejected the idea of viscous drag between crust and mantle, and proposed that ocean crust and the upper mantle move away from ridge axes as a unit (see III, §3.16 and §3.19 for Fisher's and Hess's previous views).

If the downward-moving limb [of a convection cell] follows the course comparable to the stitching on the tennis ball, the upward-moving limbs would be two lines each median in the two dumbbell-shaped pieces of material from which the ball is put together ... If the overriding continent or island arc moved forward at the same rate that the current descends then a plane dipping at 45° will represent the zone along which the motion is taking place. Actually the plane dipping under the continents and island arcs varies in slope from 30° to 45°, and the conclusion may be drawn that the downward motion is in some case a little slower than the forward motion (fig. 121). A tentative model for the convection system may be suggested (fig. 122). It may have no great merit but serves to focus attention on the possibilities. First, it is proposed that the convective system extends from the surface to a depth of 750 km, that is, the vertical extent of earthquake foci ... The flow is not that of a viscous fluid (strain proportional to stress) but that of hot creep (the pseudoviscosity of Griggs, 1939, or Andradian viscosity of Orowan, 1964, 1965[2], or of Elsasser, 1965[2] ... The crust and underlying mantle would have moved away from the crest at the same velocity as if bolted together (fig. 123), rather than as some system involving viscous drag (fig. 116) as shown by Menard (1964).

> *(Hess, 1965: 322–324; footnote 2 refers to lectures by Orowan and Elsasser*
> *delivered at an MIT Convocation, October 1964; my bracketed addition)*

Earlier, Hess had told Menard in June 1961 (III, §5.11), that he had no explanation for the great fracture zones. He now said that they formed from differential movement of "crustal plates" away from the Darwin Rise, which he and Menard had previously championed – yes, Hess used the terms "crustal plates" and "plates."

It is difficult to understand how long-continued movement along a fault such as the San Andreas could take place except by differential flow in the underlying mantle. The movement of crustal plates with respect to the great Pacific fracture-zones is a more clear cut example ... Menard (1964) had deduced that crustal plates are moving away from the crest of the East Pacific Rise extending under western North America. I believe the movement was in the opposite direction, away from the crest of a former mid-oceanic ridge, the Darwin Rise (fig 120); in either case, all the plates moved in the same direction and it is only the difference in movement which is recorded – a small part of the total amount of movement compared with a fixed grid on the Earth's surface. My reasons for believing the plates moved eastwards are as follows. The fracture-zones are not seismically active to-day. They form a consistent pattern approximately at right angles to the crest of the Darwin Rise. While Menard's reconstruction gives the approximate position of the crest of the rise, the rise probably once extended farther to the north-west, this portion of it having been obliterated by the forward motion of the western Pacific island arcs. The guyots terminate abruptly to-day at the line of trenches bordering the western side of the ocean. On my basis of reasoning the whole of the crust of the Pacific Ocean was formed on the crest of the Darwin Rise (except those parts behind the island arcs) and moved as far north-eastwards as the Gulf of Alaska and as far southwards as the Antarctic.

> *(Hess, 1965: 320)*

So unlike Menard, Hess did not relate the fracture zones to the East Pacific Rise, but to the much older Darwin Rise then acting as a mid-ocean ridge. Hess also thought that almost the entire Pacific floor was generated by seafloor spreading from the Darwin Rise.

Hess had high praise for both the Carey and the EBS fits of the circum-Atlantic continents.

Wegener first proposed the opening up of the Atlantic Ocean by continental drift because of the fit of the continental margins if moved back together. It took 40 years before anyone using suitable projections tried to test how good this fit really was. Carey (1958) demonstrated the almost perfect fit of Africa and South America using 100 fm. contour to define the edge of the continents, and more recently Bullard, Everett & Smith (1965) have done the entire circum-Atlantic with greater precision. The jig-saw puzzle fits back together so well in detail that the proposition that the pieces were once together appears to be as incontrovertible as any geological proposition ever can be. Supplementary evidence from the comparison of the geology where it is well known on both sides, and from palaeomagnetism, strengthen the case but need not be appealed to for proof which is sufficient without them.

(Hess, 1965: 321)

Perhaps Hess wanted to kill two birds with one stone: even if you hide your head in the sand to avoid the paleomagnetic case for continental drift, you would have to hide it even deeper to avoid the jig-saw fit of the circum-Atlantic continents.

4.9 Wilson develops the idea of transform faults

I was here with Tuzo and Harry Hess. Tuzo came to spend a year here ... I didn't know Tuzo in those days. I'm sure that Fred would have sent one [a reprint] to Harry because Harry was the father and mother of the paper ... They took a sabbatical. They were here. Harry didn't do very much in that year. He did not have much effect on us. He wrote that piece for the Colston volume. But he was around. That is how this place operates. There were only five staff members here; there were 70 people here. There must have been 25 post-docs here. They were here, and nothing much happened. They came and talked. And then Tuzo had a yacht. He was a keen yachtsman, and took his family sailing south of Turkey. It was away for about six weeks or so. And I remember sitting ... I remember him pounding up the stairs in the stables, which was a very dusty place where we lived atop of, and he said, "What are you doing?" I was trying to write up the Owen fracture zone paper; I had it, I was actually working on it. He said, "Whatever you're doing it cannot be as important as what I would like you to do. Stop!" So we were rather resentful. And he came up with these ... ideas that had occurred to him sitting on the deck of this yacht. He said if you want ideas go away and do nothing. The first was the transform fault idea. The second was the symmetry idea: that the anomalies ought to be symmetrical and we ought to be able to use the Cox and Doell timescale to compute them. We hadn't done it. Fred said the other day that he thought he knew about Cox and Doell's work in 1963. I'm sure I didn't know. If we did know in '63, it certainly didn't have any effect on us. But, by 1964 or so, I guess we knew vaguely what they were doing ... But paleomagnetism had collapsed here at about that point, and there wasn't anything going on. Tuzo came up the stairs, talked about transform faults, picked up a piece of paper, tore it the way he used to tear a piece of paper, and he did this, and it was obvious. I was very embarrassed, very cross because it was the obvious solution to the Owen fracture paper which I was writing. And I said, "I cannot use this. I wish you hadn't told me. I've nearly finished this paper; I simply cannot swipe your idea because this is the obvious solution to the earthquake problems on the fracture zone. For God's sake publish it quickly."

(Matthews, 1979 interview with author[9])

Wilson has made extremely important contributions to the sea floor spreading hypothesis – transform faults, the migration of the circum Antarctic ridge outward away from that continent, and others.

(Hess, March 19, 1968 review of Wilson's paper "A revolution in Earth Sciences; life cycle of ocean basins")

Hess recommended that the paper not be published, and sent Wilson a copy of his review.[10]

I have ... sent a letter to *Nature* reinterpreting many large features in terms of a new class of faults, and Harry has seen the paper and appears to agree. It is of course based upon the observations of many others and in the Pacific your book has been invaluable. Broadly speaking my views support many of your ideas.

(Wilson, June 1, 1965 letter to Menard[11]*)*

Tuzo sent the manuscript before 1 June. To his surprise, an editor phoned instead of writing him to say it was accepted.

(Menard, 1986: 247)

We must first determine when Wilson came up with transform faults. As the above quote shows, Matthews first learned about them from Wilson after Wilson returned from a month's holiday with his family on April 21. Vine recalled that he first mentioned the idea of transform faults to Wilson in February (Glen, 1982: 304). Coode is quite sure that Wilson gave a talk at Newcastle in early February before he developed the idea of transform faults, for it was immediately after this talk that Coode told Wilson about what later became called ridge-ridge transform faults (§4.11). The Wilsons left on their trip on March 12. In addition, Isabel Wilson recorded in her diary that her husband was writing an article for *Nature* on March 25 and 26 while they were in Antalya, Turkey, presumably the transform fault article that was published four months later. In addition, Wilson was definitely thinking about transform faults while attending the ICSU meeting from April 4 through April 7 because he drew on an empty page of the 1965 Year Book of the ICSU, which he took to the meeting (Figure 4.3).

The sketch, showing the Carlsberg Ridge joining the Ornach-Nal fault and connecting up to the Hindu Kush, sums up what he discussed with Matthews, who was then finishing his paper on the Owen Fracture Zone. Wilson's tour-de-force on transform faults was published in *Nature* on July 24, 1965, and he explained them in a seminar at Madingley Rise on May 26.[12] Finally, Matthews was finishing up his paper on the Owen Fracture Zone when Wilson came "pounding up the stairs in the old stables" and told him about transform faults. Matthews' paper was received by the Royal Society on May 12, and he wrote this in his daybook on April 29, 1965:

Time since getting back into lab has been spent writing papers for R.S: the Seychelles Bank, and the Owen Fracture Zone. During the last stages of completion of the latter I had several discussions with Tuzo Wilson who has been developing notions of fault transforms between mid-ocean ridges, great transcurrent faults and island (and mountain) arcs. In particular we sought for this one in West Pakistan ... [Matthews then drew a crude figure of the Carlsberg Ridge being joined by the Ornach-Nal fault near Karachi.] He put me on to a vitally important reference which I had missed: A. G. Jones.[13]

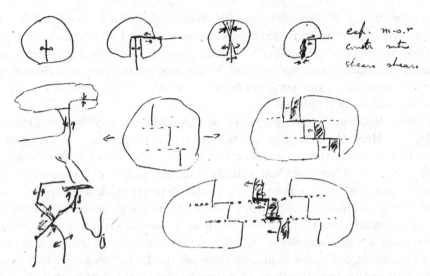

Figure 4.3 Wilson's crudely drawn figures of transform faults that he drew on a blank page of the 1965 Year Book of the ICSU. He later redrew or had them redrawn for his first paper on transform faults (Wilson, 1965b). Compare especially with Figures 4.12 and 4.13.

The "A. G. Jones" reference concerned the Ornach-Nal fault. So, all things considered, Wilson began developing the idea of transform faults, probably in late February, but definitely before mid-March. He discussed transform faults with various people at Cambridge after he returned with his family on April 21. He had developed the idea fully enough to give a seminar on May 26, and submitted it before June 1 after Hess had read and seemingly agreed with it (see above quote). This timetable is consistent with Vine's recollection (§4.10) that he told Wilson about the idea of ridge-ridge transform faults soon after Wilson's arrival. It is also consistent with Coode's recollection (§4.11) that he did likewise after Wilson gave a talk at Newcastle, a talk in which he did not mention transform faults. I hasten to add that neither Vine nor Coode believe that Wilson took their idea as his own; they think that they may have triggered Wilson to think of the idea and develop it on his own.

Before discussing Wilson's transform paper, I want to suggest that his development of his idea arose not from his attempting to work out the consequences of seafloor spreading, but from the blending of his longstanding interests in transcurrent faults, in the geometry and distribution of deformation zones (island arcs and mountain belts), and his puzzlement over the apparent abrupt termination of many of the great fracture zones at the western margins of North and Central America. His thoughts may have been brought into focus by the appearance in June 1964 of a paper by W. N. Gilliland that proposed that the Mendocino Fracture Zone, and perhaps other east–west fracture zones in the Pacific, extended under North America and reappeared as east–west fracture zones in the Atlantic; and by his plan to revise, while on sabbatical at Madingley Rise, the coauthored textbook *Physics and Geology* (Jacobs, Russell, and Wilson, 1959) now badly out of date. I want to make clear that I am not claiming that it was these two factors alone that led Wilson to come up with transform faults. As I shall explain anon, Wilson was, in early 1965, told about ridge-ridge transform faults independently by Vine and Alan Coode, and this may have been the catalyst (§4.11, §4.12).

By the 1950s, Wilson's outlook had become global like those of Wegener, Holmes, Carey, and Hess. He had longstanding interests in the geometry of deformation zones, first in North America and later globally, as he travelled widely. In the early 1950s, Scheidegger and he had offered a contractionist explanation for the arcuate shape of island arcs and young mountain belts (§1.3). He also was impressed with the two worldwide mega-belts of Mesozoic–Cenozoic mountain and island arcs that approximated great circles, and emphasized that they formed a T, meeting orthogonally at the Banda Sea (§1.3). He was interested in large transcurrent faults and he constructed a figure displaying their worldwide distribution in *Physics and Geology*; he and his co-authors argued, curiously, that transcurrent faults such as the San Andreas, Great Glen, and Alpine told against continental drift (§1.5). Once he became a mobilist, everything changed. In 1962 he argued that the convection currents which rise up along the East Pacific Rise are responsible for the San Andreas Fault, and other transcurrent faults in the Pacific such as the Alpine Fault associated with island chains (§1.9). In 1962 he argued that the Great Glen and Cabot faults had once formed a single fault when North America and Europe were united (§1.10). Nonetheless, he offered no explanation for the great fracture zones crossing the East Pacific Rise. He agreed that matching magnetic patterns indicated huge transcurrent motions along the great fracture zones, but in 1962 he was baffled by their abrupt termination, for example, of the Murray Fracture Zone at the California coast (§1.9). Wondering about transcurrent faults on a global scale was one of Wilson's longstanding concerns.

There were, I think, two factors that helped bring to the forefront his thoughts about what were to become transform faults. As I have said, Wilson had arrived at Madingley Rise planning to continue revising *Physics and Geology* (Menard, 1986: 243; Glen, 1982: 347). He had thought about doing so soon after the March 1964 Royal Society's symposium on continental drift (§3.3) (Wilson, March 26, 1964 letter to Menard). The textbook with its fixist approach and vehement attack on mobilism was in dire need of revision, although he, quite sensibly, did not complete the revisions until 1974 after completion of the restructuring of the Earth sciences required by plate tectonics. Then there were Gilliland's papers, the first of which (1962) proposed that the Mendocino Fracture Zone extended through or beneath North America at 40° N latitude, and (1962: 686) tentatively proposed that the fracture zone "is probably as old as Precambrian." Two years later, he (1964: 1276) proposed that the "world-wide distribution, in the ocean basins, of east-west trending faults with hundreds of miles of horizontal displacement ... may support the concept of an eastward-moving force below the 'crust'." Adopting a hypothesis by W. S. Jardetsky, based on the analogy

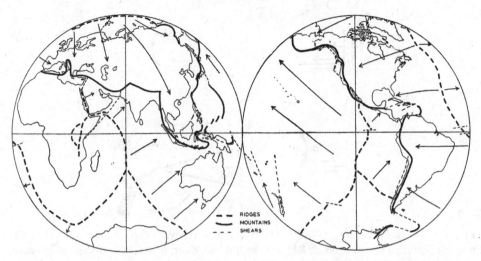

Figure 4.4 Wilson's Figure 1 (1965b: 343). His caption reads: "Sketch map illustrating the present network of mobile belts, comprising the active primary mountains and island arcs in compression (solid lines), active transform faults in horizontal shear (light dashed lines) and active mid-ocean ridges in tension (heavy dashed lines)."

with the "observed velocities of surficial materials on the Sun, Jupiter and Saturn," Gilliland proposed that Earth's rotation could cause differential zonal rotation within Earth's liquid mantle, which caused the east–west shearing of the crust above zonal boundaries (Gilliland, 1964: 1277). Even though Wilson did not mention Gilliland's work in his 1965 transform fault paper, he later recalled (Glen, 1982: 375) that he was aware of it, and was himself puzzled by the abrupt termination of the great fracture zones at the California coast.

Wilson's paper (1965b), "A new class of faults and their bearing on continental drift," was, as I have said, a blockbuster. He not only introduced the idea of transform faults, but he divided Earth's surface into about ten major and minor rigid plates. He claimed that such plates are separated by a continuous system of seismically active mobile belts, which take the form of what he called transform faults where crust is neither created nor destroyed, mid-ocean ridges where new crust is created, and trenches where old crust is swallowed up and island arc and primary mountain belts are created. The idea of transform faults, in particular ridge-ridge transform faults, like V-M, is essentially a corollary of seafloor spreading; its confirmation by Lynn Sykes at Lamont in 1967 led to the second widely recognized difficulty-free solution of seafloor spreading (§6.9).

Wilson asked this question: why do seismically active belts "which take the form of mountains, mid-ocean ridges or major faults with large horizontal movements" end abruptly? He suggested that they only *appeared* to end, they actually "are connected into a continuous network of mobile belts about the Earth which divide the surface into several rigid plates (Fig. 1)" (Wilson, 1965b: 343) (see Figure 4.4).

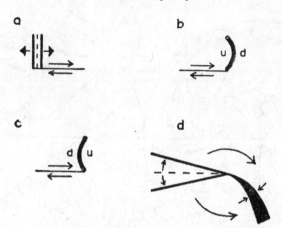

Figure 4.5 Wilson's Figure 2 (1965b: 344). His caption reads: "Diagram illustrating the four possible right-hand transforms. *a*, Ridge to dextral half-shear; *b*, dextral half-shear to concave arc; *c*, dextral half-shear to convex arc; *d*, ridge to right-hand arc."

Foreshadowing plate tectonics, he divided the Earth's surface into "several rigid plates." He also foreshadowed plate tectonics by identifying three types of borders: those where crust is destroyed (island arcs and mountain belts), those where crust is created (mid-ocean ridges), and those where crust is neither created nor destroyed (transform faults).

Wilson claimed that the relative motion across one border is transformed into that of another where they join (Figure 4.5). "A junction where one feature changes into another is here called a transform" or half-shear (Wilson, 1965b: 343).

He next introduced a type of fault that connects ridges with ridges, ridges to arcs, and arcs to arcs, and he, most importantly, appropriately named it a transform fault.

Transform faults. Faults in which the displacement suddenly stops or changes form and direction are not true transcurrent faults. It is proposed that a separate class of horizontal shear faults exist which terminate abruptly at both ends, but which nevertheless may show great displacements. Each may be thought of as a pair of half-shears joined end to end. Any combination of pairs of three dextral half-shears may be joined giving rise to the six types illustrated in Fig. 3 [my Figure 4.6]. Another six sinistral forms can also exist. The name transform fault is proposed for the class, and members may be described in terms of the features which they connect (for example, dextral fault, ridge-convex arc type).

(Wilson, 1965b: 343; my bracketed addition)

Wilson then explained and illustrated the varying growth habits of each type of transform fault.

The distinctions between types might appear trivial until the variation in the habits of growth of the different types is considered as shown in Figure 4 [my Figure 4.7]. These distinctions are that ridges expand to produce new crust, thus leaving residual inactive traces in the topography

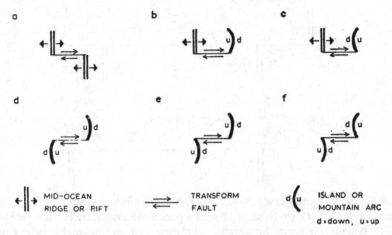

Figure 4.6 Wilson's Figure 3 (1965b: 344). His caption reads: "Diagram illustrating the six possible types of dextral transform faults. *a*, Ridge to ridge type; *b*, ridge to concave arc; *c*, ridge to convex arc; *d*, concave arc to concave arc; *e*, concave arc to convex arc; *f*, convex arc to convex arc. Notice that the direction of motion in *a* is the reverse of that required to offset the ridge."

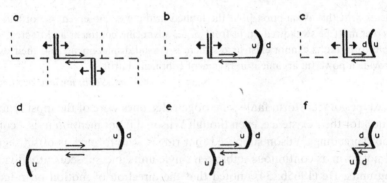

Figure 4.7 Wilson's Figure 4 (1965b: 344). His caption reads: "Diagram illustrating the appearance of the six types of dextral transform faults shown in Fig. 3 after a period of growth. Traces of former positions now inactive, but still expressed in the topography, are shown by dashed lines."

of their former positions. On the other hand oceanic crust moves down under island arcs absorbing old crust so that they leave no traces of past positions. The convex sides of arcs thus advance. For these reasons transform faults of types *a*, *b* and *d* in Fig. 4 grow in total width, type *f* diminished and the behaviour of types *c* and *e* is indeterminate.

(Wilson, 1965b: 343; my bracketed addition)

Wilson argued that the existence of transform faults implies continental drift, and therefore "their existence would provide a powerful argument in favour of continental drift."

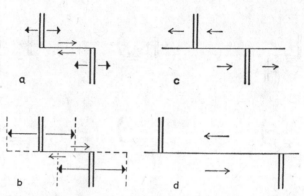

Figure 4.8 Wilson's Figure 1 (1965c: 482). (a) Dextral ridge-ridge transform fault connecting two ridges or ridge-offsets. (b) Fault in 1a after a period of movement. The distance between ridge-offsets remains the same; the direction of movement is dextral, and opposing motions are only between ridge-offsets. (c) Sinistral transcurrent fault which offsets the ridge segments further than originally offset. (d) Fault in 1c after a period of movement. The distance between ridge-offsets increases, motion is sinistral, and opposing motions continue all along the fault.

If continents drift this assumption [that the faulted medium is conserved] is not true. Large areas of crust must be swallowed up in front of an advancing continent and re-created in its wake. Transform faults cannot exist unless there is crustal displacement, and their existence would provide a powerful argument in favour of continental drift.

(Wilson, 1965b; my bracketed addition)

Of the six types of transform faults, the ridge-ridge ones were of the most immediate importance, for their existence, even though Wilson did not mention it, is a corollary of seafloor spreading. Wilson supposed that ridges were formed as offset segments; they did not form as continuous unbroken single units and subsequently divide into offsetting units. He (1965b: 343) noted that the direction of motion on ridge-ridge transform faults "is the reverse of that required to offset the ridge. This is a fundamental difference between transform and transcurrent faulting." Wilson illustrated this difference in a figure (Figure 4.8) he constructed in his next paper on transform faults, Wilson 1965c, "Transform faults, oceanic ridges, and magnetic anomalies southwest of Vancouver Island." He also emphasized another difference between ridge-ridge transform and transcurrent faults. With transform faults, seismic activity occurs only between ridge-offsets; with transcurrent faults, seismic activity occurs all along the fault. As we shall see, these predictions can be tested by seismology.

Wilson later made paper models to display differences between transform faults (Figure 4.9) and transcurrent faults (Figure 4.10). They are easily constructed and clearly display the different directions of motion of transform and transcurrent faults and restriction of seismic activity between ridge-offsets along transform faults but not with transcurrent faults.[14]

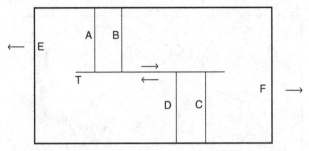

Figure 4.9 Wilson's paper model of a transform fault. Cut paper along T. Crease paper downward at B and D; crease paper upward A and C. Fold B under A, and C under D. T represents transform fault; A and C represent ridge segments. Grasp paper at E and F and slowly pull apart in direction of arrows to simulate seafloor spreading. Notice that movement along fault is right lateral as predicted by Wilson. Notice that paper moves in opposite directions along T only between ridge segments A and C simulating Wilson's prediction that seismic activity is restricted along the fault only between ridge segments.

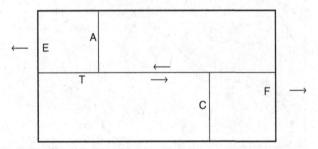

Figure 4.10 Wilson's paper model of a transcurrent fault. Cut paper into two pieces along line T. T represents a transcurrent fault; A and C represent ridge segments. To simulate transcurrent movement along T, grasp pieces of paper at E and F, and pull in direction indicated by arrows. Notice left lateral movement along fault, and that the movement is in opposite directions along entire length of T. Thus predictions based on this model differ in direction of motion and location of earthquakes.

4.10 Wilson's third trip around the world

With transform faults in hand, Wilson journeyed around the world. He had made the trip twice before, first as a fixist during the 1950s when he was guided by Scheideger's and his own explanation of island arcs (§5.3), and later in 1963 as a mobilist when he fit continents back together guided by the twin assumptions that ocean ridges lie halfway between formerly unified continents and that lateral ridges mark former points of attachment (§7.2). Before beginning, however, he acknowledged Carey's influence, but also emphasized that, unlike himself, Carey did not restrict deformation of plates to their edges (II, §6.11). He first stopped at the North Atlantic and Arctic Oceans (Figure 4.11); familiar territory for him because he already had

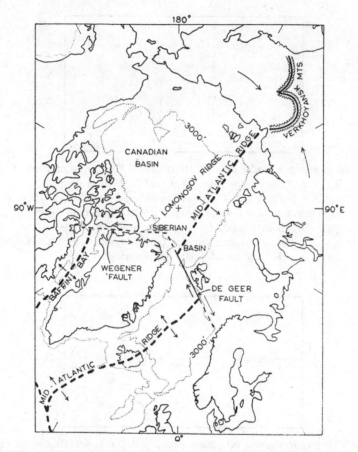

Figure 4.11 Wilson's Figure 5 (1965b: 344). His caption reads: "Sketch map of the termination of the Mid-Atlantic ridge by two large transform faults (Wegener and De Geer faults) and by transformation into the Verkhoyansk Mountains." Earthquakes no longer occur along the inactive Wegener Fault (light dashed line) and older traces of the De Geer Fault between Spitsbergen and Norway (light dashed line), but they do occur along the section of the De Geer Fault between ridge-offsets (solid line). Ellesmere Island is above the Wegener Fault opposite northern Greenland and the Spitsbergen islands are to the right of the De Geer Fault.

proposed the Wegener Fault, discussed the De Geer Fault, and described the transformation of the Mid-Atlantic Ridge into the Verkhoyansk Mountains (§3.2). He cited Wegener's *Origin of Continents and Oceans*, Wegmann's analysis of the De Geer Fault (I, §8.10), Heezen and Ewing's extending of the Mid-Atlantic Ridge into the Arctic Ocean, and mentioned the work of Harland and others in delineating the geological disjunctions between Ellesmere Island and Spitsbergen. He categorized the Wegener Fault as a non-active sinistral, ridge-ridge transform fault, the De Geer line as a dextral ridge-ridge transform fault, and the junction of the Mid-Atlantic Ridge and Verkhoyansk Mountains as a ridge to right-hand arc transform. Wilson claimed that

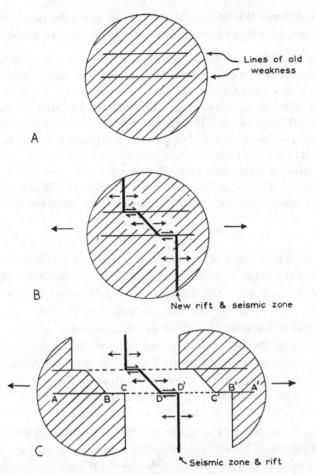

Figure 4.12 Wilson's Figure 6 (1965b: 345). His caption reads: "Diagram illustrating three stages in the rifting of a continent into two parts (for example, South America and Africa). There will be seismic activity along the heavy lines only."

the restriction of seismic activity to that section of the De Geer Fault between offsets of the Mid-Atlantic Ridge supported his interpretation (Wilson, 1965b: 344).

Turning his attention to the equatorial fracture zones, which divided the Mid-Atlantic Ridge into about ten offsets, arguing that they are not sinistral transcurrent faults but dextral ridge-ridge transform faults, Wilson prefaced his discussion by presenting an idealized rifting of a continent with the formation of a nascent mid-ocean ridge, but with the Atlantic Ocean in mind (Figure 4.12). He emphasized that the ridge formed with the initial breakup of the continents and that direction of motion along the fracture zones "is in reverse direction to that required to produce the apparent offset" for the ridge. According to Wilson, the ridge originally formed with its offsetting segments. Although Wilson did not explicitly state it, the direction of motion is that required if new seafloor is created along the ridge crests and subsequently moves horizontally away from them, eventually forming a new ocean basin.

Equatorial Atlantic fracture zones. If a continent in which there exist faults or lines of weakness splits into two parts (Fig. 6), the new tension fractures may trail and be affected by the existing faults. The dextral transform faults (ridge-ridge type) such as AA′ which would result from such a period of rifting can be seen to have peculiar features. The parts AB and B′A′ are older than the rifting. DD′ is young and is the only part now active. The offset of the ridge which it represents is not an ordinary faulted displacement such as a transcurrent fault would produce . . . It is confusing, but true, that the direction of motion along DD′ is in the reverse direction to that required to produce the apparent offset. The offset is merely a reflexion of the shape of the initial break between the continental blocks. The sections BD and D′B′ of the fault are not now active, but are intermediate in age and are represented by fracture zones showing the path of former faulting.

(Wilson, 1965b: 344)

Returning to the equatorial fracture zones of the Mid-Atlantic Ridge, he reasoned (1965b: 345) that the shape of the ridge with its offsets should match the shape of the matching edges of Africa and the Americas, if, as he believed, "the apparent offsets on the ridge are not faulted offsets, but inherited from the shape of the break that first formed between the coasts of Africa and the Americas."

Wilson moved to the Indian Ocean and Arabian Sea where he identified two large sinistral ridge-convex arc transform faults. One extended NNE from the north end of the Carlsberg Ridge to the coast of the Arabian Sea just north of Karachi, where it merged with the Ornach-Nal Fault and continued northward to the western end of the Hindu Kush (Figure 4.13). The ocean part of this transform fault is the Owen Fracture Zone, the very fracture that Matthews was writing about for the proceedings of the Royal Society's November 1964 discussion concerning the floor of the northwest Indian Ocean (Matthews, 1966a) when Wilson "pounded up the stairs" and first told Matthews about transform faults. The second transform fault extends from the northern end of the young ridge running up the Red Sea to the Jordan Valley, meets a large thrust fault in southeastern Turkey, and terminates at the Taurus Mountains. Here Wilson cited Drake and Girdler (1964) (§2.4), and a paper by A. M. Quesnell (1958) on the structure and evolution of the Dead Sea rift.

Wilson also talked to Laughton, who, like Matthews, was preparing a paper for the proceedings of the Royal Society's meeting (Laughton, 1966a). Laughton had traced the position of the young rift running the length of the Gulf of Aden, which
· extended from the Owen Fracture Zone through the Gulf of Aden where it met both the young rift running up the Red Sea and the northern end of the East African rift valleys (later identified as a ridge-ridge-ridge triple junction). Laughton argued that the Gulf of Aden formed as the Arabian block rotated counterclockwise relative to Africa, that new seafloor is being created within the rift through upwelling mantle convection. He noted Irving and Tarling's paleomagnetic work on the Aden volcanics, which independently indicated a counterclockwise rotation of Arabia relative to

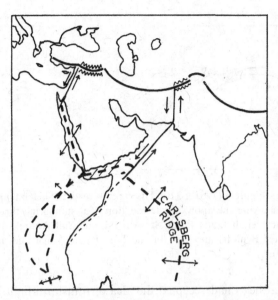

Figure 4.13 Wilson's Figure 8 (1965b: 345). His caption reads: "Sketch illustrating the end of the Carlsberg mid-ocean ridge by a large transform fault (ridge–convex arc type) extending to the Hindu Kush, the end of the ridge up the Red Sea by a similar transform fault extending into Turkey and the still younger East African rifts." The Owen Fracture Zone is the transform southeast of Arabia and it projects on land near Karachi.

Africa (II, §7.8). He also referred to Girdler's work on the Red Sea including that with Peter and Drake. Laughton (1966a: 157; my bracketed additions) also noted a strong negative central magnetic anomaly centered within the rift, and less intense "negative and positive anomalies north and south of the [rift] valley [which] may also be attributed to normally and reversely magnetized dykes." However, he said nothing about whether the pattern of magnetic anomalies supported V-M. Laughton also matched the trend of the ridge broken by faults into separate offsets with the shapes of the adjacent continental margins. Wilson (1966b: 345), referring to a private conversation with Laughton, proposed that the "many offsets described by Laughton provide another example of transform faults adjusting a rift to the shape of the adjacent coasts."

Notably, Matthews in his Royal Society paper (Matthews, 1966a: 173) did not, as Wilson would do, describe the Owen Fracture Zone as a sinistral ridge convex arc transform fault but as a dextral transcurrent fault. I say this even though Matthews, as he retrospectively noted, realized that Wilson had provided him with the "obvious solution to the earthquake problem on the fracture zone." Matthews was in the final stages of finishing his paper, but did not want to "swipe" Wilson's as yet untested idea. Matthews did acknowledge that he might well be wrong and even added that the assumption that the ridge formed as a continuous structure and later became offset by transcurrent faulting "may be incorrect" (Matthews, 1966a: 182). So I think

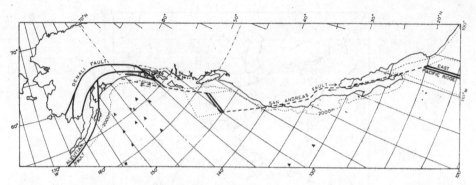

Figure 4.14 Wilson's Figure 9 (1965b: 346). His caption reads: "Sketch map of the west coast of North America showing the approximate location of a submarine thrust fault along the Aleutian trench, the Denali faults (after St. Amand) the San Andreas and another large transform fault (after Benioff) and part of the East Pacific ridge and another mid-ocean ridge (after Menard)."

it is likely that Matthews had accepted the idea of transform faults even though he did not adopt it in his paper on the Owen Fracture Zone. He also noted that the magnetic profiles were consistent with V-M, and characterized Bodvarsson and Walker's idea of "expansion of the oceanic crust by the injection of swarms of basalt dykes parallel to the mid-ocean ridges" as "very plausible." Matthews (1966a: 182–183) also briefly discussed the work that A. G. Jones had done with the Hunting Survey Corporation in Western Pakistan, work that Wilson had told him about, and mentioned that the Owen Fracture Zone might be connected "with the recently discovered Ornach-Nal dislocation which runs through West Pakistan."

Finished with the Indian Ocean, Wilson went to the Pacific, and began with a discussion of the Ridge and Trough Province, the area that would become of critical concern for Wilson and Vine. He categorized the San Andreas Fault as a dextral ridge-ridge transform fault, connecting the East Pacific Rise with a tiny ridge segment (later named the Juan de Fuca Ridge) off Vancouver Island and the Washington–Oregon coast (Figure 4.14). He (1965b: 346) also proposed that another transform fault (the Queen Charlotte Islands Fault), a dextral ridge-convex arc transform fault, extended from the tiny ridge segment to the Aleutian arc where "the Pacific Basin is sliding . . . under the Alaska Mainland, and the Bering Sea." Wilson also drew on work on trenches by Benioff, who had favored continental growth to continental drift, even though he did think that continents override adjacent denser oceanic crust which dips under continents forming trenches (III, §1.6).

The San Andreas fault is here postulated to be a dextral transform fault (ridge-ridge type) and not a transcurrent fault. It connects the termination of the East Pacific ridge proper with another short length of ridge for which Menard (1964) has found evidence off Vancouver Island. His explanation of the connexion – that the mid-ocean ridge connects across western United States – does not seem compatible with the view that the African ridge valleys are also incipient mid-ocean ridges. The other end of the ridge off Vancouver Island appears to end in a second great submarine fault off British Columbia described by Benioff (1962) as having dextral horizontal motion.

(Wilson, 1965b: 343; I have replaced Wilson's footnotes
to his references with the references themselves)

Wilson (1965b: 346) maintained that the "short length" ridge was a spreading center. Although he made it quite clear that seafloor "is being slid relatively speaking under the Alaska Mainland and the Bering Sea," he remained silent about what was happening to seafloor spreading southeastward from the opposite side of the ridge along the San Andreas Fault. Perhaps he had not sorted it out.

Wilson was unsure about what to do with the aseismic great east–west fracture zones that occur offshore in the northeastern Pacific and display such huge displacements. He was sure they are not transcurrent faults, because they are aseismic. But, if they are transform faults, then they must not be between currently active ridge-offsets, for Sykes (1963) had shown that fracture zones between ridge-offsets are active. Perhaps they are a more complex type of transform fault than any he had yet introduced; perhaps they mirror the shape of some "contemporary rift in the Pacific Ocean"; perhaps they are "related to Hess' (and Menard's) Darwin rise."

If the examples given from the North and Equatorial Atlantic Ocean, Arabian Sea, Gulf of Aden and North-west Pacific are any guide, offsets of mid-ocean ridges along fracture zones are not faulted displacements, but are an inheritance from the shape of the original fracture. The fracture zones that cross the East Pacific ridge (Sykes, 1963) are similar in that their seismicity is confined to the offset parts between ridges. An extension of this suggests that the offsets in the magnetic displacements observed in the aseismic fracture zones off California may not be fault displacements as has usually been supposed, but that they reflect the shape of a contemporary rift in the Pacific Ocean. More complex variants of the kind postulated here seem to offer a better chance of explaining the different offsets noted by Vacquier (1962) along different lengths of the Murray fracture zone than does transcurrent faulting. If the California fracture zones are of this character and are related to the Darwin rise as postulated by Hess, then the Darwin rise should be off set in a similar manner.

(Wilson, 1965b: 346)

Wilson still did not know what to do with the great aseismic fracture zones. He had not solved the problem of why the Mendocino Fracture Zone just appears to end at the coast and, unlike his usual practice, he apparently did not have an idea he thought worth mentioning. He had not envisioned the idea or thought it too speculative that much of the East Pacific Rise and its eastern limb of magnetic anomalies had been swallowed up by a former trench beneath North America as it advanced westward.

Wilson then moved south invoking transforms as he went, ending his journey in the southeastern Pacific and the ridge system surrounding Antarctica (Figure 4.15). In 1961, when first considering the implications of seafloor spreading, Wilson raised the difficulty posed by the circum-Antarctic ridge; it was this that had led him to

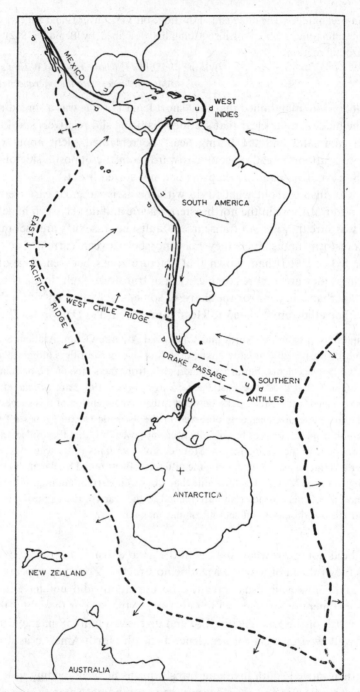

Figure 4.15 Wilson's Figure 10 (1965b: 346). His caption reads: "Sketch map of Mexico, South America, Antarctica and part of the mid-ocean ridge system (heavy dashed lines) illustrating that the great loop of the ridge about Antarctica can only grow by increasing in diameter. Transform faults are shown by light dashed lines."

propose that ridges migrate sideways (§1.9). Much later, McKenzie, then a graduate student in the Department of Geodesy and Geophysics, recalled that Wilson and he had talked about just this difficulty and "could not understand how the ridges round Antarctica could be the surface expression of convection cells."

I did have a conversation with Tuzo about exactly this issue, and I remember it well. We were standing in the entrance of the old house in 1965, and on the wall on the left as you came in was one of [G. B.] Udintsev's maps, I think of the Indian Ocean ... At that time everyone was thinking in terms of convection cells rising beneath ridges. Neither of us could understand how the ridges round Antarctica could be the surface expression of convection cells.

(McKenzie, September 16, 2009 email to author; my bracketed addition)

Wilson (1965b: 346) continued to maintain that "the great loop of the ridge about Antarctica can only grow by increasing in diameter." This solution, however, made the tie between mantle convection, or at least large-scale mantle convection, and seafloor spreading more difficult to understand. Indeed, Wilson did not mention mantle convection or even seafloor spreading per se in his transform paper, which was concerned only with kinematics. McKenzie (September 16, 2009 email to author) added that the central question that he asked in his conversation with Wilson was: "How did they [i.e. convection currents] know that they had to move away from each other at exactly the right rate to stay beneath the ridges?" Looking ahead, McKenzie continued to worry about the relationship between mantle convection and mid-ocean ridges. Perhaps it was this concern that led him to separate kinematics from dynamics and explicitly state that plate tectonics is a kinematic theory, not a dynamic one (McKenzie and Parker, 1967). Following up, McKenzie (1967a) separated mantle convection from ridge formation.

I think I was the first to sort this out, in a *JGR* paper in 1967... Ridges add to the two plates on either side at the same rate. This statement is generally true, though there are minor exceptions. A central issue then arises. Ridges cannot have any deep convective structure, and all their features (elevation, heat flow, and volcanism) must be due to passive upwelling of hot mantle between separating plates, like sea water wells up and freezes between separating ice flows. For me this was the central idea that freed us from all the difficulties that so bothered Tuzo and me in 1965. I think everyone now believes this, and I have written a number of papers, starting in 1967, showing that this idea accounts for the observations.

(McKenzie, September 16, 2009 email to author)

Wilson identified several more transform faults: a dextral ridge-convex arc transform connecting the eastern end of the East Chile Ridge with the southern end of the Northern Andes, and two dextral–sinistral pairs of convex arc-convex arc transforms at both ends of the South Antilles and West Indies arcs. In all, he identified six of the twelve types of transform faults: dextral and sinistral ridge-ridge transforms, dextral and sinistral ridge-convex arc transforms, and dextral and sinistral convex arc-convex arc transforms. He claimed that he had done enough to justify further investigation of them.

The demonstration of a few examples that at least six of the twelve types do appear to exist with the properties predicted justifies investigating the validity of this concept further.

(Wilson, 1965b: 347)

But he had another reason why further work was justified, namely "establishing the reality of continental drift."

It is particularly important to do this because transform faults can only exist if there is crustal displacement and proof of their existence would go far towards establishing the reality of continental drift and showing the nature of the displacements involved.

(Wilson, 1965b: 347)

As I shall show in the next chapter, Wilson was already heeding his own advice, teaming up with Vine, and with Hess's and Mason's help found additional support for the existence of transform faults as well as for V-M.

4.11 Vine independently proposes ridge-ridge transform faults

Vine recalls, "I had recognised what Tuzo was calling transform faults in 1964" (October 12, 2009 email to author). He was likely the first to do so. He discussed them, without naming them, during a lecture on November 4, 1964, entitled "Magnetic anomalies over the oceans," which he gave to the Geology Section of the Cambridge Natural History Society. Vine prepared a set of "skeletal notes" for the talk, reminders of what he should discuss (Figure 4.16). The notes contain no explicit statement or sketch of what later became known as ridge-ridge transform faults. They mention slides of the Indian Ocean and the east Pacific, presumably of Matthews' magnetic survey over the Carlsberg Ridge on which V-M is based, and Mason and Raff's survey of the northeast Pacific. After reminding himself to show his last slide, "Transcurrent faults of Ind. Oc.," he added "problems posed – mechanism of faulting" and "explanation of area" which suggest that he was concerned about the origin of what had been identified as fracture zones in the Indian Ocean. Perhaps he used the idea of ridge-ridge transform faults to explain movement along the fracture zone connecting two offsets of the Carlsberg Ridge. Vine may also have attempted to explain the Murray Fracture Zone in the northeast Pacific as a transform fault. His mention of the "world-wide ridge system" suggests that he also discussed other ridges.

Looking over the skeletal notes forty-five years later, Vine made these remarks.

As one might expect, this talk, probably the first one I gave on my Ph.D. work, and just over two years into the Ph.D., was essentially a summary of my thesis work. It is, therefore, a review of magnetic anomalies over the oceans. However at one point I mention the "transcurrent" faults of the East Pacific, presumably because the magnetic anomalies recorded on either side of them reveal the apparent offset on them. However I go on to refer to the "Indian Ocean," the "mechanism of faulting" and the "world-wide ridge system." It is not clear what points I made here in that they are not obviously related to the magnetic anomalies as then known.

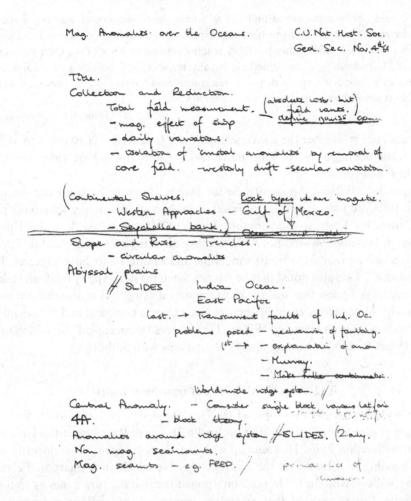

Figure 4.16 Vine's skeletal notes for his talk "Magnetic anomalies over the oceans," which he gave before the Geology Section of the Cambridge Historical Society on November 4, 1964.

It is possible that I mentioned what became known as ridge-ridge transform faults but one cannot be sure of this from these notes.

(Vine, January 25, 2010 email to author)

Vine relayed these thoughts to Cann, Hess, and Wilson. None were impressed. He recalled their reaction in an undated letter he sent to Arthur Buddington, to help him prepare his memorial for Hess (Buddington, 1970). Buddington asked Vine about Hess's time at Cambridge in 1965.

Needless to say I was somewhat preoccupied with my thesis but in passing I happened to mention to both Harry and Tuzo separately, and to anyone else who would listen at that time, that the so-called transcurrent faults which offset the Mid-Atlantic Ridge in the equatorial

Atlantic need not be transcurrent faults at all within the framework of sea-floor spreading, especially as they did not seem to have any expression on land along their strike. I cannot remember what Tuzo said to this, probably nothing because he hadn't been listening properly as usual; Harry said "yes" or something equally inspiring and reflecting his characteristic economy of words; Joe Cann, I distinctly remember, said "obviously, anyone can see that", and I, thoroughly discouraged, said nothing more about it.

(Vine, letter to Buddington)

Cann does not remember the conversation (Vine, January 25, 2010 email to author) and it was more important to Vine than it was to Cann. It was Vine's idea, and Cann thought it was obvious.

What about Wilson? Apparently he too forgot. But Vine certainly did not think Wilson stole the idea of transform faults from him. In discussing the above paragraph from his letter to Buddington, Vine (1979 interview with author) noted, "Although I certainly would not want to make anything of it ... I certainly had the idea ... which amounted to transform faults, and I think I may have triggered Tuzo into the idea." Vine also noted that he did not name the new type of fault, and added, "In many ways I guess that was the most important thing." Vine also decided not to write up his idea; there seemed little point given Cann's comment and the unenthusiastic reaction by both Hess and Wilson. He also was "preoccupied" in writing up his thesis (Vine, Buddington letter, and 1979 interview with author).

4.12 Alan Coode's idea of transform faults[15]

Alan Coode independently proposed the idea of ridge-ridge transform faults while he was a graduate student working on his Ph.D. under Runcorn at the University of Newcastle upon Tyne. He came up with the idea of ridge-ridge transform faults while listening to a lecture at the University of Newcastle in January or February 1965 by Wilson, who, in his lecture, interpreted oceanic fracture zones as transcurrent faults. Coode realized that, given the framework of seafloor spreading, this could not be correct.

Coode's father was an officer in the Royal Engineers of the British Army. Alan was born in Singapore in 1939, where his father was then stationed. Immediately prior to the fall of Singapore the family moved to India, where they stayed for three years. At the end of the war his father and family returned to the United Kingdom. The family moved from location to location as his father was reassigned to a new post, never more than three years in one place. Coode became interested in science and in geology after finding fossils in a neighbor's field. "I started becoming interested in science around fifteen when I saw belemnite fossils in a nearby farmer's field" (Coode, email to author, February 10, 2010). Although his father wanted Alan to become a mechanical engineer and study at Cambridge University as he had done, the young Coode "was interested in seeing the New World," settled on Canada, and decided to "study geology at the University of British Columbia (UBC)" preferring the climate there to that of the University of Toronto or McGill University (Coode, February 20, 2010 email to author). He enrolled at UBC in 1958, escaping Cambridge but not engineering.

In fact when it was time to register, the lady behind the counter told me that I was not going to enroll in geology, but in geological engineering. My response was "Oh am I?" to which she replied "Yes" so the future was set. When I inquired about why she did this I was told that Geological Engineering was relatively new and they were trying to populate it!

(Coode, February 20, 2010 email to author)

Coode received a Bachelor of Applied Science degree in 1962, and remained at UBC to obtain an M.Sc. studying under John (Jack) Jacobs, who had in 1958 co-authored the fixist text *Physics and Geology* with Russell and Wilson (§1.5). Jacobs was then interested in the origin and evolution of Earth's core and in micro-pulsations, very low frequency variations of the geomagnetic field that originate in the magnetosphere. Although there was some discussion of continental drift, it was not central to his education. Coode (February 20, 2010 email to author) knew about the paleomagnetic defense of mobilism and "could see no reason to doubt the work or the results. Everything seemed sound." He obtained his M.Sc. in 1963, and wanted to obtain his Ph.D. in geophysics. He almost ended up at Cambridge, but instead went to Newcastle University to work with Runcorn.

I took an M.Sc. with Jack Jacobs between 1962 and 1963. Upon completion of my M.Sc., I looked around for a different university, in which to take a Ph.D. I considered Toronto, but rejected it on the grounds that Jack Jacobs came from there and I may not have a sufficiently different exposure. I considered Sydney, Australia (Bullen), Cambridge UK and Newcastle-upon-Tyne UK. Keith Runcorn *quickly* offered me a position at Newcastle, which I accepted. Shortly after that I procured funding to go to Cambridge. I wrote to Sir Edward Bullard and asked to join him. He declined on the grounds I had accepted Keith's offer. So I went to Newcastle.

(Coode, February 20, 2010 email to author)

He began at the University of Newcastle upon Tyne in summer 1963. Within six months, he had co-authored two letters to *Nature* with Runcorn and Tozer, both as first author. The one with Runcorn, which argued against Egyed's Earth expansion, appeared on February 27, 1965; the one with Tozer, "Low-velocity layer as a source of the anomalous component of the geomagnetic variations at the coast," appeared the next month. Coode got to know Don Tarling and Ken Creer, and learned more about mobilism's paleomagnetic support. Runcorn also had him read works on continental drift and mantle convection.

Coode described what happened when he came up with the idea of ridge-ridge transform faults while listening to Wilson's talk in February 1965.

In the summer of 1963, I moved from Vancouver and the University of British Columbia (UBC) to Newcastle-upon-Tyne and the new university there. I had moved to study geophysics under Keith Runcorn and to obtain my Ph.D. A year and a half later (I have in my notes in

February 1965, but I cannot prove that), Tuzo Wilson came to the School of Physics at the University to present a review of oceanic features as he saw them and to provide his interpretation of these features. I remember I found it engaging. His interpretation of what later became known as transform faults was so geologically unrealistic that I was upset and concerned. How could so great a mind as Tuzo Wilson make such a poor interpretation? I had recently graduated from UBC with a degree in geological engineering and we covered structural geology in great detail. He was violating some basic principles. At this time I cannot remember what he had proposed in that presentation. After the talk, I waited in line to speak to him. After all I had graduated under Jack Jacobs and Jack and Don Russell and Tuzo Wilson had worked closely together in Toronto and I felt a strong loyalty towards Jack. If Tuzo Wilson had made some elementary mistakes, I should point them out and help him on his way. In talking to him, I suggested that there were simple ways to approach the issue of what later became known as transform faults. He argued with me and insisted that his view was correct, no doubt wondering what I thought I was doing challenging him with my marginal experience and lack of knowledge. After a few minutes it became clear to me that he was not flexible and truly believed that he was correct. All the hours of structural geology at UBC were of little avail. Finally, I challenged him to publish his idea and I would publish my interpretation.[16]

(Coode, February 9, 2010 email to author)

After his lecture, Wilson entertained questions from the audience, but Coode as a second-year graduate student understandably felt uncomfortable suggesting in public to such a leading figure that he had "made some elementary mistakes." So he introduced himself to Wilson, and he is absolutely sure about this, telling him that he had studied with Jacobs at UBC, and that he had taken a course in structural geology (February 9, 2010 email to author). All the very detailed enquiries that I have made of this matter (and there is only an outline here) convince me of the general correctness of Coode's testimony.

Coode challenged Wilson, wrote up his idea, and submitted it to *Nature*. He was understandably very upset when *Nature* rejected his letter, particularly by the manner in which it was done. He recalled this in a 1987 letter to Girdler.

It was rejected, almost out of hand. I asked around the Department for the reason and was told that *Nature* tends to go for someone with an established name and that as I did not have any sort of reputation, they turned it down. I believed that *Nature* must have used this policy as a safety measure to minimize the flow of material of dubious quality. I was told that if I co-authored the letter with someone like Keith [Runcorn] then it would probably be accepted and printed. In a moment of weakness, I decided that I should not co-author as the idea was entirely original and I would send it to a new and upcoming Journal, the *Can. J. of Earth Sciences*. In retrospect and in view of what you have told me, it would have made more sense to have co-authored it with Keith and had it out in March or April 1965. But I did not.

(Coode, March 7, 1987 letter to Girdler; my bracketed addition)

He later added:

I do not *think* my paper was refereed. I asked others (including Keith [Runcorn], I suspect) around the Department why it had been rejected and was told that *Nature* often did not print letters or papers unless they came attached to a good reputation name. I suspect that they looked at my name and decided that it was not prominent enough, i.e. it did not have a high enough profile. *Of course, I may have been told this to placate me, but at the time it had the ring of truth.*

(Coode, February 9, 2010 email to author; Coode's emphasis; my bracketed addition)

Perhaps *Nature* was unwilling to publish three papers by the same first author in such quick succession. Perhaps Coode's letter would have been accepted if he had co-authored with Runcorn; if so, it would have appeared before late July when Wilson's paper was published.

Coode does not remember precisely either when he submitted his letter to *Nature* or when he was notified of its rejection.[17] He thinks that he submitted it after he and Runcorn had been notified by *Nature* of the acceptance of their earlier joint letter, which was published on February 27, 1965, with no date received; Coode does not remember when they learned of their letter's acceptance. What is known, however, is that Coode's new paper was rejected by *Nature* before February 20, 1965, because that is when he resubmitted it to the *Canadian Journal of Earth Sciences*. Here is his letter to the editor, who had been Dean of Science at UBC.

Dear Dr. Gunning,

There has been a lot of speculation recently on the origin of oceanic transcurrent faults, including the quiescent faults from the northeast Pacific and those obviously generated by oceanic ridges. Because the faults extend for several thousand kilometers some authors have felt that activity must have occurred along the whole length simultaneously. Exotic and unrealistic mechanisms have been proposed which I feel are unjustified in the circumstances. I am therefore proposing a more natural mechanism and I enclose a note for publication in the *Canadian Journal of Earth Sciences*.

Yours Sincerely
Alan M. Coode [Coode signed the letter.][18]

Given this date, Wilson must have delivered his talk at the beginning of February, Coode must have submitted his note to *Nature* soon after Wilson's talk, *Nature* must have rejected the paper almost immediately, and Coode, with equal promptness, must have resubmitted to the *Canadian Journal of Earth Sciences*. Moreover, Coode must have had enough time to write or to speak with Jacobs, his old mentor, relating what happened, because it was Jacobs who suggested that Coode resubmit to the Canadian journal. Coode wrote (February 28, 2010 email to author; Coode's emphasis; my bracketed addition) "I knew Dean Gunning from my UBC days, but I had *no idea* that he was an editor of the Journal. The only person who could have told me to contact Gunning was Jack [Jacobs]." Many years later, Jacobs recalled his involvement, writing to Coode:

I am afraid that my memory of the dates of events concerning your paper on "transform" faults (1964–1965) is a little uncertain – when I returned to England I destroyed most of my early files. I do remember, however, that I was extremely excited when you first told me about your ideas and I was anxious that they should be published as soon as possible – that is why I suggested the *Canadian Journal of Earth Sciences* after it had been rejected by *Nature*.

(Jacobs, April 28, 1987 letter to Coode)

Jacobs definitely visited Newcastle in late April because he participated in a NATO conference on planetary and stellar magnetism entitled "Magnetism and the Cosmos" (Jacobs and Atkinson, 1967) and they must surely have talked there, but it is also possible that they made contact earlier in April. Coode elaborated recently:

When did Jack first talk to me? All I can say is that I have no proof one way or another that Jack and I talked until the NATO conference in April. There is evidence that we communicated many weeks before April. I knew Dean Gunning from my UBC days, but I had *no idea* that he was an editor of the Journal. The only person who could have, would have told me to contact Gunning was Jack. I firmly believe that I sent a letter to Gunning on the date, which I have on the letter, or one or two days after (not weeks or months), which was 20 February 1965. That being the case, I must have communicated with Jack in February, just after the Tuzo Wilson presentation. For me to communicate with Jack in early February, I may have done it by post, which, from a time point of view, is extremely tight, or by phone, which is possible, but at that time telephone calls were expensive and my salary was a little over £625 a year, which was my starting salary in Autumn 1963 (unless I did it using the University phone, which is possible). This leaves the last alternative, which is that Jack and I met and talked shortly after the Tuzo Wilson presentation. I am not aware that I travelled in February 1965, but I could have gone to London for an RAS meeting and met him there, but it is more likely that I met him in Newcastle when he came through to give a presentation to the School. My notes [from a later date] indicate that I met him in February in Newcastle. I wish I could substantiate this. I do not think I can.

(Coode, February 28, 2010 email to author; my bracketed addition)

Coode's paper was published in August. It attracted little attention. Wilson's paper was published in one of the leading scientific journals in the world and was widely read. Coode's paper appeared in a much lesser known journal and few noticed. Very many Earth scientists knew Wilson; few knew Coode; Coode's paper was visionary. Wilson, unlike Coode, named the new type of fault, and identified its other varieties. Would Coode's paper have received more attention if it had been published first? It is unlikely, given the relative exposure of the two journals. Would it have received more attention had it been published first in *Nature*? Decidedly yes, and he would likely have received substantial credit for being the first author to have correctly identified ridge-ridge transform faults. Coode's paper remains, as Menard (1986: 242) remarked, "one of the most remarkable documents in all of geological literature."

Coode's paper is a model of conciseness. In the first two paragraphs, he describes two types of oceanic transcurrent faults, notes the magnetic and seismic evidence for their displacement by appealing to work by Vacquier and Sykes, raises the difficulty about what happens to displacements where they abut continental margins, and introduces seafloor spreading by endorsing Bodvarsson and Walker's work in Iceland, which had attracted Matthews' attention (§4.3).

Transcurrent faults have been recorded in two different oceanic environments, off the west coast of the United States, where their presence has been inferred from studies of oceanic magnetic anomalies, and in fracture zones of ridges. Displacements of several hundred kilometers are not rare, and, when fault lines intersect a continental boundary, they disappear, and either cease to exist or continue below the continental sial (Vacquier 1962). Sykes (1963) has studied the locations of earthquakes in fracture zones and the epicenters are almost always contained within the limits of the separated ridges, indicating that the activity is by far the greatest here.

In a recent study of the geology and geophysics of Iceland, Bodvarsson and Walter (1964) concluded that the island was under east-west tension and expanding at a rate of 6 mm/year. The position of Iceland is critical. It straddles the Mid-Atlantic Ridge. One can infer that what happens to Iceland is happening on a similar scale over the whole length of the oceanic ridges. If the ridges are tension features and the crust is being separated, then new material will be introduced from the mantle below the rift.

Like Wilson, he did not refer to Hess's or Dietz's version of seafloor spreading. Did Coode know about them? In an email to me (February 9, 2010) he noted, "It is difficult to be sure 45 years after the event, but I *am* confident that I was aware of Hess and Dietz's versions of seafloor spreading; to me it was so obvious that it did not need emphasizing." I think that Coode is correct; surely he knew of Dietz's version because the 1962 paper by Vacquier, which Coode did refer to, appeared in *Continental Drift*, which contained a Dietz paper on seafloor spreading (III, §1.6, §4.9).

Coode then introduces key elements of V-M (although without naming it), and shows his preference for it over the idea that marine magnetic anomalies arise from "variable mineral content" of the ocean floor.

Oceanic magnetic anomalies suggest that production of new crust takes place at the same rate everywhere along the ridge. Obviously, either a variable mineral content or a varying magnetic field will generate these anomalies but, as the former presupposed the production of mineralogically identical extrusive material for thousands of kilometers of rift valley, the latter seems more probable.

(Coode, 1965: 400; references are the same as mine)

Is this V-M? I think so, even though he did not spell out what he meant by "varying magnetic field." Vine agrees.

I think that Alan [Coode] IS assuming the importance of remanence. There are two clues: (1) in suggesting that "a varying magnetic field will generate the anomalies" he is assuming that remanence is important and (2) later on he refers to "fossil magnetism" [see below].

(Vine, February 4, 2010 email to author; my bracketed additions)

Did Coode know about V-M? He is not sure, but his recollection is that he did not. He also believes (February 9, 2010 email to author) but cannot "guarantee" that he had in mind reversals of the field.

I feel sure that I did not know about Fred and Drum's hypothesis at this time. I similarly feel *sure* that what I meant by "varying magnetic field" was field reversals – although, again, I cannot *guarantee* it. Who is to say that I learned about it a month after or two months before writing the paper?

But there is more. Coode, in assuming "that production of new crust takes place at the same rate everywhere along the ridge," implies that the pattern of magnetic anomalies over ridge crests should be bilaterally symmetrical. So, like Wilson and the Lamont workers Talwani, Le Pichon, and Heirtzler, Coode independently proposed the bilateral symmetry of anomalies. However, unlike the others, he did not appeal to any particular magnetic profile, but arrived at symmetry solely through his vision of mantle convection and seafloor spreading; he believed that once large-scale mantle convection began it would continue at a constant rate for millions of years.

Coode next considers the positioning of fracture zones along active mid-ocean ridges, and attempts to explain them in terms of converging convection currents.

Although fracture zones occur on all ridges, they seem to be most concentrated mid-way between points of bifurcation or sharp bends. The author (unpublished) has analyzed the location of ridges and concluded that maximum tension exist under bifurcations (southeast Pacific, central Indian Ocean and south of New Zealand) and sharp bends (southeast Atlantic and north Atlantic). Consequently, fracture zones occur when crustal tension is a minimum or, if we assume some sort of localized source of tension in the mantle, where two sources overlap. As mantle convection is the only hypothesis suggested that combines all these points, one might say that fracture zones tend to occur where two upwelling currents converge.

(Coode, 1965: 400)

Most of Coode's examples are easily identified because he was partly working from Sykes' 1963 paper, the same paper that Wilson also used. Sykes (1963) discussed the restricted location of earthquakes along fracture zones extending from Easter Island to Chile, the Eltanin Fracture Zone connecting the East Pacific Rise with the Pacific–Antarctic Ridge, the fracture zone south of New Zealand extending from New Zealand to Macquarie Island, and fracture zones along the Mid-Atlantic Ridge.

In the final paragraph Coode turns to the great fracture zones in the Pacific, noting and rejecting Gilliland's solution (§4.11), and introducing his explanation of the movement along fracture zones, which is tantamount to what Wilson would soon call ridge-ridge transform faults. Implicitly he incorporated V-M without saying so specifically.

Faults the size of the Murray and the Mendocino do not mean that activity has occurred simultaneously along thousands of kilometers to produce a displacement of several hundred kilometers, as has been postulated by Gilliland (1964). Rather a model is proposed where the center line of the ridge is broken and displaced by a transcurrent fault (see Fig 1). As the new crust is generated the older material with the fossil fault and fossil magnetism is pushed to either side; in the active fracture in the southern Pacific studied by Sykes (1963) the separation

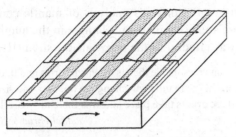

Figure 4.17 Coode's Figure 1 (1965: 401). His caption reads: "The crust on either side of the fault moves in opposite directions between the separated ridges and together, because they have the same relative speed, outside this area. Active fault generation therefore only takes place between the ridges."

is 1000 km. [This is the Eltanin Fracture Zone; see Sykes, 1963: 6001.] When the cause of the tension dies down, the linear structures, which were once parallel to the rift, and the transcurrent fault are all that remain.

(Coode, 1965: 400–4001; my bracketed addition; references are the same as mine)

His diagram (Figure 4.17) is, as Menard said, a beautifully concise description of a ridge-ridge transform fault; it clearly shows the symmetry of the magnetic anomalies and that motion along fracture zones is in opposite directions only between ridge-offsets, motion in the opposite sense from a transcurrent fault. Moreover, Coode not only believed that "production of new crust takes place at the same rate everywhere along the ridge" but he also thought that the "rate of production is roughly constant" and therefore that the observed magnetic profiles show that reversals of the geomagnetic field were not periodic but episodic.

I suspected that the rate of material production is relatively constant. The reasoning for this, in my mind, was that the convective overturn would be relatively constant (because of its high "viscosity" and moment of inertia). Therefore the variations in the width would be related to the variations in the magnetic field. So, yes, reversals vary in duration and are not periodic. It seemed relatively simple and clear to me.

(Coode, February 27, 2010 email to author)

Mantle convection was central to Coode's analysis. He thought mantle convection and seafloor spreading "obvious." Sounding like Holmes of the late 1920s (I, §5.4), Coode retrospectively remarked when I asked if he believed in mantle convection:

Of course! There was no other logical explanation. Heat produced from radioactive decay in the mantle and core would not be transmitted further than 700 km. over a time span of 4.6×10^9 years. So either the earth heats up or the heat escapes. And I do not believe that it has heated up. There are three forms of heat transmission: conduction, convection and radiation. The isostatic rebound of Fennoscandia (ship anchor rings used by the Vikings to anchor their ships in the fjords are now 50 ft up the cliff) is clear indication that the mantle has a viscosity, even if very high. So convection *has* to occur in the mantle.

(Coode, February 10, 2010 email to author)

Coode thought in terms of Runcorn's analysis of mantle convection in an Earth with an expanding core and accompanying changes in the number and dimension of convective cells, and was mindful of the timescale question (III, §1.5).

Yes, it seemed and seems quite reasonable that as the earth forms and the core separates out from the mantle, convective cells in the mantle will change in number and size. There is the question of timing, i.e. when did these events take place and was there any correlation with geology?

(Coode, February 10, 2010 email to author)

Soon after he finished his transform fault paper, Coode began fitting the distribution of tectonic features to particular spherical harmonic series, and submitted a paper to the *Geophysical Journal of the Royal Astronomical Society* in November 1965 (Coode, 1966). (As previously mentioned, it is this paper that he might have had in mind when he added a handwritten note to a copy of the letter he sent to Gunning.) Coode further developed his ideas in his 1968 Ph.D. dissertation "The evidence of convection in the Earth's mantle from the earth's surface and shape." Here is the opening paragraph of his thesis abstract.

Observational evidence of convection has been analytically investigated by fitting the distribution of tectonic features to spherical harmonic series. Assuming continental drift, then with one continent, Pangaea, analyses of ancient fold belts (compression features) show a third degree symmetry. With two continents, Gondwanaland and Laurasia, the symmetry becomes fourth degree. At the same time in the oceans, the tension features, which are now passive ridges, had and have a fourth degree symmetry. The analysis of active, modern tension and compression features each show that they are formed by the same causative mechanism. Various hypotheses have been reviewed, but the best to fit these results was a hypothesis of convection.

(Coode, 1968: ii)

Coode's and Wilson's accounts of transform faults differ in two obvious and a third less obvious respects: Coode neither gave the new type of fault a new name nor identified other types of transform faults; Wilson did both. Wilson also proposed that ridges are born with offsetting segments; offsets are part of the original ridge; ridges did not form as continuous unbroken single unities and subsequently divide into offsetting units. Coode allowed for the possibility that ridges form originally as continuous bodies without offsets, and only later split apart into offsetting segments connected by fracture zones. As he (1965: 400) noted in his paper, "a model is proposed where the center line of the ridge is broken and displaced by a transcurrent fault." Clarifying what he had in mind, Coode later remarked:

Reviewing my words and thinking back, it is more likely that the cause of the faulting was from shear stresses in the convection cells, which, with the increased heat and lower viscosity, do not fracture the material (no earthquakes), but when these forces are applied to the surface crust (lower temperature and higher viscosity) the surface manifestation is faulting in the central zone.

(Coode, February 10, 2010 email to author)

A third more obscure difference is that Coode linked ridge-ridge transform faults with what was, in all but name, V-M; Wilson said nothing about V-M. Moreover, Coode did not discuss the pattern of magnetic anomalies over specific mid-ocean ridges; his approach was less empirical more theoretical, less geological more geophysical than Wilson and Vine's. As we shall see in the next chapter (§5.2), Wilson did discuss the link with V-M in his next paper (1965c), and Vine and he (1965) in their joint paper went further, attempting to establish the rate of seafloor spreading across the Juan de Fuca Ridge during the past several million years. Very interestingly, as we shall see, Coode, the theoretician, foresaw no difficulty in supposing relatively constant spreading rates, and consequently and correctly foresaw that reversals are episodic and not periodic; if he had been questioned at the time, he would, I suspect, have considered this "obvious." Wilson and Vine, the empiricists, working initially with a very imperfect reversal timescale, opted, incorrectly, for periodic reversals and uneven spreading rates (Vine and Wilson, 1965).

Before closing this account of Coode's development of ridge-ridge transform faults, there are two other issues that need attention: the rejection of Coode's paper by *Nature*, and Wilson's submitting his paper on transform faults without acknowledging his prior discussions with Vine and especially Coode. First *Nature*'s rejection. Even though Coode had recently been the first author of two notes in *Nature*, his co-authors, Runcorn and Tozer were well known, I suspect that few noticed that Coode was the first author or even that he was an author; he was still just a graduate student working under Runcorn at Newcastle. Then there was the paper itself. Most importantly, he did not give the new type of fault a new name, and therefore did not emphasize that he was introducing an important advance in structural geology. Nor did it help that his paper was so terse. He did not help the reader understand what he was doing. Moreover, his paper was too visionary. Coode's rediscovery, as it were, of V-M is a credit to his inventiveness and theoretical understanding, but it made Coode's paper appear more prophetic than it needed to be. If he had known that Vine and Matthews had already proposed the idea, and that others had discussed it and referenced their works, thus placing his paper in clearer context, it would have been less visionary. Nonetheless, there can surely be little doubt that his paper should have been refereed. His explication of the new type of fault between ridge-offsets was of critical importance, his simple elegant figure clearly showed what he had in mind, and he clearly explained why, along fracture zones, earthquakes occur only between ridge-offsets. He explained Sykes' very important findings. A sympathetic referee may have recommended acceptance or resubmission, requiring that V-M be acknowledged and asked Coode to identify specific examples of his new type of fault. Suppose Coode had, like Wilson, made it clear that he had discovered a new type of fault by giving it a new name? Would his paper have then been accepted? I think so. It certainly would have had a much better chance of being refereed. Without the focus of a new name, the editor at *Nature* likely would not have taken the time to really understand what Coode was saying. Perhaps the editor thought that with three letters submitted within a few months he could be accused of favoritism if he published them all; had it been the first and not the last of the three, his judgment might have been different. Coode also believes his paper would have been accepted if Runcorn had been a co-author, and explained why he did not invite him.

I am sorry to report that I cannot remember what Keith said. I am sure that I showed it to him and I am also sure that it would have been published if I had co-authored with Keith. I am also sure that if I had done that, people would have remembered Keith's name and not mine. It was my first really independent idea and I wanted to see if I could float or swim. Lastly in the fabric of my mind I had made the pact with Tuzo Wilson. His idea against mine.

(Coode, March 2, 2010 email to author)

I believe that Coode is correct: it would have been accepted had Runcorn been co-author. Runcorn's name alone would have sent the paper out for review. Moreover, Runcorn could have told him to read Vine and Matthews and to rewrite appropriately.

This question raises another issue. Coode "is sure that" he "showed" his paper to Runcorn, but he "cannot remember what Keith said." Because Coode cannot remember, I assume that Runcorn did not advise him regarding changes, at least not any memorable advice. He needed sound editorial advice. Perhaps Coode should have taken more time with his paper. However, he wanted to get on with his Ph.D. work, and, once again, perhaps he thought his idea, albeit new, was so obviously correct.

I now turn to the second issue. Did Wilson pirate Coode's idea? Both Vine and Coode told Wilson of ridge-ridge transform faults; Wilson was twice primed. As already noted, Vine later stated, "I cannot remember what Tuzo said to this, probably nothing because he hadn't been listening properly as usual." Vine thinks that he "may have triggered Tuzo into the idea." Coode thinks that Wilson may have viewed what he was telling him as merely a way of "helping him sort out the problem."

It is possible that Tuzo Wilson believed that the discussions that he and I had had led him to believe that it was like the sort of discussion he had held at morning coffee or afternoon tea sessions in Cambridge. He may not have believed that it was anything more than someone helping him to sort out the problem.

(Coode, February 9, 2010 email to author)

Maybe Wilson was not really listening to Coode. Coode was just one of many who stood in line, after Wilson's presentation, to talk to him. Wilson was a dominating figure; he could and often did, take over a discussion. When he later thought of transform faults, he may have forgotten his conversations with Coode and Vine, and believed that the idea was entirely his own invention. Of course, there is the other possibility. Twice primed, once even challenged, Wilson may have remembered his conversation with Coode. Coode is quite sure that he introduced himself to Wilson, and told him about studying with Jacobs. Coode even argued with Wilson, challenged him. It seems unlikely that Wilson forgot so quickly. Coode is willing to give

Wilson the benefit of the doubt, and he bears Wilson "no ill will" (Coode, March 2, 2010 email to author).

It is most unfortunate that Coode's paper was not sent out for review. Had it still been in process when *Nature* received Wilson's paper, someone in the editorial office should have realized that Coode had independently presented the same idea of ridge-ridge transform faults and done so concisely and elegantly. Wilson developed the idea much further, but it seems to me that Coode's paper should have been accepted on merit, and both papers published in the same issue with appropriate dates of submission; Coode's had priority.

Much of Wilson's work during the previous decade and a half had been to compile and think about large-scale Earth structures, particularly linear and curved structures; he became very familiar with them. Jigsaw-like, he enjoyed piecing them together. It seems not at all unlikely that the promptings of Vine and Coode acted not so much as "a trigger" as Vine suggested, but rather as "catalysts." In Wilson's mind, during the spring of 1965, these "catalysts" quickly crystallized into comprehensive form, all the multitude of linear and curved structures that he had for so long been thinking about, linking them together, as he did in his seminal paper later that year. The acknowledgments in Wilson's article cast no light on the matter.

Notes

1 Hess kept Vine's letter, and it is included amongst his papers at Princeton University's Firestone Library.
2 Hess kept Hallam's February 15, 1965 note, which ended up in Hess's papers. He also kept Hallam's response, which is dated February 24, 1965.
3 Whittard's letter is in Hess's papers at Princeton.
4 Hess wrote out his itinerary for his speaking engagements, and listed Earth scientists along with their specialties that he might meet at each stop. (See Hess's papers at the Firestone Library, Princeton University.)
5 I want to thank Patricia Proctor and Susan Wilson, the two daughters of Tuzo and Isabel Wilson née Dickson, for their enormous help in setting me straight about Wilson's whereabouts during his sabbatical in the UK. Without their help, I would not have been able to pin down with any precision when Wilson developed the idea of transform faults. Susan Wilson kindly went through their mother's diary. Isabel Wilson often explained where she and Tuzo were on a given day. She did not always mention her husband, but nevertheless, Susan Wilson was able to definitely determine where Wilson was and even what he was doing on many of the days.
6 This letter was found in Wilson's papers that are housed at the University of Toronto.
7 This letter was found in Wilson's papers.
8 This itinerary is based on Susan Wilson's study of her mother's diary.
9 Matthews told Glen the same story during a 1979 interview (Glen, 1982: 305).
10 Hess's review and letter to Wilson are among his papers at the Firestone Library, Princeton University.
11 This letter is from Menard's papers housed at Scripps. I would like to thank Deborah Day and Carolyn Rainey for sending me requested items from Menard's papers, and Scripps for allowing me to reproduce them.
12 Hess kept an announcement of his and Wilson's talks at Madingley Rise, and it is preserved in his papers at the Firestone Library, Princeton University.

13 I thank Matthews for letting me quote this passage from his daybook, which he read to me during my 1979 interview of him.

14 Wilson made these two models for me when I met him in 1988. There is a similar figure of a paper model of a transform fault that was added as an "Editor's Note" to Wilson's published presentation he gave at the September 1965 Ottawa meeting sponsored by the International Upper Mantle Committee. He labeled the figure a "Do-it-yourself transform fault kit," or "ocean expander," and noted that it had been "prepared by Dr. Charles H. Smith in jest, but is perhaps worthy of enclosure herein" (Poole, 1966: 391).

15 I first learned that Coode's paper on transform faults had been rejected by *Nature* in 1987. Coode wrote a letter to Girdler on March 7, 1987, and Coode sent the letter to me with a short note of introduction via a third party. I neglected to do anything about the matter until 2008, but was unable to locate Coode until 2010. Coode described how his paper had been rejected by *Nature*. He did not discuss his interaction with Wilson, however, until we began to correspond in 2010. I add I knew neither that Coode's paper had been rejected by *Nature* nor that Coode had discussed what became known as transform faults with Wilson before he wrote his paper on transform faults. Menard and Coode failed to connect, and Menard died of cancer before his *Ocean of Truth* was published in 1982.

16 As Coode acknowledged, he cannot prove that Wilson gave such a colloquium in February, 1965. There appears to be no existing record of what colloquia were given in the Department of Physics at the University of Newcastle upon Tyne during this period. I asked Don Tarling and Ken Creer if they remembered such a colloquium. Tarling no longer has detailed records, but believes he was out of town at the time. In an attempt to obtain corroborating evidence of such a meeting, Coode asked Ian Wilkinson, a fellow graduate student. Wilkinson (February 10, 2010 email to author) thinks that he "talked to Tuzo about seafloor magnetic anomalies as my area of interest was geomagnetism and I'm pretty sure it was at the NATO meeting." There was a NATO meeting on planetary and stellar magnetism in the Physics and Mathematics Departments at the University of Newcastle uponTyne in April 1965, but Tuzo Wilson is not listed among the contributors. Of course, he may still have attended. Coode also remembers the NATO meeting. Indeed, Coode remembers talking with Jacobs, who did read a paper at the meeting, about his "transform" fault paper.

17 Unfortunately Coode no longer has the letter of rejection from *Nature*. He believes that it was in one of two boxes of past correspondence that he destroyed when he arrived in Saskatoon in 1983 (Coode, March 7, 1987 letter to Girdler).

18 The recorded date for reception of Coode's paper by the *Canadian Journal of Earth Sciences* is March 30, 1965, almost forty days after Coode wrote Gunning. This suggests the possibility that Coode waited after writing the letter to Gunning. Coode, however, is pretty sure that he sent the letter to Gunning either the day he wrote the letter or at most a few days later. He is also sure that he airmailed the letter. Perhaps Gunning sat on the letter or was out of town when it arrived.

There is also a handwritten two paragraph note at the bottom of Coode's letter to Gunning. Coode thinks that he wrote the note to himself after sending it to Gunning. Here is what he wrote.

I am presenting my unpublished work to the RAS on March 19th. It has been received favourably by Professor Runcorn and Vening Meinesz and I am offering it for publication in the Geophysical Journal of the RAS.

Figure 1. The fossil fault generating mechanism may extend over only a few hundred kilometers but the quantity of new crust produced may be spread over thousands of kilometers.

However, even if he added the note for Gunning, his reference to his unpublished work may refer to his paper "An analysis of major tectonic features," which Vening Meinesz and Runcorn did read, and was eventually published in the *Geophysical Journal of the Royal Astronomical Society* in 1966, and was originally submitted on November 18, 1965.

5

Continuing disagreements over the Vine–Matthews hypothesis, transform faults, and seafloor evolution, 1965

5.1 Outline

Returning to Cambridge during the first half of 1965, I discuss Wilson and Vine's interpretation of the Juan de Fuca Ridge in terms of seafloor spreading, V-M, and transform faults; this is a ridge in the Ridge and Trough Province of the northeast Pacific, north of the Mendocino Fracture Zone. I then describe the work of Heirtzler, Le Pichon, and Talwani of Lamont on the Juan de Fuca and the Reykjanes ridges, the latter being south of Iceland in the North Atlantic. Over the Juan de Fuca Ridge, the Lamont workers recognized the alternating positive and negative magnetic anomalies as Vine and Wilson did. Over the Reykjanes Ridge, they described a beautiful example of normally and reversely magnetized magnetic anomalies symmetrically arranged about the ridge. In both instances they did not support seafloor spreading and continued to interpret the magnetic anomalies as caused primarily by variations of induced magnetization in the crust and to resist V-M, which proposes that anomalies are caused by remanent magnetization of alternating normal and reversed polarity.

I then give a brief summary of the status of the premise that the geomagnetic field has often reversed in polarity, and describe the assessment given by Vine in his Ph.D. thesis (August 1965) of V-M. This is followed by an account of Matthews' attempt to explain the greater amplitude of the central anomaly compared with those on either side. I turn next to the three-day meeting sponsored by the International Upper Mantle Committee that was held in Ottawa, Canada, at the beginning of September 1965. Many papers were presented at the conference, and there were extensive discussions, transcripts of which are in the publication. I shall concentrate, not exclusively, on the presentations by Hess, Wilson, and Talwani, and the ensuing discussions. Although there was much else of interest at the Ottawa meeting, I chose these presentations because they pointed the way ahead in the mobilism debate. Concerning mobilism, seafloor spreading, V-M, and transform faults, overall, opinions varied. There was also much disagreement about the history of the northeast Pacific and western North America. In fact, confusion reigned, and I close the chapter with notes on the problems and challenges this entailed.

5.2 Wilson and Vine work in the northeast Pacific

The logic of Wilson's hypothesis predicted that there should be a short length of ridge between the two faults in this area and he named it the Juan de Fuca Ridge after the Strait of Juan de Fuca, which forms the boundary between the United States and Canada along this coastline. Tuzo was explaining this to Harry Hess and me when Harry suddenly interrupted him and said, "If you want to put a ridge there, that is one of the few oceanic areas for which there is a detailed magnetic survey, and if Fred is right, there should be a clear expression of the ridge in that survey." I dashed upstairs to the library to look at the volume of the *Bulletin of the Geological Society of America* containing the relevant article by Raff and Mason.

I cannot remember whether I took a quick look at the summary map before rushing back to Tuzo's office and setting it before Tuzo and Harry. All three of us stared at it in amazement. Not only were there linear magnetic anomalies paralleling the trend of Tuzo's putative ridge, but there was also a symmetry to the pattern of anomalies about the ridge crest.

(Vine, 2001: 59–60)

Following Wilson's transform fault paper, three others, with Wilson as author or co-author, followed in quick succession. All four, including the transform paper, were written from Cambridge. They provided further evidence for transform faults and V-M and showed how mutually supportive these two concepts were. The first, by Wilson, was entitled "Transform faults, oceanic ridges, and magnetic anomalies southwest of Vancouver Island." In the second, "Magnetic anomalies over a young oceanic ridge off Vancouver Island," Vine took the lead with substantial acknowledged help from Hess and Mason. Wilson and Vine submitted these two papers to *Science* on June 19, 1965, and they were published on October 22. On July 5, Vine sent a copy of the jointly authored paper to Hess, who had returned to the United States, wanting to give him chance to veto a quote "before it gets too far!"

I have been meaning to send you this manuscript for some time but we have been short of copies. We sent it off to 'Science' on June 17th. I am rather anxious for you to see it for, as you will see, it quotes you rather liberally and I thought you ought to have an opportunity of vetoing it before it gets too far!

(Vine, July 5, 1965 letter to Hess)

Hess vetoed nothing, replying to Vine on September 10, "The manuscript is in very good shape. I have no comments."[1] The last of the four papers, entitled "Submarine fracture zones, aseismic ridges and the International Council of Scientific Unions Line: proposed western margin of the East Pacific Rise" was published in *Nature* in late August, a month after Wilson's transform faults paper, and approximately two months before the two *Science* papers.

In the last chapter we saw that, according to Wilson, there should be a short spreading ridge segment in the northeast Pacific (Figure 4.14). Using as a base Mason and Raff's magnetic anomaly map obtained from their 1955 *Pioneer* survey, Wilson superimposed a geological sketch map of the Ridge and Trough Province, which showed various named features determined from bathymetry and earthquakes, such as the Queen Charlotte and San Andreas faults, and the Mendocino Fracture Zone. There were also one, possibly two unnamed ridges. The more northerly corresponded very well with a broad positive magnetic anomaly and this and the bathymetry and seismicity (RS1) established it as a *bona fide* ridge, likely a spreading ridge (RS1), confirming Wilson's prediction. He named it the Juan de Fuca Ridge.

The map of total magnetic field anomalies published by Raff and Mason (7) covers the region of the Juan de Fuca Ridge, and any interpretation should obviously be compatible with the magnetic observations. In order to follow the discussion, one should take the original anomaly map and superimpose on it the lines shown in Fig. 3 [my Figure 5.1].

(Wilson, 1965c: 482; reference is Raff and Mason, 1961; my bracketed addition)

By the time he had returned to Cambridge in late April after his Easter holiday still working on his transform paper, Wilson had, like Coode (Figure 4.17), come to believe that bilateral symmetry of magnetic anomalies about ridges was a corollary of seafloor spreading and V-M, but it had not yet been demonstrated. Were they bilaterally symmetrical about the newly recognized Juan de Fuca Ridge? John Sclater recalled how he attempted "to persuade Wilson that the magnetic profiles were not symmetrical."

Both Fred Vine and I spent time with Tuzo. In fact, I think Tuzo approached me in the library about the Vema-16 magnetic profile before he approached Fred. I remember copying the profile, enlarging it, and rotating one side and superimposing the other on it in an attempt to persuade Tuzo that the anomalies were not symmetric.[2]

(Sclater, September 3, 1992 letter to author; see www.escholarship.org/uc/item/4xj8c69c)

Vine had thought about profiles being bilaterally symmetrical but thought they likely were not – the process of seafloor spreading is just not that simple, geological processes are too messy. None of the few profiles they had at the time either from the North Atlantic or Indian Ocean were markedly symmetrical. But Wilson emphasized the intimate tie between seafloor spreading, symmetry of anomalies about the ridge and V-M, and sought a connection to his work on oceanic islands (§3.2); nobody talked him out of the symmetry of the anomalies over the Juan de Fuca Ridge.

The rough bilateral symmetry exhibited in the magnetic anomalies seems to follow as a corollary of Vine and Matthews' (8) original proposal that the floor of the ocean is built of strips of lava and intrusives alternately magnetized with normal and reversed polarity. Because ocean basins have a rough bilateral symmetry shown, for example, in the symmetrical relationships of some pairs of ocean islands and ridges (9), it is to be expected that the strips on the ocean floor should also be formed at an approximately uniform rate on each side of the ridge. *(Wilson, 1965c: 483; reference 8 is to Vine and Matthews (1963); reference 9 is to Wilson (1965a))*

Wilson looked for and found consilience between the width of seafloor created along the axis of the Juan de Fuca Ridge and the total displacement along the San Andreas Fault. He reasoned that if the San Andreas Fault is a ridge-ridge transform fault connecting the Juan de Fuca Ridge to the East Pacific Rise, then the total movement

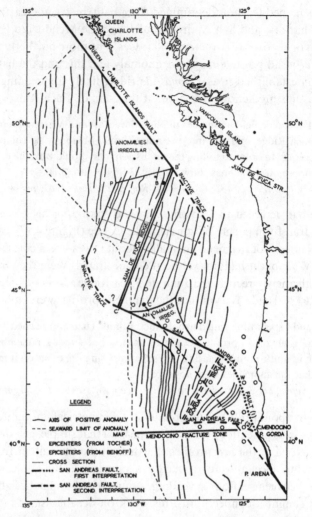

Figure 5.1 Wilson's Figure 3 (1965c: 484). Sketch map showing major geological structure in the Ridge and Trough Province superimposed on the magnetic anomalies discovered by Mason and Raff. Block PB′QRC′S encloses the magnetic anomalies parallel to B′C′, the axis of the Juan de Fuca Ridge, determined bathymetrically. The anomalies arrange themselves symmetrically about B′C′. Plotting earthquake activity placed the Juan de Fuca Ridge between B and C, which is "within the limits of error" close enough to B′C′. Cross sections a, b, and c show positions of magnetic profiles over the Juan de Fuca Ridge illustrated in Vine and Wilson's companion paper (Vine and Wilson, 1965).

along the San Andreas Fault since the inception of the Juan de Fuca Ridge should equal the width of seafloor created at the ridge. Even though, at the time, the age of the San Andreas Fault and its total displacement were matters of speculation, the best estimates of its total displacement were in rough agreement with the amount of seafloor generated along the crest of the Juan de Fuca Ridge. He summarized (Figure 5.1):

The width of the Juan de Fuca Ridge normal to B'C' is approximately 350 km but the total displacement is on the San Andreas fault since the formation of the Juan de Fuca should be measured in the average direction of the strike of the fault. According to Benioff this is a line joining Point Arena to C. This displacement is 400 km (250 miles). This estimate is neither precise nor dated, but it lies within the limits proposed by Hill and Dibblee (4), who suggested 175 miles since early Miocene with more doubtful possibilities of 225 miles since Late Eocene, 320 miles since Cretaceous, and 350 miles since Precambrian, and by Crowell (1) who gives a displacement of 175 for Oligocene rocks and perhaps "still more" for older ones.

(Wilson, 1965c: 484; references are to Benioff (1962),
Hill and Dibblee (1963), and Crowell (1960))

Mason, because of the magnetics, and Hess, because of heat flow, independently had suggested to Wilson the existence of yet another spreading ridge, just southeast of the Juan de Fuca Ridge. Wilson concurred and named it the Gorda Ridge.

R. G. Mason, for a study of magnetic anomalies, and H. H. Hess, from a consideration of measured heat flow, have both independently proposed to me in discussion that there may be another young and growing ridge between the Mendocino Fracture Zone and the Juan de Fuca Ridge. This neat solution also explains the epicenters along the Mendocino and Gorda escarpments ... It has been added as an alternative in Fig. 3 and named the Gorda Ridge, but does not affect the other arguments.

(Wilson, 1965c: 485)

On Figure 5.1, Wilson traced the Gorda Ridge by dashed lines, indicating its tentative status. He constructed what he thought at the time might be two possible northern extensions of the San Andreas Fault: (1) shows its trace if the Gorda Ridge is inactive, and (2) if it is active. Mason and Hess were correct; the Gorda Ridge does exist and it is now known to be connected to the Juan de Fuca Ridge by the Blanco Fracture Zone. This is the solid line between the south end of the Juan de Fuca Ridge (C') and the north end of the Gorda Ridge and it runs diagonally northwest from Cape Blanco on the north Californian coast. In his concluding paragraph, Wilson turned to the Mendocino and other great fracture zones. He remained undecided about whether they are related to the Darwin Rise as Hess maintained, or to the East Pacific Rise as Menard maintained. Fittingly, Vine took the lead in the second paper, concentrating as it did on magnetic anomalies – especially rates of seafloor spreading and the question of bilateral symmetry of anomalies. He realized that the discovery of the Juan de Fuca Ridge within Mason and Raff's survey area removed two difficulties that always plagued V-M (RS1). They had found a ridge that was associated with a large central magnetic anomaly and that was flanked by the zebra pattern of anomalies. They emphasized that the Juan de Fuca Ridge, originally predicted by Wilson on the basis of the concept of transform faults, had now been delineated bathymetrically and seismically (RS1).

At the time it was put forward, the Vine and Matthews hypothesis was particularly speculative in that no large-scale magnetic survey was thought to be available for an oceanic ridge ... Furthermore, it has been suggested in the preceding report that the area of detailed magnetic survey in the northeastern Pacific might include one or more short lengths of a young and active oceanic ridge (8). This suggestion was originally based on the concept of transform faults and the distribution of earthquake epicentres along the western coast of North America (9). Only subsequently was reference made to the magnetic survey (1) to find that it lends convincing additional support to the proposal.

(*Vine and Wilson, 1965: 486; reference 8 is my Wilson (1965c), reference 9, my Wilson (1965b)*)

Vine and Wilson proceeded to consider spreading rates on the basis of V-M. They first determined if spreading rates affected the amplitude of the central anomalies compared to other magnetic anomalies, reasoning that the slower the rate of spreading, the more pronounced the central anomaly should be.

The model for the 1 cm per year [per limb] rate ... suggests a possible explanation for the central high-amplitude anomaly observed over certain ocean ridges, notably the Mid-Atlantic Ridge and the northwestern Indian Ocean (Carlsberg) Ridge, as discussed previously (2). For this rate of spreading, the anomalies resulting from the normal-reverse contacts on either side of the central block reinforce each other to produce the central anomaly. For the faster rate of spreading, giving rise to a wide central block, the reinforcement is much less, and a rather broad central anomaly is produced which is not so easily distinguished from its neighbors ... This might possibly be the case over the East Pacific Rise and the new, Juan de Fuca Ridge (8).

(*Vine and Wilson, 1965: 486; my bracketed addition; reference (2) is Anderson (1951),*
reference (8) is Vine and Matthews (1963))

Vine and Wilson then determined the spreading rate that best fit the Juan de Fuca anomalies. In timely fashion and entirely serendipitously the first radiometrically dated reversal timescales for recent times had just been developed by Cox, Doell, and Dalrymple of the USGS and by MacDougall and Tarling at the ANU (II, §8.15). Vine and Wilson used the Cox *et al.* (1964) timescale to date successive magnetic anomalies of the magnetic profiles over the ridge. An average rate of 1.5 cm per year per limb gave the best fit. This implied that 120 km of new seafloor had been created along the Juan de Fuca Ridge during the past 4 million years which, they pointed out, agreed with estimates of the present rate of displacement along the San Andreas Fault and, extrapolating backwards, with its total displacement (RS1). For future reference, readers should note that they spoke of the rate as "rather irregular, as one might expect."

[The] average rate of spreading of the ridge has been rather irregular, as one might expect, but averages about 3 cm per year (1.5 cm per year per limb of the cell). This implies that the central 120 km or so of the crustal material over the crest of this new ridge has been formed within the past 4 million years. Thus if it is assumed that the rifting and associated faulting has continued without interruption and at this average rate, then the whole ridge (a total width of about

350 km) would be no more than 11 or 12 million years old. These deductions agree well with the total displacement across it, as discussed by Wilson (8).

(Vine and Wilson, 1965: 487; my bracketed addition; Wilson reference is Wilson (1965c))

No doubt pleased with their success, Vine and Wilson stepped back and noted the key advantage of V-M over competing views, as Vine had advocated from the beginning.

Clearly, the models [used] . . . are oversimplified. However, they express the basic tenet of the Vine and Matthews idea that the steep magnetic gradients so obvious from any detailed magnetic survey over the oceans might delineate the boundaries between essentially normally and essentially reversely magnetized crust, thus reproducing the observed gradients, without recourse to improbable structures or lateral changes in petrology.

(Vine and Wilson, 1965: 487)

Regarding the question of whether the source of the magnetic anomalies is distributed through Layers 2 and 3 of oceanic crust as originally proposed by Vine and Matthews (§2.13), or whether it is restricted entirely to Layer 2 as Hess had recently proposed (III, §3.15), they found (1965: 488) Hess's model "a considerable improvement." It is illustrated in Figure 5.2c. As to why the central anomaly, regardless of spreading rates, is more generally pronounced than others, they reiterated Vine's original contamination explanation.

In Fig. 3c [my Figure 5.2], the magnetic anomalies have been computed over a model in which the magnetic material is confined entirely to layer 2. As previously, the central block is assumed to be more strongly magnetized because it is the only block composed exclusively of young material which is magnetized normally, except for the minor possibility of self-reversals. Volcanism probably occurs over a wider zone than the central bock, and all other blocks will therefore be contaminated with younger material, often of reverse polarity to that of the initial block, and hence lowering or modifying its resultant magnetic effect. The serpentine of layer 3 is almost certainly riddled with basaltic feeders for the flows and intrusives of layer 2. If these feeders are taken into account they will have the effect of slightly lowering the effective susceptibility assumed for layer 2 in Fig. 3c.

(Vine and Wilson, 1965: 488; my bracketed addition)

Vine and Wilson turned to the important question of the bilateral symmetry of magnetic anomalies about spreading ridges. As we have seen, Sclater was skeptical. Vine thought it unlikely from a geological point of view; ridge processes "couldn't be that clear cut." Vine recollects:

One of the things that is difficult to remember with hindsight is that at that stage not only did we not know the reversal time scale and whether it was irregular or periodic or whatever – and I think that has a bearing on the symmetry question – but, in addition, at least I was not sure as to whether one would expect spreading to be continuous or discontinuous or regular. Drum and I were more preoccupied with geology, with what the process was at ridge crests. Even if you have got spreading which is a fantastic concept, how the hell is it accommodated in terms

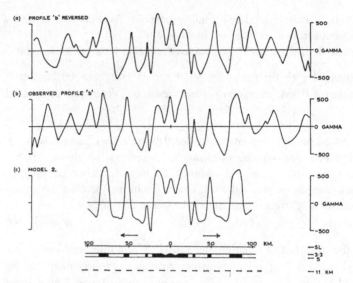

Figure 5.2 Vine and Wilson's Figure 3 (1965: 488) establishing symmetry of magnetic anomalies about the Juan de Fuca Ridge. (b) is a profile as observed over the Juan de Fuca Ridge along dotted line "b" in Figure 5.1; (a) is (b) reversed; (c) is a model of the profile based on Hess's model with the magnetic material being confined to Layer 2, and an effective susceptibility of +0.02 for the central block and ±0.01 for the other blocks.

of the way stuff is extruded and intruded and the ocean crust is formed. As it could only be messy, it couldn't be that clear cut. We were working mainly, indeed exclusively, at Cambridge with North Atlantic and Indian Ocean data which are now known, of course, to have slow spreading rates and are less clear. But, of course, at that time if you looked at a profile across the North Atlantic or Indian Ocean Ridge, it didn't look symmetrical. In fact, it looked a bit of a mess. There was no reason to suggest that the actual process by which the record would be written was that clear cut, that it would be simple enough or elegant enough to preserve that time scale very accurately or preserve symmetry. We thought it would simply be scrambled by the geological process.

(Vine, 1979 interview with author with slight changes by Vine)

Vine proceeded to demonstrate symmetry (Figure 5.2). Mirrored about the ridge the anomalies are remarkably similar. He found this "a tremendous revelation" because it had appeared to him so unlikely from a geological point of view. Twenty years later Vine still seemed surprised by the high degree of symmetry.

It was just fantastic ... the real flash in a way was the symmetry of the Juan de Fuca. It was the first thing I saw. That was '65 when Tuzo and Harry and I realized there was a ridge in that area – that is an incredible story as to why it had not been recognized before – but the magnetics over it was symmetrical, and we just stood there looking at it with our mouths open. There it was. The thing had been in the literature for four years; nobody had seen it. When you went to look for it, it just stood out. That was the first thing. The Juan de Fuca was a tremendous revelation in that it showed that the record could be sufficiently clearly written to generate symmetry and the boundaries were quite sharp and well defined. The first thing was the symmetry because, as I say, we hadn't anticipated that the record would be that clear, and suddenly realized that this was the first time we have looked at a really good survey, such an extensive survey [over both sides of a ridge]. Perhaps it is more common, you know, perhaps it is that you cannot tell from just one profile. So that was a tremendous revelation; the record was sufficiently clear to reveal the symmetry. The other thing about the symmetry: we didn't know what to expect. Are the reversals every half million years or are they every million years? Are they periodic? If it was periodic there would just be zebra stripes of the same width. But if it is an irregular pattern, then that will be much more compelling, as indeed it is.

(Vine, 1979 interview with author; my bracketed addition)

Vine and Wilson (1965: 488) remarked:

A literal interpretation of the Vine and Matthews hypothesis implies that the magnetic anomalies observed over ridges at certain latitudes and orientations should be roughly symmetrical ... but the simplicity of this model when compared with the probable complexity of the real situation makes a high degree of symmetry improbable.

Although earlier, symmetry had seemed improbable to Vine, once found, he was quick to recognize how much it increased support for V-M. Bilateral symmetry across the line of the ridge is demonstrated in Figure 5.3, which shows three profiles along the profiles a, b, and c of Figure 5.1, each 45 km apart. The central profile in Figure 5.3 is the central profile of Figure 5.2. "It is possibly significant that no active submarine volcanism has been reported despite the active nature of ridge as evidenced by recent earthquakes along it and along the bounding transform faults (8)" (Vine and Wilson, 1965: 488–489; (8) refers to Wilson (1965b)). Vine thought that the degree of symmetry may increase the more quietly a ridge grows, and proposed that the lack of any reported violent submarine volcanism associated with the Juan de Fuca Ridge independently suggested "quiet growth."

Vine and Wilson closed their paper by echoing Peter and Stewart's plea for magnetic surveys over an area as opposed to long profiles, and added that "aeromagnetic surveys would appear to be perfectly adequate."

Wilson wrote a fourth paper in which he thought about the mutual consequences of transform faults and aseismic ridges, and the generally inactive ridges that extended out from mid-ocean ridges and are occasionally topped by volcanic islands. Working from the idea that aseismic ridges are formed over hot spots, he proposed, with help from Hess, that horizontal crustal motions relative to Earth's mantle are determinable by looking at the trends of island chains with the expectation that the age of islands along each chain increases with distance from its associated spreading ridge. He also claimed that the position of ridge-offsets relative to each other should be reflected both by the magnetic pattern and by the position of aseismic ridges on adjacent blocks between fracture zones. Using these guidelines, he (1965d: 910) argued that the "floor of the Pacific is moving northwards like that of the Atlantic Ocean and several southern continents." Returning to the issue of whether the great

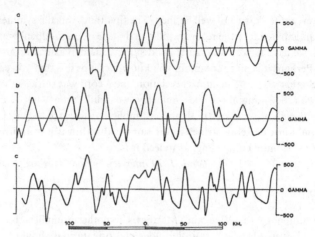

Figure 5.3 Vine and Wilson's Figure 4 (1965: 488). Observed profiles a, b, and c across the Juan de Fuca Ridge. Profile b and its mirror image are shown in Figure 5.2.

fracture zones are related to the East Pacific Rise or to the Darwin Rise, he argued that the fracture zones and aseismic ridges (Emperor Seamounts, Line Islands Ridge, Hawaiian Islands, Tuamotu Islands, Society and Austral islands, Tehuantepec Ridge, Cocos Ridge, and Nazca Ridge) are associated with the East Pacific Rise. Given the close association of all these, excepting Hawaiian Islands, with the East Pacific Rise, he claimed that their westerly extension forms the western margin of the East Pacific Rise. He named this margin the ICSU Line in honor of the International Council of Scientific Unions. He imagined the western flank of the East Pacific Rise to extend at least to the middle of the Pacific Ocean. Then (1965c: 908), assuming that spreading rates had been the same on either flank of the East Pacific Rise, inferred that "its eastern margin and northern crest" must "have been overridden by the Americas." (Much later, Menard (1986: 248) correctly criticized Wilson for this; the western margin of the East Pacific Rise extends very much further west. Wilson clung to the notion that Pacific crust west of the ICSU Line had been produced by the Darwin Rise (§4.10; III, §5.6) which was later shown never to have existed.)

Wilson concluded his paper by returning to his earlier concern over the difficulty of appealing to mantle convection as the cause of seafloor spreading in light of the fact that Antarctica is surrounded by actively spreading ridges with no significant sink within. He had already addressed this issue in his first paper on transform faults, and had discussed the difficulty with McKenzie (§4.9). In his original paper on transform faults, he did not mention mantle convection, perhaps because he was concerned about the difficulty. But he now offered a solution. He proposed that convection originating from ridges surrounding Antarctica proceeds at twice the rate away from Antarctica than towards it.

An explanation [of why ridges surrounding Antarctica move northward] can readily be offered in terms of the possible sub-crustal currents mentioned earlier. On an infinite body they might

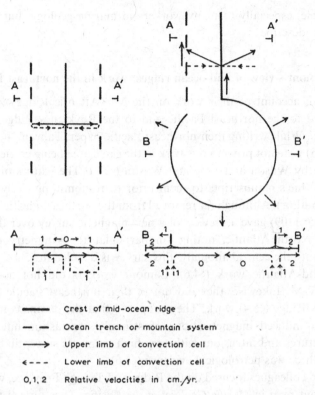

———	Crest of mid-ocean ridge
— —	Ocean trench or mountain system
——→	Upper limb of convection cell
◄---	Lower limb of convection cell
0,1,2	Relative velocities in cm./yr.

Figure 5.4 Wilson's Figure 6 (1965d: 910). Left side shows flow on an infinite body; right side, on a spherical body. The loop represents the active spreading ridges surrounding Antarctica; BB′ a cross section orthogonal to the solid line extending upward from the loop.

flow as in Fig. 6 (left), but on a spherical surface they might be constrained to flow as in Fig. 6 (right) [Figure 5.4]. This pattern, like a J folded about the Earth can be taken to be diagrammatic simplification of the real mid-ocean ridge system and can be compared with Fig. 10 in my previous article [Wilson, 1965b; Figure 4.15 above]. The stem represents the mid-ocean ridge system through the Southern Ocean to the West Chile Ridge. Antarctica would be in the centre of the loop. Because there is no sink within the looped source the diameter of the loop itself increases and the surface currents move outward with twice the velocity of the deeper counter-currents. If the sources of volcanoes lie in the counter currents (if the convecting system is shallow), or within the stagnant cores of convection cells (if the system is deep), then the differential in rates of flow will produce chevron patterns in the aseismic ridges.

(Wilson, 1965d: 910; my bracketed additions)

He (1965d: 910) recognized a similar problem with Africa, which "is ringed by mid-ocean ridges on all but the northern side." Just why the currents know to move twice as rapidly away from Antarctica than towards it remained unexplained.

These four papers (§4.9 and §5.2) on transform faults and on V-M did not provide a difficulty-free solution, but they went a long way to making seafloor spreading more acceptable, especially to many workers in marine geology, but not to all of them, as I now describe.

5.3 Lamont's view of mid-ocean ridges: work in the northeast Pacific

After giving an account of their work on the Mid-Atlantic Ridge (§4.6), Lamont workers turned to the northeast Pacific and to the Reykjanes Ridge (§5.4) in the North Atlantic. While writing their northeast Pacific paper, Talwani, Le Pichon, and Heirtzler (1965) were not privy to the work in the same area being carried on at much the same time by Wilson and Vine and Wilson (§5.2). The editors of *Science* sent Wilson's and Vine's manuscripts to them prior to resubmitting (Talwani, July 14, 1981 letter to author). Although far removed from the northeast Pacific, Talwani and company (1965: 1109) gave a preview of a new magnetic survey over the Reykjanes Ridge in the North Atlantic, noting "in particular the symmetry of the linear anomalies about" the central anomaly; this survey is detailed in §5.4.

In their Mid-Atlantic work (§4.6), Lamont workers did not accept seafloor spreading or V-M. Likewise, they did not in their northeast Pacific (this section) and Reykjanes Ridge (§5.4) work. They proposed that the oceans comprised rocks that possessed induced magnetizations of strongly contrasting intensity – with numerous fractures and intrusions with very high magnetic susceptibility for which they believed there was petrological evidence (§4.6).

Talwani and colleagues focused on the Ridge and Trough Province, working from the huge foldout map in *Marine Geology of the Pacific*. They agreed that the East Pacific Rise extends northwards beneath western North America and back out into the Pacific, into the Ridge and Trough Province, where it comprises three short segments. Talwani, Heirtzler, and Le Pichon were ahead of Wilson and Vine: they posited without reservation the existence of the Gorda Ridge, and correctly claimed that there is a third ridge (Explorer Ridge) just northwest of the Juan de Fuca Ridge (see Figures 5.6 and 5.7). Basing their case on bathymetry and seismicity, they supported it by appealing to Von Herzen's discovery of high heat flow, and to Raitt's discovery of seismically anomalous mantle beneath the Ridge and Trough Province, comparable to that beneath the Mid-Atlantic Ridge.

They presented a new magnetic profile obtained during *Vema* cruise 20, which crossed the Juan de Fuca Ridge (see Figures 5.5 and 5.6). They located their profile within the Mason–Raff survey, fine tuning its position by bathymetry and linearity of the magnetic anomalies. They commented on the "remarkable similarity" between the *Vema* 20 magnetic profile and profiles over other ridges, and noted, likely for the first time, that a strong magnetic signature ended abruptly at the outer edge of the continental margin (Talwani *et al.*, 1965: 1110). They emphasized that earthquake activity along the fracture zones connecting the ridge-offsets is restricted to the segment between ridges, and, like Wilson and Vine, noted its consistency with Sykes' 1963 findings.

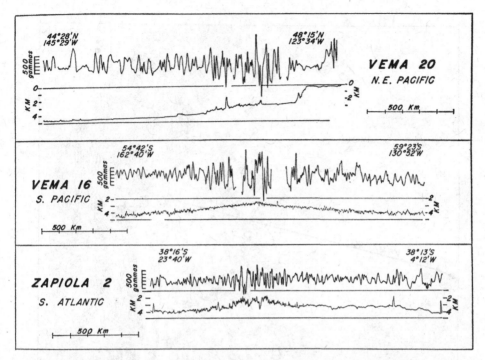

Figure 5.5 Talwani *et al.*'s Figure 2 (1965: 1111). They compare profile *Vema* 20 over the Juan de Fuca Ridge, *Vema* 16 over the Pacific–Antarctic Ridge and *Zapiola* 2 over the Mid-Atlantic Ridge in the South Atlantic. The *Vema* cruise (see Figure 5.6) is at an angle between 30 and 40 degrees to the trend of the Juan de Fuca Ridge.

The coincidence of the maximum amplitude of the axial magnetic anomalies with the topographic crest makes it possible to follow the ridge crest north of the Mendocino Fracture Zone from point A in Fig. 3 to point F in Figs. 3 and 4. The axis of the rise is clearly offset between B and C and between D and E (Figs. 3 and 4) along what appear to be fracture zones ... It is between these offset portions of the crest that the earthquake epicenters occur most frequently. Between C and D the magnetic and topographic patterns are undisturbed and earthquake epicenters are very rare. This seems to support the observation that zone of seismic activity is limited to the portions of the offsets between the crest of the rise (9).

(Talwani et al., 1965: 1111; (9) refers to Sykes, 1963)

The Lamont workers raised the same two difficulties against V-M that Heirtzler and Le Pichon (1965, §4.6) had. However, they did temper ever so slightly their rejection of the hypothesis by noting that it was "untenable at least in its present form."

In this article we do not go into details of the origin of the ridge anomalies. We merely note that the symmetry and linearity, as well as the ridge strike, indicate that these anomalies are genetically related to the formation of the ridge. We also note that the difference in character of the axial and flank anomalies implies that they do not have identical origins. The flank anomalies are not axial anomalies at greater depths. Vine and Matthews (1) explain the ridge

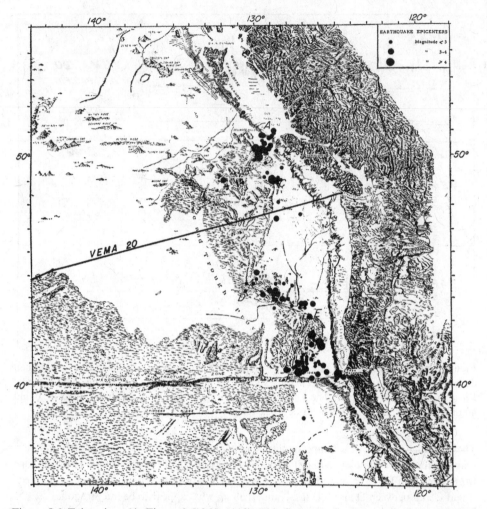

Figure 5.6 Talwani *et al.*'s Figure 3 (1965: 1112). This figure is after Menard's foldout figure "Physiography of the Northeastern Pacific" drawn by Howard Taylor and Janet Pyle and is from his 1964 *Marine Geology of the Pacific* (Menard, 1964). It shows the track of *Vema* cruise 20. The black circles mark epicenters of various magnitudes. They identified ridge segments *AB*, *CD*, and *EF*. They were later respectively named the Explorer, Juan de Fuca and Gorda ridges. They identified *BC* and *DE* as fracture zones. Menard, as they acknowledged, had already named *DE* as the Blanco Fracture Zone; *BC* was later named the Sovanco Fracture Zone. The peak to the west of the Juan de Fuca Ridge axis at approximately 47° N is the Cobb seamount.

magnetic anomalies by invoking the "spreading floor hypothesis" of Hess and Dietz. However, we believe that the flankward diminution in amplitude of the axial anomalies and the difference in character between flank anomalies and axial anomalies make their hypothesis untenable, at least in its present form.

　　　　(Talwani, Le Pichon, and Heirtzler, 1965: 1109; (1) refers to Vine and Matthews (1963))

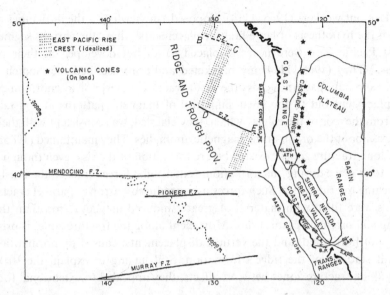

Figure 5.7 Talwani *et al.*'s Figure 4 (1965: 1113). GAR, B.P., and S.A. respectively stand for the Garlock, Big Pine, and San Andreas faults. *BC* (Sovanco) and *DE* (Blanco) are fracture zones; *CD* (Juan de Fuca) and *EF* (Gorda), ridge segments. The Explorer Ridge is cut off, its southern end is at *B*. The eastern part of the Basin and Range Province and the Colorado Plateau extend beyond the figure.

Turning to the more southern sections of the Mason and Raff survey, they proposed that offsets of the East Pacific Rise south of the Mendocino Fracture Zone extend under the continent beneath the Basin and Range Province whose topography, they claimed, is similar to that of the Ridge and Trough Province. With the northerly extension of the East Pacific Rise established, they considered and rejected Vacquier's hypothesis of large-scale movement of adjacent blocks of seafloor along the great fracture zones. Labeling Vacquier's proposal the "displacement hypothesis," they noted that "theories of continental drift and of evolution of the ocean floor have leaned heavily on" it. They raised obstacles against it, and introduced their fixist alternative. They found it difficult to explain how huge blocks of seafloor could move such vast distances with so little distortion. They pointed to the lack of evidence of huge matching east–west relative displacements of geological structures in North America above the East Pacific Rise, and the absence of earthquakes along the great fracture zones (Talwani *et al.*, 1965: 1114). The first difficulty was yet another permutation of the mechanism difficulty facing mobilism. The second difficulty does not arise if the East Pacific Rise does not extend under North America; its legitimacy depended on Sykes' excellent work and their own observation that earthquake activity along the fracture zones connecting offsets of the East Pacific Rise appeared to be restricted to those parts of the fracture zones between the offsets.

The Lamont workers (1965: 1114) proposed "in view of all the difficulties associated with the hypothesis of large fault displacements" that "the various segments of
the East Pacific Rise were never displaced at all but developed at their present
positions." They (1965: 1114; my bracketed addition) ascribed the "match in the
magnetic anomaly pattern [across the great fracture zones] not to enormous strike-
slip displacement but to the great similarity of magnetic patterns at constant distances from the axis of the ridge," which, they claimed, was consistent with their own
hypothesis about the origin of the magnetic anomalies. They maintained (1965: 1114)
that the long fracture zones "existed before formation of the rise, even though major
faulting took place during" its formation. Extensional forces cracked the crust along
what became the ridge axis. Such forces also produced narrower parallel cracks, and
all cracks were filled with material of greater induced magnetization than the surrounding and original oceanic crust. Movement along the fracture zones is primarily
vertical, not horizontal, and the vertical displacements, caused by normal faulting,
are manifestations of the ridge's formation. Still having to explain the strike-slip
faulting along fracture zones between offsets, they appealed to extensional forces.

It is reasonable to assume extensional forces away from the crest of the ridge. Where the crest is
offset, maximum shear stresses would be set up in the portion of the fracture zone between the
two crest segments. This would be the only part where the otherwise normal faulting would
have a strike-slip component, and these stresses might explain the usual earthquake activity in
this area.

(Talwani et al., *1965: 1114)*

They (1965: 1114) again noted perceptively that Vacquier's displacement hypothesis
regarding the fracture zones south of the Mendocino Fault could not explain "why
earthquakes should not occur along the entire length of the fracture zone." They
further maintained that if their hypothesis is correct the magnetic pattern should vary
"somewhat" between ridge-offsets, and added that "the disturbed zone," within the
pattern of magnetic anomalies on opposite sides of the Murray Fracture Zone, and
the greater apparent right lateral displacement at its western than at its eastern end,
also supported their view.

Vacquier's hypothesis was subject to the obstacles they had just raised. Wilson's
analysis of movement along ridge-offsets in terms of transform rather than transcurrent faulting avoided them. This the Lamont workers acknowledge in an endnote.

The recent "transform fault" hypothesis of J. T. Wilson (*Nature*, **207**: 343 (1965)), like other
hypotheses, does not require displacement of the ridge crest by transcurrent faulting. There is
an important difference in the two views, however: Wilson believes that the San Andreas fault
system is a "transform fault" connecting two different ridges and, consequently, that the East
Pacific Rise does not continue into the western part of the United States.

What they did not mention, however, is that Wilson's hypothesis was a corollary of
seafloor spreading, and that, unlike Vacquier's hypothesis, required neither huge

east–west differential displacements of north–south trending geological structures in the western United States nor earthquake activity all along fracture zones. They also did not mention that Wilson had related the great currently inactive fracture zones to the much older Darwin Rise. Talwani and colleagues offered no explanation of the San Andreas Fault.

But there is more. For, as noted above, Talwani recalled (July 17, 1981 letter to author) that after they submitted their own manuscript "the editors of *Science* sent the Wilson–Vine manuscripts to us." So they had the opportunity to revise their paper in light of Vine and Wilson's work or argue against it. Why didn't they? Talwani tried "to guess" what happened:

As far as timing is concerned ... the editors of SCIENCE sent the Wilson–Vine manuscripts to us. Why did we not pay more attention to these manuscripts? I am trying now to guess what went on in our minds sixteen years ago. Firstly, there was agreement with Wilson on one major point – the ridge crests were born in an offset position. Secondly, we, as they, related the magnetic anomaly patterns genetically to the ridge. Beyond that we must have believed that we were right and they were wrong ... At any rate, the Vine–Wilson paper was pointed out to us at a later stage – after the original submission of our paper, and at best (in our opinion) theirs was a competing hypothesis and we saw no reason not to publish our paper – or to change our paper drastically to accommodate their hypothesis.

(Talwani, July 14, 1981 letter to author)

Talwani, Heirtzler, and Le Pichon added that additional surveys were needed to determine if their hypothesis is correct, and they (1965: 1115) singled out the southeast Pacific as "a critical area of study."

Essential to our hypothesis is evidence of repetition of the general structure, and of its dependence on the distance from the rise axis, from one rise segment to another. We feel that only further geophysical surveys will establish whether or not this repetition and dependence exist. A critical area for study would be the southeast Pacific Rise, where observation from a vessel crossing the East Pacific Rise at high magnetic latitudes should reveal that the ridge has a well-developed magnetic anomaly pattern.

(Talwani et al., 1965: 1115)

Looking ahead, it is ironic that they were so absolutely correct about the eventual importance of the survey in the southeast Pacific over the Pacific–Antarctic Ridge and knew so clearly what to do next; it was precisely the results of this survey that eventually forced them to favor V-M. Indeed, Lamont workers would later find that "repetition of the general structure" of magnetic anomalies "and of its dependence on the distance from the ridge axis" was so good, even exhibiting unbelievable bilateral symmetry, that it provided greater support for the V-M hypothesis than for their own (§6.6).

The contrast between the reactions of the three Lamont scientists and Wilson and Vine to their independent discovery of ridge segments within the Ridge and Trough Province and the realization of the surprising and remarkable bilateral symmetry of the magnetic anomalies over the Juan de Fuca Ridge is arresting; they could hardly have been more different. Both groups had essentially the same data but offered opposing explanations, and the Lamont group had even seen the Wilson–Vine manuscripts before resubmitting their paper. They recognized the striking bilateral symmetry of the anomalies. Although the Lamont group still had reason to reject V-M because it offered no explanation for the increase in amplitude of the flank anomalies compared to the axial anomalies, they could have offered a more balanced comparison of it and their own concept. Bilateral symmetry was tantamount to a corollary of V-M while it was only loosely tied to Lamont's own hypothesis. More-over, the discovery of the Juan de Fuca Ridge with its associated pattern of bilat-erally symmetrical magnetic anomalies again quite spectacularly removed one of the key difficulties facing V-M, namely, the lack at first of a generally recognized ridge or ridges within the Mason–Raff survey that was parallel to some of the magnetic anomalies. Yes, V-M could not explain why the flank anomalies are of greater amplitude than the axial anomalies; however, the Lamont group offered no explan-ation at all for either the flank or axial anomalies. And, more importantly, it had no explanation of the axial anomalies that did not require appeal to unusual petrological differences between adjacent stripes of seafloor.

However, their position had softened ever so slightly, for, as noted above, they qualified their rejection of V-M by allowing that it was "untenable, at least in its present form." However, at bottom, they seemed to harbor such a deep antipathy to seafloor spreading that they were unable to muster much sympathy for V-M.

5.4 Lamont's view of mid-ocean ridges: work on the Reykjanes Ridge

During that year [1964] a scientist from the U.S. Navy, J. G. Baron, appeared on the scene and worked on magnetic anomalies south of Iceland, under the supervision of Jim Heirtzler. It soon became apparent that the magnetic patterns that emerged from the noise caused by magnetic storms were symmetrical about the ridge crest [Heirtzler *et al.*, 1966]. I had, of course, read the paper by Vine and Matthews [1963] in *Nature* and thought that it was a great idea ... When it became clear that lineated magnetic anomalies were clearly associated with the Reykjanes Ridge and were symmetrical about it, I considered that the issue was closed. Indeed, I recall saying as much to Jim Heirtzler. He reacted strongly in a negative manner, so I immediately dropped the subject.

(Opdyke, 1985: 1177; references are same as mine; for similar comments see Opdyke, 2001: 103,
and Lamont oral history project, Opdyke, March 17, 1997 interview with R. Doel)

Opdyke further added:

Jim was not happy with my comment, he also made a statement in which he said that the anomalies were so beautiful that they might be attributed to electrical currents flowing in the crust. I thought this unlikely! His strong reaction to my comment caused me to keep my mouth shut and terminate the conversation.

(Opdyke, December 8, 2009 email to author)

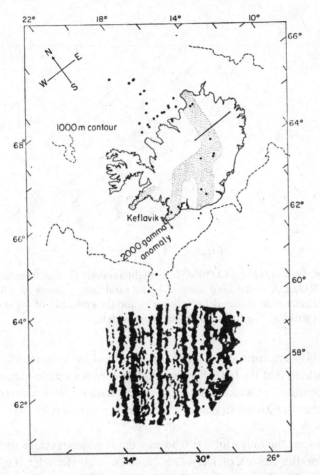

Figure 5.8 Heirtzler *et al.*'s Figure 1 (1966: 428). Their caption reads: "The magnetic survey area southwest of Iceland. Positive anomalies are shown in solid black. The belts of Quaternary volcanics in Iceland are shaded (after Bodvarsson and Walker, 1964). Epicenters north of the survey area are represented by solid dots (after Sykes, 1965). The line in eastern Iceland locates the geologic section used in Fig. 11." See my Figure 5.10 for their Figure 11.

Heirtzler, Le Pichon, and J. Gregory Baron co-authored the third paper on marine magnetic anomalies in which Lamont workers rejected V-M. It described an aero-magnetic survey carried out by Baron, who was from the US Naval Oceanographic Office, over the Reykjanes Ridge, in the North Atlantic south of Iceland. They submitted the paper in November 1965. The survey, which took two months, was originally planned in 1961 but not undertaken until October 1963. Covering an area of approximately 350 km by 350 km, the Navy made fifty-eight NW–SE flyovers, and twelve NE–SW ones. Flight paths were approximately 5 to 10 km apart. The results again showed a recognized oceanic ridge surrounded by the predicted zebra pattern of magnetic anomalies which extended well past its flanks (see Figures 5.8 and 5.9).

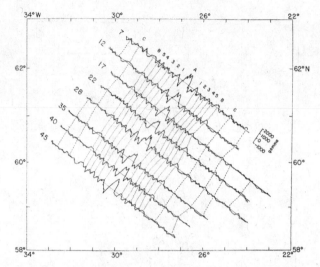

Figure 5.9 Heirtzler *et al.*'s Figure 3 (1966: 431). Eight anomaly profiles from the survey across the Reykjanes Ridge. A is the axial anomaly. The axial zone extends to BB. The thirteen anomalies in the axial zone display definite linearity and the symmetry of the anomalies within the axial zone is striking. The flank anomalies begin with C.

They (1966: 434) were impressed with the linearity and the symmetry of the anomalies, even remarking that the latter "is remarkable, even for smaller details." This was the *second* occasion on which Lamont workers had found symmetrical profiles. Nonetheless, they continued to favour their own hypothesis (§4.6) over that of Vine and Matthews.

Opdyke was not the only mobilist who saw the Reykjanes Ridge profiles prior to publication, for the editors of *Deep-Sea Research* sent Heirtzler, Le Pichon, and Baron's paper to Matthews to review. Matthews must first have been delighted when he saw the linearity and symmetry of the anomalies, and then astonished that the authors had said *nothing* about his and Vine's hypothesis. Matthews recommended that they should at least consider V-M. He also recommended to them Bodvarsson and Walker's work on Iceland, situated as it was along the crest of the Mid-Atlantic Ridge. At the time (see §4.3), Matthews had just applied their ideas about the formation of new crust in Iceland to submarine ridges, believing (like Wilson and Vine) that they neatly explained the linearity of the magnetic anomalies running parallel to the ridge axes; he thought of it as a refinement of Hess's seafloor spreading. Upon receiving Matthews' review, the authors added or altered a section entitled "Geologic Hypotheses." Heirtzler recalled:

We submitted our Reykjanes Ridge paper to *Deep Sea Research* in November, 1965. I have the vague recollection that Drum Matthews was a late reviewer of the manuscript. Drum was not the type of person who would have insisted that we agree with the V-M interpretation, but very likely that had something to do with the wording in the "Geologic Hypotheses" section. I recall that Drum had a short sabbatical at Lamont before we flew the Reykjanes Ridge. I don't think we made any other modifications in response to reviewers' comments. In fact I don't recall who the other reviewers were.

(Heirtzler, November 3, 2009 email to author)

They began the newly altered section by repeating the same two difficulties Lamont workers had already raised against V-M.

While limits can be put on the configuration and magnetization of these bodies, the origin advocated for this general structural pattern depends on the geologic hypothesis adopted. The location of the axial body under the floor of the rift valley in the Mid-Atlantic Ridge suggests that it consists of volcanic material filling a tensional crack. The existence of progressively smaller anomalies on the sides was attributed by Heirtzler and Le Pichon (1965) to a pattern of subsidiary fractures. Vine and Matthews (1963) on the other hand, following Dietz (1961), suggested that this pattern was due to a spreading ocean floor, originating at the ridge axis, and alternately normally and reversely magnetized. However, this hypothesis in its present form does not explain the characteristic change in magnetic pattern from the axial zone to the flanks and difference between the axial anomaly and the adjacent ones.

(Heirtzler, Le Pichon, and Baron, 1966: 440; the references to Vine and Matthews and Heirtzler and Le Pichon are the same as mine; the Dietz reference is to my Dietz, 1961a)

They acknowledged that Matthews had recently applied Bodvarsson and Walker's ideas about the formation of Iceland to oceanic ridges, and promised that they would investigate the matter.

More recently, Matthews (1965) has assumed a process similar to the one by which Bodvarsson and Walker (1964) explained the structure of Iceland. Crustal drift, in this case, is produced by crustal extension through injection of dykes in a central rift. It is consequently important to investigate whether the magnetic pattern found over the Reykjanes Ridge continues into Iceland and whether a structure similar to the geologic structure adopted for Iceland can produce such a pattern.

(Heirtzler, Le Pichon, and Baron, 1966: 440; the reference to Bodvarsson and Walker is the same as mine; the Matthews reference is to my Matthews, 1967)

Heirtzler and Le Pichon turned to Hans Wensink, a paleomagnetist from Holland who was visiting Lamont at the time. Wensink had worked extensively on the paleomagnetism of basalts in Iceland's central graben (Rutten and Wensink, 1959; Wensink, 1964a, b). He prepared for them "a simplified geologic section" (Figure 5.10) through the graben. Bodvarsson and Walker had proposed that the graben was an on-shore extension of the Mid-Atlantic Ridge and for which they gave two possible solutions: a mobilist one that they themselves preferred, involving crustal extension, and a fixist one involving an original syncline with a contracting central volcanic belt running down the center.

Two hypotheses appear capable of explaining the available geological data on the structure of Iceland. In one, a contracting state hypothesis, the structure of Iceland is taken to be synclinal,

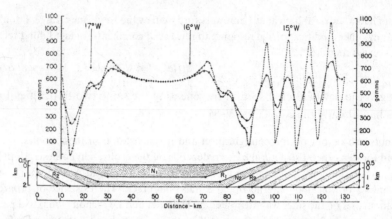

Figure 5.10 Heirtzler *et al.*'s Figure 11 (1966: 441). Their caption reads: "Calculated anomalies for geologic section of eastern Iceland. "N" and "R" refer to normal and reversely magnetized material. Anomalies are calculated for heights of 1 km (dashed line) and 2.5 km (solid line) above sealevel or 0.5 km and 2 km above the eastern plateau. The value of magnetization assumed is +0.006 c.g.s. for the normally magnetized material and −0.006 c.g.s. for the reversely magnetized material." N_1, the central graben, is the youngest normally magnetized material. R_1, N_2, and R_2 are successively older lava flows with sequentially normal and reversed polarity.

and the width of the belt of active volcanism to have contracted steadily from a maximum of several hundred kilometers in early Tertiary times to its present-day minimum of a few tens of kilometers. The dips of the Tertiary lavas are directed towards the active median belt and are due primarily to sagging below the great weight of lavas in this belt.

(Bodvarsson and Walker, 1964: 288–289)

Matthews had worked with Bodvarsson and Walker's mobilist explanation (§4.3). Heirtzler and Le Pichon instead applied the fixist solution (the one Bodvarsson and Walker rejected) and, by analogy, sought a fixist explanation of the offshore axial anomalies on the Reykjanes Ridge.

The magnetization assumed (0.006 c.g.s) was the average of magnetization measured on rock samples. The section consists essentially of a wide block of recent, normally magnetized lava with, on each side, flows of alternately reversed and normal magnetization dipping toward the graben ... The section was arbitrarily terminated at a depth of 2 km which corresponds approximately to the average thickness of flow basalts ... The interesting point is that these anomalies are not produced by vertical blocks of dykes, but by flows of lava. This fits well with the results of model studies, namely that the bodies have a much larger horizontal than vertical extent ... It is thus clear that such a model requires relatively small changes to give an anomaly pattern similar to the one observed over the Reykjanes Ridge.

(Heirtzler and Le Pichon, 1966: 441–442)

Applying, but without firmly adopting the fixist version, which Bodvarsson and Walker had rejected, to the Reykjanes Ridge, they explained the anomalies in terms of alternately reversed and normal magnetizations without saying how the reversals arose. They also noted that "the problem of the origin of the flank anomalies would still remain unsolved."

It is clear that such a model requires relatively small changes to give an anomaly pattern similar to the one observed over the Reykjanes Ridge ... In other words, the possibility, that such a geologic process is at the origin of the anomalies of axial zone of the Reykjanes Ridge, cannot be denied on the basis of the analysis of magnetic data. The arguments for or against this origin have to come from other geological or geophysical considerations. Even so, the problem of the origin of the flank anomalies would still remain unsolved.

(Heirtzler et al., *1966: 442)*

But Heirtzler and Le Pichon did not firmly adopt Bodvarsson and Walker's fixist model, taking, it seems, a wait-and-see approach.

The geologic structure of Iceland is such that a similar magnetic pattern, centered over the main graben, is probably developed there. While such a geologic structure could explain the axial magnetic pattern observed over the ridge, the differences between the Icelandic plateau and the Reykjanes Ridge are so large that it is not clear to us how their geologic structures can be similar.

(Heirtzler et al., *1966: 442)*

What about remanence? Had Heirtzler and Le Pichon changed their minds? If they had adopted Bodvarsson and Walker's model, even their fixist model, they also would have had to acknowledge that the magnetization is at least partly remanent. But they remained ambivalent about whether remanence is involved.

The lower end of the axial body cannot be deeper than the Curie isotherm depth, which, for a heat flow of $5\,\mu cal.cm^2$ per sec would be approximately 6 km. Such a body, extending from the sea-floor to a depth of 6 km, would have width of 20 km and a magnetization contrast of 0.008 e.m.u. This high value of magnetization suggests that there is an important remanent component of the magnetization in the direction of the present field. If the model is assumed to be within a reversely magnetized medium, then the magnetization has to be only one half of this value.

(Heirtzler et al., *1966: 437)*

Heirtzler (§4.6) retrospectively remarked, "It didn't make much sense to argue whether the central anomaly was induced or remanent" unless they thought that "the seafloor was moving." However, Ade-Hall (1963) had shown in a wide range of ocean basalts that Koenigsberger ratios were high, indicating that remanence far exceeded induced magnetization; they knew of this work because Opdyke had told Le Pichon about it after seeing that he and Heirtzler had failed to mention it in their Mid-Atlantic Ridge paper (§4.6). Yet they continued to disregard remanence as a potent contributor to anomalies. Talwani retrospectively told me that he believed "Lamont workers felt that remanence was involved."

Lamont workers did not specifically express a preference for whether the magnetization was induced or remanent, that is, whether it was caused by having high susceptibility or high remanent magnetization. I am sure that, implicitly, the Lamont workers felt that the magnetic anomalies were almost certainly mostly caused by bodies with remanent magnetization.

(Talwani, July 17, 1981 letter to author)

Even after discovering the bilaterally symmetrical magnetic anomalies flanking the Reykjanes Ridge, Talwani, Heirtzler, and Le Pichon remained opposed to seafloor spreading and V-M. If they did really believe that remanence was significantly involved, their resistance is very difficult to understand. Their work had shown that there were now three instances of mid-ocean ridges with bilateral magnetic anomalies extending beyond their axial zone out over their flanks, thus removing the first difficulty facing V-M (§2.13). One might have quibbled about whether or not the magnetic anomalies extended beyond the axial regions of the Juan de Fuca Ridge, but not so with the Reykjanes Ridge. Heirtzler, Le Pichon, and Talwani did not seem to recognize just how strongly their own findings actually enhanced support for V-M and, with it, Hess's seafloor spreading.

Retrospectively, Heirtzler said about their endeavors:

We really wanted to explain the magnetic anomaly pattern by some kind of geometrical effect – fractures, offsets, scarps, and things of that kind on the sea floor or buried beneath the sea floor – that might not require the involvement of the earth's magnetic field. We thought that would lead to an overall simpler theory . . . One might keep in mind that all that time in 1963, '64, and '65, the people who were measuring magnetic anomalies of rocks on land were really the ones developing the geomagnetic time scale only had a time scale that extended back something like three to possibly five million years . . . So the geomagnetic time scale was quite primitive, and to say that these magnetic anomalies were due to reversals assumed a great deal because three million years only takes you through the first couple of magnetic anomalies over the Reykjanes Ridge on either side of the ridge. And, over the Reykjanes Ridge we had many anomalies on either side.

(Heirtzler, March 1981 letter to author)

Heirtzler, however, seems to have had little awareness or appreciation of the high Koenigsberger ratios of oceanic basalts, and the ample evidence for reversals of the geomagnetic field, evidence that I shall summarize in a moment. Also, Heirtzler had been unaware of the longstanding controversy over continental drift and seafloor spreading.

I didn't become acquainted with those theories until after Fred Vine and Drum Matthews published their theory of seafloor spreading, and I dug back a little into Hess and Dietz' theory. I had quite by chance met Hess at a meeting at Columbia University. I was seated at the dinner table with him, and found him a fascinating person. At the time I didn't know who Hess was. I remembered his name at the table because he was such a very nice person, and so very well informed, and such an imaginative person. I was so involved with what he was saying, I suggested that he might come to Lamont to give a lecture.

(Heirtzler, note to author, March 1981)

Before arriving at Lamont in 1960, Heirtzler had no background in geology or geophysics. He had studied physics, had taught it in Beirut, and had worked at General Dynamics demagnetizing nuclear submarines.

5.5 Geomagnetic reversals, the dominance of remanence, and Vine's Ph.D. dissertation, August 1965

As a prelude to considering some aspects of Vine's thesis, I comment on the status in the early 1960s of two of the premises underlying V-M: the reversing geomagnetic field and the dominance of remanent magnetism as the source of marine magnetic anomalies.

Matthews and Vine never seem to have seriously doubted that, acting approximately as a geocentric dipole, *the geomagnetic field had reversed its polarity many times*, but across the geophysical community, excluding paleomagnetists, in the first half of the 1960s this was not widely regarded as a given. Vine himself referred to it as a "questionable assumption" (§2.13). As I shall describe, by the end of the decade, and central to the mobilism debate, all doubts had been dispelled following the harmonious integration of (1) the marine magnetic anomalies with (2) the reversal timescale determined radiometrically on land and (3) the reversal stratigraphy established from cores in deep ocean sediments. It is important to emphasize that this was an integration only of *the order and timing of these three records*, essentially the global correlation of geomagnetic field reversals as recorded by three independent means; it did not require or imply a full understanding either of how or why reversals occurred, or of the processes by which they were recorded, understanding of both of which was limited at the time. The resolution of the question of whether reversals of the field occurred was not hindered by these uncertainties – the remarkable agreement of three independent records carried the day. When Matthews was making his detailed two-dimensional survey of the Carlsberg Ridge and when Vine proposed V-M, .the reversal timescale was only sketchily known and then only for a few intervals of time. V-M was conceived by a first-year graduate student based on a survey carried out under the direction of his supervisor, well before geomagnetic reversals were universally accepted, or in some quarters even considered seriously. In view of the outcome, it was an astonishing achievement by any standards.

Although, by the early 1960s, there was very good evidence for believing in field reversals, at the time not many workers in the geosciences were aware of how extensive that was. Reversals did not attract general interest. To those who followed them, the most telling evidence was that reversals observed on land occurred in an orderly way – in stratigraphic sequence. They had been documented in thick volcanic and sedimentary piles, especially of Late Cenozoic age, and were independent of rock type. In particular, the most reversal had been well established, as Alexandre Roche described *"au cour du Pleistocene inférieur"* between about one million and 600 000 years ago. A very wide variety of sedimentary, volcanic, and volcaniclastic rocks were represented in the older reversed and in the younger normal polarity sequences. Polarity was not in any way correlated with rock type. It had also been established that the length of time that polarity remained constant varied very greatly. During the Cenozoic, reversals occurred every million or few hundred thousand years. By the early 1960s (see Hess's discussion in §5.4), what is now called the Kiaman Reversed Superchron had been established, lasting about 50 million years from the Late Carboniferous through Middle Permian. Sedimentary and igneous rocks of this age of a wide variety of compositions had been sampled from all continents except Antarctica; all had reversed polarity. Worldwide, rocks that acquired their magnetization during the same interval had the same polarity. Nonetheless, there were lingering doubts about the role of self-reversal; it had been claimed, in limited instances, that reversed polarity was correlated with normal polarity, and Bullard (1968a) had identified these discrepancies as in need of further understanding.

Rod Wilson of the Physics Department at Imperial College, London, was a member of one of the paleomagnetic groups there who initially thought there might be some truth to this correlation. These doubts were answered mainly by many studies worldwide of igneous bodies and their baked contacts in rocks of different ages (II, §1.11). Wilson himself answered his own doubts. Notable was his double-baked contact test and his global analysis of contact studies. For his double-baked contact test, Wilson (1962) sampled a locality in the Eocene Antrim Plateau basalts of Northern Ireland, which had been first studied paleomagnetically a decade earlier by Hospers and Charlesworth (II, §2.9). A reversely magnetized, massive basalt flow overlay baked red laterite. The baked laterite was magnetized in the same direction as the basalt – a positive contact test. A younger mafic dyke intruded both lava and laterite. The directions of magnetization of the rebaked laterite right next to the dyke were parallel to those of the dyke; both were reversely magnetized. Further away from the dyke, the magnetizations changed to the direction of the once-baked laterite. Although both lava and dyke were reversed they did not have identical direction, having cooled at different times, the difference therefore was ascribable to secular variation. As McElhinny and McFadden (2000: 143) remarked, "It is difficult to explain the observations by any reasonable mechanism other than the earth's magnetic field had changed polarity and was reversed during both reheating episodes." In his global contact test, Wilson (1962) compiled the polarities of igneous bodies and their baked contacts as they had been recorded in the literature up to the early 1960s. He argued that if the geomagnetic field frequently reversed in polarity, then igneous bodies and the rocks baked by them should always have the same polarity, whereas if self-reversals are to account for the observations, then in roughly half of the examples the polarities should be the same and in the other half opposed. Many different types of igneous and baked contact rocks were represented. In the case of the rocks deposited during the Kiaman Reversed Superchron, the polarity of all igneous bodies ought, on the hypothesis of self-reversal, to have been opposed to that of the baked contact: none were. Wilson's test was updated by Irving (1964: 176) and he found that there were, in 1963, eighty-eight examples including some from the reversed Kiaman. In eighty-five cases the polarities of the igneous body and contact agreed. This was even true in two cases of transitional directions – lavas and their baked contacts situated in transitional strata between sequences of opposed polarity had the same magnetization directions. In three cases the polarities disagreed – they were early studies in which magnetic cleaning had not been carried out and partial instability was likely the cause. Wilson's test provided very strong evidence that self-reversals are exceedingly rare, as they have proved to be.[3]

Matthews and Vine did not make clear in their paper how much of the research on reversals they had read. Wilson's tests were published in *Geophysical Journal*, their "local" journal, so the likelihood is that they saw them. They were likely aware of early work done in their department – Hospers' early work on Late Cenozoic reversals (II, §2.8), Irving's mapping of sequential reversals in the Precambrian Torridonian (II, §2.6), and Creer's observation of reversed magnetization in the Devonian and Permian (II, §2.12). Bullard, with whom they were in regular contact, favored field reversals. Laughton, with whom Matthews had close contact, had as a graduate student been a contemporary of Hospers, and thought favorably of his work on reversals (§2.4). Vine also heard Blackett lecture on paleomagnetism.

Also central to V-M was the notion that marine magnetic anomalies were caused by *remanent magnetization and the effect of induced magnetization was negligible*. From an early stage in his work, Vine seems to have settled for this. In its favour there was Matthews' earlier measurement of high Koenigsberger ratios in dredged seamount basalt (§2.7). Importantly, Vine also had shown from Matthews' detailed magnetic and bathymetric survey that a seamount on the Carlsberg Ridge was reversely magnetized and the magnetic anomaly associated with it was not caused by topography; hence, reversely magnetized rocks likely existed on the seafloor, a first requirement. And Vine learned of Ade-Hall's findings that ocean basalts from other ocean localities had high Koenigsberger ratios (§2.9). Also in their department, there was presumably an awareness of well-documented instances on land where sampling from bodies with associated negative magnetic anomalies confirmed the dominance of reversed remanence. Certainly Bullard, their mentor, was aware of them, attending London meetings of the RAS in the 1950s, where they had been discussed. There were the well-known negative anomalies found by Gelletich over the Pilansberg dykes of South Africa, which Gough later confirmed by sampling were reversely magnetized (III, §1.17). Nearer to home, there was the demonstration of reversed magnetization in samples from Tertiary dykes of northern England, which were characterized by negative magnetic anomalies (II, §1.6). Whatever he knew ahead of time, Vine was prepared, on the basis of his analysis of Matthews' survey and some ridge profiles, to clothe the entire deep oceans with basalt in which remanence strongly dominated over induced magnetization (§2.13). Now a given for Vine, this was not quickly accepted by others (§3.8).

I turn now to Vine's Ph.D. thesis. After submitting his Juan de Fuca Ridge paper with Wilson (1965), Vine returned to work on his thesis. He finished in August. Martin Bott of the University of Durham and Maurice Hill on staff at Cambridge were examiners. Bott was well informed and had already spoken in favour of mobilism at the Royal Society's symposium on continental drift (§3.3). He would later analyze the pattern of magnetic anomalies over the Juan de Fuca Ridge in terms of two-dimensional rectangular normally and reversely magnetized blocks wholly within Layer 2 (Bott, 1967).

Martin Bott was the external examiner for my thesis and Maurice Hill the internal examiner. The only things I can remember from the oral were that they thought there were too many exclamation marks in it (a fair comment), that I could not spell "benefited" and that Chapter 6 [in which Vine looked at some of the implications of V-M with regard to such things as the worldwide pattern of mantle convection] was too speculative.

(Vine, October 12, 2009 email to author; my bracketed addition)

Vine's dissertation, "Interpretation of magnetic anomalies observed at sea," was in two parts: the first dealt mainly with V-M; the second with interpreting magnetic surveys over isolated seamounts. Much of his thesis has been covered in Chapter 2. I deal only with the concluding chapter of Part I, in which he raised a certain difficulty with V-M, and speculated about the geometry of mantle convection.

Vine was concerned that V-M did not readily explain why the amplitude and wavelength of the anomalies were higher on the flank than in the axial zone of the ridge (RS2). It was essentially the same difficulty that Heirtzler and Le Pichon had raised in their Mid-Atlantic Ridge paper (§4.6). Vine recognized the difficulty, learning later that Heirtzler and Le Pichon had done so too. Their paper was published on August 15, 1965. At the time Vine, with help from his wife Sue, was putting the final touches to his thesis. Vine does not remember making contact with Heirtzler or Le Pichon. However, as he explained to me, he may have received a reprint.

I do not recall having a reprint and it seems very unlikely that I had one because I don't think I had had any contact with them by then. However, I note that the page numbers for the Heirtzler and Le Pichon reference are added in ink which suggests, perhaps, that I did have a reprint.

(Vine, May 12, 2010 email to author)

Heirtzler and Le Pichon pointed to the increase in wavelength and amplitude when moving from axial to flank anomalies; Vine pointed to the change when moving from the flank anomalies to those over the abyssal plains and it was this that he thought was serious.

However, it must be said that there is one essential point on which it would appear to break down. If the crust on the flanks of an oceanic ridge is "conveyed" passively off it, on the limb of a convection cell within the mantle, and with progressive burial under a sedimentary cover, then one would expect the anomalies observed over the abyssal plains to be comparable with those over the flanks, but slightly attenuated because of the increase in depth ... However, this is not generally the case. Although the shortest wavelengths show some attenuation with increasing depth, the amplitudes of anomalies recorded over the abyssal plains are often greater than those over the flanks and longer wavelength anomalies predominate. This has been appreciated for some time but stressed recently by Heirtzler and Le Pichon (1965).

(Vine, 1965: 112; Heirtzler and Le Pichon (1965) is the same as mine)

The intensity of the remanent magnetization of the basalts may decrease over time (RS1) – the viscous decay predicted by Néel's theory of *traînage magnétique* (Néel, 1949).

Clearly, on the face of it, this discrepancy is insuperable in the light of our present state of knowledge, but it seems quite possible to invoke mechanisms which might change or modify the magnetisation of the rocks with increasing age. Dynamic and thermal metamorphic effects produced by fracturing, shearing or intrusion due to the release of stresses in the crust would doubtless modify and probably reduce the remanent magnetisation. In particular, strongly magnetised volcanics within layer 2 might lose their remanent magnetisation to some extent due to "decay" with age, enhanced possibly by the fact that they are sitting in an alternating field (!!). As ever, the introduction of a "geological time" factor introduces possibilities which are difficult to study and almost impossible to simulate. However, this unknown should not be overlooked or underestimated. All too little is known of this aspect of magnetic properties. Konigsberger (1936) and Hospers (1954) both felt that there was evidence for a "decay" of the remanent magnetisation of volcanics with time but no further work seems to have been done along these lines. However, if such a "decay" for high remanent intensities could be established as an additional variable it would clearly play an important role in any model based on ocean-floor spreading. Over the crest and flanks of ridges the effects due to the more strongly magnetised layer 2 volcanics might predominate whereas over the abyssal plains the overall effect of layer 3 might be dominant.

> *(Vine, 1965: 113; Konigsberger reference is same as mine; Hospers*
> *(1954) is my Hospers (1954b))*

But this did not remove the difficulty. If V-M and seafloor spreading held, then the further the rocks are from the ridge axis, the older they are and the weaker their remanent magnetization should be. Appealing to viscous decay did not explain why older abyssal plain magnetic anomalies sometimes had greater amplitudes than younger axial or flank anomalies. Clearly there were many major features of marine magnetic anomalies that were not fully explained by V-M alone. It was not until later that these questions were investigated in any detail.[4]

I turn now to Vine's speculations on mantle convection. He adopted Hess's tennis-ball analogy (§4.8) with the seams representing the down-currents or sinks, and the up-currents associated with active ridges being along the long axis of the two barbell shapes outlined by the seams (Vine, 1965: 114–118). He first reasoned that, if V-M is correct, active and inactive ridges can be distinguished by whether or not they possess an axial magnetic anomaly. Given this criterion, he proposed that the Mid-Labrador Sea Ridge and "the ridge system to the south-east of Africa" are inactive (Vine, 1965: 114, 117). With these eliminated, Vine cleared the way to invoke Hess's "tennis-ball" convective pattern.

If this criterion is correct and these sections of the ridge system are inactive, it leads to an interesting speculation. It means that the active parts of the ridge system at present form two separate lengths each traversing half the circumference of the earth: one running from the New Siberian Islands in the Arctic Basin, across the Arctic Ocean, and down the whole length of the Atlantic to 40° or 50°S, and the other running from the Red Sea, via the Indian and South

(a)   (b)

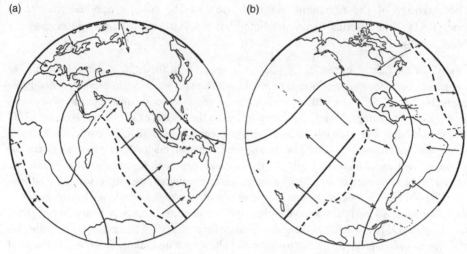

Figure 5.11 Vine's Figure 6.1 (1965: 115–116). His caption reads: "An illustration of the way in which one might fit a simple, two-cell, 'tennis-ball' convection pattern to the major features of the earth's surface if one uses the central magnetic anomaly criterion to delineate ridges which are actively "spreading" at the present time. Thick broken lines indicate the active parts of ocean ridge system. Thick solid lines indicate the up-currents of the hypothetical two-cell system, and the "seam" of the tennis ball represents the down-current or "sink." The thick broken lines indicate active transcurrent or transform shears. Arrows indicate directions of motion." (Reproduced with thanks to Jill Turner and photographer at Madingley Rise.)

Pacific Oceans to the Gulf of California (see fig. 6.1 [my Figure 5.11]). This fits well with a "tennis-ball" pattern of convection cells within the upper mantle at the present time. As can be seen from fig. 6.1, the main stumbling block to accepting this in the past has been the assumption that the ridge system is active between South Africa and Antarctica. Having omitted this, the simple two-cell "tennis-ball" system is a good approximation to the present tensional and compressional features or "sources" and "sinks" on the earth's surface. It will be seen that the continents have taken up positions over the "sinks" represented by the "seam" of the ball, and that resulting compressional features, such as mountain systems and island arcs, are associated with this locus. The centres of the cells correspond to the two active lengths of the mid-ocean ridge system, and the directions of motion agree well with the motion on active shear zones.

(Vine, 1965: 117–118; my bracketed addition)

It turned out that the ridge system southeast of South Africa is active. It appeared that it had no central anomaly, but there simply was not enough data at the time to tell, and the spreading rate is very low and spreading direction very oblique to the ridge axis. After rereading Chapter 6 forty years later, Vine recalled his appeal to Hess and the response it elicited from Hill and Bott.

The only points in the Chapter that are unsatisfactory or incorrect are my attempts to explain the amplitude differences mentioned above [i.e. higher amplitudes of the abyssal plain anomalies compared to ridge flank anomalies] and my assumption that the ridge system southeast of South Africa is inactive. (This arose from the fact that there were very few data available at that time, the spreading rate is very low and the spreading direction makes a very oblique angle with the overall trend of the ridge – hence there are numerous fracture zones.) This assumption led me to favor Harry's [Hess's] tennis ball configuration of mantle convection, and this is the one really speculative point in the Chapter. As far as Maurice [Hill] and Martin [Bott] were concerned this was going a bit too far!

(Vine, May 3, 2010 email to author; my bracketed additions)

Vine was awarded his Ph.D. He and his wife embarked on September 16 from Southampton for New York en route to Princeton.

5.6 Matthews seeks to explain the greater amplitude of the central anomaly

Matthews wrote the following entry in his daybook on August 23, 1965:

Went on striving to get maps right during w/e. I calculated a typical profile by taking means of depth of anomaly read along lines ‖ to strike. Wrote and discussed abstracts.

Idea – Calculate anomalies over a ridge formed by injection of narrow dykes injected so that probability of a dyke at distance x from centre follows a Gaussian law and in which there is a sudden reversal of the field now and then. BOMM can produce a semi random series of numbers obeying this law. As each dyke is injected co-ordinates of all others must be stepped out once. Finish when initial dyke reaches a given distance, say 100 m from centre. Then compute anomaly by adding together whole lot of series representing anomalies due to individual dykes (+ and −) at suitable lag distances. ["Idea" is underlined in red.]

When I interviewed him, he actually read the passage aloud, annotating as he went along.

Bedford [Nova Scotia] Monday, August the 23rd. Went on striving to get maps right during the weekend. I calculated a typical profile by taking means of depth and anomaly read along lines parallel to the strike: It was an idea that Bob Parker and Bosco [Loncarevic] wrote up ... Wrote and discussed abstract. We were trying to write a paper about this story. Idea: ("It's got a red line by it!") Calculate anomalies over a ridge formed by injection of narrow dykes injected so that the probability of dyke at distance x from the center follows a Gaussian law in which there is a sudden reversal of the field now and then. BOMM (that was for Bullard, Oglebay, Munk, and Miller collection of programmes for handling time series), BOMM can produce a semi-random series of numbers obeying this law as each dyke is injected coordinates of all the others must be stepped out once. Finish when the initial dyke reaches a given distance, say 100 meters from center. Then compute anomaly by adding together the whole lot of series representing anomalies due to individual dykes, plus or minus suitable lag distances.

(Matthews, 1979 interview with author)

Matthews then put the daybook down and continued.

That idea, actually, I don't know whether it occurred on Monday, August 23rd, was actually – I was having a discussion with Manik [Talwani], I think Manik was talking to us – he didn't believe anything we said – I think it was then that I said, "You know, we'll try running marbles down a board to mock up this story. I mean I always think in terms of marbles instead of computer programs. Bosco went out and got three packets of marbles, I think, and we leaned up a board and nailed another board to the corner of it, and injected marbles, and showed that one got this fall off of intensity. When we computed using a program Ron McMahon at Bedford had running by then, we computed the resulting anomaly pattern from the model, because we didn't have a random number program then that would do; random number programs were very hard to get. And, when I got back here I didn't find a random number program for ages until Bob Parker, who was then a research student now at Scripps, very good mathematician, very nice chap, who spent most of his Ph.D. time helping other people. It was from that, the new book Abramowitz and Stegun [*Handbook of Mathematical Functions with Formulas, Graphs, and Mathematical Tables*, published in 1964]. He said, "Oh, I found a random number program that will do your job, and why don't I get Jenny [Jennifer Bath] to program it." The acknowledgements of that paper were written while in hospital – I was in hospital, I'd had a stroke. It was a very little stroke, and I got very bored. I was only in hospital for three or four days, and I was sitting up and taking notice, and they said, "Are you better? What is a magnet?" I couldn't tell them what a magnet was. It is very hard to tell what a magnet is; you have to be a bit of a philosopher. If philosophers could tell you what action at a distance is, we should know much more about the world than we do. So I found that rather difficult. So I'll write a paper instead. So I wrote that paper. And I always felt extremely guilty about the acknowledgements of that paper because I didn't acknowledge Bob Parker's contribution. It was fine to acknowledge funny marbles from Bosco, but in fact what I really ought to have acknowledged was Bob, who, as he did almost for everybody in this lab, actually found the algorithm that was required, and written it out on a nice piece of paper for me so that even I could understand what was going on.

(Matthews, 1979 interview with author; my bracketed additions)

Matthews eventually co-authored a paper with Jennifer Bath who did the programming. He first performed the experiment with marbles. Playing a row of pale colored marbles, which represented non-magnetic screen rocks, along a board resting at the bottom of an inclined plane, they alternately rolled down red and blue marbles, respectively representing normally and reversely magnetized dykes. The rolling marbles, all dropped from the same place for any given run, would strike the line of pale marbles, bounce upward and sideways, and settle somewhere along the line of marbles, which represented where it was injected. The average distance of the bounced marble from the center of the line, for any particular run, represented the standard deviation from the center of dyke injections. They varied the standard deviation by changing the height or angle of the inclined plane. They found that if dyke injection occurred over a 5 km range, the resulting pattern of magnetic anomalies showed enough contamination to produce a central magnetic anomaly of approximately twice the magnetization of the surrounding anomalies. They ended the discussion of the marble simulation by showing Matthews' strong aversion to computer modeling.

The results of these early experiments indicated that a standard deviation of about 5 km would be the best. This is about one quarter or one half of the width of the median valley. Since, unfortunately, it is hard to persuade people of the truth of this kind of simple-minded experiment, it was necessary to repeat it on a computer.

(Matthews and Bath, 1967: 351)

The results were the same.

Clearly the set with a standard deviation 5 km, which is about a quarter of the width of the median valley or half the width of the other blocks, is about right. It is particularly clear that this model adequately explains the tendency of the central anomaly to be larger than the others without any requirement of any additional mechanism of magnetization ... It will be seen that only near the centre of the pattern is there an absence of oppositely magnetized material, and this is why the median anomaly is substantially larger than the others.

(Matthews and Bath, 1967: 354)

Cann and Vine (1966) had already offered this explanation of why the amplitude of the central anomaly is greater than the surrounding anomalies (§4.3). Matthews, however, thought of the idea while in Canada independently of them and added the use of standard deviation to determine the width of the median valley within which dyke injection occurs.

5.7 The Ottawa meeting, September 1965

This meeting was sponsored by the International Upper Mantle Committee and held in Ottawa September 2 through 9, 1965. There were three symposia: "Drilling for Scientific Purposes," "The World Rift System," and "Continental Margins and Island Arcs." I shall say nothing about the first and little about the third. The symposium on the world rift system was concerned with oceanic rifts, and the published volumes provide a detailed record of the status in 1965 of this central aspect of the mobilism debate. The meeting was held one month after Wilson's paper on transform faults (1965b) was published, and two months before Vine and Wilson's paper (1965) on the Juan de Fuca Ridge appeared. I suspect therefore that on arrival many participants did not know about transform faults, and very few would have known about the recent work on the Juan de Fuca Ridge: but they certainly did when they left. Vine and Matthew's 1963 paper would be well known to them, as were Hess's and Dietz's works on seafloor spreading earlier in the decade, but they were not yet widely accepted and often disbelieved. I shall concentrate on the presentations by Talwani (speaking also for Heirtzler and Le Pichon), Wilson, and Hess and the ensuing discussions, because the topics covered by them are at this juncture central to the mobilism debate.

Not many speakers directly addressed V-M, transform faults, or even seafloor spreading; several reported finding the telltale zebra pattern of magnetic anomalies without mentioning V-M, showing just how far it and seafloor spreading were from being embedded in geological discussion, as they were so soon to be. Disagreement and speculation were rampant, especially over mantle convection and whether or not the East Pacific Rise extended northward beneath North America. This leads me to conclude that without V-M and the idea of ridge-ridge transforms, controversial as they were, to tether seafloor spreading to hard data, mobilism likely would have taken many, many years before gaining acceptance. It might have happened if enough influential Earth scientists had come to their senses and recognized the difficulty-free status of the paleomagnetic case for mobilism, but that was unlikely.

I start with **Talwani's presentation**, which began with the recent Lamont/US Navy's airborne magnetic survey over the Reykjanes Ridge (§5.4). Hess, who was in the audience, saw these profiles (Figure 5.9) for the first time and promptly wrote to Vine, then about to embark for the United States and Princeton. He was not sure who gave the talk but he sure remembered its contents, singling out the "magnificent bilateral symmetry" of these magnetic anomalies, which to him signified strong support for V-M and seafloor spreading.

There were a good many references to the Vine-Matthews hypothesis at the Ottawa symposium – half favorable, half against. A magnetic survey with close spaced traverses and good position control on the ridge southwest of Iceland (presented by Bosco L. or Talwani – I forget which) showed a magnificent bilateral symmetry so striking it could be seen at a glance. I'll tell you more about the discussion when I see you.

> *(Hess, September 10, 1965 letter to Vine; Bosco L. is B. Loncarevic)*

Talwani acknowledged the "extreme symmetry" of the anomalies but did not interpret it as Hess did.

The large amplitudes, the extreme symmetry and regularity of the axial magnetic anomalies and their linearity in a direction parallel to the strike are revealed by a closely spaced aeromagnetic survey over the Reykjanes Ridge SW of Iceland. There is some evidence for the linearity of the flank anomalies over the Reykjanes Ridge and also over other segments of the Ridge-Ridge system.

> *(Talwani et al., 1966: 345)*

He and his co-authors, Heirtzler and Le Pichon, rejected V-M.

We are not sure of the ultimate origin of the ridge magnetic anomalies. However we do believe that the symmetry of the magnetic pattern and its parallelism to the strike of the ridge requires that the magnetic anomalies owe their existence to the formation of the ridge, which in turn is caused by a change of density in the upper mantle. The spreading ocean floor theory of Hess (1962) and Dietz (1962) with reversals of the magnetic field has been used by Vine and Matthews (1963) to explain the axial anomalies over the ridge crest. However we feel that the variation in amplitude of the axial anomalies as well as the completely different character of the flank anomalies argues against the Vine and Matthews hypothesis.

> *(Talwani et al., 1966: 346; Hess reference is same as mine; Dietz reference is my Dietz (1962a))*

Also he (1966) previewed their fixist interpretation of the great fracture zones of the northeast Pacific that they were to soon offer (§5.3).

The magnetic anomaly patterns in the North East Pacific have been used by Vacquier and others to deduce large scale strike slip motions along the fracture zones. The establishment of the association of the magnetic pattern with the East Pacific Rise makes it possible to give a different explanation for the matches of the anomaly patterns across the fracture zones – an explanation that does not involve large scale motions. It is suggested that (i) the different segments of the crest of the East Pacific Rise developed at their present locations, (ii) the matches in magnetic pattern are due to similarities in the pattern at equal distances from the rise crest, and (iii) differential uplift due to the formation of the rise is responsible for depth differences across fracture zones.

(Talwani et al., 1966: 345–346)

Hess was not the only one impressed by the Lamont data, but unimpressed by their interpretation. Loncarevic disagreed, offering an explanation of the difference in wavelengths between the axial and flank anomalies in terms of V-M. He proposed that the anomaly variations reflect the variations in spreading rates and reversal frequency.

I would first like to comment on a question asked yesterday and then I will offer a suggestion for the difference in the period of anomalies across the ridge. I think Hess's objections to the suggestion that dykes cause the magnetic anomalies along the Mid-Atlantic ridge were probably based on the assumption of a very long single dyke, which was not what we had in mind. We tried to emphasize that a width of one mile might be a unit size for the intrusion. You will recall that the central valley has a number of blocks, and that these blocks are normally magnetized in one section and reversely magnetized on either side. Now the distance between the normal and reversed areas is about 30 km. The last reversal of the field is a million years which gives you 3 cm per year for the growth of the ocean, which is the right sort of figure. If you go out to, say, 200 km from the centre of the rift, the wavelength of the magnetic anomalies is, as Dr. Talwani has just said, about 70 km. The possible explanation is that, 6 to 10 million years ago, there was much more dyke production along the ridge, and therefore the floor of the ocean was spreading faster and the anomalies therefore developed with a longer wavelength. An alternative possibility is that the reversals of the earth's magnetic field took longer 10 million years ago – on a cycle of two million rather than one million years.

(Loncarevic, 1966: 347)

Talwani acknowledged Loncarevic's point, "I agree – that explanation can, of course, be made." He then changed topics, and turned to recent sedimentary studies, which, he argued, raised difficulties for seafloor spreading.

I would like to point out one more problem with this theory. In a paper in press, Ewing, Le Pichon and Ewing discuss the age of sediments found on the Mid-Atlantic ridge. They find Miocene fossils in several cores less than 90 km from the crest. Dr. Saito recently examined a dredge taken 18 years ago by Ewing very close to the rift valley and examined once by Shand [1949] in which he finds – I believe – baked middle Miocene fossils covered by volcanic glass. Also the cores taken in some abyssal hill area just off the ridge flanks run into Pliocene material 5 m below the bottom. The total thickness of sediment is probably not more than 25 m. I do not know how unequivocal the evidence is, but one might make the case from the sedimentary record that there is no evidence that the ridge is younger towards its centre.

(Talwani, 1966: 347)

This evidence, as I shall explain later, was not unequivocal (§6.2).

Heezen, who within a month would abandon rapid Earth expansion (§6.10), questioned Talwani's proposal that there is no appreciable movement along fracture zones, and presented an explanation that incorporated, even though he did not say so explicitly, V-M and ridge-ridge transforms within earth expansion.

If you look at the map of the South Atlantic Ocean, you can see the same pattern repeated over and over again across each of the fracture zones – that is, a rift valley and its offsets. The offsets are variable – some of them are only a few miles; others are 120 miles or more. The fracture zones, however, are hundreds of miles long, so it is clear that, at the moment, if you can judge by seismicity, the active zone is at the crest of the ridge and along the part of the fault between the offset parts of the crest.

(Heezen, 1966a: 348)

I turn now to **Wilson's presentation and discussion**. He maintained that the fundamental differences between continents and oceans and their histories are ultimately to be explained in terms of upper mantle convection.

The geology of ocean floors differs greatly from that of continents in age, petrology, types of faulting, ore deposits, seismicity, and in the associated systems of mountains and ridges. These differences resemble those between the surface of a fluid convecting in a cauldron which is renewed like the ocean floors, and the islands of froth or slag that collect like continents. The tectonic pattern of continents is due to marginal growth distributed by the periodic breaking up and reassembly of continental blocks. The pattern on the ocean floor reveals the flow in the upper mantle which seems to take the form of a single complex cell.

(Wilson, 1966a)

Elaborating and giving Holmes his due he remarked:

It is becoming apparent from developments by Hess (1962), Elsasser (1963), Tozer (1965), and Orowan (1965) of the original convection hypothesis of Holmes (1931) that there is probably a shallow convecting layer in the mantle at a depth of from one hundred to a few hundred kilometres, and that it is likely to be convecting in rolls of great lateral extent with motions of a few centimetres per year.

(Wilson, 1966a; references are the same as mine)

He then introduced the idea of transform faults, which few in the audience had heard of. He was on home turf and he did so with characteristic flair. According to Irving, who was at the meeting:

He had brought along a very large white board, probably Bristol board, cut in two. On it he had boldly marked a ridge that was offset by a fracture zone and he had cut the board in two along them. Waving the two pieces above his head, he reassembled them and moved them apart along the fracture zone, showing that the motions were in the opposite sense to that along a transcurrent fault. It was a vivid demonstration. It was as he may well have said, "simple." Simple it might be but its repercussions were profound, allowing him to connect up many major structural features.

(Irving, September 16, 2011 email to author)

He went on to present his global vision. He had definitely settled on upper mantle convection; it was "independent of the behaviour of the core" (Wilson, 1966a: 389), and in contradiction to Runcorn who thought that the geometry of mantle convection is a function of an expanding core.

The general pattern appears to be a single cell rising under the mid-ocean ridges and sinking under the active island and mountain arcs (Wilson, 1965b). The nature of the loop in the ridge through the Southern Ocean suggests that the loop is expanding northwards and that the pattern is a self-altering one – which offers an explanation of the periodicity of mountain building independent of the behaviour of the core (Wilson, 1965c).

(Wilson, 1966a: 389; Wilson, 1965c is my Wilson, 1965d)

Elaborating on the past motions of continents and citing Carey, who, along with Lester King, was at the time the other great showman of global tectonics, he sketched them with a broad brush. Without mentioning that Köppen and Wegener had long ago suggested it (I, §3.15) and paleomagnetists had recently confirmed it (III, §1.7), he proposed northward spreading of the Gondwana continents (1966: 389). This, he argued, had led to northerly striking major rifts and shears, resulting "in dislocations along the equator," and the formation of "regions of intermediate-type crust and of ridges in the Caribbean, Scotia, and Tasman Seas." He recognized that marine magnetic anomalies are the key to unraveling the history of continental drift and seafloor spreading. In fact, he and Vine had already provided a strong link, not envisioned by Hess, between seafloor spreading, marine magnetic data, and V-M (§5.2).

Thus the pattern of flow deduced from the ocean floors, the implied spreading of the upwelled parts of the ocean floors, and reabsorption of other parts along trenches, suggests a unified explanation of the recent movement of continents and the varied behaviour of mountain belts. This will serve as a first approximation until complete magnetic surveys of the oceans enable the detailed history of the spreading to be analyzed (Vine and Wilson, 1965; Wilson, 1965d). The notion that the history of spreading of the ocean floors can be established depends upon the hypothesis advanced by Vine and Matthews (1963) that periodic reversals of the earth's main magnetic field every 10^5 or 10^6 years are recorded in the strips of positive and negative magnetic anomalies.

(Wilson, 1966a: 389–390; Wilson 1965d is my Wilson 1965c)

There was much discussion. G. B. Udintsev, not the first to be impatient with what he took to be Wilson's cavalier attitude toward facts, and getting right to point, inquired rhetorically, "Are there any data which do not fit your theory?" Wilson, deflecting the question, called on Talwani in the audience, who (1966c) in turn asked, "Have you any explanation of why the ridges are ridges? In other words, why is the whole ridge, not just the rest, higher than surrounding ocean basin?" Wilson cleverly deflected the second question, this time to Hess, adding only that convection was involved.

This has nothing to do with the geometry of transform faults and I think that the question should be directed to Professor Hess, who has contributed more to this problem than I have. I can only suppose that the crest may be uplifted over an expanded rising column.

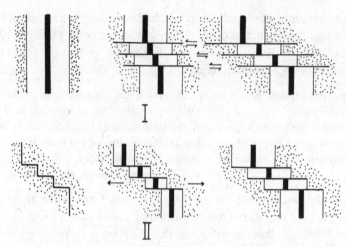

Figure 5.12 Heezen and Tharp's Figure 5 (1965: 93). The figure shows fracture zones in the equatorial Atlantic. The dashed lines indicate lines of displacement whose direction of relative motion is shown by arrows. The directions of relative motion show that Heezen and Tharp analyzed motion along the fracture zones in terms of transcurrent faulting. Compare with Figure 4.12.

Thus Wilson escaped what could have been a very technical cross-examination by one of Lamont's rising stars, who very tellingly could have called attention to the recent heat flow measurements at the crest of the Mid-Atlantic Ridge by Marcus Langseth, also from Lamont, as described elsewhere at the Ottawa meeting (§6.11). Heat flow was three times less than that expected if the ridge were caused by mantle convection; it was not emitting enough heat. This result, and others from other ridges, defied explanation for some years (§6.11).

Heezen then commented that Tharp and he, as well as Matthews, had already recognized ridge-ridge transform faults.

The idea of these transform faults, I think, is very clearly a fresh one, but if you examine the rift graben in the South Atlantic you can see that the idea wasn't exactly unknown to us. The seismic belt does go through these offsets and it has occurred to us that the Pacific anomalies could be explained in the same way. In fact, this is in press in a volume that should be out in the next few weeks from the Symposium on Continental Drift held a year ago in London.

(Heezen, 1966a: 392)

In their publication, Heezen and Tharp (1965: 93) showed fracture zones between ridge segments as transcurrent, not transform faults. This is clear in their Fig. 5 (my Figure 5.12) which shows the fracture zones in the equatorial Atlantic; the motions they mark on the fracture zones indicate transcurrent motion – the opposite of transform faults. In the same publication they showed fracture zones in the northwest Indian Ocean, which likewise are designated with transcurrent displacement (Heezen and Tharp, 1965: 97).

Matthews promptly dismissed Heezen's suggestion that he had come up with the idea of transform faults. He credited Wilson, and recalled that he had been "very embarrassed, very cross because it was the obvious solution to the Owen Fracture paper," which he had been working on at the time (§4.11).

Then Laughton explained convincingly how he had realized when talking privately with Wilson months before and as he had stated in his own presentation (Laughton, 1966), that the offsets of the ridge running down the Gulf of Aden and in the northwest Indian Ocean are not transcurrent but transform faults.

I would like to comment on a slightly smaller example of these transform faults in the Gulf of Aden and northwest Indian Ocean. Crossing the Gulf of Aden, as I described the other day, are diagonal faults which have displaced the central rift zone. One of these in particular extends from one side to the other and offsets the central zone by 15 or 20 miles. And it is precisely on this offset zone that the major concentration of earthquakes occurs. In this region the difficulty has always been that if the fault is a transcurrent one, then why don't offsets occur on the continents on either side? The theory of transform faults makes it clear why no offsets should occur, and there are no signs of young faults extending this feature onto the continents. It is evident that there are old lines of weakness considerably predating the formation of the Gulf of Aden. This proposal answers the difficulty which we had in this region, that is, that the only fracture zone, the Kossack Trench [i.e. Aluka-Fartak Trench], is a left-lateral fault, while it is apparent, according to ideas of transcurrent faults that it should be a right-lateral fault. There doesn't seem to be any particular reason why there should be a reversal in the sense of these two faults. If, however, they are both transform faults, then such a reversal is not necessary, and the transform faults are a consequence of the original lines of weakness.

(Laughton, 1966c; my bracketed addition[5])

Wilson was next asked by George Thompson (1966b) whether or not he believed the East Pacific Rise extends northward beneath North America. He prefaced his question by remarking that Wilson's transform hypothesis "has a great deal to offer," and noted that he (Thompson, 1966a), like Menard, thought the East Pacific Rise extended beneath the Basin and Range Province. Wilson did not think Menard was right, but did wonder if some early Tertiary mid-ocean ridge unrelated to the Juan de Fuca Ridge had once extended beneath western North America.

Yes, I recall Menard's idea that the East Pacific Rise passed under California and Nevada and emerged from Oregon striking in a northwesterly direction, but I think that the topography suggests that the ridge off Vancouver Island is running at right angles to that and in a northeasterly direction. Benioff's seismic data also suggest this as do the magnetic anomalies. I am wondering whether formerly, during the early Tertiary a mid-ocean ridge did not go into the Gulf of California, pass through the Basin and Range province as far as the Rocky Mountain Trench and not come out. In that case the San Andreas Fault and the ridge off Vancouver Island are younger Pliocene or Miocene additions, and the Rocky Mountain Trench might be an early Tertiary forerunner of the San Andreas Fault.

(Wilson, 1966b: 394)

Wilson's intuitions were good, but he really did not know what happened to the East Pacific Rise. He returned to the topic the next day, after Hess's presentation.

George G. Shor, Jr. of Scripps also had something to say about the Juan de Fuca Ridge off the coast of Washington State. He thought it trended northwesterly.

We have been studying the ridge off the State of Washington. There is a high heat flow determination off Cape Mendocino. Many high heat flow values have been determined northwest of there, as well as a low velocity zone off Cape Mendocino. Last spring we worked just off this ridge west of the Alaskan abyssal plain and found a low velocity mantle. As far as I can see, all the trends, except possibly the magnetics, are northwesterly. So I would much prefer to take the ridge system that way.

(Shor, 1966)

This was a serious objection, because Wilson and Vine had made much of the consilience between the magnetic, bathymetric, and seismic data; they had used them to position the ridge (§5.2). All this was news to Shor, not unexpectedly because their papers had not yet appeared. Wilson and Shor exchanged a few comments, but nothing was resolved. Shor agreed that "some of the magnetics" is inconsistent with his interpretation, and claimed "you can interpret all the topography" either his way or Wilson's way.

Runcorn then went on to claim that Wilson's position was not all that it seemed.

There are, of course, one or two points on which Professor Wilson and I agree. Perhaps, to make our position clear, I should say that I agree that he has demonstrated very clearly why the convection current in the Atlantic follows so well the coastline of South America and Africa. But it seems to me he is saying, in a way, that he is pulling himself up by his own bootstraps in talking about changes in the convection pattern. During their movements, continental blocks have lots of time to reach equilibrium. The question that the classical theory is actually asking is why the cracks develop in such places. For example, why have the convection currents parted North America and Europe, and South America and Africa, along this more or less north-trending line? Professor Wilson claims that the theory of convection is not relevant to this question. One of his reasons is that he feels it is undue simplification to talk about the mantle as if it were uniform. I am rather doubtful about his criticism ... So I believe that the theory of convection can be reasonably applied to this problem and I think that Professor Wilson is a little misleading when he talks of an enormous, single convection cell wrapped around the earth. When you analyze this by spherical harmonics, it breaks up into a set of smaller cells. The theory of convection goes on to explain why the pattern changes and why the pattern has this rather large scale, and what sort of patterns exist in, say, millions of cells. Such patterns we know are the cause of break-up of the continental masses. From the size of the fragments we can find the mode of the cells, which increase in number as the core grows and the mantle gets thinner.

(Runcorn, 1966b: 396)

Wilson responded in kind.

Professor Runcorn and I don't agree about the growth of the core. He considers the whole mantle is convection and that as the core grows slowly throughout geological time the number of cells increases and the pattern of mountain building changes, about five times. I follow Elsasser and Tozer in thinking that the lower mantle is rigid and the growth of the core – which probably happened quickly – has had no effect on mountain-building. I have examined the geology and I think the pattern I get from it is a good suggestion of what that pattern of flow is in the earth, but, on the other hand, the regular little cells which Runcorn plots from a simple interpretation of the mathematics with no reference to the geology do not look at all like the real earth to me.

(Wilson, 1966b: 396)

I now turn to **Hess's presentation**, or at least what was left of it. After submitting for publication, he withdrew most of it because he considered it "unsuitable for complete publication" by the time the symposium volume was to appear (editor's footnote within Hess's published presentation). He first considered the Darwin Rise and what he took to be its associated fracture zones "crossing the rise fall more or less on great circles and – about perpendicular to its axis." Viewed retrospectively, this is rather startling. I say this because Menard would later make the same comment about the fracture zones associated with the East Pacific Rise, and his comment would play an important role in Morgan's development of plate tectonics (III, §7.8).

The fracture system and rises in the central Pacific Ocean basin in Menard's book "Marine Geology of the Pacific" (1964), are old, probably Mesozoic or early Tertiary. They are not active today. The Darwin Rise in the Central Pacific has subsided since Cretaceous time. The fractures crossing the rise fall more or less on great circles and are about perpendicular to its axis. At one time, a mid-ocean ridge existed along the axis of the rise, and crustal material moved away from the ridge axis for 100 or 200 million years. The fractures are related to that movement.

(Hess, 1966a: 311)

Hess next considered the East Pacific Rise. He claimed that the rise was very recent, having formed during the last ten million years – he later added a footnote claiming, "Results in the past year (1966) suggest that the East Pacific Rise is much older than 10 million years." Here is what he said at the symposium.

Much later, the East Pacific Rise was formed – perhaps in the last one to ten million years. Its young age south of the equator is indicated by the existence of sediments extending across the rise without change in thickness. Furthermore, if the rise does continue into the Colorado Plateau or Basin and Range Province, there the tectonic activity is also late Tertiary.

(Hess, 1966a: 311)

What about the magnetic anomalies? He first considered those north of the Mendocino Fracture Zone that trend north–south. They were, he proposed, unrelated to the East Pacific Rise or the Darwin Rise.

The familiar linear magnetic pattern of the Pacific Ocean basin north of the Mendocino Fracture Zone extends over a large area of ocean. North of the Mendocino Fracture Zone, the anomalies extend at least 2000 kilometres out into the Pacific. If the linear anomalies grew by westward migration of crustal material (like an unrolling window blind) at the rate of 1 centimetre per year, or even 2 or 3 centimetres per year, the growth of the belt requires something of the order of 100 million years. These are old anomalies and should not be correlated with the activity on shore.

(Hess, 1966a: 311)

"If the linear anomalies grew by westward migration" they are not related to the Darwin Rise. He claimed they are old, and unrelated to the young East Pacific Rise. From what Hess said, one can only presume that these north–south magnetic anomalies were related to some ridge as yet undiscovered or that has been destroyed, but Hess remained silent on the matter – he was at sea, baffled by the Pacific Basin. Hess next turned to the northeast–southwest trending magnetic anomalies that Vine and Wilson, with Hess's help, had related to the Juan de Fuca and Gorda ridges.

The anomalies on a segment off California to Washington State on the north side of the Mendocino zone are probably young and may be forming today. It would be interesting to recontour the magnetics in the area while keeping in mind that the old anomalies trend north and the young anomalies trend east of north. Perhaps a discontinuity would appear between anomalies of the two ages.

(Hess, 1966a: 311)

Indeed, Hess, as he had indicated to Vine, was pleased with their treatment of the Juan de Fuca Ridge, and its associated magnetic anomalies. Returning to the older, more broadly spaced anomalies south of the Mendocino Fracture Zone, he proposed that the seafloor spreading had occurred more rapidly south of the fracture zone than north of it.

The older anomalies south of the Mendocino are about 70 metres apart and the younger north of the Mendocino are about 30 kilometres apart. The more widely spaced anomalies indicate a higher velocity of movement, roughly twice that of the closely spaced anomalies. If the successive positive and negative anomalies are the result of reversals in the earth's magnetic field and if the reversals occurred about every 1 million years, as calculated by Cox and Doell (1964), then the velocities of movement are 1 ½ to 3 centimetres per year of the two sets of anomalies.

(Hess, 1966a: 342)

Hess concluded by praising V-M. It provided a way out of the morass of doubt and speculation. It provided a way to figure out the history of the ocean basins.

Vine and Matthew's (1963) hypothesis is very attractive; it may be right. We should drill some of the positive and negative anomalies to see if they consist of normally and reversely polarized basalt. We could date some of the basalts isotropically to determine how much time has lapsed between the formation of successive anomalies across the belt. From this, the velocity of movement could perhaps be calculated, and eventually the structure and growth in time of the entire ocean floor could be determined. I am anxious to see more well controlled surveys of the oceans areas.

(Hess, 1966)

Hess understood what needed to be done.

As Wilson's had the day before, Hess's talk provoked considerable discussion. Runcorn began. He quite sensibly pointed out that reversals are not periodic but episodic, and that the frequency of reversals during the Mesozoic was lower than during the Tertiary.

I have one suggestion to make. It seems that the number of complete reversals in the earth's magnetic field varies considerably throughout geological time. If, for example, we consider the anomalies in the Pacific as being Mesozoic, then it's almost certain that the reversals in the Mesozoic have not been as frequent as those in the Tertiary.

(Runcorn, 1966b: 312)

Hess agreed. He was aware of Irving and Parry's (1963) recognition of what is now called the Kiaman Reversed Superchron, from the Late Carboniferous to the middle Permian (~50 million years ago), during which the geomagnetic field had continuous reversed polarity, and noted that there should be "anomaly-free areas" of the ocean floor, if the V-M hypothesis is correct and seafloor of Kiaman age still existed.

That's fine. We can sample and date the Mesozoic anomalies too, as well as examine the magnetics. We should find areas of ocean floor of Permian age if any of the oceans are that old – reversals are not observed in Permian rocks – and we should find anomaly-free areas, if old ocean floors do exist and if Vine and Matthews are right.

(Hess, 1966b: 312–313)

Although no seafloor older than Jurassic remains, Hess's observation has been applied to the Cretaceous Normal Superchron when the field remained of normal polarity for about 35 Ma. This is defined at sea by the absence of anomalies over ocean floor formed during that interval. It is often referred to as the Cretaceous Magnetic Quiet Zone.

Wilson began by emphasizing their full agreement on one matter: "I'd like to comment, if I may, that I'm glad that Professor Hess and I agree on the value and possible validity of the Vine and Matthews hypothesis and on its apparent usefulness" (Wilson, 1966b: 313). He then went on to question Hess's claim that the East Pacific Rise is at most ten million years old.

In the first part of his talk, though, there were some points which I would like to question. The extreme youth of the East Pacific Rise and its lack of connection with these great scarps and fracture zones puzzle me, because it's so obvious that the fracture zones are offset and seem to be related to the East Pacific Rise in the South Pacific. And also there is all the palaeontological data which he did not mention. In a recent paper, Reidel and Funnel found only Pliocene fossils along the centre of the East Pacific Rise in the South Pacific, and progressively farther from the centre they found sediments of all the other ages in the Tertiary as old as Eocene. This suggests that the East Pacific Rise represents all of Tertiary time.

(Wilson, 1966b: 313)

Hess was not ready to agree.

Talwani next asked Hess about the East Pacific Rise, wanting perhaps to know if he agreed with him that it extended beneath North America.

Menard has presented some very impressive arguments for the gentle slope of the East Pacific Rise north of the Mendocino escarpment and extending westward to about longitude 150°W. Using the magnetic patterns, he presented strong evidence for the offset across the Mendocino escarpment. Also the ocean is much deeper south of the escarpment; and the magnetic anomaly seems to follow the scarp. Do you have any thoughts on that?

(Talwani, 1966c: 313)

Hess (1966b: 313) replied, "I don't know. I haven't reviewed Menard's data with Wilson's hypothesis in mind. I'll do it for you though. Next time you ask, I'll have the answer." But Talwani was not the only one who thought the East Pacific Rise extended beneath North America. K. L. Cook (1966a) had already argued during his own presentation that the rise extended beneath western North America. Hess replied:

I did not mean to say that the East Pacific Rise might not extend into the Basin and Range province – it very well might, as far as I know. But I would not correlate the magnetic anomalies off California with the East Pacific rise where it joins the Basin and Range province. I think there is a fairly reasonable argument that a branch of the East Pacific Rise or the whole thing might go off into that area. And there is another, smaller branch, that Professor Wilson mentioned, off the coast of Oregon and Washington. But it is difficult to accept that the fracture zones of the Pacific cross the San Andreas Fault – they don't. So I'd make them older than most of the San Andreas.

(Hess, 1966b: 314–315)

Another honest answer from Hess: he just did not know if the East Pacific Rise extended beneath North America or not. He made a very good point: the fracture zones, which appear to be a necessary accompaniment to the rise, don't cross the San Andreas Fault. So if they don't, what happened to them? Wilson had an answer, it was not the entire answer, but he was on the right track.

I think [the East Pacific Rise] disappeared under North America – when it went down, it just vanished – and I don't see much relation between the structure of the Nevada region and that of the mid-ocean ridges. If a connection through California was found, I think it would be very difficult to find a branch from Nevada to the vicinity of Vancouver Island. So I think that that's the argument from both sides and we'll probably not be able to resolve it today. But I think it's interesting that we've brought out these differences in point of view.

(Wilson, 1966b: 315; my bracketed addition)

This differed slightly from what he had said the previous day after his own talk, and was closer to the truth. The East Pacific Rise just vanished. But just how it vanished remained a mystery. Wilson said that "it went down." But there did not seem to be a trench, that is, a zone of compression. Wilson was unable to identify one. Nor did Hess know. Vine would later suggest that a trench system had been overridden by North America as it drifted westward (§6.8). But it would take the development of plate tectonics to solve the puzzle; McKenzie and Parker (1967) and Morgan (1968) would independently identify a trench along the coast of Oregon, Washington, and south Vancouver Island (§7.11, §7.14).

I end by remarking on an acrimonious exchange on mantle convection.

Many people have drawn idealized cross-sections showing convection currents rising and spreading beneath mid-ocean ridges and descending along deep-focus earthquake belts – but it does not appear possible to relate such speculation to the earth's actual tectonic geometry. The Mid-Atlantic Ridge is the type example of the rising and diverging part of a convection cell cited by most advocates of mantle convection. There is, however, no place where this hypothetical convection cell can be postulated to descend on either the east or west sides (with the exceptions of the east ends of the Caribbean and Scotia Arcs, which represent only a trifling proportion of the length of the Mid-Atlantic Ridge). Conversely, there is no ridge to match the Kamchatka–Kuril–Japan–Marina Trench. Convection cannot explain most strike-slip faulting known on the continents, nor can it explain many other tectonic features. And it by no means established that deep-focus earthquakes are due to shear along planes dipping under continents or island arcs, although this is commonly assumed to be so ... One alternative explanation is that the ridges are raised above belts of high heat flow, and that the flanks slide gravitationally toward the deep ocean floors.

(Hamilton, 1966: 186)

Runcorn, the champion at the Ottawa meeting of whole mantle convection, rose to the bait:

Deep-focus earthquakes and the processes of ascension obviously imply some causal connection between the motion of continents at the surface and motions beneath the surface. You were quite right in drawing attention to the importance of isostasy as proof that there is the possibility of both horizontal and vertical flow beneath the rigid crust. After that point, when you get into the evidence of the hypothesis of Wegener's interpretation, you are demonstrably wrong.

(Runcorn, 1966b: 186)

Hamilton responded:

I agree that Wegener's forces won't work, but I think that his basic concept of crustal rafts moving independently over the deeper mantle is more generally correct than is the common assumption that continents float passively on top of moving mantle material.

(Hamilton, 1966: 186)

At which point, Runcorn lost patience:

Well, obviously, there has got to be some force which will move the continents, and your last statement, which pictured a raft moving independently of what's underneath, of course, is nonsense.

(Runcorn, 1966b: 187)

Such exchanges were heading nowhere. Interestingly, it was reading them later that prompted McKenzie to make sure that when he and Parker were composing their seminal paper (1967) a year and a half later, they emphasized that plate tectonics is a kinematic, not a dynamic theory. Indeed, McKenzie (May 30, 2010 email to author) recalled, "I read the two books that came from this [Ottawa] meeting with great care, especially the transcripts of the exchanges ..." Exchanges such as the one above he recalled, illustrate "why it was so important to emphasise that plate tectonics is a kinematic theory." Wrangling about mechanism was fruitless; it was a morass of doubt and speculation.

Similarly, there were disagreements about whether or not the East Pacific Rise had ever extended beneath western North America, about the relationship, if any, between the San Andreas Fault, the Mendocino Fracture Zone, and magnetic anomalies in the northeast Pacific. At this juncture, Earth scientists simply did not know what happens to the East Pacific Rise north of the Gulf of California, what its relationship to the San Andreas Fault and nearby structures is; they did not know whether the Darwin Rise ever existed, was it really an ancient spreading center? The majority were far from sure that seafloor spreading actually occurs, and that V-M was applicable; and as for transform faults, they had only just emerged. The record of the Ottawa meeting shows that these were all wide open questions.

While these disagreements were being aired and the need for more surveys emphasized, Walter Pitman was aboard research vessel *Eltanin* on its twentieth cruise in the southwest Pacific heading south. By year's end, he was able to plot magnetic records of this and the previous year's cruise (*Eltanin*-19), which, as we shall see, over the next few months, caused many Lamont workers to accept the V-M hypothesis and the seafloor spreading it entailed. In the longer term it had much wider influence.

5.8 The challenges of unraveling the Cenozoic history of the northeast Pacific

As early as 1949, Menard and Dietz discovered ridges and troughs of the aptly named Ridge and Trough Province (III, §5.5). However, the detailed bathymetric survey was classified – a Cold War requirement of US defense policy. In the paper they wrote at the time, they (1952: 272) called the Juan de Fuca and Gorda ridges "scarps," which is how they designated the Mendocino scarp; they did not (1952: 268) single them out as special; they were two among a host of ridges in the Valley and Ridge Province. Retrospectively, Menard (1986: 249) considered that he had, at the time, provided enough information for others to recognize the presence of these two ridges, and he was correct, because that is what Wilson, Vine, and Hess did; it was likely more difficult than it might have been had they had the full record, making their achievement the more notable.

In the northeastern Pacific and the adjacent parts of western North America, the very different tectonics of ocean and continent confronted one another, and seemed at the time irreconcilable. The challenge of integrating them required the development of plate tectonics. Much of the East Pacific Rise's crest and eastern flank had vanished, and there was, seemingly, no ridge that could be related to the north–south trending magnetic anomalies of the southern half of Mason and Raff's survey (III, Figure 5.5). Then, north of the Mendocino Fracture Zone, there were the north-easterly trending magnetic anomalies. What did they signify? Understanding the origin of the great fracture zones (III, Figure 5.2), and how, if at all, they were related to the San Andreas Fault, became especially challenging. Before the advent of transforms, Vacquier and others had, understandably but erroneously, concluded that transcurrent faulting along them had resulted in very large lateral displacements, which at the time were regarded as enhancing support for mobilism (§2.3).

Looking ahead, it took five years after Wilson, Vine, and Hess talked together at Madingley Rise, Cambridge, before Atwater (1970) began to reveal the Cenozoic history of the eastern Pacific and the San Andreas Fault essentially as it is currently understood. When Morgan (1968) and McKenzie and Parker (1967) proposed plate tectonics, they identified the important tectonic issues in the area partially bounded by the Juan de Fuca and Gorda ridges, and hypothesized the existence of a subduction zone to the east off the coast of Oregon, Washington, and southern Vancouver Island (§7.11, §7.14). Morgan named the area the Juan de Fuca Block, now called the Juan de Fuca Plate. They also recognized the San Andreas and Queen Charlotte faults as transform faults forming the border between the Pacific and North America plates (§7.11).

Notes

1 Both letters are from Hess's papers housed at the Firestone Library, Princeton University.
2 The *Vema*-16 profiles over the Mid-Atlantic Ridge were published in Heirtzler's technical report of the geomagnetic measurements of *Vema*'s cruise 16. According to Sclater, Wilson was looking at a profile in the South Atlantic. *Vema* crossed the Mid-Atlantic Ridge in the South Atlantic on December 5, 1959. The profile as reproduced in Figure 12 exhibits little symmetry (Heirtzler, 1961).
3 McElhinny (1973: 110–111) updated Rod Wilson's test a decade later. He found in the literature 157 examples in which the polarity of igneous bodies and their baked contacts agreed, and, as before, three that did not in rocks that had not been "cleaned."
4 After the acceptance of mobilism in the later 1960s in which magnetic anomalies at sea played a key role, many questions about them remained unanswered. It was not certain where, exactly, magnetic anomalies originate. Vine and Matthews and Hess had settled on Layer 2 (~2 km) and this served during the final stages of the mobilism debate. Shortly after, based on a geophysical study of the Reykjanes Ridge, Talwani et al. (1971) determined that the source layer for the anomalies had a thickness of 500 m and they referred to it as Layer 2A, a value now generally accepted. The question of variations in amplitude and wavelength of the anomalies has been extensively reviewed by Gee and Kent (2007).
5 I thank Laughton for explaining the misprint (June 1, 2010 email to author). Laughton named the trench, which has a "most spectacular offset."

The most spectacular offset is that at 52° E where the axis of the central rough zone is displayed by 100 mi. The offset is marked by a linear trench 150 mi. long and 10 mi. wide ... This runs between Ras Alula on the Somalia coast and Ras Fartak on the Arabian coast. It is suggested that the trench be called the Alula-Fartak trench.

(Laughton, 1966a: 154)

6

Resolution of the continental drift controversy

6.1 Outline

By the beginning of 1967, seafloor spreading and continental drift had gained acceptance among marine geologists and geophysicists. This was brought about as a result of the confirmation of the two corollaries of seafloor spreading, and also by new developments in the radiometric reversal timescale, by new work on reversal stratigraphy in deep ocean cores which confirmed and added to this timescale, and by the observation of the extraordinary *Eltanin*-19 magnetic anomaly profile across the Pacific–Antarctic Ridge.

Unlike competing views, such as M. Ewing's (III, §6.13) and Menard's fixist accounts of the evolution of ocean basins (§3.6), seafloor spreading possessed two imminently testable corollaries, the Vine–Matthews hypothesis (§2.13) and the idea of ridge-ridge transform faults (§4.9). Vine, with much help from Wilson and Hess, led the way. He and Wilson realized that in the most northern part of the Mason–Raff survey in the northeast Pacific there are two extensions of the East Pacific Rise, the Juan de Fuca and Gorda ridges, and both had the zebra pattern of magnetic anomalies as predicated by V-M. Wilson proposed that there is a ridge-ridge transform fault connecting the Juan de Fuca and Gorda ridges within the Mason and Raff survey (§5.2). This, and many other faults transversing ocean ridges, was shown to be transform by Sykes, a seismologist at Lamont. Working quickly, he completed his study near the end of November 1966. Sykes showed that, during earthquakes, movement on fracture zones occurred only between spreading centers and in the sense required for them to be transform faults; thus he established the difficulty-free status of Wilson's and Coode's concept. Vine and Wilson in their Cambridge work (§5.2) in the early summer of 1965, used an early, incomplete radiometric timescale. They calculated an average spreading rate across the Juan de Fuca Ridge of 1.5 cm/year per limb but it was erratic not constant. In November 1965 at the GSA meeting in Kansas City, Dalrymple of the USGS in California told Vine of their new improved radiometric timescale. Vine immediately realized that he could now show that spreading across the Juan de Fuca Ridge had been at a constant rate. In a seminal review in 1966, he extended this type of analysis worldwide. This was enough for him to fully accept V-M as correct: he had turned the V-M into an essentially difficulty-free solution. I shall describe this and other improvements in the reversal timescale by the Australian (ANU) and Californian (USGS) groups.

While Vine and Dalrymple were meeting at Kansas City, Pitman, a graduate student at Lamont working with Heirtzler, was digitizing magnetic profiles over the Pacific–Antarctic Ridge. He completed the analysis before the year was out, showing that one of the profiles, *Eltanin*-19, sometimes called "Pitman's magic profile," was near perfectly symmetrical about the ridge, that *Eltanin*-20 was almost as symmetrical, and that both profiles correlated with the axial profile of the Juan de Fuca Ridge. There could now be little doubt that spreading about the East Pacific Rise had occurred and was symmetrical. Some workers at Lamont, such as Heirtzler, Le Pichon, and Talwani, who individually or jointly had rejected seafloor spreading and V-M from the early 1960s through much of 1966, accepted it. With the *Eltanin*-19 profile and Sykes' confirmation of transform faults, both superb products of their own institution, Lamont workers in marine magnetics, some more readily than others, accepted V-M, seafloor spreading and continental mobilism.

Also at Lamont, while Pitman was working on his profiles, Opdyke, his graduate student Forster, as well as Glass, one of Heezen's students, were developing a reversal timescale based on marine sediment cores. By February 1966, they had produced a timescale back to the base of the Gauss normal epoch which matched beautifully the reversal timescale determined radiometrically on land and that discernable in marine magnetic profiles. This confirmation from three independent sources of the fine detail of the spectrum of reversals over the past 4 million years essentially dispelled any lingering gratuitous doubts that the geomagnetic field had actually undergone reversal and that the marine magnetic profiles reflected and thus confirmed seafloor spreading.

I shall describe a two-day conference at the Goddard Institute for Space Studies in New York City. The meeting was held in November 1966, and introduced the new support in favor of mobilism to some leading North American Earth scientists. I shall discuss why mobilism met with such strong resistance at Lamont, and conclude with M. Ewing's reluctant switch to discontinuous seafloor spreading.

6.2 Lamont workers argue distribution of Atlantic Ocean floor sediments is incompatible with seafloor spreading: Hess disagrees

M. Ewing, Le Pichon, and J. Ewing submitted a paper on the distribution of sediments in the South Atlantic in September 1965. They did so approximately six months after Heirtzler and Le Pichon had submitted their paper on magnetic anomalies in the Atlantic in which they argued against interpreting them by V-M (§4.6). Asked to revise it, Ewing and company resubmitted in early December and it was published the following year (1966). They argued that the distribution and undisturbed nature of the sediments are inconsistent with seafloor spreading. Their thickness did not increase gradually and steadily with distance from the ridge crest. The crest itself had almost no sediments, but the amount of sedimentation found beyond the crestal mountains remained constant, so the seafloor outside the ridge did not appear to increase in age. They also found sediments of Middle and Lower Miocene age at points too close to the rift valley to allow recent spreading. Reasoning that if spreading in the Atlantic had occurred at a minimum rate of 2.5 cm/year, and if South America and Africa had separated 120 million years ago, then 500 km of new crust should have been added on each side of the rift valley since Early Miocene:

Yet middle and lower Miocene sediments have been obtained at six different locations ranging from less than 10 km to about 500 km from the rift valley, and middle and upper Eocene sediments have been found at four locations ranging from 500 to 800 km from the ridge crest.

(Ewing et al., 1966: 1634–1635)

They also reasoned that seafloor spreading could not have occurred without disturbing the seafloor, any decoupling between moving seafloor and circum-Atlantic continents would have disturbed sediments at continental margins; but the sediments in the Atlantic Basin seem undisturbed. This last criticism was, I think, invalid, because with the hypothesis of seafloor spreading, circum-Atlantic continents were supposed to move along with seafloor accounting for their undeformed state. They raised the same sort of difficulty with respect to sediments within the Puerto Rico Trench.

If the convection cells are partly or totally decoupled from the continents, some kind of evidence of disturbance in the sediment layering near the continents should have been found. Yet none has been found. Even in the Puerto Rico trench, which is supposed to be held down by a downward-moving current [Fisher and Hess, 1963], the only clear evidence of disturbance is a slight tilting of the strata toward the island slope.

(Ewing et al., 1966: 1635; my bracketed addition)

Essentially they claimed to have removed from seafloor spreading's list of solved problems its ability to explain Atlantic stratigraphy, and rebranded it a difficulty (RS2).

It is becoming increasingly clear that, if continental drift occurred at all, it must have terminated in a few million years ... Or, if it took as much as a few tens of millions of years, the process of continental drift was one of adding new crust at the axis without disturbing the sediments in the basins or at the margins of the continents. In any case, all movement had ceased 20 m.y. ago at the latest. The corresponding velocity of movement is so large that convection currents, as usually understood, could hardly have been the cause of such movements. The concept of a rapid event is in sharp contrast with the steady spreading of the ocean floor. A logical conclusion is that spreading of the ocean floor cannot explain the Pleistocene tectonic activity of the crest of the mid-Atlantic ridge and the maintenance of the Puerto Rico trench out of isostatic equilibrium. Consequently, the crucial facts that were presumed to be explained by this hypothesis require a different explanation for the Atlantic Ocean.

(Ewing et al., 1966: 1635)

Table 6.1 Table showing ages and distances from ridge axis of sediments cored from the North Atlantic (Saito *et al.*, 1966a: 1076)

Cruise	No.	Source	Core Water depth (m)	Distance to axis (km)	Age
A150	RD7	30°01′N,42°04′W	4280	10	L and M Miocene
V9	32	14°10′N,45°44′W	3623	90	Pliocene
V16	206	23°20′N,46°29′W	3733	110	U Miocene
V10	91	23°23′N,46°24′W	3540	110	Pliocene
A150	RD8	31°49′N,42°25′W	3700	130	L and M Miocene
V10	89	23°05′N,43°48′W	3525	170	Miocene
V20	242	23°22′N,43°39′W	4565	170	Pliocene
V16	21	17°16′N,48°25′W	3975	200	Pliocene
V9	29	3°47′N,34°47′W	4675	280	L Miocene
V16	35	17°39′S,15°06′W	3892	300	Pliocene
V12	5	21°12′N,42°21′W	3003	330	Pliocene
V16	205	15°24′N,43°24′W	4043	330	Pliocene
V10	94	24°56′N,48°59′W	4260	370	Pliocene
V16	23	13°15′N,40°40′W	4887	420	Pliocene
V20	241	22°08′N,41°30′W	4372	420	Pliocene
V4	53	33°05′N,29°18′W	2470	480	M Miocene
V19	307	26°22′N,38°50′W	4715	500	Pliocene
V16	208	27°44′N,49°55′W	4861	560	Pliocene
V17	166	34°56′N,45°21′W	4210	570	Pliocene
A153	144	33°08′N,48°08′W	4850	660	Pliocene
V14	4	15°29′N,40°31′W	4473	670	Pliocene
A150	24	29°02′N,51°02′W	4850	670	Pliocene
V17	163	27°58′N,34°08′W	5132	700	Pliocene?
V16	38	22°59′S,06°46′W	4925	700	Pliocene
RC8	2	11°12′N,48°05′W	4614	800	U Eocene
V12	4	24°17′N,53°04′W	5009	800	M Eocene
A180	25	30°15′N,28°30′W	1280	810	Miocene
V16	209	30°00′N,51°52′W	4673	850	L Eocene
V10	96	27°52′N,54°38′W	4680	950	Eocene
A180	32	29°07′N,26°15′W	5029	1100	Pliocene
V20	238	16°28′N,36°19′W	5233	1100	Pliocene
V16	40	26°16′S,03°01′W	4790	1100	Pliocene
RC5	12	26°35′N,56°29′W	5104	1170	Cretaceous
V17	162	24°58′N,28°56′W	5480	1300	Pliocene?
V20	207	22°06′S,00°19′E	5349	1300	Pliocene

But they were not finished; they added an endnote to let readers know that T. Saito and other Lamont workers had recently found more pre-Pleistocene fossils and that there were other telling data soon to be published.

Saito, Ewing, and Burckle submitted their paper in January 1966, and it appeared at the beginning of March, actually a week and a half before Ewing *et al.* (1966). In it they maintained (1966a: 1075) that there had been no seafloor spreading since the Lower Miocene, because they had found further Miocene fossils at two sites on the Mid-Atlantic Ridge (A150-RD7 and A150-RD8 in Table 6.1) and because Miller at Cambridge

University had obtained a radiometric date of 29 ± 4 million years from a basalt from the ridge's crest – far too old for seafloor spreading as Hess envisaged to have occurred.

The facts that two rock samples of Lower Miocene age have been dredged from the Rift Valley, and an age of 29 ± 4 million years has been derived from basalt from the crest of the ridge at 45°N (25), indicate that expansion or crustal movement, if any, ceased at least 20 million years ago.

(Saito et al., *1966a: 1076; (25) refers to Baker* et al. *(1964), the larger report in which Miller (1964) is found)*

Although they acknowledged that available dates from sediments of the Atlantic floor generally show older sediments to be the more distant from the ridge axis, they stressed the need for additional cores. They also suggested (1966a: 1079), "Lower Tertiary sediments along the crest of the ridge may have been buried beneath basaltic flows, while areas distant from the crest are unaffected," a proposal that would require a special coincidence of circumstances. Nonetheless, they admitted that seafloor spreading was still a possibility.

Another explanation is that the addition of new crust, during continental drift, occurred in such a way that patches of older sediment were left behind in the crestal area rather than being completely swept away from the axis.

(Saito et al., *1966a: 1079)*

Menard was impressed with Ewing and company's findings, prompting him to comment to Hess on March 29, 1965. He had written a new paper and he asked Hess to review it and added this handwritten note.

I think that the Ewing et al. papers on Atlantic stratigraphy really eliminate significant Tertiary Atlantic sea floor spreading. One of our tentative conclusions suggests a way to retain continental drift, but the handy 1 cm/yr calculations are in trouble.

Hess was not impressed. He wrote Menard back on April 4, disagreed with Menard's assessment of Ewing's work, and noted that Vine was working on the rate of seafloor spreading over the Reykjanes Ridge.

As for your comment on ocean spreading, I don't agree at all. Ewing et al have one misplaced sample (the first one in their table [A150 RD7 in Table 6.1]) and one bad K-Ar age. If these are disregarded everything fits just as it should. Vine is presently working on the rate of spreading during the last 4 million years on the ridge south of Iceland. Will let you know his results when they are available.

(Hess, April 4, 1966 letter to Menard; my bracketed addition)

It would take more than this to dislodge Hess from his catbird seat. Hess was sitting pretty. He had come a long way since World War II when as USS *Cape Johnson* boat commander in the Pacific he had had a penchant for leaving the echo-sounder running (III, §3.5), and had, over the intervening years, thought a great deal about the oceans; he was not about to budge. In Ottawa he had, a few months earlier, seen the striking symmetry of the magnetic anomalies over the Reykjanes Ridge (§5.7). And, as I am about to explain, he knew about the "Jaramillo event" and of the use Vine (now in Princeton) was making of it to harmonize the spreading rate at the Juan de Fuca Ridge.

6.3 Vine learns of corrections to the reversal timescale and fully accepts seafloor spreading: the November 1965 GSA meeting

The turning point was the meeting of the Geological Society of America in Kansas City in November 1965. We did the work on the Juan de Fuca when still at Cambridge in early '65. We had difficulty interpreting it because we didn't have the correct timescale; the Jaramillo event was missing. So we had an irregular spreading rate. Now this is a very interesting point because one of the things it is difficult to remember because of hindsight is that at that stage not only did we not know the timescale and whether ... [reversals] were irregular or periodic or whatever ... we, or at least I, wasn't sure ... whether to expect spreading to be continuous or discontinuous or irregular – we now know that spreading seems to be remarkably continuous and that spreading rates are remarkably constant over millions of years – but then it is now perhaps difficult to realize.

(Vine, 1979 interview with author)

While Wilson and Vine were finishing their papers on the Juan de Fuca Ridge, Wilson told Vine that they were going to present them at the upcoming GSA meeting in Kansas City. Vine voiced a useless protest.

Why did I go to the GSA meeting? Tuzo's doing, he was at Cambridge then, he said, "Look," as Tuzo typically would, "You must present this ... at the GSA." I said, "I don't think I'm going to the GSA." "You must go to the GSA. I'll have a word with Harry."

(Vine, 1979 interview with author)

It was well that Vine listened to Wilson because it was at the GSA meeting in Kansas City that he met Brent Dalrymple. As we have seen (II, §8.15) it was Dalrymple who had been responsible for the very accurate radiometric dating of rock samples for work on the reversal timescale being carried out at the USGS (Cox, Doell, and Dalrymple, 1963, 1964). A notable feature of their timescale is that it gave the age of the most recent reversal as about one million years. They named (Figure 6.1) the interval of normal polarity since the last reversal the Bruhnes epoch (now called the Bruhnes Chron) and the preceding interval of reversed polarity the Matuyama epoch (Chron). At Kansas City, Dalrymple told Vine that in their more recent work they had discovered two reversals of the geomagnetic field that had occurred more recently than 1 million years ago. One was from reversed to normal, and one shortly afterwards from normal to reversed, thus defining a short interval of normal polarity within the later portion of the reversed Matuyama epoch. This they referred to as the Jaramillo normal event (now called the Jaramillo Subchron) with an age, given in their later publication, of about 0.9 million years and a duration of about

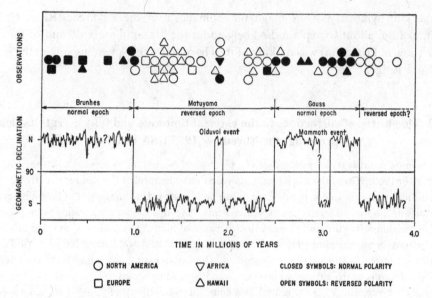

Figure 6.1 Cox, Doell, and Dalrymple's 1964 reversal timescale. Paleomagnetically observed declinations are plotted against time; anomalies are denoted by question marks. On this scale, the last (or most recent) reversal occurred about 1 million years ago. The Jaramillo normal event at approximately 0.9 million years had not yet been discovered.

100 000 years, the last reversal being 0.7 million years ago (Doell and Dalrymple, 1966: 180). The discovery of the Jaramillo normal event is of critical importance for the mobilism debate and I shall discuss it in more detail in the next section.

On this basis of the Cox *et al.* 1964 timescale, Vine and Wilson determined the average spreading rate to be approximately 3.0 cm/year but very variable (§5.2), adding that this did not bother them because it is "what one might expect"; a statement Vine retrospectively described as "bloody silly."

Assuming that the times of the field reversals are the same as those suggested by Cox, Doell, and Dalrymple (7), dates have been added to the section and an average rate of spreading deduced (Fig. 1*c*). In Fig. 1, comparison of *c* with *a* and *b* suggests that the rate of spreading of the ridge has been rather irregular, as one might expect, but averages out 3 cm per year (1.5 cm per year per limb of the cell).

(Vine and Wilson, 1965: 486–487; (7) refers to Cox et al.*, 1964)*

Vine and Wilson correlated the valleys on either side of the central anomaly with the Matuyama reversed epoch, the little hump within each valley with the Olduvai normal event, and the two outer humps with the Gauss normal epoch (Figure 6.2).

Working backwards in time and matching each of these features to points on the 1964 timescale, they estimated spreading rates of approximately 2.0 cm/year during the Brunhes; 0.6 cm/year from the end of the Matuyama until the end of the Olduvai; 2.0 cm/year during the Olduvai; 3.0 cm/year from the beginning of the Olduvai until

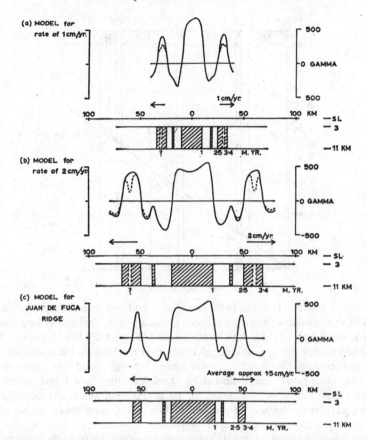

Figure 6.2 Vine and Wilson's Figure 1 (1965: 486). Their caption reads: "Models and calculated total field magnetic anomalies resulting from a combination of suggested recent polarities – for the earth's magnetic field and ocean floor spreading." The models are in accordance with Cox, Doell, and Dalrymple's 1964 timescale, and normally magnetized blocks are shaded; reversely magnetized blocks unshaded. Portions (a) and (b) assume uniform rates of spreading. Portion (c) was deduced from the gradients on the map of observed anomalies. The dashed parts of the computed profiles in (a) and (b) show the effect of including the possible reversal at 3 million years (the Mammoth event).

the end of the Gauss; 0.55 cm/year during the Gauss. These rates are represented on the distance time plot (isogram) as a dashed line in Figure 6.3; their variability is evident.

With hindsight we can see that it is because we were working with North Atlantic and Indian Ocean data where the spreading rate is only 1.5 cm per year, and the record is messy – still they have trouble and argue over North Atlantic data and how you correlate it, and what represents what period in the reversal timescale. So there was that. Then before we had the Jaramillo data, Tuzo and I tried to interpret it. Then we had trouble with the spreading rate, was it constant or irregular. It looked as if it were irregular and I said something silly like "Well you'd expect it to be irregular" which from hindsight was a bloody silly thing to say. But I was just lost.

(Vine, 1979 interview with author)

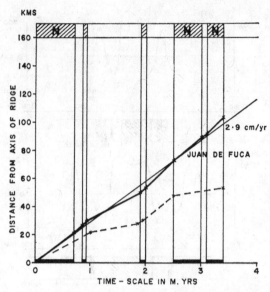

Figure 6.3 Vine's Figure 11 (1966: 1410) confirming uniform spreading across the Juan du Fuca Ridge. The dashed line shows the variable spreading rates Vine and Wilson attributed to the Juan de Fuca Ridge in 1965 using Cox, Doell, and Dalrymple's 1964 timescale. The heavy solid line shows the spreading rate Vine calculated over the Juan de Fuca once he had taken account of the recently discovered Jaramillo event. The light solid line shows the average spreading rate. Data-points are indicated by stars. For the dashed line, given the 1964 timescale, the first data-point was taken to be at 1 million years, the beginning of the Brunhes normal epoch; the second and third data-points were taken to be at 1.8 and 2.0 million years, the end and beginning of the Olduvai normal event. With the new timescale the first data-point, the end of the Brunhes, was changed to 0.7 million years; working backwards in time, the second and third data-points were now at 0.7 and 0.9 million years, the end of the Matuyama reversed epoch, and the end of the Jaramillo normal event. With the new timescale, the fourth and fifth data-points marked the end and the beginning of the Olduvai normal event, still respectively at 1.8 and slightly less than 2.0 million years.

Only "lost," temporarily; Vine saw the way ahead once Dalrymple "told him about the Jaramillo event," which he realized gave the Juan de Fuca Ridge a constant spreading rate.

Brent [Dalrymple] was there. Almost fortuitously I went to his room, which he was sharing with Sherm Grommé (C. Sherman), and I think I went to them primarily to talk about a mini-fluxgate magnetometer, which is something the USGS had developed for getting some idea of remanence in the field, in particular it was ideal for determining if something was reversed or normal, and it was a crude version of a spinner magnetometer . . . and I was keen to borrow one for work in Yellowstone, and we subsequently copied one and it went on the market a year or two later . . . So I went on a completely different matter. We were just talking, and Brent told me about the Jaramillo event, which, of course, was actually published in mid-1966. I was fascinated. And then the penny dropped because I realized that the Jaramillo was on the Juan de Fuca Ridge survey, and then you could then interpret the survey using constant spreading rates. And I have a very vivid recollection of mentioning this to Tuzo just before we gave our talk. I said, "But Tuzo it's wrong, I've got some new data." But it was too late to change our talk.

(Vine, 1979 interview with author)

The new reversal timescale differed importantly from the one they had originally used. The Brunhes normal epoch now began only 0.7 million years instead of 1 million years ago. The Matuyama reversed epoch ended at 0.7 million years instead of 1 million years ago, and had within it a short normal event, the Jaramillo, at approximately 0.9 million years ago. Vine could now correlate the valleys adjacent to the central anomaly with the end of the Matuyama reversed epoch at 0.7 million years, the tiny hump within each valley was the Jaramillo normal event not the Olduvai normal event, and the two outer humps were the Olduvai normal event not the Gauss normal epoch. With this corrected timescale Vine obtained a constant spreading rate of approximately 3.0 cm/year for the Juan de Fuca Ridge over the last 4 million years, constructing his memorable time–distance graph (Figure 6.3).

It was at this point that Vine became completely convinced of the correctness of his hypothesis. He knew immediately that the timescale based on Jaramillo would give him a constant spreading rate across the Juan de Fuca. He had all he needed: a ridge with clearly defined linear anomalies arranged in a sufficiently accurate reversal timescale. He could now confidently maintain that spreading over the Juan de Fuca had occurred at 1.5 cm/year per limb over the past 4 million years.

Vine retrospectively summarized the key steps that led him to accept fully V-M, capturing the exhilaration he felt along the way and the excitement he felt in realizing that one could "trace the footsteps of continents" through tracing the sequence of magnetic anomalies on the seafloor.

It was just fantastic ... the real flash in a way was the symmetry of the Juan de Fuca. It was the first thing I saw. That was '65 when Tuzo and Harry and I realized there was a ridge in that area – that is an incredible story as to why it had not been recognized before – but the magnetics over it was symmetrical, and we just stood there looking at it with our mouths open. There it was. The thing had been in the literature for four years; nobody had seen it. When you went to look for it, it just stood out. That was the first thing. The Juan de Fuca was a tremendous revelation in that it showed that the record could be sufficiently clearly written to generate symmetry, and the boundaries were quite sharp and well defined. The first thing was the symmetry because, as I say, we hadn't anticipated that the record would be that clear, and suddenly realized that this was the first time we have looked at a really good survey, such an extensive survey [over both sides of a ridge]. Perhaps it is more common, you know, perhaps it is that you cannot tell from just one profile. So that was a tremendous revelation; the record was sufficiently clear to reveal the symmetry. The other thing about the symmetry: we didn't know what to expect. Are the reversals every half million years or are they every million years? Are they periodic? If it was periodic there just would be zebra strips of the same width. But if it is an irregular pattern, then that will be much more compelling, as indeed it is. But we had no idea at the time. But then at the Kansas City meeting I realized that you could fit a uniform spreading rate ... The spreading was constant, and ... one could see that the reversal timescale was laid out on the ocean floor incredibly clearly, and then it was pretty well over. I realized that if this is typical of the ocean floor, my God, the whole things falls into place. If it is that clear cut, we will be able to do all those things in terms of putting isograms across the ocean floor, tracing the footsteps of the continents or whatever you want to call it. It is all there, in a sense it was all over.

(Vine, 1979 interview with author; my bracketed addition)

Vine's new analysis of the Juan de Fuca Ridge was published in December 1966 after he saw Pitman's *Eltanin*-19 profile (§6.6) and learned of Opdyke's independent discovery of the Jaramillo event (§6.5). I shall examine his account in §6.8; however, I want to note what Vine said about his and Wilson's previous analysis, and what they might have done if they had "had more faith in" V-M and "a more constant rate of spreading."

In this earlier time scale [used in Vine and Wilson's first analysis of the Juan de Fuca Ridge] the Jaramillo event ... had not been differentiated, and the most recent reversal was placed at 1 million years ago. Consequently the narrow peaks on either side of the central positive anomaly ... were correlated with the Olduvai event. This correlation implied a very erratic rate of spreading and a much looser average rate of 1.5 centimeters per annum (8).

It will be seen that, had the authors had more faith in the idea and the probability of a more constant rate of spreading, for inertial reasons, they could have predicted the Jaramillo event. The recent detailing of this event by Doell and Dalrymple (7) and its independent discovery by Opdyke *et al*. (20) are therefore of considerable interest and importance in interpretation of the magnetic anomalies.

(Vine, 1966: 1408; (8) refers to Vine and Wilson (1965); (7), to Cox, Doell, and Dalrymple (1964) and to Doell and Dalrymple (1966); (8), to Opdyke et al. *(1966))*

From the magnetic anomalies, they could have predicted the Jaramillo normal event in 1965. That would have been quite a coup.

6.4 Improvements in the reversal timescale during 1966

I want now to describe the development of the reversal timescale immediately following its initial use by Vine and Wilson in their 1965 study of the Juan de Fuca Ridge (§5.2). The work was carried out mainly on radiometrically dated lava flows by groups at the USGS at Menlo Park in California and the ANU in Canberra. Knowing something of the history of the reversal timescale helps in understanding how V-M, and with it seafloor spreading, came to be confirmed.

Hitherto, geological time had usually been defined as the time required for the accumulation of certain standard stratigraphic sections. Because time can never be completely recorded in one place there are very many standard sections spread over several continents. Time was divided into periods which were given such names as Cambrian or Cretaceous. Periods were then assembled into the geological timescale;

time was built up by piecing together the spans of time represented by these standard sequences using classical methods, mainly the principle of superimposition, fossil sequences, and cross-cutting igneous relations. Strategically placed radiometric ages were then used to calibrate the geological timescale in millions of years (I, §5.2). In the 1960s the radiometrically dated reversal timescale began being created in a very different way. It is based on only two measurements on samples from the same geological body, often the same sample. These measurements are the polarity of remanent magnetization and the radiometric age. The sequence of geomagnetic field reversals is determined solely by the absolute measurement of age. As we have seen (II, §8.15), samples do not have to be in geographical proximity; reversals recorded from rocks worldwide can be compiled directly into a single timescale. In the context of the mobilism debate, the importance of the reversal timescale is that it provided students of marine magnetic anomalies, such as Fred Vine and Walter Pitman, with the means to break new ground determining the tectonics of ocean floor, using V-M to test Hess's concept of seafloor spreading. It is also important to remark that, as I have just explained, they were also breaking new ground stratigraphically. Early workers (e.g. Hospers, 1954a) commented on the potential stratigraphic importance of reversals; now there was the opportunity of compiling reversal timescales back through time and making global correlations of improved accuracy (Opdyke and Channell, 1996).[1]

Without in any way detracting from the splendid achievement of the USGS group at Menlo Park in establishing the radiometric reversal timescale for the past few million years, I want to remark on the substantial contributions by the ANU group, contributions that have been overshadowed by those of the USGS group. Although less well known, the contributions of the ANU group were also splendid, if somewhat muted. Menlo Park's discovery of the Jaramillo simply overshadowed their contributions. Serendipitously, this discovery came at just the right time, being critical (§6.6, §6.8) for the work of Vine (1966), Pitman and Heirtzler (1966), none of whom acknowledged the work by the ANU group in their papers, papers which within a few years, would be read worldwide.

I begin with Cox, Doell, and Dalrymple's June 1964 timescale, the one that Vine and Wilson used at Cambridge when first attempting to determine the spreading rate over the Juan de Fuca Ridge (Figure 6.1). Although McDougall and Tarling (1963: 56) concluded that reversals are not periodic, it was Cox and Doell who introduced the term "events" (now called subchrons) to identify brief intervals of opposed polarity within longer epochs (chrons). The Menlo Park trio incorporated this terminology into their June 1964 timescale (Figure 6.1) and identified and named two events. The most recent, which they named the Olduvai normal event, was dated at approximately 1.9 million years; the older, which they named the Mammoth reversed event, was dated at approximately 3.0 million years (Figure 6.1). C. S. Grommé and R. L. Hay (1963) at the University of California, Berkeley, were actually the first to find the evidence in Tanganyika (now Tanzania) for the Olduvai event. Cox and company (1964: 1542) remarked, "This event is now supported by data from both Africa and North America, and may be considered well established." And the Menlo Park group soon provided supporting data in Alaska (Doell and Cox, 1965: 3404). They were less confident about the timing and duration of the Mammoth event.

Less securely documented is the Mammoth event (Fig. 3), based on two points of reversal near 3.0 million years ago; the time interval between these points is about equal to the precision of the potassium-argon dating method; thus; additional data will be needed to determine whether these two points represent one or two events.

(Cox, Doell, and Dalrymple, 1964: 1542; Fig. 3 is my Figure 6.1)

Bearing in mind experimental error, McDougall and Tarling (1963) may have been the first to observe what Cox and Doell named the Mammoth event, finding reversely magnetized rocks which they dated as 2.95 ± 0.06 million years from the Hawaiian Islands (McDougall and Tarling, 1963); however, McDougall (1964) had expressed doubts about its age, and therefore, Doell, Dalrymple, and Cox (1966: 539) decided against using it to define the timescale although it fitted well enough (Figure 6.1). Moreover, Doell and company (1966: 534) had a finding of their own: a reversed sample dated at 3.04 ± 0.09 million years near Mammoth Mine in the Sierra Nevada, hence the name.

Cox and Doell placed interrogation marks in their diagram (Figure 6.1) indicating uncertainty in the declination observed paleomagnetically. There were two associated with the Mammoth event signaling uncertainty in its duration, and one at approximately 7 million years within what they then considered the Brunhes normal epoch. Their concern was well founded; they soon discovered reversed samples with an age of approximately 0.7 million years within the Bishop Tuff, a rhyolite ash flow exposed between Bishop and Mono Lake, California (Dalrymple, Cox, and Doell, 1965). This was news. The Bishop Tuff had been previously dated at about 1 million years old and the youngest known reversed samples had been dated at approximately 1 million years, observations used by Cox, Doell, and Dalrymple in their June 1964 timescale to set the boundary between the Brunhes and Matuyama epochs. So Dalrymple and company (1965: 671) concluded that the "boundary between the Brunhes normal- and Matuyama reversed-polarity epochs is now uncertain within the limits of 0.68–1.0 million years." As it stood, and assuming that the reversals were not self-reversals, there could be either a reversed event within the Brunhes or the Brunhes/Matuyama boundary could be approximately 0.7 million years old.

Doell and Dalrymple (1966: 1060) next discovered normally, reversely, and intermediately magnetized rocks from nineteen "volcanic units from the Valles Caldera, Sandoval County, New Mexico." The intermediate magnetizations presumably reflected geomagnetic field directions during a transition from one polarity to another. The youngest ranged in age from 0.71 to 0.73 million years and were reversely magnetized. The intermediate rock was dated at 0.88 million years, a normally magnetized rock was dated at 0.89 million years, and another reversely magnetized rock was dated at 0.89 million years; hardly statistically significant. There could be three possibilities: the dates are wrong, the normally magnetized rocks had undergone a self-reversal, or there is "a short polarity event near the Brunhes–Matuyama boundary" (Doell and Dalrymple, 1966: 1060). The latter could be a normal event towards the end of the Matuyama or a reversed event late within the Brunhes. After giving strong reasons to rule out the first two possibilities and also for stratigraphic reasons, they turned to the third, and reset the Brunhes–Matuyama boundary at 0.7 million years. This done, they proposed a short normal event in the later part of the Matuyama which they called the Jaramillo normal event because of the proximity of the normally magnetized rocks to Jaramillo Creek.

The placement of the Brunhes–Matuyama boundary is more or less arbitrary in view of the present data. It could be placed between 0.9 and 1.0 million years, in which case the three reversely magnetized domes with ages between 0.71 and 0.73 million years would represent a reversed polarity event in the Brunhes normal epoch; or the boundary could be placed at 0.7 million years ago with 4D057 [intermediately magnetized] and 3X187 [normally magnetized] representing a normal event in the Matuyama reversed epoch at about 0.9 million years. For purposes of stratigraphic correlation, the last transition of polarity will undoubtedly be the most useful and we therefore prefer to assign the epoch boundary at 0.7 million years. Accordingly we here name the normal event near 0.9 million years the "Jaramillo normal event," after Jaramillo Creek, which is approximately 3 km south of the locality of unit 3X187.

(Doell and Dalrymple, 1966: 1061; my bracketed additions)

This was a most notable achievement with far-reaching consequences, both for marine work and for Quaternary stratigraphy. Their paper appeared on May 20, 1966. In it they made clear that they could not tell whether they had pinned down the beginning or end of the Jaramillo event.

From the present data it is not possible to tell whether the intermediate direction represents the transition to or from the Jaramillo normal event, nor, therefore, whether the event occurred just before or just after 0.9 million years ago. ,

(Doell and Dalrymple, 1966: 1061)

This question was answered later by the ANU group.

After the appearance of the Doell and Dalrymple paper (DD), four important papers by the ANU group were published in quick succession. McDougall and Chamalaun (1966) "Geomagnetic polarity scale of time," was submitted in October 1966, five months after DD appeared. McDougall and Wensink (1966) "Paleomagnetism and geochronology of the Pliocene–Pleistocene lavas in Iceland," was received by *Earth and Planetary Sciences Letters* on June 21, a month after DD appeared. The two others were submitted before DD was published; they were McDougall, Allsopp, and Chamalaun, (1966) on the Newer Volcanics of Victoria, Australia, originally received by *JGR* on March 24, revised and resubmitted in August, and Chamalaun and McDougal (1966), "Dating geomagnetic polarity epochs in Réunion," published in *Nature* on June 18, just 29 days after DD appeared. The latter had no date received; however, in it Chamalaun and McDougall referred to MacDougall, Allsopp, and Chamalaun's forthcoming paper on the Newer Volcanics of Victoria as "McDougall, I. and Allsopp, H.A. in preparation." So Chamalaun and McDougall submitted their paper on Réunion volcanics before they knew that Chamalaun would be a co-author of the Newer Volcanics paper. Surely therefore, the "Réunion" paper was submitted before DD appeared. Indeed, it may have been submitted in January when DD was submitted. I consider the papers in their chronological order of submission.

Chamalaun and McDougall analyzed lava flows from Réunion, an island south of the equator in the Indian Ocean located at the southwest extremity of the Mauritius–Seychelles Ridge. They analyzed approximately forty samples from two volcanoes, Piton de la Fournaise and Piton des Neiges. All samples from Piton de la Fournaise gave ages of 0.04 to 0.36 million years, and had normal polarity. Following previously established practice, they divided the lava of Piton des Neiges into the Differentiated and Oceanite series. All samples from the Differentiated Series were normally magnetized and gave ages between approximately 0.1 and 0.35 million years. They then turned to the Oceanite Series of the Piton des Neiges.

A second group of lavas from the oceanite series have measured ages lying between 0.97 and 1.18 m.y., based on samples from seven flows. Five of the specimens yield dates in the range 1.07–1.18 m.y., and have reversed polarity. These results are consistent with previously published data; the rocks can confidently be assigned to the Matuyama reversed epoch ... The remaining lava of this age group is distinctive columnar olivine basalt that crops out on the road to Takamaka. Field observations suggest that the lava occurs adjacent to the edge of a valley eroded in the oceanite series; the valley was afterwards filled by lavas of the differentiated series. The directions of magnetization measured on three different samples from this particular olivine basalt agree to within a few degrees and indicate normal polarity. Two of the specimens were collected by one of us, and third was obtained at a later time by Dr. M. W. McElhinny. All three samples were dated ... and the results agree closely at 1.01 ± 0.05 m.y. Because of the consistency of the results we are very confident of both the polarity and age measurements.

(Chamalaun and McDougall, 1966: 1213)

Chamalaun and McDougall were faced with roughly the same situation as Doell and Dalrymple after they had found the reversely magnetized rocks near the Jaramillo River. Counting backwards in time, they (1966: 1213) claimed "that a change from normal to reversed polarity occurred at about 1.0 m.y." but declared that there was not enough data "to decide whether this is the boundary between the Brunhes normal and Matuyama reversed epochs." Then they brought in results from their study of the Newer Volcanics of Victoria, which they were working on at the same time. They reported that they had found samples of reversed polarity dated at 0.81 ± 0.03 million years. Comparing them with the normally magnetized samples from the Bishop Tuff in the Sierra Nevada that Dalrymple (Dalrymple *et al.*, 1965) had re-dated at about 0.7 million years, Chamalaun and McDougall (1966: 1213), still counting backwards in time, proposed that "a change from normal to reversed polarity occurred at 0.75 ± 0.07 m.y." After discussing some additional and "conflicting evidence," they remained undecided about how to resolve the situation.

These data suggest that either the boundary between the Brunhes normal and Matuyama reversed epochs occurred at 0.75 ± 0.07 m.y., with a short interval of normal polarity (that is an event) at close to 1.0 m.y., or that the boundary lies at 1.01 ± 0.03 m.y, and the reversed polarity of rocks dated at 0.81 m.y. records an event in an otherwise normal epoch. Another possibility is that self-reversal is partly responsible for the somewhat confused picture. Clearly many additional data are needed in the age range 0.7 - 1.0 m.y. to distinguish between the alternative explanations.

(Chamalaun and McDougall, 1966: 1214)

What was the confusing evidence? The most troubling finding was one of their own: an apparently normally magnetized rock dated at 0.81 ± 0.04 million years from the Kula series of East Maui, Hawaii. If this finding from Hawaii held up to further scrutiny, they would have to propose that there had been a change in polarity from reversed to normal between 0.81 ± 0.03 and 0.81 ± 0.04 million years, that their finding from Réunion of reversed samples dated 0.81 ± 0.03 was mistaken, or that there had been a self-reversal. They (1966: 1213) actually had reason to doubt the polarity of their Hawaiian find because they had not measured the polarity of the same sample they dated: they inferred that it was normal "because other samples from the same volcano" were normally magnetized. This situation arose because Tarling the paleomagnetist and McDougall the geochronologist had made separate collections unbeknownst to each other (II, §8.15). Leaving the Brunhes–Matuyama boundary question, they turned to the Olduvai event. They found no rocks with normal polarity at 1.9 million years as there are in Olduvai Gorge. Instead they found three normally magnetized samples and five reversely magnetized samples, all dated at 2.0 ± 0.1 million years. Also they (1966: 1214) found reversely magnetized rocks of the same age overlying rocks with normal polarity. They correlated their normally magnetized samples with the Olduvai, and used their finding to reset the time of the Olduvai from 1.9 m.y to 2.0 ± 0.05 million years. In a later revision Cox (1969: Figure 4) recognized the Olduvai as double; two normal events within the range 1.95–2.13 million years.

In their study of the Newer Volcanics of Victoria, Australia, McDougall and Chamalaun, and their co-author H. L. Allsopp reconsidered the Brunhes–Matuyama. Allsopp was on leave at ANU from the Bernard Price Institute of Geophysical Research, University of Witwatersrand, Johannesburg, South Africa. The Newer Volcanics had been studied paleomagnetically by Irving and Green when developing their APW path for Australia (II, §5.3). McDougall and colleagues did not know of Doell and Dalrymple's discovery of the Jaramillo event before they submitted their paper in March 1966; however, they knew of it before their own paper was published in December. Although (working backwards) they were strongly committed to a change from normal to reversed polarity at 0.75 ± 0.07 million years, they remained undecided about whether to designate this as the Brunhes–Matuyama boundary and have a normal event within the Matuyama, or to keep the Brunhes–Matuyama boundary at about 1.0 million years and have a reversed event within the Brunhes. They found in the New Volcanics of Victoria that samples with consistent reversed polarity gave an age of 0.81 ± 0.03 million years. They claimed that these samples were securely dated and definitely reversed, noting:

The dates are in excellent agreement and are considered to provide an accurate measure of the age. Green and Irving [1958] collected 34 oriented samples from this particular quarry and found that the measured paleomagnetic directions are tightly grouped and show consistently reversed polarity ... Demagnetization experiments on several samples indicated stability. Further confirmation that at about this time the polarity was reversed is provided by a basalt from France that has reversed polarity and is dated at 0.81 ± 0.03 m.y.

(McDougall et al., 1966: 6113)

This led them, as in their previous paper, to claim (1966: 6133), "Together with the Bishop tuff results, these data suggest that a change in polarity from normal to reversed occurred 0.75 ± 0.07 m.y. ago." They again appealed to their discovery in Réunion of three normally magnetized samples dated at 1.01 ± 0.05 million years and five reversely magnetized samples dated in the range 1.07–1.18 million years, and claimed that a reversal from normal to reversed occurred at 1.00 ± 0.05 million years.

As it stands, they had yet to add any significant new finding about where to place the Brunhes–Matuyama boundary; however, they now were prepared to dismiss findings that they had allowed in their previous paper, ones that they had previously categorized as "confusing." These included the Menlo Park's sample from Devils Postpile, which has normal polarity, and was dated at 0.94 ± 0.16 million years, and their own, purportedly normal sample dated at 0.84 million years from Hawaii. They wisely rejected their own sample because as noted already it had not been measured paleomagnetically and may actually be reversed, because they knew that reversals are not uncommon in the Kula series. They rejected Menlo Park's sample because it "is too imprecise to be of much value." Together with the dismissal of their early finding from the Kula series, they no longer had to deal with the "confusing" factor of a purported change of polarity from reversed to normal at 0.84 million years. Nonetheless, they did not take the plunge.

Hence the results suggest that changes in polarity occurred at 0.75 ± 0.07 m.y. and 1.00 ± 0.05 m.y. ago. One of these changes probably records an event, and the other the boundary between the Brunhes normal and the Matuyama reversed polarity epochs. At present we have too few data to decide which is the event, and which is the boundary between the epochs; the choice may prove to be an arbitrary one. The possibility that self-reversal is responsible for some of the apparent difficulties should also be borne in mind. Clearly, additional data must be obtained on rocks whose ages lie within the range 0.7 to 1.0 m.y. in order to distinguish between the alternative interpretations.

(McDougall et al., 1966: 6114)

If the "apparent difficulties" refer to the "confusing" findings, which they had eliminated, and had been the reason for concern about self-reversals, I believe that the only reason they had for their continued caution was, as they stated, the need of "additional data." In retrospect, it can be seen that Chamalaun and McDougall's samples from the second group of lavas from the oceanite series on Réunion found on the road to Takamaka – the five reversely magnetized samples dated in the range 1.07–1.18 million years, and three normally magnetized ones dated at 1.01 ± 0.05 million years – defined the older boundary of the Jaramillo event. If McDougall and his co-workers had not been so cautious, they could have at least claimed that there had been a reversed event within the Brunhes or a normal event within the Matuyama and decided to set the Brunhes–Matuyama boundary accordingly. Or, they could have independently decided, like Doell and Dalrymple, that it made better stratigraphic sense to set the Brunhes–Matuyama boundary at about 7.5 million years. They then could have claimed that there had been within the Matuyama a normal event whose older boundary was at 1.01 ± 0.05 million years and whose younger boundary was no older than 0.84 million years. They could have been confident of the older boundary but have left open the younger boundary, and claimed, just as Doell and Dalrymple (1966: 1061) had done, that the first normal event within the Matuyama occurred "near 0.9 million years." Doell and Dalrymple, just like McDougall and his co-workers, had to leave open one of the boundaries. Doell and Dalrymple had to leave the older boundary; McDougall and company, the younger one. Doell and Dalrymple took the plunge and with Cox got almost all the acclaim for developing this critical part of the radiometric reversal timescale; McDougall and his coworkers did not, and received little acclaim even though it was they who established the older boundary of the Jaramillo.

Doell and Dalrymple's "Jaramillo" paper was published before McDougall and his co-workers' paper. Doell and Dalrymple provided the additional data that the ANU group required. Adding an addendum to their paper they set the older boundary of the Jaramillo and suggested this reversal sequence.

Since this paper was submitted for publication Doell and Dalrymple [1966] have reported polarity determinations on rocks whose measured ages lie between 0.7 and 0.9 m.y. Their results are fully consistent with those reported in the present paper. The combined data provide a much clearer picture of the behavior of the geomagnetic field in this age range and suggest the following time scale: present to 0.70 m.y. ago, normal polarity; 0.70 to 0.88 m.y. ago, reversed polarity; 0.89 to 1.00 m.y. ago, reversed polarity.

(McDougall et al., 1966: 6115; my bracketed addition)

The ANU group also did important work on other events. In their next paper, on Icelandic lavas, McDougall and Wensink uncovered a new normal event in the Matuyama. They found two normally magnetized lava flows separated by a reversed flow within the Matuyama. Samples from the upper flow had a date of 1.60 ± 0.05 million years; those from the lower flow fell within the Olduvai event, dated at about 1.9 to 2.0 million years. They proposed a new normal event, which occurred at about 1.6 million years, some 0.3 to 0.4 million years younger than the Olduvai event.

Clearly the upper normally magnetized flow gives a date too young to be correlated with the Olduvai event, and may indicate another short interval of normal polarity at about 1.60 m.y. in the R1 (Matuyama) reversed epoch. We tentatively propose the name Gilsá for this event, after a small tributary of the Jökulsa River in northeast Iceland.

(McDougall and Wensink, 1966: 235)

They also found strong confirmation for the Mammoth event, which Doell, Dalrymple, and Cox had recently identified (Doell *et al.*, 1966).

In their final 1966 paper, McDougall and Chamalaun summarized results over the past year by Doell and Dalrymple in New Mexico and their own results in Réunion, Iceland, and the Australian state of Victoria (McDougall and Chamalaun, 1966). To this end, they reviewed the recent history of work on the Jaramillo, fairly describing the contributions by both groups.

Originally the most recent change in polarity, the boundary between the Brunhes normal and Matuyama reversed epochs, was thought to have occurred about 1.0×10^6 yr ago ... The age of this boundary was controlled by the normally magnetized Bishop Tuff dated at 0.98×10^6 yr and a number of samples dated at $1.0–1.4 \times 10^6$ yr with reversed polarity. The age of the Bishop Tuff was later revised, however, to about 0.67×10^6 yr ... so that the age of the Brunhes–Matuyama boundary was then known to lie between 0.7×10^6 and 1.0×10^6 yr. Recent results ... from the Valles Caldera, New Mexico indicated reversed polarity of rocks dated at 0.71×10^6 to 0.73×10^6 yr and normal polarity for a sample with a measured age of 0.89×10^6 yr. From these results Doell and Dalrymple ... concluded that the Brunhes–Matuyama boundary was dated at 0.70×10^6 yr, and that a normal event, termed the Jaramillo event, occurred at about 0.89×10^6 yr. Because the identification of the Jaramillo event was based on one of two samples dated at about 0.89×10^6 yr the older and younger limits for this event could not be defined with any precision. Results recently obtained in this laboratory ... however, not only confirm the occurrence of the Jaramillo event, but allow its boundaries to be determined with some accuracy. The date of 1.01×10^6 yr on normally magnetized lava from Réunion ... together with numerous results on reversely magnetized rocks dated at $1.0 \times 10^6–$ 1.4×10^6 yr, strongly suggests that the older boundary of the Jaramillo event is well controlled at $1.00 \pm 0.04 \times 10^6$ yr. Several reversely magnetized rocks from Victoria, Australia, were dated by us ... at 0.81×10^6 yr and indicate that the younger limit for the Jaramillo event lies between 0.81×10^6 and 0.89×10^6 yr ago. Doell and Dalrymple ... reported an intermediate direction of magnetization from one of the Valles Caldera rocks which was dated at 0.88×10^6 yr. If this intermediate direction records the transition of the magnetic field as it changed polarity then the younger limit of the event is accurately dated to close at 0.89×10^6 yr.

(McDougall and Chamalaun, 1966: 1416–1417)

They also proposed another reversed event (the Kaena) within the Gauss normal epoch, which they dated at 2.80 ± 0.06 million years. They had Tarling to thank for this because he had found and dated the sample, and discussed it in his 1964 Ph.D. thesis.

Within the Gauss epoch, from 2.6×10^6 to 3.0×10^6 yr ago, few acceptable data are available. A previously unpublished result on a reversely magnetized lava from a quarry on Kaena Point, Oahu, Hawaii, suggests the possibility of yet another event (Kaena) in the Gauss normal epoch at $2.80 \pm 0.06 \times 10^6$ yr. The rock is petrographically of excellent quality for dating and the paleomagnetic results appear to be satisfactory.[23]

(McDougall and Chamalaun, 1966: 1417; note 23 refers to Tarling's 1963 Ph.D. thesis)

McDougall and Chamalaun were less reticent when proposing the Gilsá and Kaena events than they been for the Jaramillo. In both cases they had one sample to define the event, and neither of its boundaries.

6.5 Opdyke and others at Lamont develop a reversal timescale based on the study of deep-sea cores

By February 1966, they had produced a reversal timescale back to the base of the Gauss normal epoch, and had even discovered the Jaramillo event, which Opdyke had named the Emperor event. Opdyke then invited Vine to Lamont, and Vine came in either February or March, 1966. Vine saw *Eltanin*-19; he was impressed, and even more confident that V-M was correct. Opdyke also showed Vine the new marine sediment-based reversal timescale. Vine told Opdyke that Dalrymple and company had already discovered the same event, had named it the Jaramillo, and had presented their finding at the GSA meeting in Kansas City.

I arrived [at Lamont] in January '64 ... It was a very, very dynamic place, and filled with very dynamic people. I hadn't – I'd been abroad so long that I hadn't really fit in to the American academic scheme, although I had been originally at Lamont, or Columbia, but I didn't know personally any of the people at Lamont. My only problem was that I had differing scientific views to almost everybody at Lamont, so I got into a lot of arguments with people.

(Opdyke, interview with R.E. Doel, July 1977; my bracketed addition)

When Neil Opdyke arrived at Lamont, "the citadel of continental fixism," in January 1963 he already was a confirmed drifter. He had worked on ancient wind directions with Runcorn (II, §4.2, §5.12–§5.14), had learned paleomagnetism from Irving (III, §1.16) at ANU, and from Ian Gough and colleagues at the UCRN (now the University of Zimbabwe) where he had obtained or helped obtain several key paleomagnetic poles from the Karroo System (III, §1.17). Soon after arriving at Lamont, he witnessed Worzel's attack on the ethics of paleomagnetists (see III, §1.17 for how Opdyke got a job at Lamont). He later argued with Heirtzler and Le Pichon regarding their dismissal of V-M in their interpretation of magnetic anomalies over the Mid-Atlantic Ridge generally and its northern portion, the Reykjanes Ridge. He objected especially to their resistance to considering the possibility that remanence was the causal factor in the generation of the marine magnetic anomalies (§4.6, §5.4).

This spurred him to join forces with R. Hekinian to examine the magnetic properties of dredged basalts in Lamont's collection, in which they found that remanent magnetization dominates (§4.6).

Opdyke, however, had still not begun to work on Lamont's vast collection of deep-sea sediment cores, the principal reason he had been hired.

In January of 1964, I moved from Africa, where I had been studying the Paleozoic paleomagnetism of central Africa in the hope of demonstrating that continental drift was a reality, to Lamont Observatory, which was a citadel of continental fixism. I was hired by Jim Heirtzler to set up a paleomagnetic laboratory, one principal impetus being the desire of Maurice Ewing to have someone study the paleomagnetism of the large and growing collection of deep-sea cores available at the observatory.

(Opdyke, 1985: 1177)

In March 1965, the paleomagnetics laboratory became operational. Opdyke first worked on a land-based paleomagnetic project; he and Wensink studied the Jurassic White Mountains Series (Opdyke and Wensink, 1966). Maurice Ewing wanted Opdyke to tackle the cores; he did not think too highly of Opdyke's work "with his old rocks."

I had tried to get the stuff [Lamont's deep-sea cores] studied for about ten years. First one man was on it; his conclusions were inconclusive. Then another man. Then Neil Opdyke came; he was doing paleomagnetism, but of old rocks. I gave him a long talk, said you have here an unequaled opportunity; and he played with his old rocks some more.

(Ewing as quoted in Wertenbaker, 1974: 200; my bracketed addition)

All credit to Ewing, his instincts were on the mark. Opdyke later explained his initial reservations about working on the cores.

In the summer of 1965, we began, in a rather desultory fashion, to study the reversals in marine sediments. At the time, I was aware that others had studied reversals in cores, and indeed, I had heard a lecture by John Belshé (of Cambridge University) in Australia in 1960 on reversals in brown clay cores from the Pacific. This work was carried out as a part of Chris Harrison's Ph.D. [1964] research at the Scripps Institution of Oceanography in La Jolla, Calif., and I also knew that he had had much trouble with the study [Harrison and Funnell, 1964]. At the time the problems looked difficult, since the cores were unoriented, and I could not understand how the cores could become magnetized, since many were intensely burrowed, and possible sources for magnetized material were distant.

(Opdyke, 1985: 1177; reference same as mine)

Nonetheless, he assigned John Foster, his first graduate student, to work on them. Foster had designed a slow-speed, highly sensitive fluxgate spinner magnetometer that was just right "to measure the paleomagnetism of marine cores." Until then, spinner magnetometers had used the principle of induction. A weakly magnetized specimen was spun at a very high rate sufficient to induce a voltage in enclosing fixed coils. Uncemented sedimentary cores were likely to disintegrate. Foster's innovation was critical for the study of Lamont cores.

I had acquired my first graduate student ... John Foster, who was an excellent instrument designer and very helpful in setting up the laboratory. He developed a slow-speed spinner magnetometer of high sensitivity that used fluxgates to measure the direction of magnetization in rock specimens. This instrument and its derivatives were widely used in paleomagnetic research until the advent of the cryogenic magnetometer in the late 1970s.

(Opdyke, 2001: 103; see Foster (1966) for a detailed description of his magnetometer)

In summer 1965, Foster started with two internally oriented cores.

[O]n the Vema 21 leg between Hawaii and Guam, W. Broecker collected samples at 10-cm intervals from two brown clay cores that were internally oriented. These were analyzed by John Foster, who, at the time, was my first graduate student. The change of polarity was actually observed, but the reversal was not 180°, so the results were interesting but not compelling.

(Opdyke, 1985: 1177; see Dickson and Foster (1966) for discussion of work on the
two cores; my bracketed addition)

Real progress began in January 1966 after Foster talked to Billy Glass, one of Heezen's graduate students.[2] Importantly, Glass proposed that they study magnetic cores from high latitudes because at such latitudes the inclination of the geomagnetic field is steep and reversals would be at their most evident; accompanying changes in declination need not be ascertained, and so azimuthal orientation is unneeded. Another indicator of a reversal is a drop in intensity, which, as Opdyke *et al.* (1966: 352) noted, is expected with the dipole field decreasing to zero during a reversal. Opdyke later described what happened, and what they found within just a few months.

John Foster ... was talking about his research with other students, in particular with Billy Glass, a student of Bruce Heezen. During the course of their discussions, Billy suggested to Foster that perhaps it would be a good idea to study higher-latitude cores, in particular those around the Antarctic that had been studied by James Hays for his Ph.D. thesis. I told them to go ahead and give them a try, which they did. Sure enough, they were quite highly magnetic and changes in polarity were soon observed [Opdyke, Glass, Hays, and Foster, 1966]. Progress was very rapid, and by early 1966 research on reversal stratigraphy of cores spread to sediments of the North Pacific on the suggestion of D. Ninkovich, a colleague of Heezen, who was interested in correlating volcanic ash layers in cores south of the Aleutian Islands. By February 1966, we had magnetic stratigraphy from fossil-bearing marine sediments to the base of the Gauss normal period 2.6 million years ago, and we had identified a new normal event older than the Brunhes–Matuyama boundary, which we called the Emperor event, after the Emperor Seamount chain. This event was also observed on the profiles from the East Pacific Rise (*Eltanin*-19 profile). By March 1966, we had all of this information at our disposal, but it was known only to the members of the Magnetics Department, and to Bruce Heezen, who was very enthusiastic about the application of magnetic stratigraphy to marine geology [Ninkovich, Opdyke, Heezen, and Foster, 1966].

(Opdyke, 2001: 104–105; my bracketed additions; see also Opdyke, 1985: 1177, and
Opdyke, Session 2 (p. 132), interview with R. E. Doell, July, 1977)

The results were spectacular (Figure 6.4). Basing their results on classical stratigraphy – the law of superposition – they identified sedimentary layers corresponding

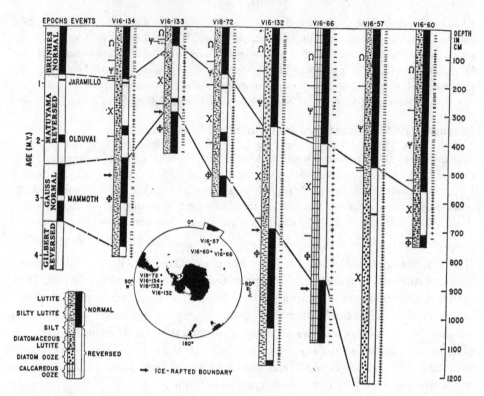

Figure 6.4 Opdyke *et al.*'s Figure 1 (1966: 350). Their caption reads: "Correlation of magnetic stratigraphy in seven cores from the Antarctic. Minus signs indicate normally magnetized specimens; plus signs, reversely magnetized. Greek letters denote faunal zones (from Ericson *et al.*, 1963)." All cores are from the Antarctic (see inset figure). Epoch and events are from Cox, Doell, and Dalrymple (1964), Doell, Dalrymple, and Cox (1966), and McDougall and Tarling (1963). V16-134 is continuous. All cores show boundary between Brunhes normal and Matuyama reversed series. They identified the short normal interval near the top in V16-134, V18-72, V16-66, and V16-57 with the Jaramillo, the next normal interval in cores V16-134, V16-133, and V18-72 with the Olduvai, and the short reversed interval within the Gauss found in core V16-134. Only core V16-134 passes through the Gauss normal series to Gilbert reversed series. The faunal boundaries were determined by Hays (1965).

to periods of the reversal timescale based on paleomagnetic and radiometric studies of lava flows by the ANU and Menlo Park groups (§6.3). One core gave them "a continuous record of the paleomagnetic field back to about 3.5 million years" (Opdyke *et al.*, 1966: 356); it was considered "continuous" because it contained no evidence of breaks in deposition. They identified a Brunhes normal series, a Matuyama reversed series with two short reversed intervals corresponding to the . Jaramillo and Olduvai events, a Gauss normal series with a short reversed interval corresponding to the Mammoth event, and a Gilbert reversed series. They (1966: 356) emphasized the "striking agreement between the continuous record of deep-sea cores

and the earlier discontinuous land record." Notable is their identification of the
Jaramillo event by procedures independent of those on land. The consilience was
further enhanced by estimates of the ages of two faunal boundaries. Using radiometrically determined ages of the changes in polarity as determined by lava studies,
they (1966: 353) calculated rates of sedimentation, and used these rates to calculate
the age of faunal boundaries found in their cores, which "agreed well with estimates
for these [faunal] boundaries from Th^{230} estimates." They noted (RS2), "The fact
that two independent methods of dating give similar ages for the faunal zones lends
credence to both methods." They closed their first paper with a promise, which they
made good during the next few years. However, I interrupt discussion of their work
and turn to the magic profile over the Pacific–Antarctic Ridge obtained by Pitman
and colleagues, whose analysis was enhanced by the above independent discovery of
the Jaramillo.

6.6 Pitman's "magic" profile over the Pacific–Antarctic Ridge, December 1965: Pitman, Heirtzler, and Talwani accept V-M[3]

At the time that this intense activity [of working on reversals in marine sediment cores] was
taking place, a significant scientific breakthrough was occurring in the office next to mine,
where two graduate students Walter Pitman and Geoff Dixon, resided. The previous fall,
Pitman had gone to sea on the *R.V. Eltanin* as chief scientist after deciding that he would do a
thesis in marine magnetics. One night, Walter ran out the magnetic profile of the *Eltanin*-19
crossing of the East Pacific rise. The next morning, when I arrived at my office, he had thumb
tacked the profile to my door. There, in fantastic detail, was the magnetic time scale to the base
of the Gauss Chron [Cox *et al.*, 1964; McDougall and Tarling, 1963] and beyond, to terra
incognita, as well as the event in the Upper Matuyama that we had seen in cores which would
later become known as the Jaramillo event. As might be expected, the excitement was intense.
Acceptance of the implications of what came to be known as Pitman's Magic Profile was by no
means instantaneous or universal [Pitman and Heirtzler, 1966].

> *(Opdyke, 1985; Cox* et al. *is my Cox, Doell, and Dalrymple (1964);*
> *other references are the same as'mine)*

Walter Pitman was born in 1931 in Newark, New Jersey.[4] Educated at Morristown
School for Boys, where he did well in mathematics and enjoyed reading fiction, in
1949, he entered Dartmouth College, planning to study philosophy and literature.
After leaving Dartmouth halfway through his second year, he attended Lehigh
University, where he remained for five years, earning, in 1956, undergraduate degrees
in arts and in engineering physics. Still not interested in pursuing a career in science,
he took an administrative position with Hazeltine Corporation, an electronics firm.
Bored completely out of his mind, he realized he had to do something else and finally
decided, in 1960, to go into the sciences. He first thought of physics, got accepted in
physics at Columbia University, but changed his mind because he thought he would
never do anything first-rate in physics. Having developed a fondness for the ocean

while growing up, he decided to study physical oceanography. This interest led him in September 1960 to Lamont, where he was told by John E. (Jack) Nafe that he would not be admitted outright, but probably would be after working as a technician aboard a Lamont ship. Nafe called Jim Heirtzler, who was looking for a technician to take magnetic measurements aboard *Vema*. Pitman went to see Heirtzler, whom he thought very nice, a good person with whom to work, and he soon found himself in December 1960 aboard *Vema*, trying to get a magnetometer properly working.

Pitman became a graduate student the next year working under Heirtzler for his Ph.D. – Charles Drake was his official advisor. He got tremendous help from Neil Opdyke. Originally planning to study micropulsation (short variation of intensity) of Earth's magnetic field, in 1965 he decided instead to study marine magnetic anomalies, and thus in early September 1965 found himself once again, five years after first meeting Heirtzler, aboard another Lamont ship RV *Eltanin* on Leg 20 "to 'pay my dues' to be part of the data-collecting process" and to gain "access to other *Eltanin* data" (Pitman, 2001: 88). He planned to work on the Eltanin Fracture Zone, but magnetic profiles collected during Legs 19, 20, and 21 over the southern extensions of the East Pacific Rise (Pacific–Antarctic Ridge) proved more fruitful (Pitman, July 31, 2011 email to author).

Before embarking on *Eltanin*, Pitman knew little about continental drift.

For me it was a time of great excitement but also wonderment. I had started out my thesis studying time variations of the earth's magnetic field, in particular diurnal variations and micropulsations, and switched over to marine geology and geophysics only at the beginning of 1965. So the problems were all new to me. We had been fed some notion of the idea of continental drift sometime back in grammar school geography class ("see how South America and Africa seem to fit together like pieces in a puzzle?"), but it was not until 1965 that I began to be aware of the immense controversy this idea had provoked. In effect, I was learning about the problem as I was helping to solve it. It would take some time for the full magnitude of what had happened to sink in.

(Pitman, 2001: 94)

Unsurprisingly, Pitman had heard nothing more about continental drift, nothing at Dartmouth or Lehigh, nothing while taking classes at Lamont (Pitman, July 31, 2011 email to author). He was unfamiliar with the classical arguments for continental drift, its strong support based on land-based paleomagnetic studies, and the evidence in favor of field reversals before the ANU and Menlo Park groups began their timescale research. Pitman was also unfamiliar with the idea of seafloor spreading and V-M before embarking on *Eltanin* in September 1965.

Again, this should not come as a surprise because he had planned to study the Eltanin Fracture Zone, not marine magnetics, and Vine and Wilson's work on the Juan de Fuca Ridge and Wilson's work on transform faults was not published until October 22 (§5.2).

Pitman returned to Lamont in November 1965, and soon read Vine and Wilson's work. He also thinks he read Dietz on seafloor spreading at the same time (Pitman,

July 31, 2011 email to author; Glen, 1982: 333). Herron, another graduate student working on marine magnetic anomalies, and he soon began reducing the magnetic data. They began with Leg 20 because they had been on the ship together (Glen, 1982: 332). When they got the *Eltanin*-20 profile, Pitman "didn't know what [it] really meant," had only a superficial understanding of V-M, and still did not know that Heirtzler and Le Pichon had opposed it in print. Here is what he told Glen:

At that point in December 1965 Ellen Herron and I were just getting data out and had not yet selected our doctoral dissertations and didn't know what the *Eltanin*-20 profile really meant. I'd read Dietz's paper casually; I didn't know the Vine-Matthews hypothesis very well at all. I was not aware of it in detail. I had not read the paper and studied it; it was not something that seemed important to me at the time. I was unaware that Le Pichon and Heirtzler had made that strong comment against Vine and Matthews in print.

(Glen, 1982: 334)

Pitman realized within "maybe two weeks at the most" the similarity between *Eltanin*-20 and the profile of the Juan de Fuca Ridge and the one he had seen from reading Vine and Wilson's October paper. So he showed the profile to someone he thought would be interested. Pitman explained to Glen what happened.

I took that *Eltanin*-20 profile down to [a staff member's] office, because I knew he had written on the subject and was interested in the problem. He was talking to [another researcher], and when they came out I said, "Look at this, this is very puzzling; these magnetic anomalies that we got on the *Eltanin*-20 look almost exactly like the same as the profile over the Juan de Fuca Ridge." He said, "Ha! Ha! I suppose you're going to prove Vine and Matthews are right." I questioned him about that and he said something about seafloor spreading. I had been doing broad reading while searching for a thesis problem and remembered Dietz. They also said something about seafloor spreading and Harry Hess. That was the first time Hess's name came into this thing. That was December of 1965. The conversation and laughter about Vine-Matthews, Dietz, Hess, and seafloor spreading stuck in my mind. His laughter didn't bother me at all; I must say I began to smell something at the time. I didn't understand it and I didn't know what it was. I went back and continued to process the data with Ellen Herron, but what problems we were mutually or exclusively interested in I had only the faintest idea.[5]

(Glen, 1982: 334)

Pitman (Wertenbaker, 1974: 205) also recalled that the remark about Vine and Matthews made him "go back and read Vine and Matthews." So the sarcastic response did not deter Pitman – just the opposite, it spurred him on. Not a new graduate student fresh out of college, but a mature thirty-three year old who had spent a half-dozen years working after college, he was able to stand up to initial ridicule. Le Pichon also recalled a conversation he had with Pitman at about this time; he credits Pitman as the first among the marine magnetics group to question their "previous fixist ideas."

Walter Pitman clearly expressed his dissent with our conclusions at the time. He told me then: "You are going too far, I would be afraid to publish such conclusions." Walter was the first of us to have entered this grey domain where we knew our previous fixist ideas were not right, but were not yet sure that seafloor spreading could work.[6]

(Le Pichon, 2001: 210)

Le Pichon did not mention the objections to their "fixist ideas" that Opdyke had raised earlier (§4.6). I guess Opdyke had passed through the "grey zone" and abandoned fixism five years earlier so, strictly, he would not qualify.

With a group of Lamont technicians to digitize the data, Pitman and Herron ran out the *Eltanin*-19 and -21 profiles. Pitman might still not have fully understood the implications, but the *Eltanin*-19 profile "hit [him] like a hammer."

It hit me like a hammer. In retrospect, we were lucky to strike a place where there are no hindrances to sea-floor spreading. There's no other place we get profiles quite that perfect. There were no irregularities to distract or deceive us. That was good, because people had been shot down an awful lot over sea-floor spreading. The symmetry was extraordinary.

(Wertenbaker, 1974: 205)

He also showed the results to Opdyke and Dickson. He told Glen their reaction was decidedly different from the others to whom Pitman had shown the *Eltanin*-19 profile.

One of the very important factors in all this was the presence of Neil Opdyke and a graduate student named Geoffrey Dickson, who ... was ... receptive to the idea of continental drift. Opdyke was most important; he believed in drift and was a very important guy to have around. Because he was our age he was a colleague; he was encouraging and kept on pushing. He talked about the paleoclimatological evidence for drift and believed in field reversals and that the polar wander paths could be explained by drift. As we processed all those data, the similarity between the Juan de Fuca and those profiles became obvious. I remember staying there at night alone one day, running out magnified projected profiles, simply taking the magnetic anomaly data and making slight adjustments to them so they look as though you've run perpendicular to a ridge axis. I pinned up all the profiles of *Eltanin*-19, 20, and 21 on Opdyke's door and went home for a bit of rest. When I came back the guy was just beside himself! He knew that we'd proved seafloor spreading! It was the first time you could see the total similarity between the profiles – the correlation, anomaly by anomaly. The bilateral symmetry of *Eltanin*-19 was the absolute crucial thing. Once Opdyke saw that he said, "That's it – you've got it! Opdyke was a very important catalyst at that point. He knew immediately what it meant and became our advocate. Having him around was a very important thing. Heirtzler took a while to come around; at first he tried to explain it away by electrical currents. Opdyke knew the significance of the profile immediately; he knew all the ramifications.

(Glen, 1982: 394)

Opdyke, who had found paleoclimatic and paleomagnetic support for mobilism, and Dickson, who had worked with F. Stacy at ANU, had come up on his own with an explanation of marine magnetic anomalies very like that of Vine and Matthews (§2.19), and knew well mobilism's paleomagnetic support based on APW paths, educated Pitman about mobilism's classical support, especially its paleoclimatic and paleomagnetic support (Glen, 1982: 354). When I asked Pitman if Opdyke had told him about Irving and Green's APW path of Australia, the APW path of India, about how they could not be reconciled with each other or APW paths from other landmasses without continental drift, about Hospers' classic work on reversals, about support of the GAD hypothesis, about drift's paleoclimatic support, he said:

Yes, all of the above!! In spades! Neil arrived at Lamont as a continental drifter. Once he saw the *Eltanin*-19 and 20 data he became a most ardent supporter. We talked about it all with him, distribution of flora and fauna, paleoclimate, paleomagnetism, etc., etc., etc.

He added, "Dickson was very helpful. He was a colleague, we worked and published together." I add Dickson must not have told Pitman that he had worked with Stacey. "I was not aware of this. Dickson was very smart and clever so I was not surprised that Dickson knew much paleomagnetism and discussed it with us" (Pitman, July 31, 2011 email to author). With their help, Pitman became a mobilist, accepting V-M, seafloor spreading, and drifting continents. Concerning Opdyke, Pitman (July 31, 2011 email to author) added, "We all talked to him a lot. He started in on paleomagnetism of deep sea sediments, he pushed us and pushed us; he was a great sounding board for ideas, a great listener, talker and advocate."

The *Eltanin*-19 profile, the one later called "Pitman's magic profile," led many at Lamont to become mobilists. Pitman [May 22, 1997 interview with Levin] believes that Heirtzler changed his mind within a week or so. Opdyke told Glen that it took Heirtzler about a month.

Opdyke recounted: "Heirtzler said the *Eltanin*-19 profile was too perfect and caused by electrical fields in the upper mantle in order to get out of the Vine-Matthews hypothesis. It was a process of about a month getting Heirtzler to change his mind about it.

(Glen, 1982: 355)

Pitman (July 31, 2011 email to author) also recalled that he did not argue with Heirtzler about the interpretation of the *Eltanin* profiles: "He changed his mind before we could talk at it." Pitman, who was still a graduate student working on his Ph.D., did not know the extent of Opdyke and Heirtzler's discussions. Regardless of whether it took a week or a month, to Heirtzler's credit, he had come a long way in the few months since November 1965 when he, Le Pichon, and Baron had submitted their paper on the Reykjanes Ridge (§5.4).

With Heirtzler on board, or at least mostly on board, progress quickened. Pitman had tacked up the profiles outside Opdyke's office, and, before Vine showed up at Lamont in late February or early March and told Opdyke the hot news from the GSA meeting in Kansas City about the discovery of the Jaramillo event by Dalrymple and company (§6.3), Pitman had done three really important things. He had demonstrated the symmetry of the *Eltanin* profiles about the crest of the East Pacific Rise, he had matched anomalies on *Eltanin*-19 with those from profiles over the Juan de Fuca and Reykjanes ridges, and he had figured out the spreading rates for the

Pacific–Antarctic, Juan de Fuca and Reykjanes ridges based on the magnetic profiles and Opdyke's new timescale with its new Emperor event back at least to the base of the Gauss normal epoch (Opdyke, July 31, 2011 email to author).

Next came the April meeting of the AGU in Washington, DC. Heirtzler and Pitman had sent in an innocuous abstract well before they knew what the *Eltanin* profiles would reveal. Opdyke recalled what happened in deciding what they should present at the meeting.

At this time, all of us were preparing for the AGU Spring Meeting, which at that time was traditionally held in Washington D. C., every April. Jim Heirtzler had spent some time agonizing over the results of the *Eltanin* profile and was reluctant to accept its implications, which would, of course, support seafloor spreading and therefore continental drift. The question arose as to how to present the data. Jim originally wished to present the profile and allow the members of the audience to form their own opinions. I recall discussing it with Walter, and I advised him and Heirtzler to bite the bullet. To do otherwise would make us look foolish, since the implication of the data was so clear.

(Opdyke, 1985: 1181)

Heirtzler agreed. "So we did bite the bullet" (Pitman, September 8, 2011 note to author). Opdyke cancelled his scheduled talk on the White Mountains and presented his work on deep-sea cores. When he got to the meeting, he told everyone who would listen to go to Pitman's talk. "It caused a stir, as did the data I presented, but the news was really spread by word of mouth" (Opdyke, 2001: 406).

After the meeting, Opdyke collected *Eltanin* cores in Tallahassee where they had been stored. He returned to Lamont and began processing them. Opdyke recalled what happened:

It was an extraordinarily exciting time. I recall running on the magnetometer the samples from *El*[*tannin*] 13–17 that I had collected in Tallahassee, and the known magnetic stratigraphy was rapidly reproduced to the base [going backwards in time] of the Gauss magnetic chron at 3.5 m.y. The magnetic stratigraphy of the Gilbert magnetic chron was essentially unknown from continental studies, but Pitman's Magic Profile predicted the existence of four events within the Gilbert, which he affectionately called Four Fingers Brown [see Figure 6.5]. When studies on the core progressed to the point that these events should be expected, Walt Pitman joined me in the laboratory, where he prepared the samples for me to measure. As predicted the events appeared one after another. It was one of the most thrilling experiences that I have ever had in science. We were overjoyed and totally convinced that the observations had proved that the magnetic anomalies were accurate records of the reversals of the field.

(Opdyke, 1985: 1182; my bracketed addition)

Pitman described to Wertenbaker what happened.

For a few months Neil's group and ours practically lived in each other's laps. His pattern that he was getting from sediments and our pattern were always the same. By spring, he'd gotten back to the beginning [as seen from going backwards in time] of the Gilbert reversed epoch, three and a half million years. That was as far as Cox and his group had gotten with the lavas.

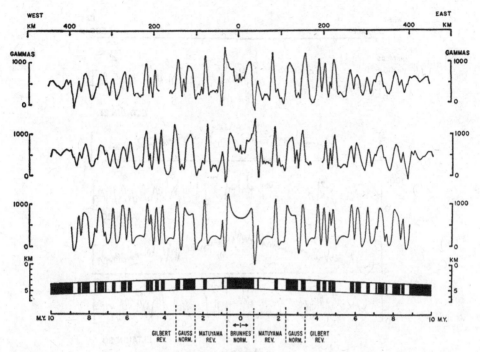

Figure 6.5 Pitman and Heirtzler's Figure 3 (1966: 1166). Their caption reads: "The middle curve is the *Eltanin*-19 magnetic-anomaly profile; east is to the right. The upper anomaly profile is that of *Eltanin*-19 reversed; west is the right. On the bottom is the model for the Pacific-Antarctic Ridge. The time scale (millions of years ago) is related to the distance scale by the spreading rate of 4.5 cm/yr. The previously known magnetic epochs since the Gilbert epoch are noted. The shaded areas are normally magnetized material; unshaded areas, reversely magnetized material. Above the model is the computed anomaly profile." Four Fingers Brown comprises the four positive anomalies within the Gilbert reversed epoch, located approximately 200 km from the ridge axis. They are easily seen on both sides of the computed anomaly profile, on the right side of the top profile, and left side of the middle profile, but are disturbed on the left side of the top profile and right side of the middle profile.

But just a little farther on in our profiles there was a conspicuous group of four high, narrow anomalies [see Figure 6.5] that I had called Four-Fingers Brown after a famous gangster in Newark, near where I grew up. One afternoon in June, Neil and I together ran the samples from an entire core through the spinner, and at the bottom of the core the reversals came booming in, Four-Fingers Brown. That was the first confirmation we had of our sequence. Until then, we'd been showing that our pattern matched someone else's. This time, it was the other way around.

(Wertenbaker, 1974: 205–206; my bracketed addition)

By the time Pitman and Heirtzler submitted their paper in early September to *Science*, they had taken the reversal timescale back approximately ten million years, based on *Eltanin*-19 and extrapolating a spreading rate of 4.5 cm/year of the Pacific–Antarctic ridge. It is time to look at their paper.

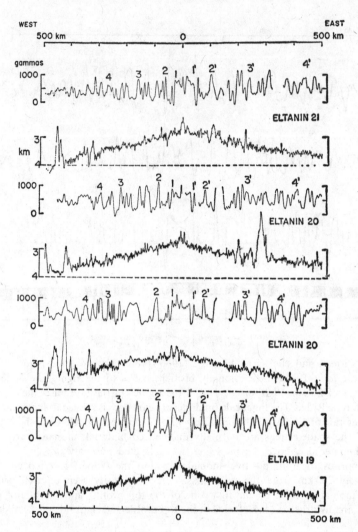

Figure 6.6 Pitman and Heirtzler's Figure 2 (1966: 1165). Observed magnetic profiles and bathymetry across the Pacific–Antarctic Ridge projected along a zenith normal to the ridge axis. Lineation demonstrated by matching anomalies from profile to profile facilitated by numbering of some positive anomalies.

They presented the four *Eltanin* profiles of bathymetry and magnetic anomalies (Figure 6.6), stressing their symmetry and linearity of patterns. "The two most striking features of the magnetic profiles are the linearity of the pattern from profile to profile, and the symmetry of the anomaly pattern about the axis of the ridge" (Pitman and Heirtzler, 1966: 1164). So no one was in doubt about what they meant, they spelt out the linearity by matching numbered anomalies from profile to profile (Figure 6.6) and illustrated the symmetry of *Eltanin*-19, the most symmetrical profile, by comparing it with its mirror image (Figure 6.5). "The most striking instance of

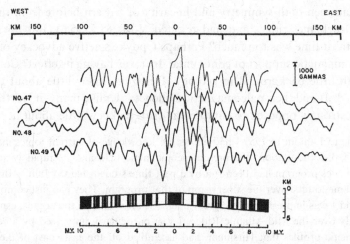

Figure 6.7 Pitman and Heirtzler's Figure 4 (1966: 1166). Their caption reads: "Pacific-Antarctic model applied to Reykjanes Ridge, with a time scale related to distance by a spreading rate of 1 cm/yr." The top profile is computed; the bottom three profiles are abstracted normal to the ridge axis from the aeromagnetic survey of the Reykjanes Ridge (Heirtzler, Le Pichon, and Baron, 1966).

symmetry is illustrated in Fig 3 [Figure 6.5] which shows the *Eltanin*-19 profile both as in Fig 2 [Figure 6.4], with east on the right-hand side, and reversed, with west on the right-hand side; the bilateral symmetry is revealed in the smallest detail" (Pitman and Heirtzler, 1966: 1165; my bracketed additions). They also emphasized what Pitman saw very soon after seeing *Eltanin*-20, namely the similarity between its axial pattern as well as the other *Eltanin* profiles with that of the Juan de Fuca Ridge. Taking Cox, Doell, and Dalrymple's most recent timescale that included the Jaramillo, and Opdyke and company's timescale based on deep-sea sediment cores with the Emperor, and the *Eltanin*-19 profile, they established that the seafloor had been spreading at a constant rate of 4.5 cm/year over the Pacific–Antarctic Ridge since the end of the Gilbert reversed epoch, basically for the last three and a half million years. They also extended the reversal timescale back approximately ten million years, based on *Eltanin*-19 and extrapolation of the 4.5 cm/yr spreading rate over the Pacific–Antarctic Ridge (Figure 6.5). Turning to the Reykjanes Ridge, they applied their model of the Pacific–Antarctic Ridge, and were able to show that it had been spreading at a constant rate of 1 cm/year for the last ten million years (Figure 6.7).

There is a good indication that the spreading rates at the Eltanin-19 profile and at the Reykjanes Ridge have both been constant during the last 10 million years; if not, one would have to assume that the rate changed nearly simultaneously and similarly at both locations. If neither of these hypotheses has been true, it would have been difficult to compare the Pacific-Antarctic-body model with observed data from Reykjanes Ridge. It is more reasonable to assume constant spreading rates.

(Pitman and Heirtzler, 1966: 1166)

Heirtzler had seen both symmetry and linearity of pattern before from the Juan de Fuca and Reykjanes ridges and had rejected V-M (§5.5, §5.6). But not this time! Perhaps a third time was too much? Perhaps Opdyke's active advocacy of the land-based paleomagnetic support of continental drift was having its effect? Coming from physics with little background in geology, Heirtzler "knew little about theories of continental drift and their paleomagnetic evidence." Unlike many in North America, he was not strongly opposed to continental drift, he knew little about it.

When I came to Lamont in 1959 I had almost no knowledge of current topics in geology or geophysics. I did know about the present geomagnetic field and its rapid variations. The marine magnetics program had been run by a part time student named Julius Hirshman and Ewing asked me to take over the supervision of that program. They had just completed Vema Cruise 16 and I was impressed by the great magnetic anomalies in the oceans, especially that high anomaly over the Mid-Atlantic Ridge. I was particularly impressed by a collection of marine magnetic profiles that Hirshman had assembled off the east coast of the US which showed a distinct magnetic anomaly running all along the east coast from Canada to Georgia. Nothing like this had ever been found on the continents and suggested a process with which we were unfamiliar. I soon learned that Scripps people had found impressive linear anomalies running off the Pacific coast. They suggested that these might be due to differences in heat flow, although such had not been measured. I knew little about theories of continental drift and their paleomagnetic evidence. If there had been large linear magnetic anomalies on land I would have given it more attention. I don't know that I ever had any strong feelings against continental drift; it was just a subject that didn't come up. I knew that the paleomagnetic evidence for continental drift was confined to a relatively closed group associated with Runcorn and his camp, Irving and Hospers and some Europeans. Enter Neil Opdyke ... Neil had measured the magnetic properties of continental rocks. (Neil developed a magnetometer at Lamont, with important input from technician John Foster that allowed him to measure the low magnetization of marine sediment cores, which was extremely important later.) A big move for us was the survey of Reykjanes Ridge south of Iceland in 1963, We had no money for this survey but did it by begging time on a Navy aircraft and I, personally, reducing the data on my own time. It was while we were flying that survey that the Vine-Matthews paper appeared in print. Le Pichon and I had looked at a number of magnetic profiles over the Mid-Atlantic Ridge and continental drift did not come into consideration. It was the mirror symmetry of anomalies over the Reykjanes Ridge and the fantastic symmetry along the *Eltanin*-19 profile that showed that continental motion had occurred, at least in those two oceans.

(Heirtzler, August 4, 2011 email to author)

Pitman and Heirtzler now made it abundantly clear that they strongly supported the V-M hypothesis and seafloor spreading.

We feel that these results strongly support the essential features of the Vine and Matthews hypothesis and of ocean-floor spreading as postulated by Dietz (16) and Hess (17). The very rapid apparent-spreading rate in the South Pacific permits one (using a constant spreading rate) to date reversals of the geomagnetic field back to 10 million years ago.

(Pitman and Heirtzler, 1966: 1171; (16) referenced Dietz (1966a); (17) referenced Hess (1962))

They added the anomalies associated with the Gorda Ridge to their list of ridges showing "great similarity to the *Eltanin*-19 profile," and hinted at what they had begun to see from looking at Lamont data from elsewhere.

The bodies of the Pacific-Antarctic, when adjusted for slower spreading rate, fit well with the magnetic anomalies observed over the Reykjanes Ridge in the North Atlantic. Preliminary examination of magnetic data from Gorda Ridge in the northeast Pacific shows (especially on the western side) great similarity to the *Eltanin*-19 profiles over the Pacific-Antarctic Ridge in areas other than those discussed by us. Other magnetic profiles from over the Pacific-Antarctic Ridge, the Indian-Antarctic Ridge (south and southwest of Australia), and the Mid-Atlantic Ridge in the south Atlantic indicate the possible application of this model to those areas.

(Pitman and Heirtzler, 1966: 1171)

Pitman and Heirtzler also considered one of the difficulties that Heirtzler, Le Pichon, and Talwani had raised against V-M, namely why was the central anomaly so much more pronounced than the other anomalies?

To obtain the proper magnitude for the central anomaly, the intensity of the magnetization of the central block has been doubled; essentially the same effect could have been obtained by doubling the thickness of the central block, but this would have presented mechanical difficulties as the central block moved toward the less-thick off-axial positions. Why the central block is thus different from the adjacent blocks is not yet clear. Near the edges of the observed profile the anomalies are smaller than those of the model, although both are of the same form; this could mean that the bodies there are slightly less than 2 km thick, that the magnetization has decreased because of remanent viscous decay, or that chemical alteration has occurred.

(Pitman and Heirtzler, 1966: 1165)

Vine and Cann (§4.3), and Matthews and Bath had already proposed contamination (§5.6). Pitman and Heirtzler said nothing about the other difficulty that Heirtzler and Le Pichon had raised against V-M (§4.6), namely the increase in intensity and wavelength of the flank anomalies. Their new results (Figures 6.5 and 6.6) however show that these difficulties do not apply to the new results from the East Pacific Rise and the segmented ridges of the northeast Pacific, where there is little if any difference between the magnitude of the central anomaly and those immediately to the side. Nor do the flank anomalies show the same features as in the Atlantic. These former difficulties were not enough to stop them from accepting "the essential features of V-M." In essence, they left them as unsolved problems or minor difficulty, not of sufficient moment to prevent acceptance of V-M's "essential features."

What about Talwani and Le Pichon? When did they change their minds? Both accepted seafloor spreading after seeing the *Eltanin* profiles. Opdyke recalled, "Once the *Eltanin* profiles came along in 1966, he [Talwani] changed his mind pretty fast ... [It] was, as far as he was concerned ... definitive, a definite answer" (Opdyke, July 11, 1997 Doel interview; my bracketed addition).[7] Pitman talked only briefly with Talwani about the *Eltanin* profiles and V-M and Pitman is not sure how long it took

Talwani to change his mind, but thinks that Heirtzler did before Talwani (Pitman, July 31, 2011 email to author). Le Pichon did not change his mind until late April 1966, but he did change it, he retrospectively believed, within hours after seeing *Eltanin*-19. He recalled:

On February 13, 1966, I left Lamont for Recife, Brazil, to participate as chief scientist in a South Atlantic cruise that would lead me to Buenos Aires and then to Cape Town. I then joined the faculty at Strasbourg, where I defended my thesis. It was April 26 when I returned to Lamont, where many of my colleagues were now "converted" to sea floor spreading. Walter Pitman showed me the "magic" magnetic anomaly obtained over the South Pacific ridge crest, the *Eltanin*-19 profile that had been presented by Jim Heirtzler at the American Geophysical Union (AGU) meeting in Washington, D. C. on April 27. My wife still remembers that on my way back from the laboratory, I asked her to get me a drink and told her: "The conclusions of my thesis are wrong: Hess is right."

(*Le Pichon, 2001: 211–212*)

Pitman (September 8, 2011 note to author) however retrospectively believes that he showed Le Pichon the *Eltanin*-19 profile "in late 1965, shortly after we processed the data and recognized it."

6.7 Cox and Doell become mobilists

Cox and Doell sat on the fence in their 1960 and 1961 reviews of paleomagnetism and its case for mobilism. I argued that their response was unreasonable and that they should have accepted continental drift or at least given it more credence (II, §8.7). I think they were still sitting on the fence until they saw *Eltanin*-19. Certainly, seeing it profoundly affected them, and their responses imply that they had accepted V-M, seafloor spreading and, necessarily, drifting continents.

   Cox chaired the session at the 1966 AGU at which Pitman presented *Eltanin*-19. He first saw it while having a beer with Heirtzler before the session. Retrospectively, Cox told Glen (1982: 337) that seeing *Eltanin*-19 was a "truly extraordinary experience." Further describing the experience, he continued:

The *Eltanin*-19 profile really has everything on it that we found in all our work on reversals. When it came out, it had things I knew were there and things I thought were probably there, including short polarity intervals slightly older than the Olduvai event. The potassium-argon dates at the beginning of the Olduvai even are more inconsistent than they should be, in view of what we knew of the dating errors. This led me to suspect that one or more short events, slightly older than the Olduvai, were fuzzing up the boundary. *Eltanin*-19 shows a big event, the Olduvai, and then slightly older than that on both flanks of the Rise, two little blips come in – I think they're both real. I said so in an article shortly afterwards [Cox (1969); McDougall and Chamalaun (1966) named it the Gilsá event (§6.4)]. There was so much happening all at once. That was the most exciting year of my life, because in 1966, there was just no question any more that the seafloor idea was right.

(*Glen, 1982: 337–339; my bracketed addition*)

Heirtzler (August 8, 2011 email to author) independently recalled, "I particularly remember meeting with Allan Cox at that meeting and it seemed to be no question in his mind that we had evidence of continental drift."

Around the same time, Doell saw *Eltanin*-19 and some of Opdyke and company's deep-sea core results. Opdyke told Glen what happened.

I saw the stunned look on Dick Doell's face; he was sitting in the lab outside my office that April in 1966 when we gave those talks in Washington [at the AGU meeting]. Doell looked at the magnetic stratigraphy in the cores . . . and at the *Eltanin*-19 profile and said, "It's so good it can't be true, but it is."

(Glen, 1982: 339; my bracketed addition)

So Cox and Doell finally, a half-dozen years after their *GSA* review (II, §8.4), had become mobilists.

6.8 Vine turns the Vine–Matthews hypothesis into a difficulty-free solution

Dear Dr. Vine,

. . . I present my belated compliments on the Vine-Matthews hypothesis. The corroborative evidence which you and others have assembled presents a compelling argument to me and is even shaking the skeptics (I seem to detect weakening in Maurice Ewing). I have cherished a belief in mantle currents sweeping the ocean floors clean ever since my 1939 work (copy enclosed). I had expected that oceanic heat flow measurements would clearly reveal the existence and outlines of such mantle currents. Here I was disappointed, since my analysis of the data encountered the same difficulties as Langseth, Le Pichon and Ewing (JGR, 71, 5321 – (1966)) . . . Now your hypothesis together with Tuzo Wilson's transform fault analysis seems clearly to show the current activity and recent history of such mantle motions.

Your demonstration of the reality and details of ocean floor spreading serves as a spur to all of us who are interested in the mechanism by which these motions come about. Yet I, at least, am way behind you, for I have only recently come to appreciate the full grandeur and power of your hypothesis. I am most anxious to learn your thoughts on the nature and sources of these mantle motions, insofar as you have something ready to release into the public domain . . .

I agree with Teddy Bullard, who on his last visit said to me, "Vine and Matthews' idea is the most important development in earth science in this century" (or words to that effect).

Sincerely,
D. T. Griggs (January 9, 1967 letter to Vine; from Hess's papers at Princeton)

In February 1966, when Vine at Opdyke's invitation visited Lamont, he, unlike those at Lamont, already knew of the discovery of the Jaramillo event by Doell and

Dalrymple (§6.2). Vine recalled what happened when he first walked into the room where Opdyke was drawing the first diagram for Opdyke *et al.* (1966), his first paper on the deep-sea cores (Figure 6.4) and "the *Eltanin* profiles were stacked up on the opposite wall."

When I walked into the room Neil [Opdyke] was poring over a light table and he was drawing up the diagram – literally drawing the diagram – which appeared in his 1966 paper. Although I think the first thing we talked about was that, I distinctly remember that all Walt's profiles, the *Eltanin* profiles, were stacked up on the opposite wall, and so I looked from one to the other. Neil said, "Look, Fred, fantastic, we just discovered a new event. We call it such-and-such event." I said, "Oh, yes, I hate to tell you this Neil, but Cox, Doell and Dalrymple have discovered that event and they have named it and presented it." He was just astounded. And I said, "Yes, Neil it is called the Jaramillo. Moreover, here it is on the *Eltanin*-19 profile." They [Opdyke and Pitman] both looked at *Eltanin* and looked back at me. They said, "My God!"
(Vine, August 1979 note to author; confirmed by Opdyke; see also Vine (2001: 60–61) and Opdyke (1985: 1177, 1181; 2001: 105). Opdyke, however, believes that Vine visited Lamont in March; Pitman told Glen that it was in February (Glen, 1982: 336).)

Surprises all around! Vine learned of Opdyke's independent discovery of the Jaramillo event, saw the *Eltanin* profiles and could pick out the Jaramillo; Opdyke and Pitman, and Heirtzler, who was also there, learned that the Emperor event had already been discovered and named by the Menlo Park group. Heirtzler gave Vine the *Eltanin* profiles and details about the Reykjanes Ridge survey either before he left or shortly thereafter (Pitman, comment to Glen (1982: 336); Vine, 2001: 63). Vine incorporated them into his "review article" that appeared in December 1966 in *Science*.

In the months that followed my memorable meeting with Neil, Walter, and Jim, I prepared my review article and made a further visit to Lamont to give a talk. In May or June, Walter and Jim visited Princeton and were surprised, I think, to discover that my review article was rather wide ranging and essentially complete.

(Vine, 2001: 63)

Vine (2001: 63) recalled, "Jim [Heirtzler] very generously suggested that we should try to arrange to publish simultaneously in *Science*, and I was happy to agree to this." Pitman and Heirtzler's paper actually appeared on December 2, 1966, two weeks before Vine's paper, which "struck [Vine] as entirely reasonable" (Vine, 2001: 63).[8]

Vine's paper was a blockbuster. He presented strong and abundant support for V-M, showed that the hypothesis had become essentially difficulty-free, tied seafloor spreading with continental drift, and speculated about the evolution of the Northeast Pacific. He first noted that the longstanding controversy over continental drift had been "enlivened" by paleomagnetism where he meant land-based paleomagnetism, referring to Bullard's first paper in which he argued for continental drift because of its strong paleomagnetic support, and by marine geology, where he singled out Hess's seafloor spreading idea, which, he claimed, feasibly answered the longstanding mechanism difficulty, accounted for "many features of the ocean basins and continental margins," and was "perhaps first and independently suggested by A. Holmes" (Vine, 1966: 1405, and endnote 1). Introducing V-M, he (1966: 1406) said of its field reversals, the importance of remanent magnetization, and seafloor spreading, its three basic assumptions, that the first two had "recently become more firmly established and widely held," and therefore the establishment of V-M "might provide virtual proof of the third assumption: ocean-floor spreading, and its various implications."

He (1966: 1406) identified the three difficulties that V-M faced when first proposed: no recognized ridge in the northeast Pacific that seemed parallel to the magnetic anomalies, no good example of a mid-ocean ridge with "linear anomalies paralleling" its central anomaly, and the shift from low-amplitude, short-wavelength crestal anomalies to high-amplitude, long-wavelength flank anomalies (§4.2). The first impediment had been removed by the identification of the Juan de Fuca and Gorda ridges; the second, by the pattern of linear anomalies surrounding the Reykjanes Ridge as well as the *Eltanin* profiles over the southern extension of the East Pacific Rise (Pacific–Antarctic Ridge). Vine now had an answer to the third difficulty, the one that Lamont workers had raised (§4.6, §5.3, §5.4) and Vine had previously unsuccessfully attempted to remove (§5.5).

Of the three points originally unexplained by the Vine-Matthews hypothesis, two have been answered; the remaining difficulty concerns the change in character of anomalies as one moves from the axial zone of a ridge to the flanks ... If the Vine and Matthews hypothesis is applicable beyond the central, axial zones of ridges, this change in character may reflect a change in the intensity or frequency, or both, of reversals of Earth's magnetic field. If the frequency of reversals is high, the resulting "blocks" of material of a particular polarity will be narrow; their width will depend on the rate of spreading, but if they are a few kilometers in width they will give rise to a considerably reduced anomaly ... Narrower blocks may well have no obvious individual expression in the magnetic anomaly but will tend to lower the bulk resultant magnetization of the surrounding block. Thus this boundary between the flank and axial-zone anomalies may reflect an increase in the frequency of reversals of Earth's field, together possibly with a decrease in its intensity ... Clearly, if this is the case, the boundary should occur at different distances from ridge axes according to the average rate of spreading in that region. A preliminary investigation of many ridge profiles suggests that this change may have occurred approximately 25 million years ago. Changes in the frequency of reversal seem quite probable when one bears in mind that for the whole of the Permian and part of the Upper Carboniferous the field appears to have been a single polarity (41).

(Vine, 1966: 1414; (41) referenced Irving, 1966)

By this argument he rendered V-M essentially difficulty-free (RS1).

Vine (1966: 1406) identified what he took to be "the two most obvious corollaries of a literal interpretation of the hypothesis: (i) linear magnetic anomalies should parallel or subparallel ridge crests, and (ii) for many latitudes and orientations the anomalies should be symmetric about the axis of the ridge." Selecting surveys over the Reykjanes, Juan de Fuca, and Gorda ridges as providing "convincing confirmation" of the corollaries, and adding the southern extension of the East Pacific Rise (Pacific–Antarctic Ridge), he showed reversed profiles over the Juan de Fuca Ridge and East Pacific Rise to emphasize their incredible symmetry (Figure 6.8).

Using the corrected reversal timescale, he obtained almost constant spreading rates over Juan de Fuca (as opposed to the fluctuating rate he and Wilson had previously calculated without the Jaramillo (§5.2, Figure 6.3)), the Pacific–Antarctic, and the Reykjanes ridges (Figure 6.9). Moreover, he (1966: 1408) found slight deviations from constant spreading rates (imperfect linearity of plots of distance from the ridge axis against time) over the East Pacific Rise and Juan de Fuca, and "rather remarkably, the deviations from linearity for the East Pacific Rise and Juan de Fuca Ridge – 11 000 kilometers apart – are exactly analogous." He did, however, find one discrepancy which led him to suggest that this might mean that the reversal timescale was in need of further, minor revision, and indeed, such plots using more recent reversal timescales imply essentially constant rates of spreading. In particular he proposed that the Mammoth event "may include a short period of normal polarity."

The only discrepancy occurs in the region of the Mammoth event, concerning which there is a suggestion, from the profile resulting from the very fast rate of spreading in the South Pacific (Fig. 7 [Figure 6. 8]), that this event is multiple; that is, it may include a short-period of normal polarity (see Fig. 13 [Figure 6. 11]).

(Vine, 1966: 1408–1409)

Earlier, Vine had written, regretfully, that he and Wilson could have predicted the Jaramillo event from the Juan de Fuca profiles if they had had more confidence that spreading rates were constant. Not about to make the same mistake again; he went ahead and split the Mammoth event on the basis of marine magnetic anomalies. Vine's prediction was essentially correct for McDougall and Chamalaun (1966) had, as already noted, just proposed the Kaena event, another reversed event within the Gauss normal epoch, which occurred very shortly after the Mammoth event (§6.4) – they submitted their paper in October 1966, and it was published eight days after Vine's paper. What we are seeing here is a growing realization that the best record of reversals of the geomagnetic field is to be found not on land, where they were first discovered, but at sea.

He also found constant spreading rates over the Carlsberg Ridge situated in the Northwest Indian Ocean (1.5 cm/year) and the South Atlantic portion of the Mid-Atlantic Ridge (1.5 cm/year) (see Figure 6.8). Moreover, he linked continental drift and seafloor spreading by finding support for Backus's 1964 suggestion (§3.8) that the width of anomalies in the South Atlantic should increase in width

reflecting an increase of spreading, because of the rotation of South America relative to Africa (21) and the resultant southward increase in separation. The increase southward, in the width of the envelope of the central magnetic anomalies indicated by Heirtzler and Le Pichon (15, fig. 3) may well be an expression of this phenomenon.

(Vine, 1966: 1408; (21) refers to Bullard, Everett, and Smith (1965);
(15) refers to Heirtzler and Le Pichon (1965))

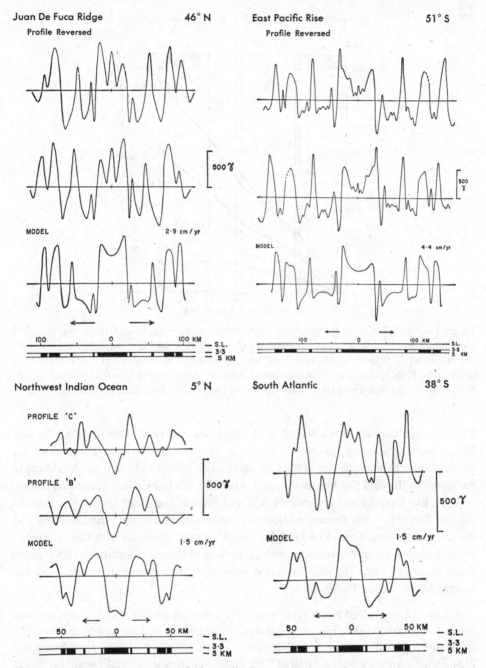

Figure 6.8 Vine's Figures 6–9 (1966: 1409). Reversed observed, observed, and modeled magnetic-lineation profiles over the Juan de Fuca Ridge (from the Mason and Raff survey), East Pacific Rise (*Eltanin*-19), Carlsberg Ridge (Matthews, Vine, and Cann, 1965), and Mid-Atlantic Ridge in the South Atlantic (Heirtzler and Le Pichon, 1965). Simulated profiles based on Cox, Doell, and Dalrymple's timescale with the Jaramillo, Olduvai, and Mammoth events. S.L., sea level. Note discrepancy between the two profiles: only the profile of the East Pacific Rise suggests a short period of normal polarity within the reversed Mammoth event.

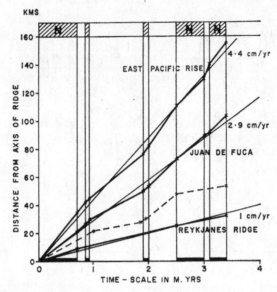

Figure 6.9 Vine's Figure 11 (1966: 1410). "Inferred normal-reverse boundaries within the crust plotted against the reversal time scale based on" Cox, Doell, and Dalrymple's timescale with the Jaramillo, Olduvai, and Mammoth events. "The dashed line represents a similar plot for the Juan de Fuca Ridge, if one assumes the earlier time scale – as did Vine and Wilson" (§5.2). "Note the similar deviations from linearity for the East Pacific Rise and Juan de Fuca Ridge."

Vine also successfully applied V-M to the Red Sea, the very place that Girdler and Peter years earlier had adopted remanence as the source of the anomalies, but appealed to self-reversals not field reversals (§2.4). Vine thought that he should apply his model to the Red Sea because he and others took it to be a new spreading center.

   Vine, like Opdyke and company (§6.5) and Pitman and Heirtzler (§6.6), accented the consilience between the new radiometric reversal timescale with the Jaramillo and the reversal timescale based on Opdyke's work on deep-sea cores with their Emperor event, and their further agreement with the widths of magnetic lineations (RS1) given linear spreading rates. He also correlated anomaly patterns between profiles from the Northeast and South Pacific (RS1).

New data ... have enabled Peter (37) to trace a particular pattern of flank anomalies approximately from north to south from the Aleutian trench to just south of the Murray fracture zone [see Figure 6.11], a distance of 2800 kilometers ... Christoffel and Ross (38), working south of New Zealand, have similarly correlated flank anomalies on adjacent north-south profiles at approximately 173°E ... It is suggested that the two patterns are the same except for difference in the rate of spreading that formed them. The two areas are 11 000 kilometers apart.
(Vine, 1966: 1412; references are equivalent to my Peter (1966b) and Christoffel and Ross, 1965)

He then unearthed another link between seafloor spreading and continental drift: agreement between the rate of seafloor spreading based on the reversal timescale and

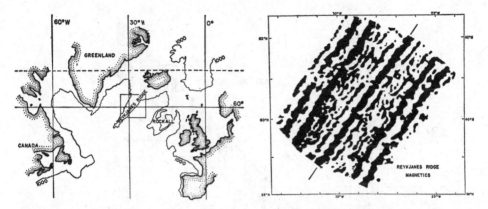

Figure 6.10 Vine's Figures 2 and 3 (1966: 1407). "The location of Reykjanes Ridge, southwest of Iceland" showing the area of the Lamont and US Naval Oceanographic Office aeromagnetic survey. The results are shown in the box on the right; black (white) areas have positive (negative) magnetic anomalies in which the regional field is enhanced (diminished). There is a larger scale anomaly map shown in Figure 5.8.

width of the anomalies extrapolated back to initiation of seafloor spreading along the Reykjanes Ridge (1 cm/year for each half-limb) and the full rate of continental drift (2 cm/year) based on preliminary potassium-argon dates of rocks from northwestern Scotland and eastern Greenland (Figure 6.10) (RS1).

When one applies the concept of continental drift to this region, it seems reasonable to assume that Rockall Bank, southeast of the ridge [Figure 6. 10], is a continental fragment, as was assumed by Bullard, Everett, and Smith (21) in reconstructing the fit of the continents around the Atlantic. In this instance the deep to the southeast of Rockall may represent an initial abortive split; the oceanic area to the northwest, centered on Reykjanes Ridge, a subsequent and more persistent site of spreading of the ocean floor ... This area, therefore, 1200 kilometers in width, may well record a comparatively simple and straightforward example of drifting and spreading. The oldest rocks in the Thulean or Brito-Arctic Tertiary Igneous province occur in northwestern Scotland and eastern Greenland. Preliminary potassium-argon dates from Arran, Mull, and other centers in the British Isles suggest an age of approximately 60 million years (perhaps slightly greater) ... If it is assumed that this igneous activity indicates the initiation of drift in this area, then the implied average rate of spreading from Reykjanes Ridge (half width, 600 km) is approximately 1 centimeter per annum – that is, the rate of "drifting" is approximately 2 centimeter per annum.

(Vine, 1966: 1407)

Vine also took a crack at explaining the evolution of the North Pacific and what had happened to the East Pacific Rise, a topic of speculation at the Ottawa meeting (§5.7). Assuming a constant rate of spreading of the East Pacific Rise of 4.4 cm/year, as determined over the crestal region from the width of the magnetic lineations and the reversal timescale and its agreement with the rate Pitman and Heirtzler determined for the entire length of the *Eltanin*-19 profile, he deduced that the profile showed the

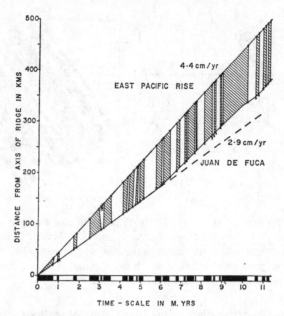

TIME - SCALE IN M. YRS

Figure 6.11 Vine's Figure 13 (1966: 1411). His caption reads: "Magnetic boundaries across the East Pacific Rise deduced onto 500 kilometers from the crest and plotted on a line representing a constant spreading rate of 4.4 centimeters per annum. Similar boundaries from Juan de Fuca Ridge are plotted out to 150 kilometers on the assumption of a constant spreading rate of 2.9 centimeters per annum. The time scale out to 5.5 million years is based on both plots." Note that, based on East Pacific Rise profiles, Vine inserted a short normal interval within the reversed Mammoth event at just over three million years.

reversal timescale for approximately the last 11.5 million years (Figure 6.11). He then matched the magnetic anomalies from the *Eltanin*-19 profile with composite profiles of the Juan de Fuca and Gorda ridges based on the Mason and Raff survey (Figure 6.12), and deduced (1966: 1410) that the rate of spreading "in the Juan de Fuca area may have decreased within the Pliocene from a rate of 4 or 5 to 2.9 centimeters per annum for the last 5.5 million years" (Figure 6.11). Vine then asked what could cause the decrease in spreading rate? Vine found his answer in D. U. Wise's 1963 "An outrageous hypothesis for the tectonic pattern of the North American Cordillera." Wise had proposed that the east–west shift in the pattern of north–sound trending magnetic lineations between the great fracture zones and their change in trend from north–south between the great fracture zones to northeast–southwest by the Gorda and Juan de Fuca ridges reflect "the anomalous width and unique features of the American Cordillera in the western United States." Vine (1966: 1411) speculated:

To a first-order approximation, the more recent geological history and structure of the western United States can be ascribed to the progressive westward drift of the North American continent away from the spreading Atlantic Ridge, and to the fact that the continent has overridden and partially resorbed first the trench system and more recently the crest of the East Pacific Rise.

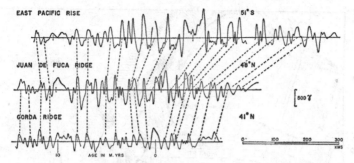

Figure 6.12 Vine's Figure 14 (1966: 1412). His caption reads: "The East Pacific Rise profile *Eltanin*-19 (23) compared with a composite profile across and to the northwest of Juan de Fuca Ridge, and with a profile normal to the strike of the anomalies across and to the west of Gorda Ridge." Reference 23 is to Pitman and Heirtzler (1966).

He (1966: 1411) further proposed, "the northeast Pacific basin represents a flank of the East Pacific Rise, the crest of which has been modified and arrested by the encroachment of North America." The modified section is the ridge system of the Explorer, Juan de Fuca, and Gorda ridges, connected by the Sonvanco and Blanco transform faults. Modification included a change in direction of spreading during the Pliocene from west–east to northwest–southeast as reflected in the changing trend of the magnetic anomalies.

The former east-west direction of spreading from the East Pacific Rise, reflected in the north-south magnetic anomalies of the northeast Pacific, has apparently been replaced within the Pliocene so that the present direction of motion is northwest-southeast, paralleling the San Andreas faults, as Wilson (30) has suggested. In the area off Washington and Oregon and to the north of the Mendocino Fracture Zone, this change in direction has been accommodated by faulting and a gradual stifling and reorientation of the ridge crest to form Juan de Fuca and Gorda ridges. This stifling is illustrated by profiles in Fig. 14 [Figure 6.12], but is most graphically shown by coloring the anomaly bands of [Figure 5.5, Volume III]. [Note shift in trend from older north–south trending bands, especially on northwest flank of Juan de Fuca Ridge, to younger northeast–southwest trending anomaly bands.]

(Vine, 1966: 1411; Wilson (30) is my Wilson (1965b); my bracketed additions)

Continuing southward, Vine agreed with Wilson (1965c) that *current* crustal spreading is along the San Andreas Fault, itself a transform fault. He then speculated about *former* crustal spreading from the East Pacific Rise before it was overridden by North America. He even attempted to reconstruct the ridge.

In this area, between the Mendocino and Murray fracture zones, the former ridge crest has presumably been overridden and damped out, perhaps not without attempts at modification as suggested by the northeasterly trending anomalies, near the continental margin, in the magnetic survey of this area ... However, it is interesting to reconstruct the ridge crest as it would be had it not been overridden and modified. If one calculates the position of the ridge crest

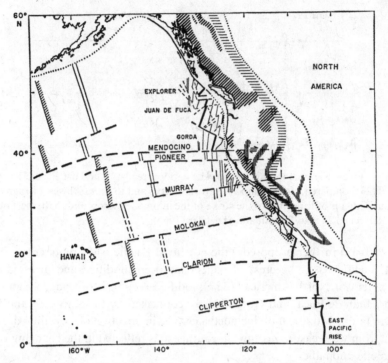

Figure 6.13 Vine's Figure 15 (1966: 1412). His caption reads: "The East Pacific Rise in the North Pacific. Solid black lines indicate the present crest and active transform faults [the crest south of the Gulf of California, from Menard (34).] Thin lines represent key magnetic anomalies: in particular, the pair of anomalies traced by Peter (37) between 160° and 140° W. The half-herringbone pattern to the west suggests a possible boundary of the rise and associated north-south magnetic anomalies. Broken lines indicate inactive faults or fractures. Dotted lines enclose the circum-Pacific cordillera within which the tectonic belts of Wise (28) are shown by ruled shading." Reference (34) refers to Menard (1964); (37) to Peter (1966b); and (28) to Wise (1963).

north of the Mendocino (had it not been stifled) and assumes the offsets measured further west on the Mendocino and Pioneer fractures (12), the reconstructed ridge crest lies beneath Utah and Arizona – the areas of the Colorado Plateau uplift (see Fig. 15 [Figure 6.13]).

(Vine, 1966: 1411–1412; (12) refers to Vacquier, 1965; my bracketed addition)

Thus Vine correctly related the north–south magnetic anomalies of the northeast Pacific with the overridden East Pacific Rise and viewed at least the Mendocino Fracture Zone as a fossilized ridge-ridge transform fault. I believe he was the first to view the great fracture zones as fossilized transform faults. Tying his account to Wise's 1963 speculation, he went one step too far, suggesting that "quasi-transform faults may have existed along the continental extension of the Mendocino and Pioneer fracture zones, producing the right lateral offset of the various tectonic belts indicated by Wise (28) (see Fig. 15 [Figure 6.13])" (Vine, 1966: 1412; my bracketed addition). Vine (July 19, 2011 email to author) recently recalled that his suggestion of "quasi-transform faults" and oceanic crust "being subducted" but "retaining its integrity" was "rubbish." Looking ahead, he also noted that explaining the evolution of the Northeast Pacific could not be done until McKenzie and Morgan presented their account of triple-junctions (§7.21) and McKenzie showed that ridges "do not have to be associated with convective upwelling" (§7.5).

Yes, the interpretation of the Mendocino fault as a formerly active right-lateral transform fault is correct, but with the benefit of hindsight, and the formulation of plate tectonics and the evolution of triple junctions in particular, the statement that "earlier ... quasi-transform faults may have existed along the continental extension of the Mendocino and Pioneer fracture zones ..." is rubbish. Yes, at the time I think I did envisage the oceanic crust being "subducted" beneath western North America and retaining its integrity for a while before being thermally resorbed into the mantle. The other conceptual block at the time was the fact that, because the Holmes–Hess model associated ridge crests with convective upwellings in the mantle, I thought that the upwelling would continue to impact on western North America after "subduction." However Dan [McKenzie] elegantly showed that ridges are passive rather than active features where plates separate and do not have to be associated with convective upwellings.

(Vine, July 19, 2011 email to author; my bracketed addition)

Vine also attempted to determine when the East Pacific Rise first formed. Constructing an east–west profile immediately north of the Mendocino Fracture Zone running from the crest of the Gorda Ridge to the surveyed boundary of the north–south magnetic lineations at 168° W, and using a spreading rate of 4.5 cm/year, he calculated

that East Pacific Rise, at least in the north and south, and perhaps throughout the length of the Pacific, was initiated in the late Cretaceous, possibly at the time of the extinction and beginning of subsidence of the Darwin Rise (2).

(Vine, 1966: 1413; (2) refers to Hess (1962))

Vine's estimate was not far off. He later explained why he had slightly underestimated the ridge's age, and acknowledged that he probably had been influenced by Hess and Menard, who both argued in favour of an earlier Darwin Rise, now regarded as a phantom.

In the South Pacific the East Pacific Rise (EPR) or Pacific-Antarctic Ridge was initiated between the New Zealand (i.e. Campbell) Plateau and Antarctica in the Late Cretaceous and records the complete sequence of Late Cretaceous/Cenozoic anomalies/reversals which follow the Cretaceous Quiet Zone (i.e. period of no reversals). (However, in 1966 the Cretaceous Quiet Zone had not been recognized.) The same/complete sequence of Late Cretaceous/ Cenozoic anomalies occurs in the northeast Pacific with the "Quiet Zone" to the west of them. Thus the ages I assigned to these anomalies in Figure 17 of the 1966 article are essentially correct. Because of the situation in the South Pacific I suggested that the EPR throughout the Pacific may have been initiated in the Late Cretaceous. I think I may also have been influenced by the idea that the crust to the west of the anomalies in the north Pacific may have been formed about the "Darwin Rise" (a pet concept of Harry Hess and Bill Menard at the time). However we now know that the Darwin Rise concept is incorrect and that the EPR in the North Pacific was initiated earlier than in the south, i.e. in the Jurassic.

(Vine, August 18, 2011 email to author)

Not finished, Vine (1966: 1414) proposed that V-M "may provide the best criterion for distinguishing between active and inactive ridges. Ridges that have actively spread during the last 1 million years should be characterized by a central magnetic anomaly of appropriate sign and shape." He (1966: 1414) cited two definite examples out of "approximately 100 available crossings of the worldwide ridge system:" a ridge in the Labrador Sea and another to the southeast of South Africa. Wilson (§3.2) had argued that such a ridge had existed in the Labrador Sea to explain the separation of Ellesmere Island and Greenland, and Godby and company had found magnetic lineations (§4.4), but Drake had argued that the area was currently seismically inactive (§4.4). So Vine basically removed a counterexample (RS1). The other instance, profiled by Heirtzler (1961) aboard *Vema* cruise 16, was over the southwest Indian Ridge. Vine (1966: 1415) was also pleased to remove both from the list of actively spreading ridges because they did not fit his (§5.5) and Hess's (§4.8) "'tennis ball' pattern of convection within the upper mantle."

In all, Vine had done a masterful job showing that the positive support of V-M was very strong, and that it faced no more outstanding difficulties. He also sprinkled his "virtual proof" with speculation about the evolution of the Northeast Pacific and the "tennis ball" pattern of mantle convection, but they could be taken with a grain of salt because the positive evidence was so strong and conceptually independent of them. Unless a new finding led to a serious difficulty, rejection of V-M was no longer reasonable.

6.9 Sykes confirms ridge-ridge transform faults

During the early and middle 1960s, three Lamont seismologists, Jack Oliver and his two terrific graduate students, Bryan Isacks and Lynn Sykes, played significant roles in furthering mobilism's support. In 1966, Sykes confirmed the existence of ridge-ridge transform faults; in 1967, Oliver and Isacks found strong support for subduction of rigid blocks of seafloor and upper mantle (lithosphere) beneath trenches (§7.4), and in 1968 all three co-authored a paper relating the seismological studies with the new global tectonics suggested by continental drift, transform faults, seafloor spreading, and the just-introduced theory of plate tectonics (§7.18).

Sykes was born in 1937 in Pittsburgh, Pennsylvania; two years later his family moved to Virginia, just outside of Washington, DC. In 1955, with the help of a Procter and Gamble scholarship, he began studying at MIT.[9] Participating in a joint-degree five-year program, he earned, in 1960, B.S. and M.S. degrees, studying geology with an emphasis on geophysics. In 1959 he was a Summer Research Fellow at WHOI, participated on two cruises, and wrote his master's thesis under Brackett

Hersey. What Sykes learned most of all at WHOI was that marine geophysics and WHOI were not for him. He wanted to specialize in seismology, and after an encouraging Saturday afternoon meeting with Oliver on a visit to Lamont, decided to study there under him for his Ph.D. Oliver was equally impressed with Sykes.

When Lynn Sykes arrived at Lamont ... he had no deficiencies ... He arrived, in every respect, ready to go into meaningful research. In classes, he quickly demonstrated the strength of his training and his outstanding ability and competence. He also demonstrated strong personal attributes as well. He had those intangible qualities that quickly made him outstanding as a researcher. In particular, he had a decided knack for finding and exploiting opportunities in science overlooked by others. Such a knack, cultivated or inherited, is a critical asset of outstanding scientists.

(Oliver, 1996: 53)

What Sykes also had, like most North American geology and geophysics students, was a low opinion of continental drift. He had not read Wegener, du Toit, or Holmes on continental drift, but only "derivative things."

I had read ... derivative things that were in textbooks about continental drift and pole fleeing forces. I can remember that one of the classical things ... I was told as a student at MIT as an undergraduate in the 1950s ... that continental drift is impossible. The continents can't plow through the strong rocks of the oceanic crust and that bright and serious young scientists shouldn't work on such a stupid idea.

(Sykes, January 4, 1997 Ron Doel interview)

He knew little about mobilism's paleomagnetic support, even after he had been at Lamont, and, I suspect, did not read any of the key papers. He certainly did not have a deep understanding of the subject. I also suspect from what he said below that he did not read Cox and Doell's 1960 or 1961 reviews.

And I did not know that in the late 1950s that Ted [Edward] Irving had left England, his thesis was turned down, and that he went to Australia to work on paleomagnetism of rocks from Australia. He was able to clearly show that the polar wander path [of Australia] differed extremely from that of the northern continents. He published it in an obscure Italian journal. So I was not aware of it until, probably the late 1960s ... I was aware of some of Runcorn's data showing a difference in polar wander paths for Europe and North America. At the time I wasn't very convinced because most of the opening had been one of longitude and not of latitude which is what paleomagnetism senses. I now know in retrospect, in fact, that his curves went back far enough in time and the position of the poles for those two continents had moved enough and there was enough of a difference in latitude. It was something that I was skeptical about, including when I heard it at Runcorn's Newcastle meeting.

(Sykes, January 4, 1997 Doel interview; my bracketed additions)

He thought the existence of field reversals "contentious," and did not know much about arguments in their favor. He did not know of Hospers' work (II, §2.8).

I think it [field reversals] was something that I felt was contentious. It was not something that I knew that much about. And clearly in North America there was this impression that there might be self-reversals, even though I now know in retrospect there's only one example that was ever found of a self-reversal in one rock. So I think I was just of the standard North American predisposition to be skeptical.

                    (Sykes, January 4, 1997 Doel interview; my bracketed additions)

So Sykes, like most marine geologists and geophysicists in North America, was unfamiliar with paleomagnetism's defense of continental drift, earlier work on field reversals by Hospers and others, and had not read key papers in paleomagnetism.

At Oliver's suggestion, Sykes for his Ph.D. thesis studied high-frequency surface waves generated by shallow earthquakes. Oliver and he thought that such waves were associated with the low rigidity of unconsolidated sediments on the ocean floor. To do the work, Sykes had to get a good fix on the location of the earthquakes, and solving this problem turned out to be essential for his later, more important work in locating, with much more accuracy than before, first earthquake epicenters and then both epi- and hypocenters. He benefited from having available reliable data from the USCGS, the setting up of first-class seismic stations in Antarctica during IGY (1957–8), improvements in echo sounders, and from an increase in the number of scientific cruises (Sykes, 1963). He also benefited from a computer program that Bruce Bolt of the University of Sydney, Australia, had written at Lamont just before Sykes arrived.

One thing that did come out of my thesis just by serendipity was that, in order to get good measurements of the dispersion, or the speed with which these various seismic waves propagate at certain frequencies across the oceans, it was necessary for me to have earthquakes that were in the oceans and for them to be well located. A person that had done a post-doc at Lamont, either the first or, I guess it was he was just finishing up when I arrived at Lamont, Bruce Bolt (1960), had written a program for computer relocation of earthquakes.[10]

                              (Sykes, January 4, 1997 Doel interview)

Oliver's characterization of Sykes as having "a decided knack for finding and exploiting opportunities in science overlooked by others" was quickly evident. Sykes knew immediately how to take advantage of the situation. Mark Landisman, a recent Lamont Ph.D., and he improved Bolt's program, and could process earthquake data more quickly. Sykes soon realized that former positioning of earthquakes had sometimes been off by hundreds of kilometers (Sykes, January 4, 1997 Doel interview). In 1963, Sykes determined the location of mostly shallow earthquakes in the South Pacific and discovered that seismic activity along fracture zones was restricted to segments between ridge-offsets. It was to this seminal work that Wilson appealed in 1965 as the basis for his hypothesis of ridge-ridge transform faults (§4.6).

I now turn to Sykes' work on the seismicity and deep structure of island arcs, in particular his work on locating the hypocenters of earthquakes associated with the Tonga/Fiji region, where Oliver and Isacks, found, in 1967, strong support for subduction of 100 km thick slabs of lithosphere. Work on the region arose from a seminar Oliver offered, which both Sykes and Isacks took.

[While] I was a graduate student, Jack [E.] Oliver decided to run a seismology seminar for credit on deep earthquakes. He thought that the subject had really not been explored much in about twenty years. At that time, we were able to read twenty or twenty-five papers that were the totality of papers written on deep earthquakes. It did become evident that the Tonga/Fiji region had more deep earthquakes that occurred per year than any other region. South America had some of the largest ones, but it didn't have as many moderate size earthquakes ... Brian Isacks and I were two of the members. I think it was a fairly small seminar. I can't remember who else was in there. But I think at the end of the seminar, Jack Oliver had decided to try and mount a field program to study deep earthquakes. Brian Isacks and I were finishing up our theses about the same time. Brian was quite interested in doing a field project on deep earthquakes. Jack Oliver felt that picking Tonga/Fiji was a better area, both because moderate size earthquakes were more numerous, it was an area that was reasonably easy of access, and it didn't involve crossing as many international boundaries as in South America. Logistics would be easier. And in South America you didn't have as frequent moderate size deep earthquakes. Fiji and Tonga are malaria free. It was an area in which you didn't have to go prepared to deal with tropical diseases. So the decision [was made] – Jack Oliver then wrote a proposal to the National Science Foundation, and Brian was involved in that – of doing a field study in that area.

(Sykes, May 14, 1997 Doel interview; first bracketed addition is mine)

Isacks and Oliver took the lead on the project, and Isacks went to Fiji with his wife and three small children at the beginning of summer 1964. He set up seismological stations on Fiji and Tonga.

Meanwhile Sykes began calculating hypocenters of over 1500 earthquakes in the Arctic, Alaskan, East African, Tonga–Fiji, Kermadec, Kuril–Kamchatka, and Caribbean regions, presenting results in a series of single and co-authored papers (Sykes, 1964, 1965, 1966; Sykes and Landisman, 1964; Sykes and Ewing, 1965; Tobin and Sykes, 1968). Sykes and Landisman checked the accuracy of their new program by computing the position of the nuclear explosion Longshot in the Aleutian Islands (Sykes and Landisman, 1964). With the added help of more than thirty new seismological stations in the southwest Pacific, Sykes determined the hypocenters of over 850 earthquakes in the Tonga–Fiji and Kermadec regions. He found that they lined up on a plane dipping approximately 45° beneath the inner (landward-side) margin of the trenches. The dipping plane often extended 650 km vertically and 500 km horizontally (Sykes, 1966). He also found that the entire tectonic ensemble of the trench, deep and shallow earthquakes, and volcanoes fish-hooked toward the west at the northern end of the Tonga Trench (Figure 6.14).

While Sykes was making progress on his new project, he spent July to October in 1965 helping Isacks install two more seismological stations (Sykes, email to author, July 30, 2011; Sykes checked his old passport to confirm dates). It was in Fiji that Sykes learned that Wilson had appealed to his work as major evidence for his idea of transform faults. Still viewing mobilism with disfavor, he wondered if he could use first motion studies of earthquake mechanisms to show that Wilson "was wrong."

But one of the very important things that happened while I was in Fiji was that I received a letter from Jim [Henry James] Dorman – who was on the Ph.D. staff at Lamont – telling me

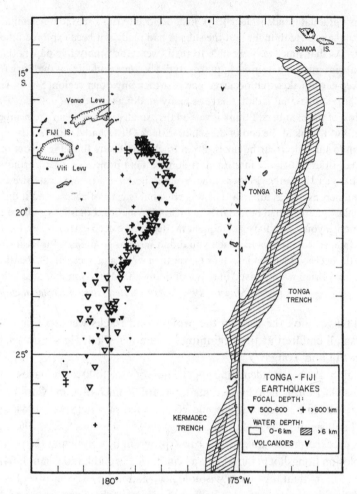

Figure 6.14 Sykes' Figure 3 (1966: 2985). His caption reads: "Deep focus earthquakes in the Tonga-Fiji region. Focal depths greater than 500 km." The severe bend at the northern end of the Tonga Trench is echoed by the deep earthquakes and volcanoes. The shallow earthquakes, which were plotted on a different figure, do the same.

about this paper that had been published by Tuzo [J. Tuzo] Wilson – his famous [July 14] 1965 transform fault paper. And that Wilson had utilized my locations of the zigzag patterns of earthquakes. So the germ of the idea was there for me, but ... maybe I [could] do something more to try and prove or disprove Wilson. And I think probably at that time my sense was that [Wilson's idea] sounded like a rather woolly idea ... and that ... [by] using the mechanisms of earthquakes, it might be possible to show that he was wrong.

(Sykes, May 14, 1997 Doell interview; last four bracketed additions are mine)

On his return journey to Lamont, Sykes stopped in Kansas City to attend the November 1965 meeting of the GSA. Needing to finish his present work on locating earthquake hypocenters in the Tonga–Fiji, Kermadec, and New Hebrides regions before he could begin testing Wilson's ridge-ridge transform faults, he finished his manuscript by mid-December, revised it and resubmitted in mid-March, 1966. His results gave qualitative support to the idea that seafloor was being carried downward back into the mantle, as envisioned by Fisher and Hess (1963). He was, however, bothered by the pronounced hook (Figure 6.14) as well as by various details (RS2). Knowing of Orowan's work at MIT, he referred only to Orowan's pro-drift views, not mentioning Hess (1962), Fisher and Hess (1963), or any of Dietz's papers on seafloor spreading.

Many theories of mantle convection currents attempt to relate the spatial distribution of earthquakes to the configuration of the convection cells. If this correlation is assumed, several facets of the spatial distribution of earthquakes should be considered. These include the abrupt curvature of the earthquake belts in some regions and their marked linearity in others, the constancy of the dip of the focal surface for depths of about 0 to 650 km, changes in the depth of maximum deep activity from about 625 km to about 475 km at the southern end of the Tonga arc, the abrupt decrease in seismic activity at that depth of about 650 km, and the existence of a discontinuity in seismic activity between the Tonga and Kermadec arcs. The concentration of seismic activity in zones less than 50 to 100 km thick also is a factor that must be considered in theories of large-scale deformation of the earth. Orowan [1964] concluded that the instability of creep appears to be the fundamental cause of the concentration of deform-ation in the circum-Pacific zone. The existence of focal surfaces of nearly opposite dip in the Tonga and New Hebrides arcs also seriously constrains hypotheses on tectonics.

(Sykes, 1966: 3004)

Sykes was invited to present his findings on the worldwide distribution of deep and shallow earthquakes at a conference organized by the School of Physics at the University of Newcastle upon Tyne. The conference, whose general theme concerned mantles of Earth and terrestrial planets, was held from March 30 to April 7, 1965. Other participants whom we have already or soon will encounter included Coode, Creer, Girdler, Hospers, Jeffreys, Runcorn, and McKenzie. Creer, Hospers, and Runcorn presented papers summarizing the paleomagnetic support for mobilism, and Creer and Runcorn also discussed the movement of Australia and India relative to each other and to other landmasses. Sykes spoke in the same session as Creer and Hospers. He toed the Lamont line defended by Ewing that continental drift was probably not correct. It was in Newcastle that Harold Urey told him that he needed "to take continental drift more seriously."

And so I merely remarked that there was this major structure that extended all the way down to six hundred kilometers, and that ideas of earth mobility would have to take that into account. I was certainly of the standard Lamont opinion that [W. Maurice] Ewing held that probably a lot of the current ideas of earth mobility weren't right, like continental drift, but that something else would emerge from that ... Harold [C.] Urey, Nobel Prize winner in chemistry ... was there at the meeting. He said to me, young man, you need to take continental drift more seriously.

(Sykes, May 24, 1997 Doel interview)

Sykes added during the above interview that Urey's remark "was probably not one of the things that I wanted to hear, but it turned out to be good that I heard it ... It certainly made an impression on me. It did not 'convert' me right away."

Sykes, still a fixist, next saw Pitman's magic profile in June. Heirtzler asked Oliver to come over to see the profile, Oliver asked Sykes and they walked over to Heirtzler's office, and Heirtzler and Pitman showed them the profile. Here is what Sykes told Wertenbaker.

It was beautiful. They'd made transparencies of a thousand-mile length of the *Eltanin*-19 profile. By taking two of them, turning one around and putting it over the other, you saw vividly the symmetry of the profile. Some other people's work had been suggestive, but it was the Pitman profile that really made people believe Vine and Matthews, which hit you that this really was so ... At the time I had been trying to think of ways to test Wilson's idea of transform ... faults, but hadn't got started ... I saw the profile in June. The next morning I went to work. Essentially I knew what I had inside a week. I looked at twenty earthquakes and they all worked ... I was a little hesitant about presenting my results. It was a matter of trying them out on people first; and the more I tried, the better things sounded. Sea-floor spreading was the only thing that could produce those earthquakes.[11]

(Wertenbaker, 1974: 209–210)

Once he saw *Eltanin*-19, Sykes accepted V-M, or at least thought it was probably correct, and he began to work on fracture zone earthquakes. This was the moment that his splendid training and solid work had prepared him for. Within a week, based on first motion studies of earthquakes along fracture zones between ridge-offsets, he began to find support for ridge-ridge transform faults. He showed through fault plane solutions that the sense of motion of the earthquakes on fracture zones between ridge-offsets was consistent with the idea of ridge-ridge transform faults and inconsistent with the motions inferred by matching features across them – transcurrent faulting. This important distinction is made clear in Figures 4.8–4.10. Fault plane (focal mechanism) solutions provide the orientation of the plane along which the motion caused by the earthquake occurs (fault plane) and the direction of that motion (slip vector). The strike (φ) of the fault plane is the intersection of the fault plane with the horizontal plane, and its dip (δ) is the steepest angle of descent relative to the horizontal. There are three types of faults: strike-slip, normal, and thrust. With strike-slip faults (transform, transcurrent, wrench, lateral faults), δ is near 90°; with normal faults (dip-slip, normal slip, tensional faults) $45° < \delta < 90°$; with thrust faults (dip-slip, reverse-slip, compressional fault) $0° < \delta < 45°$. Fault plane solutions are obtained by studying first motions of the first P-wave from an earthquake at seismological stations surrounding the position of the earthquake. If the recorded P-wave is compressional (positive, up) it means the direction of motion was toward the station; if it is dilational (negative, down), the direction of motion was away from the station.

In the 1950s and early 1960s, J. H. Hodgson of the Dominion Observatory, Ottawa, Ontario, attempted to make first motions studies, but inadequate distribution of stations and unreliable reports of first motions gave inconsistent results – a few seismographs were even improperly connected and the polarity of P-waves and compressional waves were recorded as negative; dilatational, as positive (Sykes, August 2, 2011 email to author). Sykes used data obtained from the World-Wide Standardized Seismography Network (WWSSN) operated by the USCGS. Data from a few seismological stations became available in 1962 and by the mid-1960s there were sufficient stations collecting reliable data for Sykes to tackle the task. He also knew that "film clips of the seismograms" from the stations were available to him right there at Lamont. But there was another problem: Hodgson misidentified thrust faults as strike-slip (transcurrent) faults. At the time, nobody realized his mistake (Sykes, August 1, 2011 email to author).[12] Sykes avoided the problem by plotting first motion data in a certain way. He imagined a small sphere centered on the earthquake's focus (focal sphere), and plotted, on an equal-area projection of the lower hemisphere of the focal sphere, the directions of the S-waves. Sykes recounted how he obtained reliable fault plane solutions for earthquakes on the Mid-Atlantic Ridge, and what he did to get reliable solutions.

Here was all of this data that we were getting from the World Wide Standard Seismic Network that were right there at Lamont, and which you could go down and read the film clips of the seismograms and determine the first motions. You could get them better from the long period instruments for the moderate to large earthquakes. This was the first time, in fact, that this network presented the opportunity of looking at original seismograms, a whole bunch of them, in a relatively short amount of time. Before that, people might have taken years to collect the records, just from one earthquake and then to study it. [John H.] Hodgson in Canada had a project from the 1950s, in which he would just send out questionnaires to seismic observatories, [asking] please tell me for the following earthquake whether the P-wave motion is up or down. He always had something like twenty percent inconsistencies, which turned out to be pretty fatal for focal mechanism studies. He didn't realize that … the problem comes that particularly in an island arc with thrust earthquakes as in Tonga that there is one of the nodal planes where you get the transition from upward to downward motion that is quite well defined. The other plane is nearly horizontal, and with stations at a distance, you usually don't have good control on it. So if you have a whole lot of stations at a distance that are largely sampling a cone oriented straight downward [at the source], plus or minus thirty degrees, you got one nodal plane through them. If you have twenty percent inconsistencies, you will try and draw another plane through there which also will be a steep plane. Hence, you will come out with a strike slip mechanism, even though in reality, it's a thrust earthquake. So he was misled. I realized this pretty soon with my work. So I sat down with some earthquakes from the mid-ocean ridges, particularly from the mid-Atlantic ridge, and right away started going through the records for a couple of them. I thought about how to portray the data. One of the things that came back to me from taking structural geology from Bill [William] Brace at MIT was a standard method, was an equal-area projection. Use of it does not bias data that leave in various directions from a small sphere surrounding an earthquake.

(Sykes, May 24, 1997 Doell interview)

When Sykes began, he soon learned that William V. Stauder (1922–2002) and some of his colleagues at St. Louis University had already started making fault plane solutions using the same projection method, but had not yet begun to test the idea of transform faults (Stauder and Bollinger, 1964).[13]

I had worked out a number of mechanisms in 1966, before I realized they were using the same projection in the 1966 paper [Stauder and Bollinger, 1966a]. Their project was mainly aimed at the use of S waves to ascertain if earthquake mechanisms were of the single or double couple type. They did not concentrate much on tectonic interpretations until I sent Stauder a preprint of what was later published in 1967 (Sykes, August 1, 2011 email to author).

They had not made the connection between Wilson's work on transform faults and first motion studies. They, as Sykes later pointed out (Sykes, August 1, 2011 email to author), "did not have the interaction with colleagues in marine studies that I had." Indeed, Stauder and Bollinger said nothing about transform faults and tectonics, and did not relate their fault plane mechanisms to tectonics. In their 1966 paper, which they submitted in June 1966, they noted three earthquakes that Sykes also analyzed:

Three earthquakes (May 19, August 3, November 17) occurred along the mid-Atlantic ridge ... The mechanisms of the three shocks are similar, and in all three cases the tension axis is nearly horizontal and normal to the local trend of the mid-oceanic ridge.

(Stauder and Bollinger, 1966a: 1371)

If the tension axis is "nearly horizontal and normal to the local trend of the mid-oceanic ridge," then it would appear that they were describing normal (tensional) faults in the ridge itself. But they classified all three faults as strike-slip, and as Sykes retrospectively emphasized, "For strike-slip earthquakes along transform faults the tensional axis is horizontal but is not normal to the local trend of the mid-ocean ridge along ridge segments" (Sykes, August 4, 2011 email to author). The tension axis for strike-slip faults is horizontal, bisects the compressional quadrants, oriented 45° to the fault plane. Thus Sykes noted in his paper:

Stauder and Bollinger [1966a] concluded that the inferred axes of maximum tension were approximately normal to the trend of the ridge. They apparently failed to realize, however, that in each of the solutions dominated by strike-slip motion the epicenter was located on a fracture zone and one of the two nodal planes for *P*-waves nearly coincides with the strike of the fracture zone. Stauder and Bollinger's analysis indicated that all their solutions were of the double-couple type; hence, from *P*- and *S*-wave data alone it would not be possible to choose which of the two nodal planes was the fault plane.

(Sykes, 1966b: 2133)

As I shall explain below, the ambiguity in determining which nodal plane (see Figure 6.18) is the active fault plane can be settled by observing the trends of the relevant geological features (ridges and fracture zones). Stauder and Bollinger deserve much credit for having used the very same projection method as Sykes, but were not interested in applying their improved solutions to tectonics.

Sykes worked quickly. "Within a month or less," he knew that he "really had something." He had results from the Atlantic and East Africa, "which clearly agreed with the direction of motion that Wilson had proposed."

I found that in a matter of two or three days I could do an earthquake. At the same time, I pretty quickly wrote this computer program. So probably within about three months, I had done about a dozen earthquakes – several from the Atlantic, some from East Africa. They are most of the earthquakes that are in my 1967 paper, Mechanism of Earthquakes and the Nature of Faulting on the Mid-Oceanic Ridges. So, certainly within a month or less, I knew that I really had something. When I plotted up the first couple of strike slip earthquakes, they clearly agreed with the direction of motion that Wilson had proposed for transformed faulting. They were exactly opposite the direction as if the seafloor across the fracture zone had originally been together and later offset.

(Sykes, May 24, 1997 Doell interview)

Within six months, Sykes had gone from thinking that Wilson's transform faults "sounded like a rather woolly idea" to showing that Wilson "was right." In April, Urey told him he should take drift more seriously; in June, he saw *Eltanin*-19, which further spurred him on to get cracking, and by about late August, he had shown, despite his previous skepticism, that Wilson "was right."

So very quickly I knew that I had something really big, and that I was not going to prove Wilson wrong. I was going to prove him right. [Laughter] I turned out to be, in fact, a good person for doing this. I was skeptical about his proposal, but I did a test that showed that he was right.

(Sykes, May 24, 1997 Doell interview)

Sykes discussed his progress with Isacks, once he returned from Fiji. Oliver also was encouraging.

I certainly also was having talks with Bryan Isacks. I forget exactly when Bryan came back from Fiji, but it was probably early in that time interval. I had already started working on the mechanisms of mid-ocean ridges. But I was certainly talking to him when he returned from Fiji, pretty early on. So he was the main person I talked to. Jack Oliver was generally encouraging. And so, let's see what else happened there. The other thing was that I also wrote a computer program for plotting out all hundred and twenty-five of the world wide plus Canadian stations. I could into the computer enter the coordinates and depth of the earthquake, and the program then would plot out all the stations on an equal-area projection.

(Sykes, May 24, 1997 Doell interview)

Pitman, Opdyke and others in marine geophysics at Lamont began to hear about Sykes' work and he was invited by Lamont geochemist Paul Gast to present his results at the upcoming November 11–12 Goddard conference.

Some of my colleagues at Lamont started to hear about it in the summer of '66 – Pitman and Opdyke, and a lot of the other people in marine geophysics. I was then invited by Paul Gast, who was a geochemist, who heard about it and asked me to give a talk at an upcoming invitational meeting that was to be held at the Goddard Space Flight Center in New York.

(Sykes, May 24, 1997 Doell interview)

Although he did not have a completed manuscript, and given the short notice did not have time to finish one, he accepted, and presented at the Goddard Conference the seventeen focal mechanisms he had managed to analyze. However, rather than wait for the proceeding to be published, he wisely decided to write up his results immediately and submit them to *JGR*. The paper was received on November 28, 1966, and published in April, 1967.

Sykes wanted to let Wilson know that he had confirmed transform faults. He wrote Wilson on August 19, 1966, and included an abstract of the paper he thought he was going to present at the upcoming November 14–16 GSA meeting in San Francisco.

I have been trying to determine focal mechanism solutions for a number of earthquakes on the mid-oceanic ridges. I will report on this work at the G.S.A. meeting on November 16, 1966. Since this is quite pertinent to your work on transform faults, I am enclosing an abstract of the paper.

(Sykes, August 29 letter to Wilson)

Sykes also asked for some reprints. Wilson sent them. On November 22, Sykes sent him a copy of the manuscript he had just sent off to *JGR*. Wilson gave a talk at Lamont on October 28. They hoped to meet, but could not. They finally met in late 1966 or early 1967. Wilson gave a talk at Queens College in New York City, and Sykes went to hear him. During the talk Wilson mentioned that Sykes' work had confirmed the transform fault idea (Sykes, August 30, 2011 email to author).

Sykes also showed Ewing his mechanism solutions. Sykes believes that Ewing found the results "interesting," the data "really good," was "encouraging" but nonetheless, "was skeptical" and thought that he should delay and get more solutions.

But at about the time that I obtained the focal mechanisms – probably just before the Goddard conference – I went to see Ewing to show him them, which was rather customary for everybody to do. He said, "This is really interesting and some really good data. But I would suggest that you get two or three times as many solutions." I don't think that he said before publishing, but it was rather clear that he was skeptical and yet encouraging. Fortunately, I decided to go [publish] – and I think Jack Oliver encouraged me – with what I had.[14]

(Sykes, May 24, 1997 Doell interview)

Sykes was right not to wait; his results were very good, and Stauder might suddenly have become interested in transform faults.

Sykes (1967b) presented fault plane solutions to seventeen earthquakes in his *JGR* paper (Figure 6.15). Ten were associated with the Mid-Atlantic Ridge.

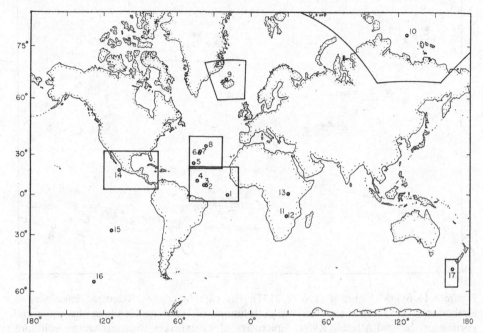

Figure 6.15 Sykes' Figure 3 (1967b: 2136). His caption reads: "Earthquakes on mid-ocean ridge system for which mechanism solutions are provided in this paper." Earthquakes are numbered 1–17. Earthquake 13 is from the East African Rift Valley thought to be an extension of the mid-ocean ridge system. Earthquakes 11 and 12 were induced by dam loading.

Earthquakes 1–4 were from fracture zones between ridge-offsets along the equatorial region of the ridge (Figure 6.16). Earthquakes 5–8 were from the North Atlantic: 5 and 8, from fracture zones between offsets; 6 and 7, from the ridge itself (Figure 6.17). Earthquake 9 was from a fracture zone just north of Iceland; earthquake 10 was from the northward extension of the Mid-Atlantic Ridge in the Laptev Sea, near the Arctic shelf of Siberia. The earthquakes 11–13 were from East Africa: 11 and 12 were produced by loading of the artificial Lake Kariba in Southern Rhodesia (now Zimbabwe); 13 from the Western branch of the East African rift system. The remaining earthquakes were from the Pacific: 14–16, from fracture zones between segments of the East Pacific Rise; 14, from the Rivera Fracture Zone, 15, from the Easter Fracture Zone, and 16, from the Eltanin Fracture Zone. The last earthquake was from the Macquarie Ridge, south of New Zealand.

The results were spectacular. The earthquakes had now been accurately relocated and the mechanism solutions were highly precise. Less than 1 percent of the data was inconsistent with the quadrant distribution of first motions as compared to the 15 to 20 percent from previous studies (Sykes, 1967b: 2131). Every focal mechanism of an earthquake from a fracture zone between ridge-offsets was strike-slip

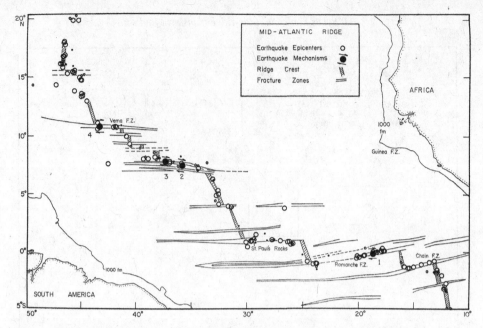

Figure 6.16 Sykes' Figure 4 (1967b: 2137). His caption reads: "Relocated epicenters of earthquakes (1955–1965) and mechanism solutions for earthquakes along the equatorial portion of the mid-Atlantic Ridge. Epicenters of earthquakes with mechanism solutions shown by solid circles; other epicenters shown by clear circles; large circles denote more precisely determined epicenters; smaller circles, poorly determined epicenters."

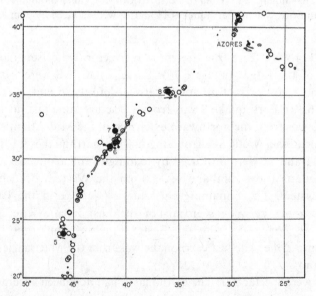

Figure 6.17 Sykes' Figure 9 (1967b: 2141). His caption reads: "Relocated epicenters of earthquakes (1955–1965) and mechanism solutions for earthquakes within rift valley of the mid-Atlantic Ridge. Rift valley in mid-Atlantic Ridge is denoted by diagonal hatching ... Other symbols same as Figure 4 [Figure 6.16]." Earthquakes 5 and 8, on fracture zones are strike-slip; earthquakes 6 and 7, on ridge segments, are dip-slip (tensional).

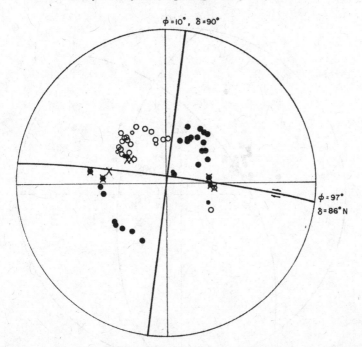

Figure 6.18 Sykes' Figure 7 (1967b: 2141). Mechanism solution for earthquake 3 within fracture zone between ridge-offsets of the Mid-Atlantic Ridge (Figure 6.16). His caption reads: "Diagram is an equal-area projection of the lower hemisphere of the radiation field. Solid circles represent compressions; open circles, dilatations; crosses, wave character of seismograms, indicating station is near nodal plane. Smaller symbols represent poorer data. φ and δ are strike and dip of the nodal planes. Arrows indicate sense of shear displacement on the plane that was chosen as the fault plane [see also Figure 6.16]" (my bracketed addition). The mechanism is strike-slip, indicative of a transform fault. Sykes changed the strikes of the two planes when reviewing the proofs to 8° and 98° from 10° and 97°.

(e.g. Figure 6.18); with one exception, every mechanism solution of an earthquake from a mid-ocean ridge system and its extension in East Africa was entirely or predominantly dip-slip, and indicative of a normal fault (e.g. Figure 6.19). The exception, an earthquake over the Macquarie Ridge, was dip-slip and indicative of a thrust fault (Figure 6.20). He suggested that its tectonics "may be more similar to the tectonics of New Zealand than to the tectonics of the mid-oceanic ridges" (Sykes, 1966b: 2148). But the most important result was his confirmation of ridge-ridge transform faults.

The study of the mechanisms of earthquakes is in agreement in every case with the sense of motion predicted by Wilson [1965a, b] for transform faults. The results support the hypothesis of ocean-floor growth at the crest of the mid-oceanic ridge. The sense of displacement indicated from these studies of earthquakes is opposite to that expected for a simple offset of the ridge crest.

(Sykes, 1967: 2132; Wilson references are my Wilson 1965b, c)

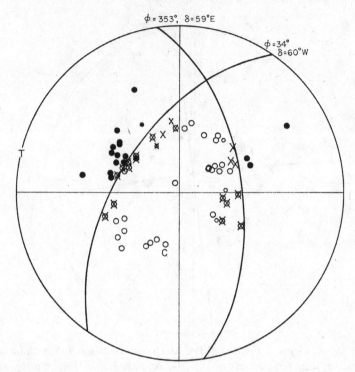

Figure 6.19 Sykes' Figure 11 (1967b: 2142). Mechanism solution for earthquake 6 located within rift valley of mid-Atlantic ridge segment (Figure 6.17). "*T* and *C* are inferred axes of maximum tension and compression, respectively." Inferred tensional axis normal to trend of ridge segment; see Figure 6.17. Other symbols same as in Figure 6.18. The mechanism is dip-slip and indicative of a normal earthquake.

Sykes had identified ten transform faults, eight dextral and two sinistral. Earthquake 8, on an unnamed facture zone on the Mid-Atlantic Ridge, and earthquake 16 on the Eltanin Fracture Zone were sinistral, consistent with the fracture zones being transform faults.

Before leaving this account of Sykes' work on focal mechanisms, I need to discuss an ambiguity endemic to fault plane solutions based on fault first motion studies and explain why it presented no significant hindrance to Sykes. Look at the fault plane solution to the earthquake 3 on an unnamed fracture zone in the North Atlantic (Figure 6.18). Given the distribution of the compressional and dilational quadrants, either nodal plane may represent the actual fault plane. It could strike east and move dextrally, which is what Sykes thought, or it could strike north and move sinisterly, for in either case, the movement would be toward the compressional quadrants and away from the dilational ones. So how did Sykes decide? And, was his choice warranted? Absolutely, the geology made it clear. Here is what he wrote:

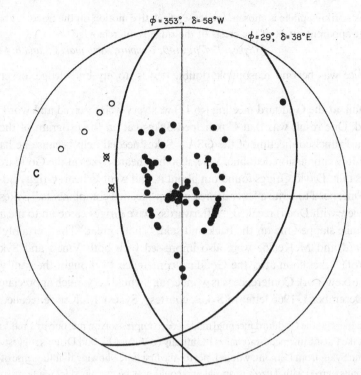

Figure 6.20 Sykes' Figure 24 (1967b: 2148). Mechanism solution for earthquake 17 on the Macquarie Ridge (Figure 6.15). Symbols same as Figure 6.19. The mechanism solution is strike-dip and indicative of thrusting. Inferred axis of compression is roughly normal to the trend of the ridge.

Choice of the fault plane. Although a unique choice of the fault plane cannot be made from the first motion data or from an analysis of S waves, the east-striking plane seems to be overwhelmingly favored for the following reasons:

1. For each earthquake the epicenter is located on a prominent fracture zone; the strike of one of the two nodal planes nearly coincides with the strike of the fracture zone [e.g. Figures 6.16 and 6.18].
2. Earthquake epicenters are aligned along the strike of the fracture zones.
3. The linearity of fracture zones suggests a strike-slip origin ... The bathymetry and other morphological aspects are similar to the morphology of the great strike-slip zones on continents.
4. Many of the rocks from St. Paul's Rocks [from a fracture zone] ... are described as dunite-mylonites ... Rocks dredged from other fracture zones ... as well as samples from cores taken in fracture zones ... exhibit a similar petrology and provide evidence for intense shearing stress in the vicinity of fracture zones.
5. The choice of the north-striking plane would indicate strike-slip motion nearly parallel to the ridge axis. On the contrary, earthquakes on the ridge axis but not on fracture zones are characterized by a predominance of normal faulting.

6. If the east-striking plane is chosen, the sense of relative motion on the fracture zone reverses when the apparent offset of the crests of the ridge is interchanged.

(Sykes, 1967b: 2140; compare earthquakes 5 and 8, Figure 6.17)

Sykes' choice was beyond reasonable doubt. It was, to my knowledge, never seriously questioned.

Sykes' talk at the Goddard meeting (§6.12) was very well received and word of it soon got around. One result was that Cox helped Sykes get on the program of the fall 1966 annual San Francisco meeting of the GSA – Sykes needed help because he had already missed the July submission deadline. Vine, who had heard Sykes at the Goddard meeting, plugged his talk. David Griggs found out about it, and went to hear. Griggs did not, as he had with Vine, need to write a letter expressing his "belated compliments." Sykes (May 24, 1997 interview with Doell) recalled, "Afterwards, Dave Griggs came up to me and with a great big smile slapped me on the back and said, 'That's great!' That certainly made my day." Bullard and McKenzie were also impressed with both Vine's and Sykes' work. Bullard wrote Sykes soon after the Goddard conference, "I thought the stuff you talked about at the New York Conference was wonderful; I would very much appreciate reprints" (Bullard, December 1, 1966 letter to Sykes; courtesy, Sykes). McKenzie recalled:

Two of the papers [at the Goddard meeting] made a deep impression on me: one by Fred Vine . . . who showed that the Pacific magnetic anomalies beautifully confirmed his and Drum's suggestion, and the other by Lynn Sykes from Lamont, whom I met for the first time, showing that the sense of motion on transform faults agreed with Tuzo's proposal and could only be explained by sea floor spreading.

(McKenzie, 2001: 176; my bracketed addition)

Just as the work of Vine and Heirtzler and Pitman converted the first key corollary of seafloor spreading, V-M, into a difficulty-free solution, so Sykes' work converted the second key corollary of seafloor spreading, transform faults, into a difficulty-free solution. Sykes, by establishing a firm connection between seismology and tectonics, gave seismology a vital role in turning the tide toward mobilism, a role Isacks and Oliver were already beginning to strengthen (§7.4).

On November 28, 1966, Father Stauder wrote Sykes a gracious letter. Stauder began with strong praise, and then elegantly discussed his and Bollinger's apparent failure to recognize that the strike-slip earthquakes were "located on a fracture zone and one of the nodal planes for P waves nearly coincides with the strike of the fracture zone."

Dear Lynn:

Thank you for the preprint of your paper, "Mechanism of Earthquakes and Nature of Faulting in the Mid-Oceanic Ridges." I regard it as a most significant contribution and wish to congratulate you on a very careful and thorough investigation. The evidence in support of transform faulting is exciting indeed, and explains what is otherwise a puzzle, the confining of present seismicity of the region of offset between ridge crests.

I had to smile a little, ruefully, on reading on page 6 that Bollinger and I "apparently" failed to realize that in each of the solutions dominated by strike-slip motion the epicenter was located on a fracture zone and one of the nodal planes for P waves nearly coincides with the strike of the fracture zone. It is correct that "apparently" this is the cases, for we gave no other indication. Actually I have a map of Menard's with two epicenters of '63 spotted on it. I was puzzled, however, by the sense of motion and not too reliant on the accuracy of the epicenters. I figured a good deal more data and careful examination were needed. You have supplied both very well. Also at the time I was inclined to an opinion that stress fields from groups of mechanism solutions were a more basic conclusion of first motion studies than individual fault planes. I am not so sure of this any longer.

(Stauder, letter to author, courtesy of Sykes)

Stauder's intuition that he may have been wrong about stress fields being more important than fault planes was correct. For plate tectonics, fault plane solutions that indicate relative plate motion and yield slip vectors are what are important, not stress fields. Stauder, however, did not immediately change what he was doing. McKenzie later mentioned that he heard Stauder talk at the spring 1967 AGU meeting in Washington, DC, and that he was still more concerned with "directions of greatest and least stress," with mechanics rather than kinematics.

The other paper [which made a deep impression on me at the April 1967 AGU meeting] was by Father William Stauder, from St. Louis University, concerned with the mechanisms of earthquakes along the Aleutian Trench, which I misunderstood. Earthquakes are generated by movement on faults, when one side of the fault slides past the other. The direction of this motion is known as the slip vector. I thought Stauder had plotted these slip vectors, whereas in fact he plotted what he believed to be the direction of greatest and least stress. This rather technical issue sounds trivial but is not: plate tectonics is concerned with slip vectors, not with stresses, and the relative motion between plates produces earthquakes whose slip vectors show the direction of their relative motion.

(McKenzie, 2001: 180)

In closing, Stauder (November 28, 1966 letter to Sykes) noted that he "enjoyed reading the paper" and "anticipate[d] that this will be a much referenced work."

6.10 Menard accepts seafloor spreading; Heezen renounces rapid Earth expansion

At the end of March 1966, when he wrote Hess (§6.2), Menard still opposed seafloor spreading or at least continuous seafloor spreading. At the time he was in Washington, DC, serving as a science advisor in the Johnson administration. He left in July 1966, returning to Scripps in La Jolla, California "to do what counts." He wanted to examine the Mason and Raff magnetic anomaly maps to see if the symmetry of anomalies about the Juan de Fuca Ridge that Vine and Wilson had claimed, held up under his own scrutiny. Here it is, in his own words:

There were two things on my mind when I returned to La Jolla in July 1966: the Nova expedition and the reality of magnetic anomalies.... As to what counts, all I had, other than hearsay, was the profiles published by Vine and Wilson. I had found them too small, too short, and too few to be convincing. I hauled out my copies of the drafts of the Mason and Raff maps. The drafts were three times as large as the published versions. I made numerous profiles across the full width of the anomalies at this large scale. The profiles covered a much wider area than Vine and Wilson had discussed. As I did the work it became obvious what was coming. Even so, when I laid out all the profiles I could only shake my head and think, "Well, I'll be damned."

(Menard, 1986: 272)

Menard received a letter from Wilson on August 11, 1966 (partially reproduced in Menard, 1986: 272) saying, "I find submarine geology in a very exciting state." Menard recalled, "I could not have agreed more," and wrote Wilson back, "I have been looking further into the problems of seafloor spreading and magnetic anomalies and I must say I am getting increasingly convinced. The symmetry of the magnetic anomalies in a number of places is really overwhelming" (Menard, October 27, 1966 letter to Wilson partially reproduced by Menard (1986: 273)). He wrote this *before* either Pitman's magic profile (Pitman and Heirtzler, 1966) or Vine's review (1966) had appeared.

Actually, Menard had already changed his mind. He had already written his talk for the Goddard meeting, which was then just ten days away, in which he accepted the confirmation of V-M. In the 1968 published version of his talk, this is how he began, making it very clear that he had accepted seafloor spreading:

The newly discovered symmetry of magnetic anomalies and seismicity of transform faults presents convincing evidence of sea floor spreading (Hess, 1962; Wilson, 1963a; Vine, 1966). However in this forward thrust in our understanding a few puzzles have been left behind, and these are the subject of this paper. I shall deal with problems of the terminations of fracture zones and apparent complications in sea floor spreading.

(Menard, 1968: 109)

One of the complications Menard examined was the outstanding difficulty of how could seafloor spreading occur when Antarctica entirely, and Africa partly, are surrounded by spreading ridges. The confirmation of V-M and ridge-ridge transform faults meant that seafloor spreading was beyond reasonable doubt. But its proponents still had to deal with the "ridge" issue, which, although no longer an impediment to its acceptance, still had to be explained. Menard proposed an ingenious way for ridges to migrate, and still generate bilaterally symmetrical magnetic anomalies. Suppose one-sided convection. Suppose seafloor is generated on only one side of a ridge crest at rate $2V$. If the ridge itself migrates at a velocity V in the same direction as the generated seafloor, then the ridge will bisect the block of newly generated seafloor, and the pattern of magnetic lineations will be bilaterally symmetrical about the ridge. This solution required inexplicable coordination between the rates of ridge migration and mantle convection; it received little attention. A more radical solution would be needed, involving rejection of the idea that ridges form where they do as a consequence of uprising mantle convection.

Heezen, as already described (III, §6.15), renounced rapid Earth expansion in a 1966 co-authored encyclopedia entry, and nonetheless accepted continental drift "largely due to the convincing paleomagnetic evidence of large continental displacements" (Heezen and Fox, 1966: 515). Thanks to Ursula Marvin, Heezen's rejection of rapid Earth expansion can be further pinned down. Marvin heard Heezen abandon rapid Earth expansion at a one-day conference called "What's New On Earth?" held at Rutgers University in October 1966. Summarizing festivities at the meeting, she wrote:

Intimations of change were very much in the air, and the meeting was called to discuss them. J. Tuzo Wilson spoke and held up colored cardboard sheets, which he pulled apart to produce a very nice illustration of ridge-spreading and transform faults. Bob Dietz ended the sessions with a review of ideas for and against drift, using his cartoon of torn newspapers fitted together, mammoths galumphing over meter-wide land bridges, and weeping creatures being separated by drifting blocks. It was a great day. I attended and took notes. I was most impressed to see Bruce [Heezen] announce his abandonment of radical expansion right there in public – in real time, as it were.

(Marvin, January 26, 1987 letter to author)

Although Marvin could not find her notes, she kindly guided me to a report that she had written in December 1966, just two months after the meeting. She quoted Heezen directly from her notes:

It [rapid Earth expansion] has recently been abandoned, however, by one of its main advocates (Heezen, 1966), because it would require a radial expansion of 4 to 8 mm year^{-1} for the past 200 million years to account for the Atlantic Ocean alone.

(Marvin, March 24, 1987 letter to author quoting Marvin, 1966: 60; my bracketed addition; Heezen, 1966 (my Heezen, 1966b) refers to Heezen's unpublished talk at the Rutgers meeting)

She (1966: 60) also noted that Heezen had stated, "During the Tertiary an estimated $120 \times 10^6 \, \text{km}^2$ of area was added to the ocean floors (Heezen, 1966)."

6.11 Lamont workers argue that heat flow over the Mid-Atlantic Ridge is too low for seafloor spreading; Hess disagrees

The most illuminating example of our dilemmas at the time is the interpretation of the oceanic pattern of distribution of heat flow. With Lamont's heat flow specialist, Marcus Langseth, in 1965, we were trying to analyze and interpret the numerous heat flow measurements he had made in the Atlantic. In particular, I made the first numerical computation of the heat flow pattern that should be produced by Hess's seafloor spreading. While the overall pattern was consistent with Hess's model, quantitatively, the disagreement was obvious. The computed heat flow was three times larger than the measured one. In contrast, deeper convection currents, those that could not reach the seafloor (as proposed by Ewing), would produce a heat flow pattern in good agreement with the measurements. I was convinced then that we had obtained the quantitative demonstration that the Hess model did not work. This is the conclusion we published in 1966.[28] Sea-floor spreading should leave a clear heat flow signature – but it was not present. Our computations were correct, our measurements were correct, but our conclusion was wrong.

(Le Pichon, 2001: 210–211; reference 28 is to Langseth et al. *(1966))*

Hans Petterson, leader of the 1947–8 Swedish Deep Sea Expedition took the first heat flow measurement at sea in 1947.[15] Bullard and Revelle, recognizing the promise of Petterson's work, took steps to develop a heat flow group at Scripps and Bullard designed a heat probe instrument (III, §2.21). It did not take long for workers there and at the Department of Geodesy and Geophysics at Cambridge, Bullard's home institution, at Lamont where Ewing designed his own heat probe, and at other institutions to take heat flow measurements in the Atlantic, Pacific, and Indian oceans.[16]

Based on this very substantial body of heat flow data, Von Herzen and Langseth (1965) wrote an extensive review. Investigators had found appreciable differences between measurements taken in close proximity to each other. They had found unexpected and, at the time, inexplicable low values on ridge crests. Overall they found that the average heat flow through mid-ocean ridges in the Atlantic, East Pacific, and Indian oceans was significantly higher than the oceanic mean. It was the sporadic high heat flow results at ridge crests that led Bullard to adopt mantle convection (III, §2.12) and that were instrumental in Hess's development of seafloor spreading (III, §3.12–§3.14); they also played a role in Menard's claim that ridges are ephemeral and in his adoption of seafloor thinning (III, §5.7, §5.12), and were instrumental in the Ewing brothers' adoption of small-scale mantle convection as an explanation of ridge formation (III, §6.12). But did the high heat flow measurements at ridge crests support the large-scale mantle convection thought necessary for seafloor spreading or the small-scale convection the Ewing brothers preferred? This question was more easily asked than answered. Heat flow experts had to develop a model that predicted how much excess heat would be produced by large-scale mantle convection, and then compare it with mean heat flow observed on ridges and in the deep ocean basins.

Because of the wide scatter of the heat flow values at ridge crests and the many peculiarly low values, heat flow experts in the mid-1960s believed (as it turned out rightfully) that excess heat produced at ridges tends to be underestimated. Consequently the overall mean over ridges is too low, but they really had no idea by how much. There were many uncertainties. Some heat may be quickly dissipated before it can be measured. Some chemical reactions at ridge sites are endothermic. There also was a problem in determining the "normal" heat flow of the deep ocean basin, how much was produced by processes other than current mantle convection; some of the heat flow through the deep ocean floors may be from residual heat produced by earlier mantle convection. There were uncertainties in assigning values to key parameters of the model such as the thickness of the moving seafloor and geothermal gradients.

In 1965, Lamont researchers Marcus Langseth, Le Pichon, and M. Ewing decided it was time to forge ahead and attempt to test seafloor spreading quantitatively through heat flow measurements (Langseth *et al.*, 1966). Le Pichon did the modeling. They (1966: 2348) set the thickness of the moving seafloor at 100 km with a bottom temperature of 1500 °C. They also set other parameters such as conductivity, heat capacity, and density, and computed the distribution of temperatures and surface over a width of 1000 km centered on the ridge axis for spreading rates of 1 cm and 2 cm per year. They ran the figures for the Atlantic; their results did not, they argued, support seafloor spreading.

In conclusion, the following points can be made: Horizontal drift of the oceanic crust and upper mantle, according to the spreading-floor hypothesis, results in a peaked distribution of surface heat flow. The [predicted] average heat flow is at least 2 or 3 times as high as the [observed] normal basin heat flow of 1.2–1.4 cal over a width of 1000 km, if the velocity is of the order of 1 cm/yr or more. Partial fusion of the upper layer is to be expected in the axial zone, and the deficit of density, created by the increase in temperature, will be less than 50 km deep. The measurements reported in this paper do not support the existence of geologically significant continuous continental drift in the Atlantic Ocean during the Cenozoic. We cannot, however, rule out the possibility of Jurassic or Cretaceous continental drift, and the spreading-floor hypothesis could account for the narrow zone of high heat flow at the axis of the mid-Atlantic ridge by a recent (less than 10 m.y.) rejuvenation of spreading.

(Langseth et al., *1966: 5352–5353)*

Even though the measured mean heat flow over the Mid-Atlantic Ridge was higher than that over the basins on either side, the difference was not large enough and the region of higher heat flow was not wide enough for seafloor spreading to have occurred.

The average heat flow over the mid-Atlantic ridge is within 20% of the heat flow of the basins, the high values being confined to a narrow axial zone. The absence of a wide heat-flow maximum over the mid-Atlantic ridge precludes the possibility of continuous continental drift of the spreading-floor type during the Cenozoic area.

(Langseth et al., *1966: 5353)*

Results from the Pacific were ambiguous.

On the other hand, the total amount of heat released over the East Pacific and Galapagos rise system is consistent with a rate of drifting of the order of 1 cm/yr. However, the flat maximum of surface heat flow (except for narrow anomalies probably due to intrusions) is not consistent with the theoretical pattern expected. In any case it would be very difficult to account for the huge amount of heat released over the equatorial East Pacific Ocean without admitting the existence of some type of convective transfer.

(Langseth et al., *1966: 5353)*

Langseth and company argued that the Mid-Atlantic Ridge "is in an advanced state of evolution," which was Ewing's view.

Hess wrote a response which he thought important enough to keep – I found it among Hess's papers at Princeton University. Titled "Comments on a paper by Langseth, Le Pichon and Ewing entitled: Crustal Structure of the Mid-Oceanic Ridges, No. 5: Heat-flow Through the Atlantic Ocean Floor and Convection Currents," it reads like a review that Hess might have written for the editors of *JGR*. But there is no accompanying letter, the review is not dated, and no recommendation as to whether it should be published. There are several small discrepancies between Hess's review and the published version of Langseth *et al.*; Hess referred to a Table 4, which became Table 5 in the published version suggesting that it was a review of an earlier version. Even if it is a review intended for *JGR*, I have no idea if it was ever sent to the authors. The original paper was received by *JGR* on May 30, 1966; the revised version was received on August 9, 1966 – several months after Le Pichon had accepted V-M and seafloor spreading (§6.6).

Whatever the status of Hess's comments, he clearly thought the authors had failed to show that "the amount of excess heat-flow over the ridge is incompatible (too low) with the concept of rising convection currents." Here is Hess's review. He had six objections (RS2):

I will restrict my comments to the part of the paper which tries to show that the amount of excess heat-flow over the ridge is incompatible (too low) with the concept of rising convection currents.

First, it seems amusing to point out that we have had the situation that the oceanic heat-flow was too high to be compatible with convection currents and continental drift (MacDonald); now it appears, according to the Lamont authors, that the heat-flow under the oceans is not high enough to sustain the idea of convection currents. Truth might, as usual, be somewhere in the middle.

There are two kinds of objections against the Lamont calculations; the first kind being that in a number of assumptions and numerical simplifications they seem to be systematically on the side of lowering the so-called excess heat they finally arrive at; the second kind being that it is easy to show that their answer is critically dependent on the model chosen. It can be shown that no numerical value of excess heat can preclude the assumption of horizontal drift at the usual rates.

We will first enumerate the objections of the first, not fundamental type.

1: The authors make clear that the rough topography over the ridge tends to make heat-flow measurements systematically lower, yet in their final calculation they do not incorporate such a correction term of 20%, which even if applied only to the part 0–100 km from the ridge would make a correction of 12 cal per second per 1 cm strip across the ridge.

2: In most models of convection currents a basaltic layer is formed together with the rifting and drifting; at a rate of 1 cm/yr this releases about 30–40 cal/sec per 1 cm strip across the ridge, which will not be measured by the usual heat-flow measurements, as this heat of the cooling extrusive rock will be quickly dissipated. This amount must then be subtracted from the amount of excess heat which is brought up by the rising convection current.

3: This same argument will hold in a general way for most transformations the rising matter will go through; as all the possible solid-solid transformations have a positive slope, a pressure release without large cooling will cause the high temperature/low pressure forms to become stable. These reactions are endothermic so that again part of the heat brought up will not be measured as heat-flow. If the rising convection current passes through an equilibrium curve of, e.g.

Jadeite + 2 enstatite → albite + forsterite

.this reaction will take up about 15 cal/sec for a 200 km strip across the ridge rising at a rate of 1 cm/yr (calculated from enthalpy data in Robie, Table 20, Handbook of Physical Constants).

4: Geothermal gradients of 6°/km at 100 km depth and of temperatures of 1500 °C at these depths seem to me rather on the high side; this again increases the heat-flow which should be measured, according to the authors, if rising convection currents existed under the ridge.

These four objections together make their estimate of the excess heat-flow over the ridge ·area low by a factor of 2–3.

More fundamental are the following two objections.

5: On page 28–29 [5341 in published paper], the authors define the heat-flow in the basins as being the average heat-flow. This completely arbitrary assumption defines thereby the level of excess (or for that matter, deficit) heat over the ridge area. I think that in any model of continental drift having taken place in the last few hundred million years, it can be shown that the heat-flow values over the whole ocean should be still in excess of what would be the steady-state situation (Schuiling[17], [June] 1966). If, instead of values of 1.12, 1.47, 1.3 and 1.3 (Lamont Paper, Table V, basin values [Table 4 in the published paper]), we would adopt as normal values 0.7, we would see that the excess heat would be much more. As an example, a strip across the ridge of 4000 km long with an average heat-flow of 1.3 would have an excess heat of 240 cal, instead of no excess heat at all.

6: On page 38 [5347 in published paper], the authors state that "the creation of new oceanic crust at this rate (i.e. 1 cm/yr – R.D.S.) requires that the underlying upper mantle be moving at a speed at least as great." This is certainly true for the horizontal movement. It depends entirely on the ratio of the thickness of the vertically rising column to the thickness of the horizontally spreading layer, what velocity we must attribute to the rising column, and therefore, how much "excess heat" is brought up. If the horizontal movement is confined to a layer of, let us say 10 times thinner than the thickness of the rising column, the vertical movement will be only 1 mm/yr, and the "excess heat" will be only 1/10 of what the authors calculate. This unknown ratio of thicknesses precludes in principle any use of a numerical value of "excess heat" as an argument against the possibility of drifting, as long as there is no "excess heat" at all.

In summary then, objections 1–5 show that the numerical value of the excess heat over the ridges and oceans as a whole, at which the authors arrive, is too low by probably an order of magnitude, and objection 6 shows that it is fundamentally impossible to use an argument from the numerical value of excess heat against the possibility of convection currents occurring now, or having occurred in the geologically recent past.

(Hess review; my bracketed additions)

Hess's was an appropriate strategy. His objections raised just the sorts of problems encountered in attempts to model numerically scattered data. Were his objections correct? Were they fair? As I do not have the expertise to judge, John Sclater kindly assessed Hess's comments.

I will make an attempt to analyze Hess' comments on Langseth *et al.* (1967).

His general comments starting "there are two kinds of objections" are absolutely correct. (1) They did attempt to reduce the observed heat flow as much as they could and (2) [Dan] McKenzie (1967) showed that by thinning the plate to 50 km you could get a much better fit to the data. (Dan's concept is correct but his application is wrong as he used too thin a plate.)

1: It was known that rough topography at the ridge axis led to highly variable heat flow with a predominance of low values. The 20% figure for the correction is totally *ad hoc*. Hess is correct here. I was already doing the same thing when thinking about heat flow over the ridge axis for the same reason – I think I said something in Sclater and Francheteau (1970). [He did; see p. 529.]

2: This is correct but too small an effect to change the conclusions in Langseth *et al.* (1966).

3: Again this is correct but the heat involved is too low to be important at the ridge axis, and would not change the conclusions in Langseth *et al.* (1966).

4: This is the most interesting part of the whole criticism as Langseth *et al.* are correct here and Hess is fishing to make the data fit better his model. I suspect that if Xavier read this review it must have set him really thinking. The 1500 °C temperature is too high but the plate thickness is very close to modern ideas. The temperature for the best fitting plate model is around 1350°C. Rising convection currents under the ridge axis do not heat up the plate enough to make much difference at the surface. The high heat flow comes almost all from the intrusion of the magma. Xavier's model (he did the modeling for the paper) is basically the now generally accepted quantitative model for plate tectonics. No one believed that the ocean floor could act as simply as Langseth *et al.* (1966) suggested. They proposed such an outrageously simple model because they knew that it was likely not to fit the data. However, it was brilliant and a totally original relook at the data without having to deal with convection cells.

5: Again here Hess is trying to fiddle with the data to make it fit his own ideas. The background heat flow is around 1 μcal/cm^2 sec not 0.7. Langseth *et al.* are right here. Once hydrothermal circulation is taken into account the values at the ridge axis are almost an order of magnitude higher than the average measured not 20%.

6: Hess is right here. You cannot argue from the surface values about what is happening at depth.

Hess's last paragraph is right on. The heat flow over the ridges is about 10 times that background value on 1 Ma crust. Hess really knew he was right and he was trying to get Langseth *et al.* to see that their data could just as easily be modified to fit with the creation of a 100 km thick plate. Noise in the heat flow data enabled them to fudge things so that the model did not fit. However, it was not so much the noise as not taking into account the fact that heat lost by advection was not being measured. Lister (1972) showed that hydrothermal circulation that gave rise to this advection is what caused the noise in the heat flow, and reduced the mean heat flow at the ridge axis by almost a factor of ten.

(Sclater, March 31, 2009 email to author; slightly amended, August 24, 2011 email to author; my bracketed additions)

Sclater is saying that Hess's objections 2, 3, and 6 were correct and that it was objection 6, one of Hess's fundamental objections that essentially blunted the Lamont researchers' argument against seafloor spreading. As with Menard's query to Hess about Ewing's one misplaced sample and potassium-argon age (§6.2), it was very likely that Hess was by now quite confident that the basic idea of seafloor spreading was correct. Unless Hess was sent a preprint of Langseth *et al.* before it was received by *JGR* on May 20, 1965, which is highly unlikely, he already knew of the Jaramillo event (§6.3) and of Opdyke's independent confirmation of it (§6.5); he already knew of Vine's recalculation of the spreading rate over the Juan de Fuca Ridge, of the bilateral symmetry of the magnetic anomalies over the Reykjanes Ridge, which he had seen at the Ottawa meeting, and most likely had also been shown the *Eltanin*-19 profile by Vine. As far as Hess was concerned, the game was over.

As for objection 1, Hess took the idea of a correction term of 20% (Objection 1) from Langseth and company's paper. His complaint was that they neglected to make such a correction "in their final calculation."

On the other hand, the large scatter in the data obtained over the midocean ridge and the possibility of a systematic error due to environmental disturbances do not permit conclusions based on variations of heat flow smaller than about 20%. However, if there has been in the recent geological history significant continental drift with injection of new material along the axis of the midocean ridge, the resulting distribution of temperatures within the crust and the upper mantle should produce large systematic variations in the surface heat flow.

(Langseth et al., 1966: 5346)

Sclater, looking ahead, briefly described why the 20% correct was *ad hoc*, and why "the large scatter in the data obtained over the midocean ridge" brought about primarily by the peculiar low readings, turned out, once correctly attributed to hydrothermal circulation, to provide mobilists all the excess heat they needed.

Heat flow measurements require soft sediments to enable the probe or corer to penetrate far enough in to bottom for reliable measurement. The soft sediments occur in sediment ponds at the base abyssal hills. The cold water flows down though these sediments and then out through the rocks of the abyssal hills in jets (black smokers) or slowly moving upward vents off the ridge axis. The high values that would be observed in these areas cannot be measured because the advection of heat is occurring over bare rock where measurements could not be taken. The only place where one can get reliable estimates of the heat loss is over extensive well sedimented areas where the outward loss of heat by advection is removed by the sedimentary blanket. Clive Lister suggested this in 1972 and I and my student John Crowe were able to prove it in

1979 by a heat flow survey across the well sedimented Reykjanes ridge. We showed that when measurements were taken in such an extensive well sedimented area the heat flow values were high like those predicted by plate models that matched the topography.

(Sclater, March 31, 2009 email to author)

Of course, this is not to say that Langseth, Le Pichon, and Ewing should have known about hydrothermal circulation, nor is it to say that they should not have done the numerical modeling, for as Sclater (2004: 27) noted, Pichon's model "became a cornerstone for quantitative plate-tectonic plate theory," and was a "brilliant and a totally original relook at the data." But they should have tempered their conclusion that "the absence of a wide heat flow maximum over the mid-Atlantic ridge precludes the possibility of continental drift of the spreading-floor type during the Cenozoic era." This is not correct and was not correct at the time because as Sclater (2004: 6) retrospectively noted, the feeling among members of the heat flow community was that "great scatter in the measurements prevented a confident quantitative interpretation of the results."

6.12 The Goddard conference: selling continental drift and seafloor spreading to the establishment

Word began to spread about Pitman and Heirtzler's magic profile and Opdyke's work on sediment cores at the April AGU meeting in Washington, DC. Those in attendance were impressed; recall, for example, Cox's reaction (§6.7). But the most important 1966 meeting was a two-day by-invitation-only meeting held at the Goddard Space Center in New York on the mornings of November 11 and 12, immediately before the GSA meeting. It was sponsored by Columbia University and the Goddard Institute. Here is what three attendees, Opdyke, Le Pichon, and Menard had to say.

The acceptance of seafloor spreading at Lamont set the stage for what turned out to be an exceedingly important meeting, organized in August of 1966 by Paul Gast in the Geochemistry Division of LDGO and Robert Jastrow, director of the Goddard Space Center in New York, where the meeting was held. The power structure of North American Earth Science was invited to this meeting. Attendance was by invitation only. It was a carefully arranged confrontation, with summations at the conclusion, pro and con, by Sir Edward Bullard and Gordon MacDonald. I was invited at the last moment because Allan Cox was unable to attend.

Most of the data presented at the meeting was unpublished at the time, so that the amount of data that could be brought to bear on the subject of seafloor spreading was surprising to many. However, as the meeting wore on, it became apparent that the data clearly supported H. H. Hess' ideas of seafloor spreading [Hess, 1962]. Indeed, at the end of the meeting, Bullard gave a meaningful summation pro continental drift and seafloor spreading, and Gordon MacDonald, who had been such a vehement opponent of continental drift, discovered that he had pressing business elsewhere. It's my opinion that this one meeting fostered the rapid acceptance of the seafloor spreading hypothesis ... This contributed to the acceptance of crustal mobility envisioned by Wegener and led directly to the modern theory of Plate Tectonics.

(Opdyke, 1985: 1182)

It was during a conference organized by NASA in New York, on November 11 and 12, 1966 that the victory of mobilism was clearly established. Teddy Bullard, who presided, could not find a single scientist to defend fixism.

(Le Pichon, 2001: 213)

The [Goddard] meeting was fortunately timed because the scientists from nearby Lamont were ready to display their work, but nothing was published [until 1968]. The list of invited participants included only 39 people, but more than half of them were or would be members of the National Academy [US] or Royal Society [London]... Maurice Ewing did not give his scheduled paper on "Sediment Cover of the Deep-Sea Floor," although he attended the meeting. Gordon MacDonald did not publish his portion of the review, although I don't recall whether or not he was mute at the time of Review. [He left the meeting before his scheduled talk.] The withdrawals give a clue about the way the meeting went.

(Menard, 1986: 273; my bracketed additions)

Paul Gast, a geochemist at Lamont who helped organize the meeting, McKenzie, and R. McConnell, Jr. spoke at the first session on the upper mantle (Gast, 1968; McKenzie, 1968a; McConnell, 1968). Opdyke, Vine, Heirtzler, Menard, Le Pichon, and Sykes spoke at the next session on evidence from the ocean basins (Opdyke, 1968; Vine, 1968a; Heirtzler, 1968; Menard, 1968; Le Pichon, 1968b; Sykes, 1968). Cox, Doell, and Dalrymple were asked to participate, but could not do so although they did contribute to the proceedings (Cox, Doell, and Dalrymple, 1968). Speakers at the third session on evidence from the continents, included P. M. Hurley and J. R. Rand, John Dewey and Marshall Kay, E. Irving, J. C. Briden, F. Stehli, and J. Boucot (Hurley and Rand, 1968; Dewey and Kay, 1968; Irving and Robertson, 1968; Briden, 1968; Stehli, 1968; Boucot, Berry, and Johnson, 1968), Gordon MacDonald, whose attacks on mobilism were well known, and E. C. Bullard were scheduled to debate at the closing session (Bullard, 1968b). However, MacDonald who did attend the first day sessions, canceled. Maurice Ewing, who was scheduled to speak on sediment cover of the deep-sea floor, also canceled. Menard's remark above about withdrawals giving a clue about what happened at the meeting is correct. Leading Earth scientists were also invited. They included D. L. (Don) Anderson, Anton Hales, Bruce C. Heezen, R. (Spike) Hide, John Imbre, Frank Press, Walter H. Munk, and Jerry C. Wasserburg.

Except for Stehli, longtime fixist foe of Irving, Runcorn, Briden, and Brown (II, 7.10; III, 1.9, 1.18), and J. Boucot, another North American paleontologist, everyone who discussed mobilism favored it – those who spoke on the mantle did not directly address mobilism. Yes, Marshal Kay, once a firm fixist (I, §7.3), who had begun to pay attention to mobilism's land-based paleomagnetic support a few years before

(III, §1.14), had by then become a mobilist.[18] Even Le Pichon, who spoke on heat flow, and included only an abstract in the conference proceedings published two years later, did not speak against mobilism, if his published abstract reflects what he said. In his opening statement, he acknowledged the strong support accorded to seafloor spreading by Heirtzler and Vine.

The contributions of Heirtzler and Vine have shown that the pattern of magnetic anomalies over the mid-ocean ridges can be interpreted according to the spreading floor hypothesis, thus providing strong support for the horizontal mobility of the oceanic crust.

(Le Pichon, 1968a: 119)

He then proceeded to summarize his arguments against seafloor spreading based on the apparent lack of heat released over the Mid-Atlantic Ridge; however, there is no mention, for example, that the difficulty he had raised in his paper with Langseth and Ewing (§6.11) "precludes the possibility of continuous continental drift of the spreading-floor type during the Cenozoic era." Given the strong support of seafloor spreading accrued through confirmation of V-M, Le Pichon no longer viewed the difficulty arising from heat flow data as a reason for rejecting seafloor spreading.

Indeed, it was Vine's presentation, and, I think, to a lesser extent those by Heirtzler, Opdyke, and Sykes that made the most impact on those in attendance. Menard recalled their reaction to Vine's presentation:

My memory is that the audience was stunned, but certainly there was a discussion. Don Anderson asked why the crust was spreading. Because of convection. Jerry Wasserburg asked why, if spreading was symmetrical, the central anomaly was not. It was symmetrical, but it didn't always look that way because it was also a function of latitude and orientation. Frank Press asked if the symmetry had been tested statistically, Vine replied,

I never touch statistics. I just deal with the facts [laughter].[41]

Walter Munk felt that the statistically determined orientation of anomalies might indicate whether spreading was influenced by the rotation of the earth. It was an old question, said Vine. He had easily fielded questions from what were normally some of the toughest hitters in geology.

(Menard, 1986: 274; endnote 41 refers to Vine, 1968a)

Vine also showed age-colored figures of the magnetic anomalies over the Juan de Fuca and Gorda ridges (reproduced at www.cambridge.org/frankel4), and the Reykjanes Ridge.

Apparently there were a few who were not pleased. Maurice Ewing was one. In his biography of Ewing that he wrote for both the US National Academy of Sciences and the Royal Society (London), Bullard (1980: 153–154) recalled, "Just before the meeting started Ewing came up to me, looking, I thought, a little worried, and said: You don't believe all this rubbish do you?" McKenzie (March 18, 1990 letter to author) also remembered "Teddy [Bullard] teasing Ewing, who did not like what he was hearing, that all the best evidence came from his lab." Pitman, who attended the meeting but did not speak, told Wertenbaker (1974: 218):

One man who had been violently against continental drift just got up and walked out. But I remember that Menard from Scripps, who had opposed it, sat and looked at Eltanin-19, didn't say anything just looked and looked and looked.

Pitman did not identify the person. I suspect that Worzel was unhappy with the pro-drift results. Pitman told Glen (1982: 335) what happened at Lamont when Heirtzler first showed Worzel the profiles:

Heirtzler later called in Joe Worzel and showed him the profiles, reversing them to emphasize the bilateral symmetry. Worzel looked at it for a while and finally said, "Well, that knocks the seafloor-spreading nonsense into a cocked hat." I said, "What do you mean, Joe?" He said, "It's too perfect," and walked out of the room.

With MacDonald a no show, Bullard entertained the audience by imitating MacDonald's arguments. MacDonald heard from some who attended that Bullard's imitation was "delightful and comical."

Nevertheless [despite the time demands of the President's Science Advisory Committee] I still accepted an invitation to participate in what turned out to be the second memorable meeting [after the Royal Society meeting (§3.3): the "History of the Earth's Crust," convened by the Goddard Institute]. Because of my new commitments, I had little time to think about the recent developments. I let the organizers know that I could not be present at the summary session and arranged for Teddy Bullard to present my arguments. Having listened to me make the same case a number of times, he imitated my presentation in a delightful and comical way, according to those who heard him.

(MacDonald, 2001: 125; my bracketed additions)

Bullard, putting the new work from the ocean basins in perspective, reminded his audience of the importance of the continental paleomagnetic work, emphasized that these new developments had not helped us further our understanding of continental geology, and suggested problems in need of further investigation. He began by noting what the new work on the ocean floor had and had not done.

This conference has been about the history of the crust of the earth but has only covered part of the subject. The crust has many features which have a history and which we might have discussed but have not. We have concentrated on the great division between continents and oceans and especially on the history of the oceans and of the mid-ocean ridges. It is in these matters that the great advances have been made in the last year or two, advances that seem to me to herald a new precision in our knowledge about the Earth and a real revolution in geology. The main subject we have left out is the structure and history of the fold mountains of the continents. The change of emphasis is remarkable; until a few years ago the fold mountains were at the center of thought on structural geology, and it would have been inconceivable that a conference on the crust of the earth could have left them out.

(Bullard, 1968b: 231)

Next came his acknowledgment of the importance of land-based paleomagnetism, in particular the great discrepancy of the Australian APW path with those of Europe and North America, the previous paleomagnetic and paleoclimatic support of the GAD hypothesis, which had not made the GAD hypothesis much more attractive than its denial, and an acknowledgment of the relevance of Briden's contribution at the meeting.

Our central theme has been the demonstration the continents have moved large distances in the last 200 million years. Like many people I had been uncertain about such movements ever since I was an undergraduate. The thing that convinced me was the work of E. Irving on the magnetization of Australian rocks. The discrepancy with the European and American results was so great that there were only two possibilities: either Australia had moved by 40° or so relative to the northern continents or the earth's magnetic field during the Mesozoic was grossly different from that of a dipole. There is some direct evidence that the field was not far from a dipole field, and it should be possible to put the matter beyond doubt; in the meantime the assumption of a non-dipole field seems a quite arbitrary and unattractive hypothesis; J. C. Briden's contribution is relevant to this question.

(Bullard, 1968b: 231)

Turning to the new work, he (1968: 231) emphasized that "N. Opdyke's measurements ... have confirmed every detail of Cox, Doell and Darymple's" reversal time scale "during the last 2½ million years," claimed, referring to earlier findings which he did not reference, that the "reality of reversals of the earth's field is put beyond doubt," noted that magnetostratigraphy has become an important sub-discipline, and challenged theorists, perhaps including himself, to explain the origin of field reversals.

Bullard turned to Vine and Heirtzler; seafloor spreading was a reality. Giving credit to the magnetics program at Scripps, he summarized the situation.

But this is not all: Fred Vine has shown that we have the history of magnetic field spread out horizontally on the sea floor in the pattern of magnetic anomalies. The first of the great linear patterns of magnetic anomalies was discovered by Mason, Raff, and Vacquier off the west coast of North America, but their widespread occurrence and true nature has only very recently become apparent, largely through the work of J. Heirtzler. The pattern runs parallel to the mid-ocean ridges and is very accurately symmetrical about them. It is not a reflection of either the topography of the sea floor or of a buried basement beneath the sediments. The hypothesis of Matthews and Vine that it is associated with the spreading of the ocean floor, postulated by Hess, Dietz, and others, seems very likely to be correct.

(Bullard, 1968b: 232)

Bullard (1968b: 232) turned to Sykes' work and seafloor spreading's other key corollary, transform faults.

These ideas are strikingly confirmed by L. Sykes's study of the first motion in earthquakes along the mid-ocean ridges. These occur either beneath the central valley or on the transverse fracture zones. The one along the valley shows that it is in fact cracking open, and those on the fracture zones show that they are, in Tuzo Wilson's nomenclature, "transform faults."

Moving to the continental drift itself, he began (1968b: 232) by noting, "It is possible to believe in the spreading of the ocean floor without believing in continental drift; the two ideas do, however, go very naturally together." He then mentioned, without singling out the "Everett/Bullard/Smith" fit, the "striking" geometric and geological fits of the opposing coast lines on the two sides of the Atlantic, and briefly described Hurley's and Dewey and Kay's contributions. Not forgetting Ewing's presence, he cited the importance of drilling "holes through the ocean through the sediments beneath the ocean floor to help settle some uncertainties about the age-distribution of ocean floor sediments. But as far as Bullard was concerned the controversy over continental drift had been resolved. Everything he had already mentioned was fact: it was time to accept them as fact and figure out what to do next: "All the things I have mentioned so far are facts or near facts, at any rate they are things about which one can plan rational programs" (Bullard, 1968b: 233).

Looking to the future, Bullard discussed mechanism issues, and, perhaps having in mind an issue he might have raised with Gordon MacDonald had he been present, contrasted the factual nature of what he had discussed with the speculative nature of figuring out the mechanism behind the opening of the Atlantic. Rejecting rapid Earth expansion because it requires an unreasonable "increase by over 10% in the last 2% of geological time, he (1966: 233) thought mantle convection more reasonable. He then raised a difficulty that MacDonald had previously raised against continental drift.

The principal difficulty arises from the equality of the mean continental and oceanic heat flows. The rocks at the surface of the continents are about an order of magnitude more radioactive than those beneath oceans. If the continents do not move the explanation is simple. The radioactivity from the mantle beneath the continents has worked its way upward into the continental rocks, but under the oceans it is still in the mantle; the total amount beneath oceans and continents is the same, but that beneath the oceans is spread over a greater range of depth. If the continents move this account cannot be correct; the heat flow through a continent advancing over an ocean should be that derived from the over-ridden ocean plus that from the continental crust and should be greater than normal. The situation is complicated by the thermal effect of the convection current and by the very long thermal time constants of crust blocks, but there seems no reason to expect equal flows on land and at sea.

(Bullard, 1968b: 234)

So what was to be done given that continental drift and seafloor spreading were now firmly established? What was needed? "Detailed calculations on various hypotheses are required; the work of X. Le Pichon [on heat flow] provides an example" (Bullard, 1968b: 234; my bracketed addition). In closing, Bullard cited Menard's presentation about using seafloor spreading to explain details connected with the great fracture zones in the Pacific.

The other 1966 meeting at which these new ideas were beginning to be spread was the GSA fall meeting in San Francisco, which began just two days after the Goddard meeting. As already noted, Sykes' talk impressed those who heard it; recall, for example, Griggs' reaction (§6.9). But, as Menard (1986: 279) noted, "the big event of the meeting was the awarding of the Penrose Medal to Harry Hammond Hess." William W. Rubey gave the citation.

Hess's contributions in geology, mineralogy, and geophysics are known internationally. His detailed studies of the pyroxene and plagioclase groups of minerals, of serpentine and ultra-basic rocks, and of the petrology of layered igneous complexes are widely recognized as classics. His bold and sweeping hypotheses about the tectonic significance of peridotite belts, the role of island arcs and ocean deeps in mountain-building, and the spreading floor origin of ocean basins – these hypotheses are perhaps not characterized by the same detail as the mineralogic studies, but they are no less renowned – and on the whole, favorably renowned. These two almost antithetical aspects of his work – a meticulous attention to the fourth decimal place in refractive indices, on the one hand, and provocative geopoetry about the origin of ocean basins, on the other – these two contrasting aspects have characterized Hess's scientific interests and contributions from the beginning of his career ... It is difficult to say, and perhaps not even profitable to ask, which of the two contrasting aspects of his work has been more valuable to science. The detailed descriptions ... certainly stand as solid foundation on which others have built. Yet the imaginative invention of what some may consider outrageous hypotheses has also inspired excellent new field and laboratory investigation – some of it very probably aimed at trying to prove Harry wrong.

(Rubey, 1968: 83–84)

6.13 Making sense of why Le Pichon, Heirtzler, and Talwani tried so hard to prove Hess wrong

When Le Pichon, Heirtzler, and Talwani raised difficulties against V-M, they, being good scientists, were acting in accordance with RS2. Their objection, for example, concerning the increased amplitude and wavelength of flank anomalies was important. Heirtzler and Le Pichon had reason to reject V-M based on their spring 1965 analysis of Lamont's profiles over the Mid-Atlantic Ridge (§4.6). They also had reason not to *accept* it in their analyses in the summer and fall 1965 of the Juan de Fuca (§5.3) and Reykjanes ridges (§5.4). However, even their second discussion of the Juan de Fuca and Reykjanes ridges gives the impression that they thought consideration of V-M hardly worth their bother. Although they recognized the striking bilateral symmetry of the magnetic lineations over both ridges, their opposition between the two studies lessened, at most, only a minute amount. Moreover, when the editors of *Science* received their paper discussing the East Pacific Rise and Juan de Fuca Ridge and sent them Wilson and Vine's forthcoming papers on the same area, Talwani, Heirtzler, and Le Pichon did not rework their manuscript (§5.3). Le Pichon and Heirtzler did not mention V-M in their (and Baron's) originally submitted manuscript to *Deep-Sea Research* on the Reykjanes Ridge, and only discussed V-M in their revised manuscript because Matthews, who refereed the paper, suggested they do so (§5.4). Le Pichon and Heirtzler (and Baron) also continued to ignore remanence as a causative factor (§5.4), although Opdyke had previously told them of Ade-Hall's work showing the dominance, in oceanic basalts, of remanent magnetization (§4.6).

Why, after seeing the striking bilateral symmetry of the magnetic lineations over the two ridges, should they be so unsympathetic to V-M, even ignoring it in some of their papers? Talwani, on July 17, 1981, sent this letter to me.

To appreciate my view (and I suppose those of most others at Lamont), about the Vine-Matthews hypothesis, you have to realize that we regarded the hypothesis of sea floor spreading as *the* important entity that was to be proved or disproved. We had a number of objections to sea floor spreading which influenced our thinking. When the Vine-Matthews hypothesis was first postulated, the evidence for it was not tremendously convincing. However, evidence piled up – and the three most important in terms in the evidence were: (i) the reversal chronology explained the crestal anomalies assuming nearly uniform rates of spreading. (ii) Assuming the same reversal chronology (actually obtained from magnetic profiles themselves) anomalies in most of the oceans could be explained by spreading at uniform rates. (iii) The strong symmetry of the *Eltanin*-19 and to a lesser extent the symmetry of the magnetic pattern over the Reykjanes Ridge. We were persuaded to accept sea floor spreading because the above evidence showed that the V-M hypothesis had to be correct.

The important reason why we initially thought that sea floor spreading was wrong was: (i) the supposed presence of Miocene fossils (discovered by Saito and Ewing?) on the crest of the mid-ocean ridge. Later on it turned out that while the fossils were indeed of Miocene age, the position of the ridge crest was poorly known. Because of an offset in the position of the ridge crest which was not known at the time of the collection of the fossils, the collection site was actually some distance away from the ridge crest. (ii) Thermal calculations by Langseth, Le Pichon and Ewing which "proved" that the observed heat flow pattern over mid ocean ridges was incompatible with spreading. (iii) A new instrument called the seismic reflection profiler was able to measure sediment thickness in the oceans and in particular on the ridges. While the general pattern of increase of sediment thickness away from the ridge crest favored the hypothesis of sea floor spreading, important exceptions to this pattern, especially the near absence of sediments in the abyssal hill areas of the N. Atlantic, cast doubt on the sea floor spreading hypothesis. (iv) The presence of undeformed old sediments (also discovered by the seismic reflection profiler) demanded the transport of the crust with the overlying sediment completely undisturbed through distances of several thousand kilometers. We considered that difficult.

It is certainly true that the Reykjanes Ridge survey supported the V-M hypothesis. However, even with this survey, we felt that the bulk of the other evidence against sea floor spreading appeared so strong that we were unwilling to accept it. However, as the evidence in support of the V-M hypothesis piled up (see three points noted earlier) we became convinced that the V-M hypothesis had to be correct. Hence, sea floor spreading had to be correct. Hence, other explanation for the difficulties that stood in the way of sea floor spreading had to be found.

(Talwani, July 17, 1981 letter to author)

In essence, they initially rejected V-M because they felt the evidence against seafloor spreading was so very strong, therefore they could not see how V-M could possibly be correct. Their minds were set against seafloor spreading.

Until they saw *Eltanin*-19, until the Jaramillo (Emperor) event had been discovered, until spreading rates were shown to be essentially uniform, not only across
the Juan de Fuca, Pacific–Antarctic, and Reykjanes ridges but "most of the oceans,"
until they had seen *all these things* they did not accept V-M and break their strong
conviction against seafloor spreading. It is not easy changing your world-view. *It is
likely the most difficult thing a researcher ever has to do.* Given the last condition,
("most of the oceans") it is surprising that they did not accept V-M until after hearing
Vine speak at the Goddard conference. As for the difficulties they raised against
seafloor spreading, they were at the time legitimate, but questionable. As Hess told
Menard in April, 1966, the difficulties disappear and "everything fits as it should" if
the bad potassium-argon aged sample and one misplaced sample on the ridge crest
"are disregarded" (§6.2). As Hess maintained, the parameters chosen by Le Pichon in
his thermal model were arbitrary (§6.11). Nonetheless, these were the reasons they
did not accept seafloor spreading until after they were reconciled to V-M. Retrospectively, Le Pichon (1986, 2001) said much the same as Talwani. In essence, they
had to balance their objections to seafloor spreading and V-M against the striking
bilateral symmetry and linearity of the magnetic lineations over the Juan de Fuca and
Reykjanes ridges. The difficulties they raised against seafloor spreading arose from
their detailed analysis of the distribution of sediments on the seafloor and their
numerical modeling of the heat flow data, in which Le Pichon showed the measured
values to be three times lower than the computed ones. As for V-M, it could not, they
argued, explain the increased intensity and wavelengths of the flank anomalies. He
also recalled how he had noticed "the remarkable similarity in the magnetic anomaly" patterns over the Juan de Fuca and Reykjanes ridges, and raised the issue with
Talwani and Heirtzler when they were working on their East Pacific Rise paper, but
they decided to keep their fixist interpretation.

In late 1964 or early 1965 I noticed, at about the same time as, but independently from Tuzo Wilson
and Fred Vine, the remarkable similarity of the magnetic anomalies over what is now known as the
Juan de Fuca Ridge, off western North America, with the anomalies above the Reykjanes Ridge,
south of Iceland. I pointed it out to Manik Talwani and Jim Heirtzler. It was obvious that a portion
of active mid-ocean ridge crest was present to the north of Mendocino off western North America.
But how could one explain the remarkable similarity in the magnetic anomaly pattern over any
portion of the mid-ocean ridge crest, in both the Atlantic and Pacific. After considerable debate, we
considered that too many observations remained unexplained by the sea floor spreading model,
and our conclusions were published in *Science* with a fixist interpretation.

(Le Pichon, 2001: 210)

Again, this does not explain why they did not temper their conclusion: they acknowledged the support for V-M but then did nothing about it – made no serious attempt
to give alternative explanations. They were sent Wilson and Vine's papers by the
editors of *Science* prior to making their revisions, they knew about Ade-Hall's work
on oceanic basalts, but still decided not to consider Wilson and Vine's interpretation
of the Juan de Fuca in terms of transform faults and V-M (§5.2).

Le Pichon, who "has often been asked why the Lamont people had such a late conversion to seafloor spreading," offered one answer that I think hits the mark in his 2001 retrospective essay.

I have insisted earlier in this essay on the quasi-absence of serious awareness of paleomagnetic results in Lamont and on the prevalent fixist culture of the geophysicists there. I know that, as far as I am concerned, these were two very serious obstacles. When I was a Ph.D. student at Lamont, my supervisors never seriously exposed me to paleomagnetics and continental drift.

(*Le Pichon, 2001: 221–222*)

Heirtzler has also acknowledged that he was, to use Le Pichon's phrase, not "seriously aware" of the land-based defense of paleomagnetism or of the long-standing controversy over continental drift. Given Heirtzler's background, his attitude is not at all surprising. Heirtzler in this frank statement made clear just this point:

When I came to Lamont in 1959 I had almost no knowledge of current topics in geology or geophysics. I did know about the present geomagnetic field and its rapid variations. The marine magnetics program had been run by a part time student named Julius Hirshman and Ewing asked me to take over the supervision of that program. They had just completed Vema Cruise 16 and I was impressed by the great magnetic anomalies in the oceans, especially that high anomaly over the Mid-Atlantic Ridge. I was particularly impressed by a collection of marine magnet profiles that Hirshman had assembled off the east coast of the US which showed a distinct magnetic anomaly running all along the east coast from Canada to Georgia. Nothing like this had ever been found on the continents and suggested a process with which we were unfamiliar. I soon learned that Scripps people had found impressive linear anomalies running off the Pacific coast. They suggested that these might be due to differences in heat flow, although such had not been measured. I knew little about theories of continental drift and paleomagnetic evidence for that. If there had been large linear magnetic anomalies on land I would have given it more attention. I don't know that I ever had any strong feelings against continental drift, it was just a subject that didn't come up. I knew that the paleomagnetic evidence for continental drift was confined to a relatively closed group associated with Runcorn and his camp, Irving and Hospers and some Europeans.

(*Heirtzler, August 4, 2011 email to author*)

I suspect they did not know, as Sykes did not know (§6.9), of the large differences in Australia's APW path compared with those of North America and Europe, just the thing that convinced Bullard of continental drift. In fact, I suspect that except for Opdyke, Dickson, Dunn, and Heezen few others at Lamont knew. If Opdyke explained the paleomagnetic evidence for drift to Heirtzler and Le Pichon, it fell on deaf ears; they did not seem to recognize the implications of what he told them about the dominant effect of remanence in ocean basalts (§4.6).

When at first Le Pichon and Heirtzler, and, perhaps, Talwani balanced the pros and cons of V-M and seafloor spreading, they did not include on the pro-side the strong paleomagnetic support of continental drift and the previous decade of work establishing episodic field reversals. They also did not consider the paleoclimatic support of continental drift. Even Heirtzler, given the climate of opinion at Lamont, may have come to see continental drift as something not to take seriously (see Vine's August 17, 2011 email to author below).

At the end of Volume I (§9.7), I mentioned the insidious effects of "tunnel vision" and "group-think." At the beginning of Volume I (§1.15) I stressed how regionalism and specialization helped shaped the controversy. I also stressed the importance of keeping an open mind, quoting Feynman's eloquent description of what is involved in being open-minded. Here again is what he wrote:

Now, in order to work hard on something, you have to get yourself believing that the answer's over *there*, so you'll dig hard there, right? So you temporarily prejudice or predispose yourself – but all the time, in the back of your mind, you're laughing. Forget what you hear about science without prejudice. Here in an interview ... I have no prejudices – but when I'm working, I have lots of them.

But the thing that's unusual about good scientists is that while they're doing whatever they're doing, they're not so sure of themselves as others usually are. They can live with steady doubt, think "maybe it's so" and act on that, all the time knowing its only "maybe." Many people find that difficult; they think it means detachment or coldness. It's not coldness! It's a much deeper and warmer understanding, and it means you can be digging somewhere where you're temporarily convinced you'll find the answer, and somebody comes up and says, "Have you seen what they're coming up with over there?", and you look up and say *"Jeez! I'm in the wrong place!"* It happens all the time.

(Feynman, 1999: 199–200)

Most at Lamont were not aware of what paleomagnetists were coming up with, and those who thought they did, often from listening to someone who had not actually read the papers himself, had a misapprehension of what it was they had accomplished. Recall Worzel's comment about paleomagnetists fudging their data. Most had only a superficial knowledge of paleomagnetism's strong support of mobilism and the overwhelming evidence in favor of geomagnetic field reversals. As for the latter, many thought that the study of reversals began with Cox and Doell. Speaking of Cox and Doell, Heirtzler (August 4, 2011 email to author) mentioned "Cox and his colleagues" while discussing why he did not consider pro-drift paleomagnetic work when working on marine magnetic anomalies, "but just let the evidence speak for itself."

Allan Cox and colleagues had a large collection of rocks and could only develop a geomagnetic time scale back to about 4 million years and seemed to know much about the ancient geomagnetic field and didn't dwell on continental drift. We had as many as four marine magnetic surveys going at one time and soon provided the evidence for motion of the major continental plates. I did not have to be preoccupied with the arguments of Runcorn, Hospers, or even Allan Cox. I just let the evidence speak for itself!

(Heirtzler, August 4, 2011 email to author)

At Lamont, pro-drift paleomagnetic work was not a factor when weighing the evidence for and against V-M. It helps to keep a lookout for work that is contrary to your own point of view, and not always listen only to those who agree with you. It helps to read works by people who support opposing views. Otherwise you will not recognize group-think and bias when it smiles at you.

But this story gets a little more complicated. Le Pichon, Talwani, and, likely, Heirtzler heard Irving lecture on paleomagnetism at Lamont in summer 1965, and Le Pichon had lunch with him afterwards, at the same table. Irving recalled what happened.

In 1965 (it was summer because our dated family photographs show everyone in summer clothes) we visited for a week or so, Sheila's brother in Leonia, not far north of Lamont. It is my recollection that I drove to Lamont for the day. Ewing was away and Neil was teaching for him part of what I think was a graduate course. He had asked me to give a lecture. I was, as it were, a stand-in once removed for Ewing. It was then my habit always to talk about new results and I recollect that I spoke on work that was about to be published (Irving, 1966) which described my new results from the Late Carboniferous glacial sequences of New South Wales that placed Australia in high southern latitudes. I would also review other Late Paleozoic results from Australia as I did in that paper. Very likely I would show a slide [Figure 1.9, Volume III] tracing the motion of the south geographical pole path from Silurian to Permian as we had determined from Australia projected onto Lester King's reconstruction of Gondwana. Silurian and Devonian poles were in West Gondwana and migrated close to Australia in the Late Carboniferous in the sense that Wegener, du Toit and King had proposed on geological grounds. As you know from my publications, that since my 1956 foray into paleoclimatology, it was usual for me to stress the good intra-continental agreement between paleomagnetic latitudes and paleoclimatic indicators. It is unlikely that I failed to point this out to the Lamont audience, and argue that the above was another good example – consilient with the GAD time-averaged model of the field and hence with continental drift. As reward, Neil very appropriately presented me with a blue shirt on which was printed the reassembled southern continents and entitled "Gondwanaland Reunited." I still have it. Afterwards we lunched in the cafeteria. Xavier Le Pichon, whom I had not met previously, was at our table. He was unhappy with my news. He said, "There must be something wrong with paleomagnetism." I can vouch for that. It was a long time ago, but I shall never forget being told with such confidence that, working on paleomagnetism, as I had for the past fifteen years, I had been barking up the wrong tree.

(Irving July 16, 2011 letter to author)

One lecture, just like one field trip to southern Africa, was for the then fixists Wilson (§1.2) and Oliver (§7.2) an insufficient antidote to tunnel vision and group-think.[19] Le Pichon apparently immediately rejected even toying with the idea of picking up the crustal mobility end of the stick, and asking, for example, if there was some way to reconcile the heat flow data with seafloor spreading? This is what McKenzie (§7.5) did soon after leaving the Goddard meeting, having been convinced by Vine's and Sykes' talks that seafloor spreading was essentially correct. Le Pichon commented on McKenzie's move:

It is interesting to note that Dan McKenzie, a young scientist from Cambridge who was then working at Scripps and who would soon play a significant role in the elaboration of the plate tectonic model, made the same computations one year later, arguing that the heat flow was compatible with sea floor spreading. To obtain the correct results, he chose a temperature inside the mantle three times smaller – 550°C instead of the 1500°C that we had chosen. This latter temperature was then and is still considered to be much closer to the actual mantle temperature. But McKenzie was already convinced of the validity of the sea floor spreading model, and he preferred to adjust the parameters rather than arrive at obvious discrepancy.[30] At the time, whether the fixist or the mobilist model was adopted a certain number of observations did not agree with the predictions. The choice made was heavily influenced by the environments, the working philosophy, and discipline in which one worked.

(Le Pichon, 2001: 210; endnote 30 refers to McKenzie (1967b))

I shall consider McKenzie's solution later (§7.5). My point here is that Le Pichon was confident that he had considered all the relevant factors. But he had not. Unbeknownst to him, and in fairness to everyone at the time, was the huge cooling effect of hydrothermal circulation (§6.11). If Le Pichon had paid more attention to what paleomagnetists were digging up elsewhere, he might have wondered if the measured heat flow over the crest of the Mid-Atlantic Ridge, excluding the particularly low readings, was more in line with his predicted value given seafloor spreading. He might have predicted that there was some unknown process causing these peculiar low values, some hidden parameter, that immediately removed much of the rising heat. Had he done so, he would have been hailed as a prophet for having had such foresight to foresee Lister's 1972 proposal concerning the cooling effect of fluid advection in the crust, later confirmed in 1979 (§6.11); he would have imagined prophetically an auxiliary hypothesis that would have reconciled or lessened the clash between the measured and calculated heat flow. In describing his emotional state after he saw *Eltanin*-19 and realized that "Hess is right," he touched on just this issue, and the lesson he learned.

This extremely painful "conversion" experience has been crucial in shaping my own vision of what science is about. During a period of 24 hours, I had the impression that my whole world was crumbling. I tried desperately to reject his new evidence, but I had an extraordinary predictive power! Why then was the heat flow three times smaller than expected for sea floor spreading? Why were the magnetic anomalies so different over the flanks of the ridge? Why was the sediment fill in the trenches undisturbed? I did not know, but I was progressively forced by the convincing power of the magnetic anomaly profiles to assume that in all these unexplained observations, there must have been hidden parameters that had not yet been taken into account. Since that time, I know that good data and correct models do not guarantee that our conclusions are definitive: the possibility of hidden parameters is always present.

(Le Pichon, 2001: 212)

Yes. Hess was right about hidden parameters too, having said as much about the very analysis in question, that by Le Pichon (§6.11). Geology is messy compared to physics.

Ewing's strong fixism and sometimes heavy-handed leadership, has also been suggested as a reason for the tardiness of Lamont's acceptance of mobilism. How responsible was Ewing for this strong preference toward fixism? Opdyke (December 8, 2009 email to author) believes that Ewing's negative attitude toward crustal mobility "absolutely" influenced the views of others at Lamont. He believes that Ewing and "all the heavies at the major research centers" resisted "crustal mobility."

The prevailing opinion at the Observatory was opposed to crustal mobility at all and Doc Ewing was the towering figure and he opposed it. He was not alone: all the heavies at the major research centers were of the same opinion, both geologists and geophysicists alike.

(*Opdyke, December 8, 2009 email to author*)

If Opdyke is correct, most at Lamont and other major research centers, at least in North America, were opposed to "crustal mobility." I think this *was* the case. Heezen and Menard were exceptions. But Menard flirted with paleomagnetism when it was in his interest, and Heezen stopped talking about paleomagnetism once its results told against rapid Earth expansion. Moreover, as far as Lamont was concerned, following Heezen's example after his and Tharp's rift with Ewing probably did not seem a good idea.[20] According to Le Pichon, "At Lamont, two schools of thought prevailed. There was Heezen's rapid Earth expansion, and there was Ewing's fixism."

But Maurice Ewing thought that the idea of such a fast expansion ... was physically absurd. He remained a fixist; he preferred to explain the tectonic activity of the rift by deep convection currents that did not reach the surface but were the cause of extension and volcanism, without wholesale movement of the crust. For Ewing, such speculations were premature. What did they bring to science? New facts were within reach of our dredges, corers, cameras, and magnetometers. With his younger brother, John, he was inventing marine seismic reflection, a technique that would continuously record the thickness of the sedimentary layers. This technique was soon to reveal the very thin ocean sediment cover, and its total absence near the rift. But Ewing could not stop Heezen from developing his ideas, and ultimately, conflict between the two men could not be avoided. Ewing could not accept that one of *his* scientists would act in a completely independent way, without any control of the director of the laboratory. In 1967, it would lead to an open and painful split.

(*Le Pichon, 2001: 204; my emphasis*)

Le Pichon (2001: 204–205) maintained that students at Lamont did disagree. There were those who were "mobilist and expansionist." They "were more geologically inclined and included the students of Heezen." "The other, which was more geophysically inclined, and to which I belonged, was fixist and believed in long-standing ocean–continent distribution." Vine also believes that Ewing's presence and attitude toward mobilism influenced the views of Lamonters, and he cites a conversation he had with Heirtzler as an example. Vine speculates that Ewing's negative attitude toward seafloor spreading may have been enhanced by his poor personal relationship with Hess.

I am pretty sure that Jim asked me at that February 1966 meeting whether this all meant that we had to accept continental drift, and I said yes. He seemed rather worried by this answer and

I just felt that he was thinking "How the hell am I going to get this past Maurice Ewing?" Clearly the linear anomalies further away from ridge crests, as in the N.E. Pacific for example, might have been formed by older phases of spreading, but I think that Harry and I were convinced that discontinuous spreading was unlikely, in part for theoretical reasons, but also because the rates deduced for the North Atlantic, coupled with the geological evidence for the timing of the initiation of drift around the margins of the North Atlantic, were consistent with continuous spreading and drift.

I repeat that my conjecture that Maurice Ewing did not want to believe in sea floor spreading because Harry had proposed it is pure speculation on my part, but a distinct possibility. As I understood it, the personal relationship between them was poor and that this derived from time spent together in a submarine in the late '30's with Vening Meinesz. However I could be completely wrong – it needs corroboration.

(Vine, August 17, 2011 email to author)

There is no doubt about the frosty relationship between Ewing and Hess. Annette Hess even remarked about it when, in August 1979, she kindly talked to me about her late husband. Whether his poor relationship with Hess affected Ewing's judgment of seafloor spreading, is, as Vine notes, a matter of speculation.

Talwani and Oliver, on the other hand, emphatically deny that Ewing pressured Lamonters to not favour mobilism. They see Ewing as wanting to "prevent people from going off half cocked" and to "justify thoroughly and in the best scientific fashion what they proposed to publish." Talwani put it this way:

Ewing did, of course, have a tremendous positive influence on scientific thinking at Lamont. But it would be absolute nonsense to suggest that he would try to suppress any papers that expressed views contrary to his own. The influence that he exerted did prevent people from going off half-cocked and presenting ideas without having thought them through.

(Talwani, July 17, 1981 letter to author)

Oliver finds such statements "contrary to what" he "observed."

Ewing's early opposition to sea floor spreading and then nascent plate tectonics has somehow led some to infer that he tried to suppress support for these concepts by controlling the work or the publications of scientists at Lamont during the period when those ideas were being tested and developed. Of course, I cannot claim to have monitored all of Ewing's actions, but I would like to go on record here as noting that I find these statements or inferences surprising and contrary to what I observed. During that critical period, I was a senior staff member at Lamont as well as a faculty colleague of Ewing. As head of the earthquake seismology program and one of the founding group of Lamont, I was clearly a part of the inner circle of leaders ... At no time, however, did that work or the corresponding publications encounter any interference, or barriers, or even negative comment from the Director of the Lamont Geological Observatory. Ewing was a strong and sometimes partisan leader who did not hesitate to demonstrate his opinions and the depth of his conviction, but he was too much the solid scientist driven by the basic truth of observation in science to oppose those truths once attention was drawn to them and the case made ... I thus find myself in complete opposition to, and often somewhat astonished over, some statements that have been made and sometimes published about Ewing's supposed negative reaction and obstruction to certain developments in the plate tectonics revolution. Of course, it might be that Ewing treated some of *his* scientists in one way, others in another. However, I, and some others, think it is more likely that the conflicting views of Ewing are a consequence of different interpretations of his motives as he called upon Lamont scientists to justify thoroughly and in the best scientific fashion what they proposed to publish. Some saw this practice as unjustified and prejudicial opposition to a pet theory, others saw it as merely tough, hard, but nevertheless reasonable and good, science.

(Oliver, 1996: 44–45; my emphasis)

Most Lamonters were opposed to mobilism and Ewing's anti-drift attitude was well-known. I leave readers to judge if and to what extent Ewing's anti-drift views influenced those at Lamont. In addition, Ewing may have thought that any paper supportive of mobilism had to be half-cocked, that is until listening to Vine and his own scientists at the Goddard meeting. Talwani also maintained that the issue at Lamont by 1965 was not whether or not continental drift had occurred, but when it had stopped occurring.

By 1965 most of the scientists at Lamont had accepted the concept of continental drift. The principal question was how late in the history of the earth had it occurred and whether it was still going on. Ewing together with Vening Meinesz had postulated the concept of Ur continent which implied that continents had drifted but drift had stopped no later than the Pre Cambrian (I am guessing the exact time). Others believed that the continents had been in their present positions since some time in the Tertiary. I believe that the imprecision of the pole positions determined from paleomagnetic studies was such that it was not clear whether N. America and Europe had had much relative motion late in the Tertiary. So the question was not really whether the continents had ever moved, but how late the last motion had occurred and whether it was still occurring through the process of sea floor spreading.

(Talwani, July 17, 1981 letter to author)

If believing that fragmentation of a primordial continent, Vening Meinesz's Ur continent, is enough to be a mobilist, then Jeffreys was a mobilist, not the arch-fixist he was. The huge highly significant differences in the APW paths of Australia and India, with each other and with North America and Europe, indicated continental drift since the Tertiary – something Talwani did not mention. Their personal testimony shows that they were not well informed about the evidence for mobilism – either the classical evidence or the more recent paleomagnetic evidence.

I believe that my review of the 1965 publications of M. Ewing, Heirtzler, Le Pichon, and Talwani himself (§4.6, §5.3, §5.4, §6.2, §6.11), their performance at meetings (§5.7, §6.12), and their personal testimony shows that they did not believe continental drift had occurred, and did not consider it seriously until the very end. Regarding the oceans, they did not believe seafloor spreading had occurred. They preferred fixism to mobilism.

6.14 Maurice Ewing reluctantly accepts discontinuous
seafloor spreading

Ewing and company continued to ask whether Atlantic stratigraphy is consistent with seafloor spreading. They wrote three papers. Saito *et al.* (1966b) was received by *Science* in August 1966. It and J. Ewing *et al.* (1966) appeared in the same December issue of *Science*. Ewing and Ewing (1967) was received by *Science* in April 1967, approximately five months after the Goddard conference. In particular, they discussed the age and·nature of Horizon A, the strongest and most continuous sub-bottom reflective horizon they had found in the Atlantic (J. Ewing *et al.*, 1966). In Saito *et al.* (1966b), a more technical paper, they discussed the lithology and paleontology of an outcrop of Horizon A, located approximately due east of the Florida Keys and due north of the most eastward tip of Cuba. They (Saito *et al.*, 1966b: 1173) claimed to have recovered Late Cretaceous aged turbidites interspersed with red clays of deep-sea origin. In J. Ewing *et al.* (1966), the more expansive paper, they claimed, from their work on the outcrop, that the Horizon A is "a buried abyssal plain of Upper Cretaceous age." They also identified two other sub-bottom reflecting surfaces, Horizon β and Horizon B. They claimed that Horizon β, which is directly below Horizon A, is another turbidite layer, and fossil identifications suggested that it was of Middle Cretaceous age. They proposed that Horizon B, which underlies Horizon β, "must be well over 200 million years old," an estimate based by extrapolating downwards from estimating depositional rates of sediments beneath Horizon A. They admitted, however, that it may be as young as mid-Jurassic in age. Given that their localities marked the eastward limits of the three horizons, they agreed (1966: 1131) that Atlantic stratigraphy is consistent with continuous and constant seafloor spreading of the Atlantic at a rate of "about 1 centimeter per annum."

If we accept a late-Jurassic age for horizon B, Cenomanian [Middle Cretaceous] for horizon β, and Maestrichtian [Upper Cretaceous] for horizon A, and if we accept the apparent eastward limits of each as representing the position of the base of the mid-Atlantic ridge at the time of deposition, we can plot the rate of expansion of the North America basin . . . The plot indicates a spreading rate of about 1 centimeter per annum, and extrapolation back to zero basin width gives an age of about 190 million years for the earliest deep basin. The initial break-up of the continents would have occurred about 60 million years before this date, if a constant rate of spreading is assumed.

(J. Ewing et al., 1966: 1131)

They were not convinced, and raised difficulties with the idea of such extensive and constant seafloor spreading; instead they proposed fixism or intermittent seafloor spreading (RS2). They first questioned the key assumptions of uninterrupted constant seafloor spreading.

Such reasoning clearly can be used to support the spreading-floor concept if the assumptions on which it is based are correct. Neither the eastward limits of reflector B and β, nor their ages, are exactly established. Horizon β may well be older than Cenomanian, and the eastern limits of both B and β may extend farther than indicated; thus the line in Fig. 6 probably can be considered to represent the maximum possible rate of expansion, as movement of the Jurassic or Cenomanian points either upward or to the right would steepen the slope. Even if we ignore the Jurassic point, which is certainly the more questionable, the line could still be valid for continuous spreading since Cenomanian.

(J. Ewing et al., 1966: 1131)

They again raised the theoretical difficulty that the undisturbed nature of horizon A was difficult to explain, if seafloor spreading had occurred. "The relatively small distortion of horizon A is remarkable if spreading of the sea floor has indeed transported the basin ... westward during the 70 million years since its deposition" (J. Ewing *et al.*, 1966: 1131). Moreover, they (1966: 1131) thought that there was some "rather good evidence" that horizon A extends out to the base of the ridge in some places. If their "suspicion" were correct, then "post-Cretaceous spreading ... would be restrict[ed] to the ridge province."

They then noted that what was known of the stratigraphy of the Atlantic oceanic crust was also consistent with fixism.

So far we have assumed that the sediments at each stage of the deposition flowed out from the North American continent and covered essentially the entire basin floor that existed at the time. We should point out that the observed pattern of eastern limits versus age of sediments might be expected if the width of the basin had not changed at all.

(J. Ewing et al., 1966: 1131)

Given a fixist interpretation, the lack of sediments on the crestal region of the Mid-Atlantic Ridge did not indicate a youthful seafloor but burial of sediments by lava. They also let their readers know of the difficulty that Langseth, Le Pichon, and Ewing had just raised against massive mantle convection because of the limitation of high thermal gradient to a "narrow strip along the crest."

The observation that the thermal gradient is high only in a narrow strip along the crest of the mid-Atlantic ridge also poses serious objections to a long-sustained convective flow (20); the objections would be less serious if the convection had been stopped for a long time and only recently resumed.

(J. Ewing et al., 1966: 1131; (20) refers to Langseth et al. (1965) as forthcoming; it appeared in 1966)

Not finished, they played their trump card, the Miocene fossils recovered from the rift valley. Of course, these may have been left behind, as they had previously acknowledged.

It seems possible, therefore, that a recent resumption of convection, after a period of quiescence extending back possibly to the end of Cretaceous time, can explain the sediment distribution, the pattern of heat flow, and the pattern of magnetic anomalies (21), but there is serious conflict between recent convection spreading and the recovery of Miocene fossils from the rift valley unless we allow the possibility that "patches of older sediment were left behind in the crestal area rather than being completely swept away from the axis (11).

(*J. Ewing* et al., *1966; (21) refers to Vine and Matthews (1963); (11); to Saito* et al. *(1966a))*

Perhaps Ewing was still hoping the *Eltanin*-19 profile would disappear.

Ewing and his brother sent a note to *Science* in April 1967; the paper appeared in June. They accepted seafloor spreading but only intermittently. They rejected continuous seafloor spreading. They began by describing their previous attack on seafloor spreading as merely an attempt to reconcile Atlantic stratigraphy with seafloor spreading, and noted that they had proposed recent spreading after a lengthy hiatus of no spreading.

In an earlier paper (1) that describes certain features of the sediment distribution in the North Atlantic Ocean, an attempt was made to reconcile the pattern of accumulation with the hypothesis of spreading sea floor ... It was noted that although the virtual absence of sediment in a strip 160 to 240 km (199 to 150 miles) wide along the crest of the mid-Atlantic ridge supported the hypothesis of spreading, a relatively uniform sediment cover on most of the ridge flank seemed to require a uniform age of the flanks rather than a progressively greater age away from the crest. Accordingly, the suggestion was made that the spreading that has produced the strip of thick sediments on the crest is relatively recent and that it has been preceded by a long period of quiescence during which the flank sediments had accumulated.

(*Ewing and Ewing, 1967: 1590; (1) refers to J. Ewing* et al. *(1966))*

They admitted that Pitman and Heirtzler (1966) and Vine (1966) "have given strong support of the hypothesis of spreading by demonstrating close agreement between the patterns of the magnetic anomalies of ridge crests and the known reversals of the earth's field" (Ewing and Ewing, 1967). But they still rejected continuous seafloor spreading. They reproduced sedimentary profiles over the Mid-Atlantic Ridge in both the North and South Atlantic, over one of the active ridges in the Indian Ocean, over the East Pacific Rise or Antarctic Ridge in the South Pacific, and over the East Pacific Rise near the equator. They (1967: 1591) argued that changes in sediment thickness as a function of ridge-distance showed that seafloor spreading during the Late Mesozoic or Early Cenozoic had broken up Gondwanaland and Laurasia and moved their fragments "approximately to their present locations"; that this early spreading cycle was "followed by a period of quiescence, during which most of the observed sediments were deposited on a static crust; that a new spreading cycle began approximately 10 million years ago simultaneously along all active ridges," and added that "the ridge axes of the latest cycle follow those of the preceding one with remarkable accuracy."

The Ewing brothers were nothing if not ingenious. They had found a way to interpret the magnetic anomaly patterns in such a way that the magnetic tape recorders over all active spreading ridges ceased at the *same* time, and, after a hiatus of approximately 60 million years, simultaneously restarted in the *same* places.

But Heirtzler and co-workers at Lamont would soon begin identifying anomaly patterns among different oceans extending over 1000 km from ridge axes (§7.7).

The nature and age of Horizon A was taken up again several years later by S. Gartner, Jr. He (1970: 1077) based his arguments on cores obtained during leg 2 of the Deep Sea Drilling Project. They indicated that the North Atlantic had been spreading at a constant rate of 1.1 cm per year for the last 80 million years "without any major episodes of quiescence or rapid spreading." He agreed with Ewing and company that Horizon A is widespread in the North Atlantic. Paleontological dates from the five cores taken at different distances from the Mid-Atlantic Ridge, along a line roughly normal to the ridge, indicated that the Horizon A formed during early and middle Eocene time and that the age of the seafloor increased as its distance from the ridge increased. He explicitly noted:

This conclusion is in contrast to the view, advanced by Ewing and Ewing (4), that sea-floor spreading is episodic, with the present episode of spreading beginning about 10 million years ago and preceded by a long period of quiescence (4).

(Gartner, 1970: 1078; (4) refers to Ewing and Ewing (1967))

Gartner also argued against Ewing and company's claim that Horizon A was formed entirely of turbidites. Turbidites were restricted to cores taken near continental margins; cores farther from the continents showed "little or no evidence of turbidite origin of the horizon A chert." He argued that Horizon A is a chert, initially a siliceous ooze, and that turbidites are found only near continents.

6.15 Why seafloor spreading was rapidly accepted by most marine geologists and geophysicists

Confirmation of V-M and the idea of ridge-ridge transform faults led to the overnight acceptance of seafloor spreading by marine geologists and geophysicists. Field reversals and remanence of oceanic basalts were no longer in question, and Vine had credible answers to all the difficulties that had been raised against V-M (RS2). Profiles across ridges showed bilateral symmetry – they matched magnetic lineations across the Juan de Fuca Ridge, and especially the wonderful profiles by Lamont across the Reykjanes and Pacific–Antarctic ridges (RS1). Vine had also linked seafloor spreading and continental drift by showing agreement between independent determinations of when the Atlantic opened based on the rate of seafloor spreading across the Reykjanes Ridge and based on radiometric dating of continental rocks (RS1). Work by Opdyke and colleagues on deep-sea sediment cores independently confirmed the reversal timescale as determined by the USGS and ANU groups, and matched the reversals displayed by the magnetic profiles (RS1). Consilience abounded (RS1). Reversals as manifested through seafloor spreading over several ridges, through magnetization of ocean basin sediments, and through magnetization of dated lava flows, basically gave the same reversal timescale. As for transform faults, the second key corollary of seafloor spreading, they were splendidly confirmed at Lamont by Sykes (§6.9, §6.12). With much improved data, and stereographic projection techniques, his study of first motions of earthquakes either on ridge-offsets or on fracture zones between ridge-offsets confirmed the existence of ridge-ridge transform faults. His result was never questioned. Sykes' work was so clean; the evidence was so "black and white" as McKenzie later (March 18, 1990 letter to author) put it. With the acceptance of these two difficulty-free solutions came the acceptance of seafloor spreading. There also were, as Bullard reminded those at the Goddard meeting (§6.12), the long-known geological and geometric disjuncts recently buttressed by computer fits of continents surrounding the Atlantic (§3.4), Hurley's matching radiometrically determined ages of rocks from Nigeria and Brazil (§3.3, §6.12), restated by Kay and Dewey (§6.12). And, there was the independent confirmation of continental drift by land-based paleomagnetic work beginning in the early 1950s that led to mobilism's first difficulty-free solution (II, Chapter 1). This was much less readily accepted by most former fixists than the two difficulty-free solutions related to seafloor spreading.

I think there were several reasons why this happened. Understanding these solutions was much less of a stretch than understanding the land-based continental paleomagnetic support of mobilism. Few North American geologists knew enough geomagnetism to understand and appreciate drift's paleomagnetic support, few read the actual papers, and if they read anything in paleomagnetism it likely was Cox and Doell's fence-sitting *GSA* review (II, §8.4; II, §8.7). Second, most of those who followed from afar the debates over seafloor spreading, transform faults, and V-M, would recognize Lamont as a fortress of fixism. So when workers at Lamont not only obtained many of the key results, but changed their minds, and joined the mobilists leading the charge, they figured that everything was aboveboard; pre-existing bias could not be involved. Again, this stands in contrast to the land-based paleomagnetic work where many fixists, ignorant of what these paleomagnetists had done, conveniently assumed that their data were unreliable, their analysis flawed, or, worse, that they had fudged the data. That never happened in the debate about the mobility of ocean floors. Third, those who had formerly questioned continental paleomagnetic work now realized that it was correct after all, given the new consilient findings in marine geology and geophysics. Ironically, those who worked on marine magnetic anomalies were actually doing paleomagnetism because the anomalies were produced by the fossil magnetization of the basalts that clothed the ocean floor. They were doing paleomagnetic work remotely, doing it by other means. Uniquely at the time, because of Lamont's unrivalled core collection, it was only Opdyke, Glass, and Foster who were doing marine-based paleomagnetism by the usual means, that is measuring directly the magnetization of partially oriented deep ocean cores (oriented vertically but not in azimuth and sufficient to determine polarity in higher latitude

(§6.5)). By whatever means, the role of paleomagnetism within the mobilism controversy, whether over continents or oceans, was front and center.

I want to emphasize that although the two corollaries with their difficulty-free solutions warranted the acceptance of seafloor spreading it still faced difficulties. There was the major conflict between predicted and measured heat flow over ridge crests and flanks, minor issues over the distribution of ocean floor sediments, and questions about the thickness and rigidity of the material beneath the seafloor that moved along with it. Just what happened beneath island arcs and trenches was also not well understood. During the 1950s, Benioff had argued that oceanic crust downthrusted beneath trenches, and, of course, Fisher, Hess, Vine, and Wilson believed that seafloor descended into mantle beneath trenches, but none of the details were known satisfactorily. There was also Earth expansion, and its defenders argued that confirmation of V-M and ridge-ridge transforms confirmed it (Carey, 1976). But it continued to fail paleomagnetic tests (§3.10) and did not gain support from Wells' study of Devonian and Carboniferous rugose corals (§3.10), and was thought by most to be physically impossible. Heezen, one of its two most influential supporters, abandoned it. Carey, its other major proponent, held fast, but became more and more isolated. But Earth expansion was not entirely out of the picture; getting a better fix on what was happening beneath trenches was needed. The most significant difficulty, I believe, was the apparent impossibility of seafloor spreading, given that Antarctica and partially Africa are surrounded by ridges. This problem, initially raised in 1957 by Heezen and Ewing in regard to continental drift and mantle convection (§6.8), remained. Although it was no longer an impediment to accepting seafloor spreading, it still had to be explained. Wilson (§5.2) and Menard (§6.10) toyed with the possibility of migrating ridges and one-sided convection. However, a more radical solution was needed, namely a separation of upwelling mantle convection and the formation of ridges. Most thought that the ridges were formed by active convective upwelling, and that the position of a ridge was determined by where mantle convective currents ascended and broke the crust. This problem was eventually solved by viewing ridges as passive structures. With the breaking apart of a lithosphere slab, upwelling isothermal basalt simply fills in the gap.

Finally the mechanism question remained unsolved. I suppose most everyone who accepted seafloor spreading thought it was somehow caused by mantle convection. However, what sort of mantle convection was involved and how it worked were outstanding questions, as Bullard emphasized at the Goddard meeting. Although seafloor spreading provided the process by which continents drifted, it did not ultimately explain what forces and how such forces caused drifting continents. So mobilism was accepted without a mechanism. Earth scientists had showed that continental drift and seafloor spreading are possible by demonstrating that they occur. This was entirely different from attempting to prove that they are possible through some theoretical argument as a response to Jeffreys' and MacDonald's arguments that because no satisfactory dynamic model had been proposed continental drift cannot happen. As Le Pichon learned to appreciate the hard way, Earth's complexity often surprises those who argue that something has been shown to be impossible through mathematical physics; hidden parameters are sometimes lurking in the abyss. An elegant kinematic model, plate tectonics, would soon be proposed, but it was geometrical only and not dynamic.

Notes

1 This paragraph was contributed by E. Irving.
2 See Glen (1982: 328) for an account of the conversation, Foster's construction of his magnetometer, and Foster and Glass's first measurement of one of Hays' cores.
3 Frankel (1982), Glen (1982), and Wertenbaker (1974) offered early accounts of Pitman's and Heirtzler's work on *Eltanin*-19. Pitman (2001) later presented his own account, and Opdyke (1985, 2001) also recounted events surrounding *Eltanin*-19. I highly recommend Glen (1982), which contains interesting details based on interviews, especially with Pitman, Opdyke, and Heirtzler. I shall draw on all the above in this account, as well as on new information from Heirtzler, Opdyke, Pitman, and Vine.
4 This biography of Pitman is based almost entirely on the three-part interview of Pitman by Doel (December 14, 1995; January 9, 1997), and by Levin (May 22, 1997). Niels Bohr Library & Archives, American Institute of Physics, www.aip.org/history/nbl/oralhistory. html.
5 Pitman "would rather not identify the researchers, so as not to embarrass them. They had been a bit arrogant about it" (Pitman, July 31, 2011 email to author).
6 Pitman does not remember actually making the statement, "You are going too far, I would be afraid to publish such conclusions" (Pitman, July 31, 2001 email to author).
7 Opdyke (1985: 1181) also remarked that Talwani was "still opposed to seafloor spreading in early summer." However, when asked about his conflicting answers, he said that he thought what he said in his interview with Doel was correct (Opdyke, August 9, 2011 email to author).
8 Vine recalled an amusing story about apparent resistance by some at *Science* to publish so many papers on Earth science.

> There were many rumors circulating at the time regarding the lobbying and discussions that were going on in relation to these two papers. The only hard evidence I ever had of this, apart from the delay in publication, was a comment from Harry Hess, who was a good friend of Phil Abelson, the editor of *Science* at the time. Abelson had asked Harry whether he thought that my paper was worth publishing; apparently he felt that *Science* was carrying too many earth science articles at the time.
>
> *(Vine, 2001: 63)*

9 This biography of Sykes is based almost entirely on Ron Doel's excellent interview of Sykes, Niels Bohr Library & Archives, American Institute of Physics, www.aip.org/history/ nbl/oralhistory.html). Sykes is refreshingly open, admitting, for example, his ignorance of the classical arguments for continental drift, and, more interestingly, his unfamiliarity with paleomagnetism's strong support of continental drift and field reversals.
10 See Bott (2001) for an account of his development of the program and for a more general discussion of the state of seismology in the 1960s, and work on fault mechanisms.
11 In his interview with Doel in 1997, Sykes thought that he had seen *Eltanin*-19 in February 1966, and that he had attended a symposium in 1965 in Newcastle and saw Urey before he visited Isacks in Fiji, but in his comments to Wertenbaker in 1974 or before, he said that he did not see *Eltanin*-19 until June 1966. I believe that what he told Wertenbaker is correct. After almost twenty-five more years, I think he simply mixed up the sequence and when

he saw *Eltanin*-19. Both Sykes and Urey are listed as having participated in a 1966 symposium in Newcastle that began on March 30 and ended on April 7 (Sykes, 1967a; Urey, 1967). There also was a symposium in Newcastle in April 1965, but neither Urey nor Sykes are listed as having attended and neither contributed anything to the proceedings. Sykes was definitely visiting Isacks in Fiji after July 26, the date when Wilson's first paper on transform faults appeared. Thus I think Sykes returned from Fiji, finished his paper "Seismicity and deep structure of island arcs," spoke at the Newcastle meeting, saw *Eltanin*-19 in June after his return to Lamont, realized V-M was correct, and immediately began to test the idea of ridge-ridge transform faults.

12 Sykes' point that nobody realized Hodgson's mistake of thrust faults as strike-slip (transcurrent) helps explain, I believe, why Hess, relying on Hodgson's analysis of earthquakes associated with island arcs and trenches as strike-slip, proposed a vertical plunge of seafloor into the mantle at trenches, even though it was contrary to Benioff's and Fisher's belief that the seafloor was thrusting back into the mantle along descending planes that extended underneath trenches and continued underneath island arcs or continental margins (III, §3.20).

13 The Dutch seismologist A. R. Ritsema also developed the same projection method. Ritsema's 1964 paper appeared in *Geofisica Pura e Applicata* and was probably missed by many working in North America (see Bolt, 2001).

14 Sykes later added that he does not know if Ewing made the connection between his focal plane solutions and Wilson's idea of ridge-ridge transform faults and seafloor spreading.

I don't know if Ewing made the connection or not. He must have seen the Pitman-Heirtzler results. [He did.] He made a lot at that time and for at least a year later about the sudden increase in sediment thickness away from mid-ocean ridges as an argument against seafloor spreading.

(Sykes, August 3, 2011 email to author; my bracketed addition)

15 My understanding of developments in measuring and analyzing heat-flow during the 1960s has been greatly enhanced by discussions with John G. Sclater. I could not have written this section without his aid. See Sclater (2001) for his views about the importance of Richard Von Herzen's contributions to the heat-flow community, and why its members did not have a greater impact on the outcome of the debate over seafloor spreading. See Sclater (2004) for a concise early history of the heat-flow community, and the impediments in figuring out why the measured heat flow over ridges was lower than expected, given most versions of seafloor spreading.

16 Bullard (1951, 1954a, b), Bullard and Day (1961), Vacquier and Von Herzen (1964), Lister (1962), Lee (1963), and Langseth and Grim (1964) made measurements in the Atlantic; Bullard (1952), Revelle and Maxwell (1952), Bullard, Maxwell, and Revelle (1956), Von Herzen (1959, 1960, 1963), Von Herzen and Maxwell (1959), Von Herzen and Uyeda (1963), and Langseth, Grim, and Ewing (1964) made measurements in the East Pacific; Von Herzen and Vacquier (1966) and Sclater (1966) made measurements in the Indian Ocean.

17 Dr. R. D. Schuiling, Chair Geochemistry (emeritus since March 1997) was a NATO Research Fellow at Princeton during the 1965–6 academic year. He received his Ph.D. at Utrecht under Nieuwenkamp. Concerning his attitude toward continental drift, he wrote (July 25, 2011 email to author), "I was trained between 1950 and 1957 not to believe in continental drift. When I returned to the Utrecht University in 1961 continental drift was generally accepted. Remember that people at our Paleomagnetic laboratory had significantly contributed to it. Despite the negative attitude before 1957, my supervisor had a map of the continents along the Atlantic Ocean in his room, where you could not only see the congruent coastlines, but also how the general geology on one side of the ocean fitted the geology on the other side." Schuiling did not know about Hess's review.

18 Dewey and Kay (1968) discussed the Appalachian-Caledonian disjunct. They mentioned F. B. Taylor, but apparently thought there was insufficient reason to cite Argand (I, §8.7), Collet (I, §8.9), Holtedahl (I, §8.11), and Bailey (I, §8.13).

19 An interesting aside is that Ewing decided to teach a course in paleomagnetism. Opdyke explained what happened.

> Ted [Irving] did give a lecture and Ewing did take note. In fact the next semester he taught a course using Ted's book as the text. I think that he figured that he could learn paleomagnetism by teaching it. I taught some of it as you might expect.
>
> *(Opdyke, December 8, 2009 email to author; my bracketed addition)*

20 Even if the rift between Heezen (and Tharp) and Ewing was not based on Heezen's promotion of rapid Earth expansion (III, §6.16), it was, at least by people such as Le Pichon, viewed as such.

7

The birth of plate tectonics

7.1 Outline

Once the drift controversy was resolved, progress was rapid as seafloor spreading and continental drift were cast into the mold of plate tectonics, which was further confirmed and conceptually articulated within two years after its discovery. During the period immediately after the Goddard meeting, McKenzie, who accepted seafloor spreading because of Vine's and Sykes' work, treated the heat flow question as an unsolved problem for seafloor spreading rather than as a difficulty raised against it, and offered a solution. In doing so, he jettisoned any close tie between upward mantle convection and mid-ocean ridges, essentially making them rootless, and paving the way for the development of the purely kinematic theory of plate tectonics. Heirtzler and his group of geophysicists mined Lamont's huge marine paleomagnetic data-set and mapped out the histories of several ocean basins, and with the help of land-based paleomagnetic results extrapolated the reversal timescale back 70 million years. In early spring 1967, Jason Morgan discovered the theory of plate tectonics. This monumental discovery turned seafloor spreading and continental drift into a more powerful theory. He essentially took Wilson's ideas of transform faults and rigid blocks and applied them to a sphere. This allowed him to describe their relative motion in terms of Euler's Point Theorem. Morgan first presented his theory at the spring 1967 AGU meeting where few understood or appreciated what Morgan had done. Only Le Pichon pursued Morgan's idea, further developing and testing it. Meanwhile, in fall 1967, McKenzie, not knowing of Morgan's work, discovered plate tectonics on his own while at Scripps. His paper, co-authored with Parker, was published in December 1967, but Morgan has priority for discovering plate tectonics. Isacks, Oliver, and Sykes (1968) soon showed, using Lamont's huge data-set of seismological earthquake data, that plate tectonics explained the origin and character of earthquakes, and Isacks also developed an explanation of deep foci earthquakes associated with subduction zones. Menard and Atwater (1968), Pitman and Hayes (1968), and Vine and Hess (1970) applied plate tectonics to new problems. This highly progressive period was finished off by McKenzie and Morgan (1969), explaining the evolution of triple junctions, the points where three plates meet. They

437

provided a formal procedure for determining the stability of triple junctions and their work was crucial for applying plate tectonics to past relative motions between plates. The story is continuous, yet lines have to be drawn somewhere, and in ending this book there is a need for me to relate in an important way the first serious challenge to fixism in the mid-1950s (continental paleomagnetism) with the final coup a dozen years later. So, except for discussing the complementary and natural relationship between land-based paleomagnetism and plate tectonics, which is essential to paleo-geography and which is necessary to link the volumes together, I have ended the account of the emergence of plate tectonics with triple junctions. As a result, I have not considered Tanya Atwater's magnificent application of plate tectonics to the Cenozoic history of the northeastern Pacific and its interaction with the western margin of North America (Atwater, 1970), Richard Hey's propagating rift model (Hey, 1977; Hey et al., 1980), perhaps the last theoretical (geometric) addition to plate tectonics; Morgan's influential development of the idea of mantle plumes (Morgan, 1971, 1981, 1983), which arose from Wilson's (1963c) idea of hotspots; and John Bird and John Dewey's application of plate tectonics to mountain building (Bird and Dewey, 1970; Dewey and Bird, 1970).

7.2 Bryan Isacks and Jack Oliver at Lamont

Although John (Jack) Ertle Oliver (1923–2011) and Bryan Isacks have been intro-duced (§6.9), it is time to become better acquainted with them and to describe their analysis of Benioff–Wadati zones in terms of subducting slabs of lithosphere. Oliver was born in Massillon, Ohio.[1] For Oliver, high school meant mathematics, science, and football – not necessarily in that order. Oliver's high school football coach was Jim Brown, the famous coach of The Ohio State University, and later of the Cleveland Browns. Through football with Brown, he learned discipline and gained confidence. Graduating in 1941 from high school, he thought of attending Case Western Reserve, Harvard, Cornell, and Columbia, but chose Columbia. Columbia gave him a football scholarship but he still had to work two jobs. He concentrated on mathematics and science during his first two years, before being drafted into the Navy in mid-1943, spending two years in Hawaii and one in the Philippines working as a naval civil engineer. In summer 1946, he returned to Columbia, concentrated on physics, received his undergraduate degree in 1947, and began doing graduate work at Columbia in physics. He also got a job working for Maurice Ewing.

At the end of the academic year, I made a late decision to forego my senior year of college in order to begin graduate study in physics at Columbia during the following semester... Early in the fall of 1947, I confided to a friend that I needed money and was seeking a job. As luck would have it, the conversation was overheard by Dick Edwards, and he kindly volunteered some help. Dick led me to an unfamiliar part of the campus and introduced me to his employer, Professor Maurice Ewing. In less than five minutes Ewing interviewed me, offered me a job, and I accepted with delight. As I left Ewing's office, I noticed the title "Professor of Geology" on the door. It struck me that he must have thought that I was a graduate student in geology, not physics. I reentered his office and said, "Professor Ewing, I'd like that job very much. However, you're a professor of geology. You didn't ask me about it, but I feel I must point out that I've never studied geology in my life." I was certain he would withdraw the offer and I would once again be job seeking. Instead Ewing's response was my introduction to the real world of science, a world that extends far beyond the strict organization and regimentation of classrooms, curricula, and books. "Well," he said, smiling and shaking my hand, "that'll be two of us!" Ewing would become one of the greatest of earth scientists, and of course, he knew a great deal of geology, but it was all self-taught or learned informally from others. His Ph.D. was in physics and he never studied geology in a formal way.

(Oliver, 2004: xi–xii)

By 1950 Oliver had earned an M.A. in physics. Inspired by Ewing, he switched to geophysics, finding it more interesting, and liking the idea of doing research outside. He took only one course in geophysics; most were in geology, including ones from Bucher and Kay. Settling on earthquake seismology, he worked with Ewing and Press and, in 1953, earned his Ph.D. in geophysics. In 1955, Press left Lamont for Caltech, and Oliver took over his vacated position as head of Lamont's earthquake seismology program, soon becoming an assistant professor of geology.

With Bucher and Ewing as his guides, it is no wonder that Oliver became a staunch fixist. He witnessed the 1951 debate at Columbia between Bucher and King over continental drift (I, §6.10). Like everyone else who has recounted what happened, he agreed that King won the debate; however, he believed that Bucher deliberately lost the debate, sacrificing himself for students in order to stimulate discussion of continental drift. Regardless of whether Bucher planned it, students did discuss the merits of continental drift but at the end all remained fixists.

At one point during the 1950s [October 31, 1951], Bucher debated Lester King, the prominent South African geologist, on the subject of continental drift. Bucher was a staunch opponent of drift. In the debate, held in the era preceding the arrival of plate tectonics, Bucher's arguments against drift were noticeably weaker than those he normally presented in class. King won the debate easily and the student body at Columbia was stimulated to discuss continental drift heatedly for many months (unfortunately we then discarded it and had to revive it later). I think Bucher planned it so as to startle us and sacrificed himself in the cause of stimulation of thinking by those students. . . That's my view because of the way it sounded to me. But maybe other people had a different idea. He never told me that. Anyway for several months after that, the whole student body, graduate student body, was stimulated to talk about continental drift. And this must have been in the, oh gee, in the 1950s some time. And I would guess maybe mid-1950s. That was a time when we all looked into this matter of continental drift, semi-seriously at least. Well at the end of that period, almost every last one of us, I think every last one of us, all went back to becoming fixists, saying well there must be something wrong with this idea of continental drift, because everybody else around us says it's incorrect. But nevertheless that was a good thing for us – memorable thing.

(Oliver, 2004: 109–110; Oliver, March 8, 1996 interview with R. E. Doel)

Again, the power of group-think!

Oliver had a second chance to be persuaded by King about continental drift. In 1954, he met King while installing a seismograph in South Africa, but by his own admission was too close-minded to take him seriously.

I became acquainted ... in South Africa, and in Durban with Lester King, a well-known geologist and a leading proponent of continental drift. Favoring continental drift was not a popular position for an earth scientist, at least one in the US, at that time. It was some years before the coming of the plate tectonics revolution of the 1960s that incorporated continental drift, so King was considered a rebel. However, he was on the right track. I should have learned from him then rather than forgiving him for an idiosyncrasy as I did at the time. Later, of course, as I played a role in the plate tectonics revolution, I would come over to his side of that debate.

(Oliver, 1998: 55)

Oliver also recalled that everyone in Columbia's geology department, except for the South African petrologist Arie Poldervaart, was a fixist and that nobody at Lamont besides Heezen, who Oliver mentioned in a different context, was a mobilist until Opdyke arrived.

Now you'd have to remember in the geology department, at the time before Lamont was formed or shortly thereafter, there was a South African. It was Arie Poldervaart. He was a petrologist. And he I think was a drifter. I never took his course, and he was pretty quiet about it. But I believe it's the case that he was a drifter. . . I didn't know Arie too well. And at least we had one guy with the right kind of a background to be a drifter there. Then the next one that showed up at Lamont . . . was Neil Opdyke. Opdyke had been trained in England under a drifter, Runcorn, and they worked in Africa and so on. And so he was a drifter when he came to Lamont.[2]

(Oliver, March 8, 1996 interview with R. E. Doel)

Bryan Isacks was born in New Orleans, Louisiana, on July 25, 1936.[3] An excellent student, he was accepted at Harvard, Princeton, and Columbia, and chose Columbia because he "got the best scholarship from Columbia and was attracted by its location in New York City," and in fall 1954 entered Columbia College, where he concentrated on geology and physics, and in 1958 received an A.B. degree in physics and geology (Isacks, September 9, 2011 email to author).

According to Isacks, continental drift was "rarely discussed" until after Vine and Matthews (1963) appeared.

During my undergraduate and much of my graduate school time at Columbia, issues related to continental drift were rarely discussed or thought about. I was busy learning basic physics and the classical fields of geology – mineralogy, structural geology, etc. At Lamont, pre-1964, the geophysical research centered around seismology and marine geology and geophysics. Interpretation of the magnetic anomalies obtained by the Lamont research vessel *Vema* was the only activity in magnetics that I remember at Lamont pre-1964. I remember vividly talking with the graduate student struggling in vain with the interpretation of the large positive and negative magnetic anomalies. He became so frustrated that he eventually left Lamont and joined a mining exploration company. I can't remember his name. This was before the famous Vine & Matthews paper appeared.

(Isacks, September 9, 2011 email to author)

His "earliest brush with continental drift was listening to a talk given by Heezen at the International Oceanographic Conference given at the UN in New York in 1959 [III, § 6.9]" (Isacks, September 9, 2011 email to author; my bracketed addition). Isacks, like others I have asked, and like Le Pichon, believes that the "paleomagnetic work going on during the 1950's and early 1960's didn't seem to have much impact at Lamont before the mid 1960's, probably because there were no paleomagnetists there then [until Opdyke arrived in December 1963]" (September 9, 2011 email to author; my bracketed addition). He also recalled that before he left for Fiji, "I knew little about [seafloor spreading] really, nor had yet read the Dietz and Hess papers, or little if any of the emerging paleomagnetic work" (September 9, 2011 email to author). He also recalled that he became a mobilist after seeing *Eltanin*-19. "Like Sykes and Oliver I was bowled over by the *Eltanin*-19 profile" (September 9, 2011 email to author). So Isacks, Oliver, and Sykes had unfavorable opinions of continental drift before they learned of *Eltanin*-19, were unfamiliar, at least in any detail, with the land-based paleomagnetic support of continental drift, and, except for Oliver, knew little about drift's longstanding arguments.

In summer 1957, Isacks met Oliver, having just finished his junior year of undergraduate studies. Oliver needed an immediate replacement to run Alpha, a scientific station on T3 (Fletcher's Ice Island), a former iceberg which used to drift clockwise in the Arctic Ocean. Oliver called Isacks to see if wanted to go.

For the summer of 1957, I had hired a scientist for that field position well in advance of the departure date, but as that date approached he reneged, and I had to find a last minute replacement. With no other applicants for the Arctic job, I chose the best, as yet unhired, candidate for summer work elsewhere at Lamont. He was a junior in Columbia College, majoring in geology and physics. He had spent three summers doing seismic reflection work in the steamy swamps of the Gulf Coast of Louisiana, but had no previous Arctic experience and so was marginally qualified on that score. Nevertheless, a professor who had had him in class spoke highly of him, so I called him at home in Louisiana where he had gone following the end of spring semester. It was a Thursday afternoon. In a few words, I described the project and the job, the trying conditions and the remote locations, and the enforced isolation, and then asked if he wanted the job. There was a pause of a very few seconds while he pondered this situation that was completely new to him, and then, decisively he responded, "Yes!" "All right," I said, "be here on Monday morning, ready to go." He arrived on schedule, adjusted quickly to his new lot, departed shortly thereafter for T3, and, fortunately for both of us, spent a successful and rewarding field season on the ice island.

(Oliver, 1996: 63)

Isacks also remembered the phone call. It was a welcome one.

Jack's story is correct as I remember it. I talked with him on the phone. Jack did not teach undergraduate classes, so this call was my first contact with him. I had applied to Lamont to go on the *Vema*, but did not get accepted for that. Instead, I was planning a trip out west, combining working on farms or whatever with hiking and climbing in the Rocky Mountains. The details of this plan were not clear, so Jack's call was quite welcome. It seemed like a great outdoors adventure. My family was away on vacation, so I called on my grandmother for money to fly up to New York.

(Isacks, September 9, 2011 email to author)

He also recounted his time on T3. He became adept at running equipment at remote stations, which put him in good stead to equip and to run seismological stations in the Tonga–Fiji region.

When I arrived at T3, after an 8 hour flight from Thule, Greenland on a twin engine WWII "goony bird," I learned that I was replacing the man who was to be the scientific chief of the station. The team at that point consisted of several Air Force support staff and meteorological observers. My job was to get the ocean bottom sounding and coring equipment working. The equipment was located in a tent on the ice pack just off the edge of the ice island, where the ice was thin enough to have a hole through which hydrophones and coring equipment could be lowered into the ocean beneath. I found the seismic equipment completely submerged beneath a melt water pool and useless, so I spent much of the summer learning how to do ocean bottom coring and helping with the meteorological observations. During the following summer, after graduating, I spent another summer and most of the fall of 1958 on Ice Station Alpha, located on the ice pack itself. There, I and Gary Latham worked under Ken Hunkins on seismic sounding of ocean depths and seismic experiments to study wave propagation in the ice and ocean beneath. Gary and I were both entering graduate school at Lamont ... One of the highlights of that summer of 1958 was the visit of the nuclear submarine Skate. The submarine surfaced in one of the open leads and some of the crew came "ashore" for the visit.

(Isacks, September 9, 2011 email to author)

After getting to know Oliver, Isacks decided to continue his education at Lamont. He liked Oliver and Lamont's open atmosphere. Deciding to remain at Station Alpha during fall 1958, he delayed graduate work until January 1959.

During my senior year in college I worked part time at Lamont for Jack Oliver and Ken Hunkins, and would often talk to Jack informally. Lamont then was a kind of classless society where everyone was on a first name basis and shared a love of geophysics and an enthusiasm for the work at hand. Although I applied to a number of grad schools, I chose Columbia after getting to know Jack Oliver and the atmosphere at Lamont during my senior year. I stayed on Station Alpha for the fall of 1958 and so deferred entering graduate school until January 1959. In November the ice floe on which the station was located started breaking up, with a crack right through the station itself, which led to a successful but rather dramatic rescue effort by the air force.

(Isacks, September 9, 2011 email to author)

Isacks worked under Oliver on his Ph.D. thesis, which he wrote in 1964, and, in 1965, received his Ph.D. Oliver discussed Isacks' Ph.D. work on high-frequency (short wavelength) P waves propagated by earthquakes in the Caribbean through the upper mantle to a seismic station located in an old zinc mine in New Jersey. He stressed Isacks' penchant "not to follow the crowd."

Lamont's aggressive program in seismology had a strong observational slant that led, among other things, to the first seismograph on the moon and to an unusual observatory in a deep mine in New Jersey. The mine was chosen especially for operation of the low-frequency seismographs that were Lamont's specialty. However, another exceptional graduate student, Bryan Isacks, elected not to follow the crowd and initiated a study there on waves at the opposite, high-frequency end of the spectrum. As part of that study, Isacks in the early 1960s studied waves like those investigated earlier by Ewing, i.e., the high-frequency shear waves traveling in the upper mantle between the West Indies and the New York–New Jersey area. Isacks, partly as a result of those efforts and that experience, would later become a major factor in the discovery [of subduction zones].

(Oliver, 2004: 171; my bracketed addition)

I want now to describe Isacks and Oliver's work that led to the confirmation of the idea that seafloor returns to the mantle beneath trenches, an idea put forth by Holmes in the late 1920s and early 1930s (I, §5.4), middle 1940s (I, §5.8), and middle 1960s (III, §2.16), by Benioff in the 1950s (III, §3.16) but later rejected by him in the early 1960s (III, §1.6), by Fisher in the late 1950s and early 1960s (III, §3.18), and by Hess and Fisher in 1962 (III, §3.16).

7.3 Isacks and Oliver launch their study of deep earthquakes

Yes, in hind sight, one could say that it is obvious that from the concentration of earthquakes along planar zones in the mantle that of course there must be something peculiar in mantle properties in such a zone. However, no one focused on this. The work of Benioff and others trying to relate this to some kind of mantle convection might be taken to imply that here was a zone where concentrated shearing was occurring. The key idea in our paper was instead to propose that the Wadati–Benioff zone is located within the descending Pacific lithosphere the plate itself, i.e. the subduction model.

(Isacks, September 14, 2011 email to author)

Isacks and Oliver's identification of a descending slab of lithospheric plate was based on results from Lamont's seismological study of the Tonga–Fiji region. This study grew out of a seminar on deep earthquakes offered by Oliver and taken by Isacks and Sykes. It was funded by the National Science Foundation as part of the US Upper Mantle project. Sykes has already described the seminar, accenting that Oliver decided to study deep earthquakes because he thought they had been ignored during the previous twenty years (§6.9). Here is what Oliver later said about it.

I gave the matter [of what to study] very careful and prolonged thought and eventually hit upon the subject of deep earthquakes as one that looked promising. There were two reasons for this decision, one strategic, one practical. Strategically, deep earthquakes seemed a prime and timely target. That deep shocks occurred and that they occurred in zones dipping beneath arcs to depths of about 700 km was known, mostly from Japanese seismologists led by Wadati. Except for some speculative hypotheses however, almost nothing was known about the nature of deep shocks, or why deep shocks occurred where they did. Furthermore, deep shocks were so prominent and so frequent that they had to be important components of the seismological and tectonic stories of earth dynamics. In addition, past observations of deep shocks were relatively sparse and spotty so that it seemed, the subject was ripe for an observational program. In typical Lamont style, we could set out to observe them thoroughly, even though there was no particular hypothesis to test. On the more pragmatic side, there was good reason to think that a sound project to study deep earthquakes could be funded.

(Oliver, 1996: 65; my bracketed addition)

At Lamont, we bandied the idea of a project on deep earthquakes about and then held a graduate seminar on the subject to explore it further. The style of the seminar was intensive, each student was required to read all of the assigned papers and come to class prepared to lead the discussion and make a hard and tough evaluation of the literature and to formulate and select ideas for future research. I cannot recall the names of all of the handful of attendees, but both Lynn Sykes and Bryan Isacks were among them . . . Bryan Isacks was also enthusiastic about the project and, indeed, could devote himself to it, and, it would turn out, did.

(Oliver, 1996: 65)

Isacks agreed with Oliver's description. He tried to remember what they read.

As I remember, the Tonga–Fiji project proposal was funded probably in early 1964 or late 1963. I think the seminar was in 1963. Jack's description that you quote is what I remember too. His motivation then was not to test the then barely emerging ideas of seafloor spreading (at Lamont), but was classical Jack Oliver strategic thinking: what phenomena are lurking out there which are being largely ignored but are dramatic and quite puzzling, and therefore might be good targets for new and fertile observations. The papers we assembled were primarily from the ground breaking investigations of Gutenberg, Richter and Benioff at Caltech, the prolific Japanese literature (I remember Honda in particular, but, as it turned out, we missed a key paper by Utsu – this deserves a special discussion [Utsu, 1966]).

(Isacks, September 14, 2011 email to author; my bracketed addition)

When further asked about Benioff's work, Isacks recalled:

Benioff was a very brilliant scientist, but I don't remember that during the seminar that his 1954 paper made any particular impression. Many ideas were bouncing around then, and the dominant research philosophy at Lamont then was to seek new and startling observations which would speak for themselves.

(Isacks, September 9, 2011 email to author)

This insular research philosophy at Lamont, also shown by Heirtzler and company's initial ignoring of V-M in their paper on the Reykjanes Ridge (§5.4), is also reflected by Oliver's offhand remark to Fisher when he, surprised that Oliver and Isacks had not referenced any papers by Scripps workers on the Tonga trench, asked him why, and Oliver quipped, "Why should we read what Lamont doesn't write?" It may also help to explain why Oliver insisted that they planned to study deep earthquakes "even though there was no particular hypothesis to test." The seminar was held, probably in 1963, and the work began in 1964, well after seafloor spreading had been proposed. When I asked Isacks about the decision to study deep earthquakes being independent of hypotheses about their origin, he mentioned Jeffreys' view that they were "theoretically impossible" but emphasized the "inductive" approach to research so prevalent at Lamont.[4]

[This decision] wasn't simply the neglect of deep earthquakes, but, as I said above, they were of considerable interest in themselves. The famous British geophysicist Harold Jeffreys thought them to be theoretically impossible, although the seismological evidence for their existence was quite solid. We certainly hoped that new observations would have something important to say about the ideas of Benioff and others about what happens at island arcs, but our focus was on obtaining the new observations and understanding what they were trying to tell us. You have to understand how pervasive the inductive philosophy that Jack [Oliver] gets across in his book [*Shocks and Rocks*] was at Lamont. This was something that he absorbed from Doc Ewing, as did many of the researchers at Lamont.

(Isacks, September 14, 2011 email to author; my bracketed additions)

So Oliver and Isacks were not interested in testing seafloor spreading or any other particular hypothesis; they wanted to study deep earthquakes because they had been neglected, were dramatic, and not understood. They definitely accomplished what they set out to do, and also, albeit unplanned, found support for seafloor spreading and mantle subduction. Given that they had no intention of testing seafloor spreading, their finding support in its favor was, as Oliver (1996) has claimed, serendipitous not only in Isacks getting to know Oliver, not only in how Isacks' Ph.D. work helped him make the discovery, but also in finding support for it and subduction.

Once Oliver had decided to pursue the project, he needed to find someone to take charge. Isacks was a perfect choice.

I next tried to assemble the elements of a project that would observe deep earthquakes with some thoroughness. The key element was recruitment of a young scientist who could devote full time to the project. (I had other time-consuming continuing responsibilities.) My first choice for this position, and in retrospect it was an inspired one, was Bryan Isacks. Bryan had recently completed his doctorate. He had exceptional ability, including a strong intuitive sense for science, training in both physics and geology, and appropriate and relevant experience with instrumentation and analysis from his field studies in the mine. He was enthusiastic, determined, eager to see and explore the earth, and anxious to make a contribution to science. It was my good fortune, indeed, when Bryan joined the nascent project.

(Oliver, 2004: 173)

They first had to decide where to do the study. They wanted a place where deep earthquakes occur frequently and where they could set up and manage a network of seismological stations. There were two obvious options: the South American Andes, and the Tonga–Fiji region of the South Pacific. Isacks successfully argued in favor of the Tonga–Fiji region:

Yes, I remember our discussions very well. I did push for the Pacific site, primarily because of the enormous number of deep earthquakes in the Fiji–Tonga region, comparable to the total number in the rest of the world. This seemed a definitive argument assuming that field work was feasible in the Fiji–Tonga region. That led to the second argument: to cover the South American deep earthquake zone properly, we would have to operate in a number of countries, none of which were English speaking and some of which were politically dicey. The multiplicity of governments that we would have to deal with for importing equipment, logistics and site arrangements presented formidable challenges, while the Fiji–Tonga region then was a quite stable UK colony (Fiji) or a UK Protectorate (Tonga). A researcher in marine geophysics at Lamont then actually worked as the head geologist in Fiji previously and gave us confidence that Fiji and Tonga would be a feasible place to run the project. The choice seemed quite clear.
(Isacks, September 14, 2011 email to author; see also Oliver, 2004: 174)

Isacks left for Suva, Fiji in September 1964; his wife and three daughters joined him three months later, and they remained there over a year, returning in December 1965 (Isacks, September 9, 2011 email to author). Unbeknownst to Isacks at the time, there was resistance by the head of Fiji's branch of the Colonial Geological Survey. Fortunately, the head of the meteorological service in Fiji came to his aid.

This leads to an interesting story. When I arrived in Fiji in the Fall of 1964, I approached the Colonial Geology Survey of Fiji for help, especially a place to work, and help with logistics and siting. The head of the survey there was quite disinterested in our project. I found out later that he then sent out a memo to all the government agencies "warning" that an American scientist was looking for help and beware of getting commitments that might drain resources away from their proper work. This message came across the desk of the head of the meteorological service in Fiji, a New Zealander who disliked the idea of "a limey b..." handing down orders. He got in touch with me and offered excellent space in the meteorological building and valuable help with contacts for sites and logistics. The success of the project owed much to him.[5]
(Isacks, September 14, 2011 email to author)

7.4 Isacks and Oliver pin down subduction

The basic strategy was to put stations above and around the deep earthquakes [see Figure 7.1]. Geography and logistics dictated the rest. Stations were not placed on the Fijian Lau islands because they were too difficult to get to and maintain. Jack did have the idea that the planar zone of deep and intermediate depth earthquakes, the Wadati–Benioff zone, might have different physical properties than the surrounding mantle at comparable depths. He promoted getting a station in Tonga to explore that idea, enabling comparison of paths through the zone (deep earthquakes to Tonga) with paths through the non-seismic mantle (deep earthquakes to Fiji). We in fact found a dramatic difference in the amplitudes of high frequency waves. Differences in velocity also exist, but are harder to pin down, because they are partially masked by the mis-location of the deep earthquakes. The anomaly can be detected only by a careful

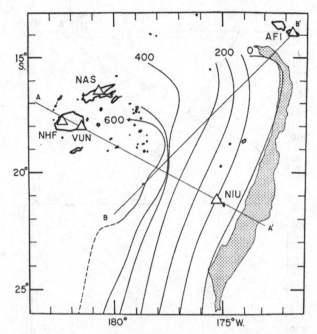

Figure 7.1 Oliver and Isacks' Figure 1 (1967: 4262; my bracketed additions). Their caption reads: "Index map showing locations of stations Niumate, Tonga (NIU); Afiamalu, Samoa (AFI); and Vunikawi (VUN), Nasangga (NAS), Nausori Highlands (NHF), Fiji. Shaded area indicates water depths greater than 6 km and shows location of the Tonga trench. Contours express the configuration of the seismic zone and are drawn at intervals of 100 km. AA' and BB' show location of sections of Figure 4 [Figure 7.2 below]."

study of the times of arrival at many stations, both near and worldwide, while the differences in the amplitudes and frequencies of the seismic waves are directly and dramatically displayed on the seismograms.

(Isacks, September 14, 2011 email to author)

As the data began to come in, Bryan, who was the first to see them, was continually alert for any sign that the seismic zone was somehow different from its surroundings. We hoped for a detectable effect, probably in the form of a difference in velocity, but were concerned that such a velocity effect might be small and obscured by errors in location of the earthquake foci. It eventually turned out, to our delight, that the earth was good to us, much better than we had ever hoped or anticipated. There was a velocity difference of a few percent and it could be detected. It was some time, however, before that effect was resolved or even given much attention. Something bigger was in store. From some of the early data, Bryan quickly recognized that there was a huge effect in attenuation that far outstripped anything we had dreamed of. The amplitudes of seismic waves traveling up the inclined seismic zone to Tonga were sometimes more than three orders of magnitude larger than those traveling a comparable but aseismic path to Fiji (figure 4) [Figure 7.2]. The differences in velocity were secondary. The differences in attenuation were predominant and astonishing. It was a startling result, and we knew, or at least suspected,

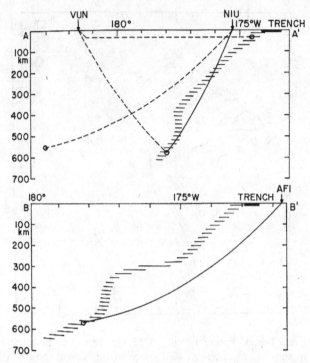

Figure 7.2 Oliver and Isacks' Figure 4 (1967: 4263; my bracketed additions). Their caption reads: "Sections showing the seismic zone, seismic foci, and paths to stations. Paths for which high-frequency waves are observed are indicated by solid lines; those for which only low-frequency waves are observed, by dashed lines. The seismic zone is indicated by the hachures, seismic foci by open circles. Section AA′ (see Fig. 1 [Figure 7.1] for location) shows paths and wave types at Fiji and Tonga after typical deep and shallow shocks in the main seismic zone and for one deep shock in an anomalous location. Section BB′ (see Fig. 1 [Figure 7.1] for location) shows path from a deep focus to Samoa."

that we were on to something important. But what? We had to develop some understanding of the effect as it related to other earth features and phenomena to make it meaningful.

(Oliver, 2004: 175–176; my bracketed addition)

Isacks began examining the results, and noticed a dramatic difference in the seismic waves recorded at Fiji and Tonga, especially the S waves, generated by deep shocks. Those recorded at the Tonga station "were predominately of high frequencies" while waves of "such high frequencies" were not found at the Fiji stations (Oliver and Isacks, 1967: 4261) (see Figure 7.2). Fortunately, Isacks had installed instruments that recorded a wider band of frequencies than "is usual for short-period instruments," without which they would not have seen the difference in amplitude of the high-frequencies waves. He does not claim that he installed them because they anticipated "looking for frequency differences."

I chose the recording instruments to have a broader band of frequency response than is usual for short-period instruments. This turned out to be a good choice for looking at the dramatic difference between the Fiji and the Tonga recordings of deep earthquakes, although at that point in the planning, I don't think Jack or I thought specifically about looking for frequency differences in the way it turned out.

(Isacks, September 14, 2011 email to author)

Isacks noticed these differences in amplitude of the high-frequency waves while still in Fiji. He was puzzled by them, but with the demands of field work did not spend a lot of time trying to understand them before he returned to Lamont in December 1965 (Isacks, October 20, 2011 email to author).

At the time neither he nor Oliver, nor Sykes for that matter, felt any urgency to make sense out of their data. Sykes, recall, felt as such only after he and Oliver saw *Eltanin*-19. Recall also that Isacks "knew little about" seafloor spreading when he left for Fiji and that originally there were no plans to test any hypothesis. When he returned to Lamont, he was "bowled over" by *Eltanin*-19, and also learned that Sykes had confirmed the idea of ridge-ridge transform faults. Isacks is unclear just when all this happened. It was during the first half of 1966, and he thinks that he learned of Sykes' work first.

All this happened so rapidly for me at Lamont after I returned from Fiji in the first half of 1966 that I cannot remember the sequence, but since I talked with Lynn mostly then, I think that I heard about that confirmation of Wilson's idea before *Eltanin*-19.

(Isacks, September 9, 2011 email to author)

When Sykes visited Isacks in Fiji from June to October 1965, they talked about how "the advent of the World Wide Standardized Seismograph Network (WWSSN)" could "revolutionize the study of focal mechanisms." Isacks later used Sykes' program to obtain focal sphere projections of deep earthquakes in the Fiji–Tonga region (§7.16).

Our conversations about focal mechanisms started in during field work in Fiji. We were sitting looking out over the ocean, when Lynn asked me what did I think were the most important problems that we could be addressing. I remember talking about the possible impact of the new long period seismic recordings that had started coming available with the advent of the World Wide Standardized Seismograph Network (WWSSN). Jim Brune, a gifted researcher at Lamont then working on long period surface waves, had shown (Chander, R. and J. N. Brune, 1965) that the body waves from earthquakes recorded by the long period seismograms were often very simple and pulse like, permitting one to see clearly the polarity of the P-wave first motion.[6] We talked about how this could revolutionize the study of focal mechanisms. and get us out of the mess that Hodgson encountered when trying to do that using reported first motions from short period instruments. As Lynn showed in his 1965 paper, and Stauder showed slightly earlier, this is the secret of getting reliable and robust focal mechanisms, and I took the lead in exploiting this for the deep and shallow earthquakes in the Fiji–Tonga region.

(Isacks, September 9, 2011 email to author)

Isacks remembers that the Goddard meeting, which he and Oliver both attended and "where Lynn [Sykes] was already moving on the transform fault story . . . definitely turned

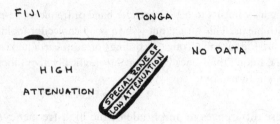

Figure 7.3 Oliver's Figure 5 (2004: 178). This is Oliver's redrawn figure of the one he drew on the blackboard in his office after Isacks and he decided that the difference in amplitude between the seismic waves recorded at Tonga and Fiji was best explained by supposing an anomalous zone of low attenuation angled downward beneath the Tonga trench.

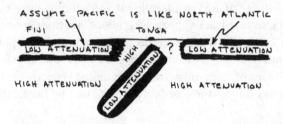

Figure 7.4 Oliver's Figure 6 (2004: 179). This is Oliver's redrawn figure of the one he drew on the blackboard in his office when Isacks and he decided that the difference between the seismic waves recorded at Tonga and Fiji was best explained by supposing that the anomalous low-attenuation zone was composed of the same material as that which allowed the propagation of high-frequency waves from earthquakes in the Caribbean to the seismological station in the New Jersey zinc mine.

up the fire beneath Jack [Oliver] and myself to get our story out" (Isacks, September 9, 2011 email to author). Once the burner was turned up, Isacks and Oliver hit upon their solution and wrote it within about five months, for their paper was received by JGR on April 13, 1967. Although it is unclear just when they came up with their solution, it is clear that it happened in Oliver's office after they had accepted seafloor spreading.

Then the great moment of the discovery, what some have described as the "Eureka phenomenon," arrived. The setting was my office at the Lamont Observatory. Bryan and I sat at a blackboard where I had drawn a cross-section through the Tonga–Fiji area (Fig. 5 [Figure 7.3 below]). It showed, in simple fashion, the deep seismic zone that exhibited the anomalous seismic wave propagation. It showed no detail for the shallow mantle of the region outside our seismic network in the Tonga–Fiji area. In particular, we had no information on the shallow mantle east of Tonga, that is, east of where the west-dipping seismic zone lay near the surface. Recalling the unusually efficient propagation of high-frequency shear waves over Atlantic paths between the West Indies and Lamont, I said, "The efficient shear wave propagation in the inclined seismic zone of Tonga–Fiji is something like shallow horizontal shear wave propagation in the western Atlantic. Why don't we assume that shallow horizontal propagation in the Pacific mantle is also similar? Then we could draw it this way." And I sketched figure 6 [Figure 7.4 below] and then drew the now familiar picture of the slab that is horizontal beneath normal

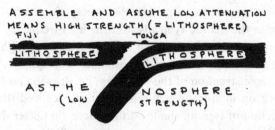

Figure 7.5 Oliver's Figure 7 (2004: 180). This is Oliver's redrawn figure of the one he drew on the blackboard in his office after Isacks and he figured out that the difference between the seismic waves at Tonga and Fiji was best explained by supposing the slab of lithosphere beneath the Pacific angled beneath the Tonga Trench as it was being subducted back into the mantle.

sea floor and that bends downward and descends in island arcs (figure 7) [Figure 7.5 below]. Almost before I had completed the picture, Bryan, conscious of the developing sea-floor-spreading story and the accompanying enigma, said, "Of course. It's underthrust!" How simple. How delightfully simple. We knew at once that we had the answer. We knew it was an important discovery. We were elated beyond description. It did not matter that we had been on the verge of taking the simple step for months and had somehow failed to do so. It only mattered that we had found something new and important that no one had recognized previously . . . Our elation was reinforced almost immediately by a burst of enthusiasm for investigating the additional meaning of the simple but elegant concept that had appeared on that blackboard.

(Oliver, 2004: 179–180)

Isacks and Oliver entertained four hypotheses to explain the differences in amplitude of the waves recorded at Tonga and Fiji. The alternatives to the propagation path, the one they eventually accepted, were as follows:

(1) Amplification effects on the incoming seismic waves due to layered structures at shallow depths (10's to 100's of meters) – i.e. strong amplification effects at Tonga compared to Fiji; (2) Spatial radiation pattern from the same earthquake; (3) Instrumental differences.

(Isacks, October 20, 2011 email to author)

The third was easy to eliminate. The instruments at different stations were also the same type, and they were all working properly. Isacks recalled why they eliminated the two other alternatives.

The second refers to any specific earthquake. The seismic waves radiated from an earthquake vary in amplitude and polarity in a systematic way, and the variations in amplitude can be quite large. However, the frequency content of the waves and the consistency of the Tonga–Fiji difference for numerous earthquakes with differing radiation patterns argues fairly strongly against this too. The first was the one that Jack and I struggled with most. However, nothing in the literature that we looked into indicated that the very large magnitude of the differences in amplitudes could be produced by shallow layering.

(Isacks, October 20, 2011 email to author)

This left them with one candidate, namely, the propagation path (Oliver and Isacks, 1967: 4266). Because the mantle typically absorbs high-frequency waves from deep earthquakes, they reasoned that the zone through which the high-frequency waves traveled to Tonga is anomalous and of low attenuation (Figure 7.3).

Given Oliver's reconstruction of the discovery, he, drawing on Ewing's work and Isacks' Ph.D. work on high-frequency seismic waves, realized that the anomalous zone was like a section of typical, shallow lithosphere. So Oliver drew Figure 7.4.

According to Oliver, Isacks then immediately said, "Of course. It's underthrust!" The diagonal lithosphere slab beneath the Tonga Trench and the horizontal, shallow lithosphere slab beneath the seafloor were attached; they were one and the same huge slab of lithosphere, which had been created through seafloor spreading, angled downward beneath the Tonga Trench and was being subducted back into the mantle (Figure 7.5).

Isacks thinks that Oliver's memory of what happened is incorrect in one important respect.

My only disagreement with this (memory is unreliable, mine as well as Jack's) is that I remember being the one who jumped on the western Atlantic observations, because I had been working on the attenuation story to try and understand what the dramatic difference in the high frequency waves were telling us.

(Isacks, September 14, 2011 email to author)

Yes, memories are unreliable, and there is now no way to prove whether Isacks or Oliver "jumped on the western Atlantic observations." However, I'm inclined to believe that Isacks was the one who first made the leap for precisely the reason he gave. Given Isacks' memory of what happened, he made the suggestion, and Oliver then drew the figure on the board. As Oliver acknowledged (§7.2), Isacks' dissertation work prepared him to make the leap.

Just what is the relation between Isacks' Ph.D. work and the Fiji–Tonga work? Isacks explained, also giving his recollection of their epiphany.

Here is the connection. In my Ph.D. thesis I used the mine site to record and study high frequency seismic waves (approximately in the band from 0.5 to 20 cycles/sec). The seismic waves included background noise, small local earthquakes, and more distant earthquakes. The seismic waves from most distant earthquakes (say 1000's km distant) do not have frequencies this high, because they travel beneath the lithospheric plates in the part of the mantle that absorbs high frequency waves. However the waves from the Caribbean are able to travel within the cooler lithospheric plate to northeastern US where the mine is located and thus arrive with their high frequencies not absorbed. At the time that I wrote the thesis (in early 1964), we were not thinking of this in terms of a transmitting lithospheric plate over an absorbing mantle asthenosphere. But in 1966, with the seafloor spreading ideas beginning to explode around us as we were trying to make sense of the Tonga–Fiji observations, the eureka moment that Jack mentions happened. We realized that the high frequency waves that were traveling up to the Tonga station but not to the Fiji stations had the same significance as the high frequency waves that travel from the Caribbean to the zinc mine: they were both traveling along a path in the mantle that did not absorb them. The next step was the eureka moment where we imagined that the paths were both within lithospheric plates: in the case of the Caribbean earthquakes to the New Jersey mine, the lithospheric plate was the western Atlantic/North American plate while the Tonga high frequency waves were going through the Pacific plate which had bent downwards and was descending into the mantle. On the surface, lithospheric plates have good high frequency transmission because the mantle part of the plate has become cooler than the deep mantle (which is responsible for its strength and behavior as a plate). The descending plate west of Tonga and beneath Fiji retains its ability to transmit high frequency waves because it remains cool, although it is slowly heating up as heat flows from the hotter surrounding mantle. The waves recorded in Fiji travel through that hot mantle and lose their high frequency content. Although at that time we had no observations of high frequency waves beneath the Pacific crust east of the Tonga Trench, the concept of the Pacific plate simply bending down beneath the Tonga Island arc (the bend is directly manifested by the bathymetry of the eastern side of the Tonga Trench) and moving down into the mantle where it becomes seismically active, producing the zone of deep and intermediate depth earthquakes, just jumped out to both of us and Jack drew the picture on the blackboard. We realized that this is where the plates created by seafloor spreading along the ocean ridge system go back down into the mantle.

Besides my Ph.D. work on the Caribbean earthquakes, I was particularly sensitive to the significance of the high frequencies because as we struggled to understand the dramatic differences between the recordings of waves in Tonga and Fiji, I had been exploring the effects of absorption on the frequency content of seismic waves. I realized that, of the competing hypotheses, only significant differences in absorption along the travel path could account for the observations. The reason for the frequency selectivity of absorption is simple. Seismic waves are composed of cycles of stress and strain. In real materials, there will always be a small loss of energy in the cycle due to non-elastic processes such as frictional heating. This loss results in a diminution of the amplitude of the wave. Since for a given travel path there are more cycles of stress/strain for high frequency waves than for low frequency waves, there will be a diminution of high frequency waves relative to low frequency waves. This effect is quite dramatic, and practically wipes out high frequency waves above a certain value of frequency, that value being dependent upon the fraction of energy lost to absorption per cycle of elastic energy. For the type of waves we were considering, which involve cycles of shear stress and shear strain primarily, the very strong high-cut filtering effect on the frequencies associated with increased absorption per cycle is most simply associated with the temperature of the material.

(Isacks, September 9, 2011 email to author)

In their paper they (1967: 4272–4273) introduced Q, the reciprocal of the specific attenuation factor, and following Anderson *et al.* (1965), assumed that Q is "a measure of the strength of the material." This grounded the specific links between the anomalous, low-attenuation zone and the strong lithosphere, and the typical, high-attenuation zone and the weak asthenosphere, and thereby introduced subduction and seafloor spreading. They borrowed the terms "lithosphere" and "asthenosphere" from

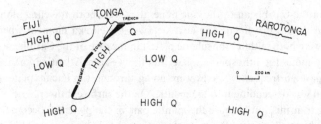

Figure 7.6 Oliver and Isacks' Figure 13 (1967: 4272). Their caption reads: "Hypothetical section through Fiji, Tonga, and Rarotonga based on data of this paper. Boundaries between high Q and low Q zones are not well determined but can be taken as a first approximation."

Reginald Daly. They also found, as should be expected, that S and P waves both traveled with slightly greater velocities through the less attenuated, stronger lithosphere than through the more attenuated, weaker asthenosphere.

The evidence described above provides important new information on the structure of the earth's mantle, particularly on the part of the mantle that must be directly involved in major tectonic processes ... Even from the most conservative view point, it seem clear that the data presented here demonstrate that zone of the upper mantle, which is more or less defined by the seismic foci, differs markedly from that of parts of the upper mantle at comparable depths elsewhere, not only because the seismic foci occur there but also because the characteristics of the seismic waves, particularly of the S wave, propagating through the anomalous zone are different for those of waves traversing more normal parts of the mantle, and, less certainly, because the velocities of P and S waves appear to be high in the zone. From this it is a minor and reasonably safe step to explain the pertinent character of the seismic waves as an effect of attenuation along all, or a substantial part of the propagation path. When this is done, the anomalous zone is found to be a region of low attenuation that, if the semiquantitative estimates are correct, is comparable to the attenuation in the uppermost layer of the mantle ... and lower than that of "normal" mantle at comparable depths in other regions. "Normal" mantle with regard to Q is here taken to be similar to that described by Anderson and Archambeau [1964], Anderson *et al.* [1965], or Anderson [1966] ... With this evidence on the distribution of anelasticity, it is possible with some assumptions on the continuity of the various zones and their configuration to draw a generalized section through the Tonga arc. Such a section is shown in Fig. 13 [Figure 7.6]. In addition to illustrating the anomalous zone of the upper mantle, Fig. 13 shows continuity between that zone and the uppermost layer of the mantle with regard to attenuation properties in the media. In fact, the configuration of Fig. 13 could clearly be reconciled with an underthrusting, or dragging, or settling of the uppermost mantle layer on the outward side of the arc beneath the arc itself. Such a process is in general agreement with the hypothesis relating to seafloor spreading under the influence of mantle convection currents as proposed by Hess [1962], Dietz [1961d], and others. These hypotheses call for rising convection currents beneath the ocean ridges and sinking currents in the vicinity of the ocean deeps associated with island arcs. An important point is suggested by the present work if Q, the reciprocal of the specific attenuation factor, is assumed to be a measure of the strength of the material, as suggested, for example by Anderson *et al.* [1965]. This assumption would imply that the upper regions of high Q in Fig. 13 correspond to the

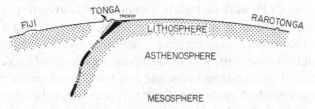

Figure 7.7 Oliver and Isacks' Figure 14 (1967: 4273). Their caption reads: "Hypothetical section through Fiji, Tonga, and Rarotonga, assuming *Q* correlates with strength. The lithosphere and mesosphere are zones of significant strength, and the asthenosphere is a zone of vanishing strength on appropriate time scale. The terminology is that of *Daly* [1940]."

lithosphere; those of low *Q*, to the asthenosphere (Fig. 14 [Figure 7.7]), a conclusion that receives some support from the general distribution of seismic velocities with depth. The low-velocity zone for S waves would, within the uncertainties involved, coincide with the low *Q* zone of Fig. 13. If the high *Q* zone of Fig. 13 does indeed represent the lithosphere, a layer of significant strength, then it is apparent that theories of tectonics based on convection currents must take this layer into account... The concepts of lithosphere, asthenosphere, and meso-sphere are, of course, long standing ones in geology [Daly, 1940] and need not be elaborated here. The new feature of this paper is the evidence that suggests a new configuration for, and perhaps the extent of the mobility of, the lithosphere.

(Oliver and Isacks, 1967: 4272–4273; my bracketed references to Figures 7.6 and 7.7)

Interestingly, they referenced Hess (1962) but not Fisher and Hess (1963), the very paper in which they proposed that seafloor descends back into the mantle beneath trenches.

Oliver and Isacks (1967: 4273) speculated about the thickness of the high-*Q* layer. Acknowledging that it had not been well defined by their study, they reported that they had first thought it coincides with the seismic zone. Noting that Sykes' ongoing but unpublished work on positioning hypocenters of deep earthquakes suggested that the seismic zone beneath the Tonga arc is less than 20 km thick, they proposed, on the basis of "various ray path locations," that the thickness of the high-*Q* layer, the lithosphere, is "perhaps of the order of 100 km," and that the seismic zone is "near the upper boundary" of the high-*Q* layer. They also related but did not reference Isacks' Ph.D. on Caribbean earthquakes and Ewing's work, retrospectively mentioned by Oliver, to the thickness of their proposed shallow high-*Q* layer beneath the seafloor at the top of the mantle.

The approach developed in the present study coupled with such as, for example, the high-frequency S (or *Sn*) phases recorded at Rarotonga from shallow shocks in Tonga, or similar phases recorded at Bermuda and the eastern United States from shocks in the West Indies, or high-frequency *Sn* phases recorded for paths crossing the Canadian shield and the eastern United States indicate that for these areas the base of the lithosphere is clearly below the surface of the mantle, at least some tens of kilometers below.

(Oliver and Isacks, 1968: 4275)

Even though Oliver and Isacks did not start out planning to test seafloor spreading, they did just that. Their discovery was serendipitous, from Oliver asking Isacks to man

T3 and his ensuing Ph.D. work on high-frequency seismic waves, through Isacks using broad-band recording seismographs, to the key observations not being of differences in seismic wave velocities but in high-frequency amplitudes. Within a year, Isacks, Oliver, and Sykes would combine their work in identifying regions of subduction, creation of new seafloor, and along transform faults, add to it Isacks' work on first motion studies of deep earthquakes which explained that they occur *within* the anomalous high-Q descending lithospheric slab either through compression or tension, and summarize the seismological support of the new theory of plate tectonics (§7.18).

7.5 McKenzie, the making of a geophysicist

Dan McKenzie was born on February 21, 1942, in Cheltenham, England. His father, like his father's father, was a doctor on Harley Street in London; his mother Nan, née Fairbrother, was noted for her work on garden design and its history (Fairbrother, 1956, 1970, 1974). She also wrote two books about raising Dan and his younger brother; the first, *An English Year*, recounts her time spent in Buckinghamshire raising her two sons during World War II.[7] The family moved to London when McKenzie was seven. He attended Westminster School, one of England's oldest public schools. He was a "hopeless student" until puberty, when he "discovered physics and chemistry and to some degree biology and after that never looked back" (McKenzie, May 11, 2007 interview by Alan Macfarlane). After Westminster, in fall 1960 he went to King's College, Cambridge. He described his education at Westminster, and circumstances surrounding his admission to King's College.

I was a tolerably good mathematician, and good at physics and chemistry at school. I greatly enjoyed understanding the world in a precise way. I decided that I wanted to be a scientist almost as soon as I started to do it seriously at about 15... My physics teacher [Austin Stokoe] had been at King's, and it was he who sent me to talk to the senior tutor [John Raven] at King's, who was a classicist but was interested in British wild flowers. We talked about orchids and about Dostoevsky, on the strength of which he offered me a place, even though I had not at that stage taken any exams at all! This has changed, which is perhaps sad. I must have been interviewed in spring 1959, taken A and S levels and won a state scholarship in the summer of 1959, taking maths for scientists, physics and chemistry, then won a major scholarship to King's in the same subjects in December 1959. I decided to stay at school for the following two terms, because I thought the maths teaching [by Adolf Prague, a German refugee, who was terrific and taught pure maths and Fisher who taught applied maths] at the school I was at, Westminster, was probably better than at Cambridge (it was!) and won another State Scholarship in pure and applied maths in the summer of 1960.

(McKenzie, March 18, 1990 letter to author; my bracketed additions)

McKenzie needed a third science at Cambridge for the Natural Sciences Tripos. He had physics and chemistry, and decided on geology. Earth science has Lyell and Geikie to thank.

My career as a geologist really began at school where I was principally interested in Physics and Chemistry. After I got a scholarship at King's I discovered that I had to start a new science subject. [The rules, then as now, about taking science (called the Natural Sciences Tripos) at Cambridge are that you have to take three science subjects. Maths you also take, but it does not count as a science.] My director of studies would not let me do Physical Chemistry and strongly recommended me to do Physiology. So I took out all the books from the school library on both subjects. There were only two on Geology: Lyell's *Principles* and Geikie's *Ancient Volcanoes of Great Britain*, and I thought they were marvelous. The Physiology books were modern textbooks for medical students and were extremely dull. So my mind was made up. But I am sad to say that Geology at Cambridge in 1960 was not a patch on the Nineteenth Century subject, and I gave it up after a year in favor of Theoretical Physics.

> *(McKenzie, February 18, 1983 letter to Janet Watson; bracketed addition from McKenzie, March 18, 1990 letter to author)*

He later elaborated on the drudgery of his geology classes, adding, however, that he became friends with the geology students.

I found the geology lectures awful. I have no memory of any concern with the sort of issues you may raise [about the drift controversy]. I learnt the names of 600 zone fossils and what zones they came from, I can still give an extremely detailed account of British stratigraphy, and could at a pinch draw detailed cross sections of crinoids, echinoderms, sutures of ammonites and so on. The tectonics we got was principally concerned with using very small scale features to determine the stress field. I passed the exam, but gave up the subject at the end of my first year. But I ended up with a tremendous liking for geologists and most of my undergraduate friends were either at King's or were geologists. After the first year I dropped everything except maths and physics, but went to Spitzbergen in the summer of 1961 as a field assistant (to David Gee who now works for the Swedish Geological Survey), and talked my way onto the Easter geology field trip to south west Wales, to spend two weeks in the field with my friends.

> *(McKenzie, March 18, 1990 letter to author)*

Excepting geology, he found science "wonderful" during his first two years at Cambridge, which is much better than he found undergraduate life and politics.

It was somewhat of a shock to go up to Cambridge, which I found very much a backwater. The undergraduate life was to me rather like children playing compared with what my parents and their friends were doing in London, and I found the whole enterprise dull.[8] But the science was wonderful. I went to some of the lectures I was supposed to, but also went to listen to Dirac talk about quantum theory and the Dirac equation, to Hoyle talking about stellar structure, and to a number of other people who I thought might be interesting.

> *(McKenzie, March 18, 1990 letter to author)*

He became bored with physics in his last year, but not with science. Thanks to Hill, Matthews, and Bullard, he decided to switch to geophysics.

My last year, doing physics fulltime, was not a success. I was bored by the lectures. It was the coldest winter I can remember, and I spent most of the time between Christmas and Easter skating on the Cam, which was frozen for the whole term. Nothing I heard about in the Cavendish really interested me, but I was good enough to do research, though not at all convinced that I was one of the very few people who could do something fundamental in physics. I got to know Fred Hoyle, who had been a walking friend of my mother's when she was at the University, and went out with his daughter for some time. I talked to him about what I should do. I remember that I met Willie Fowler at their house, who impressed me tremendously by sitting on the floor with a coffee pot full of ice and whiskey (nearly a whole bottle), and drinking the lot with no obvious effect while talking about nuclear reactions in stars with Fred. I though all American scientists were like this until I went to the U.S.! Fred thought I should do biophysics. The other person who took an interest in me was Maurice Hill, who looked after me in King's. He introduced me to Drum Matthews, who then was unmarried and lived in rooms in King's next to Morgan Forster. I went and talked to Drum often in my last year, and it was he who sent me to meet Teddy at Madingley. Drum and Maurice were both quite clear that I should do geophysics and be supervised by Teddy. I liked the place and the people, who were like geologists, but more intelligent.

(McKenzie, March 18, 1990 letter to author)

In October 1963, McKenzie began graduate work in geophysics working under Bullard. Like Vine the year before, he was given a Shell Scholarship, selected by Hill and Bullard as the most promising new graduate student (McKenzie, March 18, 1990 letter to author).[9] Bullard set out his problem. He wanted McKenzie to use "atomic force laws to calculate the seismic velocities in the mantle." The problem was then not tractable, but he was awarded a fellowship at King's College for his efforts, and managed to impress Don L. Anderson at Caltech's Seismology Laboratory. He also asked his mother to teach him how to write.

Teddy [Bullard] started me off on a problem he had worked on, of trying to use the atomic force laws to calculate the seismic velocities in the mantle. It was far too soon to try this: it is now starting to look possible with modern computers. I wrote a fellowship dissertation for King's on this at the end of my first year, in the autumn of 1964, persuaded my mother to teach me to write and got Teddy to go through the science. One of the external referees was Francis Birch, and Don Anderson read it, either as a referee or later, and was impressed. He tried to get me to publish it when I was at Caltech [during the first half of 1967], but I was not happy with it. I got a King's Fellowship after 4½ terms of research, principally on promise and because Maurice Hill was on the relevant committee.

(McKenzie, March 18, 1990 letter to author; my bracketed additions)

McKenzie recalled that it was about this time, the beginning of 1966, that he became interested in continental drift and debates about it that were then going on at Cambridge.

By this time I wanted to do something more central to the debates about continental drift which were going on at Madingley, and decided to look at Munk and MacDonald's argument about the shape of the Earth (this became my Ph.D. [The Shape of the Earth] and led to [the following] publications: McKenzie (1966a, b, 1967a, 1968a, b, 1969a).

(McKenzie, March 18, 1990 letter to author; my bracketed additions)

What exposure and attitude did he have toward continental drift up to this time? He knew nothing about it before going to Cambridge (McKenzie, March 18, 1990 email to author). Recalling what he may have learned about continental drift from his course work in geology during his first year at Cambridge, and touching on several other related issues, including a lovely story about his mother, McKenzie wrote:

Harland, Hughes and Bill Black lectured to me in my first year geology course. Fred Vine was the year ahead of me, though I got to know him a little in my first year. I don't remember any particular concern with continental drift, though I am sure Harland must have talked about it. What I liked about Lyell and Geikie was their breadth and openness to new ideas. I certainly read Holmes: I have a copy of the first edition that I bought second hand, probably while I was in my last year at school, and used the second edition in as an undergraduate text. I lent it to my mother to read: she loved it but found it too heavy to read in bed. So she took the carving knife and cut it into four volumes! I made her buy me a new copy. [Holmes' second edition weighs 2.56 kg.] I never met Holmes. Your question assumes that I took a particular interest in continental drift in particular and geology in general. I did not: I read hugely across a great range of science. I was also very busy with course work.

(McKenzie, March 18, 1990 letter to author; my bracketed additions)

McKenzie did hear Hess speak about seafloor spreading at the 10th Inter-University Geological Congress in January 1962 (§2.8), but he does not remember Hess's talk. He did not attend Vine's May, 1962 Presidential Address to the Sedgwick Club where he talked about HypothHESSes (§2.8).

I do not remember any particular time when I suddenly became interested in continental drift. I had attended the conference at which Hess talked in Cambridge, which must I think have been in January 1962, but what I principally remember is how everyone kept referring to papers by the author's name and date of publication, which seemed to me a very odd way to talk, and that Teddy [Bullard] could not remember the authors of the papers he talked about, and was helped out by Harry... [It should be remembered that I] was no longer formally doing geology and went along because the meeting was in Cambridge, out of term, and my friends were involved... I did not hear Fred [Vine] talk at the Sedgwick club.

(McKenzie, March 18, 1990 letter to author; my bracketed additions)

He attended the March 1964 Royal Society's symposium on continental drift (§3.3) because he was "particularly interested in meeting David Tozer and Gordon Mac-Donald" (McKenzie, March 18, 1990 letter to author). Both shared McKenzie's interest in mantle convection, and McKenzie discussed their work in his Ph.D. thesis. McKenzie was not at this time particularly interested in continental drift, seafloor spreading, or V-M, even though, as he retrospectively noted, he (2001: 175) "wrote a Ph.D. thesis on mantle convection, which taught me fluid mechanics and enough materials science to know that all materials creep at high temperatures and low stress." But, he was not alone, for as he later remarked:

Although it now seems strange, neither I nor the other graduate students changed thesis topics to work with Fred [Vine] and Drum [Matthews] ... As graduate students we were not especially stupid, and three of us (Bob Parker, John Sclater, and myself, in addition to Fred

Vine) have since been elected to the Royal Society. At the time it simply was not obvious to us
that what Fred and Drum were doing was important.

(McKenzie, 2001: 175; my bracketed additions)

But as he also remarked, the magnetic anomaly data from the North Atlantic
collected by Cambridge workers did not reveal magnetic anomaly patterns strongly
supportive of V-M (McKenzie, March 18, 1990 letter to author; see also McKenzie,
2001: 173–175). McKenzie remembered that even Matthews, in 1963, began to think
that V-M might not be correct.

After Fred [Vine] and Drum [Matthews] wrote their paper they collected all the magnetic data
from Madingley Cruises in the N. Atlantic (most of which I and Carol [Williams] later
published in 1971 [Williams and McKenzie, 1971]), cut out the profiles and stuck them onto
huge boards in the stables here at Madingley, where they both worked. They could not see any
convincing correlations, and Drum in particular became convinced that this good idea was
wrong. At Madingley there was no hostility to the idea, but nor was there a conviction that it
was correct.

(McKenzie, March 18, 1990 letter to author; my bracketed additions)

Matthews' doubt, I add, was short-lived. After reviewing results from the Indian
Ocean survey which extended his original survey, Matthews defended V-M in a paper
(Matthews *et al.*, 1965) that was submitted in June 1964 (§4.3). But, there is no
question that some such as Sclater and Hill thought V-M was highly speculative and
most likely incorrect (§4.3). Asked about what he thought of the paleomagnetic
support of continental drift while a graduate student and whether or not he discussed
it with Runcorn, Irving, Creer, Blackett, or Clegg, he wrote:

As a graduate student I talked to Teddy [Bullard] about these issues, and my views were
formed by him. I disliked the statistical nature of the arguments, and their complete dependence on assumptions about the core field that we had no real way of checking and did not
understand theoretically. Like Teddy, I was impressed with Irving's Australian results and
by the faults of the western U.S. I found Brian Harland tiresome, and was very put off by
Keith Runcorn's missionary zeal for drift and convection. I got to know Runcorn as a
graduate student, and have never got to know Creer well. I met Blackett several times with
Teddy (they were close friends), but he was too grand and old for me to get to know well.
He came to talk at Cambridge to the Physics Society when I was an undergraduate, presumably about paleomagnetism, but I can remember nothing about what he said. But I clearly
remember his appearance, which was very striking: tall and thin with a long thin and very
deeply lined face ... I also went up and talked to him at the end of his lecture (but I don't
remember what about!). I don't remember when I first met Ted Irving. I knew his name and
the story of his Ph.D. before I met him ... members of Mad Rise form a sort of worldwide
mafia. So we probably met for the first time at some conference and once when out for dinner.

(McKenzie, March 18, 1990 letter to author; my bracketed addition)

Apparently McKenzie's views were "formed by" Bullard, but they did not agree
with Bullard's. In June 1963, Bullard accepted continental drift on the basis of its

paleomagnetic support (§2.14). I suspect that McKenzie's attitude about Runcorn's zeal in defending both continental drift and mantle convection did affect his attitude toward paleomagnetism's support of mobilism. In his Ph.D. dissertation, McKenzie argued against Runcorn's view of mantle convection. McKenzie typically does not engage in polemics in his published papers. Instead, he presents his view, provides the evidence, and leaves it at that. As for continental drift's paleomagnetic support, he later retrospectively remarked:

The paleomagnetic work on continental drift, which was carried out by Keith Runcorn, Ted Irving, Ken Creer, Jan Hospers, and others with such energy and success at Cambridge in the early 1950s, stopped with Keith's departure [in January 1955], and had surprisingly little influence on later developments – even at Cambridge, where it had been so prominent. Later discoveries have clearly shown that the conclusions based on paleomagnetic measurements were correct, and their failure to convince the wider community of geologists and geophysicists remains to me the most interesting part of the history of plate tectonics.

(McKenzie, 2001: 173; my bracketed addition)

On this score, I certainly agree with McKenzie, and have devoted over one volume to paleomagnetism's support of mobilism. As for why paleomagnetism's support of mobilism had little or no effect on McKenzie himself and most other graduate students who arrived after Runcorn left, I think the primary reason is that no one of Runcorn's, Creer's or Irving's caliber in paleomagnetism, excluding marine magnetists, remained in the Department of Geodesy and Geophysics after (or since) McKenzie arrived. I also doubt whether McKenzie and most other graduate students at Madingley Rise were reading the key papers in paleomagnetism supportive of mobilism or had then worked through Irving's 1964 *Paleomagnetism and Its Application to Geological and Geophysical Problems*. Although McKenzie heard Bullard's 1963 address to the Geological Society of London in which he explained why paleomagnetism's support of mobilism warranted its acceptance (Bullard, 1964), he still did not accept his mentor's assessment of paleomagnetism as confirming continental drift. Of course, McKenzie and the other graduate students were working on other things, and their interests lay elsewhere. McKenzie (March 18, 1990 letter to author) recalled, "There were a number of problems that I had a go at in 1964 and 1965. I remember working on convection in layers and dynamo theory both with no success." His lack of success in working on dynamo theory probably did not increase his confidence in the paleomagnetic evidence for drift; however, it did not prevent him from believing, like most at Cambridge, in field reversals.

McKenzie, ready to leave Cambridge for a spell, set off in summer 1965 for the United States. He learned new skills and met new people; most, I add, excepting George Backus, were not at the time believers in continental drift.

By this time I had been in Cambridge for 5 years and wanted to go away. Freeman Gilbert was on sabbatical during 1964 and 1965, and I persuaded him to invite me to IGPP at Scripps for six months. I came in early summer and stayed until the late fall, learned to compute and met people like George Backus, Bill Menard and Walter Munk. I went back on a Greyhound to New York, through the Deep South and visited Cecil Green in Texas, and Francis Birch in Harvard. I don't remember going to Lamont. I also went to Caltech from Scripps, and met Don Anderson and Jerry Wasserburg, and to UCLA to meet Gordon MacDonald. This trip made a huge impression on me. I went back home to write my Ph.D., which I did in the summer of 1966 (*The Shape of the Earth*).

(McKenzie, March 18, 1990 letter to author)

McKenzie passed his oral exam in October 1966. Dai Davies was McKenzie's internal examiner and David Tozer his external. Davies later sent him his October 26, 1966 report.

Report of Mr. D. P. McKenzie's dissertation:– The Shape of the Earth

Mr. McKenzie's dissertation deals with a subject of increasing interest to geophysicists concerned with properties of the deep interior of the Earth. Apart from elasticity information from seismology which is increasingly abundant, very little else has yet been inferred about the mantle, and particularly the lower mantle. Mr. McKenzie however takes one significant piece of information which has been revealed repeatedly by satellite geodesy – that the Earth has a shape which is not quite that of a rotating fluid sphere – and arrives eventually at valuable results about the viscosity of mantle rocks. He uses the argument that a decelerating viscous earth will have a figure which is a function both of the rate of angular deceleration and of the viscosity.

His discussion of the problem is very full – separate chapters deal with the basic equations of continuum mechanics, Laplace transform techniques of importance to the solution of the problem of an earth on which forces change with time, numerical investigations using the Laplace technique and an invaluable survey of literature practically unknown to the geophysicist on the behavior of ceramics (which are not unlike mantle rocks) under high pressures and temperatures. It is unfortunate that Mr. McKenzie has to spend so long setting up the problem, but he assumes, rightly, I believe, that few readers of his dissertation will have the relevant equations and data at their finger-tips. Only when a complete background to the subject has been established does he venture to draw any major conclusions, and these are all the more impressive for his previous careful groundwork.

Mr. McKenzie writes with a lively skepticism and conviction which makes the dissertation readable and most thought provoking.

I have no hesitation in recommending that he be awarded the degree of Ph.D.

McKenzie himself described (October 9, 2010 email to author) what he concluded about the bulge in his Ph.D. thesis, and in a fast-forward touched on current views.

Various people determined the Earth's ellipticity from satellite orbits, especially that of Sputnik, and found it was a bit greater than the hydrostatic value (which is true). Munk and Macdonald argued that the explanation was that the rotation rate of the Earth was decreasing because of tidal friction (which is true) and that the ellipticity lagged behind, because it did not adjust to the rotation rate sufficiently fast. They used this observation to calculate the viscosity of the mantle, and found a value that was too great to allow convection. What I did was to show that this constraint need not apply to the whole mantle: a highly viscous lower mantle could account for the observations.

The general view now is that the non-hydrostatic bulge of the Earth is maintained by mantle convection, and does not require a special explanation. I think this view is correct, though

I have never seen how to do the problem properly. We also believe that the lower mantle is more viscous than the upper mantle, but not as viscous as I required.

Another complication is that the Earth's shape is still recovering from the glaciation, which also produced a non-hydrostatic bulge when the ice melted.

So I believe the central observation I used in my thesis was wrongly interpreted. But what I did learn was a lot of materials science and fluid dynamics, neither of which I knew anything about before I started and both of which stood me in good stead later. As Drum used to observe, I am one of those people who are still working on their Ph.D. problem, even though I am about to retire! In my defence I would point out to him that I had worked on a lot of problems other than my Ph.D.!

Davies was correct to emphasize the need for McKenzie to provide the background. In particular, few geophysicists were familiar with the behavior of ceramics under especially high temperatures. They were more familiar with the behavior of metals under high temperatures and pressures. As already noted (§6.13), Bullard got McKenzie invited to the Goddard meeting, and he presented his work on the behavior of ceramics under high temperatures such as found in the mantle. He did not prove the existence of mantle convection, but showed that its existence was not an unreasonable assumption given Earth's bulge.

7.6 McKenzie interprets heat flow data in terms of seafloor spreading

I think the two most important aspects of the paper are that it used the idea of dimensionless variables, and that it argued against the ridges having any deep structure. The first I learnt from Adrian Gill when I was at Scripps in 1965, and used extensively in my thesis. The approach has the great virtue that you can solve the equations in general, irrespective of what thickness the plate has. The approach was well known in fluid dynamics, but this is its first use in tectonics I think. The second point is even more important. The real problem with coupling the ridges to convective upwellings was that it was hard to understand how the ridges around Africa and Antarctica could all move away from the continents. But if all the features of a ridge are produced by upwelling of isothermal material, such problems disappear. People are still muddled on this point, but this separation was central to my development of a kinematic model of plate motions. It also meant that much of what Harry [Hess], Tuzo [Wilson] and Keith Runcorn was saying was unnecessary.

(McKenzie, March 18, 1990 letter to author; my bracketed additions)

The first paper I wrote on this [rootless ridges] was published in *JGR*: Some remarks on heat flow and gravity anomalies, volume *72*: 6261–6273. It is widely used and referenced, because I obtained an analytic solution to the cooling plate problem. But no-one took any notice of what was for me the central issue, which was that I produced a model of a ridge which was in agreement with the observations but which had no deep structure.

(McKenzie, October 5, 2008 email to author)

McKenzie returned to Cambridge from the Goddard meeting convinced that seafloor spreading was correct; the papers by Vine and Sykes had done the trick

(§6.9). But he was well aware of three difficulties facing seafloor spreading. The first two pertained to peculiarities in the geometry of the proposed convection cells. Africa and Antarctica were surrounded by ridges which, as McKenzie and Wilson had previously discussed in 1965 (§4.10) and Heezen and Ewing had raised in 1957 against earlier mobilist theories that invoked uprising mantle currents beneath mid-ocean ridges with the continued generation of new seafloor (III, §6.8), made the geometry of mantle convection absurd. The second difficulty was that if rising convection currents are centered beneath ridges, then uprising convection currents should be offset as are ridges (McKenzie, 1967a: 6263). The third difficulty, which had been raised by Le Pichon (§6.11) and mentioned at the Goddard meeting by him (§6.12), was that heat flow beneath ridges was not high or wide enough to warrant the idea of rising limbs of convection cells centered beneath ridges. McKenzie found a common solution to these difficulties by modifying seafloor spreading (RS1). He made ridges rootless, rejecting Hess's idea that upward-moving mantle convection currents cause the lithosphere to split apart and ridges to form. He proposed that ridges passively form with the upwelling of basalt in gaps created where lithosphere splits apart. This "obvious solution occurred" to him while waiting for Davies and Tozer to read his Ph.D. thesis after he returned to Cambridge from the Goddard meeting.

The obvious solution occurred to me while I was waiting for my examiners to read my Ph.D. thesis, which is that ridges have no deep structure and are simply passive features. So, instead of being on top of the rising limbs of convection cells. . .the isotherms are instead horizontal beneath the ridges. Though. this idea immediately solves all the geometric problems, it is not obvious that it is satisfactory, because it requires all features of ridges to be produced by passive upwelling.

(McKenzie, October 5, 2008 email to author)

He did the calculations in a few days while still in Cambridge waiting to take his oral exam for his Ph.D., did the needed programming during the first half of 1967 while at Caltech before he went to the spring (April 17–20) 1967 AGU meeting. He submitted the paper after the meeting. It was received by *JGR* on June 30, 1967, and published in December 1967.

I did the thermal calculation in a few days in England after returning from the NASA conference. It is not at all hard, and I now use it as an example question in handouts I give to physics undergraduates. Bob [Parker] was at Cambridge and told me to use the Eulerian form of the temperature equations. I did none of the programming until I got to Caltech. There I also did the analysis for plate flexure and the gravity anomalies. I spent the period at Caltech doing this and learning seismology, and went to the AGU from Caltech. I did not give a paper.

(McKenzie, March 18, 1990 letter to author; my bracketed addition)

Given McKenzie's retrospective comments, he believes that for him "the central issue" was his production of a model of mid-ocean ridges that agreed with observation but in which ridges have no deep structure. He also believes that the capacity of his new model of mid-ocean ridges to avoid the difficulty facing seafloor spreading arising from Africa and Antarctica being surrounded by ridges was more important than its capacity to explain the anomalous high heat flow readings over ridges. With Hess's version, seafloor spreading is initiated by upwelling mantle convection currents which split Earth's upper layers, and pull them apart as unmoving currents diverge, move along the horizontal, and carry the newly created seafloor on their backs given sufficient viscous drag. With McKenzie's version, seafloor spreading is initiated by the splitting apart of the lithosphere. As it splits apart, upwelling mantle material fills the gap, becomes new lithosphere as it solidifies above the appropriate isotherm, and attaches itself onto the trailing edges of the diverging lithosphere slabs, creating new lithosphere and making room for more upwelling material. McKenzie, like Hess, attributed the constancy in thickness of the newly created layer to the depth of the appropriate isotherm. With McKenzie's model, however, there is no notion of the depth of the relevant isotherm becoming less because of rising convective currents as with Hess's model. With McKenzie's version, the newly generated lithosphere is much thicker than the seafloor generated with Hess's version. In both, the layer of material containing the magnetic anomalies is simply carried along with the rest of the newly generated material. McKenzie's model still depends on mantle convection in the sense that the intrusion of uprising mantle material into the gap is driven by thermal buoyancy forces; however, such convection will occur after the gap forms due to splitting of the lithosphere; it does not cause Earth's surface layers to split apart creating a gap. In addition, looking ahead, the separating lithosphere slabs, once they become sufficiently cool, sink back into the mantle driven by thermal buoyancy forces. The moving lithosphere is itself part of the convection system. With Hess, the moving seafloor is carried along given sufficient viscous drag by mantle convection. Thus there are significant differences between their versions of seafloor spreading, and we can see why McKenzie views his introduction of this new model of ridges and seafloor spreading as "the central issue" of his paper. Why didn't anyone take "any notice of" his new version of seafloor spreading? Because, as I shall suggest below, McKenzie did not make clear that he was presenting a new model. Moreover, nobody noticed that his new model of rootless ridges removed the difficulty facing Hess's version of seafloor spreading arising from Africa and Antarctica being surrounded by ridges because he never stated that it did.

McKenzie introduced his new model of rootless ridges within the context of his analysis of "the strength and thermal properties of the crust and uppermost mantle, of what is often called the lithosphere." He did not explicitly state that he was introducing a radically new model of ridges, a model in which they have no deep structure. He did not explicitly state that he also was introducing a new version of seafloor spreading.

It is the principal purpose of this paper to demonstrate that both gravity and oceanic heat flow are probably controlled by the strength and thermal properties of the crust and uppermost mantle, of what is often called the lithosphere ... Any discussion of the significance of gravity and heat flow observations requires some mechanical and thermal model for the crust and upper mantle. At temperatures (less than $\approx$ 700 °C) the mechanical behavior of hard high-melting point minerals, like olivine, is principally elastic until the yield stress is reached, and then the material fails by brittle facture. At higher temperatures various diffusion-controlled processes permit mineral to creep at all stresses and therefore any stress deep within the mantle must be maintained by some active process against dissipation by creep. These remarks suggest that the outer perhaps 50–100 km of the earth must behave as a rigid layer and that the places where this is not true will be marked by shallow earthquakes ... The early measurements of heat flow at sea ... show that high values were common on the mid-Atlantic ridge. Many measurements have since been made, and the average heat flow through the ridges in the Atlantic, East Pacific, and Indian oceans is considerably above the oceanic mean ... These anomalies are larger and more consistent than any others in the heat flow field. Two explanations have been suggested. The first produced the high flow by intrusion of hot basalt and peridotite in dykes along the axis of the ridge. As seafloor spreads away from the ridge, the hot rock slowly cools. This idea suggests that the width of the anomaly should be controlled by the spreading rate. If this theory is correct, the high heat flow is a result of tension in the lithosphere and is not the surface expression of temperature anomalies in the upper mantle.

(McKenzie, 1967: 6261–6262)

He now introduced the standard form of seafloor spreading, the one in which heat flow at ridges is caused "by narrow upwelling limbs of convection cells." However he did not cite Hess (1962) or Dietz (1961a), but a paper by Turcotte and Oxburgh (1967) within which they developed a quantitative model of seafloor spreading in which heat flow beneath ridges is caused by upward-moving convection currents. Moreover, he wanted to stress the fact that their model, unlike his own, "relates the surface observations directly to flow in the mantle."

The other suggestion ignores the lithosphere and produces the high heat flow by narrow upwelling limbs of convection cells, which are required to be beneath the axes of the ridges. Turcotte and Oxburgh [1967] have used boundary layer theory to analyze this idea and have neglected the variation of viscosity with temperature. The difference between the two suggestions is in the nature of the boundary layer ... The lithosphere is a mechanical layer produced by the rapid variation of creep rate with temperature; its existence does not depend on a high Rayleigh number. The thermal boundary layer required by the second hypothesis is governed by the convection process and, hence, relates the surface observations directly to flow in the mantle.

(McKenzie, 1967a: 6262)

He then raised the difficulty that with their model (and by the same reasoning with Hess's model), convective cells beneath ridges would themselves be split into individual segments, offset from each other wherever mid-ocean ridges are segmented by transform faults.

If the first hypothesis is correct, the heat flow anomaly is a consequence of seafloor spreading and cannot be used to examine the temperature distribution in the mantle. The anomaly should follow the axis of the ridge and be offset by transform faults, as *Vacquier and Von Herzen* [1964] believe. With this hypothesis, however, there is no reason to suppose that the axis of the mantle convection cell is offset in the same manner.

(McKenzie, 1967: 6262; see §3.8 for discussion of Vacquier and Von Herzen's paper)

McKenzie next mentioned his "simple model for a spreading seafloor" but said almost nothing on ridge structure, his interest was in explaining how his model accounts for heat-flow measurements, offering a counter to Le Pichon's model (§6.11), and introducing his analytic solution to the cooling plate problem in which he used dimensionless variables, which allowed him to "solve the equations in general irrespective of [the moving slab's] thickness." He viewed this advantage as methodologically important.

A simple model for a spreading sea floor in a slab of constant thickness moving with a constant velocity... The complicated intrusion mechanisms by which the sea floor is produced along a mid-ocean ridge are neglected. Instead, a slab is produced at $x = 0$ at constant temperature. Though this boundary is certainly unrealistic, the agreement between the heat flow anomaly from this mode and the anomaly from the model used by Langseth *et al.* [1966] shows that, except in the central region, the shape and size of the anomaly are little affected by the details of the intrusion. If the thickness of the lithosphere is taken to be 50 km, there is good agreement with the observed width and shape of the ridge anomalies... The difference between this conclusion and that of Langseth *et al.* is caused by the difference in thickness of the spreading layer. In the model used here the temperature is constant at and below 50 km, and therefore horizontal isotherms below this depth are not in conflict with the surface heat flow. Clearly, this model is not the only possible one, and a choice between it and the model of Turcotte and Oxburgh cannot be made on the basis of heat flow observations alone. Since only the thickness of the lithosphere is variable, however, the model used here is probably the simplest that can account for the observations and that does not involve any assumptions about the Rayleigh numbers within the mantle.

(McKenzie, 1967: 6262–6263)

There was of course a huge advantage of McKenzie's rootless ridge model over that of Turcotte and Oxburgh and over any model of seafloor spreading that required upward-moving convection currents beneath mid-ocean ridges, namely, that it avoided the difficulty arising from Africa largely, and Antarctica entirely being surrounded by ridges. McKenzie, as we have seen, was well aware of this advantage, but said nothing about it in the paper. (Morgan also introduced a rootless ridge model of seafloor spreading and incorporated it into his own version of plate tectonics (§7.11).)

I think it is therefore quite understandable why no one acknowledged that McKenzie was introducing a new model of the structure of mid-ocean ridges, one without any deep structure, and a new model of seafloor spreading, one in which ridges formed as consequence of seafloor spreading, not as one in which the formation of ridges initiates seafloor spreading. I think we can also see why he developed his new model while waiting to take his orals for his Ph.D., for it was in working on his Ph.D. that he learned about the relationship between creep and temperature, so essential to his account of the development how lithosphere is formed and thus fundamental to his new model of seafloor spreading and the decoupling of uprising convection currents and ridges. Moreover, his new model was also important, I think, for his development of plate tectonics. First, he already was thinking in terms of thick, rigid lithosphere slabs. Second, his account of seafloor spreading and passive formation of ridges helped him to separate kinematics and dynamics, put dynamics aside and concentrate on kinematics. Finally, I think that many, not realizing that plate tectonics incorporated a version of seafloor spreading that did not involve a direct link between upwelling mantle convection and formation of ridges, failed to appreciate that plate tectonics is not a dynamical theory.

What now about his explanation of heat flow in terms of his version of seafloor spreading? He found "good agreement with the observed width and shape of the ridge anomalies" by setting the thickness of the lithosphere at 50 km. As already noted (§6.11), Le Pichon quite correctly later noted that McKenzie, "already convinced of the validity of the seafloor spreading model," adjusted the parameters and "chose a temperature inside the mantle three times smaller – 550° instead of the 1500 °C that we had chosen," and that their estimate "is still considered to be much closer to the actual mantle temperature" (Le Pichon, 2001: 211).

Here is McKenzie's own retrospective response.

I was not at the time a marine geologist, and knew little about data collection at sea. I agree with what Xavier [Le Pichon] has to say about the disagreement of my model with the bathymetry... What Xavier missed was the importance of the depth at which the lower boundary condition was imposed, and his insistence on coupling the convection to the plate motions. I had no clear idea of how good the agreement between theory and experiment should be, and have been astonished by how far the plate model has been taken by John Sclater and Barry Parsons (1977). Their success over the next ten years convinced me of the need to go very deeply into the analysis of relevant data when I had ideas. But in 1967 I had never worked with any data, whereas Xavier had for years ... The question of the lower boundary condition was (and still is to some degree) unclear. Obviously the basalt comes out at 1200 °C, so the mantle interior must be at this temperature. But it is not obvious that this is the temperature of the base of the layer that moves with the stripes. I took 550° in 1967, and this is too cold. We now believe it is between 900 °C and 1000 °C. This part of the argument is complicated and subtle, and not understood by many people even now. You will find a recent attempt to get it over in McKenzie and Bickle (1988).

(McKenzie, March 18, 1990 letter to author; my bracketed addition)

Both McKenzie's and Le Pichon's retrospective comments, I believe, are essentially correct. Le Pichon did not believe that seafloor spreading was correct; McKenzie did. McKenzie adjusted the parameters, but Le Pichon did not think of the possibility of decoupling seafloor spreading and mantle convection. As it turned out (§6.11), there was a key hidden factor, the cooling effect of fluid advection in the crust, proposed by Lister (1972) and confirmed by Sclater and Crowe (1979). McKenzie recognized that seafloor spreading's inability to explain the heat-flow data, given the confirmation of V-M and ridge-ridge transform faults, should be treated as an unsolved problem instead of a difficulty so severe that seafloor spreading should be rejected. So he developed an alternative version of seafloor spreading with rootless mid-ocean ridges. He didn't get it all right, but he did introduce, basing his work partly on Le Pichon's model, a model that formed the basis of all future work on the problem.

7.7 Heirtzler and company extend the reversal timescale

I wish here to acknowledge the firm leadership of Jim Heirtzler, who bulldozed us into producing this impressive collection of published work in a short amount of time.

(Le Pichon, 2001: 220–221)

The resulting timescale became known as the Heirtzler scale, after the first author, Jim Heirtzler. It has turned out to be surprisingly accurate, good to a few percent in most parts – one of those great strokes of genius or luck or both – but of course, at the time, no one knew if they were even close.

(Atwater, 2001: 254)

With regard to JOIDES [Joint Oceanographic Institutions for Deep Earth Sampling] and our comments about gaps in spreading [which Vine had opposed]. JOIDES Leg 3 across the South Atlantic confirmed the magnetic time-scale in an unbelievable way. The plot of age of oldest sediment in the core (i.e., immediately above or incorporated within pillow lava) against distance from the ridge axis is fantastically linear and implies a uniform rate of spreading of, I think, 1.3 cm/yr at this latitude in the South Atlantic for the past 75 m.yr. This is remarkable in that Lamont assumed a constant spreading rate of approximately this value in the S. Atlantic in calibrating their reversal time-scale, consequently the ages predicted by the magnetic anomalies are never more than 2 m.yr different from those deduced from the micropalaeontology. The resolution on the micropalaeo. dating has still to be investigated – it's pretty good but we are now having to talk about details of the order of 1 m. yr!

(Vine, February 6, 1969 letter to McKenzie; my bracketed additions)

On October 20, 1967, *JGR* received four papers by Heirtzler and members of his group at Lamont who were working on marine magnetic anomalies (Pitman, Herron, and Heirtzler, 1968; Dickson, Pitman, and Heirtzler, 1968; Le Pichon and Heirtzler, 1968; Heirtzler, Dickson, Herron, Pitman, and Le Pichon, 1968). The papers appeared in the March 15 issue, accompanied by Morgan's (1968b) blockbuster paper on plate tectonics. In fact, Le Pichon included some of his own forthcoming work on plate tectonics in the paper co-authored with Heirtzler. Heirtzler and company had wasted little time exploiting Lamont's huge data-set of marine mag- netic anomalies, submitting four key papers just about a year after the appearance of Pitman and Heirtzler's "*Eltanin*-19" paper (§6.6). Their task was made much more doable by Ewing's insistence on digitalizing Lamont's data-sets, neither Scripps nor Woods Hole were digitalizing their data at this time.[10]

Dividing the oceans amongst themselves, Pitman, Herron and Heirtzler took the Pacific; Dickson, Pitman and Heirtzler, the South Atlantic; Le Pichon and Heirtzler, the Indian Ocean. Beginning with the radiometrically dated timescale now extending back 3.35 m.y. (§6.4), they determined relative timescales for the widening of these

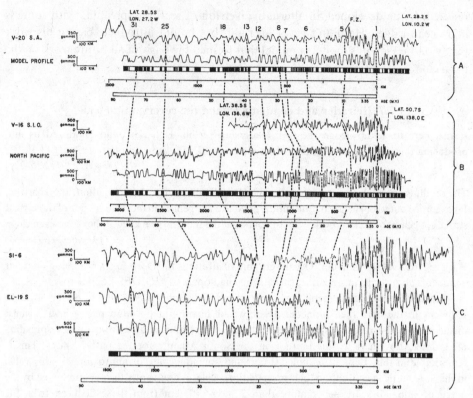

Figure 7.8 Heirtzler *et al.*'s Figure 1 (1968: 2120). Their caption reads: "Sample magnetic profiles from various oceans. The South Atlantic (S.A.) profile is from the paper by Dickson *et al.* [1968], the South Indian Ocean (S.I.O.) profile is from Le Pichon and Heirtzler [1968], and the remaining profiles (North Pacific, SI-6, EL-19S) are from Pitman *et al.* [1968]. Beneath each of the observed profiles is a theoretical profile calculated from the normally magnetized (black) and reversely magnetized (white) bodies shown. Each body is 2 km thick. With each model is a time scale constructed by assuming an age of 3.25 m.y. for the end of the Gilbert reversed epoch. Dashed vertical lines connect similarly shaped anomalies identified by the numbers at the top of the dashed lines."

oceans by matching magnetic anomalies and for ease of correlation, numbered prominent ones outwards from the spreading ridge. This Lamont numbering of characteristic anomalies identifiable in different oceans is the framework within which ocean floors are now dated: it was a huge step forward. They traced back the anomalies to number 16 in the Indian Ocean, and to number 31 in the North and South Pacific and South Atlantic (Figure 7.8). The question then became: which ocean gave a timescale that is "most reasonable"?

In Figure 1 [Figure 7.8] are displayed observed magnetic profiles from the North and South Pacific, the South Indian, and the South Atlantic oceans and model bodies that could account for the observations. The latitude and longitude of the end points of each profile are noted on

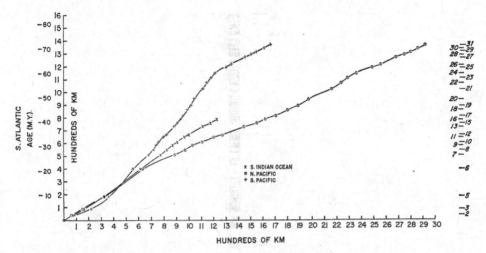

Figure 7.9 Heirtzler *et al.*'s Figure 2 (1968: 2121). Their caption reads: "The distance to a given anomaly in the South Atlantic (V-20 S.A.) versus distance to the same anomaly in the South Indian, North Pacific and South Pacific oceans. Numbers on the right refer to the anomaly numbers."

the figure. Study of the theoretical profiles for the three models shows that these models can account for the observed profiles immediately above them. With each model an age scale is given. Each scale is based on the date of 3.35 m.y. for the beginning of the Gauss normal polarity epoch [Doell *et al.*, 1966]. The time scales obviously differ and are not linearly related, since the models for the South Pacific and the South Atlantic were derived from the basic model for the North Pacific by adjusting the set of model blocks according to the curve for the relative spreading rate. The relative spreading rate curves (Figure 8, Pitman *et al.* [1968] and Figure 2 of this paper [Figure 7.9]) though nonlinear, are continuous. The problem is to choose the most reasonable time scale.

(Heirtzler et al., *1968: 2120; my bracketed figure numbers)*

They easily eliminated the Indian Ocean because it extended back only to anomaly number 16. Paleontological studies led to the elimination of the South Pacific. They appealed to studies by Saito and Ewing *et al.* (1968).

T. Saito (personal communication) has paleontologically determined the age at the bottom of a core (V 20-80) taken at 46°30′ N, 135°W as Lower to Middle Miocene (13–26 m.y.). The core location is on anomaly 6. By the North Pacific and South Atlantic standards the magnetic basement at anomaly 6 should be 20–22 m.y. old. The South Pacific time scale suggests an age of 14 m.y. for this basement, much younger than the age of the core. The core was taken on an outcrop where no sediment was detected by seismic reflection methods. The seismic reflection technique employed, however, was unable to resolve reflectors less than 50 meters in thickness. Thus, allowing for some sediment beneath the core, we can accept the older date for anomaly 6 as more reasonable.

(Heirtzler et al., *1968: 2121)*

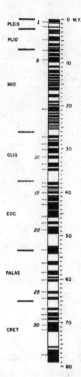

Figure 7.10 Heirtzler *et al.*'s, Figure 3 (1968: 2123). Their caption reads: "The geomagnetic time scale. From left to right: Phanerozoic time scale for geologic eras, numbers assigned to bodies and magnetic anomalies, geomagnetic field polarity with normal polarity periods colored black."

Work by Ewing *et al.* (1968) also ruled against the South Pacific.

Ewing *et al.* [1968] have found the eastern boundary of layer A (as detected by reflection techniques) in the North Pacific to be just west of anomaly 32. They have proposed an Upper Cretaceous age (70 m.y.) for this layer. The South Pacific time scale would suggest an age of 47 m.y. for anomaly 32. Unless there has been a major discontinuity in spreading rates, the South Pacific time scale is in error at this point by a factor of 2.

(Heirtzler et al., 1968: 2120)

This left them with a choice between the North Pacific and the South Atlantic. They (1968: 2122) dismissed the North Pacific because Vine (1966) had "suggested that the spreading rate at the Juan de Fuca Ridge has not been constant over the past 10 m.y. but has decreased from 4.4 to 2.9 cm/yr" (see Figure 6.11). With the South Atlantic as their standard, they presented their geomagnetic timescale based on the V-20 profile (top profile in Figure 7.8) with its average spreading rate of 1.9 cm/yr, which gave the timescale in Fig. 7.10. This timescale is of first importance and testifies to the good judgment of the Lamont team.

They identified anomalous situations and discussed three of them; eliminating the first two, but not closing the door on the third. The first pertained to a questionable dating of 27 m.y. to a boulder from the Cobb seamount in the north-east Pacific; the second, to a Miocene dated boulder supposedly dredged only 10 km from the Mid-Atlantic Ridge, which, according to Ewing, Le Pichon, and Ewing (1966) and Saito *et al.* (1966a) eliminated recent seafloor spreading (§6.14); the third, to Ewing and Ewing's (1967) discovery of an apparent buildup of 10 m.y. old sediments in the North Atlantic near anomaly 5, which led them to propose that seafloor spreading had begun again 10 m.y. ago after having stopped for millions of years (§6.14). Recent dating by Dymond *et al.* (1968) of two basalt boulders from the Cobb seamount as younger than 2.0 m.y. made highly dubious the earlier date of 27 m.y.; Hess had already explained to Menard why the Ewing–Saito sample should be ignored (§6.2), and Heirtzler and company appealed to Sykes.

L. R. Sykes (personal communication) has pointed out that although the boulder was located within 10 km of the axis of the north side of the fracture zone, it is about 60 km from the axis on the south side. Hence, it is impossible to predict the age of such a boulder if we assume the existence of sea floor spreading.

(Heirtzler et al.*, 1968: 2125)*

This left the Ewing brothers' discovery of the buildup of 10 m.y. old sediment near anomaly 5, which had led them to propose a cessation of seafloor spreading 10 m.y. ago, and which was later shown in 1971 to be based on a mistaken identity (§6.14). Of course, Heirtzler and company did not know that, and they admitted (1968: 2115) that "their data cannot exclude such an interruption" but added (my italics), *"provided that it occurred simultaneously everywhere."* If it occurred everywhere, it would have to have gone undetected on every marine magnetic profile. Nonetheless, this notion of episodic seafloor spreading with the most recent episode beginning 10 m.y. ago was reintroduced by Isacks *et al.* (1968) and Le Pichon (1968a). Indeed, even Le Pichon and Heirtzler (1968: 2114) gave it considerable credibility in their companion paper on the later history of seafloor spreading in the Indian Ocean.

I have argued that before *Eltanin-19* was revealed, Heirtzler protested too much against V-M and seafloor spreading. I also argued that he ignored land-based paleomagnetic work supportive of continental drift. However, he had now changed. Certainly he and the Lamont group deserves credit for this huge accomplishment, based mainly on their own data, of establishing the geomagnetic polarity reversal time scale of Fig.7.10, and for using land based paleomagnetic results in combination with those from seafloor spreading to get a more complete picture of seafloor spreading and continental movements in the southern hemisphere since the early Mesozoic.

The magnetic evidence for ocean floor spreading presented in the companion papers places several restrictions on the allowed paths of continental movements since the breakup of Gondwanaland.

It is thus possible to synthesize a picture of continental movement in the southern hemisphere since the early Mesozoic. Paleomagnetic reconstructions by Irving [1964] and Creer [1965] have the continents of Africa, Antarctica, Australia, India, and South America united and close to the south geographic pole in the Early Permian. Gough *et al.* [1964] show that there has been little change in the paleomagnetic latitude of Africa since the middle of the Mesozoic. They also show evidence of rapid northward movement of Africa during the early Mesozoic and Permian. Studies by Creer [1965] suggest that there has not been appreciable change in the paleolatitude of South America since Triassic-Jurassic times but that large northward movements did take place in the Lower and Middle Permian. Investigations on the Triassic-Jurassic lavas that appear in all the Gondwanic continents indicate that the breakup of Gondwanaland had already occurred by this time.

From the study of the magnetic pattern in the Indian Ocean the initial northward movement of the Africa-South America block in Mesozoic-Permian times occurred as a result of spreading about the southwest branch of the mid-Indian ridge. The break from the initial mass started at the Horn of Africa and then opened down the east coast of Africa. At present there is little or no spreading about this branch of the mid-Indian ridge, and, on the basis of paleomagnetic evidence, it appears that it was active during the lower Mesozoic and Permian but that by Jurassic the spreading had mainly ceased. The split on the South American and African continents may have begun in early Mesozoic almost simultaneously with the break away of the African-South American block. Thus, the Argentine and Cape basins were born. The cessation of spreading about the southwest branch of the mid-Indian ridge, in Cretaceous and the development of a major part of the Argentine and Cape basins ended the first major phase of spreading in the southern Indian and South Atlantic oceans.

(*Heirtzler* et al., *1968: 2134–2135*)

With its vast bank of data this could only have been done at Lamont.

7.8 Morgan discovers plate tectonics

Born in 1935, William Jason Morgan earned his undergraduate degree in 1957 from Georgia Institute of Technology, majoring in physics. After spending two years in the US Navy, he began graduate work in 1957 at Princeton University in physics under Robert Henry Dicke. He soon co-authored a paper with Dicke (Morgan, Stoner, and Dicke, 1961). In 1964 he obtained his Ph.D. (Morgan, 1964), which he described (October 1989 interview with author) as "looking for geophysical and astronomical effects that might be sensitive to fluctuations in the strength of gravity." Recall Dicke's interest in Earth expansion (§1.7). While working on his Ph.D. thesis, he became interested in geophysics. "The geophysics was very data rich; relativity was very theory rich but data starved – I was more attracted to the data rich field" (Morgan, February 8, 2012 email to author). He had not, however, ever taken a course in geology or geophysics. But he had done some reading on his own, and enjoyed *Physics and Geology*, the 1959 textbook by Jacobs, Russell and Wilson. He recalled that other books did not compare in their sweep, and specifically mentioned discussion of trenches as an example of its greater sweep. This text was fixist in approach and strongly anti-drift. Morgan also recalled that "the paper which probably had the most profound effect" on him was Wilson's (1963d) paper "Continental drift," which appeared in *Scientific American,* and which contained a discussion on hot spots (Morgan, presentation at 25th Anniversary Celebration of Plate Tectonics, May 15, 1992).

In September 1964, one month before finishing the first draft of his Ph.D., thesis, he accepted a post-doc in the Department of Geology at Princeton.

Hess and the department had received the O.K. from the Dean of Faculty to expand into geophysics – to my knowledge the number four new positions had been approved (to be filled over a period of several years). Elsasser arrived in 1963.

(Morgan, February 8, 2012 email to author)

Elsasser wanted two post-docs, and he and Hess hired Morgan "and another post-doc, Bernard Durney, who left after a year and went on to solar physics" (Morgan, February 8, 2012 email to author).

My salary came from a new NASA "Pre-moon-landing, built up of corps of scientists ready to capitalize on the moon data" grant. I never worked on the moon with Hess, but in 1960 I did work on a 1st graduate student project on lunar craters with Dicke. This work was my real introduction to the computer facilities at Princeton. I was one of the few graduate students spending much time in the computer center – I had taken a course on computers at Georgia Tech in 1957 – and to learn about the moon was also my introduction to the Geology department's library.

(Morgan, February 8, 2012 email to author)

Vine arrived in October 1965, and Morgan and he shared an office for two years. Morgan later credited Vine with helping him to understand the petrology involved with Hess's seafloor spreading.

I didn't need help with the seafloor spreading part, but Hess was basically a petrologist and his explanations of how things worked always had a lot of petrology included. That was the part with which Fred [Vine] was indispensable.

(Morgan, February 8, 2012 email to author)

Morgan remembers when he first learned of seafloor spreading; Hess gave him a copy of his 1960 preprint on seafloor spreading just after he passed his Ph.D. oral exam.

My introduction to seafloor spreading was the day of my final public oral examination of my Ph.D. There were two physicists and since several of my chapters concerned geophysics Hess on was on my committee. I had essentially used Jeffreys' interpretation of *The Earth* (a great book) for the mathematics and for the model of the Earth (all elastic, not fluid). Right after my exam, Hess gave me a copy of his 1960 pre-print on seafloor spreading and said to effect, "You ought to read this."

Elsasser had let me work entirely on finishing the thesis my first two months in the department. He then directed me to a project on viscous flow at a trench. This was pretty much an extension of my 1st year project with Dicke where I was calculating the viscous-flow/changing-shapes of craters to see if a minimum viscosity of the moon could be estimated from the height/shape of central peaks and crater rims. (Essentially the glacial rebound problem.)

(Morgan, February 8, 2012 email to author)

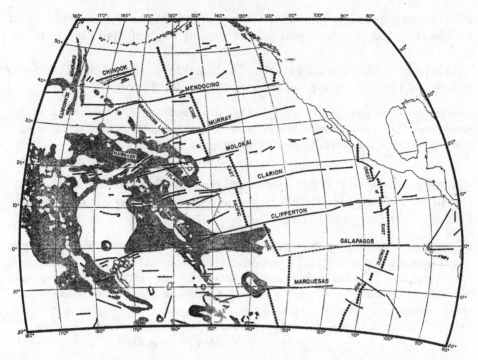

Figure 7.11 Menard's Figure 1 (1967: 73). Principal fracture zones of the northeastern and central Pacific: equal-area projection. Stippling indicates smooth archipelagic aprons and the belt of equatorial pelagic smoothing; white areas within archipelagic aprons are volcanoes and volcanic ridges. Individual lineations are troughs, asymmetric ridges, or regional changes in depth: broken-line bands have fracture-zone topography, but information is not adequate for tracing of individual lineations.

Vine also got him interested in computing spreading rates, and Morgan, a quick learner, received Vine's acknowledgement for valuable discussions in Vine's 1966 paper confirming V-M (§6.8). Morgan began working on mantle convection, and planned to read a paper on trenches at the April 1967 AGU meeting. His abstract read as follows:

Convection in a Viscous Mantle and Trenches. The effects of a concentrated cylinder with a density excess of 0.3 g/cm^3 sinking in a viscous mantle are examined. A model was chosen in which the viscosity decreased with depth. It is shown that sinkers above this low-viscosity layer pull down on the surface and produce trenches. Sinkers below this layer are decoupled from the surface and produce broad gravity maxima. An evolutionary model of trenches in which the crust above the sinker thickens with time is also considered.

(Morgan, 1967a: 217–218)

But all this changed once he met with success in initially testing his idea of what would later become known as plate tectonics.

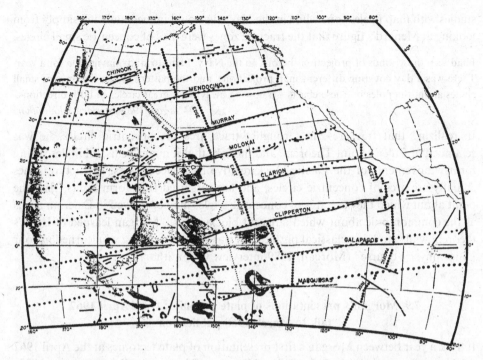

Figure 7.12 Morgan's Figure 11 (1968: 1968). Old fracture zones in the Pacific are compared with circles concentric about a pole at 79° N, 111° E. (Figure adapted from Menard (1967).)

During the first four months of 1967, he began developing his new idea after reading a paper by Menard that was published in January on the great fracture zones of the eastern Pacific. Menard (1967: 74) had no explanation to offer for their origin and exclaimed that they "are not becoming more obvious as additional facts accumulate." But Menard's pessimism is not what caught Morgan's attention. Wanting to show that the great fracture zones are nearly straight lines for long distances, Menard used a great-circle projection to construct his Fig. 1 (Figure 7.11). With such a projection, straight lines on a spherical surface appear as straight lines on a flat surface. What excited Morgan was not that the fracture zones almost follow great circles, but that they "are all not great circles" (Morgan, October 1989 interview with author). It was the "not quite" rather than the "almost" that intrigued him. This was the catalyst that got Morgan thinking, and thinking about spherical rather than planar surfaces, thinking, in particular, about Euler's theorem. He later redrew Menard's figure for his April presentation and for his first paper on plate tectonics, where it appeared as his Figure 11 (Figure 7.12), and noted, "Figure 11 shows the great fracture zones of the Pacific block. Menard's [1967] demonstration that these great fracture zones are not all great circles initiated the present investigation" (Morgan, 1968b: 1968; Menard reference is same as mine). From his navigational studies with map projections while in the Navy, he could immediately tell simply from looking at Menard's figure that the fracture zones were arcs of concentric small circles.

I had seen those kinds of projections before. In the Navy, I had to study navigation for a year. There was a day covering different map projections, and the pattern is like the lines of small circles about the poles on these charts – a bowing away from great circles in both directions.

(Morgan, 1989 interview with author)

In realizing that fracture zones roughly trace arcs of concentric circles, he was reminded of Euler's Point Theorem, and thought that it might be useful in describing former movements of the Pacific block relative to adjacent ones. If fracture zones actually are arcs of concentric circles, and are fossilized transform faults, then the point about which they are arcs of concentric circles may be viewed in Eulerian terms as the common pole about which adjacent blocks rotate. Morgan learned of Euler's theorem while studying classical physics: "It was the sort of thing you learned in your junior physics course" (Morgan, 1989 interview with author).

7.9 Morgan's presentations of plate tectonics: from April 1967 talk to March 1968 publication

It took a year between Morgan's first presentation of plate tectonics at the April 1967 AGU meeting and the March 1968 publication of his paper. What happened? He began writing up the extended notes for his talk a week before the meeting, finishing at 2 am, the night before (Le Pichon, 1991: 1). "I then caught the early morning train to Washington" (Morgan, February 8, 2012 email to author). After Le Pichon heard Morgan's talk, he asked him for a copy of the notes, and many years later had the good sense to have them published with Morgan's permission in a special issue of *Tectonophysics* in celebration of the silver anniversary of plate tectonics (Le Pichon, 1991).[11] Morgan thinks that he gave the outline to ten people at the meeting (Morgan, February 8, 2012 email to author). Besides Le Pichon, he is "quite sure" that he gave copies to Menard, Tuzo Wilson, Sykes, Vine, and Carl Bowin and/or Joe Philipps (Morgan, February 8, 2012 email to author). The extended notes are just over seven pages of double-spaced typed text, eleven figures, and a page of references. Figures *not* in the original outline but in the published paper include Figure 1 of the twenty or so rigid blocks making up Earth's surface, the two figures (Figures 8a and 8b) showing attempts to position the Eulerian rotational pole by constructing circles normal to the transform faults associated with the Mid-Atlantic Ridge or normal to the strike of earthquake mechanism solution and finding their point of intersection; all these figures are related to the determination of the Eulerian rotational pole of relative motion between the Pacific and North American blocks (Figures 13 and 14a–14f), between the Pacific and Antarctic blocks (Figures 16 and 17), and the figure he later used to show the anomalous Juan de Fuca block (Figure 12). The extended outline, which has the same title as the published paper, included a copy of Vine's Figure 15 (Vine, 1966: 1412) (my Figure 6.13). In particular, it included all the key theoretical ideas of plate tectonics and determination of the pole of relative rotation between the South and North American blocks, which he treated as a single block, and the African block. Morgan also briefly discussed the relative motion between the North American and Pacific blocks, and included what became Figure 11 in his 1968 paper (Figure 7.12 above).

Morgan spent July and August 1967 at WHOI, where he expanded and refined his paper, submitting it at the end of August 1967 to *JGR*.[12] The editor, Orson Anderson, wrote Morgan on September 5 acknowledging receipt of the manuscript and promising that his "paper will receive prompt attention." On November 14 Anderson wrote Morgan telling him that his paper had been "recommended for publication . . . after minor revision."

The paper which you submitted has been reviewed and recommended for publication in the *Journal of Geophysical* after minor revision. Therefore we are returning your typescript with the reviewers' comments. I strongly suggest that you avail yourself of the detailed comments provided.

(Anderson, November 14, 1967 letter to Morgan)

Reviewer A was impressed. He recommended publication, thought the paper contained significant ideas, but added that Morgan should remove the section "Pacific-North America." He also recommended a change in one of the figures, which again was connected with one of Menard's longtime concerns, the great fracture zones of the eastern Pacific.

This paper deserves publication in some form because of the significant ideas it contains. However I believe that the ideas are diffused rather than strengthened by the discussion of "Pacific-North America" and I recommend that this section be deleted. It might well serve as the nucleus of a second paper after the author has had more time to sharpen the analysis.

I am doubtful about one aspect of the manuscript and figures. If a continent is split by spreading convection I do not see why any transverse fracture zones would extend into the continental blocks. Such extensions are dashed in the figures but I believe that it would be more logical to remove them.

Reviewer B was clearly underwhelmed with Morgan's paper but still recommended publication. Here are B's comments. They are answers to a standard set of review questions. I do not know the questions specifically, but the answers suggest their general nature: Does the paper break new ground? Is it clearly written? Are the figures needed, and do they help to understand the material? Does the abstract provide an adequate summary of the central thesis and explain its importance? Should *JGR* publish the paper? If so, is it in need of revision? If so, please elaborate.

1. To me, it was a new treatment; but I am not thoroughly informed on relevant literature.
2. I found that the paper read well and that I could follow his reasoning.
3. As far as I know – see first comment.
4. I cannot judge for I am not familiar with policy. The figures seem necessary, and are adequately used.

5. The abstract seems adequate. I do not find that the paper contains new information that is of great impact – it is rather an exercise in explanation of described data.

6. The paper is not one that impresses me as being of great urgency, but it is one that brought to my attention some relationships in a way in which I had not seen them presented heretofore. Thus it seems to me to be a paper warranting publication in being instructive. It does not introduce new data, nor does it impress one with the significance of the interpretations that are presented. In specific answer, I do not recommend publication in its present form, but with some further emphasis on the significance of the interpretations given.

I have not made specific criticisms, nor could I do so without appreciable study of references. I have no objection to the writer's having my comments, but I believe that others more familiar with rifts should be better able to give specific criticisms.

I wonder how much Reviewer B understood Morgan's paper. Obviously, he did not understand its far-ranging implications.

Menard, I believe, was Reviewer A. He (1988: 290) recalled, albeit tentatively, that he "reviewed the paper for an editor."

Jason Morgan sent me a preprint of his manuscript in its early draft, probably in the late spring of 1967. I believe I also reviewed the paper for an editor. In any event the manuscript certainly circulated among my students, and we discussed it. The original draft however was difficult for me to fathom, and it did not have the impact of the final publication. Moreover, I had other things on my mind. I spent July through September on the Nova expedition in the southwestern Pacific. Teddy Bullard was among the old hands to participate, and there were many students including those who knew about Morgan's papers.

(Menard, 1988: 290–291)

Given that Menard reviewed the paper, I think he was Reviewer A because of his concern over representation of the great fracture zones and discussion of Pacific–North America, Menard's home waters. I also think Menard, unlike Reviewer B, realized that the paper was highly significant; he was at the time listening to McKenzie, who was visiting Scripps (§7.12), explain to him his own version of plate tectonics. It is also clear that Morgan did not send him the version of his paper that he sent to _JGR_, but did send him as he later recalled, a copy of the extensive notes he prepared for his AGU talk. What Menard (1988: 293) quoted as the opening lines of the "preprint" are those of the extended outline, while Morgan's submitted paper to _JGR_ begins the same way as the published version: "A geometrical framework with which to describe present day continental drift is presented here."

Morgan resubmitted his paper on November 29, 1967. In his cover letter to Anderson, he wisely ignored Reviewer B, but answered Reviewer A.

Enclosed is the revised version of my paper "Rises, trenches, great faults, and crustal blocks". One comment by Reviewer A concerned my use of dashed lines for the inactive extensions of the transform faults in my figures. This convention has been used by other authors, e.g., Sykes _JGR_ 72, p. 2131, 1967, and I think it requires no explanation in my text. The reviewer's other comment is more substantial, "the ideas are diffused rather than strengthened by the discussion of 'Pacific–North America'." I concur with his comment but would like to leave this portion in for the following reasons. First, the data listed in the tables in this section are used by and referred to in a manuscript by Xavier Le Pichon (Sea floor spreading and continental drift), and my conclusions regarding the anomalous nature of the Juan de Fuca region are referred to in a manuscript by Tobin and Sykes (Seismicity and tectonics of the Northeast Pacific Ocean). If these tables and comments are not included here, they would have to be included in these other authors' papers. Second, I have added two short sections following the 'Pacific – North America' section which I think strengthen the main argument more than enough to counteract the diffuse effects of this section. The sections 'America – Africa', 'Pacific – America', and 'Antarctica – Pacific' are inter-related in the section 'Antarctica – Africa' and the one section could not be eliminated without weakening the final argument.

(Morgan, November 29, 1967 letter to Olson Anderson)

Except for the additions, he made few changes to the originally submitted manuscript. Morgan also removed Vine's Figure 15 (Vine, 1966: 1412), and replaced it with his Figure 12. Morgan received notice from Anderson on February 14, 1968, two and a half months after resubmission, that his paper was scheduled for publication on March 15, 1968.

7.10 Morgan's April 1967 AGU presentation

Morgan wasted no time stating the key elements of what became known as plate tectonics during his talk at the AGU meeting. He began his presentation (extended outline) by assuming that the moving blocks making up Earth's surface, which have three types of boundaries, are rigid enough that the motion (rotation) of one block relative to another may be described in terms of Euler's Point Theorem. Here is how he opened his talk (began his extended outline).

Consider the earth's surface to be made of a number of rigid crustal blocks, and each block is bounded by rises (where surface is being destroyed), and great faults. Assume that there is no stretching, folding, or any distortion within a given block. Then on a spherical surface, the motion of one block (over the mantle) relative to another block may be described by a rotation of one relative to the other. This requires three parameters, two to locate the pole of relative rotation and one for the magnitude of the angular velocity. If two adjacent blocks have as common boundaries a number of great faults ... these faults must lie on "circles of latitude" about the pole of relative rotation.

(Morgan, 1967b: 1)

With hindsight, what could be clearer? After describing his idea as applied to a plane, he returned to a sphere. He repeated the key ideas, stated without naming the relevant "theorem of geometry," and identified the "three parameters needed" to describe the relative motion between blocks.

We now go to a sphere. A theorem of geometry proves that a block on a sphere can be moved to any other conceivable orientation by a single rotation about an appropriate pole. We may use this to prove that the relative motion of two rigid blocks on a sphere may be described by

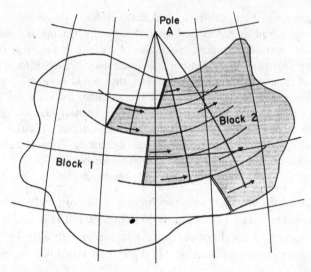

Figure 7.13 Morgan's Figure 4 (1968: 1962). The motion of block 2 relative to block 1. The transform faults (fracture zones segments between ridge-offsets) cut out arcs of concentric circles. The common center, Pole A, is the common pole of relative motion for blocks 1 and 2, if it is at the point of intersection of great circles, which are perpendicular to the common transform faults.

an angular velocity vector. We need three parameters, two to specify the location of the pole and one for the magnitude of the angular velocity. Look at Figure 4 [also Figure 4 in Morgan, 1968; my Figure 7.13], and consider the left block to be stationary and the right block to be moving as shown.

(Morgan, 1967b: 3; my bracketed addition)

Once Morgan came up with the idea that the kinematics of seafloor spreading and continental drift could be viewed as rotations of rigid blocks around Eulerian poles, the first order of business was to test the idea using the geometrical implications of Euler's theorem and the kind of available data. "The idea was certainly," recalled Morgan (1989 interview with author), "to look at the kinematics, to see how close an agreement you got, assuming that everything was rigid, without assuming any distortion." Once he began getting promising results, he decided to present them rather than his model of trench formation.

My first paper was just a uniform-viscosity-upper-Earth. And the sea floor going back in was supposed to pull down the trench: it was supposed to dynamically make the trench. And, what I was planning to talk on at the meeting was the effect of having increased viscosity near to the surface – I was working with layered viscosities and having increased viscosities near the surface had little effect on how concentrated the trench was. It did have an effect but not a great effect. That was what I was planning to talk about until I went off on this other track.

(Morgan, 1989 interview with author)

He also knew that Elsasser, who was working on the dynamics of seafloor spreading, had begun viewing Earth's outer region as rigid. "I was so exposed to Elsasser's concept of a strong outer skin that I believed everyone knew this and only gave Elsasser a passing reference" (Morgan, February 8, 2012 email to author).

Having introduced the key elements of his idea, he devoted the remainder of his talk to justifying his assumption, to determining if the moving blocks were sufficiently rigid to explain their relative motion in terms of Euler's Point Theorem. Treating North America and South America as fixed relative to each other (treating them as a single block), he determined the position of their common rotational pole of relative rotation with Africa by two methods that were based on independent datasets. If each method fixed the common pole of relative rotation, and if the positioning of the pole as determined by both methods roughly agreed, then his assumption of rigidity, if not confirmed, certainly was worth further investigation. He first determined if some of the transform faults forming part of the common boundary between the African and American blocks "lie on 'circles of latitude' about the pole of relative rotation."

Figure 5 [essentially Figure 7 in Morgan (1968b); my Figure 7.15] shows the offsets of the ridge in the equatorial Atlantic. Sets of concentric circles about various poles were computed and plotted on this figure of Heezen and Tharp (1965). The case for the pole at 65° N, 35° W is shown, but pole positions at 60° N, 30° W and 70° N, 65° W give almost equally good fits. There is fair control of the longitude of the best fitting pole but larger uncertainty in the latitude of the best pole. That is, we can draw lines perpendicular to the faults and they intersect up north at grazing angles giving good control in one direction but poor control in the other.

(Morgan, 1967b: 4; my bracketed addition)

Morgan then introduced his second method of determining whether the moving blocks are sufficiently rigid for their relative movement to be described in Eulerian terms. Taking the Eulerian pole to be positioned at 65° N, 35° W, as determined by the transform fault method, he calculated spreading rates as angular velocities along the Mid-Atlantic Ridge. Given Euler's Point Theorem, the velocity of one block relative to another varies along their common boundary, with the velocity being at its maximum at the Eulerian equator, and vanishing at the Eulerian poles. In other words, given that the angle of rotation of two blocks is the same, the linear velocity of one block relative to another will vary being maximum at the equator of relative rotation and decreases along a sine curve towards the poles.

Figure 10 [essentially Figure 9 in Morgan (1968b); my Figure 7.17] shows several observed spreading rates in the Atlantic compared to the model. Since the ridge runs almost monotonically north-south with only a minimum of doubling back at the equator, latitude is a convenient coordinate against which to plot the rates. Knowledge of the latitude, longitude, and strike of the ridge were needed at each point; these values were taken from Figure 2 of Talwani, Heezen, and Worzel [1961] for the northern part, from Figure 7 shown here [Heezen and Tharp, 1965] for equatorial region, and Figure 3 of Heirtzler and Le Pichon [1965] for the southern region. The solid line in Figure 10 was calculated using these values and with the choice of PLONG = 65°N, PLAT = 35°W, VMAX = 1.8 cm/yr. The dashed line ignores the strike correction; it gives not the half-velocity perpendicular to the strike of the ridge but rather the half-velocity parallel to the direction of spreading.

(Morgan, 1967b: 7; references are same as mine; PLONG, longitude of Eulerian pole; PLAT, latitude of pole; VMAX, maximum spreading rate = rate at the equator of spreading)

If this move of Morgan's sounds familiar, it should because it is what Backus suggested as a way to test V-M over the very same ridge (§3.8). Morgan did not mention Backus or reference his forward-looking paper because it is what Backus suggested as a way to test V-M over the very same ridge (§3.8). Morgan did not mention Backus or reference his forward-looking paper because he only later learned of it.

I learned of the Backus paper in the summer of 1969. I worked at Scripps for the summer; a summer job arranged through Menard. I was located in the IGPP building a few doors down from George [Backus]. Sometime mid-summer he gave a reprint of his Nature paper to me with a comment something like, "You probably haven't seen this." After I had read it I walked back to his office (shocked) and he said to the effect, "Don't think about it, almost no one read that paper." He told me he was at Cambridge while Hess and Wilson were talking about seafloor spreading and Vine and Matthews, talking of the magnetic stripes, and he wrote the note to Nature. He told me when he returned to Scripps he wrote an NSF proposal to hire an assistant to digitize the paper strip-recorder charts of the magnetic records of ship tracks crossing the South Atlantic – he wanted digital data of magnetic value versus position so he could make Fourier transforms of the cross-Atlantic tracks to see if the spectral peaks shifted in frequency as the sine-of-the-distance from the Euler Pole. He thought only statistically via a Fourier transformation method could you detect the "faster-broader" from the "slower-narrower" magnetic anomalies – no one yet had any idea that you could match the individual peaks to a magnetic timescale age. He told me that the proposal was not funded by the NSF; he didn't want to digitize the analog records himself; and besides he and Freeman Gilbert had begun having great success with their inversion theory techniques, and so he dropped the project.

(Morgan, February 8, 2012 email to author)

Once Morgan did the calculations based on Euler's theorem, he compared his results with those based on spreading rates over the Mid-Atlantic Ridge as "calculated at locations near each crossing of the ridge."

The observed values were obtained in the following manner. The two points marked "Phillips" were obtained by J. D. Phillips (private communication, to be presented at the Washington meeting, A.G.U., 1967). The magnetic profiles used here may be found in Heirtzler and Le Pichon [1965], Talwani, Heezen, and Worzel [1961], and Vacquier and Von Herzen [1964]. The strike of the ridge at the crossings of Vema 4, 17, and 10 assumed to be 38°, 38°, and 30° respectively. Zero strike was assumed at the crossings of Argo, Zapiola, and Vema 12. No error is introduced by this assumption if the ship was heading due east of west; the error is larger the more the ship's course departs from this ideal. In the case of Vema 12, if the strike of the ridge were −30° instead of the assumed 0°, the spreading rate would be 20% less than the 1.8 cm/yr shown here. Expected magnetic anomalies were calculated at locations near each crossing of the ridge using the normal-reversal time scale (and program) of F. J. Vine.

(Morgan, 1967b: 7; references are same as mine)

The results were very good as shown in Figure 9 from Morgan (1968) (Figure 7.17), which is very similar to his Figure 10 in his extended outline – the estimated rate of seafloor spreading is the same (1.8 cm/year), there is an additional data-point (*Vema* 18) in 1968, and the rotational poles differ by 3° latitude and 1° longitude. The results, as he later noted (1968: 1959), "roughly" agree with those determined from the transform fault method. However, at his talk, or at least in his extended outline, he just let the results speak for themselves and made no direct comparison to those obtained by the transform fault method.

7.11 Morgan's 1968 paper "Rises, trenches, great faults, and crustal blocks"

Morgan began his paper rather modestly, describing it "as an extension of the transform fault concept [Wilson, 1965] to a spherical surface." What, however, made Morgan's presentation so impressive was how he, in transferring the idea of transform faults to a spherical surface, was able to show that "each crustal block is perfectly rigid" or at least sufficiently rigid that Euler's theorem provides a useful and fairly accurate framework to describe and predict the relative motions of rigid crustal blocks (Morgan, 1968: 1960). Even if seafloor spreading, or at least some of its versions, suggested rigidity, and even if the idea of transform faults explicitly introduced the idea of rigid plates, which it did, it was not until Morgan tested the idea of transform faults on a spherical surface via Euler's theorem that a rigorous test of the rigidity of Earth's outer surface, of Elsasser's tectosphere, of Daly's lithosphere was achieved.

Morgan identified boundaries of his rigid blocks and determined their common rotational poles. He divided Earth's surface into twenty rigid blocks on the basis of present-day tectonic activity. Extending the idea of transform faults to a sphere, he recognized, as suggested by the title to his paper, three types of block boundaries.

The surface of the earth is divided into about twenty units, or blocks, as shown in Figure 1 [Figure 7.14] ... The boundaries between blocks are of three types and are determined by present day tectonic activity. The first boundary is the rise type at which new crustal material is being formed. The second boundary is the trench type at which crustal surface is being destroyed; that is the distance between two landmarks on opposite sides of a trench gradually decreases and at least one of the landmarks will eventually disappear into the trench floor. Other compressive systems in which the distance between two points decreases and the crust thickens, e.g., the folded mountains north of the Persian Gulf, are considered to be of this second type. The third boundary is the fault type at which crustal surface is neither created nor destroyed. Each block in Figure 1 is surrounded by some combination of these three types of boundaries.

(Morgan, 1968: 1959; my bracketed addition)

He determined poles of relative rotation between adjacent blocks by using rise-type and fault-type boundaries. He did not use trench-type boundaries. He also

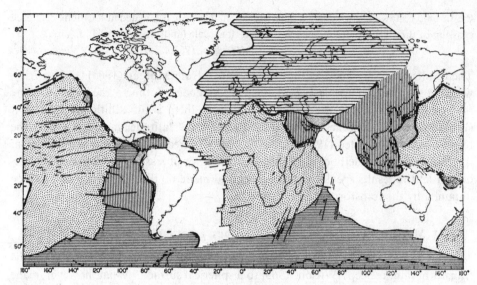

Figure 7.14 Morgan's Figure 1 (1968: 1960). His caption reads: "The crust is divided into units that move as rigid blocks. The boundaries between blocks are rises, trenches (or young fold mountains), and faults. The boundaries drawn in Asia are tentative, and additional sub-blocks may be needed."

concentrated on active spreading ridges and active transform faults, leaving behind, except for illustrative purposes, the great inactive fracture zones of the Pacific as they represented fossilized transform faults, not currently active ones. He eventually used his testing procedure at three rise-type boundaries, determining the motion of the African block relative to South America and North America treating the Americas as a single block, the North American–South America block relative to the Pacific block, and the Pacific block relative to the Antarctic. He then closed the loop by computing the motion of the Antarctic block relative to the African from the other three.

Beginning with the motion of the African block relative to the South American block, Morgan turned to the equatorial region of the Mid-Atlantic Ridge. Using the transform fault method he had introduced at his AGU talk, he constructed great circles perpendicular to the trend of the fracture zones to see if they had a common point of intersection (Fig. 8a; my Figure 7.16). They did, more or less, and the common point represented the pole of relative rotation as determined by the transform fault method.

Figure 7 [my Figure 7.15] shows the offsets of the ridge in the equatorial Atlantic Ocean. A set of circles concentric about a pole at 58° N, 36° W is plotted on a figure of Heezen and Tharp [1965]. Figure 8a shows how this pole position was obtained [my Figure 7.16]. Great circles were constructed perpendicular to the strike of each fracture zone offsetting the crest of the ridge ... The intersections of the great circles define the pole of rotation: the great circles are analogous to meridians and the fault lines are analogous to lines of latitude about this pole.

   (Morgan, 1968b: 1963–1964; my bracketed additions about Figures 7.15 and 7.16)

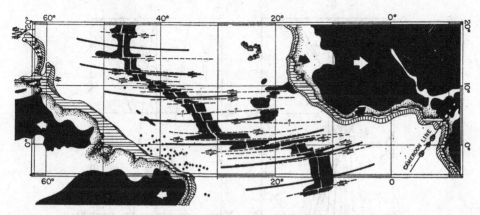

Figure 7.15 Morgan's Figure 7 (1968: 1964; reference same as mine). In this figure, adapted from Figure 5 of Heezen and Tharp (1965) [my Figure 5.3], the "strike of the transform faults in the equatorial Atlantic are compared with circles concentric about a pole of 62° N, 36° W. These circles indicate the present motion of Africa relative to South America." The pole position of 62° N, 36° W is the position found through the seafloor spreading rate method.

Morgan now introduced a new method to fix the rotational pole. He used the strike of strike-slip earthquakes occurring along fracture zones between ridge-offsets of the Mid-Atlantic Ridge as determined by Sykes (§6.9) from first motion studies. His Figures 8a and 8b (Figure 7.16) show the determination of the pole by intersection of great circles perpendicular (a) to the trend of the fracture zones and (b) to strikes of strike-slip earthquakes occurring along the active fracture zones. Because the common intersection of the great circles perpendicular to the strike of the seven earthquakes was poor compared to the one formed by intersection of great circles perpendicular to the trend of the fracture zones, he rejected the former in favor of the latter, and located the pole at 58° N (±5°), 36° W (±2°). Instead of reasoning that a lack of common intersection counted against block rigidity, Morgan questioned the accuracy of first motion studies in determining fault planes or the accuracy of the fault planes in tracking fracture zone trends.

The common intersection in Figure 8b [Figure 7.16] is not as good as the intersection in Figure 8a [Figure 7.16]. It is interesting that a circle of about 6° radius can be used both here and in the Pacific-North America case (see Figure 14f) [my Figure 7.18] to illustrate the departure from a point intersection. The bulk of the epicenters are about 45° (5000 km) from the point of intersection in both of these cases. This suggests that the accuracy to which the fault planes are determined, or the accuracy to which the first motions represent the strike of a long fault, is the cause of the scatter in Figures 8b and 14f.

(Morgan, 1968: 1964; my bracketed additions)

Noting, as he had done at his AGU talk, that the fracture zone method of determining the position of rotational poles provides a good fix on their longitude but a poor

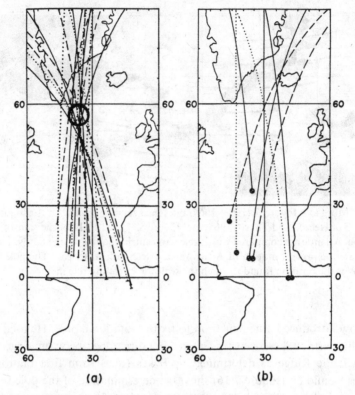

(a)   **(b)**

Figure 7.16 Morgan's Figure 8 (1968: 1965). His caption reads: "Great circles perpendicular to the strike of offsets of the mid-Atlantic ridge are shown in (a). With one exception, all of these lines pass within the circle centered at 58° N, 36° W. Great circles perpendicular to the strike determined by earthquake mechanism solutions are shown in (b)."

one on their latitude, he used seafloor spreading rates to better fix the latitude of the rotational pole. Treating the Americas as a single block and setting PLAT at 62° N, PLONG at 36° W, and V_{max} at 1.8 cm/year (where he meant what we now mean by half-spreading rate[13]), he displayed the results in his Figure 9 (my Figure 7.17).

Figure 9 shows several observed spreading rates in the Atlantic Ocean compared with the model ... The solid line in Figure 9 was calculated with these quantities and with the choice of PLAT = 62°, PLONG = to 36°W, V_{max} = 1.8 cm/yr [half-spreading rate]. The dashed line was calculated without the strike correction.

(Morgan, 1968: 1966; my bracketed addition)

The position of the pole based on using spreading rates differs from the one obtained by using the trends of transform faults. This is not surprising because the pole position determined by the trends of the transform faults gave the position of the pole between Africa and South America, while the seafloor spreading rate method

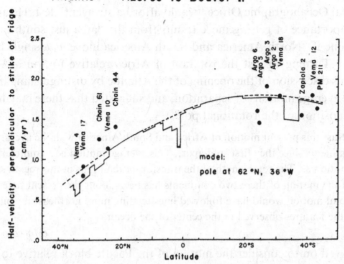

Figure 7.17 Morgan's Figure 9 (1968: 1967). "[Half-]Spreading rates determined from magnetic anomaly profiles are compared with the values calculated with the model. The solid line shows the predicted rate perpendicular to the strike of the ridge; the dashed line shows the rate parallel to the direction of spreading." Chain 44 and Chain 61 were labeled "Phillips" in the earlier version of Figure 9 that he presented at his AGU talk. Data-point Vema 18 was included in his earlier figure. His earlier calculations gave a best fit at 65° N, 35° W; he now obtained a best fit at 62° N, 36° W.

gave that between the African blocks and the North and South America block. In addition the seafloor spreading rate method is sensitive to latitude but not longitude while the fracture zone method is sensitive to longitude but not latitude. Morgan found the best overall fit. He also had to select a pole position and spreading velocity which allowed him to find the best fit between the spreading rates of the northern and southern Atlantic.

This figure was originally calculated with PLAT = 58° N, the latitude of the center of the circle in Figure 8a [Figure 7.16]. With this pole position the computed curve does not satisfactorily fit the observed points: the points south of 20° S alone fit V_{max} = 1.8 cm/yr [half-spreading rate]; the points north of 20° N better fit V_{max} = 2.2 cm/yr [half-spreading rate]. If the pole is chosen further north, say at 62° N as shown in the figure, a single curve apparently fits both the northern and southern portions of the data within the scatter of the points. The velocity pattern is sensitive to the latitude but not the longitude of the chosen pole, whereas the intersection of the perpendiculars in Figure 8 was just the opposite. A pole at 62° N (±5°), 36° W (±2°) with a maximum velocity of 1.8 (± 0.1) cm/yr [half-spreading rate] satisfies both of these criteria.

(Morgan, 1968: 1966; my bracketed additions)

Morgan (1968: 1967–1968) recommended analysis of more magnetic profiles "to ultimately test" his hypothesis as applied to the Atlantic. He thought results of the aeromagnetic survey of the Reykjanes Ridge jointly sponsored by Lamont and the US Naval Oceanographic Office "would afford a stringent" test. He also emphasized the importance of getting more results from the "area just south of 20° N" to help determine if "North America and South America move as a single block or as two blocks." Turning to just the rotation of Africa relative to South America, he concluded his discussion of the opening of the Atlantic by distinguishing between the present and average motions of separation, and suggested that there may have been a shift in the position of the rotational pole.

We may contrast this present motion of Africa and South America with the average motion of these two continents since they first split apart. This average motion is ... quite different from the present motion ... The total length of the transformation faults in this region suggests that about half of the motion of these two continents has been about the present pole. The earlier half of this total motion would have followed lines tending more northeast to southwest than the strike of the features observed in the center of the ocean.

(Morgan, 1968: 1968)

Morgan moved on to consider the motion of the Pacific block relative to the North American block. He determined both coordinates of the common pole of relative rotation by the transform-strike method, drawing primarily upon the trends (and thus strikes) of thirty-three continental transform faults extending from about 65° N (Fairweather–Queen Charlotte group) to 23° N (Gulf of California group) and, to a lesser extent, by the strike of earthquakes determined from first motion studies. Because there were no marine magnetic anomaly profiles he could not compare his transform fault method of pole placement with that determined by seafloor spreading rates. (I add here, parenthetically, that he did not use the profiles over the Juan de Fuca Ridge because he treated the Juan de Fuca block as a discrete block separate from the Pacific and North American blocks. Indeed, he removed a difficulty in the application of plate tectonics to this region by hypothesizing a separate block. As we shall see, McKenzie and Parker treated the area in the same way.) However, because of the 42° latitudinal spread of the transform fault between the Fairweather–Queen Charlotte and Gulf of California groups of transform faults, the great circle lines constructed perpendicular to the strikes of the north and southern transform faults intersected at sufficiently large angles for him to fix both coordinates of the pole (see Morgan's Figure 14e, my Figure 7.18a). He obtained the same pole position by using first motion studies of earthquakes (see Figure 14f, my Figure 7.18b). By estimating the motion along the San Andreas Fault to be 6 cm/year, he calculated the relative spreading rate of the two blocks to be 4 cm/year.

He then turned to the relative motion of the Pacific and Antarctic blocks. Using the transform-strike method and basing his placement of the strikes only on the trends of the transform faults (first motion studies were unavailable), he found that all great circles passed within 2° of 71° S 118° E, where he placed the pole (Figure 7.19). He arrived at such a pole position by using a rotational half-spreading velocity of 5.7 cm/year based on nine marine magnetic profiles from the Pacific–Antarctic

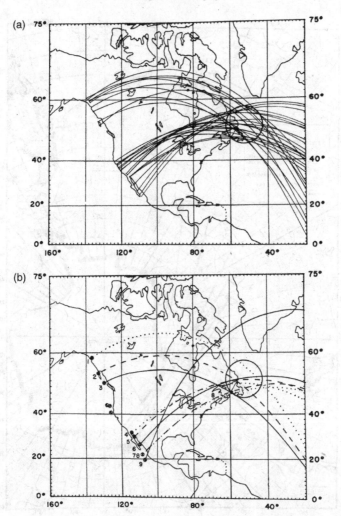

Figure 7.18 Morgan's Figures 14e and 14f (1968: 1976). His determination of the pole of relative motion of the Pacific block relative to the North American block. (a) shows the pole position as determined by the construction of great circles perpendicular to the strike (trend) of the common transform faults, where the strike is determined on the basis of the trends of the faults; (b) utilizes first motion studies. Because the angular separation between the northern and southern-most faults is 42°, the intersection of the great circles is sufficiently pronounced to fix both the latitude and longitude of the pole.

Ridge, including Pitman's *Eltanin*-19 profile. The produced fit was excellent, and he displayed it in his Figure 16 (my Figure 7.20). Thus Morgan found either the transform-fault or seafloor-spreading method independently gave basically the same position of the Euler pole, and thereby was able to apply a more severe test of plate rigidity than by using only one method or having to combine both methods.

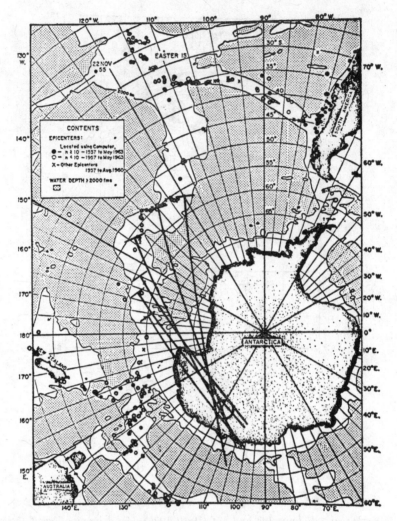

Figure 7.19 Morgan's Figure 15 (1968: 1978). Great circles constructed perpendicular to the strikes of fracture zones offsetting the Pacific–Antarctic Ridge are constructed from Sykes' (1963) seismic map of the region. The great circles all pass within 2° of the pole at 71° S, 118° E.

Once Morgan had poles and relative spreading rates for the American relative to the African, the Pacific relative to the American, and the Antarctic relative to the Pacific blocks, he developed an additional test for his hypothesis: he closed the loop around the world by determining the motion of the Antarctic block relative to the African by summing the relative rotational vectors of the three relative rotations he had already determined. This gave him a rotational half-spreading velocity of 1.6 cm/yr about a common rotational pole at 25° S, plus or

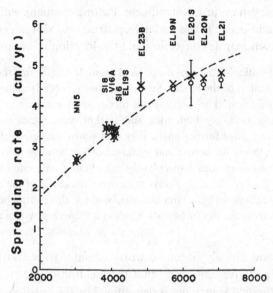

Figure 7.20 Morgan's Figure 16 (1968: 1979). Spreading rates on the Pacific–Antarctic Ridge are compared with a model with $V_{max} = 5.7$ cm/year about a pole at 71° S, 118° E. The circles are spreading rates measured perpendicular to the strike of the ridge; the crosses are these rates projected parallel to the direction of spreading.

minus 30° and 35° W, plus or minus 20°, which was consistent with the available but scanty and somewhat unreliable data.

Morgan closed his paper with a strong statement of what he thought he had shown. Moreover, there is little difference between what he wrote in his submitted and published versions. I quote from the originally submitted version, which can be compared with the published version. First are his conclusions concerning the existence of rigid 100 km thick blocks.

The evidence presented here favors the existence of large "rigid" blocks of crust. That continental units have this rigidity has been implicit in the concept of continental drift. That large oceanic regions should also have this rigidity is perhaps unexpected. The required strength cannot be in the crust alone; the oceanic crust is too thin for this. We instead favor a strong tectosphere, perhaps 100 km thick, sliding over a weak asthenosphere. Theoretical justification for a model of this type has been advanced by Elsasser [1967]. In the simple two-dimensional picture of a rise and a trench with a continent between them, we imagine a conveyer-belt process in which the drifting continent need have no great strength. In the model considered here, we may have local hot spots on the rise and faster sinking at some places on the trenches. The crustal blocks should have the mechanical strength necessary to average out irregular driving sources into a uniform motion; the tectosphere should be capable of transmitting even tensile stresses. The crustal block model can possibly explain the median position of most oceanic rises and the symmetry of their magnetic pattern.

(Morgan, 1968: 1982)

Now comes what is perhaps just as significant. Parting company with Hess, and those who thought that mid-ocean ridges had a deep structure, Morgan essentially severed any direct tie between mantle convection and the location of oceanic ridges.

We assume that the location of the rises is not fixed by some deep seated thermal source but is determined by the location of the blocks. Suppose a crustal block is under tension and splits along some line of weakness. The forces which tore it apart continue to set and the blocks move apart creating a void, say 1 km wide and 100 km deep, which is filled with mantle material. As the blocks move further apart, they split down the center of the most recently injected "dike," since this is the hottest and weakest portion between the two blocks. Even if one block remains stationary with respect to the mantle and only one moves, we will have a symmetric pattern if a new "dike" is always injected up the center of the most recent "dike." If the initial split was entirely within a large continental block, this control of mantle convection by boundary conditions at the top surface will result in a ridge crest with a median position.

(Morgan, first submitted version of Morgan (1968))

Perhaps Morgan had already begun to worry about Africa and Antarctica being surrounded by ridges; given his view "that the location of the rises is not fixed by some deep seated thermal source but is determined by the location of the blocks," he no longer had to worry. As we shall see, McKenzie and Parker made precisely the same move.

7.12 McKenzie discovers plate tectonics

McKenzie came up with the idea of plate tectonics after Morgan, but independently of him. He made the discovery, probably in mid-August, while at Scripps from July through November 1967. Improbable as it may sound, McKenzie, because of a lucky set of circumstances, did not know of Morgan's work until after discovering plate tectonics on his own and running a preliminary test based on first motion studies of earthquakes along the common boundary of the Pacific and North American plates. After developing his version of plate tectonics, he teamed up with Robert Parker, and their paper became the first published paper on plate tectonics. Morgan's presentation of plate tectonics is more complete than McKenzie and Parker's, but theirs is conceptually tighter.

Just what were these lucky circumstances? I begin with a chronology of the relevant events. This chronology is based on hard historical evidence and retrospective works by Menard (1986), McKenzie (2001), and Le Pichon (2001) as well as McKenzie's March 18, 1990 letter to me and John Sclater's retrospective comments. I include some relevant historical documents and sources.

1. On April 19, 1967, McKenzie leaves the morning AGU session before Morgan speaks. Morgan is scheduled to talk about trenches but instead discusses plate tectonics.
2. McKenzie visits Lamont, Yale, and Harvard after the AGU meeting; he hears nothing of Morgan's talk.

3. Morgan hands Menard, who co-chaired the spring AGU session at which Morgan spoke, a copy of the extended outline of his AGU talk on plate tectonics. Menard found the outline "difficult ... to fathom" (Menard, 1986: 290). I add that Menard never told McKenzie that he co-chaired the session or that he had a copy of Morgan's extended outline.

4. On June 5, McKenzie resigns his research fellowship at Caltech, effective at the end of June. He writes Dr. Lee A. DuBridge, President of Caltech, informing him of his decision.

> Dear Dr. DuBridge:
>
> At the end of June I would like to resign from my Research Fellowship in the Division of Geology. I am very grateful to you and to the Institute for the opportunity you gave me to carry on my research here, and will always remember my visit here with pleasure.
>
> Yours sincerely,
> Dan McKenzie Research Fellow

5. On June 30, *JGR* receives McKenzie's paper on heat flow (§7.6).
6. At the beginning of July, McKenzie arrives at Scripps, having fulfilled his obligations at Caltech. He accepts a postdoctoral research position at the Institute of Geophysics and Planetary Physics. He is reunited with Parker, Sclater, and John Mudie who were graduate students with him at Cambridge in the Department of Geodesy and Geophysics. He shares an office with Parker and a house with Sclater, although Sclater is at sea on the Nova Expedition aboard R/V *Argo* on Leg III (June 16 to July 8) when McKenzie arrives. Sclater remains at sea on the Nova Expedition aboard R/V *Horizon* on Leg IV (July 12 to August 8) where he joins Menard.
7. At the end of June or very beginning of July, Menard leaves Scripps to participate in the Nova Expedition and embarks on R/V *Horizon* in Suva, Fiji. He is Chief Science Officer aboard the ship during Leg III (July 3 to July 11), stays aboard R/V *Horizon* as Chief Science Officer during Leg IV (July 12 to August 8) and remains at sea as Chief Science Officer aboard R/V *Argo* during Leg V (August 11 to September 12). He returns to Scripps from Auckland, arriving around September 15 (14 below). (See Nova Notice #11 "Changes in programs and people to 1 June 1967" by Menard, UCSD Libraries, online).
8. During the first two months (July and August), McKenzie reads all of Hess's papers, Menard's *Marine Geology of the Pacific* (III, §5.12), and rereads Bullard, Everett, and Smith (1965) in which they use Euler's Point Theorem to reassemble the continents surrounding the Atlantic Ocean (§3.4). McKenzie "disliked the method" they had used to fit the continents together (McKenzie, 2001: 180). In particular he found their method in determining the amount of misfit wanting.
9. While reading Bullard, Everett, and Smith (1965), McKenzie thinks about Euler's theorem and realizes that if finite displacement can be described by finite rotations

about finite poles, which is what is being done in their paper, instantaneous rotations require angular velocities. This point seems obvious to him but not useful.

10. At some point, definitely before the second half of August, McKenzie realizes that the point above about finite and instantaneous rotations is *very* useful. He realizes that he can use slip vectors of earthquakes to find instantaneous Euler poles. He uses a National Geographic globe with a cap to find the pole of instantaneous relative rotation between the North American and Pacific plates that best fits the slip vectors. He determines the pole by finding the common intersection of great circles perpendicular to the slip vectors. He works this out in two days; he is convinced that he is right.

11. McKenzie also talks to a lot of people at Scripps, especially to Bob Parker, and to John Sclater about plate tectonics, once Sclater returns to Scripps within a few days after the August 8 completion of Leg IV on *Horizon*. Sclater remembers that McKenzie began talking to him about his work on plate tectonics when he returned to Scripps from the Nova Expedition.

> I was much more interested in wondering if my then girlfriend would become engaged to me when I returned than any science. She did but somewhat later. Dan told me about the idea he had of using the Bullard, Everett and Smith paper and the Euler poles to check Tuzo's ideas when I got back after the Nova Expedition which was in late July or early August.
>
> (Sclater, September 30, 2011 email to author)

12. McKenzie wonders how to display his results. He talks to Parker who has developed SUPERMAP, a map projection program. McKenzie suggests a Mercator projection.

13. On September 5, Morgan's paper is received by *JGR*.

14. Morgan presents his paper at a two-day conference on September 7 and 8 at Woods Hole. Mudie attends the conference, and returns to Scripps where he tells people, including McKenzie, about Morgan's talk. From what Mudie says to McKenzie, McKenzie is not sure whether Morgan is simply doing little more than what Backus did in his 1964 *Nature* paper (§3.8) or, like himself, has discovered plate tectonics. Regardless, this is the first time that McKenzie has heard anything about Morgan's work pertaining to plate tectonics, and it is after McKenzie has himself discovered plate tectonics.

15. On September 12, or a day or two later, Menard leaves Auckland and returns to Scripps.

16. Anderson, editor of *JGR*, receives Morgan's paper on September 5 and decides to ask Menard to serve as a referee. He sends Morgan's paper to Menard (§7.9). (I do not know when Anderson writes Menard – the letter has not been found in Menard's papers, and Menard does not give the date in *Ocean of Truth*, where he recounts what happened.)

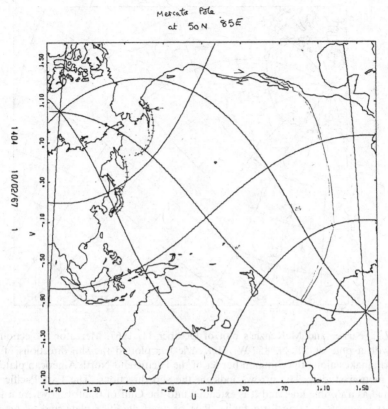

Figure 7.21 McKenzie and Parker's October 2, 1967 plot. Mercator projections of the Pacific with a pole of 50° N, 85° W. The arrow off the coast of Alaska represents the average slip direction of aftershocks of the 1964 Alaskan earthquake in the Kodiak Island region. It will become 2b in later plots and in Figure 3 of their published paper (McKenzie and Parker, 1967: 1278). The arrows along the San Andreas Fault indicate, as was well known, that it is a dextral transform fault. The East Pacific Rise is illustrated as a double line. Morgan also represented ridges as a double line (see Figure 7.13). Trenches are represented as lines with short lines at right angles. McKenzie would later use the same convention in his paper with Morgan on triple junctions (McKenzie and Morgan, 1969).

17. McKenzie tells Menard about his work on plate tectonics. Menard still finds plate tectonics difficult to understand. McKenzie explains and reexplains it to him. (Menard, recall found Morgan's extended outline difficult to understand.) He also asks Menard about the South Pacific, especially Tonga and the New Hebrides.

18. On October 2, McKenzie and Parker produce at least one inky pen plot (rapidograph) (see Figure 7.21). These plots, requested only in the likelihood of publication, are made after ballpoint pen plots have been made. The plot shows that McKenzie and Parker already have decided on a Mercator projection, that

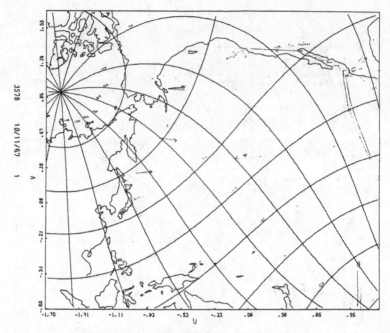

Figure 7.22 Parker and McKenzie's plot of October 11, 1967. Mercator projection of the Pacific with a pole of 50° N, 85° W. McKenzie has plotted the slip directions of several more earthquakes along the common border of the Pacific and North American plates based on first motion studies. The arrows indicate the slip directions. The East Pacific Rise is represented as a double line, and it is extended into the Gulf of California where it is offset into short segments by transform faults. Part of Juan de Fuca plate and accompanying transform fault is shown.

Parker has had enough time to write the program for producing Mercator projections, that McKenzie has determined slip directions of a transform and thrust fault, and that McKenzie already has found slip directions of enough earthquakes to position the instantaneous rotational pole of the Pacific and North American plates at 50° N, 85° E. The arrow indicating the direction of relative instantaneous rotation between the Pacific and North America plates shows that McKenzie plans to describe their motion in terms of rotating the Pacific plate relative to a fixed North American plate. The plot is an early precursor of what will become Figure 3 in their published paper, but they have not yet decided that it will become Figure 3.

19. On October 11, Parker and McKenzie plot slip directions of three earthquakes along the common boundary of the Pacific and North American plates from Japan through the Gulf of California (Figure 7.22). The directions of the slip vectors indicate that McKenzie is toying with the idea of showing the instantaneous rotation of North America relative to a fixed Pacific plate instead of keeping the North American plate fixed and rotating the Pacific plate. The East

Pacific Rise is represented as a double line. He shows the rise extending into the Gulf of California where it is offset into short segments by transform faults. They still have not decided that the plot will become Figure 3.

20. In Mid-October, probably just before or soon after October 11, McKenzie receives a referee's report of a lengthy and highly mathematical paper entitled "The influence of the boundary conditions and rotation on convection in the Earth's mantle," which he had submitted to the *Geophysical Journal of the Royal Astronomical Society*. The report by Nigel Weiss, a Cambridge friend, is a highly detailed analysis of what McKenzie should do to improve the paper.

21. McKenzie begins to rewrite the lengthy mathematical paper, deciding to use writing the plate tectonics paper as a reward for revising the convection paper. He works on the revision through the first few days of November. But the rewrite is taking much longer than first expected; it is a long slog, and it is not received by the *Geophys. J.* until February 6, 1968.

22. Menard, who, I presume, has been working through the mountain of mail awaiting him after being at sea since early July, finds Anderson's request to referee Morgan's paper. He now understands Morgan's paper because McKenzie has explained plate tectonics to him.

23. Menard tells McKenzie about Morgan's paper that he has been asked to referee. McKenzie asks Menard what he should do about his own paper on plate tectonics. Menard strongly urges him to write it up immediately for publication, and to send Morgan a preprint. McKenzie accepts Menard's advice, stops revising the lengthy mathematical paper, and returns to the plate tectonics paper.

24. McKenzie and Parker produce at least four more plots on October 30, including two which clearly show an anomalous region that they have isolated as a small plate – the same plate that Morgan also has identified named the Juan de Fuca block. McKenzie also discards depicting North America as the moving plate, and returns to his original idea of showing the North American plate as stationary and the Pacific one as moving. He has by this point designated the plot as Figure 3, so he evidently has begun writing or at least mapping out his paper (see Figure 7.23).

They plot (Figure 7.24) slip directions of five more earthquakes. With this plot, McKenzie and Parker have now plotted slip directions of the eight earthquakes they show in Figure 3 of their paper (see Fig. 7.23).

They also plot a Mercator projection of the South Pacific. The Euler pole's position is not given. McKenzie no longer remembers "which pair of plates" he "intended to use." He decided not to pursue the project. McKenzie recalls as best he can what happened.

I don't remember how I chose the S. Pacific pole, or which pair of plates I intended to use. Probably Pacific-Australia, since I thought then (and still do) that the whole of the boundary between Australia and Pacific from Tonga to New Guinea was strike slip,

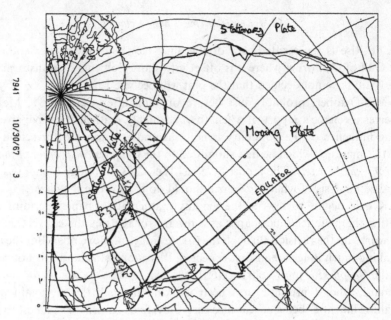

Figure 7.23 One of McKenzie and Parker's October 30, 1967 – a forerunner of Figure 3 in their paper – Mercator projections of the Pacific with a pole of 50° N, 85° W. The arrows show slip directions based on first motion studies of three earthquakes that occurred along the common boundary of the Pacific and North America plates. They also show the direction of motion of the Pacific plate relative to the North American plate. Possible boundaries of other plates are shown. Mistakes in boundaries are crossed out.

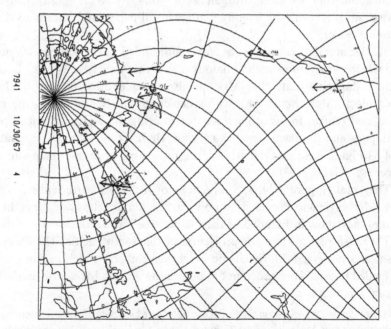

Figure 7.24 One of McKenzie and Parker's October 30, 1967 – a forerunner of Figure 3 – Mercator projections of the Pacific with a pole of 50° N, 85° W. The arrows show slip directions based on first motion studies of five earthquakes that occurred along the common boundary of the Pacific and North America plates. They also show the direction of motion of the Pacific plate relative to the North American plate.

and that this would give me enough of an angle at intersection to determine the pole position reasonably well. What worried me, and what I talked to Bill about extensively, was the nature of the structures west of Tonga, in the New Hebrides. These are not nearly as clearly accounted for by the relative motion between Pacific and Australia as are those in the N Pacific. Why became clear only much later, when GPS showed how fast the back arc spreading was in the Lau Basin.

(McKenzie, November 23, 2011 email to author)

25. On November 1, McKenzie and Parker make a Mercator plot of the Euler pole for the separation of South America and Africa. McKenzie places the pole at 44.0° N, 30.6° W based on one of the estimates by Bullard, Everett, and Smith (1964). However, he abandons the project. Even if he were, for example, to have used Sykes' mechanism solutions for earthquakes along transform faults and within the Mid-Atlantic Ridge, they could give at best only the longitude and not the latitude of the pole. To pin down the pole's latitude, he needed spreading rates for the South Atlantic based on profiles of magnetic anomalies, but he did not have them. Indeed, Morgan had them, and used them to pin down the pole's latitude.

26. On November 3, McKenzie and Parker plot what will become Figure 4 in their publication, an orthogonal projection of the North Pacific centered on the Mercator Pole. The plot does not include slip directions of earthquakes that already have been determined along the common boundary of the Pacific and North American plates. The plot also does not include grid lines, and McKenzie writes by the plot, "Can you draw me one with grid lines?"

27. Menard shows McKenzie Morgan's paper, decides to recommend its publication, discusses it with McKenzie, and writes his referee's report. I do not know the order of these events. I suspect that he tells McKenzie that he is going to recommend publication of Morgan's paper, shows and discusses the paper with him, listens to McKenzie raise difficulties with Morgan's treatment of the common boundary of the Pacific and North American plates, and then writes his referee's report (§7.9).

28. On November 7, McKenzie and Parker revise their November 3 orthogonal projection putting in grid lines, and including slip directions of earthquakes along the common boundary of the Pacific–North American plates (Figure 7.25).

29. Menard sends his referee's report to Anderson at *JGR*. (I do not know when Menard writes Anderson – the letter has not been found in Menard's papers, and he does not give the date in *Ocean of Truth*. However, he certainly wrote Anderson early enough for Anderson to write Morgan on November 14 telling him that his paper had been accepted for publication with minor revisions.)

30. McKenzie writes the paper in one or two days during the first week of November. Parker makes few changes, and on Saturday November 10 they mail it to *Nature*.

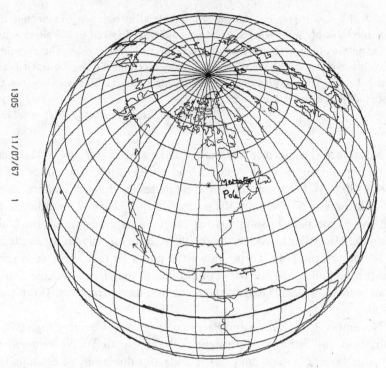

1305   11/07/67   1

Figure 7.25 McKenzie and Parker's November 7, 1967 plot of a forerunner of their Figure 4: "Orthogonal projection of the North Pacific centered on the Mercator Pole." Slip directions are tangents to concentric circles about the center. Figure 4 in their publication does not show the grid lines.

Dear Sir,

   We would like to submit this paper for publication in Nature, since we believe it to be of importance and general interest. We would be grateful if unnecessary delays could be avoided, and will not object if minor changes (for instance, insertion of 'that' and commas) are made without consultation. All correspondence should be addressed to R. L. Parker at the address above.

Yours sincerely,
Dan McKenzie

31. On November 14, Orson Anderson, editor of *JGR*, writes Morgan telling him that his paper has been accepted pending minor revisions and recommends that he remove the section on the boundary between North American and the Pacific plates. (Anderson's letter and the referee reports are reproduced in §7.9)

32. On the same day, McKenzie writes Vine at Princeton. He includes a preprint of his paper on plate tectonics, and asks Vine to give it to Morgan. He also explains

what happened, and asks Vine not to tell Morgan that Menard refereed Morgan's paper. Here is McKenzie's letter to Vine.

Please note that if McKenzie meant that he began working on the overall project in mid-October, he is mistaken as evidenced by the fact that he and Parker were already producing publishable plots at the beginning of October. I assume that he meant that he had begun to lay out the actual paper in mid-October before setting it aside to rewrite the convection paper.

Dear Fred,

Here is a paper which I believe will interest you, but is a cause of some embarrassment. I started work on the project in Mid-October, but had to stop for three weeks while I rewrote a convection paper. I started up again and got everything sorted out and written as a rough draft when Menard passed on to me Jason Morgan's paper of J.G.R, which put me in a spot, so I asked him what to do and he suggested I send a preprint to Morgan. Could you arrange this without letting him know Menard was a referee? As you will see our approaches are very different, especially our use of first motions of earthquakes, but I do want to avoid any nastiness if at all possible. It is after all a genuine case of two people having the same idea at the same time, and I don't see why either of us should be troubled.

Yours ever,
Dan McKenzie

33. McKenzie leaves Scripps for a holiday, drives across the continent, and arrives at Lamont in late November in time for a goodbye party for Le Pichon, who is returning to France.

34. McKenzie meets Le Pichon for the first time, and finally learns that Morgan first presented his version of plate tectonics at the spring AGU meeting. Recall that Menard never told McKenzie. He decides to try to delay publication of his and Parker's paper. Parker agrees. McKenzie telegraphs *Nature* and sends a letter asking if publication may be delayed until Morgan's paper is published in *JGR*. Understandably, he says nothing about Menard's role as a sounding board or Menard's having refereed Morgan's paper. Instead, he mentions but does not name Mudie as the person who first told him about Morgan's work.

Dear Sir:

... I would like to hold publication of "The North Pacific: an Example of Tectonics on a Sphere" by McKenzie and Parker so that it can appear simultaneously with that of Morgan's in the *Journal of Geophysical Research*. His article is due to appear in the issue dated March 15, 1967. After I had thought of the various ideas expressed in the paper I sent you I was told that Morgan was working along the same lines. My informant had not understood what Morgan had talked about, and I saw no reason to change my publication plans. What

I did not know was that Morgan had talked about his work at the A.G.U. meeting in March. The reason for this ignorance was that his published abstract was not related to the talk he gave. Thus the paper Parker and I sent you contains work done independently of Morgan, but for the reasons I have given, I would be grateful if it could appear simultaneously with Morgan's if this is still possible. I have spoken to Parker, who agrees with me.

Yours sincerely,
Dan McKenzie

35. On November 29 Morgan returns his revised paper to Anderson at *JGR*. (Letter reproduced in §7.9.)
36. On December 8, Morgan writes McKenzie. Morgan shows no animosity toward McKenzie and raises questions about their differing treatments of the Pacific–North American boundary. (Letter reproduced in §7.14.)
37. On 29 December, John Maddox responds, apparently unable to resist pointing one of *Nature*'s great advantages over most journals:

> We must regret to say that your paper has already appeared in this week's issue of NATURE (December 30th). This, I am afraid, is one of the penalties of dealing with a really rapid journal.

38. On February 14, Morgan is informed by *JGR* that his paper is to be published in the March 15, 1968 issue.
39. McKenzie and Morgan meet in Princeton for the first time. They discuss their work and learn that they been working on several of the same problems. Morgan is friendlier than McKenzie expected. They are joined by Tanya Atwater. They realize the need to figure out the evolution of triple junctions, but are unsuccessful. McKenzie and Morgan decide to write a joint paper on triple junctions.

This chronology of events concerned with the development of McKenzie's ideas and their temporal relationship to Menard's activities is more complete than previous ones and is based on historical documents as well as recollections. Although an improvement over previous ones it is far from perfect. As noted above, I am not sure about the sequence of events surrounding Menard's learning that he had been asked to review Morgan's paper, his deciding that it should be published, his telling, showing, and discussing it with McKenzie and his sending Anderson his positive report. Also I am not sure whether McKenzie and Parker made their October 30 and November 1 plots before or after Menard told McKenzie about Morgan's paper. However, the ordering of these particular events does not throw doubt on McKenzie's discovering plate tectonics independently of Morgan. Menard surely would not have advised McKenzie to continue working on his paper, and send a preprint to Morgan, if McKenzie had not discussed plate tectonics in some depth with him before he learned that he had been asked to referee Morgan's paper. Nor would he have shown McKenzie or discussed Morgan's extended outline of his AGU talk with him, and later recommended that McKenzie publish his own paper. It is, of course, possible that McKenzie could have added something to his paper after seeing Morgan's paper, but because their approaches differ and their positioning of the Euler pole for the rotation of the Pacific plate relative to the North American plate do not agree, I do not think that McKenzie took anything from Morgan's paper. I do however think that McKenzie helped Menard see the importance of Morgan's work, and ironically helped Menard see that his paper should be published, even if McKenzie also helped convince Menard that there were difficulties with Morgan's determination of the Euler pole common to the rotation of the Pacific plate relative to the North American plate.

What about the events themselves making up the chronology? Do any seem unbelievable? Some may find it incredulous that McKenzie left the AGU session before Morgan spoke, and that he heard nothing about Morgan's talk while visiting Lamont, Harvard, and Yale after the meeting, before resigning his research fellowship at Caltech and moving to Scripps. Here is McKenzie's explanation.

At the AGU, I was in the session in which Morgan was to speak. I had read the abstracts (I always try to do so at big meetings, otherwise you miss so much) and had read Jason's paper. I remember all the Lamont people sitting together waiting to tear him apart (he had reinterpreted their data from the Puerto Rico trench!). I found the whole affair rather nasty and thought I would learn nothing from it. I had already talked to some of those involved from Lamont before the session. So I left and went elsewhere (my tastes were very catholic). His paper made no stir at the meeting, nor at Lamont, Yale or Harvard where I was afterwards. I only discovered what had happened when I got to Lamont in December [1966]. I then went back to Caltech, finished [the heat-flow paper] and moved to Scripps.

(McKenzie, March 18, 1990 letter to author)

Was McKenzie's expectation about Lamonters "waiting to tear ... apart" Morgan reasonable? It was. Here is how Le Pichon remembered it:

I have retained a precise memory of that morning of April 19, 1967, at the spring AGU meeting during which Fred Vine and H. W. Menard, from Scripps Institution of Oceanography presided over a "sea floor spreading" special symposium. The large amphitheater was full and expectations were very high. "Sea floor spreading" was the subject of most discussions: 70 abstracts on the topic had been submitted to this AGU meeting. At the end of the session, the program announced that Jason Morgan would present a paper which, according to its title, concerned the formation of oceanic trenches by viscous convection [see §7.8 for Morgan's abstract of his planned talk]. Manik Talwani and I were preparing to listen very attentively because we had had a vigorous argument with Morgan on this subject. Morgan assumed for his model the absence of any long-term rigidity even at the surface, and we considered this assumption incompatible with the gravity data. But to our great surprise, Morgan announced that he would present a different paper, entitled "Rises, Trenches, Great Faults and Crustal Blocks." Thus the talk he made did not correspond to the abstract he had sent.

(Le Pichon, 2001: 214; my bracketed addition)

So when did McKenzie leave the session and where did he go? Here is the schedule of the session, which began at 8:30, as published in AGU's synopsis of the meeting:

1. James R. Heirtzler: Marine Magnetic Anomalies and a Moving Ocean Floor (20 min)
2. Lynn R. Sykes: Seismological Evidence for Sea Floor Spreading and Transform Faults (21 min)
3. Maurice Ewing and John I. Ewing: Deep Sea Sediments in Relation to Island Arcs and Mid-Ocean Ridges (20 min)
4. H. W. Menard: SeaFloor Spreading, Topography, and the Second Layer (20 min)
5. Dennis E. Hayes and James R. Heirtzler: Magnetic Anomalies and Their Relation to Island Arcs and Continental Margins (20 min)
6. Manik Talwani and Dennis E. Hayes: Continuous Gravity Profiles over Island Arcs and Deep Sea Trenches (20 min)
7. Russell W. Raitt: Marine Seismic Studies of the Indonesian Island Arc (20 min)
8. John C. Rose and Alexander Malahoff: Marine Gravity and Magnetic Studies of the Solomon Islands (20 min)
9. A. S. Furumoto, George H. Sutton, and John C. Rose: Results of the Seismic Refraction Survey in the Solomon Islands Group (20 min)
10. Jason Morgan: Convection in a Viscous Mantle and Trenches (20 min)

(Anonymous, 1967: 32–33)

McKenzie is not sure just when he left, which is not unexpected after forty-four years, especially because he had no inkling that his leaving would someday be of any historical interest. He thinks he may have left before the Ewings' presentation.

You do expect me to have total recall! My guess is that I left after Lynn Sykes's talk, since Maurice was still trying to claim that something happened at 10 Ma, I knew what Bill [Menard] would say, and Malahoff was making his name by opposing sea floor spreading.

(McKenzie, November 3, 2011 email to author; my bracketed addition)

Given McKenzie's catholic interests, he may have gone almost anywhere; but he might have found particularly enticing a session on general seismology, where he would have heard talks on the upper mantle, which would have been of interest to him given his Ph.D. and early papers, including the one he presented at the God-dard meeting (§6.12) and one on mantle convection that he would later rewrite at Scripps. He may also have wanted to hear Don Anderson, whom he had got to know at Caltech, and who spoke at the seismology session. But there is an added factor. Although Morgan's session was supposed to end at noon, it ran over, and Morgan believes that he did not *begin* talking until noon (Morgan, February 8, 2012 email to author). Morgan (February 8, 2012 email to author) also thinks, "As best I remember, Dan told me ('68) that he left early in order to have lunch before the afternoon sessions began." Given that the session went over, McKenzie may well have been eating lunch while Morgan spoke. But the key point is that McKenzie would not have heard Morgan's talk.

I also do not think it is surprising that he heard nothing about Morgan's talk while at Harvard, Yale, or Lamont after the AGU meeting before returning to Caltech, for as Le Pichon remembers, "Very few people, if any, actually paid attention to what" Morgan "said."

Morgan has a special gift for disorienting his listeners. This gift was especially well displayed on that occasion, and very few people, if any, actually paid attention to what he said... On the basis of this document [Morgan's eleven-page extended outline of his presentation] it seems extra-ordinary that, in the hall packed with the best geophysicists and geologists in the United States, nobody got excited by or even interested in the implications of Morgan's ideas. They were too new, too different from anything that had been done. Even among those who received an extended outline and had time to digest the new concepts, I apparently was the only one to have considered it sufficiently important to drop everything else and start working along these new lines.

(Le Pichon, 2001: 214–215; my bracketed addition)

So who would have told McKenzie? Le Pichon is the only one who took any interest in Morgan's talk. But McKenzie did not meet Le Pichon until November 1967, when he revisited Lamont after he and Parker had submitted their paper to *Nature*. Moreover, Menard, given all his later involvement with Morgan, his paper, and McKenzie, did not remember ten years afterwards that he had co-chaired the session where Morgan first spoke of plate tectonics.

That session incidentally provided me with an excellent example of how unreliable memory can be. A decade later Dick Hey, then a recent Princeton Ph.D., conceived of celebrating the tenth anniversary of the session at another Geophysical Union meeting. Fred Vine was unavailable, but Dick invited me to be co-chairman as at the original session. I not only did not remember hearing Jason's famous talk, I didn't remember presiding over the session.

(Menard, 1986: 286)

What about when McKenzie got to Scripps? Would anyone there have told him about Morgan's talk? Menard, of course, is the one who comes to mind. Menard not only co-chaired the session, but had a copy of Morgan's extended outline. Menard surely did not tell McKenzie before McKenzie talked to him about his own discovery of plate tectonics. Did Menard discuss Morgan's talk and extended outline with his colleagues or students at Scripps? And, if so, did they talk to McKenzie? Indeed, Menard later claimed that Morgan's ideas were known to others at Scripps.

John Mudie and John Sclater joined the staff. Mudie had heard Jason Morgan's talk on plate tectonics and returned full of enthusiasm to spread the word. He was in Fred Speiss's Deep-Tow Group, which included many students with Tanya Atwater among them. The basic ideas of plate tectonics became familiar to this group. It was also known to others on campus. When I returned from Washington in the summer of 1966 a few students including Clem Chase and Dan Karig asked to work with me on marine tectonics. Clem received a desk in my laboratory, and I saw him constantly. Jason Morgan sent me a preprint of his manuscript in its early draft, probably in the late spring of 1967. I believe I also received the paper for an editor. In any event the manuscript certainly circulated among my students, and we discussed it. The original draft, however, was difficult for me to fathom, and it did not have the impact of the final publication. Moreover, I had other things on my mind. I spent July through September on the Nova expedition in the southwestern Pacific. Teddy Bullard was among the old hands to participate, and there were many students including those who knew about Morgan's papers.

(Menard, 1986: 290–291)

Did Menard misremember? If he did not understand Morgan's original eleven-page outline, it is somewhat surprising that he would have discussed it with his students. Moreover, if Morgan's ideas were discussed on board R/V *Horizon* while Menard was on board, it is surprising that Sclater did not hear about them while he was with Menard and Bullard on *Horizon* from July 12 to August 8. I asked Daniel (Dan) E. Karig and Clement (Clem) G. Chase if they remembered discussing Morgan's extended outline. Karig does not remember discussing Morgan's work while participating on Nova. Karig was with Menard aboard *Argo* on Leg V from August 11 to September 12 and Chief Science Officer aboard *Horizon* on Leg VI from September 17 to October 16.

Good Lord; it's been almost half a century since those days and I hardly remember what I did yesterday. Unfortunately, events during those seminar years are all blurry in my mind. Jason and Dan and Xavier Le Pichon all came through Scripps but I'm quite sure all this was post Nova. We certainly talked about sea floor spreading and subduction during the cruise, but damned if I can recall talking about plate tectonics before or even during the cruise. My personal objective on that cruise was to investigate what was happening in the basin behind the island arc (Tonga in that case) because Doc Ewing had sent Bill copies of his primitive seismic profiles across that area that showed anomalously little sediment and a very rough sea floor. This didn't make sense, given the subducting trench so close, which tells me that we were certainly aware of seafloor spreading and convergence but it didn't imply knowing anything about plate tectonics.

I have a gut feeling that there was some discussion as to who did what when concerning the "discovery" of plate tectonics, which would have been only to be expected, but I can't remember anything more.

(Karig, November 27, 2011 email to author)

So Karig first discussed Morgan's paper in October after returning to Scripps. Chase accompanied Menard aboard *Horizon* on Legs III and IV from July 3 to August 8, and then remained on *Horizon* during Leg V, and most of Leg VI, returning to Scripps after disembarking on September 26. He did see Morgan's preprint, but not until after he returned to Scripps. I suspect he is right about this, given what else happened, his memory is probably good.

I did see Jason's preprint, but at Scripps. Probably that was after coming back from Nova. I can visualize the preprint laid out on the big map table on Bill's outer office. There was a lovely flap at Noumea [New Caledonia] because all the magnetic and navigation records went missing in LAX, whole leg's data lost. Fortunately, the suitcase showed up again after a lot of travel agent effort; it had been sent right back to New Caledonia.

(Chase, November 27, 2011 email to author)

Seeing Morgan's preprint after returning from Nova makes perfect sense. What is at issue is whether Chase, Karig, and others saw and discussed Morgan's extended outline of his AGU talk before participating in Nova, and told McKenzie, Parker, or Sclater about it. I suspect that Menard misremembered. I think he mixed up showing Karig, Chase, and other students Morgan's extended outline with the preprint of Morgan's paper that he was asked to referee.

What about Mudie? Mudie heard Morgan discuss plate tectonics on September 7 or 8 during the two-day conference at Woods Hole, returned to Scripps, and told McKenzie about Morgan's talk. However, McKenzie, as witnessed by Sclater, came up with the idea of plate tectonics and "understood how to use slip vectors" before he talked to Mudie. McKenzie has also claimed that Mudie did not really understand what Morgan did.

John Mudie . . . went to Woods Hole, and came back, I would guess in September, with a muddled version of what Jason had said there, which sounded little different from George Backus's paper. John had not understood Jason very well. By this time I had understood how to use the slip vectors, so thought I should get a move on since I was leaving Scripps in November.

(McKenzie, March 18, 1990 letter to author)

I think McKenzie is correct. If he had thought that Morgan had independently discovered plate tectonics, then, if he had been anxious to publish his own ideas, it makes no sense that he would have stopped working on the plate tectonics paper and begun rewriting his paper for the *Geophysical Journal of the Royal Astronomical Society* when he received Weiss's review. Menard also claimed that Mudie, who was in Speiss's Deep-Tow Group, talked to others in the group about Morgan's ideas. I asked Tanya Atwater if she remembered such conversations because Menard singled her out as among those in the group. She is "quite sure" that she did not.

I spent the summer and fall of 1967 data processing and putting out our first paper about the workings of sea floor spreading centers. Even if someone had told me about rigid tectonics on a sphere, it wouldn't have stuck. I'm quite sure I didn't discuss this subject with Mudie.

(Atwater, November 11, 2011 email to author)

There is, I suppose, the question of why he just happened to reread the paper by Bullard, Everett, and Smith in which they fit together the continents surrounding the Atlantic by using Euler's fixed Point Theorem. He was rereading their paper because he "had disliked the method" they had used to fit the continents together (McKenzie, 2001: 180). More specifically, he did not think their method worked when transform faults were involved.

Teddy [Bullard] used the misfit in longitude $\Delta\varphi$ at a given latitude as the quality to be minimised. If part of the margin is a transform fault, this procedure can give a large value for the misfit even when the fit is good. I thought that it would be better to use the area of gap + overlap, without regard to sign, and used this in McKenzie, Molnar, and Davies, 1970 and Fisher, Sclater and McKenzie, 1971. It does not in fact make much difference.

(McKenzie, March 18, 1990 letter to author; my bracketed addition)

Thus, bizarre as it may seem to some at first glance, closer examination of what happened, even given what cannot be completely confirmed, shows that McKenzie discovered plate tectonics after, but independently of, Morgan.

I leave the last word to Jason Morgan.

I personally have no doubt but that Dan [McKenzie] independently came up with the idea of plate tectonics. As I said, the session ran very long – my talk began after the scheduled lunch-break had already started and many people had already left the meeting room. (However, I am surprised that Bill [Menard] didn't take knowledge of my talk back to Scripps; he was the chair, I gave him a rough copy of my talk & figures at the end of the session, and in my talk I had made the point that the starting point of my idea had come from his Jan 6, 1967 paper in Science on the great fracture zones of the Pacific. Maybe he had dozed off.)

I think the time was just ripe for plate tectonics to be discovered. Euler's (Fixed Point) Theorem was a core piece of classical mechanics – in all junior-senior level textbooks on rigid body motions. Euler's Theorem had been used explicitly (E. Bullard, J.E. Everett, and A.G. Smith, The fit of the continents around the Atlantic, Phil. Trans. Roy. Soc. London, 258A, pp. 41–51, 1965) and implicitly via a cartographic technique (S.W. Carey, Wegener's South America-Africa assembly, fit or misfit?, Geological Magazine, v. 92(3), pp. 196–200, 1955; and pages 218–223 of S.W. Carey, A tectonic approach to continental drift, in the book "Continental Drift: a symposium" (March, 1956, Hobart), pp. 177–355, 1958) to make the case that continents deform very little after rifting. With the expanding interest in seafloor spreading/magnetic anomalies/fracture zones/focal mechanisms, if Dan's and my papers had not been written in 1967, I think there would have been several papers written in 1968 discovering the basic rigidity of plates and hence plate tectonics.

(Morgan, March 15, 2012 email to author; my bracketed additions)

7.13 The keys to McKenzie's discovery

McKenzie was primed to come up with plate tectonics when he decided to re-read Bullard, Everett, and Smith (1965). As already described (§7.6), McKenzie, through his work on marine heat flow, had already developed a new model of seafloor spreading which included a thick, rigid lithosphere and rootless ridges. The former gave him the rigidity he needed for applying Euler's theorem to blocks of lithosphere; the latter allowed him to accept seafloor spreading without worrying about Africa and Antarctica being surrounded by ridges, and to separate the kinematics of seafloor spreading from its dynamics. He also was familiar with Euler's Point Theorem.

I did not know about Euler's theorem until I came to Madingley [Rise]. I remember well having it explained to me at tea by George Backus, who was on sabbatical and was therefore not as busy as Teddy [Bullard], when he was writing his *Nature* paper. I knew of George's work on magnetic anomalies, which was done at Cambridge, and of course knew about Teddy's use of Euler's theorem. But it seemed simply a convenient parameterization, not a central idea.

(McKenzie, March 18, 1990 letter to author; my bracketed additions)

McKenzie was well prepared to discover plate tectonics. What he needed was a stimulus, and that was provided by his rereading Bullard, Everett, and Smith's 1964 paper. They had used Euler's Point Theorem as a basis for opening the Atlantic and their method made sense only if the continents rotating about their common Euler pole of rotation, or I suppose, strictly speaking, their trailing margins, were assumed to be rigid bodies. Nor did it hurt that McKenzie's concern about their paper was about their way of estimating misfit in longitude if "part of the margin is a transform fault," for as McKenzie and Parker (1967: 1277) noted after introducing their version of plate tectonics: "These remarks extend Wilson's concept of transform faults to motions on a sphere, the essential additional hypothesis being that individual seismic areas move as rigid blocks on the surface of a sphere."

McKenzie's breakthrough was two-fold. Reading Bullard, Everett, and Smith (1964) got him thinking about Euler's theorem, thinking that it could be used to describe rotations of lithospheric plates relative to each other, that it was more than a convenient parameterization useful for putting continents back together again. It was during this period that I realized that, if finite displacement could be described by finite rotations about finite poles, instantaneous ones required angular velocities. It seemed obvious and not very useful at first.

(McKenzie, March 18, 1990 letter to author)

Once McKenzie began thinking about instantaneous Euler poles and the need to find a way to determine instantaneous angular velocities, he hit on the idea of using slip motions of earthquakes. After all, what could be more instantaneous? Impressed with Sykes' work on using slip motions to confirm the idea of ridge-ridge transform faults, it was quite natural that he thought of using slip vectors. However, McKenzie did not restrict himself to pure strike-slip earthquakes; he realized that he could obtain slip directions from looking at the horizontal slippage of slip-dip earthquakes, and it was this move that allowed him to work with earthquakes all along the common boundary of the North American and Pacific plates. He then thought of a talk by Stauder on earthquake mechanisms along the Aleutian arc that he heard at the spring AGU meeting.

For me the critical step was realizing that I could use the slip vectors of earthquakes to find the instantaneous poles ... I did not have access to Lynn's [Lynn Sykes'] solutions, but remembered I had heard Stauder talking about mechanisms along the Aleutian arc at the AGU in the spring. [Stauder spoke on the afternoon of April 19, at a session entitled "Island Arcs, Mid-Ocean Ridges, and Sea-Floor Spreading, 2"; papers were also given by Fisher; Vacquier; Isacks, Sykes, and Oliver; Oliver and Isacks; and Hess.] I found his papers and was very surprised that he thought the P and T axes were important, rather than the slip vectors. But the slip vectors made complete sense. I used a National Geographic globe with a cap to find the pole that best fitted the slip vectors. The whole theory was then obvious. It was this that took two days.

(McKenzie, March 18, 1990 letter to author; my bracketed additions)

McKenzie's retrospective comment about slip directions, and P and T axes is important. Stauder and other seismologists were more interested in determining the compressional (P) and extensional (T) axes than in determining the slip vectors. But it is slip directions that are important for positioning instantaneous Euler poles.[14] McKenzie was the first to recognize that slip directions are what are needed to test plate tectonics *and* to correctly determine them from first motion studies. Moroever, use of slip vectors distinguishes his use of fault plane solutions from Morgan's. As explained above (§7.11), Morgan did use first motion studies, but he used them to determine the strike of the fault plane of the earthquake in question, not to determine the horizontal component of its slip vector. Once McKenzie determined a few slip directions and found that great circles perpendicular to them intersected roughly at just two points on Earth's surface 180° apart, which is what should be expected, he was sure that he was right, and he then talked to his officemate Bob Parker about how to best display his results.

I remember Dan explaining to me how the initial motions of earthquakes were used to get the slip directions. We discussed how to calculate relative velocities from a rigid plate model on a sphere. I am pretty good at this three-dimensional spherical geometry.

(Parker, March 15, 1990 letter to author)

Importantly they also talked about how to best display McKenzie's results. It is time to introduce Parker.

Parker was born in 1942 in Reading, England. Showing an aptitude for science and mathematics, he did his undergraduate work at Downing College, University of Cambridge, where he was supported by a State Scholarship Grant (1960–2) and a Downing College, Major Open Scholarship (1960–2). It was expected that he would study at Cambridge. "The reason I went to Cambridge for my bachelor's degree was simply that almost everyone thought Cambridge was the best University in the country for science and the schools all pushed their best people to take the entrance exams" (Parker, March 15, 1990 letter to author). He studied Natural Science as an undergraduate, concentrating on physics. He took no geology classes.

I studied Natural Science at Cambridge as an undergraduate; this was mostly straight physics, but I did some mineral physics because of the strong department in this field at Cambridge at the time. I had no geology courses and as an undergraduate I had no idea that I would end up as an Earth scientist of any kind.

(Parker, March 15, 1990 letter to author)

A summer job working with Alan Cook, a former Bullard student, eventually led Parker to become a graduate student in the Department of Geodesy and Geophysics at Cambridge, studying under Bullard.

Every summer like most students I found a job. At the end of my second year at Cambridge I worked in the Standards Division of the National Physical Lab in Teddington (the British equivalent of the National Bureau of Standards in the US). The man I worked for was Alan Cook, who was planning an experiment to measure G, Newton's universal gravitational constant of gravitation. He put me onto doing the calculations for this experiment. Cook was a former student of Teddy Bullard's ... and when I had finished the summer job, Cook asked me if I liked the sort of classical physics I had been doing for him, "because, if so, you should consider going into geophysics." He told me to make sure to look up Bullard and to go to his lectures, which were optional evening classes that were not part of the mainstream physics curriculum. Teddy was a wonderful lecturer; I was looking around for an area to do research in, and so after his talk on the Earth's magnetic field I asked him if he was interested in recruiting research students. His response was very encouraging and I was hooked.

(Parker, March 15, 1990 letter to author)

Bullard and Parker were an excellent fit. Parker was interested in theoretical problems and Bullard provided them for him.

I began working on the dynamo theory; Teddy had an idea there was some universal anti-dynamo based on topology, but neither he nor any of his students were ever able to make that idea work – he tried it out on subsequent students too. The first thing I did that worked out was on the reconnection of lines of force in a moving fluid. Nigel Weiss, who was a former student of Teddy's and was working at a government lab at the time, had computed numerical solutions of the evolution of magnetic fields in a moving, electrically conducting, fluid. Teddy thought the same phenomena revealed in the movies Nigel made could be captured in a much simpler system that could be solved analytically, and he was right. The issues centered on technical questions about the breaking of field lines, which [J. W.] Dungey had shown to be a necessary precursor to dynamo action. The other thing I did was to work on a simplified model to explain the electric currents observed flowing in the oceans; again some numerical work had been done, but Teddy thought I could solve a simple system by analytic means, which I did.

(Parker, March 15, 1990 letter to author; my bracketed addition)[12]

Parker did not work in geology while a graduate student. What he learned, he learned primarily from tea and coffee breaks at Madingley Rise. With nothing to unlearn, he found V–M "perfectly acceptable as a description for the creation of the seafloor," and could not understand Vacquier's resistance (§3.8) or his preference for his original explanation of the magnetic anomalies in the Pacific (III, §1.6, §5.6).

Not having had any geological training, I found Vine and Matthews' ideas perfectly acceptable as a description for the creation of the seafloor. Vacquier, as I recall, did not accept the V-M explanation for the magnetic anomalies so he had to have some way of getting them all formed in a nice long line and then have them torn apart by a subsequent fault. While I didn't have a detailed explanation for the old Pacific anomalies, it didn't seem necessary to me to have a huge subsequent displacement if you could form the same magnetic anomaly pattern anywhere on the Earth's surface where there was a ridge. I remember thinking how perverse Vacquier was being in not accepting the V-M model.

(Parker, March 15, 1990 letter to author)

He also was impressed with Wilson's idea of transform faults. He did not get to know Wilson or Hess while they were on sabbatical at Madingley Rise, but heard Wilson's seminar on May 26, 1965 (§4.9).

As a junior graduate student working in electromagnetic induction, I really had little contact with Hess or Wilson. I recall Wilson's seminar at Madingley Rise where he revealed the idea of transform faults for the first time. It seemed so natural and obvious at the time, and I remember thinking that this was a truly monumental discovery.

(Parker, March 15, 1990 letter to author)

So Parker was in a position to appreciate what McKenzie would present to him when asked about a way to display his results. More importantly, however, he was also in a position to help McKenzie because he had his map-projecting program, which grew out of his work on electric currents in the ocean.

I worked mostly on electric currents in the ocean, moving from the very simple models of my Ph.D. work to more realistic ones in which the true shape of the ocean basins was accurately represented in the computer. To display the results I had to have a plotting program capable of drawing coastlines as well as current streamlines, so I wrote one. In those days there was virtually no computer software, except stuff you wrote yourself. I was and still am quite a good programmer. This means that the programs I write are easy to use, easy to understand and easy to add improvements to. When the importance of such programs to plate tectonics became apparent, it was simple to augment my display programs to add continental rotations and so on.

(Parker, March 15, 1990 letter to author)

Parker later described the program, and rightly accented the fact that he designed it in such a way that he could adapt it to McKenzie's request.

The program was originally named SUPERMAP after Harold Macmillan, Prime Minister of Great Britain (1957–1963), known in the press as SuperMac. SUPERMAP was written in Fortran-63 and recorded on punched cards; the database comprised a primitive coastline of about 5000 points, digitized by hand by an undergraduate student as part of a summer job from a large Mercator projection map provided to me by Bill Menard. . . I have always believed it is more efficient in the long run to build programs that can be used repeatedly and that are easily used and upgraded. . . So, even though I needed only one kind of map projection for my electromagnetic induction problem, I made SUPERMAP a general-purpose program, running under an easily used common language . . . Thus SUPERMAP was ready for application when plate tectonics came along; all it needed was three or four lines to implement the oblique Mercator map projection, to generate one of the figures in our *Nature* paper.

(Parker, 2001: 196)

Parker obtained his Ph.D. in fall 1966.[15] Nigel Weiss and V. C. A. Ferraro were his examiners. Parker recalled what happened when he took the train with Weiss down to London from Cambridge to meet Ferraro.

I took my oral examination in the Fall of 1966. Nigel Weiss was one of my examiners and the other was a Professor Ferraro, a plasma physicist at Queen Mary College in London. Ferraro was rather ill at the time and so Nigel and I traveled down to London by train to have the exam. Nigel had just recently been appointed to the Applied Maths and Theoretical Physics Department in Cambridge. Nigel talked to me so loudly on the journey, that a woman in the compartment complained that she was unable to concentrate on her newspaper. Nigel turned and boomed at her: "Madam, if you listen closely you might learn something more valuable than you will find in your paper."

(Parker, March 15, 1990 letter to author)

I suspect that Parker paid more attention to Weiss's review of his lengthy mathematical paper than did the woman in the compartment.

After obtaining his Ph.D., Parker accepted a postdoctoral fellowship in January 1967 at the IGPP at Scripps. Parker wanted to work with George Backus (§3.8) and Freeman Gilbert, the two founders of modern inverse theory. Parker recalled how it came about.

In my last year as a student, George Backus and Freeman Gilbert, both from IGPP, La Jolla, were spending their sabbaticals in Cambridge. They are both mathematical types and I spent some time talking to them about my work. Freeman Gilbert asked me if I would like a postdoc in La Jolla. In January 1967, I arrived in La Jolla.

(Parker, March 15, 1990 letter to author)

Parker is still at Scripps. He has become a leading figure in inverse theory (Parker, 1994).

I return to McKenzie's development of plate tectonics. When he talked to Parker about how to best represent his results, Parker suggested an oblique Mercator projection. He also helped McKenzie work out "for the spherical system how to represent the instantaneous velocities through angular velocity vectors and how those vectors were combined at the points where three plates meet."

Sykes' work at Lamont (where Dan had been and was going back to in 1968) convinced him that the first motions contained important information for tectonics – he started to look at the earthquake mechanisms in a systematic way, not just at the ocean ridges. Dan quickly realized that it might be possible to treat the interior aseismic (i.e. seismically inactive) regions of the earth as rigid bodies. This meant that the regions under consideration became so large that pictures based on a flat-earth were no longer adequate. Dan wasn't sure how the geometry of the plane velocity vectors that he was used to would translate into a spherical setting. This is where I came in: during his visit to Scripps he told me about the problem, and I worked out for the spherical system how to represent the instantaneous velocities through angular velocity vectors and how those vectors were combined at the points where three plates meet. Furthermore, as I have already described, I had on hand my computer mapping program SUPER-MAP, which we immediately put to work displaying the amazingly compelling results. It was my idea to use an oblique Mercator map projection, which made such a dramatic graphical demonstration in the 1967 *Nature* paper.

(Parker, 2001: 197)

Parker has also acknowledged, however, that McKenzie was the creative force behind their work on plate tectonics, and accented how McKenzie realized the importance of first motion studies: "Without false modesty, I must make it clear that Dan McKenzie was the creative force behind the 1967 *Nature* paper" (Parker, 2001: 196).

7.14 McKenzie and Parker's version of plate tectonics

McKenzie and Parker introduced their version of plate tectonics, which they named the paving stone theory, by describing the kinematics but not the dynamics of seafloor spreading, by relating it to the Vine–Matthews hypothesis, by accenting the importance of rigid plates, by defining the three types of plate boundaries, and by emphasizing that transform faults being "lines of pure slip ... are always parallel ... to the relative velocity vector between two plates."

The linear magnetic anomalies[1, 2] which parallel all active ridges can only be produced by reversals of the Earth's magnetic field[1] if the oceanic crust is formed close to the ridge axis[3]. Models[4] have shown that the anomalies cannot be observed in the North Atlantic unless most dyke intrusion, and hence crustal production, occurs within 5 km of the ridge axis. The spreading sea floor[3] then carries these anomalies for great horizontal distances with little if any deformation. The epicenters of earthquakes also accurately follow the axis and are offset with it by transform faults[5, 6]. The structure of island arcs is less clear, though the narrow band of shallow earthquakes suggests that crust is consumed along a linear feature. These observations are explained if the sea floor spreads as a rigid plate, and interacts with other plates in seismically active regions which also show recent tectonics activity. For the purposes of this article, ridges and trenches are respectively defined as lines along which crust is produced and destroyed. They need not also be topographic features. Transform faults conserve crust and are lines of pure slip. They are always parallel, therefore, to the relative velocity vector between two plates – a most useful property.

> (McKenzie and Parker, 1967: 1276; references are respectively equivalent to my Vine and
> Matthews, 1963; Vine, 1966, Hess, 1962; Matthews and Bath, 1967; Sykes, 1963, 1967b)

They next introduced Euler's Point Theorem, reintroduced the three types of plate boundaries in terms of the theorem, noted the usefulness of transform faults in determining the position of Euler poles, and, to avoid confusion over reifying Euler poles, explained (p. 1277) that "the pole position itself has no significance, it is merely a construction point."

The movement of blocks on the surface of a sphere is easiest to understand in terms of rotations. Any plate on a sphere moves by two successive rotations, one of which carries one point to its final position, a second about an axis through this point then produces the required orientation. These two rotations are equivalent to a single rotation about a different axis, and therefore any relative motion of two plates on the surface of a sphere is a rotation about some axis. This is Euler's theorem, and has been used to fit together the continents surrounding the Atlantic[7]. If one of the two plates is taken to be fixed, the movement of the other corresponds to a rotation about some pole, and all relative velocity vectors between the two plates must lie along small circles or latitudes with respect to that pole. If these small circles cross the line of contact between the two plates the line must be either a ridge or a trench depending on the sense of rotation. Neither of these structures conserves crust. If the line of contact is itself a small circle, then it is a transform fault. This property of transform faults is very useful in finding the pole position and is a consequence of the conservation of crust across them.

> (McKenzie and Parker, 1967: 1277; reference 7
> refers to my Bullard et al., 1965)

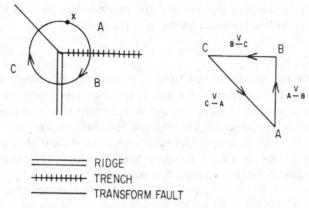

RIDGE
TRENCH
TRANSFORM FAULT

Figure 7.26 McKenzie and Parker's Figure 1 (1967: 1277) captioned "The circuit and its vector diagram show how a ridge and a trench can meet to form a transform fault." The boundary between plates A and B is a trench, between B and C, a ridge, and between C and A, a transform fault. In the vector diagram, the velocity of B relative to A ($_A v_B$) is in direction AB, of C relative to B ($_B v_C$) in direction BC, and of A relative to C ($_C v_A$) in direction CA. The length of side BC represents the magnitude of spreading, and angles ABC and BCA the direction of spreading with respect to the common boundaries of plates B and C, and C and A. Knowing these three parameters allows for determination of the sides CA and BA of the vector triangle, if the circuit is instantaneous.

McKenzie and Parker, as mentioned above, also described their theory as an extension of Wilson's idea of transform faults to a spherical surface, and emphasized as an "essential additional hypothesis ... that individual aseismic areas move as rigid plates on the surface of a sphere" (McKenzie and Parker, 1967: 1277).

Continuing to present the fundamentals of their paving stone theory, they introduced the notion of a triple junction, a point where three plates meet, and explained how, if the plates are rigid, the instantaneous velocity vector of the third plate relative to the two others can be determined if the instantaneous velocity vectors of the other two plates relative to each other are known. They explained why this holds on a plane and then extended it to a sphere. Because plates are rigid, instantaneous relative velocity vectors can be added, and because a complete circuit around the triple junction is closed, their sum around the circuit is zero. They also provided an example, referring to their Figure 1 (my Figure 7.26).

There are several points on the surface of the Earth where three plates meet. At such points the relative motion of the plates is not completely arbitrary, because, given any two velocity vectors, the third can be determined. The method is easier to understand on a plane than on a sphere, and can be derived from the plane circuit in Fig. 1. Starting from a point x on A and moving clockwise, the relative velocity of B, $_A v_B$ is in the direction AB in the vector diagram.

Similarly the relative velocities $_Bv_C$ and $_Cv_A$ are represented by BC and CA. The vector diagram must close because the circuit returns to x. Thus:

$$_Av_B +_B v_C +_C v_A = 0 \qquad (1)$$

The usual rules for construction of such triangles require three parameters to be known, of which at least one must be the length of a side, or spreading rate. Transform faults on both ridges and trenches are easy to recognize, and they determine the directions, but not the magnitude, of the relative velocities. The magnetic lineations are one method of obtaining $_Bv_C$ though this value must be corrected for orientation unless the spreading is at right angles to the ridge. Then the triangle in Fig. 1 determines both $_Av_B$ and $_Cv_A$. This method is probably most useful to determine the rate of crustal consumption by trenches.

(McKenzie and Parker, 1967: 1277)

Applying equation (1) to a sphere in terms of relative instantaneous rotational velocities (ω), which "behave like vectors," gave them equation (2): $_A\omega_B + {}_B\omega_C + {}_C\omega_A = 0$.

They also emphasized that equation (1) (and, I add (2)) should not be used with finite velocities – with rotations over durations that are not infinitesimally short.

Equation (1) must be used with care, because it only applies rigorously to an infinitesimal circuit round a point where three (or more) plates meet. If the circuit is finite, the rotation of the plates also contributes to their relative velocity, and therefore these simple rules no longer apply.

(McKenzie and Parker, 1967: 1277)

Thus McKenzie and Parker definitely recognized the difference between instantaneous and finite rotations, and that indicators of plate rotations which suggest a change in the rotational velocity of one plate relative to another, such as changes in the trends of transform faults, even of active ones, or peculiarities in the pattern of magnetic anomalies along common plate boundaries, should, if used at all, be used with extreme caution in determining the position of their common instantaneous rotational (Euler) pole. Of the three ways of determining the position of such poles then available – marine magnetic anomalies, trends of transform faults, and slip vectors of earthquakes – McKenzie and Parker preferred slip vectors, as they are the most instantaneous. Nowadays, space geodetic methods that directly and reliably measure positional changes of one plate relative to another, such as very long baseline interferometry, are used to determine instantaneous Euler poles (Argus *et al.*, 2011). McKenzie later recalled why he did not use the trend of transform faults associated with the San Andreas Fault when determining the Euler pole common to the rotation of the Pacific plate relative to North America.

You cannot really test the idea [of plate tectonics] with transform faults unless you map the active trace or unless you have slip vectors. Transform faults are produced by finite rotations and cannot in general be small circles. This I understood in 1967. I also knew that the San Andreas was not purely strike-slip, as the Loma Prieta [of October 18, 1989] earthquake has recently shown with its combined strike-slip and reverse dip motion. I had travelled almost the whole length of the fault in the summer of 1967 with Bill Farrell. The photograph I gave to Cox was taken on this trip.

(McKenzie, March 18, 1990 letter to author; my bracketed additions)

McKenzie knew that marine magnetic anomalies could be used with caution to determine instantaneous rotational velocities and thus instantaneous Euler poles. As he later put it, "I did indeed realize I could use velocities (though in a strict sense you cannot use magnetic anomalies because of finite rotations) ... But I had no access to profiles in the Atlantic, nor had I then written the programs to produce synthetic profiles" (McKenzie, March 18, 1990 letter to author). Nor did he have spreading rates for the Pacific, for as he also recalled, "I had no data [to fix the pole of relative rotation for the Pacific and North American plates] such as spreading rates from magnetic anomalies because Scripps data had not been digitalized or interpreted" (McKenzie, 2001: 180; my bracketed addition).

McKenzie realized that if he restricted himself to slip directions in testing his idea of plate tectonics he should look at the relative rotation of North American and Pacific plates because he needed slip directions from widely spaced earthquakes to fix both the longitude and latitude of the instantaneous Euler pole. In addition, McKenzie, perhaps unlike Morgan, was quite willing to use earthquakes all along the common boundary of the two plates because he realized he could use the slip component of slip-dip earthquakes, and therefore did not restrict himself to using just strike-slip earthquakes.

If all the earthquakes between the Gulf of California and Japan are produced by a rotation of the Pacific plate relative to the continental one, any pair of widely spaced slip directions can be used to determine the pole of relative rotation.

(McKenzie and Parker, 1967: 1278)

Choosing the slip direction of the June 28, 1966 Parkfield earthquake, a pure strike-slip earthquake along the San Andreas Fault south of San Francisco (McEvilly, 1966) (2a in their Figure 3, my Figure 7.27), and the average slip vector for a number of aftershocks in the Kodiak Island region of the March 27, 1964 Alaska earthquake (Stauder and Bollinger, 1966b) (2b in Figure 3), McKenzie located the instantaneous pole of relative rotation at 50° N, 85° W. He then plotted the slip directions of over eighty available fault plane solutions of earthquakes that had occurred during and after 1957 to determine whether they cut out arcs of small concentric circles around the estimated position based on Sykes (1967b), Stauder (1960), Stauder and Udias (1963), Udias and Stauder (1964), Stauder and Bollinger (1964, 1966a, b).

Rather than present their results in tabular form, they plotted them on a Mercator projection (Figure 3) whose projection pole coincided with the estimated pole of relative rotation. With this projection, they had a most perspicuous way of presenting the data, because if both plates are rigid then "all slip vectors must be parallel with each other and with the upper and lower boundaries of the figure" (McKenzie and Parker, 1967: 1278).

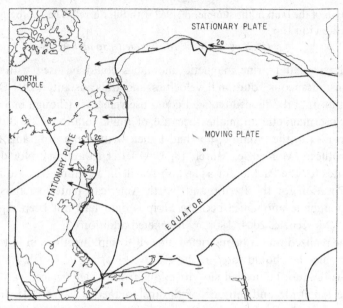

Figure 7.27 McKenzie and Parker's Figure 3 (1967: 1278). Their Mercator projection of a moving Pacific plate relative to a bordering plate containing North America and Kamchatka allowed for a visual presentation of their paving stone theory. The arrows show the slip vectors, obtained from first motion studies of earthquakes occurring along the plate boundary between the two plates. With this projection and a pole placement at 50° N, 85° W, the slip vectors should be parallel to each other and the upper and lower boundaries of the figure. The bicuspid-shaped area by arrow 2a represents their hypothesized small plate, the Juan de Fuca plate, between the Pacific and North American plates. The Mendocino triple junction is at the southeast corner of the Juan de Fuca plate where it meets the North American and Pacific plates. The kite-shaped area with its northernmost corner where the Kurile, Ryukyu, and Mariana trenches meet in central Japan was later named the Philippine plate. The junction of the three ridges, where the Philippine, Pacific, and Eurasian plates meet, was later named the Bosco triple junction, the only one where three ridges meet. The triple junction that McKenzie and Parker proposed at the entrance to the Gulf of California is on the upper right where the Middle America Trench running along the western margin of Mexico meets the East Pacific Rise and what they thought to be the southern extension of the San Andreas Fault. It is where the North American, Pacific, and what became known as the Cocos plates meet. The V-shape on the right margin of the figure below the western margin of Mexico is part of the Cocos plate. The common border of the Pacific and India plates extends northward through New Zealand, turns westward toward the southern corner of the Philippine plate, and meets the common boundary of the Indian and Eurasian plates.

McKenzie and Parker claimed (1967: 1279), "The agreement with the theory is remarkable over the entire region. It shows that the paving stone theory is essentially correct and applies to about a quarter of the Earth's surface." Of the more than eighty published fault plane solutions (from 1957 through 1967) for shallow earthquakes occurring along the common boundary of the two plates, "about 80% had slip vectors with the correct sense of motion and within ±20° of the direction required by Fig. 3" (McKenzie and Parker, 1967: 1278–1279). Given that McKenzie did not restrict himself to strike-slip earthquakes along transform boundaries but considered the slip component of the dip-slip overthrust earthquakes where the Pacific plate dips beneath the North America plate, and even analyzed a strike-slip earthquake where the strike of the common consuming boundary is parallel to earthquake's slip direction, he was able to show that regardless of the type of earthquake, if it had any horizontal component to its slip, it was within ±20° of the direction required by Figure 3. Referring to Figure 3 (Figure 7.27), they accented the variety of earthquakes in the following:

The fault systems of the San Andreas, Queen Charlotte Islands and Fairweather form a dextral transform fault joining the East Pacific rise to the Aleutian trench ... In Alaska the epicentral belt of earthquakes changes direction[10] and follows the Aleutian arc. The fault solutions also change from strike slip to overthrust (for example, 2b), and require that the islands and Alaska should override the Pacific on low angle (≈7°) faults. Though the direction of slip remains the same along the entire Aleutian arc, the change in strike changes the fault plane solutions from overthrusting in the east to strike slip in the west (2c). A sharp bend occurs between the Aleutians and Kamchatka. Here the fault plane solutions change back to overthrust (2d). This motion continues as far as Central Japan, where the active belt divides. Thus the North Pacific contains the two types of transform faults which require trenches[6] and clearly shows the dependence of the fault plane solutions on the trend of the fault concerned.

> (McKenzie and Parker, 1967: 1279; reference 10 is to
> Gutenberg and Richter, 1954; reference 6, to Wilson, 1965b)

Just how good were these results? When McKenzie and Parker were later asked, McKenzie replied as follows, thinking back to the different ways he and Le Pichon dealt with the heat-flow data:

This is a complicated and technical question, and one that sounds like Popper! I can only really answer it by explaining how you do fault plane solutions. The two planes you draw on the focal sphere are only constrained by the observations at discrete points, and how strong the constraints are depends on the details of the geometry of the seismic stations when mapped into the focal sphere. Island arcs are generally awkward, because one side is sea with no stations. In the Aleutians things are not bad because the strike of the auxiliary plane is well controlled (that of the fault plane is not). But 20° is not a large error for a fault plane solution. The other assumption you have to make in ray tracing is that there are no lateral velocity variations. This we already knew was wrong in island arcs. These cause systematic errors in the strike of the auxiliary plane, due to refraction out of the vertical plane containing the station and the source. There is a good paper on this by Toksoz, Minear, and Julian in *JGR* in 1971 [see Minear and Toksoz, 1970 and Toksoz, Minear, and Julian, 1971]. But I had thought this through in 1967. There were also complications due to the fact that the plate boundary is not always a single fault [see McKenzie and Parker, 1967: 1280 in the discussion about the Kuriles]. So I thought the agreement was good, considering all of this. But remember the contrast between my attitude and Xavier's [Le Pichon's] about heat flow [see §6.11, §7.6]. I tend to look for simple and obvious results in good data, whereas Xavier has, I think, a tendency to over interpret the observations (this is obviously a theoretician's view!). We did not do a reliability study then or when I used the same technique later. It is not obvious how to do it, because of the discrete nature of the observations.

(McKenzie, March 18, 1990 letter to author; my bracketed additions)

Parker thought that the results were so much better than what seismologists had previously found.

What impressed us as much as anything was the agreement in the sense of motion. As I said earlier, most seismologists at the time were using the arrival-time seismic data to give them the directions of the principal axes of stress. Their results were utterly chaotic and made no sense at all even within a relatively small area. The seismologists had every right to expect to learn something about tectonics by examining the forces acting to produce the earthquakes, but in a medium like the crust that is broken up and nonuniform, the picture is just too complicated. The (relatively) coherent picture emerging on a global scale from the analysis of slip vectors was "remarkable" by comparison.

(Parker, March 15, 1990 email to author)

With this appeal to different kinds of earthquakes and their known association with transform faults and trenches, McKenzie and Parker noted the following correlation, which, they argued, gave further support to their theory.

The variation of trend also controls the distribution of trenches, active andesite volcanoes, intermediate and deep focus earthquakes. All these phenomena occur in Mexico, Alaska, the Eastern Aleutians, and from Kamchatka to Japan, but are absent where the faults are of a strike slip transform nature. This correlation is particularly obvious along the Aleutian arc, where all these features become steadily less important as Kamchatka is approached, then suddenly reappear when the trend of the earthquake belt changes.

(McKenzie and Parker, 1967: 1279)

They also raised several difficulties (RS2), which they then removed (RS1). I consider only the most important.

The most obvious of these [difficulties] is the complicated region of the ocean floor off the coast between northern California and the Canadian border. The difficulties begin where the San Andreas fault turns into the Mendocino fault. Figure 5 [Figure 7.27] "shows that the change in trend of the epicenters is possible only if crust is consumed between C and A . . ."

(McKenzie and Parker, 1967: 1279; my bracketed additions)

Let me explain the difficulty in a slightly different way. The trends of the San Andreas and the Queen Charlotte Island transform faults, both common to the North American and Pacific plates, are parallel to each other but are not parallel to several other neighboring fracture zones or transform faults. If, however, these other transform faults are part of the common boundary between the North American and Pacific plates, the paving stone theory is not correct. McKenzie's solution to the difficulty was tantamount to supposing the existence of another plate, a small plate

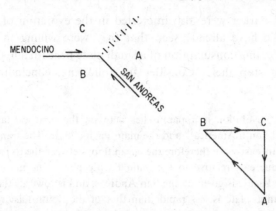

Figure 7.28 McKenzie and Parker's Figure 5 (1967: 1279). The triple junction of plate A, the North American plate, B, the Pacific plate and C, the small hypothesized plate. The boundary between A and C is a hypothesized trench. The triangular velocity vector diagram shows that if B moves relative to A along the Sun Andreas, and C moves relative to B along the Mendocino, then crust is consumed along a trench that separates C and A. This triple junction became known as the Mendocino triple junction.

A between the North American and Pacific plates and bounded by a trench, small ridge segments and transform faults.

The earthquakes along the coast of Oregon and the presence of the volcanoes of the Cascade range, one of which has recently been active and all of which contain andesites, support the idea that crust is destroyed in this area. In the same area two remarkable seismic station corrections . . . also suggest that there is a high velocity region extending deep into the mantle similar to that in the Tonga-Kermandec region. These complications disappear when the ridge and trench structures join again and become the Queen Charlotte Islands Fault.

(McKenzie and Parker, 1967: 1279)

In other words, McKenzie and Parker's Figure 5 (Figure 7.28) is of a triple junction where plates *A*, *B*, and *C* meet at the point of intersection of a hypothesized trench and the Mendocino and San Andreas transform faults. Plate *A* is the North American plate, Plate *B* is the Pacific plate and Plate *C* is the hypothesized plate. Plate *C* is consumed along the hypothesized trench that separates *C* from the North American plate. Thus the neighboring transform faults, such as the Mendocino and those between the small ridge segments (such as the Gorda and Juan de Fuca ridges, and those that connect them with the Queen Charlotte Island fault), do not form a common boundary between the Pacific and North American plates. Consequently, the fact that these neighboring transform faults are not parallel with the San Andreas and Queen Charlotte Island transform faults is not a problem for plate tectonics, and McKenzie was able to turn the difficulty into a solved, but unconfirmed, solution for his new paving stone theory.[16]

McKenzie and Parker were also interested in the evolution of plates and their past rotations. We have already seen that they were willing to hypothesize the previous and continuing consumption of a small plate beneath a hypothetical trench, and they did not stop there. Consider the following, concluding paragraph of their paper.

One area where the evolution is apparent lies between the plate containing the Western Atlantic, North and South America[10] and the main Pacific plate. The transform faults in the South-East Pacific are east-west: therefore the ocean floor between the rise and South America is moving almost due east relative to the main Pacific plate. The motion of the Atlantic plate relative to the Pacific is given by the San Andreas, and is towards the south-east. If the motion of the Atlantic plate is less rapid than that of the South-Eastern Pacific north of the Chile ridge, then the crust must be consumed along the Chile trench. The faults involved must have both overthrust and right-handed strike of the North-Eastern Pacific if there was originally a plate of ocean floor between North America and the main Pacific plate joined to that which still exists to the west of Chile. This piece of ocean floor has since been consumed, and therefore the direction of spreading in the Pacific appears to have changed in the north but not in the south. This explanation requires changes in the shape of the plates but not in their relative motion, and therefore differs from those previously suggested[2, 6]. This study suggests that a belief in uniformity and the existence of magnetic anomalies will permit at least the younger tectonic events in the Earth's history to be understood in terms of sea floor spreading.

(McKenzie and Parker, 1967: 1280; reference 10 refers to Gutenberg and Richter, 1954; reference 2, to Vine, 1966; reference 6, to Sykes, 1967b)

They also were definitely interested in triple junctions and their evolution. Besides hypothesizing about the existence of the triple junction, later called the Mendocino triple junction and classified by McKenzie and Morgan (1969) as a transform-transform-trench triple junction (FFR – see Figure 7.51), they identified what became known as the Bosco triple junction, the point where the Philippine, Pacific, and Eurasian plates meet. It is the only triple junction where three trenches meet (see Figure 7.27); McKenzie and Morgan (1969) later classified such junctions as trench-trench-trench (TTT(a) – see Figure 7.51). They also proposed the existence of a triple junction at the mouth of the Gulf of California, where they thought that the San Andreas Fault, East Pacific Rise, and Middle America trench meet.[17]

The two ends of the North Pacific belt may also be discussed with the help of vector circuits. The end in Central Japan gives the trivial results that two trenches can join to give a third. The other end at the entrance to the Gulf of California is the circuit in Fig. 1, and shows how the East Pacific rise and the Middle America combine to become the San Andreas transform fault.

(McKenzie and Parker, 1968: 1280)

The latter triple junction, illustrated in Figure 7.26, is where a ridge, a trench, and a transform fault meet. McKenzie and Morgan (1969) later classified it as type-a ridge-trench-transform (RTF(a) – see Figure 7.51). Although McKenzie and Parker had not figured out how triple junctions evolve, and if there is a general procedure for determining whether or not a particular triple junction is stable, they were interested in both issues. McKenzie would soon return to both, and would classify possible types of triple junctions on the basis of their geometry and present procedures for determining whether or not a particular junction is stable (McKenzie and Morgan, 1969). They would show, for example, that the Mendocino triple junction (or any FFT triple junction), is stable if the San Andreas Fault and the proposed trench formed a straight line; that any RTF(a) triple junction is stable if the trench and transform fault form a straight line; and that any TTT(a) triple junction is stable if two of the trenches form a straight line (§7.21).

I find their discussion of the evolution of triple junctions and their application of plate tectonics to the past, insightful and forward looking; however, I also find it in tension with their categorization of the theory as "only an instantaneous phenomenological theory."

The North Pacific shows the remarkable success of the paving stone theory over a quarter of the Earth's surface, and it is therefore expected to apply to the other three quarters. It is, however, only an instantaneous phenomenological theory, and also does not apply to intermediate or deep focus earthquakes. The evolution of plates as they are created and consumed on their boundaries is not properly understood at present, though it should be possible to use magnetic anomalies for that purpose.

(McKenzie and Parker, 1968: 1280)

They assumed that the rotation between two plates could be treated as a rotation about a fixed pole. They were willing to apply their paving stone theory to the evolution of the Pacific, to apply it to finite rotations of the Pacific and North American plates, and to suggest "that a belief in uniformity and the existence of magnetic anomalies will permit at least the younger tectonic events in the Earth's history to be understood in terms of sea floor spreading," even though they thought their theory was at that point "only an instantaneous phenomenological theory." However, they were confident that it could be applied to the past.

It was perfectly obvious in 1967, from the coastline fits and from the preservation of marine magnetic anomalies, that the plates behaved rigidly when they underwent finite rotations. We also knew perfectly well (as did Teddy) how to describe such motions: the relevant theory was developed by Euler, who used what are now called Euler angles to describe the finite rotations rather than poles of rotations and angles. He used Euler angles because they are more convenient for dynamical calculations than are poles. The standard first year mathematics course here uses all this theory for instance to analyse how spinning tops behave. Quaternions are simply yet another parameterization of finite rotations (in addition to poles and rotation angles, and to Euler angles). They are convenient if you want to combine two finite rotations to get the parameters describing the total rotation. There is yet another parameterization that uses Lie groups.

(McKenzie, January 8, 2012 email to author)

But, at the same time, they were, rightfully so, cautious about whether or not plates continue to rotate around the same instantaneous poles over time. Indeed, McKenzie and Morgan (1969: 131–132) would later show that it is impossible for three plates "to rotate through finite angles about their instantaneous relative rotation axes" unless all rotational vectors "lie along the same axes."

This clash between applying plate tectonics to present and to past relative movements of plates, I believe, continued throughout the development and articulation of plate tectonics. When determining instantaneous rotations, "instantaneous" will come to mean anything from an instant to several million years (Fowler, 2000: 11), and, as Morgan already has done, magnetic lineations and the trends of transform faults, at least active ones, will be used to determine instantaneous Euler poles. When workers attempted to explain evolution of plates and to explain their creation, destruction, and relative rotations over time, they proposed instantaneous Euler poles that existed for 50 million years or longer. Looking ahead and in most cases beyond the scope of this book, I quote Le Pichon, Francheteau, and Bonnin on the matter.

In practice all workers in the field had to assume that for some finite time interval, the motion between two plates could be described by a single pole of rotation: as much as 70 m.y. for Le Pichon (1968), 40 m.y. for McKenzie and Morgan (1969), Atwater (1970) and Phillips and Luyendyk (1970), 36 m.y. for McKenzie and Sclater (1971), and 10–75 m.y. for Pitman and Talwani (1972). A study of the pattern of relative motion from facture zone and spreading-rate data in the northeastern Pacific suggests that changes in the relative motion of the plates involved have occurred at least every 10-20 million years (Francheteau *et al.*, 1970a).

> (*Le Pichon* et al., *1973: 34; all references are the same as mine except
> for Francheteau* et al. *(1970a), which is my Francheteau* et al. *(1970), and
> Le Pichon (1968), which is my Le Pichon (1968a)*)

Parker, reflecting on his and McKenzie's approach as compared to Morgan's, later described the tension they felt about appealing to transform fault trends or magnetic anomalies to determine instantaneous rotational poles.

Yes, it was obvious that the rest of the world had to be looked at. But I think Dan and I were more cautious than Jason Morgan about the ability of plate tectonics to describe motions over geological extended time periods. It is a mathematical fact that instantaneous motions on a sphere can always be described by rotations about a pole, but once the motion has proceeded for a time, the pole itself must shift relative to the plates (except in very special circumstances). At the time we did not know whether plate motions were steady enough to allow one to use the transform arcs or the magnetic anomalies. The seismic data were obviously good approximations to the concept of instantaneous motions.

> (*Parker, March 15, 1990 email to author*)

In addition, there is also the realization that finite rotations cannot be represented by vectors, and therefore they cannot be added like vectors. Finite rotations have to be represented by quaternions, for which there are simple rules for "adding" them.

It is time to compare Morgan's presentation of plate tectonics with McKenzie and Parker's.

7.15 Comparison of Morgan's and McKenzie and Parker's presentations of plate tectonics

Morgan and McKenzie independently discovered, developed, and tested their versions of plate tectonics. Morgan discovered plate tectonics approximately five months before McKenzie, and worked on refining and testing it for six months before submitting his paper. McKenzie and Parker's paper appeared at the end of December, 1967; Morgan's paper, in April, 1968. Morgan's testing of plate tectonics was more extensive than McKenzie's; McKenzie's presentation of plate tectonics was conceptually cleaner. Despite the differences between their presentations, both made essentially the same monumental discovery: a precisely formulated kinematic theory that subsumes the kinematics of continental drift and seafloor spreading, the Vine–Matthews hypothesis and the idea of transform faults, the two key corollaries of seafloor spreading. By dividing Earth's outer surface into plates, most comprising both oceanic and continental lithosphere, and of sufficient rigidity to apply Euler's Point Theorem, they could explain the movements of Earth's outer surface in terms of relative rotational velocities, why almost all earthquakes occur in typically long and narrow seismic zones, why most earthquakes in the same type of seismic zones have the same mechanism, and why slip directions of earthquakes along the common boundary of two plates are roughly parallel to each other. There simply was no competing theory that solved so many important problems without creating serious difficulties. Despite differences in presentations, each made a monumental discovery.

I briefly compare their backgrounds and presentations. (1) Both came to geophysics from physics and were immediately able to see the relevance of Euler's Point Theorem. Both were convinced of seafloor spreading, and each, familiar with Wilson's idea of transform faults, extended his idea to a sphere, accenting the importance of rigidity. They differed, however, in their knowledge of map projection and their assessment of the rigidity of Earth's outer regions. McKenzie already believed in its rigidity before he thought of plate tectonics, while Morgan may not at the outset have been so convinced. Morgan, however, had Elsasser. Morgan was quite familiar with different types of map projections from his experience in the Navy. Indeed, it was this knowledge that led him to think of Euler's theorem when looking at Menard's figure of fracture zones in the northeast Pacific. McKenzie did not have such background knowledge. McKenzie, however, had Parker, who knew all one needed to know about different map projections to give McKenzie what he needed. McKenzie thought about Euler's theorem while rereading Bullard, Everett, and Smith's reassembly of the continents surrounding the Atlantic Ocean. (2) In order to test their idea kinematically, to see whether Earth's outer surface is sufficiently rigid to apply profitably Euler's theorem, both recognized that fault plane solutions, transform fault trends, and lineations of marine magnetic anomalies could be used to determine Euler poles. Morgan used all three; McKenzie restricted himself to results of fault plane solutions. McKenzie, however, used first-motion studies to determine the slip direction of strike-slip faults and the horizontal component of dip-slip faults, but Morgan, as described above (§7.11), used fault plane solutions to determine the strike of the fault on which the earthquake in question occurred. Morgan also ended up restricting his use of fault plane solutions to earthquakes along transform faults. But McKenzie used fault plane solutions to determine the horizontal component of dip-slip earthquakes along the entire common boundary of the Pacific and North American plates, and thereby determined, by using only slip directions, both coordinates of the pole of instantaneous relative rotation for the Pacific relative to North America. Morgan, however, unlike McKenzie, exploited the databases of marine magnetic profiles and trends of transform faults, and Morgan used them together to determine the latitude and longitude of various Euler poles. (3) Both recognized the anomalous situation in the northeastern Pacific. There were a few small fracture zones which were not parallel with those that separated the Pacific and North American plates. Morgan hypothesized the existence of a small plate; a plate he identified as the Juan de Fuca block, and suggested the existence of a compressive region along the coast of Oregon and Washington. Similarly, McKenzie hypothesized the existence of a trench along the coast of Oregon and into the Cascade Range, which requires that the anomalous region is itself a small plate. Their treatments of the anomalous region differed somewhat, but the differences were more a question of emphasis indicative of their having utilized different databases. Morgan developed his analysis primarily on the basis of the marine magnetic surveys and profiles in the area. McKenzie concentrated on the slip vectors of earthquakes and their apparent anomalous directions at the junction of the San Andreas and eastern part of the Mendocino transform faults. Thus, Morgan's treatment placed more emphasis on the postulation of a small rigid block; he even gave it a name. McKenzie, on the other hand, began his analysis of the area by looking at the junction of the two fracture zones and viewing it as an incompletely analyzed triple junction, a point where boundaries of three plates come together. This led him, through vectorial analysis, to hypothesize a third boundary, a trench and, therefore, to postulate another plate. (4) Morgan's testing of plate tectonics was considerably broader than McKenzie's. McKenzie restricted his discussion almost entirely to the motion of the Pacific relative to the North American plate. Morgan, on the other hand, also determined Euler poles for the rotation of Africa relative to South America, of Africa relative to both Americas, of Antarctica relative to the Pacific, and Antarctica relative to Africa by closing of the Africa–North America–Pacific–Antarctica–Africa circuit. (5) McKenzie's presentation of plate tectonics was conceptually tighter because he explicitly distinguished between instantaneous and finite Euler poles when discussing triple junctions (McKenzie and Parker, 1967: 58–59). With the distinction in mind, he and Parker correctly rejected the use of the trends of transform faults and magnetic lineations to position instantaneous Euler poles. Admittedly, McKenzie has stated that he also did not have any magnetic anomaly profiles that he could have used, but my guess is that he would have used them with extreme care because he recognized the difference between the instantaneous and finite rotational poles, and that it was an open question as to whether magnetic lineations indicated separately distinguishable instantaneous and finite Euler poles. Moreover, McKenzie had available but rejected the use of the trends of transform faults. In contrast, I believe that Morgan did not explicitly distinguish between instantaneous and finite rotations. This is not to say that he did not recognize the difference, which I suspect he did, it is only to say that he did not explicitly distinguish between them. Perhaps he thought the difference was obvious.

Both presentations were highly creative and masterful. In retrospect, I think most geologists, unaware or unconcerned with the differences between instantaneous and finite rotations, found Morgan's treatment more informative and helpful because it allowed them to see how Euler poles, regardless of whether they were instantaneous or finite, can be determined by using the trend of transform faults and magnetic lineations. In addition, Le Pichon, who certainly came to understand differences between finite and instantaneous rotations, basing his work on Morgan's presenta-tion of plate tectonics, used Lamont's huge databases of marine magnetic lineations, of fracture zones on the seafloor, and of earthquake epicenters worldwide, to develop a consistent and complete six plate model. This model allowed him to determine the relative motion of plates along consuming plate boundaries based on measurements of plate motions at producing plate boundaries and to try to determine the history of seafloor spreading during the Cenozoic (§7.17).

In closing this discussion of their versions of plate tectonics, I concentrate on their differing treatments of the relative motion of the North American and Pacific plates featuring an exchange of letters between them. I begin with two letters that Morgan wrote to McKenzie on December 8 and 9, 1967, approximately three weeks after Vine gave Morgan the preprint of McKenzie and Parker (1967) that McKenzie sent him on November 14, 1967.

Before considering the scientific content of their letters, I want to emphasize that throughout, Morgan and McKenzie behaved quite admirably toward each other. As noted above, McKenzie when he found out from Le Pichon that Morgan had presented plate tectonics at the April AGU meeting, wrote to *Nature* asking that publication of his and Parker's paper be delayed so that their paper and Morgan's appear around the same time. He also wrote Vine, asking him, albeit at Menard's suggestion, to give Morgan a copy of Parker and his paper. As we shall see, Morgan's two letters to McKenzie show that even though he was surprised about being scooped, he expressed no anger about it, even if he had been upset. After saying, "I surmise that you did not attend the AGU meetings in Washington last April," which was the obvious but incorrect inference, and which must have puzzled McKenzie because he still did not know that Morgan had presented an earlier version of his paper at the meetings, he kindly told McKenzie about relevant work by Sykes and Tobin, by Stauder, and by Le Pichon, even providing him with copies of his own paper and Stauder's unpublished preliminary results. He also closed his second letter by kindly telling McKenzie, "I look forward to your coming to Princeton next spring." McKenzie also later recalled that when he met Morgan in Princeton, "He was more friendly than I expected after what had happened, and we talked about what we had been doing" (McKenzie, March 18, 1990 letter to author).

Here are the letters. The first three paragraphs of each are almost identical. The first letter is unsigned; perhaps Morgan waited a day before posting the first letter, and mailed both in the same envelope. He also repeated the first three paragraphs in the second, perhaps to remove several typographical errors and a parenthetical expression that were in the first letter.

December 8, 1967

Dr. Dan McKenzie
Institute of Geophysics and Planetary Physics
University of California at San Diego
La Jolla, California

Dear Dan,

I surmise that you did not attend the AGU meetings in Washington last April: three papers presented there are relevant to the work you and Parker are engaged in. The first is my own paper, I have enclosed a copy of the manuscript I submitted to JGR in September. The second is the work of William Stauder, he presented several new solutions of earthquake mechanisms from the Rat Islands. I have included a copy of the list (unpublished as yet) he sent me last May; I have not communicated with him since then but he most likely has solutions for more earthquakes in this region now. He wrote that he would have a preprint available in mid-summer. I think that these mechanism solutions are most relevant to your study as they have better internal consistency than solutions published for other sections of the Aleutian, Kamchatka, Kurile boundary. Don Tobin and Lynn Sykes presented a paper, "The seismicity and tectonics of the northeast Pacific Ocean" in which several earthquake mechanism solutions from the Fairweather, Queen Charlotte, and Juan de Fuca region are presented. The location and strikes of these solutions are listed in Table 4 of my manuscript; they now have preprints available which show the closed and open dots plotted on the lower hemisphere for each of their solutions.

Xavier Le Pichon at Lamont has also worked on this problem. He has prepared transverse Mercator projections of all of the ocean basins. Each map is projected about a pole of relative rotation found for each pair of blocks; the transform faults in the appropriate ocean then plot as horizontal lines (lines of "latitude") and lines connecting a chosen magnetic anomaly feature, say feature #13, then plot as a series of vertical lines (meridians). He has determined rotation rates for all of these poles and has predicted closing rates for the boundaries that are not rises. His predicted closing rate in the Kurile region is 9 cm/yr; his pole position for the motion of the Pacific relative to North America is at 53°N, 47°W. He has also has run the clock backwards and projected where various blocks of continents were at say 40 to 80 my ago by matching chosen magnetic features on both sides of each ridge. He has not yet completed this manuscript, but a summary of this will soon appear in JGR as part of a series of Lamont papers on magnetic anomalies.

I think you are imposing upon your readers when you state that you have looked at over 80 mechanism solutions but then present only 4 as evidence in favor of rigid block motion in the Pacific. I think you are demanding that the reader place too much faith in your judgment that these four solutions are representative of the 80: this is much like presenting mechanism solutions with the open and closed circles shown in the lower hemisphere and allowing you to make your own decision as to the reliability of the result and presenting only the coordinates of the planes found – take it or leave it. I have looked at many of the 80 solutions you mention, the five figures included here show great circles plotted perpendicular to the direction of thrust of selected earthquakes. If there were absolutely no scatter and the blocks were rigid, the great circles would intersect at the "pole." The selection of earthquakes was based on four criteria: (1) the earthquakes had to be on the landward side of the trench, this eliminated a few tensional type solutions, (2) the earthquakes had to be shallower than 40 km (or was it 50? I forget), the thrust type pattern tended to break down at greater depths, (3) the dip of the thrust plane had to be less than 30° – this requirement merely reinforced the depth requirement above, and (4) the null axis could not be inclined more than 30° (?) from the surface. I now realize that this last requirement is inappropriate (it eliminates solutions like your "c" of June 14, 1962), what I should have required instead was that the fault plane of the solution was parallel to the plane determined by the location of hypocenters within a certain amount. From my figures I can attest that your four solutions are a fair representation of all of the solutions, but I think you would have a better paper if you established this fact.

I see three ways by which your pole in Ontario can be reconciled with my pole in Newfoundland. (1) My pole in Newfoundland is based on the intersection of perpendiculars from two groups of faults: the Fairweather group and the California group. If the Fairweather region is not in pure transcurrent motion – if some compression and mountain building is going on there – then the perpendiculars constructed here have no value and the pole in Ontario is correct. (2) There may be some systematic shift in the earthquake first motion if the axis of the trench is not perpendicular to the direction in which the Pacific floor is moving. As the Aleutians bend around more oblique to the direction of ocean floor motion, there does seem to be some systematic shift in earthquake first motion. Perhaps the percentage of slip parallel to the trench axis relative to the amount of thrust changes with this obliqueness. In this case the first motion solutions are of no value until this empirical correction is found and applied, and the pole is near Newfoundland as established by the faults in the Fairweather region. This choice was my decision last June when I abandoned this project to return to the more straightforward study of topographic expression of faults. (The much large scatter in the pole found by first motions as compared to topography – see figures 8(a) and (b) and 14(e) and (f) – was also a factor in this decision.)

[Morgan here put in a hand drawn sketch]

I ignored the solutions in Kamchatka since I felt they were of poorer quality. If the earthquake fault planes were initially fixed (parallel to the trench axis and with a dip indicated by the plane of hypocenters), perhaps with this restriction the motion determined by the poor distribution of receiving stations would give a better indication of the direction of motion of the Pacific floor here. (3) The two possibilities above allow a "rigid" crust, but perhaps both the Fairweather directions and the earthquake studies in the Aleutians and Kamchatka correctly show the local relative motion and the Pacific floor stretches like rubber.

I now suspect that this third alternative is the correct choice, although some correction to the earthquake solutions as outlined in (2) above may be necessary. The pole position determined from motions in California-Fairweather, California-Kodiak, California Rat Is., California-Kamchatka, and California-Kuriles appears to shift uniformly from about 53°N, 53°W to 45°N, 100°W as we move down this sequence. This pole shifts about 30°, this corresponds to about a 50% change in the angular velocity vector. This 50% change occurs over about 10,000 km, we could say the plate is not rigid at about 5% per 1000 km. With this condition on the "rigid" plate scheme, we may use it in a local region to interpret motion with great precision, but we cannot precisely predict motion in Japan based on measurements made in California.

At each point on the earth's surface we can define a velocity relative to some fixed coordinate system. If we see how the direction and magnitude of the velocity change in one direction and the direction perpendicular to this in a region around a point, we may assign an angular velocity vector to this point to describe the motion of this small region. Within any of the crustal blocks, this angular velocity slowly changes from point to point – if the block were rigid it would be constant. Discontinuous changes in angular velocity occur only at rises, trenches, and faults; the amount of discontinuity is nearly constant along the boundary between two blocks the more rigid they are. I expect to present something along this line at the next AGU meeting.

Here is the second letter.

December 9, 1967

Dr. Dan McKenzie
Institute of Geophysics and Planetary Physics
University of California at San Diego
La Jolla, California

Dear Dan,

I surmise that you did not attend the AGU meetings in Washington last April: three papers presented there are relevant to the work you and Parker are engaged in. The first is my own paper, I have enclosed a copy of the manuscript I submitted to JGR in September. The second is some recent work of William Stauder, he presented several new solutions of earthquake mechanisms from the Rat Islands. I have included a copy of the list (unpublished as yet) he sent me last May; I have not communicated with him since then but he most likely has solutions for more earthquakes in this general region now. He wrote that he would have a preprint available in mid-summer. I think that these mechanism solutions are most relevant to your study as they have better internal consistency than solutions published for other sections of the Aleutian, Kamchatka, Kurile boundary. Don Tobin and Lynn Sykes presented a paper, "The seismicity and tectonics of the northeast Pacific Ocean", in which several earthquake mechanism solutions from the Fairweather, Queen Charlotte, and Juan de Fuca region are presented. The location and strikes of these solutions are listed in Table 4 of my manuscript; they now have preprints available which show the closed and open dots on the lower hemisphere for each of their solutions.

Xavier Le Pichon at Lamont has also worked on this problem. He has prepared transverse Mercator projections of all of the ocean basins. Each map is projected about a pole of relative rotation found for each pair of blocks; the transform faults in the appropriate ocean then plot as horizontal lines (lines of "latitude") and lines connecting a chosen magnetic anomaly feature, say #18, then plot as a series of vertical lines (meridians) in these transverse Mercator maps. He has determined rotation rates for all of these poles and has predicted closing rates for the boundaries that are not rises. His predicted closing rate in the Kurile region is 9 cm/yr; his pole position for the motion of the Pacific relative to North America is at 53°N, 47°W. He has also run the clock backwards and projected where various blocks of continents were at say 40 and 80 my ago by matching chosen magnetic features on both sides of each ridge. He is still working on the implications of rigid block motion, but a summary of this will soon appear in JGR as part of a series of Lamont papers on magnetic anomalies.

Last June I looked at mechanism solutions of thrust earthquakes in the Alaska – Kurile region, but then shelved this project to return to the more straightforward study of topographic expression of faults. I think the motion of the Pacific Block is more complicated than you indicate with the four mechanism solutions; there is a systematic change in the strike of earthquakes and the motion predicted by rotation all along the Alaska to Kurile boundary. A student and I are using this variation to measure the 'non-rigid' or rubber-like properties of the crustal blocks.

I look forward to your coming to Princeton next spring.

Sincerely,
Jason Morgan

Morgan's mentioning of the work by Stauder, Tobin and Sykes, and Le Pichon (§7.17) was indeed pertinent. Sykes and Tobin, who knew of Morgan's but not McKenzie's work on plate tectonics, not only allowed Morgan to use their own Figure 2 (Tobin and Sykes, 1968: 3834) of epicenters of earthquakes along the Denali and Queen Charlotte Islands faults, the Juan de Fuca and Gorda ridges, and the Blanco and Mendocino fracture zones to the west of the Gorda Ridge as the basis of his own Figure 2 (Morgan, 1968b: 1969) showing the Juan de Fuca as a separate block, but also argued that their analysis of approximately 260 earthquakes, supported Morgan's (and by the same reasoning McKenzie's) view "that earth's surface is composed of a number of rigid blocks."

In the marine areas adjacent to the northwestern United States most of the relocated epicenters are associated with several of the large fracture zones and ridges (Figure 2). Most of the remaining marine areas (with the exception of the region between the Gorda ridge and the coast of northern California) are exceptionally quiet seismically, a fact that argues in favor of Morgan's [1968] contention that the earth's surface is composed of a number of rigid blocks with relatively little or no deformation within the individual blocks.
        (Tobin and Sykes, 1968a: 3832; the reference to Morgan is the same as mine)

McKenzie, as seen above, already had made good use of Stauder's previous results, and Stauder's new results were also pertinent. Stauder submitted the paper in question just in time to argue that his results not only supported seafloor spreading, the idea of transform faults, Sykes work confirming transform faults, and Isacks and Oliver's work showing subduction of rigid plates, but also supported the ideas of McKenzie and Parker, and Morgan about "the tectonics of the entire North Pacific" (Stauder, 1968a: 3853).

Stauder also wrote McKenzie on January 29, 1968, letting him know that his results are "in rather good agreement" with his rotational pole.

January 29, 1968

Dr. Daniel P. McKenzie
Institute of Geophysics and
Planetary Physics
University of California
La Jolla, California 92037

Dear Dr. McKenzie:

I am enclosing a copy of a paper I have completed on the Rat Island earthquake of 1965. I received your note and the preprint of the paper by yourself and Ron Parker just as I was completing my own work. I have included a reference to your paper. The motion in the Rat Island sequence is in rather good agreement with your pole.

Best wishes.
Sincerely yours,
William Stauder, S. J.

In a second paper submitted in July 1968, Stauder concluded that additional focal mechanism studies of earthquakes which occurred under the Aleutian trench were in "satisfactory" agreement McKenzie and Parker's pole.

Comparison of the motion expected for this pole of rotation [proposed by McKenzie and Parker] with that indicated by the focal mechanisms here examined throws further light on the relation between the motion inferred for individual foci and the moving plate hypothesis. Figure 4 [which I have not reproduced] presents this comparison. The large arrows positioned to the south of the trench on the map (Mercator projection) indicate local directions of motion corresponding to the pole at 50°N, 85°W... The single arrows in Fig. 4 drawn at the representative foci on the inner portion of the arc indicate the direction of motion of the underthrust oceanic plate as determined by the focal mechanisms. The agreement with that expected for the indicated pole of rotation is satisfactory. It is also noteworthy that while the tectonic motion becomes more and more oblique to the trend of the arc as one progresses from the eastern to the western extremity of the Aleutian Islands, the motion remains that of a plate bending and being thrust under the islands as a tabular body.

(Stauder, 1968b: 7700–7701; my bracketed additions).

I now return to Morgan's letters to McKenzie. Morgan's complaint in the first letter that McKenzie should have shown more of his mechanism solutions is fair and correct. McKenzie, I suspect, probably agreed with Morgan upon reading his letter. As already noted (note 14), he later came to regret that he had not shown the dip directions of many more of the earthquakes he had analyzed. It must have been encouraging to both McKenzie and Morgan that Morgan agreed that the solutions McKenzie did show "are a fair representation of all of the solutions." They also had no significant disagreement about the criteria used to select earthquakes. But, as we know, their poles differed: McKenzie and Parker's pole for the rotation of the Pacific relative to North America was 50°N, 85°W, in Ontario just south of Hudson Bay; Morgan's pole was further east at 53°N, 53°W, just east of Newfoundland. Morgan offered three explanations for the discrepancy. The trends of the faults in the Fairweather group that Morgan used to position the pole do not reflect pure strike-slip movements because they are modified by compression and mountain building which alters their trend slightly too far to the west of northwest. As a result, perpendiculars to them do not intersect further west in Ontario but just east of Newfoundland (see Figure 7.18). Indeed, given the density of intersections of the perpendiculars of the trends of the California transform faults, this reason for the discrepancy, I believe, makes sense. However, I am not sure if it is correct. Regardless, Morgan rejected it. Morgan's second reason is more complicated, and he included a sketch to help explain what he had in mind. Again, I am not sure if I fully understand what he had in mind. It might be that Morgan wanted to determine the strikes of earthquakes from focal mechanism studies while McKenzie used them to determine slip directions. Morgan did use the strikes of earthquakes as one way to determine the pole position for the motion of the Pacific plate relative to the North American one, while McKenzie used the slip directions to position the pole. In general, the horizontal component of the slip direction of any earthquake is normal to the strike of the auxiliary plane. So the strike of the horizontal component of the slip direction can be determined by adding 90° to the strike of the auxiliary plane (Fowler, 2000: 101). Thus, with pure strike-slip earthquakes the slip direction is parallel to the fault plane because the fault plane is perpendicular to the auxiliary plane. Consequently, with pure strike-slip earthquakes Morgan, functionally speaking, determined the same result as McKenzie. With pure thrust earthquakes, perpendiculars to both the auxiliary and fault planes plot parallel to each other, and therefore the horizontal component of the strike-slip vector is perpendicular to the auxiliary plane *and* the fault plane (Le Pichon *et al.*, 1973: 69). However, in oblique situations, those with both slip and dip movements, the slip direction is normal to the auxiliary plane but not to the focal plane. Thus, Morgan's attention to the strike of fault planes rather than to the slip direction of thrust faults led him to conclude mistakenly that the fault plane solutions in Kamchatka were unreliable and to abandon analyzing earthquakes beneath trenches and island arcs common to the North American–Pacific plate boundary. Although this explanation makes sense, it is not correct. Morgan actually wanted to get slip directions; he just did not know how to get reliable ones from dip-slip earthquakes.

I knew what the slip directions meant but I thought when it was on a dipping plane [the downgoing slab] that a correction needed to be made or else systematic errors would result and I didn't see how to make the correction.

(Morgan, February 8, 2012 email to author)

Morgan wrote to Stauder ten days after he wrote to McKenzie, and claimed quite rightly that unreliable results are worthless. Besides kindly including a copy of his own paper, two figures from Le Pichon's forthcoming paper, and two figures from McKenzie and Parker's paper, and noting the discrepancy between their rotational pole and his and Le Pichon's pole, which was much closer to his own, he restated his idea that fault plane (first motion) solutions of the thrust faults in question "are of no value" until empirically corrected.

There appears to be two possible ways to reconcile the directions obtained from thrust fault earthquakes and surface faults. There may be some systematic shift in the earthquake first motion if the axis of the trench is not perpendicular to the direction in which the Pacific floor is moving. Perhaps the percentage of slip motion parallel to the trench axis which is radiated in first motion compared to the amount of thrust radiated changes with this obliqueness. In this case the first motion solutions are of no value until this empirical correction is found and applied.

(Morgan, December 19, 1967 letter to Stauder).

In contrast with Morgan, McKenzie, who realized that he could reliably determine the horizontal component of the dip-slip vector of thrust earthquakes from their auxiliary plane, was able to determine the slip vector of earthquakes along most of the entire boundary of the North American and Pacific plates because the auxiliary planes of such earthquakes were well determined. With a good read on the auxiliary plane, he calculated the direction of horizontal slippage of the earthquakes, found them parallel to each other, and knew he was right.

Morgan's third way of reconciling the difference between his and McKenzie and Parker's positioning of the Euler pole was to accept the difference as real and to give up the idea that the Pacific plate is perfectly rigid, proposing in his first letter to McKenzie that the "Pacific floor stretches like rubber," that "the plate is not rigid at about 5% per 1000 km." At the time, this was clearly his first option, for he restated it in his December 19 letter to Stauder, even doubling his estimated reduction to 10%.

I now suspect that the directions from the Fairweather fault and the earthquake studies in the Aleutians and Kuriles correctly show the local relative motion and that the Pacific floor stretches like rubber . . . [The enclosed] Xerox #5 shows great circles constructed perpendicular to "typical" directions from earthquakes in this region: no single pole fits this data. The pole positions determined from motions in California-Fairweather, California-Kodiak, California-Rat Is., California-Kamchatka, and California-Kurile Is. appear to shift uniformly from about 53°, 53°W to 45°N, 100°W as we move down the sequence. This pole shift of 30° (a 50% change in the angular velocity vector describing the relative motion of the two blocks) occurs in about 5000 km; we could say the blocks are not rigid about 10% per 1000 km. With this condition on the "rigid" plate scheme, we may use it in a local region with precision, but we cannot precisely predict motion in Japan based on measurements made in California.

(Morgan, December 19, 1967 letter to Stauder)

Morgan also sent Le Pichon a copy of McKenzie and Parker's paper. Le Pichon replied on December 13, 1967. After expressing his surprise that McKenzie and Parker "did not seem to know about" Morgan's paper, he agreed with Morgan that plates may not be perfectly rigid but, unlike Morgan, suggested that the North American plate was less than perfectly rigid.

December 13, 1967

Dr. W. J. Morgan
Geology Department
Princeton University
Princeton, New Jersey 08540

Dear Jason,

Thank you for your letter and the McKenzie-Parker paper. I am a little bit surprised that they do not seem to know about your paper. The main objection I have to their work is that it may be a dangerous assumption that the whole continental system eastern Asia, Aleutian, North America, is perfectly rigid. If there is slow deformation of this system, then you might expect the results they get. It seems to me that the spreading floor evidence suggests that the North Pacific was larger in Cretaceous than it is now or that several thousand km of shortening have occurred in Asia since this time. I prefer the first solution. Maybe the actual pole is somewhere between McKenzie's position and yours.

I have a second version of my paper typed now. It has been reviewed within Lamont. My problem is that it has grown out of proportion. I will send you a copy when it is ready, probably before Christmas.

I am returning to France before the Spring, so I am rather anxious to have it cleared before I leave. I intend it for "Review of Geophysics."

Sincerely,
Xavier Le Pichon

Looking ahead, it turns out that McKenzie and Parker's positioning of the *instantaneous* Pacific–North American Euler pole is closer than that of Morgan to the current estimate of 49°N, 72°W (Argus, Gordon, and DeMetes, 2011: Table 3). There also has been no need to ease up on the rigidity of the Pacific plate, which is almost entirely without continental components. Le Pichon, Francheteau, and Bonnin, looking back at Morgan's and Le Pichon's positioning of the pole and why they differed so much from McKenzie and Parker's, suggested a reason reminiscent of Morgan's first explanation that he proposed in his letters to McKenzie and Stauder.

The America/Pacific pole was determined by Morgan (1968b) on the basis of azimuths of strike-slip faults in the complex ridge-transform fault system extending from the Gulf of California, through the San Andreas system, to the Queen Charlotte Island and Denali systems. The intermediate Gorda, Juan de Fuca and Explorer accreting ridges define a small Juan de Fuca plate which probably underthrusts the North American continental margin (Tobin and Sykes, 1968) in a direction and at a rate not yet properly determined. Consequently, the transform faults between these three portions of accreting ridges were not taken into account in the determinations of the Am/Pa pole. Even so, it is not clear whether the movement along these continental faults is pure strike-slip and a component of thrust faulting may exist in places. This would partly explain the difference between the original Morgan and Le Pichon's determination. The latter authors used the fault-plane solution method along the whole Pacific-American boundary from California to the Kurile Trench. Stauder (1968a, b) has shown that the fault-plane solutions along the Aleutian Trench are in general agreement with McKenzie and Parker's pole.

(Le Pichon et al., 1973: 86–87; references are equivalent to mine)[18]

Despite their different views of how fault plane solutions should properly be used to position instantaneous Euler poles, which itself adds support to the independence of McKenzie's discovery of plate tectonics, both had essentially the same, great overall idea. Their next contribution to the theory, the evolution of triple junctions, would be a joint effort (§7.21).

7.16 Isacks spearheads discovery of how deep earthquakes beneath island arcs are caused

I begin with three comments: the first is by Isacks, the second, by McKenzie and the third, by Peter Molnar.

The Tonga Fiji focal mechanism story developed rapidly in 1967, although this is not reflected in the published literature until my 1969 paper [Isacks, Sykes, and Oliver (1969)]. The key results of that paper I had in hand in 1967 and I gave at a talk in at the XIV General Assembly of the International Union of Geodesy and Geophysics held in Zurich in the fall of 1967. The two key ideas that I presented there included (1) the interpretation of deep earthquake focal mechanisms as indicating the stress orientation within the descending plate and (2) the shallow earthquake focal mechanisms beneath Tonga indicate the direction of slippage (given by the "slip vector") between the descending plate and the upper plate.

(Isacks, October 25, 2011 email to author; my bracketed addition)

A more important application [of Anderson's account of faulting] is to deep earthquakes. Isacks (Isacks *et al.*, 1968 and personal communication) has demonstrated that the P axes of intermediate and deep focus earthquakes in the Tonga-Fiji and Kermadec regions are approximately parallel to the dip of the plane containing the earthquake. Such clustering is more obvious for the P axes than for any other axes, though there is a weak orientation of T at right angles to the dipping plane. This result would not be expected if the earthquakes were caused by slip on pre-existing fault planes. It is best explained by failure of a homogeneous material with little internal friction. The observed fault-plane solutions therefore require the greatest principal stress to lie in the plane containing the earthquakes and to be directed down the dip. The weak orientation of T is also explained if the intermediate stress lies along the strike of the plane.

(McKenzie, 1969c: 600)

Anyhow, when I got my first good idea, to use fault-plane solutions of intermediate and deep-focus earthquakes to study the "driving mechanism of plate tectonics," I promptly learned that Bryan [Isacks] had already been pursuing this. A kind man, he allowed me to share the study with him, and I think I did contribute to it, although not as much as he did.

(Molnar, 2001: 294, my bracketed addition)

Isacks took the lead on this project to determine focal mechanism solutions of deep and intermediate earthquakes. During the first stage of the project, he undertook first motion studies of earthquakes, determined focal mechanisms for deep and shallow earthquakes in the Tonga–Fiji and Kermadec regions, combined his results with those obtained by seismologists working on focal mechanism solutions from other island arcs, especially from the region near Japan, and established that deep and shallow earthquakes occur *within* the descending lithospheric slab. Finding that the *compressional* axes of every non-shallow earthquake but one were parallel to the dip of the seismic zone, Isacks and company argued that "deep earthquakes occur within the downgoing slab in response to a compressional stress within it (Isacks *et al.*, 1969: 1443). During this first stage, Isacks worked with Oliver and Sykes. During the second stage, he worked with Peter Molnar, a graduate student at Lamont. They first extended the original study to over a dozen trench–island arcs and trench–continental margins (Isacks and Molnar, 1969), and later added about a dozen more (Isacks and Molnar, 1971). Besides confirming that the compressional axes of all deep earthquakes and some intermediate earthquakes were parallel to the dip of the seismic zone, they also found that the *extensional* axes of most intermediate earthquakes were parallel to the dip of the seismic zone. Plotting the down-dip extensional and compressional stresses of each descending lithospheric slab in terms of depth, they found that compressional axes of deep earthquakes were parallel to the dip of the seismic zone below 300 km, while extensional axes of intermediate earthquakes were prevalent from 70 to 300 km. Appealing again to Elsasser's view that lithospheric plates acted as stress guides, they explained the resulting pattern by proposing that extensional stresses accrue within descending slabs as they sink through the non-resisting asthenosphere, and that compressional stresses accumulate once the descending slab reaches denser lower regions of the asthenosphere or even the mesosphere, where further sinking is greatly hindered or even prohibited.

Soon after Oliver and Isacks began working in the Tonga–Fiji and Kermadec regions and identified the anomalous zone as a downgoing lithospheric slab, they (1967: 4273) realized that "the anomalous zone coincided with the zone of seismic foci," and began to speculate about the cause of the non-shallow earthquakes whose foci were in the seismic zone. The cause of deep earthquakes was a longstanding problem, which became even more puzzling given Isacks and Oliver's hypothesis that the ductile asthenosphere of the upper mantle offers little resistance to the sinking of lithospheric slabs. Drawing on Sykes's (1966) and ongoing work in locating precisely the foci of earthquakes in both regions, they speculated that the crustal material dragged down on the upper surface of the descending lithosphere might in some way be involved.

Early in this investigation it appeared that the anomalous zone coincided with the zone of seismic foci. When the most recent data of L. R. Sykes (personal communication) are considered, however, it appears that the seismic zone, which was thought to be 50 to 100 km in thickness, may, in the case of the deep shocks of the Tongan arc, be less than 20 km in thickness . . . The data further suggest that the foci are located near the upper boundary of the anomalous zone, i.e., near the upper surface of the lithosphere. This is also the region in which crustal materials would be found if they were carried down into the mantle. Thus, the presence of such materials in the mantle may be the key to understanding the nature of the focal mechanism of deep earthquakes. In fact, it is possible to speculate that *all* earthquakes are confined to, or associated with, materials that are normally a part of the crust or, perhaps, the uppermost mantle . . . In any event, it is a general conclusion of this paper that deep earthquakes occur where they do at least partly because the material of the mantle is different there than it is at comparable depths elsewhere. Just what process or processes are involved is not clear at present. A detailed study of radiation patterns and focal mechanisms of shocks in the Tonga arc is currently under way [Isacks and Sykes, 1967].

(Oliver and Isacks, 1967: 4273–4274; Isacks and Sykes, 1967 morphed into Isacks, Sykes, and
Oliver, 1969)

From this it is thus clear that when Oliver and Isacks submitted their paper in April 1967, they already had become interested in discovering the cause of deep earthquakes, had no solid idea about what it was, and that Isacks and perhaps Sykes had already begun determining focal mechanisms of deep earthquakes in the Tongan and Kermadec regions.

Isacks recalled that "much work was done during the summer and early fall of 1967," that he first presented his results at the 14th IUGG meeting in Zurich held during the last week of October and first week of November, and that "the key ideas" were that the slip vectors of shallow earthquakes "reflect the directions of slippage" *between* "adjacent plates," while "deeper earthquake mechanisms reflect the directions of stress *within* . . . descending" plates (my added emphasis).

The IUGG meeting in Zurich in the fall of 1967 is the main pin. I'm supposing that much work was done during the summer and early fall of 1967. The key ideas of this paper were that shallow mechanisms reflect the directions of slippage across the plate boundary (the "slip vectors") and thereby indicate the directions of relative motion of the adjacent plates, while the deeper earthquake mechanisms reflect the directions of stress within the descending plate.

(Isacks, October 25, 2011 email to author)

Unfortunately the meeting's proceedings were not published and his abstract quite understandably did not reflect what he said. He may have seen an announcement of the meeting in the December 1966 issue of the *Bulletin of the Seismological Society of America*. In any case, he submitted his abstract before he had "pinned down" the cause of deep earthquakes.

I submitted the abstract before I had the story pinned down. It was a common practice to submit abstracts which are often due many months in advance of the meeting as kind of place holders if the results were not pinned down well (no one wanted to give a talk basically negating what was in the abstract!). The abstract outlined the area of study and purposes, etc. I don't remember what was in the IUGG abstract, but I think it fell into the "place holder" category.

(Isacks, October 7, 2011 email to author)

Isacks' fault plane solutions did not appear until 1969 when they were presented in Isacks, Sykes, and Oliver, 1969, which was not submitted for publication until October 1968, more than a year after the IUGG meeting. Presumably, given the mention of "Isacks and Sykes, 1967" in the above quoted passage from Isacks *et al.*, 1968, Isacks and Sykes originally planned to publish the results in 1967. Why the change in plans and long delay? Because, in part, they and Oliver decided soon after Isacks returned from the meeting to write their monumental paper "Seismology and the new global tectonics," which they submitted in May 1968. With this change of plans, they included an initial interpretation of Isacks' findings. In addition, even though Isacks started working on the project before he knew of Morgan's and McKenzie and Parker's discovery of plate tectonics, and Le Pichon's extension of Morgan's ideas, he knew of them while he, Oliver and Sykes were working on Isacks *et al.*, 1968. As result, they interpreted his results in terms of plate tectonics.

Isacks determined focal mechanisms for eighteen shallow, six intermediate, and fifteen deep-focus earthquakes in the Fiji–Tonga–Kermadec region. Here is his summary of his findings as of October 1968 when Isacks *et al.*, 1969 was submitted:

The orientations of the mechanisms of deep and shallow earthquakes appear to be systematically and fundamentally different in respect to the orientation of the seismic zone. Whereas the shallow mechanisms all appear to accommodate movements between the adjacent sides of the seismic zone, the slip planes of the deep earthquake mechanisms are systematically nonparallel to the deep seismic zone. Hence the deep zone of activity does not appear to be a simple thrust fault. The P [compressional], B [null] and T [extensional] axes of the double-couple solutions tend to parallel the dip, strike, and normal directions, respectively, of the portions of the Tonga seismic zone deeper than about 80 km. The P axes tend to be more stable in orientation than the B and T axes. Large variations in the orientations of some of the deep mechanisms may reflect contortions of the deep seismic zone in a simple geometrical fashion. The shallow mechanisms indicate that thrust faulting is occurring beneath the inner (islandward) margins of the Tonga and Kermadec Trenches and that transform faulting is occurring at the northern end of the Tonga Arc . . . These results, which are also in agreement with mechanism data for other regions such as Japan where both numerous and reliable data are available, are most simply interpreted by a tectonic model of an island arc in which (a) shallow earthquakes occur between a segment of lithosphere that moves downward into the mantle and the segments of lithosphere on the surface, and (b) deep earthquakes occur within the downgoing slab in response to a compressional stress within it.

(*Isacks*, et al., *1969: 1443*)

Except for several shallow earthquakes at the northern end of the Tonga arc where it bends abruptly westward, all shallow earthquakes were beneath the inner margin of the Tonga and Kermadec trenches and all their fault-plane solutions were dip-slip. They resolved the fault plane-auxiliary plane ambiguity by selecting as the fault plane the one more parallel to the dip of the seismic zone, which also turned out to be the less steep nodal plane. This choice indicated that the Pacific plate is being overridden by the Australian plate. The trends of the slip vectors, roughly parallel with each other, were "remarkably consistent with one another," which was consistent with the existence of a common pole of instantaneous rotation between the two plates.

The orientations of the mechanism of the shallow earthquakes are such that a simple interpretation of the zone as a large-scale zone of thrust faulting can be made. In most cases, one of the nodal planes is approximately parallel to the seismic zone; in all cases, one nodal plane is clearly more parallel than the other nodal plane. These nodal planes can therefore be assumed to be the slip planes . . . these slip planes are the nodal planes which are more nearly horizontal than vertical; they also tend to dip westward. The auxiliary planes so chosen are nodal planes

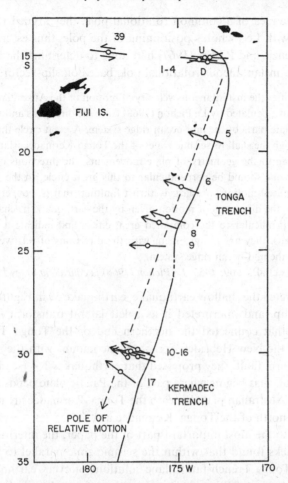

Figure 7.29 Isacks *et al.*'s Figure 5 (1969: 1453). "Interpretation of the nature of the faulting at shallow depths in the Tonga–Kermadec region. The arrows through the epicenters of earthquakes 5–17 are azimuths of the slip vectors, and they show the motion of the eastern side of the island arc relative to the western side. The strikes of the nearly vertical nodal planes are shown for earthquakes 1–4, and the vertical motions inferred are shown by U for up and D for down. The strike of the inferred fault plane and sense of strike-slip motion are shown for earthquake 39. The solid line marks the axis of the Tonga and Kermadec . . . ."

that tend to be more nearly vertical than horizontal; they tend to dip steeply eastward. The poles of the auxiliary planes give the directions of the slip vectors, that is, each slip vector is located in the shallow dipping plane and is perpendicular to the auxiliary plane. In each case, the sense of motion is such that the eastern or oceanic side, is thrust beneath the western or islandward side of the shallow seismic zone. The plunges of the slip vectors vary from about 0° to about 45°. Thus slip vectors are nearly parallel to the dips of the shallow parts of the inclined seismic zone. The trends of these slip vectors are shown in Figure 5 [Figure 7.29 below]; they are remarkably consistent with one another.

(Isacks, 1969: 1453; my bracketed addition)

Turning to the idea of a common rotational pole, they argued that their results were consistent with Le Pichon's positioning of the pole, thus essentially using slip vectors as McKenzie and Parker (1967) had done to determine the longitude of the Australia–Pacific instantaneous rotational pole based on slip-vectors.

The pole that specifies the instantaneous velocity of motion of the Australian plate relative to the Pacific plate was calculated by Le Pichon (1968) from the directions and rates of spreading along the appropriate parts of the mid-ocean ridge system. A great circle that passes through this pole and through the shallow seismic zones of the Tonga-Kermadec island arc is shown in Figure 5. According to the geometry of plate movements, the directions of relative motion along the seismic zone should be perpendicular to this great circle. In the Tonga-Kermadec region, these computed motions are largely thrust faulting (that is, convergent motion). As shown in Figure 5, the directions of motion given by the earthquake mechanisms are indeed approximately perpendicular to the computed great circle, and indicate a predominance of thrust faulting. Thus, they are in agreement with the directions of relative motion that are extrapolated from the mid-ocean ridge system.

(Isacks et al., 1969: 1455; Le Pichon (1968) is equivalent to my Le Pichon (1968a))

They also interpreted the shallow earthquake, earthquake 39 in Figure 7.29, as strike-slip, not strike-dip, and interpreted it as a left lateral transform fault an arc-arc transform fault that connected the northern end of the Tonga Trench with the southern end of the New Hebrides Trench. Concomitant with their postulation of an arc-arc transform fault, they proposed that earthquakes 1–4 (see Figure 7.29) are along a hinge fault that begins where part of the Pacific plate bends downward as it underthrusts the Australian plate beneath the Tonga–Kermadec arc and is torn from the Pacific plate north of the Tonga–Kermadec arc.[19]

Turning now to the most important part of the paper, the intermediate and deep earthquakes, Isacks found that within the seismic zone parallel to the main linear segment of the Tonga Trench fault plane solutions of every earthquake showed P axes parallel to the dip of the seismic zone with horizontal or vertical nodal planes (see Figure 7.30). The P axes of earthquakes in the deep zone associated with the sharp westward bend of the Tonga Trench were likewise parallel to the dip of the seismic zone. Thus, focal mechanism solutions of all deep earthquakes associated with the Tonga Trench had slip planes that were non-parallel to the dip of the seismic zone and axes of compression that were parallel, indicating, they argued, that deep earthquakes occurred within the descending plate in response to internal compressional stresses, not to stresses between the descending and overriding plates or between descending plate and ductile asthenosphere (Figure 7.30). They found further support for their interpretation by examining thirteen reliable fault plane solutions by Japanese seismologists Honda and others (1956, 1967), Ichikawa (1966), and Aki (1966a, b) of earthquakes in the Izu-Bonin and North Honshu arcs in the region near Japan; except for an earthquake located at the intersection of the two arcs, all solutions were just like Isacks' solutions of deep

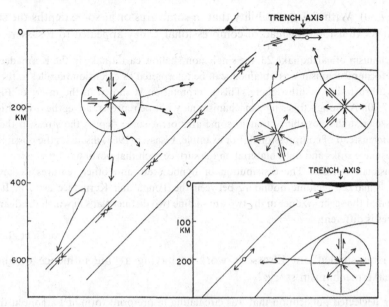

Figure 7.30 Isacks *et al.*'s Figure 10 (1969: 1460); see also identical figure (Figure 11) in Isacks *et al.* (1968: 5874). "Vertical sections perpendicular to the strike of an island arc showing schematically typical orientations of double-couple focal mechanisms. The horizontal scale is the same as the vertical scale. The axis of compression is represented by a converging pair of arrows; the axis of tension is represented by a diverging pair; the null axis is perpendicular to the section. In the circular blowups the sense of motion is shown for both of the two possible slip planes. The features shown in the main part of the figure are based on results from the Tonga arc and the arcs of the North Pacific. The insert shows the orientation of a focal mechanism that could indicate extension instead of compression parallel to the dip of the zone [earthquake 23, the only non-shallow earthquake Isacks found in the Kermadec arc]" (Isacks *et al.*, 1969: 1460; my bracketed addition). The top far right circular blowup shows the mechanism of an earthquake below the trench axis. It has a dip-slip orientation and was interpreted as a normal (extensional) fault type of earthquake. Such earthquakes sometimes occur seaward of the thrust fault type earthquakes (whose mechanism is shown in the neighboring top circular blowup). Stauder discovered such extensional earthquakes beneath the Aleutian trench (§7.15). Isacks did not find them in the Tonga–Kermadec region, but kept them in Figure 10 because it was supposed to represent a "typical" arc area (Isacks *et al.*, 1969: 1456). Stauder (1968a, b), Isacks *et al.* (1968: 5869–5870) and Isacks *et al.* (1969: 1456) suggested that such earthquakes are caused by tensional stresses on the convex side of lithospheric slabs at the point where they bend downward beneath the trench.

and intermediate earthquakes associated with the main part of the Tonga Trench (Isacks *et al.*, 1969: 1459–1460).

   Isacks did find, however, one true exception. He found that the focal mechanism solution of the only non-shallow earthquake in the Kermadec arc, earthquake 23, had, as expected, either vertical or horizontal nodal planes, but its *extensional* axis, not its compressional axis, was parallel to the dip of the seismic zone (see insert,

Figure 7.30). With this "possibility that in some arcs or at some depths the stress in the slab is extensional instead of compressional," they appealed to Elsasser.

The mechanism of earthquake 23, the single non-shallow earthquake in the Kermadec arc for which a focal mechanism was obtained, can be interpreted as extension parallel to the dip for the downthrust slab of lithosphere. This interpretation is shown in the insert of Figure 10 [Figure 7.30]. Although this single mechanism may not be representative, the orientation does at least suggest the possibility that in some arcs or at some depths the stress in the slab is *extensional* instead of compressional. For example, Elsasser (1967) discusses the possibility that the lithosphere sinks and pulls material on the surface such that portions of the slab are subject to extensional stresses. The distribution of hypocenters and other features (Sykes, 1966) indicate a marked tectonic boundary between the Tonga and Kermadec arc . . . It is thus possible that the seismic zones in the two arcs define two distinct slabs in which the distribution of stresses is different.

(Isacks, et al., 1969)[20]

Isacks later singled out Elsasser's work as having a "big influence on the deep earthquake mechanism story."

An often neglected publication that was circulating in pre-print form at Lamont at the time, and which had a big influence on the deep earthquake mechanism story, was Elsasser's 1967 paper (Elsasser, W. M., 1967, Convection and stress propagation in the upper mantle: Technical Report No.5, June 15, 1967, Princeton University, Princeton, New Jersey, 65 p.)
   (Isacks, email to author; his parenthetical addition; his Elsasser, 1967, is my Elsasser, 1967b.)

Elsasser (1967b) maintained that the lithospheric slabs (his "tectosphere") act as stress guides because stresses (forces) act on them over large horizontal distances, and because they are rigid. Moreover, these slabs can only act as stress guides "because [they] can slide readily on the underlying aesthenosphere." Here is how Elsasser put it in one of his Princeton technical reports mentioned above.

On the whole, the existence of strong long-distance correlations . . . is a patent fact. It is obvious that such correlations must be the expression of a system of forces; in a solid this takes the form of stresses. These stresses act over very large distances such as the width of the Atlantic and apparently even across the Pacific. But stress must somehow propagate . . . We can hardly conceive of the stresses inferred as propagating through the deep interior of the mantle . . . We may conclude safely that stresses are transmitted horizontally, along the Earth's curvature, in the topmost layers, that is, in the tectosphere. The tectosphere acts as a stress guide. This is possible only because it can slide readily on the underlying aesthenosphere.

(Elsasser, 1969b: 237)

Speculating about the nature of stress next to trenches, stresses caused by gravitational forces, Elsasser continued as follows:

Compressive stress will occur at the flanks of ridges and tensile stresses adjacent to trenches; it is only when the two act simultaneous on a piece of the tectosphere that motion of the latter can occur . . . Little can be said about the tensile stresses since little is known. Material, which by various process has been made denser and slides down under about 45° along the

Gutenberg fault zones [Wadati–Benioff zones], can either be "pushed" down or it can sink down on its own by having acquired a sufficiently greater density than surrounding mantle material. Several arguments suggest that the second of these alternatives is applicable. The motion along the 45° zones then leads to a tensile "pull" in the adjacent horizontal part of the tectosphere by simple mechanical means which we shall not analyze here.

(Elsasser, 1969b: 243)

Moreover, Elsasser was himself impressed with Lamont's seismologists, for by 1969 he praised Sykes' work on confirming the idea of ridge-ridge transform faults, and Oliver and Sykes for having "given impressive seismological evidence to the effect that the slanting descent under about 45° . . . is a realistic picture" (Elsasser, 1969b: 235). During the second stage of Isacks' work on fault plane solutions of non-shallow earthquakes within subduction zones, he and Molnar would find many more examples of intermediate earthquakes whose extensional axes are parallel to the dip of the seismic zone, and would continue appealing to Elsasser.

Molnar attended Oberlin College, where in 1965 he received an undergraduate A.B. degree in physics. Deciding to study geophysics, in particular seismology, he decided to go to Columbia, hoping to study with Oliver. Molnar's father, an experimental physicist and administrator at Bell Laboratories had been impressed with Oliver while both served on the Berkner panel, engaged with nuclear test ban talks (Molnar, 2001: 291). Arriving in 1965, it took him a year before he "was ready to take my field of seismology seriously" (Molnar, 2001: 291). In fall 1966, he took a seminar from Oliver and Isacks on deep earthquakes, heard Heirtzler discuss his group's work confirming V-M and seafloor spreading, and, more importantly for him, heard Sykes discuss his work confirming ridge-ridge transform faults. During the next year he worked with Oliver on determining that discontinuities of the lithosphere occur only beneath accretion plate boundaries and behind subduction zones, and with Sykes on determining the boundaries of the Cocos and Caribbean plates and their motion relative to surrounding plates (Molnar and Oliver, 1969; Molnar and Sykes, 1969). Both papers were submitted in 1968. The paper with Sykes definitely prepared Molnar to work with Isacks on intermediate and deep earthquakes. Molnar and Sykes determined fault plane solutions for many earthquakes, including four of intermediate depth that appeared to have their extensional axes parallel to the dip of the descending plane: Two (events 144 and 154) were in the Mid-America Trench and they (1969: 1662) said of one of them, "Unlike most intermediate-depth earthquakes in Tonga and Japan, the axis of tension is parallel to the inclined zone of earthquakes." The two others (events 125 and 126) were in the Bucaramanga nest of Colombia, and they (1969: 1660) said of them, "If a seismic zone dips east here, these mechanisms imply that the axis of tension is parallel to the dip of the zone." Thus, they not only found possibly four intermediate earthquakes with their axis of tension parallel to the inclined zone of earthquakes, but they also noted that such earthquakes were unlike those found in Tonga and Japan. They also found hinge faulting, similar to that which Isacks *et al.* (1969: 1659) had discovered at the northern end of the Tonga Trench. Isacks thinks that Molnar began working with him near the end of 1968, about the time when Molnar was finishing up his work with Oliver and Sykes. Isacks also believes that he was primarily responsible for the key ideas, "although the many interactions with Peter greatly helped focus and sharpen the story and get it out in a timely fashion."

Peter Molnar, a graduate student then, worked with me to expand the focal mechanism study world-wide. We divided the various regions up to get the focal mechanisms done. I'm supposing I started working with him the latter part of 1968. The basic idea of that paper was to relate the global pattern of predominantly down-dip tensional stresses at intermediate depths and compressional stresses at greater depths to the varying effects of gravity on the denser descending lithosphere, mantle viscosity, and mantle phase changes. I take major credit for these ideas, although the many interactions with Peter greatly helped focus and sharpen the story and get it out in a timely fashion.

(Isacks, October 10, 2011 email to author)

Molnar, I suspect, agrees, for as noted at the beginning of this section, he recalled, "he [Isacks] allowed me to share the study with him, and I think I did contribute to it, although not as much as he did."

Isacks and Molnar wrote two papers. The first, a short 1969 paper in *Nature*, in which they presented a simple two-dimensional model that accounted for most of their key findings, and a lengthy 1971 paper in which they presented their data showing many of their fault plane solutions, and introduced a more complete three-dimensional model which allowed them to introduce additional factors and thereby account for findings that they could not explain with their two-dimensional model. The *Nature* paper was received on July 9, 1969; the second paper was not submitted until a year later. I consider the first paper in some detail, but deal only briefly with the second because it is beyond the scope of this volume.

They summarized their results as follows:

The main result of this survey is that the stress in those portions of the mantle seismic zones that have a relatively simple inclined planar configuration, that is, in those portions removed from remarkable contortions or changes in trend, is often oriented such that either the axis of maximum compressive stress or the axis of minimum compressive stress is approximately parallel to the dip of the inclined seismic zone. We find clear evidence for down-dip extensional stress within slabs at intermediate depths in at least six regions. These regions seem to be also characterized by prominent gaps in the seismicity at depths between about 300 and 500 km or by an absence of earthquakes deeper than about 300 km. On the other hand, down-dip compressional stress is predominant at depths greater than 300 km in all regions studied.

(Isacks and Molnar, 1969: 1121)

The most important of the new findings was that extensional stresses characterized the intermediate section of six out of the fourteen reported island arcs and continental margins studied. Molnar had already discovered that the intermediate portion of Middle America's seismic zone was extensional at intermediate depths. This

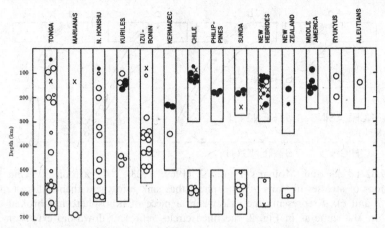

Figure 7.31 Isacks and Molnar's Figure 2 (1969: 1123). "Down-dip stress type plotted as a function of depth for fourteen regions . . . A filled circle represents an orientation such that the axis of tension T is within 20–30 degrees of the local dip of the zone, that is, down-dip extension; an unfilled circle represents an orientation such that the axis of compression P is within 20–30 degrees of the local dip of the zone, that is, down-dip compression; and the Xs represent orientations that satisfy neither of the preceding. Smaller symbols represent less reliable determinations. The enclosed rectangular areas approximately indicate the distribution of earthquakes as a function of depth by showing the maximum depths and the presence of gaps for the various zones. In addition to the references listed in the caption to Fig. 1 [Isacks *et al.*, 1969; Stauder, 1968b; Sykes *et al.*, 1969; Adams, 1963; Katsumata and Sykes, 1969; Hamilton and Gale, 1969; Honda *et al.*, 1956; Hirasawa, 1966; Ritsema, 1966; Katsumata, 1967; Molnar and Sykes, 1969; Engdahl and Flinn, 1969] data from Barazangi and Dorman [1969] and Guttenberg and Richter [1954] were used for each of the fourteen regions. The zones are grouped (from left to right) according to whether the zone is continuous to depths of 500–700 km, discontinuous with a gap between intermediate depth and deep earthquakes, or continuous but reaching depths less than 300–400 km" (my bracketed additions).

discovery of extensional-type mechanisms at intermediate depths, which had not been observed by Isacks in the Tonga Trench or Honda, Masatsuka, and Ichikawa (1966), or Aki in North Honshu, is what allowed them to adopt with more justification than previously Elsasser's proposal that lithospheric slabs behave like stress guides, that the negative buoyancy of the descending part of a slab causes it to sink, creating tensional stresses as its hangs in the asthenosphere supported from above by the rest of the slab.

Isacks and Molnar introduced two exceptionally clear figures, one summarizing their results (Figure 7.31), and another illustrating their two-dimensional model (Figure 7.32).

Once a downgoing slab reaches a depth of about 650–700 km, the increasing strength of the mantle impedes progress and it stops sinking. Its support transfers from above to below and it becomes subject to gravitational compressional stress. If the slab breaks, the severed part keeps sinking until it is supported from below and

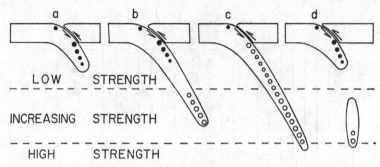

Figure 7.32 Isacks and Molnar's Figure 3 (1969: 1123). "A cartoon showing possible distributions of stresses in slabs of lithosphere that sink into the asthenosphere (*a*) and hit bottom (*b* and *c*). *d* represents the case where a piece of lithosphere has broken off. The symbols are the same as in Fig. 2; the filled circles represent down-dip extension and the unfilled circles represent down-dip compression. The size of the circles qualitatively indicates the amount of activity at the respective depths. In *b* and *d* gaps in the seismicity would be expected. Also shown is the underthrusting and the extensional stresses near the upper surface of the slabs due to the bending of the slab beneath the trenches. These features are inferred from the mechanisms of shallow earthquakes [Isacks *et al.*, 1968]. The lower boundary where the slabs hit bottom might correspond to the transition region or discontinuity near 650–700 km [Anderson, 1967; Engdahl and Flinn, 1969]" (my bracketed additions).

becomes subject to compressional stress. Its top part is left hanging and remains subject to gravitational extensional stress. Elegant.

The model offered explanations for every region but the Aleutians, Ryukyus, and Marianas. Middle America is explained in terms of scenario *a*; Tonga, N. Honshu, and Izu-Bonin, in terms of scenario *c*; Kermadec, Chile, the Philippines, Sunda, New Hebrides, the North Island of New Zealand, in terms of scenario *c* or *d*, depending on whether the gap in seismic activity arises because no earthquakes occurred from 1962 through part of 1968, the time span of the experiments, or because the bottom part of the downgoing lithosphere broke away from the top part. The two-dimensional model did not explain all the data.

Further comparisons of Figs. 2 and 3 show that the correlation breaks down in certain regions. In particular, the down-dip compressional stresses in the Aleutians and Ryukyus, the presence of both compressional and extensional stresses at intermediate depths in the Kuriles and New Hebrides, and the solutions denoted by "X" are not explained by the two-dimensional models of Fig. 3. These discrepancies . . . indicate the unsurprising result that the two-dimensional model of Fig. 3 does not explain all the data. Certainly, contortions and disruptions of the lithospheric plates would be expected to affect the stresses, although a search for such effects does not yield any simple relationship between the mechanism orientations and the contortions of the zones.

(Isacks and Molnar, 1969)

Moving to three dimensions, they gave several examples of what they had in mind, one offered an explanation of the compressional stresses characteristic of the

Ryukyus and Aleutians, and also found in the New Hebrides; another, for the "mechanisms represented by Xs."

The down-dip compressional stresses present in the Aleutians, the New Hebrides and the Ryukyu arc (Fig. 2) are located where the curvature of the arc is appreciable. In each case the T axis, instead of the B axis [axis of intermediate compressive stress] is nearly parallel to the strike of the seismic zone; the orientation of the extensional stress might result from the type of deformation suggested by Stauder [1968b]. In the less arcuate portion of the New Hebrides the T axes are predominantly parallel to the dip of the seismic zone and the B axes are parallel to the strike. Many of the mechanisms represented by Xs in Fig. 2 may reflect other unresolved contortions of the downgoing slab or other sources of stress such as thermal gradients within the plates or local density variations.

(Isacks and Molnar, 1969: 1124; my bracketed additions)

They closed their discussion with what amounted to an endorsement of Elsasser's sinking lithospheric slabs.

Nevertheless, the consistent pattern shown in Fig. 2 for zones that are relatively uncontorted offers support of the interpretation that gravitational body forces are a major source of stress in the lithosphere. If we make this interpretation, then the results require that in many regions the slabs are sinking and are exerting a downwards pull. The results also require the complications of Fig. 3 in which the sinking slabs are supported from above or below, or both. Thus if we use the results to extract information on the forces that drive the global plate movements, we can infer two important effects: (1) the downgoing slabs can exert a pull on the surface portions of the plates, and (2) as the downgoing slabs "hit bottom" beneath the lithosphere the downward pull on a surface plate is significantly decreased. These interpretations suggest that the pull of the descending slabs may be an important contribution to the driving forces of global tectonics.

(Isacks and Molnar, 1969: 1124)

In their second paper, which was not published until February 1971, they did what they said had to be done: they examined regions that did not fit or entirely fit their two-dimensional model on a case by case basis. They also presented many of their 204 reliable focal plane solutions, and expanded their coverage to twenty-four regions, adding "regions in the Alpide belt such as Burma, Hindu Kush," Rumania, Aegean, Calabria, and island arcs such as the Sandwich Islands, Solomon Islands, New Britain, Mindanao, and the trench-continental by Peru. Moving to three dimensions allowed them to consider the variations in strike of the descending plate, and they introduced factors such as a bent edge, a contortion, and a lateral extension of the descending tongue to explain stresses at depths inconsistent with the basic two-dimensional model. They also documented extensively the work of others, again giving special thanks to work of two other Lamont seismologists, Muawia Barazangi and James Dorman, whose 1969 "maps of epicenters within of 100-km depth were very helpful" (Isacks and Molnar, 1971: 126). Barazangi and Dorman (1969) sharpened Gutenberg and Richter's maps so that much that could not be seen with Gutenberg and Richter's maps of seismic activity worldwide maps became clear.

This work by Isacks, both with Oliver and Sykes and with Molnar, finally explained how non-shallow earthquakes occur. In addition, they showed that even though Benioff (1949, 1955) and Raitt *et al.* (1955) were convinced that crustal material descended into the mantle beneath ridges, their account of a huge mega-thrust zone where intermediate and deep focus earthquakes presumably occur between the downgoing material and the medium through which it descends was mistaken. They also gave an additional reason to reject rapid Earth expansion (§7.18).

7.17 Le Pichon loops plate tectonics around the world

I apparently was the only one to have considered . . . [Morgan's 1967 AGU spring talk] sufficiently important to drop everything else and start working along these new lines . . . I decided immediately to test this kinematic approach, in spite of the skepticism of my colleagues, who considered it more important to continue to decipher the magnetic anomalies. I had to elaborate a rather complex methodology and a system of computer programs, which kept me busy until July. I could then verify that each of the different rift openings behaved according to spherical geometry: plates (as they were later to be called) were indeed rigid, and Morgan was right. Part of my work got incorporated into the 1968 paper by Heirtzler and colleagues on magnetic anomalies and crustal motion [Heirtzler *et al.* (1968: 2129–2134)].

(*Le Pichon, 2001: 215; my bracketed additions; Heirtzler* et al. *(1968) is discussed in §7.7*)

This has never been mentioned. But I wish to emphasize that it was accomplished alone, without the help of anyone, except for the extended outline of Morgan and the suggestion by Heirtzler to use Mercator plots. And that this was very unusual at Lamont. Mine was really a solitary work, between the AGU and October, mostly at night, alone at the computer, followed by the write up of this long paper in November and December. In particular I had to do all the programming to play with both instantaneous and finite rotations and to plot the different projections. And this was in a context where none of my colleagues were really interested in what I was doing and in which they were thinking, at least in the beginning, that I was wasting my time. As if I should have been working on magnetic anomalies instead! The fact that I did not find anybody who wanted to collaborate was most unusual at Lamont. There is no other major paper with a single author coming from Lamont at that time, at least in this general domain. Thinking back of that time, I still marvel that I had sufficient drive to carry this project to the end in six months.

(*Le Pichon, January 20, 2012 email to author*)

Le Pichon is proud of what he did. Indeed, he should be. He was the only one in attendance at Morgan's talk who understood it, recognized its importance, and most importantly, did something about it. Moreover, he did it without, at least initially, the encouragement of his colleagues at Lamont. Le Pichon, as we have seen, had tremendous confidence in his abilities, and did not seem to doubt the conclusions he reached. He behaved as such while a fixist, and again once he had become a mobilist. In this regard he was much like Tuzo Wilson. Both vehemently rejected mobilism, but, once they changed their minds, vehemently accepted mobilism, and immediately produced work of fundamental importance which extended mobilism's explanatory power.

To understand what Le Pichon aimed to do and what he did do in taking Morgan's version of plate tectonics around the world, let me review what Morgan did because it is particularly relevant for understanding Le Pichon. Morgan used three accreting plate boundaries to determine the relative motions of three pairs of plates: the southern part of the Mid-Atlantic Ridge from the Azores to Bouvet Island to determine the rotational velocity of America relative to Africa; the western coast of North America from the entrance to the Gulf of California to the northern extension of the Fairweather–Queen Charlotte system of transform faults to determine the motion of the Pacific plate relative to the North American, and the Pacific–Antarctic Ridge to determine the motion of the Antarctic block relative to the Pacific block. He then closed the loop by summing the rotational velocities so obtained to determine the motion of the Antarctic block relative to the African block – what he called predicting the Antarctica–Africa pole from closure of the Africa–North America–Pacific–Antarctica–Africa circuit (see his Figure 4; Morgan, 1968b: 1979). He (1968: 1980) deduced that Africa and Antarctica are "moving apart," which was consistent with observations. This result, of course, provided additional support for Morgan's model. Nonetheless, Morgan did not determine the relative rotational velocity of plates at consuming plate boundaries. Moreover, Morgan used spreading rates and the strikes of transform faults determined topographically or from the strikes of strike-slip earthquakes occurring along them. He needed to combine both the transform-fault and spreading-rate methods to position the Africa–America pole, used the transform faults to determine the Pacific–America pole, and found basically the same Antarctic–Pacific pole using either method.

Le Pichon wanted, as he later put it, "to test whether [Morgan's] model was internally coherent" (Le Pichon *et al.*, 1973: 80; my bracketed addition). He needed to obtain a unique solution; he needed to close every loop, he needed to determine rotational poles between all adjacent plates, which required finding the instantaneous relative rotation vectors between plates at every common boundary. Once done, he could then compare his results with observations of the present relative movements of his selected plates at their common boundaries. He found he could obtain a unique solution by using six large plates: *Eurasia* (which also included the Philippines and China), *India* (which also included Arabia), *Pacific*, *Antarctica* (which also included Cocos and Nazca), *America* (which included both Americas and the Caribbean), and *Africa* (which included Somalia) (see Figure 7.35). This is not to say that he thought there were only six plates. He knew there were more. Le Pichon and company later explained the criteria he used in constructing his "six-plate model".

One of the most important assets of plate tectonics is that it can provide a precise global kinematic model of the surface of the earth. The model can integrate the deformations at the various plate boundaries, as expressed in geological phenomena which have occurred in the last few millions of years (Neotectonics).

The first problem is to define the number of plates necessary to describe a useful global model. The question is not how many plates presently exist at the surface of the earth. This question probably will never be answered satisfactorily, as the answer depends on the criteria one adopts for the minimum size of a plate and minimum relative movement along plate boundaries. The question rather is what is the minimum number of plates necessary to define a satisfactory global model. It is clear that it is possible to insert additional smaller plates between parts of the boundaries of two larger plates. Yet, the general pattern of motion elsewhere will not be affected by this modification. This suggests that the minimum number of plates should be such that all pairs of adjacent plates have in common a boundary along which their relative movement can be measured with precision. Consequently, the movement of one plate with respect to any other of the chosen set of plates will be accurately described. In addition, for the model to be realistic, the plates chosen should cover a major proportion of the surface of the earth.

(Le Pichon, et al., 1973: 80)

Le Pichon followed Morgan's basic procedure. Like Morgan, he used spreading rates and the strikes of transform faults as determined topographically, to position Euler poles. Unlike Morgan, he did not use the strikes of earthquakes occurring along transform faults, and unlike McKenzie and Parker, he did not use slip vectors of earthquakes. Like Morgan, he considered the poles so derived to be instantaneous rotational poles, thus treating "instantaneous" as extending to several million years. Using Lamont's database, he began with five instead of three accreting plate boundaries. Like Morgan, Le Pichon used the Pacific–Antarctic Ridge to determine the motion of Antarctica relative to the Pacific block; the Mid-Atlantic Ridge south of the Azores to determine the motion of the American block relative to Africa; the North Pacific, from the entrance to the Gulf of California through the northern extension of the Fairweather–Queen Charlotte system of transform faults to determine the motion of the Pacific block relative to the American block. Unlike Morgan, Le Pichon also used the Carlsberg Ridge to determine the motion of the Indian relative to the African block, and the Mid-Atlantic Ridge north of the Azores and Arctic Ridge (Gakkel Ridge) through the Lena River delta to determine the motion of the Eurasian relative to the American block. When determining the instantaneous Euler poles between the respective blocks, Le Pichon, like Morgan, was able to use *both* fracture zones and spreading rates (magnetic lineations) to determine the poles between the Pacific/Antarctica blocks, and, unlike Morgan, between the America/ Africa blocks. Because the methods are independent and gave consilient results, he provided an excellent confirmation of plate rigidity. He determined the other poles, like Morgan, based on the trends of transform faults alone (see Figure 7.33). Just as McKenzie and Parker were doing, Le Pichon displayed his results by using oblique Mercator projections, a suggestion Heirtzler made to him (see Figure 7.34). Le Pichon finished this first stage of his project by the end of August. At the time, neither Morgan nor he knew about McKenzie and Parker, and Morgan did not know what Le Pichon was doing.

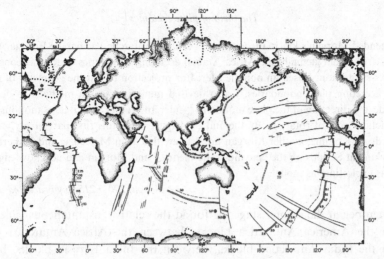

Figure 7.33 Le Pichon's Figure 1 (1968a: 4663). "Available data on seafloor spreading. The axes of the actively spreading mid-ocean ridges are shown by a double line; the fracture zones by a single line; anomaly 5 ($\approx$ 10 m.y. old) by a single line; the active trenches by a double dashed line. The spreading rates are given in centimeters per year. [See his Figures 7.5 and 7.6 for position of anomaly 5 magnetic lineations of SA, NP, SP, IO; see Figure 7.6 for its position in the geomagnetic timescale.] The locations of the centers of rotation obtained from spreading rates are shown by X; those obtained from the azimuths of the fracture zones by +. NA stands for North Atlantic; SA for South Atlantic; NP for North Pacific; SP for South Pacific; IO for Indian Ocean; A for Arctic. The ellipses drawn around the NA, NP, SP, and A centers of rotation obtained from the fracture zones are the approximate loci of the points at which the standard deviation equals 1.25 times the minimum standard deviation. The ellipse around the IO center of rotation is too small to be shown. These ellipses indicate how fast the least-squares determination converges" (my bracketed addition).

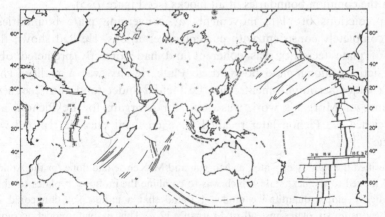

Figure 7.34 Le Pichon's Figure 2 (1968a: 3666-3667). "The pole projection is . . . the South Pacific center of rotation (69° N, 157° W). The continent outlines, ridge crests and fracture zone locations, and the location of anomalies 5, 18, and 31 were digitized from the original of Figure 1, which was traced on the World Map of the U.S. Navy Oceanographic Office (scales of 1/ 39 000 000 at the equator). [As noted in caption to Figure 7.33, see Figures 7.5–7.7 for specified anomalies in different contexts.] The outlines of the continents north of 80° N or south of 70° are not as accurate, being tracked by hand in each case from an atlas" (my bracketed addition).

I first extended Morgan's kinematic analysis . . . to the Antarctic/Pacific, the Eurasia/America, and the Africa/India [actually the Africa/Arabia] accreting boundaries to test his concept. On Heirtzler's suggestion, I used an oblique Mercator projection to test the geometry of opening of these accreting plate boundaries. I also devised numerical search methods to define the magnitude and direction of the plate motion as "Eulerian vectors" – that is, as motions around a hypothetical pole of rotation. By the end of August 1967, this first part of my work was completed. At the time, neither Morgan nor I knew that Dan McKenzie and Robert Parker were working at Scripps on the "paving-stone theory," and Morgan had no knowledge either that I was exploiting his model.

(*Le Pichon, 2001: 215; my bracketed addition*)

He then began the second stage: he found the relative instantaneous Euler poles between the America–Antarctica blocks, between the Africa–Antarctica blocks, between the India–Antarctica blocks, between the India–Eurasia blocks, between the India–Pacific blocks, between the Eurasia-Pacific blocks, and between the Africa– Eurasia blocks.

The resulting instantaneous rotational vector between two blocks can be obtained from the geometrical or vector sum of the rotational vectors of the blocks. These can be obtained from the five rotational vectors previously determined. For example, the America-Antarctica vector is the sum of the Pacific-Antarctica and America-Pacific vectors, and the India-Antarctica vector is the sum of the four vectors: Pacific-Antarctica, America-Pacific, Africa-America, and India-Africa [see Figure 7.35]. The more vectors that are involved in the sum, the greater is the probable inaccuracy.

(*Le Pichon, 1968: 3676; my bracketed addition*)

He had his unique solution; he had determined the twelve relative rotational poles between the common boundaries of all blocks (see Figure 7.35).

His predictions of plate movements at converging plate boundaries were almost completely consistent with current observations. He had shown that the theory of plate tectonics was coherent and had correctly predicted observed seismic activity along plate boundaries. Plate rigidity was a reality. This was indeed a major accomplishment. That Le Pichon understood and recognized the importance of Morgan's work was good for Morgan, for Le Pichon, and for Earth science. Le Pichon later recalled his excitement once he realized what he had done.

Once I verified the rigidity of plates, as Morgan and McKenzie had done for the Atlantic and Pacific, I moved to the next state, which was to combine the motions of plates to obtain the first predictive global quantitative model. I found that a unique solution could only be obtained by using six plates instead of Morgan's 12 ... This six-plate model accounted for most of the world seismicity, as Bryan Isacks and his colleagues would later show [Isacks *et al.*, 1968; see §7.18]. Even now it is difficult for me to forget my extraordinary excitement the day I realized that my six-plate model worked, and that it could indeed account as a first approximation for the broad geodynamic pattern. I remember coming home early in the morning for breakfast after a night at the computer and telling my wife: "I have made the discovery of the

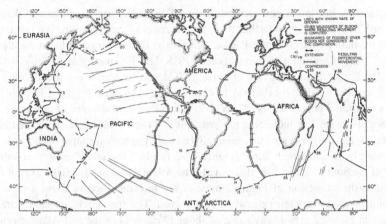

Figure 7.35 Le Pichon's Figure 6 (1968a: 3675). "The locations of the boundaries of the six blocks used in the computations." The number vectors represent computed differential movements. 1–6 are on a common consuming Eurasia–Pacific boundary, calculated at a converging rate of 8.15×10^{-7} deg/yr with their rotational pole at 67.5° S, 138.5° E; 7–11 are on a common consuming India–Pacific boundary, calculated at a converging rate of 12.3×10^{-7} deg/yr with their rotational pole at 52.2° S, 169.2° E; 12–19 are on a common consuming America–Antarctica boundary, calculated at a converging rate of 5.44×10^{-7} deg/yr with their rotational pole at 79.9° S, 40.4° E; 20–22 are on a common consuming America–Pacific boundary, calculated at a spreading rate of 6.0×10^{-7} deg/yr with their rotational pole at 53° N, 47° W; 23 and 24 are on a common consuming America–Eurasia boundary, calculated at a spreading rate of 2.8×10^{-7} deg/yr with their rotational pole at 78° N, 102° E; 25 and 26 are on a common spreading Africa–Antarctica boundary, calculated at a spreading rate of 3.24 $\times 10^{-7}$ deg/yr with their rotational pole at 42.2° S, 13.7° W; 27 and 28 are on a common spreading India–Antarctica boundary, calculated at a spreading rate of 5.96×10^{-7} deg/yr with their rotational pole at 4.5° S, 18.1° E; 29–32 are on a common consuming Africa–Eurasia boundary, calculated at a converging rate of 2.46×10^{-7} deg/yr with their rotational pole at 9.3° S, 46.0° E; 33–37 are on a common consuming India–Eurasia boundary, calculated at a converging rate of 5.50×10^{-7} deg/yr with their rotational pole at 23.0° N, 5.2° W.

century." Well, I was young and my enthusiasm carried me too far. But this statement is a good indication of how we felt during those days of frantic discoveries.

(Le Pichon, 2001: 216)

Le Pichon also realized that he had an elegant refutation of Earth expansion. If Earth is expanding and keeps its basic shape, it must expand at roughly the same rate along every radius. Every great circle (circumference) should increase at roughly the same rate. Given that the axes of the rotational poles of the South Pacific (Antarctic–Pacific), the North Pacific (America–Pacific), and Atlantic (America–Africa) are roughly the same, the great circles of their spreading equators are roughly the same. Given the combined rates of spreading along the three ridges, Earth, if there is no subduction, is expanding at an approximate rate of 17 cm/yr. However, the rates of expansion along any other great circle vary between 0 and 7 cm/yr. Thus Earth expands only if its shape changes.

Without entering into other considerations, we will consider first the pattern of deformation of the Earth's surface, assuming that there are no zones of compression. We can then easily find with the data [Antarctica and Pacific rotate away from each other at an angular rate of 10.8×10^{-7} deg/yr around a rotational pole at 70° S, 118° E; America and Pacific rotate away from each other at an angular rate of 6.0×10^{-7} deg/yr around a rotational pole at 53° N, 47° W; America and Africa rotate away from each other at an angular rate of 3.7×10^{-7} deg/yr around a rotational pole at 58° N, 37° W] how much the circumference of any great circle on the surface of the earth is increasing per year. Ideally, each great circle should expand at the same rate, so that the earth's surface maintains its nearly spherical shape. Such an idea can be tested rapidly by considering Figure 2 [Figure 7.34], in which the axis of projection is the axis of rotation of the South Pacific; it is also close to the Atlantic and north Pacific axes of rotation. Note that in the projection of Figure 2 most of the spreading occurs about axes which run north-south, whereas very few ridge axes run east-west. The great circle formed by the equator of rotation in Figure 2 expands at a rate of about 17 cm/yr (4 at the mid-Atlantic ridge, 12 at the East Pacific rise, and 1.5 at the Carlsberg ridge.) On the other hand, the great circles of longitude of Figure 2 are parallel to the crests of the Atlantic and Pacific ridges and would at most intersect the Arctic and South Indian Ocean ridges. The resulting rates of expansion vary between 0 and 7 cm/yr. There is no evidence of important pole migration during the last 10 m.y., yet the equator of rotation [of the three axes of rotation along the spreading ridges of North and South Pacific and Atlantic] would have increased its circumference by 1700 km and some of the great circles of longitude would not have expanded. This implies a differential expansion of as much as 270 km between the average radius along extreme great-circles circumferences and possibly as much as 500 km between individual radii. It is unacceptable, and it becomes even more so when we recognize that the present pattern of spreading has prevailed during the whole Cenozoic era. Consequently, in the expansion hypothesis, we have to assume some compensating large-scale process of earth's surface shortening by compression or thrust to maintain the nearly spherical surface of earth. The expansion hypothesis then loses most of its appeal. Other strong arguments have previously been advanced against the expansion hypothesis [e.g., Runcorn, 1965], and it will not be considered further here.

(Le Pichon, 1968: 3674; my bracketed additions, except for reference to Runcorn's paper, which is equivalent to my Runcorn, 1965b)[21]

Once Le Pichon had fixed the common instantaneous Euler poles of his six plates, he, as Morgan put it in his December 8, 1967 letter to McKenzie, ran "the clock backwards." When trying to figure out what to do, he also learned that finite rotations do not behave like instantaneous ones: finite rotations are not commutative. Vine (1966), Pitman and Heirtzler (1966), Pitman *et al.* (1968), Dickson *et al.* (1968), and Le Pichon and Heirtzler (1968) had already unraveled with varied success the history of the ocean basins in terms of seafloor spreading, using marine magnetic anomalies and, as an auxiliary aid, land-based paleomagnetic studies. In addition Heirtzler *et al.* (1968) had also developed a geomagnetic timescale back 80 million years (see §6.6, §6.8, §7.7). Le Pichon now went through the same exercise of tracking magnetic anomalies, and matching them from ocean to ocean, but he now performed the exercise in terms of Euler's Point Theorem, as Bullard *et al.* (1965) had done, and in terms of the Euler poles that he already had determined during the first two stages of his overall project.

Finally, I made the first kinematic reconstruction of the evolution of the surface of the earth based on magnetic anomalies. To do this, I had to fit the magnetic anomalies that had identical ages on both sides of the rift in the same way as Bullard and his co-authors had done to fit the continental margins on both sides of the Atlantic [Bullard *et al.*, 1965]. This was the beginning of a paleogeographic method, which has proven to be especially powerful. The fit of the anomalies was done on the computer and involved combining rotations that were no longer small but could reach several tens of degrees. Small rotations can be treated as vectors, whereas this is not true of large ones, which must be treated as matrices. Not knowing that, it took me some time to discover the origin of large discrepancies in my early computations. The rules of spherical geometry were poorly known at the time among geophysicists.

(Le Pichon, 2001: 217; my bracketed addition)

Le Pichon argued in favor of episodic spreading, agreeing with the Ewing brothers that seafloor spreading had been episodic (§6.14).

On the basis of seismic reflection studies, Ewing and Ewing [1967] suggested that the sediment distribution could be explained much more easily in terms of an episodic spreading [see §6.14]. A discontinuity in sediment thickness near anomaly 5 (10 m.y. in the Heirtzler *et al.* time scale) was interpreted as indicating an interruption of spreading possibly covering all of Miocene time. Ewing *et al.* [1968b] further suggested that the sedimentary structures in the Atlantic basins and the western Pacific implied some important reorganization of spreading, possibly accompanied by a large interruption, most probably at the Mesozoic-Cenozoic boundary [see § 7.7]. Thus the study of sediment distribution led Ewing *et al.* to hypothesize three main episodes of drifting: (1) Mesozoic, during which the basins were formed; (2) early Cenozoic, during which most of the mid-ocean ridge area was created; and (3) latest Cenozoic, during which the crestal regions appeared. Langseth *et al.* [1966], to explain the distribution of heat-flow values, had also suggested that spreading was episodic [see §6.10].

(Le Pichon, 1968; his references are respectively equivalent to my Ewing, J. and Ewing, M., 1967;
Heirtzler et al., *1968; Ewing, J.* et al., *1968; Langseth* et al., *1966); my bracketed additions to*
sectional references)

Thus he definitely rejected Heirtzler's timescale based on the assumption of a constant spreading rate in the South Atlantic. Heirtzler and company had acknowledged that their timescale would have to be altered if the Ewing brothers were correct about episode spreading (§7.7). Le Pichon (1968: 3680) now argued that it should be so altered, even though Heirtzler and company had not recognized a systematic change in anomaly patterns and had thought it unlikely that seafloor spreading had simultaneously and completely stopped worldwide. He now thought that the Ewing brothers' work plus his own examination of spreading during the Cenozoic indicated episodic spreading which he explicitly proposed

by giving to anomaly 32 an age of 60 m.y. (early Paleocene) instead of 77 m.y. and by placing an interruption of spreading about 10 m.y. long at anomaly 5, preceded by a general slowing down of the movement. This time scale would not violate any of the core or sediment distribution data. It would also explain the change in magnetic pattern without advocating some over-all change in the average reversal periodicity of the geomagnetic field between early and late Cenozoic times. This adjusted time scale does not disagree with the time scale of Heirtzler *et al.* by more than 17 million years.

(Le Pichon, 1968: 3681)

Assuming as later noted (Le Pichon *et al.*, 1971: 34) that the motion between two plates could be described by a single pole of rotation for as long as 70 m.y., he proposed three episodes of seafloor spreading.

We have attempted to apply this concept [of plate tectonics developed by Morgan (1968b)] to obtain a reconstruction of the history of spreading during Cenozoic time. Three main episodes of spreading are recognized – late Mesozoic, early Cenozoic, and late Cenozoic. The beginning of each cycle of spreading is marked by the reorganization of the global pattern of motion. A correlation is made between slowing of spreading at the ends of the two previous cycles and paroxysms of orogenic phases. This history of spreading follows closely one advocated by Ewing *et al.* [1968] to explain the sediment distribution.

(Le Pichon, 1968: 3663; my bracketed additions)

Later, after Maxwell *et al.* (1970) showed that results from the Joint Oceanographic Institutions for Deep Earth Sampling (JOIDES) vindicated Heirtzler and his time-scale and demonstrated continuous spreading in the South Atlantic, Le Pichon rejected episodic spreading (Le Pichon *et al.*, 1971:63). As we shall see, Le Pichon was not the only one at Lamont who continued to entertain the Ewing brothers' idea that the most recent episode of seafloor spreading began 10 m.y. ago, for Isacks, Oliver, and Sykes also thought there was something to the idea (§7.18).

Once Le Pichon finished the paper he wrote Morgan on December 13, 1967,

"I have a second version of my paper typed now. It has been reviewed within Lamont. My problem is that it has grown out of proportion. I will send you a copy when it is ready, probably before Christmas. I am returning to France before the Spring, so I am rather anxious to have it cleared before I leave." By this time, I had just decided to go back to France and I was on leave in early February 1968. This would have the consequence that I would be cut off from the Lamont data bank and would not be able to work on the development of plate tectonics for the next two years.

(Le Pichon, 2001: 219–220)

He also asked M. Ewing if he want to be a co-author. Recalling what happened, Le Pichon later wrote:

It was in this context [of my having been offered a position at MIT] that I had shown my paper to Maurice Ewing and asked him whether he wanted to be an author, as was usually the case in Lamont for papers based on data collected there. He declined and told me: "This is your work; publish it alone." He may have done this because he wanted me so much to stay at Lamont. Alternatively, he may have refused to go against his fixist ideas. In any case, this is how

I became the sole author of the most important paper of my career, which was very unusual at Lamont at the time, especially for a young scientist.

(Le Pichon, 2001: 220)

He waited to submit it until after Morgan's paper had been accepted.

I waited to submit it until Morgan's paper was accepted, in order to respect his priority. Morgan's paper was delayed three months by Menard's review and could have been published in December 1967 instead of March 1968 if Menard had immediately accepted it as Wilson and Oliver later did mine. It also could have been published in abbreviated form in June or July, had Morgan decided then to publish a cleared-up version of his extended outline. I had more luck than he had. My reviewers, Wilson and Oliver, recommended immediate acceptance of my paper. The title would be "Sea floor Spreading and Continental Drift."

(Le Pichon, 2001: 220)

Le Pichon also asked Morgan to review the paper. Morgan sent his review, again as requested by Le Pichon, directly to Anderson at *JGR*. On January 9, 1968, he wrote a strongly positive review to Anderson. He made two major suggestions.

Dear Dr. Anderson,

Xavier Le Pichon telephoned me about two weeks ago and asked me to review this manuscript. He said I should correspond directly with you as he had already discussed this paper with you. Please forward a copy of my comments to him.

I think this is a very good paper which should be published right away. I have several suggestions for major changes, but these could as well be incorporated in a later paper.

1. I don't like the use of the assumption "The direction of spreading generally occurs perpendicular to the ridge axis." Then in the South Atlantic (page 16) where the model and data disagree, it is claimed that this disagreement is partly artificial (resulting from this assumption). I think it would be much better to use the strike of the ridge and the direction of spreading (from a trial pole) in all of the calculations, so that exceptions like this would not need to be made and so that the data would generally better agree with the model.
2. It is not clear to me exactly how the least square fit of spreading rate and fracture zone data was made. A more explicit description (perhaps in an appendix) would be helpful.
3. I have several questions of style, there seem to be too few capitals in "south Pacific ocean" – I would prefer "South Pacific." Also the names "Atlantic-Indian Rise" and Southeast Indian Rise" as used on maps by the Oceanographic Office and National Geographic seem preferable to "southwest mid-Indian ocean ridge" and "southeast mid-Indian ocean ridge". What are JGR's editorial styles for these names?
4. Other comments appear in the margins of the manuscript.

Sincerely,

W. J. Morgan

Le Pichon appreciated Morgan's suggestions, adopting (1) and (2). On January 24, 1968, he wrote Morgan, letting him know what he had done, and thanking him for his "helpful criticisms."

Dear Jason:

O. Anderson gave me last Monday the copy of your review of my paper. In view of your criticisms, I decided to revise my method of computing center of rotation from spreading rates and insert a small appendix as you suggested. I join a copy of this appendix and of the corrected tables I and II. In case you want to compare with your result, I have inserted the values of the assumed azimuths along which the spreading rate was measured.

As you will see it makes no difference for the South Pacific but makes a large difference in longitude (not in latitude) for the Atlantic. The main advantage of this method is to provide an independent estimate of the maximum spreading rate. The main problem, for me, was that I used other people's determinations of spreading rates so that I am not completely sure of the azimuth of the line along which they measured it and of the consistency of the method of measurement. This is why I always used the determinations from the fracture zones. However, it is clear that in the Atlantic the center of rotation of the part south of the equator is about 10° farther north than the part north of it. This is independently shown by the distance to anomaly 18 and 31 in the South Atlantic which indicates a center of rotation near 70° N (My figure 2). I do not know what to make out of it as this difference is not between North and South America but between the northern and southern part of South America.

I had to rush the whole thing as I am leaving for France next Wednesday. Thank you for your helpful criticisms.

Yours sincerely

Xavier Le Pichon

NB I had no comments from F. Vine or H. Hess. Do you know of any reactions from them to my paper?

Morgan (1968b) and McKenzie and Parker (1967) laid out most of the key elements of plate tectonics, and Le Pichon further tested Morgan's version, and attempted to apply it to the whole of the Cenozoic. McKenzie and Morgan joined forces and explicated the evolution of triple junctions, the next major conceptual element of plate tectonics (§7.20). Meanwhile, Isacks, Oliver, and Sykes wrote a masterful defense of the "new global tectonics" by showing how it was so strongly supported by seismology. I now turn to their paper.

7.18 Isacks, Oliver, and Sykes integrate seismology and plate tectonics

Isacks, Oliver, and Sykes' 1968 paper "Seismology and the new global tectonics" helped tremendously in conveying what later became known more or less as plate tectonics to Earth scientists who did not work in areas that played a significant role in establishing seafloor spreading, V-M, and plate tectonics. The three Lamont seismologists thoroughly and clearly presented the rapidly growing earthquake seismology data supportive of what they called "the new global tectonics," by which they (1968: 5856) meant "current concepts of large-scale tectonic movements and processes within the earth, concepts that are based on hypotheses of continental drift . . . seafloor spreading, and transform faults and that include various refinements and developments of these ideas."

Oliver recalled when and why they decided to write the paper.

The early success of Lynn Sykes' seismological paper on transform faults and the widespread attention given to the paper by Bryan Isacks and me on the down-going slab, plus the excitement generated by the 1967 AGU meeting and the news of assorted advances elsewhere, stimulated Lynn, Bryan and me to join together for intensive further action . . . It was easy to see that seismology was almost certain to be a major contributor . . . We also recognized that we had been fortunate enough to get a temporary, and almost certainly short-lived, lead on other seismologists. Therefore we felt that, at Lamont with its archives of data, supporting facilities, and colleagues in related fields, we were in a strong and probably leading position to make a truly major advance both in global dynamics and seismology by relating those two subjects thoroughly and comprehensively. In fact, we convinced ourselves that we had a once-in-a-lifetime opportunity and set out with great enthusiasm and urgency to make the most of it . . . Lynn, Bryan and I made an ambitious plan with the ultimate plan of writing a grand paper that would report a comprehensive study of every aspect of the field of earthquake seismology that bore on the developing story of global tectonics. We intended to test the new ideas on tectonics against all relevant seismological observations, to develop and enhance the story of global tectonics with new ideas derived from the seismological perspective, and to call attention to problems and opportunities in the field of seismology revealed through the perspective of the new tectonics.

(Oliver, 1996: 75–76)

Oliver (1996: 77–78) also recalled that they decided to share equally the burden of writing the paper, thinking it was the best way to complete such an extensive review and still "stay at the forefront of seismological research." This decision led them to add a footnote on the first page of their paper – "the order of authors by lot" – to indicate that the ideas and writing were shared equally among them (Sykes, January 30, 2012 email to author).

I think there were six principal reasons why the paper was so successful. First, they had a tremendous amount of their own data and that of other Lamont workers. There was Sykes' work on transform faults confirming ridge-ridge transform faults, Oliver and Isacks' work on establishing the existence of downgoing slabs of lithosphere beneath island arcs, and Isacks' ongoing and still unpublished first-motion studies of shallow, intermediate, and deep earthquakes associated with downgoing lithospheric blocks, but there was also work by other Lamonters such as Barazangi and Jim Dorman's 1969 plots of the worldwide distribution of earthquake epicenters that so nicely delineated plate boundaries, improving and updating, with the WWSSN of the USCGS, works by Gutenberg and Richter (1954) and by Rothé (1954). Like Isacks and Molnar (§7.16) and Atwater (§7.19), Oliver thought highly of Barazangi and Dorman's work. He recalled the background story and explained its importance:

As Lynn, Bryan and I were preparing the NGT [New seismology and global tectonics] paper, other seismologists at Lamont were catching the plate tectonics fever. Jim Dorman had been the first Lamont seismologist to capitalize on the advent of digital computing. It was the time of the IBM 650, a major technological marvel then but no match for a desktop of today. Nevertheless, that early computer opened many new horizons in geophysics. Having long been familiar with the relatively crude maps of seismicity found in Gutenberg and Richter's book, and then noticed how J. P. Rothé in France had improved the detail and information content of such maps by using more modern data, I suggested to Jim and he agreed, that he might use the computer to go one step further and plot the new hypocentral data of the WWSSN era. Perhaps, we thought, a still more revealing set of maps would result. Jim and Muawia Barazangi took on that job, and with considerable effort produced a set of global maps of seismicity that also became icons of the plate tectonics story.

   The maps showed far more epicenters and more precisely and accurately located epicenters than previous maps. And some maps showed epicenters for shocks in only certain ranges of hypocentral depths. All of the maps were published in a separate paper by Barazangi and Dorman (1969), and the one showing shocks at all depths was also reproduced in the NGT paper (NGT, Fig. 14). The map was a dramatic demonstration of how the seismic belts define the plate boundaries and in turn how most earthquakes occur at those boundaries. Everyone who saw it grasped that fundamental relationship at first glance and the paper became widely-known.

(Oliver, 1996: 79; my bracketed addition)

Isacks, Oliver, and Sykes also had pre-publication access to Le Pichon's six-plate model, which they basically adopted. Second, plate tectonics explained the overall pattern, variety, slip directions, and origin of most earthquakes in terms of tectonics. This was a huge advantage of plate tectonics and huge advance in our understanding of earthquakes, which were both, I believe, easily appreciated by most readers. The three seismologists were not shy about letting their readers know. Here is how they began their abstract.

A comprehensive study of the observations of seismology provides widely based strong support for the new global tectonics which is founded on the hypotheses of continental drift, sea-floor spreading, transform faults, and underthrusting of the lithosphere at island arcs. Although further developments will be required to explain certain parts of the seismological data, at present within the entire field of seismology there appear to be no serious obstacles to the new tectonics. Seismic phenomena are generally explained as the result of interactions and other processes at or near the edges of a few large mobile plates of lithosphere that spread apart at the ocean ridges where new surficial materials arise, slide past one another along the large strike-slip faults, and converge at the island arcs and arc-like structures where surficial materials descend. Study of world seismicity shows that most earthquakes are confined to

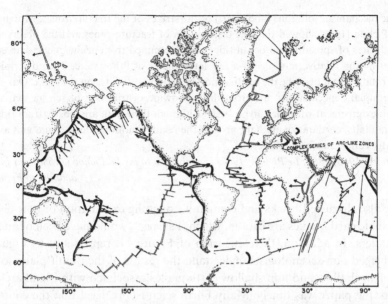

Figure 7.36 Isacks *et al.*'s Figure 3 (1968: 5861). "Summary map of slip vectors derived from earthquake mechanism studies of relative motion of block on which arrow is drawn to adjoining block. Crests of world rift system are denoted by double lines; island arcs and arc-like features, by bold single lines; major transform faults, by thin single lines. Both slip vectors are shown for an earthquake near the western end of the Azores-Gibraltar ridge since a rational choice between the two could not be made. Compare with directions computed by Le Pichon (Figure 2 [Figure 7.34 above])" (my bracketed addition).

narrow continuous belts that bound large stable areas. In the zones of divergence and strike-slip motion, the activity is moderate and shallow and consistent with the transform fault hypothesis; in the zones of convergence, activity is normally at shallow depths and includes intermediate and deep shocks that grossly define the present configuration of the down-going slabs of lithosphere. Seismic data on focal mechanisms give the relative direction of motion of adjoining plates of lithosphere throughout the active belts. The focal mechanisms of about a hundred widely distributed shocks give relative motions that agree remarkably well with Le Pichon's simplified model in which relative motions of six large, rigid blocks of lithosphere covering the entire earth were determined from magnetic and topographic data associated with the zones of divergence. In the zones of convergence the seismic data provide the only geophysical information on such movements.

(Isacks et al., *1968: 5855)*

Becoming more technical, they later added:

Evidence for motions of lithospheric plates. One of the most obvious features in Figure 3 [Figure 7.36] is that the slip vectors are consistent with the hypothesis that surface area is being created along the world rift system and is being destroyed in island arcs. Along the mid-Atlantic ridge, for example, slip vectors for more than ten events are nearly parallel to one another and are parallel to their neighboring fracture zones within the limits of uncertainty in either the mechanism solutions (about 20°) or the strikes of the fracture zones. Morgan [1968] and Le Pichon [1968] showed that the distribution of fracture zones and the observed directions and rates of spreading on ocean ridges as determined from geomagnetic data could be explained by the relative motions of a few large plates of lithosphere. They determined the poles of rotation that describe the relative motion of adjacent plates on the globe. Our evidence from earthquake mechanisms and from the worldwide distribution of seismic activity is in remarkable agreement with their hypothesis. Although their data are mostly from ridges and transform faults, earthquake mechanisms give the relative motions along island arcs as well as along ridges and transform faults.

(Isacks et al., *1966: 5882; Le Pichon (1968) is equivalent to my Le Pichon (1968a); my bracketed addition to Figure 7.36)*

They also discussed Isacks and company's ongoing explanation of how deep and intermediate earthquakes are caused, which, I suspect, would have been of interest to most readers. In addition, their inclusion of Figure 11, reproduced as Figure 7.30 above, helped non-seismologists understand the results of their fault-plane solutions and explain shallow and non-shallow earthquakes associated with island arcs.

Third, the paper was timely. Many Earth scientists realized that the controversy over continental drift had been resolved. Even if the vast majority of Earth scientists, especially those working on continental interiors, saw no use for mobilism in their own work, many had seen and understood the confirmation of V-M and ridge-ridge transform faults, and their role in confirming seafloor spreading. In addition, their paper appeared from three to nine months after Morgan (1968b), McKenzie and Parker (1967), and Le Pichon (1968a), the three key papers on plate tectonics, so readers who knew that the papers were important but found them difficult to understand were eager to learn more about the new ideas and welcomed Isacks, Oliver, and Sykes' paper. Fourth, they organized the earthquake seismological data around plate tectonics, not around classical divisions within seismology. This made it easier for readers to see just how the data supported plate tectonics, helped them to think in terms of Earth's surface being made of rigid plates whose movements are described relative to each other, and to make their paper more accessible to readers who did not work in seismology.

The organization of the NGT paper and of the studies on which it was based was the result of rather careful consideration. We chose to organize the efforts and the text according to the principal effects of the new global tectonics, and not according to the classical divisions of seismology. Thus major subdivisions of the paper have titles like "ridges," "island arcs," and "movements on a global scale," rather than "travel time curves," "focal mechanisms," or seismicity," although these later topics are discussed in minor subdivisions. This type of organization seemed to work out well.

(Oliver, 1996: 78)

Fifth, they kept the jargon of seismology at a minimum. Sixth, several of their figures were simple but very informative enabling the general reader who may not

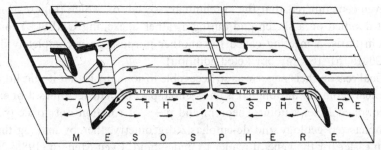

Figure 7.37 Isacks *et al.*'s Figure 1 (1968: 5857). "Block diagram illustrating schematically the configurations and roles of the lithosphere, asthenosphere, and mesosphere in a version of the new global tectonics in which the lithosphere, a layer of strength, plays a key role. Arrows on lithosphere indicate relative movements of adjoining blocks. Arrows in asthenosphere represent possible compensating flow in response to outward movements of segments of lithosphere. One arc-to-arc transform fault appears at left between oppositely facing zones of convergence (island arcs), two ridge-to ridge transform faults along ocean ridge at center, simple arc structure at right."

have a deep appreciation of the ideas of Morgan, McKenzie and Parker and Le Pichon to understand the main features of plate tectonics: that Earth's surface is divided into rigid lithospheric plates and that there are three types of plate boundaries, those where plates separate, those where they slip past each other, and those where they converge (Figure 1, reproduced as Figure 7.37 below). Their paper helped the reader to visualize the vertical orientation of axes of compression, tension, and slip vectors of earthquakes at island arcs (Figure 11, reproduced as Figure 7.30 above); to see at a glance the directions of slip-vectors obtained from studies of earthquakes at plate boundaries indicating the relative movement of adjacent blocks (Figure 3, reproduced as Figure 7.36 above); to visualize downward movement of the lithospheric slabs and crustal sediment beneath island arcs and the formation of volcanoes (Figure 7, reproduced as Figure 7.39 below); and to see the worldwide distribution of earthquake epicenters nicely outlining the plate boundaries of large plates (Figure 15). Addressing the fifth and sixth points, Oliver retrospectively remarked:

We hoped that the NGT paper would appeal to a large and diverse audience so we tried to limit jargon, to make the introduction and most other sections readily readable by non-seismologists, and to include some figures that were cartoon-like and designed to illustrate concepts rather than to present data. One figure, the block diagram (NGT Fig. 1) that illustrates the basic elements of what is now plate tectonics, became very popular and has been reproduced frequently and in a wide variety of places. It appears in many textbooks, and it is considered one of a set of what some (see Le Grand, 1988) have called the "icons" of plate tectonics. The "icons" in this sense are a few figures that, taken together, communicate graphically and almost at a glance the essence of plate tectonics to the uninitiated.

(Oliver, 1996: 78)

Oliver even remembered that the "preliminary sketch on which [Figure 1] is based was first doodled on a sketch pad during a rather unexciting UNESCO committee meeting in Paris, a meeting called for another purpose on a completely different topic" (2001: 161–162; my bracketed addition).

The final reason their paper was so accessible to so many is that it said little about the spherical geometry although it is an essential part of plate tectonics. For example, they mentioned Euler's Point Theorem and poles of rotation only once (p. 5882). They emphasized geology and de-emphasized geometry, thereby making their presentation easier for the general reader to understand. Even Menard (1986: 293), for example, found Morgan's original outline of his AGU talk difficult to understand. He found Morgan's published paper easier to understand because, he claimed, "the style had been transformed from geometry to geology," even though, as we have seen, there was little change between the original outline and paper (§7.10). The absence of discussion of Euler's Point Theorem and instantaneous versus finite rotational poles made their paper accessible to most readers. Their paper was a primer on plate tectonics for those challenged by the application of Euler's Point Theorem.

I now turn to the paper itself, but shall keep the discussion brief because the key works of the authors and the ideas of plate tectonics have already been dealt with. I begin with their Figure 1 (Figure 7.36), which illustrates both the general movements of lithospheric blocks relative to one another and more specifically "a part of the Pacific basin including the New Hebrides, Fiji, Tonga, the East Pacific rise, and western South America" (Isacks *et al.*, 1968: 5857). It illustrated of the relationship between lithosphere, asthenosphere, and mesosphere; the division of the lithosphere into rigid, "relatively thin blocks, some of enormous size"; that "major tectonic features" such as ridges, transform faults, trenches and island arcs "are the result of relative movement and interaction of these blocks"; their Fig. 1 showed that following Elsasser, the lithospheric blocks act as stress guides, are driven by "instability resulting from surface cooling" and that movement of the lithosphere brings about "a compensating return flow . . . in the asthenosphere" (Isacks *et al.*, 1968: 5857–5858). They also used it to highlight their own work, and justifiably so. Sykes' first motion studies of earthquakes along transform faults and within ridges (§6.9); Oliver and Isacks' recognition of the downgoing lithospheric slab through their study of deep earthquakes in the Tonga–Fiji region (§7.4), and Isacks and company's ongoing study of first-motion studies of earthquakes associated with the Tonga–Fiji and Kermadec regions (§7.16); it even shows the hinge fault and arc-arc transform fault connecting the northern end of the Tonga Trench with the southern of the New Hebrides Trench. Indeed, they got a lot of use out of their Figure 1.

Turning their attention to what transpires where the lithosphere bends beneath island arcs, they used their Figure 7 (Figure 7.38) to illustrate two hypotheses concerning the cause of thrust-fault earthquakes, and upward movement of crustal material previously carried downward into the mantle while being trapped in tensional graben-like structures on the convex (upper) side of the downgoing

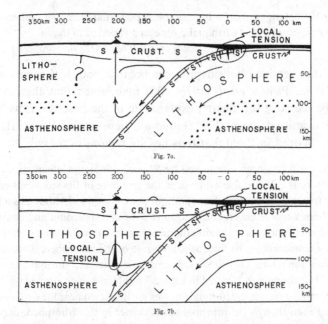

Figure 7.38 Isacks *et al.*'s Figure 7 (1968: 5869). "Figure 7 shows vertical sections through an island arc indicating hypothetical structure and other features. Both sections show down-going slab of lithosphere, seismic zone near surface of slab and in adjacent crust, tensional features beneath ocean deep where slab bends abruptly and surface is free. (In both sections, S indicates seismic activity.) (a) A gap in mantle portion of lithosphere beneath island arc and circulation in mantle associated with crustal material of the slab and with adjoining mantle [Holmes, 1965]. (b) The overriding lithosphere in contact with the down-going slab and bent upward as a result of overthrusting. The relation of the bending to the volcanoes follows Gunn [1947]. No vertical exaggeration."

lithosphere as it was bent downwards. With the first hypothesis, illustrated in Figure 7a, there is little or no well-defined lithosphere beneath the island arc. Petrologically altered formerly-trapped crustal material rises to the surface forming volcanoes (volcanic arcs). With the second hypothesis, illustrated in Figure 7b, there is a well-defined lithosphere beneath the island arc which is in contact with the downgoing lithospheric slab. Thrust earthquakes are caused by friction between the underthrusting downgoing lithospheric slab and the overthrusting lithosphere beneath the island arc, and petrologically altered formerly-trapped crustal material forms volcanoes (volcanic arcs) once it rises to the surface through extensional cracks in the lower surface of the overriding lithosphere formed by its being bent upward as it overrides the descending slab. With the first hypothesis, explaining shallow thrusting-type earthquakes was not straightforward.

One implication [of the first hypothesis] . . . is that at least part of the shallow earthquake zone might not result from the contact between two pieces of lithosphere but might instead indicate an embrittled and weakened zone formed by the downward moving crustal materials [Raleigh and Paterson, 1965; Griggs, 1967]. In this the exponential decay in activity might reflect changes in the properties of the earthquake zone as a function of depth.

(*Isacks* et al., *1968: 5877; first bracketed phrase is mine*)

Nonetheless, they retained both hypotheses because both had some empirical support. In 1973, Le Pichon *et al.* (1973: 220) maintained that the first hypothesis "appears to be the most satisfactory explanation of the observed facts."

They also dismissed the earlier "Lamont" view of the 1950s (III, §3.17, still maintained by Worzel in 1965) that trenches are regions of tension.

Thinning of the crust has been interpreted by Worzel [1965] and others as an indication of extension and there is considerable evidence in the structure of the sediments on the seaward slopes of many trenches supporting the hypothesis of extension . . . Although such evidence for extension has been cited as an argument against sea-floor spreading and convection on the basis that down-going currents at the sites of the ocean deeps would cause compression normal to the arcs, the argument loses its force when the role of the lithosphere is recognized. All the evidence for extension relates only to the sediments and crust, i.e., the upper few kilometers of the lithosphere. For the models pictured in Figure 7 in which a thick strong layer bends sharply as it passes beneath the trench, extensional stresses are predicted near the surface on the convex side of the bend even though the principal stress deeper in the lithosphere may be compressional. Earthquake activity beneath the seaward slope of the trench is, in general, infrequent and apparently of shallow depth. The focal mechanisms that have been determined for such shocks indeed indicate extension as predicted, i.e. normal to the trench, the axis of bending (Stauder [1968] and T. Fitch and P. Davis, personal communication). Stauder demonstrates this point very well in a paper on focal mechanisms of shocks of the Aleutian arc.

The extensional features also suggest a mechanism for including and transporting some sediments within the down-going rock layers. As implied by Figure 7a, sediments in the graben-like features may be carried down to some depth in quantities that may be very significant petrologically, as suggested by Coats [1962].

(*Isacks* et al., *1968; Stauder, 1968 is my Stauder, 1968a; Worzel, 1965 is equivalent to my Worzel,*
1965c; Coats reference is the same as mine).

With this, I believe, except for Carey and his loyal followers, the idea that trenches are extensional features where no subduction occurs was laid to rest. Indefatigably, Carey (1975: 125), attempted to rebut the Lamonters' interpretation of first-motion studies of earthquakes associated with island arcs characterizing it as "astonishing." Only his few loyal followers, I believe, were listening. At this point, as readers will no doubt recognize, Carey was abusing RS2.

As already noted, Isacks, Oliver, and Sykes found Le Pichon's "six-plate model" attractive because his calculation of the directions of the slip component of dip-slip earthquakes along consuming plate boundaries was consistent with reliable first-motion solutions along all three types of boundaries. They reproduced his Figure 6 (Figure 7.35 above) as their Figure 2 (Figure 7.40 below), which showed his six plates, and their relative movements, mostly compressional, that "he derived from rates of spreading determined from magnetic data and from orientations of fracture

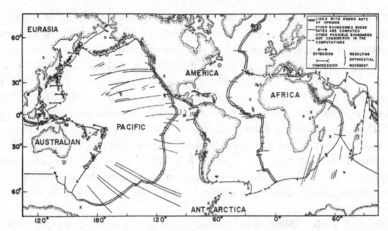

Figure 7.39 Isacks *et al.*'s Figure 2 (1968: 5859). "Computed rates of compression and extension along boundaries of six lithospheric blocks [after Le Pichon, 1968]. Computed movements were derived from rates of spreading determined from magnetic data and from orientations of fracture zones along features indicated by double lines. The extensional and compressional symbols in the legend represent rates of 10 cm/yr; other similar symbols are scaled proportionally. Symbols appearing as diamonds represent small computed rates of extension for which the arrowheads coalesced. Historically active volcanoes [Gutenberg and Richter, 1954] are denoted by crosses. Open circles represent earthquakes that generated tsunamis (seismic sea waves) detected at distances of 1000 km or more from the source."

zones." They added positions of active volcanoes and designated dip-slip earthquakes on the landward side of trenches that had generated tsunamis.

They also appreciated Le Pichon's argument against rapid Earth expansion (§7.17). Referring to their Fig. 2 (Figure 7.39), they characterized his argument as "a strong argument."

Le Pichon used data from ocean ridges to infer the direction of motion in island arcs. His predicted movements (Fig. 2) [Figure 7.39 above] which are based on the assumption of conservation of surface area and no deformation within the plates of lithosphere, compare very closely with mechanism solutions in a number of arcs. This agreement is a strong argument for the hypothesis that the amount of surface area that is destroyed in island arcs is approximately equal to the amount of new area that is created along the world rift system. Thus, although modest expansion or contraction of the earth is not ruled out in the new global tectonics, rapid expansion of the earth is not required to explain the large amounts of new materials added at the crests of the world rift system. This approximate equality of surface area is, however, probably maintained for periods longer than thousands to millions of years, but minor imbalances very likely could be maintained for shorter periods as strains within the plates of lithosphere.

(*Isacks* et al., *1968: 5882; my bracketed addition*)

Still refering to their Fig. 2 (Figure 7.39), they hinted at "new possibilities for explaining certain seismological observations, particularly the configuration of the deep earthquake zones."

Figure 2, adapted from Le Pichon [1968] with additions, shows the plan of blocks of lithosphere as chosen by Le Pichon for the spherical earth and indicates how their movements are being accommodated on a worldwide scale. The remarkably detailed fit between this scheme, based on a very small number of rigid blocks of lithosphere (six) and the data of a number of fields, is very impressive. The number and configuration of the blocks of lithosphere is surely larger than six at present and almost certainly the pattern has changed within geologic time, but the present pattern must, in general, be representative of at least the Quaternary and late Tertiary. The duration of the current episode of sea-floor spreading is not known. Some evidence suggests that it began in the Mesozoic and has continued rather steadily to the present. Other evidence [Ewing and Ewing, 1967] indicates that the most recent episode of spreading began about 10 m.y. ago. This suggestion is considered here because it opens new possibilities for explaining certain seismological observations, particularly the configuration of the deep earthquake zones.

(Isacks, et al., 1968: 5858; Ewings reference is equivalent to mine; Le Pichon, 1968 is my Le Pichon, 1968a)

More specifically, they thought they could explain the length of lithosphere subducted at various island arcs. To understand how and why the appealed to the Ewing brothers' proposal that the most recent episode of seafloor spreading began 10 million years ago, I begin with Oliver and Isacks's earlier estimate that the average annual half-velocity of seafloor spreading over the entire ridge system is 1.3 cm/yr. Oliver recalled how they made such an estimate.

Bryan and I took my desktop globe, shaded the geographical locations of deep seismic activity, measured the shaded areas with a plainmeter, corrected for dip, divided the total by the length of the world rift system and the ten million year time constant, and so got a tentative figure for the average annual rate of sea floor spreading along the entire rift system. The half-velocity was 1.3 cm/yr. What that number meant was subject to the assumptions we had made of course, but the very fact that we could casually pick up a nearby globe, do a few very unsophisticated things, and come out in an hour or two with an answer that seemed reasonable and consistent with other data such as spreading rates based on magnetic anomalies, left us with the euphoric feeling that we really were on the right track to understanding the earth.

(Oliver, 1996: 88)

Isacks *et al.* also described Oliver and Isacks previous estimate, again mentioning the Ewing brothers' idea of episodic spreading. Curiously, they also attributed episodic spreading to Vine; I say curiously because Vine (1966) always favored continuous spreading.

Length of seismic zones in island arcs. The systematic changes in the seismic zones in the Aleutians and in the southwest Pacific suggest that the lengths of zones of deep earthquakes might be a measure of the amount of underthrusting during the last several million years. Using the maps of deep and intermediate-depth earthquakes prepared by Gutenberg and Richter [1954], Oliver and Isacks [1968] estimated the area of these zones and divided the total area by the length of the world rift system (about two great circles) and by 10 m.y., which is the duration of the latest cycle of spreading based on data from ocean-floor sediments and from magnetics [Ewing and Ewing, 1967; Vine, 1966]. They obtained an average rate of spreading for the entire rift system of 1.3 cm/yr for the half velocity. This value is reasonable for the average spreading along the world rift system.

(Isacks et al., *1968: 5884; references are equivalent to mine)*

Examining "the hypothesis that the lengths of deep seismic zones are a measure of the amount of underthrusting during the past 10 m.y.," they rounded off the average half-spreading rate of 1.3 cm/yr to 1 cm/yr and introduced Figure 16 [Figure 7.40] with the following:

Figure 16 illustrates the lengths of the seismic zones in various arcs and the corresponding rates of underthrusting as calculated by Le Pichon [1968] from observed velocities of sea-floor spreading and the orientation of fracture zones along the world rift system. In nearly all cases the regions with the deepest earthquakes (and hence the longest seismic zones as measured along the zones and perpendicular to the arcs) correspond to the areas with the greatest rates of underthrusting; regions with only shallow- and intermediate-depth events are typified by lower rates of underthrusting. Since the calculated slip rates and some of the measured lengths may be uncertain by 20% or more, the correlation between the two variables is, in fact, surprisingly good.

Although six points fall well above a line of unit slope, which represents an age of 10 m. y., all but one of the lengths [South America] are within a factor of 2 of the lengths predicted from the hypothesis that these zones represent materials underthrust during the last 10 m.y. Three of the more discrepant points, which are denoted by crosses on the figure, represent a small number of deep earthquakes that are located in unusual locations with respect to the more active, planar zones of deep earthquakes. They include the unusual deep Spanish earthquake of 1954, three deep earthquakes in New Zealand, and a few deep events under Fiji that appear to fall between the deep zones in the Tonga and New Hebrides arcs.

(Isacks et al., *5884–5885; my bracketed addition of South America)*

They offered two hypotheses to explain this correlation, which Le Pichon *et al.* (1973: 63) later characterized as "remarkable," between the length of the seismic zone and calculated slip rate. The first was the Ewing brothers' idea of episodic spreading with the last episode beginning 10 million years ago; they obtained the second one from McKenzie. He proposed 10 million years as the time needed for a descending lithospheric slab to warm up enough for it to lose its integrity and be assimilated into the mantle. In 1997, Sykes recalled how McKenzie made such a suggestion after reading an earlier version of their paper.

So I just want to make one other comment and that there was one other person who I and we interacted with quite a bit – Dan McKenzie. Dan McKenzie spent quite a bit of time in the U.S. at Lamont, where he looked at a lot of film chips, and at Scripps several times. We showed him a version of our new global tectonics paper. One of the things that he rightly criticized was our

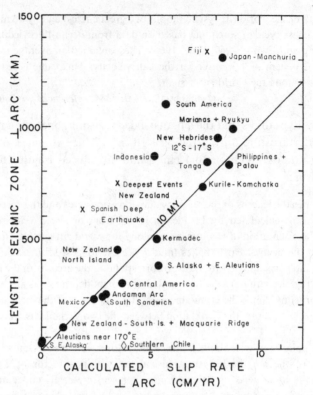

Figure 7.40 From Isacks *et al.*'s Figure 16 (1968: 5885). "Calculated rates of underthrusting [*Le Pichon*, 1968], and length of seismic zone for various island arcs and arc-like features (solid circles), for several unusual deep events (crosses), and for South Chile (diamond). The solid line indicates the theoretical locus of points of uniform spreading over a 10-m.y. interval."

interpretation of the lengths of downgoing seismic zones. The places that had the highest rates of underthrusting had the longest ones and generally those that go down the deepest. We explained that in the first version of our paper, i.e. a draft as the pattern of plate motion having changed about ten million years before that, and so that underthrusting had started in those places at that time. I think that idea came from Ewing and Ewing's statement that they thought in the North Atlantic that there had been a big change [in spreading] as indicated by the thickness of sediments, about ten million years ago. McKenzie pointed out that there was another viable hypothesis and that was that ten million years or twenty million years was about the time for a slab to warm up and that could turn off the earthquakes. So we added that as a second hypothesis in the paper. I think that that is surely the correct explanation, and not of there having been a world-wide beginning of underthrusting that was new, some ten million years ago. That was really a very catastrophic [hypothesis].

(Sykes, May 24, 1997 interview by Ron Doel, Niels Bohr Library & Archives, American Institute of Physics, www.aip.org/history/nbl/oralhistory.html)

As we now know, and as Sykes knew by the early 1970s, seafloor spreading did not begin 10 m.y. ago after a considerable hiatus. McKenzie's suggestion was in the right direction. He (1969b) presented the idea himself a year later, proposing, as Le Pichon *et al.* (1973: 212) put it, "the calculation of the thermal evolution of a single plate within an isothermal mantle is very similar to the calculation of the cooling of a newly created plate, provided the heating is only due to transfer of heat by conduction." Le Pichon *et al.* (1973) also provides an assessment of the hypothesis; see Kearney, Klepeis, and Vine (2009: 259–262) for a more recent discussion.

What Lamont seismologists did was absolutely marvelous, from Sykes's confirmation of ridge-ridge transform faults and Oliver and Isacks' identification of an anomalous zone in the mantle which they identified as a downgoing slab of lithosphere, and Barazangi and Dorman's 1969 plots of the worldwide distribution of earthquake epicenters, to Isacks' (and Molnar's) work on fault plane solutions of shallow, intermediate, and deep earthquakes. But I would be remiss not to acknowledge the excellent work by other seismologists, especially the work of Stauder and his colleagues at St. Louis University and that of Honda and other distinguished Japanese seismologists. Stauder and Bollinger (1966a) had already developed a method of reliably plotting results of first-motion studies in determining fault plane solutions before Sykes perfected what was essentially the same method. Stauder and his colleagues also determined many fault plane solutions, and their work, for example in the Aleutians (1966b), was important for McKenzie in testing his ideas about plate tectonics. Utsu (1966, 1967, and 1971) and Katsumata (1960) identified the anomalous zone of high Q and low attenuation from studying earthquakes in Japan and its vicinity. Honda (1932), Honda, Masatsuka, and Emura (1956), Ichikawa (1961), and Katsumata (1967) had studied mechanisms of deep earthquakes and had obtained fault plane solutions of intermediate and deep earthquakes before or around the same time when Isacks and company began their outstanding work in the Tonga–Fiji area. This, of course, is not to take anything away from the contributions of the Lamont seismologists, for they were the ones who deliberately set out to confirm key components of the new global tectonics (Sykes on transform faults, and Isacks and later Molnar on subduction) or later realized the importance of their work in confirming them (Oliver and Isacks on discovery of downgoing slabs). I should also emphasize that Lamont seismologists acknowledged Stauder's work and that of these Japanese seismologists. McKenzie (1969c) also deserves credit for realizing that it is the horizontal components of slip vectors and not fault planes that are important for testing plate tectonics.

In closing this section on Lamont's contributions, I want to emphasize that even though most at Lamont were reluctant to accept mobilism, once they changed their minds, they were quick to exploit the extensive databases and confirm key elements of the new global tectonics. Lamont may not have been the best place to develop new hypotheses, but it had the data to test them.

7.19 Atwater and Menard apply plate tectonics to the great
fracture zones of the northeast Pacific

With the launching of Sputnik in 1957, Tanya Atwater decided to become a scientist. In 1961, encouraged by her mother, a botanist, and her father, an engineer, she began undergraduate studies at MIT. During her junior year she "accidentally took a physical geology course" (Atwater, 2001: 243). After an enjoyable summer of field work in Montana where she "was in heaven," she decided to move west, transferred to the University of California, Berkeley, and in 1965 earned an undergraduate degree with a major in geophysics. She spent summer 1965 at WHOI as a Summer Fellow in marine geophysics. She also attended the September 1965 Ottawa meeting that was sponsored by the International Upper Mantle Committee and held in Ottawa (§5.7). She was most impressed with Tuzo Wilson and his idea of transform faults.

The whole meeting was exciting, but the presentation that made the biggest impression on me was the one about transform faults by J. Tuzo Wilson. Tuzo was a wonderful showman with a great twinkle in his eye. After he had explained this idea, he passed out paper diagrams with two mid-ocean ridges connected by a transform fault. It said "cut here," "fold here," "pull here." We all laughed, and I felt embarrassed (kindergarten games at this august scientific meeting?). But I took the paper back to the privacy of my hotel room and cut and folded and pulled; wow; the light bulbs went on in my brain. The simple geometry of the transform faults with their fracture zones holds the key to the geometry of formation of all the ocean basins – right there in that little piece of paper.

(Atwater, 2001: 245)

Deciding to join some of her siblings who were traveling around South America, she obtained, in late 1965 or early 1966, a position at the Universidad de Chile in Santiago as a Visiting Research Associate in earthquake seismology. While there she attended a symposium on Antarctic Oceanography from September 13 to 16, 1966, in Santiago. Heirtzler made an unscheduled appearance and showed his and Pitman's *Eltanin*-19 profile.

I was dozing through a series of papers full of Latin names of diatoms and foraminifera (single-celled planktonic organism) when they announced an extra paper. Jim Heirtzler was passing through from Lamont-Doherty Geological Observatory on the way to meet a ship in Valparaiso and he wanted to present some marine geophysical results. In his talk he put up the *Eltanin*-19 magnetic anomaly profile – still, to this day, the clearest, most beautiful, and symmetrical profile in the world – and with it made the case for sea floor spreading. It was as if a bolt of lightning had struck me. My hair stood on end.

(Atwater, 2001: 245–246)

Afraid that she would miss the revolution before starting graduate school, she applied to Scripps, and in January 1967, four months after she heard Heirtzler and saw *Eltanin*-19, arrived at Scripps ready to join the revolution.

John Mudie became her Ph.D. supervisor. She gleefully accepted his offer to participate in his Deep Tow group, went to sea with the group in spring 1967, and became the lead author of Atwater and Mudie (1968), which was published in February, a month after they submitted it to *Science* in January, 1968, about a year after she had arrived at Scripps. She also got to know McKenzie during fall 1967, and learned about his and Parker's version of plate tectonics.

Dan McKenzie was there [at Scripps] during the fall of 1967 and he was thinking hard about many aspects of the new theories. My advisor, John Mudie, had set up a monthly beer party at a local German dance hall to get people together for informal talk. I especially remember one of these sessions during which McKenzie and Bob Parker arrived, bubbling over about some project they were working on. I couldn't figure out what they were talking about and could barely hear them over the loud accordion music, but during a lull I asked, rather timidly, what the fuss was about. Dan took a napkin and sketched out the San Andreas and Queen Charlotte Fault systems and the Aleutian/Alaskan subduction zone. He showed me how all these features lay along the boundary between two large rigid plates, the Pacific and North American plates. "That's all very well, but what about the Mendocino fracture zone? That doesn't line up," I complained, trying to grab his pencil so I could add the offending feature to his tidy sketch. (They were acting so smug, I hoped I could trip them up.) "Easy," said Dan, and he drew a third plate, the Juan de Fuca/Gorda plate, meeting the other two at the Mendocino triple junction. Three plates! Of course. So elegant, so simple, and so powerful, I sat there, agog, my brain zooming around in all directions . . . After that evening in the beer hall, I became a McKenzie groupie, attending his seminars, dogging him with questions, making a big nuisance of myself, I'm sure. He was humorously generous and I learned a lot: about tectonics, about the scientific approach, and about tectonic passion and delight.

(Atwater, 2001: 250–251)

Little did she know at the time that she soon would become obsessed with the relationship between the San Andreas Fault and Mendocino Fracture Zone and the interaction of the Pacific and North American plates. Indeed, in early 1968, she and Menard would begin meeting in his office where they studied a map detailing magnetic lineations and great fracture zones of the Pacific Basin, attempting to explain variations in both by directional changes of seafloor spreading.

I got in the habit of dropping in at Bill Menard's lab. It was already known that the magnetic anomalies in the northeast pacific were exceptionally clear, and that they were well lineated and offset across the fracture zones, but no one had compiled them for a look at the regional pattern. Menard had his draftswoman, Isabel Taylor, transfer all the available magnetic profiles from their paper records to their ship tracks on a big map. She did it all by hand – this was before computer data processing became routine. The result was spectacular. The magnetic anomalies of the northeast pacific are especially easy to read and the emerging pattern was full of information about sea floor spreading and transform faulting. Every session that we had over the map was full of discovery and excitement. Menard and I began seeking each other out first thing in the morning to share our middle-of-the-night thoughts. Often I couldn't sleep at night, my head was so abuzz with geo-possibilities and implications. Apparently he was having the same problem, because I often arrived in the morning to find his ideas scribbled on my blackboard, "What about this . . . .?"

(Atwater, 2001: 248)

Atwater attended the 1968 April AGU spring meeting in Washington, DC, where she presented the paper she had co-authored with Mudie. After the meeting she joined a group of students visiting Lamont. She remembers two things about the visit. She was invisible, dismissed as a nonentity because of the prevailing and dismissive attitude of many male scientists toward female scientists, and she was impressed by Barazangi and Dorman's worldwide plot of earthquake foci.

After the meeting, I heard that some other students were going up to New York for a tour of Lamont Geological Observatory . . . Curious, I joined them. I remember two things from that tour. One is that I was invisible. In every lab we visited, they introduced all the young men and skipped me, every time. I guess our student guide assumed I was someone's tag-along girlfriend and therefore of no account. I introduced myself and tried to establish that I was a scientist, too, but my hints fell on deaf ears. The other thing I remember was the map of earthquake locations that student Muawia Barazangi, working with Jim Dorman, had plotted onto transparent mylar sheets and had overlain on a huge wall map. There they were: the plates of the world all outlined by the earthquakes. It was stunning, awesome, so simple and clear and full of details about the individual plates. It was oh-so-hard to pull myself away from that map.

(Atwater, 2001: 247)

Atwater left Lamont for Princeton where she vaguely remembers talking to McKenzie and Morgan (§7.11) about the Mendocino triple junction.

I suppose that a Princeton visit happened while I was east, after attending the spring 1968 AGU meeting in Washington, though I mostly remember the Lamont visit. I do remember us drawing triple junctions on a blackboard somewhere (Princeton?) sometime, me bugging them about how the Mendocino triple junction doesn't work well – it keeps drifting out of whack.

(Atwater, November 13, 2011 email to author)

Returning to Scripps, she and Menard kept meeting. Making excellent progress, in mid-June, 1968, they submitted their first of three co-authored papers on the evolution of the Pacific Basin. Identifying changes in the pattern of magnetic lineations between, especially, the Mendocino and Murray fracture zones, which they coordinated with changes in the trend of both fracture zones, they argued that both independently and together indicated changes in the direction of seafloor spreading. It is no surprise that Menard and Atwater worked so well together. Menard had already devoted much of his professional life trying to explain the great fracture zones, and

Atwater could not stop thinking about the Mendocino Fracture Zone and its junction with the San Andreas and Queen Charlotte fault systems. They also liked looking for geometric patterns and figuring out what they meant in terms of seafloor spreading and plate tectonics.

In Bill Menard, I found a soul mate, a fellow enthusiast for geometric patterns and their implications. He was constantly cutting up pieces of paper and moving them around – "What if such and such happened? How would that play out the sea floor patterns?" He had a thorough knowledge of the oceanic data sets of the time; we would predict some geometric relationship with our paper cut-outs and he would then recall examples of the same patterns from the real world. Imagine my surprise when, after a few weeks he presented me with a draft manuscript describing our conversations. I was just having fun, playing intellectual games, and it was actually serious science. Indeed, those playful sessions resulted in three early papers in prestigious journals, summarizing the magnetic anomaly patterns in the northeast Pacific and generalizing them to examine the effects of changes in direction of sea floor spreading.

> *(Atwater, 2001: 248; she identified the three papers as Menard and Atwater, 1968; 1969, and Atwater and Menard, 1971)*

Menard (1967: 73) had already noted that the Mendocino fracture zone was "The most notable exception from a great-circle trend of the other northeastern Pacific zones only in its central portion." At the time, Menard thought that the great fracture zones were somehow related to the East Pacific Rise, but had no good idea about how they were related. Menard and Atwater (1968: 464), given Morgan's identification of the great fracture zones as concentric about a former rotational pole, now knew that the great fracture zones "are parallel to the spreading direction active at the time that they were frozen into the crustal block." Moreover, Morgan (1968b: 1968) had already proposed a change in direction of the Pacific block relative to the North American based on changing trends of the Mendocino and Pioneer fracture zones.

5.06 Except for the Mendocino and Pioneer fracture zones, the concentric circles are nearly coincident with the fracture zones. The Mendocino and Pioneer fracture zones depart from the circles farther west than do the other facture zones; this departure is likely related to North America "overriding" and interfering with the flow of the northernmost end of the rise at an earlier date. These old fracture zones indicate that the Pacific once moved away from North America toward trenches of New Guinea and the Philippines. About 10 m.y. ago this pattern changed, and the Pacific now moves toward the Japan and Aleutian trenches.

> *(Morgan, 1968b: 1968)*

Menard and Atwater identified three similar changes in the trends of the Mendocino and Murray fracture zones, one along the Blanco Fracture Zone, and proposed that such changes indicated five spreading episodes. They reproduced Morgan's Figure 11 (Figure 7.12) as their Figure 2 (Figure 7.41), and showed the five episodes of spreading beneath.

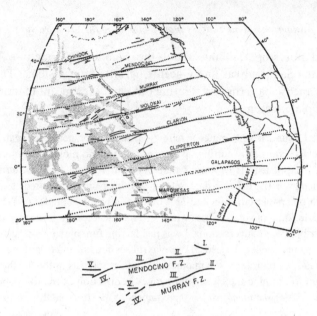

Figure 7.41 Menard and Atwater's Figure 2 (1968: 464). "Fracture zones in the north-eastern Pacific showing trends corresponding to five possible spreading episodes. Dotted lines are small circles about the pole at 79° N., 111° E. suggested by Morgan [1968]. It is the pole of rotation for episode III." The fracture zone showing the direction of spreading during episode I, the most recent episode, is the Blanco Fracture Zone, which connects the Gorda and Juan de Fuca ridges.

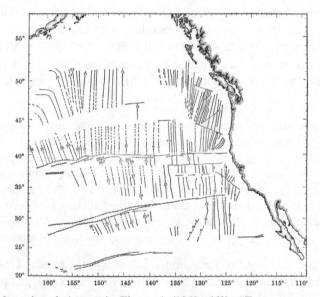

Figure 7.42 Menard and Atwater's Figure 1 (1968: 463). "Fracture zone and magnetic anomalies of the north-eastern Pacific. Numbers follow system of Pitman *et al.* [1968]. The Explorer, Juan de Fuca, and Gorda ridges are dotted. Some magnetic anomalies are numbered, and may be traced from fracture zone to fracture zone. Compare with Fig. 7.41 to identify fracture zones."

Turning to the magnetic lineations, and referring to their Figure 1 (Figure 7.42) they noted that the lineations are typically perpendicular to the fracture zones, that individual anomalies are "readily identified and classified," and that even though ridge segments offset by the fracture zones need not be parallel to the magnetic lineations and perpendicular to the offsetting fracture zones, they usually were.

They proposed that changes in spreading directions should be marked by changes in pattern of magnetic lineations, even given spreading stoppages of spreading as proposed by the Ewing brothers. Referring to their Figure 3 (Figure 7.43), they identified the change in pattern as a "Zed pattern."

Changes in directions of spreading can also be investigated using trends of magnetic anomalies; however, a criterion for identifying a change in direction by means of anomalies is required first. The chief clue is that the sequence of anomalies is continuous in all the ocean basins [Heirtzler *et al.*, 1968] despite prolonged time lapses between anomalies [Ewing and Ewing, 1967]. Thus the oceanic crust in any region breaks along any pre-existing spreading centre . . . Immediately after a change in the direction of relative motion of two separating blocks, the transform faults between them become parallel to the new spreading direction. If the spreading centres are intrinsically perpendicular to the transform faults . . . then the centres will gradually become realigned. The pattern that would emerge is a "Z" or an "N" or a mirror image, all of which may be called "Zed patterns" (Fig. 3).

(Menard and Atwater, 1968: 465)

They (1968: 465) identified Zed patterns (between anomalies 2 and 5) around the Juan de Fuca and Gorda ridges, and proposed that the distorted pattern of anomalies on the eastern side of the Gorda Ridge was probably caused by its interaction with both North America and the Mendocino Fracture Zone.

The magnetic anomalies around the Juan de Fuca ridge form just such a Zed pattern slightly split by parallel anomalies and fractured by minor adjustments. The western side of the Gorda ridge also shows the fan of anomaly trends characteristic of the Zed pattern although the eastern side is grossly distorted by interaction with North America and with the Mendocino fracture zone which has resisted the latest change of spreading direction.

(Menard and Atwater, 1968: 465)

The presence of these Z patterns marked the end of spreading episode II, and the beginning of episode I (Figure 7.41). Both Juan de Fuca and Gorda ridges were in the center of a Zed pattern. They noted that Zed patterns could also be used to identify changes in spreading directions even if the ridge had already been subducted as had the East Pacific Rise between the Gorda Ridge and the extension of the East Pacific Rise in the Gulf of California. They identified such a Zed pattern between anomalies 20 and 24, north and south of the Mendocino Fracture Zone, and claimed that it marked the transition between episode II and episode III.

It is evident that one side of the Zed pattern may be almost as diagnostic of a change in direction of spreading as the two sides together. Thus it is possible to identify an ancient change in spreading direction by means of an anomaly pattern far out on the flank of a rise or

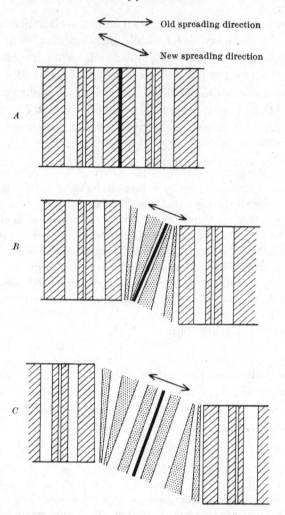

Figure 7.43 Menard and Atwater's Figure 3 (1968: 465). "Proposed mode of adjustment of spreading rises to a change in spreading direction. *A*, A segment of rise has been spreading forming symmetrical anomalies. The spreading direction changes. *B*, Spreading occurs in the new direction. The ridge centre gradually migrates around forming anomalies of a Zed pattern as it migrates. *C*, Spreading continues in the new direction. The ridge and its resulting anomalies are perpendicular to the direction. A split Zed pattern emerges."

ridge. This is particularly useful in the north-eastern Pacific because most of the eastern half of the East Pacific Rise is missing. North of the Mendocino fracture zone the the western group of anomalies have a different trend from those farther east, and between No. 21 and No. 23 they form half of a Zed pattern (Fig 1). South of the Mendocino zone, anomalies 21-23 also appear to form a transition in trend between the more regular patterns of higher and lower numbers.

(Menard and Atwater, 1968: 465)

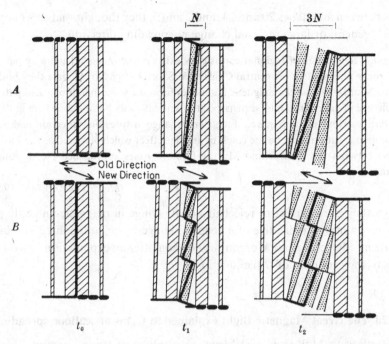

Figure 7.44 Menard and Atwater's Figure 5 (1968: 466). "Two possible modes for adjustment of a ridge to a change in spreading direction. *A*, The ridge readjusts as a unit taking on a new direction by t_2, after $3N$ km of spreading has occurred. *B*, The ridge breaks into segments, each piece becoming realigned by time t_1, after N km of spreading."

They added that changes in spreading direction were better supported if both markers were present, not just one. Moreover both markers were present between anomalies 20 and 24, and 2 and 5.

Attempting to explain the manner in which the crust adjusts to changes in spreading directions, they (1968: 466) proposed, "transform faults take on the new direction almost as soon as the change begins," which "necessarily involves fracturing of the corners of the old blocks, if the transform fault is offset in one sense or addition of new crust along them if it is offset in the other." In addition, they offered and illustrated (Figure 7.44) two scenarios (A and B) for the adjustment of ridges to a change in spreading direction: either the ridge adjusted as a single unit or broke into sections with each making the adjustment.

Further proposing that scenario *B* may occur if the ridge segment is particularly long or the change in spreading direction is large, they (1968: 466) suggested that the Pioneer fracture zone may have formed in such a way with the change in spreading "at the time of anomaly No. 23."

Menard and Atwater identified another change in both the trend of the fracture zone and magnetic anomalies "immediately off central California." Unlike the changes between anomalies 20 and 24, and 2 and 5, they thought that this one did not indicate a general or large regional change in spreading direction.

An exception to the idea of regional consistency exists in the magnetic anomaly pattern and fracture zones immediately off central California. Shortly after No. 11 time they both swing toward the south-west and the magnetic anomalies form the western half of a Zed pattern. No similar change in direction can be seen associated with anomaly No. 11 elsewhere in the north-eastern Pacific, as would be required if a general change in spreading direction had occurred. This exceptional pattern may be the result of an edge effect which occurred as the East Pacific Rise crest between the Pioneer and Murray fracture zones approached North America in No. 11 time.

(Menard and Atwater, 1968: 465)

Although they were correct to reject this dual change in magnetic anomaly pattern and fracture zones as indicative of a general change in spreading direction, its cause eluded them. McKenzie and Morgan would soon offer an explanation based on their analysis of the evolution of triple junctions (§7.23)

7.20 The Great Magnetic Bight explained in terms of seafloor spreading

Before turning to McKenzie and Morgan's analysis of triple junctions, I want to discuss Pitman and Dennis Hayes' (1968) and Vine and Hess's (1971) explanations of the origin of the Great Magnetic Bight in terms of seafloor spreading, V-M, and plate tectonics. Both explained its origin by proposing the former existence of a ridge and what McKenzie and Morgan (1969) would call a ridge-ridge-ridge (RRR) triple junction, both of which were later subducted. The Great Magnetic Bight, located at 50° N, 160° W south of the Aleutian arc, is a sequence of ten or so magnetic lineations that are bent 90° (Figure 7.42 and Figure 7.45). Explaining the origin of the Bight, which had been fully plotted and named by Elvers *et al.* (1967a, b), became a new and challenging problem for proponents of seafloor spreading. The challenge arose because there was no obvious way to generate bent magnetic lineations by seafloor spreading and V-M, and because the youngest rather than the oldest magnetic anomalies were closer to the nearest trench (Figure 7.45). I first consider Pitman and Hayes' solution.

Hayes, a former Ph.D. student of M. Ewing's had already co-authored a paper on magnetic anomalies with Heirtzler (Hayes and Heirtzler, 1968). Indeed, Hayes and Heirtzler (1968: 4643–4644) previewed Hayes and Pitman's explanation of the Great Magnetic Bight before fully presenting it in Pitman and Hayes (1968), which later appeared in *JGR* in October. They divided the North Pacific into four plates (Figure 7.46). Plate I represented the Pacific plate; Plate III, the North American plate. They supposed the Pacific plate fixed, and kept the North American plate fixed relative to the Pacific plate. They speculated that Plate II moved northward relative to Plates I, III, and IV. It was separated from Plate I by a east–west trending ridge offset by a

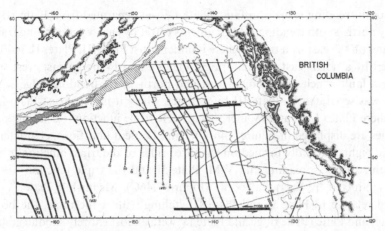

Figure 7.45 Pitman and Hayes' Figure 3 (1968: 6574) "Magnetic lineations in the North Pacific. The broad dashed lines are the lineations [constitutive of the Great Magnetic Bight] from Elvers *et al.* [1967a]; the solid lines show lineations inferred from data" collected from Lamont cruises of the *Robert D. Conrad* and *Vema*, "where dashed, the lineations are regarded as poorly controlled. The three bold east-west trending lines show the location of proposed fracture zones." (first bracketed addition mine).

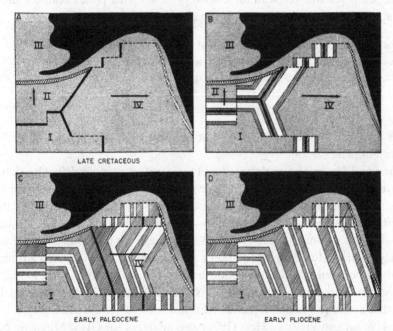

Figure 7.46 Pitman and Hayes' Figure 4 (1968: 6576). "Schematic diagrams of four stages in the development of the lineations of the north-east Pacific . . . The arrows indicate relative motions." The thick lines represent active ridges; the fine dashed lines, fracture zones; the crisscrossed areas are trenches or will soon be trenches; the alternating lined and clear strips are normal and reversed magnetic lineations.

north–south trending transform fault. Plate IV was separated from Plate III on the north by north–south trending ridge segments offset by east–west trending transform faults, and on the east by a trench where it underthrust Plate III. Plates II and IV were separated by a north–east trending ridge; Plates I and IV by an east–west trending transform fault which offset north–west and north–south trending ridge segments. Plate II was separated by Plate III by a trench where it under thrust Plate III. The point where Plates I, II, and IV met was a RRR triple junction. Resulting magnetic anomalies are displayed in Figure 7.46B, C, and D. Bent lineations, characteristic of magnetic bights naturally formed on the three sides of the triple junction in accordance with seafloor spreading and V-M. Plate II and the triple junction were completely subducted by Early Paleocene (Figure 7.46C). Meanwhile Plate IV moved eastward relative to Plate I, and Plate III, taking with it the ridge, the boundary between it and Plate I. The ridge and Plate IV were almost entirely subducted beneath Plate III by early Pliocene (Figure 7.46D). What remains is the Juan de Fuca plate.

Plate II is more or less the same plate that Atwater later named the Kula plate, choosing Kula because it means "all gone" in an Arthapaskan Indian dialect (Grow and Atwater, 1970: 3717). She, having learned from Menard the importance of giving good names to newly discovered objects or phenomena, sometimes feels embarrassed when credited with the discovery of the Kula plate because Pitman and Hayes actually made the discovery. But names are important.

Water Pitman and Denis Hayes at Lamont had already pointed out the evidence of this [Kula] plate, but they had described it and its neighbors as plates I, II, III, and IV, not exactly names that stick in the mind. Plates I, III, and IV were, in fact, the Pacific, North American, and Farallon plates.

(Atwater, 2001: 249)

Plate IV is more or less the same plate that McKenzie and Morgan would soon name the Farallon plate (§7.23).

Pitman and Hayes also anticipated and removed an obvious difficulty that probably would have been raised against their reconstruction of the Cenozoic history of the North Pacific. Realizing that their placement of ridges and trenches in close proximity clashed with classical seafloor spreading with its intimate tie between ridges with upward mantle convection and trenches with downward mantle convection, they favored McKenzie and Parker's plate tectonics which severed any straightforward tie between mantle convection and ridges. In addition, they (1968: 6577) noted that close proximity of the active East Pacific Rise and Middle America Trench proves that active ridges and trenches "can reside in very close proximity."

Vine (1968a) and Vine and Hess (1971) also analyzed the Great Magnetic Bight in terms of a formerly existing triple junction. Vine's analysis was done independently of Pitman and Hayes, but they deserve priority. Recalling what he did, when he did it, and why he did not think the Great Magnetic Bight or any RRR triple junction presented a difficulty for seafloor spreading, Vine wrote:

The paper by Harry and myself for *The Sea*, vol. 3 was written over the summer of 1968. (Presumably early summer, because I left Princeton in July for Cyprus) It was submitted in November, as noted at the head of the Chapter. I am not sure when I first heard of the magnetic bight. I think it was at a conference and during a talk by George Peter.[21] As I recall he presented it with some glee in the hope that it would be inexplicable in terms of sea floor spreading and hence lead to its downfall. From the outset I did not consider it to be a problem and that it must be the result of spreading about a triple ridge junction now subducted beneath the Aleutian arc. Why? Simply because if there are more than two plates covering the Earth's surface there must be triple junctions, and if you have a triple junction in an oceanic area then presumably it will be a triple ridge junction.

(Vine, January 25, 2012 email to author)

Vine completed his section of the paper before he left Princeton on July 12 for England then on to Cyprus, where he spent from August 12 through October 3 doing field work (Vine, January 31, 2012 email to author). So Vine had finished his analysis of the triple junction and left Princeton before Hayes and Heirtzler (1968) and Pitman and Hayes (1968) appeared in *JGR*.

In Vine (1968a), the abstract of the paper he presented at the 1968 spring AGU meeting, he argued that the Pacific plate had moved "northward through approximately 25° of latitude" since the Cretaceous based on various types of paleomagnetic evidence, including "the ratio of the amplitudes of the anomalies on north–south and east–west profiles" which "give a measure of the paleolatitude of the crust at the time of its formation." The north–south and east–west anomalies were those making up the two sides the Great Magnetic Bight.

Paleomagnetic Evidence for the northward Movement of the North Pacific Basin during the Past 100 m.y. Existing analyses of magnetic anomalies associated with isolated seamounts southwest of Hawaii and east of Japan suggest that these areas have moved through approximately 30° of latitude, with respect to the magnetic pole, since the mid-Cretaceous. A similar analysis for Midway atoll confirms this picture, and paleomagnetic results from an oriented drift core from the island indicate that Midway may have been 15° south of its present latitude in the Oligocene. The older linear magnetic anomalies of the North Pacific trend both approximately north–south and east–west south of the Aleutians. If it is assumed that the oceanic crust beneath these anomalies is essentially uniform in magnetic properties and was created simultaneously at former ridge crests, such that the anomalies reflect the same sequence of reversals of the Earth's magnetic field, then the ratio of the amplitudes of the anomalies on north–south and east–west profiles give a measure of the paleolatitude of the crust at the time of its formation. Again crust thought to have formed 60–70 m.y. ago would appear to have moved northward through approximately 25° of latitude. This movement may result from spreading from the East Pacific rise and its extension south of New Zealand and Australia, and imply the resorption of vast areas of former North Pacific crust in marginal and encroaching trench systems.

(Vine, 1968b)

Vine repeated his analysis in his joint paper with Hess, which was submitted in November 1968. He also considered the former position of the North American

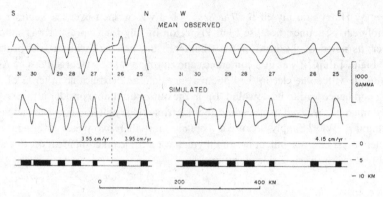

Figure 7.47 Vine and Hess's Figure 6 (1971: 602). "Observed and simulated magnetic anomaly profiles across the two limbs of the magnetic "bight" shown in Fig. 7 [Figure 7.48]. The observed profiles were obtained by averaging the four north–south and east–west profiles indicated on Fig. 7." Normal anomalies are shaded; reversed, unshaded.

plate during the Cretaceous, and argued that there "must have been considerable relative movement between the Pacific and American blocks during the Cenozoic with resorption of vast areas of oceanic crust in an intervening trench system."

The discrepancy between the predicted and observed amplitudes of anomalies on east–west profiles across the Great Magnetic Bight (Fig. 6 [Figure 7.47]) is thought to indicate its magnetic latitude during formation (Vine 1968[a]). In order to reproduce the observed ratio of amplitudes on east–west and north–south profiles it is suggested that the bight was formed at least 25° *south* of its present latitude. This contrasts with the paleomagnetic pole for the North American block, which is at 70°N in the Chukchi Sea, during the Cretaceous (Larochelle, 1968). The Alaska peninsula would then have been at 70°N; 20° north of its present position. There therefore must have been considerable relative movement between the Pacific and American blocks during the Cenozoic with resorption of vast areas of oceanic crust in an intervening trench system.

(Vine and Hess, 1971: 604; my bracketed additions)

Thus Vine disagreed with Pitman and Hayes who favored the Ewing brothers' view that seafloor spreading began about 10 m.y. ago after a worldwide cessation of many millions of years.[22]

Vine also reconstructed the RRR triple junction responsible for the Great Magnetic Bight, and argued that it is "consistent with spreading" and even constructed what amounted to a velocity triangle of the junction, what McKenzie and Morgan (1969) would do for sixteen different types of triple junctions.[23] He described the reconstruction with the following:

The rates, and the azimuths of the linear anomalies indicated in Fig. 7 [Figure 7.48], enable one to predict the geometry of the corner (see Fig. 8 [Figure 7.49]). Alternatively the azimuths of the anomalies and of the corner enable one to predict the ratio of the spreading rates.

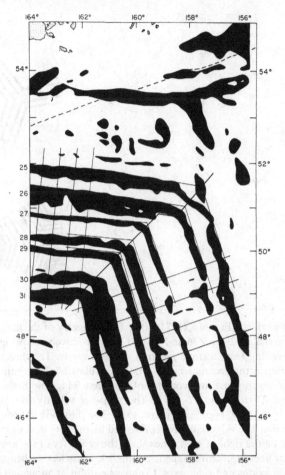

Figure 7.48 Vine and Hess's Figure 7 (1971: 603). "Summary of magnetic anomalies defining the "magnetic bight" of the northeast Pacific (after Elvers *et al.*, 1967). Positive anomaly areas are indicated in black and numbered for identification as in Fig. 6. [Figure 7.47]. . . The dashed line indicates the topographic axis of the Aleutian trench and the solid lines the profiles referred to in the caption to Fig. 6. The geometry of the bight assumed and reproduced in Fig. 8 [Figure 7.49] and 9 is also indicated by thin solid lines." Anomaly numbers are shown along the left margin (Elvers *et al.*, 1967 is my Elvers *et al.* 1967a; my bracketed additions).

Either way it can be seen from Figs. 7 and 8 that the predicted and observed directions and rates are indistinguishable, and the geometry of the bight is therefore consistent with spreading. If it is assumed that the relative movement between crustal plates that form the bight was at right angles to the anomalies . . . then one can deduce a possible plate geometry at the time of formation of the bight as shown in Fig. 8. The geometry will be slightly different during the formation of anomalies 25 and 26 in comparison with anomalies 27 through 31.

(Vine and Hess, 1971: 601–602; my bracketed additions)

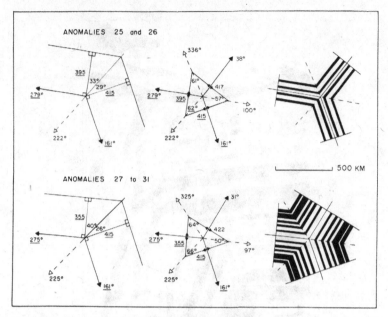

Figure 7.49 Vine and Hess's Figure 8 (1971: 604). The geometry of the magnetic bight during the formation of anomalies 25 and 26 and 27 and 31, respectively, showing that it is entirely consistent with spreading from a triple junction of ridge crests. The direction and distances underlined are assumed; the remainders have been calculated." The center diagrams in the series of three diagrams showing the geometry of the Great Magnetic Bight during formation of anomalies 26 and 27 (top) and 27 and 31 (bottom) are velocity triangles (compare with McKenzie and Morgan, 1969, Figure 3 [Figure 7.51]). The thin dashed lines with open arrows are the velocity vectors. The ridge axes are represented by the very thin solid lines with shaded arrows. The figures on the right of each series show the geometry of the respective anomalies. Solid thick and clear lines represent magnetic anomalies; arrowless dashed and thin solid lines represent velocity directions and ridge axes. Compare pattern of anomalies with Figure 7.48.

Vine also recalled that he did not think about the general issue of whether or not various triple junctions are stable.

I don't think I had thought about the stability of triple junctions in general, although my geometric analysis of the triple ridge junction, in which the ridge crests are perpendicular bisectors of the three sides of the velocity triangle, implies that RRR triple junctions are always stable, i.e. the three perpendicular bisectors always intersect at a single point, the circumcenter of the triangle.

(Vine, January 25, 2012 email to author)

In closing this discussion of the origin of the Great Magnetic Bight and the treatment of this formerly existing RRR triple junction, I would be remiss not to mention the recognition by Herron and Heirtzler (1967) and Raff (1968) of a RRR triple junction west of the Galapagos Islands where the Galapagos Rift Zone joins the East Pacific Rise.

It is time to turn to McKenzie and Morgan's formal treatment of the triple junctions.

7.21 McKenzie and Morgan explore the evolution of triple junctions

I must humbly apologise for not replying sooner to your letter of mid-November . . . I thoroughly enjoyed your manuscript on triple junctions. I think your formulation is great and very timely. I think people underestimate the importance of formulating concepts correctly and thoroughly, after all what more did Newton do? Many of us recognized "transform faults" before Tuzo but none of us bothered to name them or formulate then properly. Similarly I have only been tinkering with triple junctions as the need arose i.e., in connection with the magnetic bight, Guatemala, and the Japan, Bonin and Ryuku arcs. The diagrams I sent you in connection with the magnetic bight are from our manuscript for "The Sea" Vol. IV and probably will not see the light of day for another year.

(Vine, January 15, 1969 letter to McKenzie)

What I did was to go through the various geometric possibilities, define stability and show how to work out whether a particular triple junction was stable or not. Given what Tuzo had done with transform faults, it seemed to me to be worthwhile to consider the kinematics of all possible combinations. It is this, not the idea of a triple junction, which was new.

(McKenzie, January 25, 2012 email to author)

In early 1968, Hess invited McKenzie to Princeton. Robert Phinney wrote McKenzie with the official offer of a visiting appointment, and McKenzie provisionally accepted it on June 22, when he replied to Phinney.

Dear Bob:

I would be delighted to have a visiting appointment provided those who need to know it is only for two months or so. I have never given a lecture course, but it is high time I learnt to. Would perhaps 6 lectures on the Physics of the Earth's Interior, mainly mathematical, fit in? I don't want to read up on things I am not working on because my years of unfettered research are limited. Here they pay me $8,500 p.a., at La Jolla $11,000 and I am going back to $2,200; I suggest something in between. I will be coming to Princeton from Lamont, so that will cost nothing; as to my fare back to England we can discuss that when I arrive. I paid my own way over here, and am expecting to do the same going back.

McKenzie arrived at Princeton from Lamont in January, 1968. He stayed for about four months, and does not "remember giving a lecture course." Like Atwater, Morgan, and Vine, he gave a talk at the spring 1968 AGU meeting, for which, to his "astonishment" he needed Ewing's permission since he submitted his abstract while at Lamont – Ewing ran a tight ship.

I went to Princeton round about Jan 1968, and stayed for about 4 months. I don't remember giving a lecture course. I certainly went to the AGU and gave a paper, which I submitted from

Lamont. To my astonishment I had to get permission from Ewing to do so, and he required me to make some changes (I don't remember what these were, or what the talk was about). I met him to shake him by the hand at the conference in New York in 1966, when Teddy introduced me. But I never met him again.

(McKenzie, January 25, 2012 email to author)

As noted already, Atwater visited Princeton after the AGU meeting, and she, McKenzie, and Morgan talked about triple junctions and the Mendocino triple junction, attempting to unravel the recent history and interaction of the Pacific plate with western North America. Once McKenzie (2001:183) got to know Morgan, he was "astonished" that they "had been working on the same problems in the same way for much of the previous year." Morgan did not end up speaking about the plate rigidity at the spring AGU meeting as he had expected to do and so written in his December 8, 1967 letter to McKenzie (§7.15). Morgan instead wrote on the Juan de Fuca Ridge and the interaction of the ridge with western North America. Here is his abstract:

Pliocene Reconstruction of the Juan de Fuca Ridge. A magnetic anomaly map of the Juan de Fuca ridge area has been cut with scissors along the San Andreas, Mendocino, Blanco, and Queen Charlotte faults. The central area of positive magnetic anomaly of the Gorda, Juan de Fuca, and Explorer ridges that correspond to sea floor created in the last 0.75 m.y. has been cut out and removed. The "jig saw puzzle" has been reassembled with the Pacific Block displaced 45 km (0.75 m.y. at 6 cm/yr) southeast relative to the North America block. This process is repeated for several distinctive magnetic boundaries back to about 8 m.y. ago. No unique solution to the puzzle is found since there are too many degrees of freedom, but the best solutions require the ocean floor off the coast of Washington and Oregon to have a small component of eastward velocity, with crustal shortening occurring in either the Cascades or Coast ranges.

(Morgan, 1968a)

Morgan, seeing this abstract after many years, remembered what happened when he began coloring and cutting up the magnetic anomalies to see how they fit together.

I've always thought of this paper as my "paper doll paper." I had a stack of xerox copies of the black and white magnetic anomaly pattern of the Juan de Fuca area (yes, there was a Xerox machine in Guyot [Hall] by then) and used colored pencils to indicate the different ages of the magnetic stripes – i.e., a cheap-quick version of the beautiful figure of the Juan de Fuca that Fred had made that was on the cover to the book of the Goddard Inst. Conference [§6.12]. While I was sitting on the living room floor that Sunday afternoon, coloring and cutting out pieces to try to fit together at different ages, my 5 year old daughter watched me for a while, then went upstairs to her room and came down with a paper doll book, and sat down beside me coloring and cutting.

(Morgan, March 16, 2012 email to author).

Given Atwater's work with Menard, it is no wonder that when she met with Morgan and McKenzie they talked about the recent history of the Pacific plate and its interaction with western North America. McKenzie, who eventually provided the preliminary answer as to how triple junctions evolve, recalled how the idea came to him when waking up one morning.

I remember it very clearly. I was at Princeton at the time and Tanya Atwater came to visit. All three of us had been worried about the problem of what happens at triple junctions. I remember the three of us trying to work out on the board one day what happened. We tried to do the same thing the next day and we got different answers. None of us could see how to do this. I went back to Cambridge. I went off with John Sclater to the Indian Ocean, and studied its magnetic anomalies. I came back to Cambridge. I wasn't thinking about this at all, I was thinking about the Indian Ocean and reconstructing its magnetic anomalies. One morning I essentially woke up with the solution. I have no idea where it came from. I wrote it up quickly before I forgot it and then worked out why it worked – a most extraordinary situation.

(McKenzie, response to a question at a plate tectonics 25th anniversary celebration, 1992 spring AGU meeting).[24]

McKenzie had a good enough draft of the paper by mid-November 1968, and sent it to Morgan, asking him what he thought about it and whether he wanted to be a co-author. Morgan agreed to be a co-author. Morgan, I add parenthetically, believes (March 11, 2012 phone conversation with author) that the triple-junction paper was 90% McKenzie's work. He also thinks,

My main contribution to this paper was a demonstration that it wasn't possible for the three plates at a triple junction to continue to rotate through finite angles about fixed poles [except in the special case where all the relative motion poles (but not angular velocities) were coincident].

(Morgan March 16, 2012 email to author).

McKenzie also sent copies to Vine and Parker, and showed it to Bullard, who told him that he needed to explain just what he meant by "stability." He received some feedback from Vine, who, as noted above, told him about his own explanation of the Great Magnetic Bight and reconstruction of the now subducted RRR triple junction from which the bight was generated.[25] McKenzie and Morgan met at the AGU 1969 spring meeting, and Morgan affirmed that he wanted to co-author the paper. McKenzie, who returned to the seismological laboratory at Caltech in April and worked there through June, is sure that he submitted their paper there.[26]

The date of submission is June 1969, and I am sure I submitted it from Caltech for two reasons. One is that the diagrams were drawn by Laszlo Lenches, the draughtsman at Caltech, whose style is distinctive. The other is that Warren Hamilton came to visit the Seismological Lab at Caltech and gave both Don Anderson (who was head of the lab) and me a very nasty turn. Don had brought him to talk to me, and Warren was particularly interested in what I had to say about triple junctions, which I explained to him rather thoroughly. He was the seminar speaker that day. But, instead of talking about what had been announced, he gave the whole seminar about triple junctions

(McKenzie, January 25, 2012 email to author)

Although Hamilton's reaction provided the impetus to submit the paper, McKenzie (January 30, 2012 email to author) emphasized that "his and Anderson's concern was misplaced: Warren did not make any attempt to take credit for the ideas he talked about, and has remained a good friend of mine ever since this first meeting." As previously described (I, §7.6–7.9), Hamilton was one of the few North American geologists who had argued in favor of mobilism before the confirmation of ridge-ridge transform faults or V-M. He certainly recognized a good idea when he saw one, and surely appreciated how it could be applied to unravel the recent tectonics of western North America. He also remembers that he incorporated what essentially amounted to McKenzie and Morgan's ideas about the evolution of triple junctions, in particular as they applied to the San Andreas triple junction. Hamilton also emphasized that he does not *"deserve any credit for triple junctions."*

I did give a talk at Caltech in 1969. My big topic everywhere that year was western US plate tectonics, mostly the Cretaceous subduction story, plus a little earlier colliding arcs, but I was fully aware of the San Andreas triple junction pitch, and knew on my own that it broadly fit the onshore geology, and presumably added that as Dan McKenzie recalls. . . *I do not, however, deserve any credit for triple junctions.*

(Hamilton, January 31, 2012 email to author)

McKenzie asked Laszlo Lenches, the draughtsman at Caltech, to prepare the figures, and sent the manuscript to *Nature* once Lenches was finished. *Nature* received the manuscript on June 9, and it was published on October 11, 1969, and Atwater proceeded to use McKenzie and Morgan's analysis of triple junctions to unravel the Cenozoic history of the Pacific and how it affected the tectonics of western North America (1970). But that was in the future, beyond the scope of this book and I must return to McKenzie and Morgan's paper.

After delineating various independent ways, given what is now "often called plate tectonics," the instantaneous relative motions of any two plates may be determined. Emphasizing the "striking" agreement between such determinations (RS1), they identified

the two main reasons why plate tectonics does not yet provide a complete theory of global tectonics. The first is that the mechanism by which the motions are maintained is still unknown, though it now seems that some form of thermal convection can provide sufficient energy 13. This problem will not be discussed further. The other is that the original ideas only apply to motions at present talking place, and are not concerned with either the slow evolution of plate boundaries or with changes in their relative motion through geological time.

(McKenzie and Morgan, 1969: 125; endnote 13 refers to McKenzie (1969b))

Further narrowing their current concerns, they wanted to investigate "two causes of plate evolution" that "are the geometric consequences of the motion of ridge plates." The first concerned changes in the nature of plate boundaries, which do not involve changes in the direction or magnitude between plates be they simply along a common boundary or, more interestingly, at triple junctions, where three plates meet. The

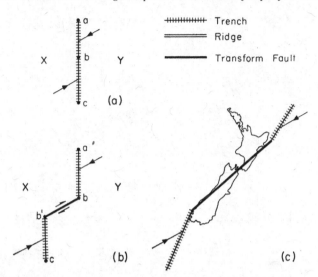

Figure 7.50 McKenzie and Morgan's Figure 1 (1969: 126). "The evolution of a trench. The arrows show the relative motion vector and are on the plates being consumed. Thus Y is consumed between a and b, X between b and c. The trench evolves to form two trenches joined by a transform fault. (c) is a sketch of New Zealand showing that the Alpine fault is a trench-transform fault of the type in (b)."

second did involve real changes in the relative motions of the three plates joining at the triple junction. But, again, the change is simply a consequence of the geometry of plate motions, of their kinematics, not their dynamics, and it derives from the fact that finite rotations, unlike instantaneous ones, are not vectors. Following McKenzie and Morgan (1969), I first consider those that do not involve changes in plate rotations.

The example they offered of a change along a common boundary (but not a triple junction) involved situations where a trench by the mere geometry of the situation, in particular because trenches "consume lithosphere on only one side," will evolve into a trench-transform fault-trench boundary, and they had in mind an actual case, namely the Alpine fault in New Zealand (Figure 7.50)

Turning to triple junctions and defining trenches, ridges, and transform faults in terms of "destruction, and creation of plates rather than in terms of topographic features," McKenzie and Morgan introduced the idea of stable triple junctions:

For the purposes of plate tectonics, the surface of the Earth is completely covered by a mosaic of interlocking plates in relative motion. There are many points where there plates meet, but, except instantaneously, none where four or more boundaries meet. The relations between the relative velocities of the plates at triple junctions have been discussed previously[3]; they are a consequence of the rigidity of the plates and do not impose any restrictions on the orientation of plate boundaries or on the relative velocity vectors. If, however, the triple junction is required to look the same at some later time, there are important restrictions on the possible

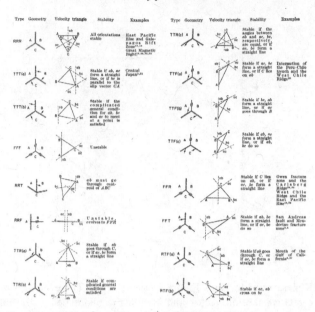

Figure 7.51 McKenzie and Morgan's Figure 3 (1969: 127). "The geometry and stability of all possible triple junctions. Representation of structures is the same as in Fig. 1 [Figure 7.50] Dashed lines *ab*, *bc* and *ac* in the velocity triangles join points the vector velocities of which leave the geometry of *AB*, *BC* and *AC* respectively, unchanged. The relevant junctions are stable only if *ab*, *bc* and *ac* meet at a point. The condition is always satisfied by *RRR*; in other cases the general velocity triangles are drawn to demonstrate instability. Several of the examples are speculative" (my bracketed addition) Their endnotes respectively refer to following papers: 3 to McKenzie and Parker (1967), 4 to Morgan (1968b), 10 to Barazangi and Dorman (1969), 17 to Herron and Heirtzler (1968), 18 to Raff (1968), 19 to Pitman and Hayes (1968), 20 to Vine and Hess, 1971, 21 to Gutenberg and Richter (1954), 22 to Peter (1966b), 23 to Elvers *et al.* (1967a), 24 to Matthews (1966a), 25 to Sykes (1968), 26 to Morgan *et al.* (1969), and 27 to Larson *et al.* (1968).

orientations of the three plate boundaries. Unless these conditions are satisfied the junction can exist for an instant only, and for this reason is defined as unstable. If evolution is possible without a change in geometry, then the vertex is defined as a stable junction. The distinction between the two types is important because movement of stable junctions permits continuous plate evolution.

(*McKenzie and Morgan, 1969: 126; endnote 3 refers to McKenzie and Parker, 1968*)

A triple junction is stable if, and only if, its geometry remains the same as the plates in question rotate relative to each other. In such cases the triple junction is able to migrate along one of its boundaries permitting continuous plate evolution. McKenzie, as already noted, wanted to give a systematic account covering all possible types of triple junctions. He isolated sixteen, and proposed a general method for determining if examples of each type are stable or unstable. They displayed their results in Figure 7.51, letting *R* stand for ridge, *T* for trench, and *F* for transform fault.

Combinations of the three types of plate boundaries yield ten types of triple junctions, three where the same type of boundary meet (*RRR, TTT, FFF*), one where each boundary is different (*RTF*), and six where two but not three boundaries are the same (*RRT, RRF, TTR, TTF, FFR, FFT*). Because trenches can consume lithosphere only on one side, the number of possible triple junctions increases to sixteen (Figure 7.51).

Thus they provided general stability conditions for all possible triple junctions. Easing the reader into their analysis, they began with *RRR* triple junctions which are always stable and perhaps the easiest to understand. Indeed, Pitman and Hayes, and Vine and Hess (§7.20) had already shown how seafloor spreading could generate the magnetic anomaly pattern associated with such triple junctions, and Vine had even constructed their velocity triangle, anticipating the velocity triangle shown in Figure 7.51.

Perhaps, the simplest example of the general method is the triple junction between three ridges, which we shall call an RRR junction (Fig. 3 [Figure 7.51]). An example of such a junction is the meeting of the east Pacific rise and the Galapagos rift zone in the equatorial east Pacific[17, 18]. The Great Magnetic Bight in the north Pacific was probably formed by another such junction which has now ceased to exist[19, 20]. In this and all other examples the relative velocity vectors at the junction are required to satisfy[3] (Fig. 3)

$$A^VB + B^VC + C^VA = 0$$

This equation must be satisfied if the plates are rigid.

(*McKenzie and Morgan, 1969: 127; endnote references are to Herron and Heirtzler, 1968; Raff, 1968; Pitman and Hayes, 1968; Vine and Hess, 1971, which they referred to as being in press*)

They then explained why *RRR* triple junctions are always stable.

In Fig. 3, the lengths *AB, BC* and *CA* are proportional to and parallel to the velocities $_AV_B$, $_BV_C$, and $_CV_A$ respectively. The triangle is therefore in velocity space, and represents the condition imposed by equation 1. Because ridges spread asymmetrically at right angles to their strike, a point on the axis of the ridge *AB* will move with a velocity $_AV_B^2$ [that is, $(_AV_B)2$] relative to *A*. This velocity corresponds to the mid point of *AB* in Fig. 3. Consider a reference frame moving with a velocity corresponding to some point on the perpendicular bisector *ab* of *AB*. *ab* is parallel to the ridge *AB*, so in this frame the ridge will move along itself and will have no velocity at right angles to *AB*. The same is true of the plate boundaries *BC* and *CA* when observed from reference frames whose velocities lie on *bc* and *ac* respectively. The perpendicular bisectors of the sides of the sides of any triangle meet at a point called the centroid, and this point *J* in velocity space gives the velocity with which the triple junction moves. It is therefore always possible to choose a reference frame in which the triple junction does not change with time. From the velocity triangle the relative velocity V of all plates relative to the triple junction is:

$$V = \frac{|B^VC|}{2\sin\alpha} = \frac{|C^VA|}{2\sin\beta} = \frac{|A^VB|}{2\sin\gamma}$$

Also the angle between *AB* and *AJ* is 90 − γ. Such a junction between three ridges is therefore stable for all ridge orientations and spreading rates.

(*McKenzie and Morgan, 1969: 126–127; my bracketed addition*)[27]

RRR triple junctions are the only ones that are always stable.

The use of velocity triangles provides an easy method for determining whether any possible triple junction is stable. If the triple junction's velocity triangle is such that its three velocity vectors intersect at a common point, the triple junction is stable. Indeed, the ease of constructing velocity triangles and this straightforward procedure for determining whether or not a triple junction is stable is a highlight of their paper, and made it immediately useful to others attempting to describe the evolution of plate boundaries. They constructed their Fig. 3 (Figure 7.15) in such a way that they could show, except for *RRR* triple junctions which are always stable, an unstable example of the triple junction. They also described at least one set of circumstances which would yield a stable example of the type of triple junction, except for *FFF* triple junctions which are never stable as shown by the fact that its velocity vectors *ab*, *bc* and *ac* can never have a common point of intersection.[28] Nine junction types are stable, if two of their boundaries are collinear (*TTT*(a), *TTR*(a), *TTR*(c), *TTF*(a), *TTF*(b), *TTF*(c), *FFR*, *FFT*, and *RTF*(a). They also identified those triple junctions presently found on ocean floors, and noted the former *RRR* triple junction that generated the Great Magnetic Bight prior to its consumption in the Aleutian Trench. Moreover, they were able to use velocity triangles representing velocities on a plane rather than having to use angular velocities because triple junctions are points; the velocity triangle shows instantaneous velocities on a plane to tangent to the "spherical" Earth at the point of the triple junction. Finally, at Vine's suggestion (January 15, 1969 letter to McKenzie), they (1968: 129) pointed out that McKenzie and Parker had diagramed two triple junctions that were unstable.

After describing all sixteen types of triple junctions and their stability conditions, they turned to evolution of the northeast Pacific and tectonics of the western margin of North America, essentially the same group of problems that they had discussed with Atwater at Princeton the previous spring, and which Atwater would solve, in part, by extending their analysis. Reproducing as their Figure 4, Figure 1 (Figure 7.42) from Menard and Atwater (1968), McKenzie and Morgan agreed with them that there was a change in spreading direction between anomalies 23 and 21, and proposed the previous existence of a plate, which they called the Farallon plate, between the Pacific and North American plates.

There are at least four examples of triple junctions at present in the north Pacific . . . Three of the active junctions occur along the west coast of North America, and their evolution demonstrates how complicated the interaction of three plates can be even without any changes in their relative motion. Fig. 4 is taken from Menard and Atwater (1968), and shows diagrammatically the magnetic lineations in the north-east Pacific. They point out that there are two striking changes in the trend of both transform faults and of magnetic lineations. The first is between anomalies 23 and 21 throughout the north-east Pacific north of the Pioneer fracture zone, and is best explained by a change in the motion direction of one of the plates. Most probably the plate between the Main Pacific plate and the American plate was the one involved. This plate is called the

Farallon plate throughout the rest of the discussion, after the Farallon Islands off the coast of central California.

(McKenzie and Morgan, 1969: 129)

Turning to the second change in trend of the transform faults and magnetic lineations, they attributed it to the evolution of a triple junction rather than to a change in spreading directions. Otherwise, they would have to propose that the plates were not rigid.

The second change in trend occurs at anomaly 10 near San Francisco, but not until after anomaly five north of the Mendocino. It is thus not possible to produce this change by a change in the motion of any of the plates at a given time. Such apparent changes in spreading direction are, however, easily explained by the evolution of the triple junctions formed at the time of anomaly 10.

(McKenzie and Morgan, 1969: 129)

Displaying the versatility of their analysis of triple junctions, they offered an explanation of the longstanding problem of the origin of the San Andreas fault and its relationship with the Mendocino Fracture Zone.

The main features of the north-eastern Pacific before the time anomaly 10 was formed have now largely vanished. Except near the Gorda and Juan de Fuca ridges, and south of Baja California, only one half of the anomaly pattern remains. Thus there must have been a trench between the ridge and the coast of North America which consumed the Farallon plate with its anomalies. This trench must have existed as a continuous feature up to about the time of anomaly 10, or the Middle Oligocene. Fig. 5a [Figure 7.52a] shows the arrangement of plates at about the time of anomaly 13. If we assume that all relative plate motions remain constant from the time of anomaly 13 onwards we can deduce the motion of all the junctions relative to any plate. In Fig. 5 [Figure 7.52] all fracture zones except the Mendocino and the Murray have been omitted. The offsets in the ridge show that it will first meet the trench just south of the Mendocino fracture zone to form two triple junctions, *FFT* in the north and *RTF*(a) in the south (Fig. 5b [Figure 7.52b]). Fig. 5c [Figure 7.52c] shows that the first of these is stable if the transform fault between *A* and *C* and the trench between *A* and *D* lie in a straight line. The triple junction is then at rest relative to *C*, and therefore *J* moves north-westward relative to *A*, changing the trench into a transform fault. Similarly the second junction is stable if the trench and the transform fault are in a straight line (Fig. 5d [Figure 7.52d]). Clearly this junction can move north-west or south-east relative to *A*, depending on the magnitude and direction of the relative velocities. Unless *bc* lies to the east of *A* in Fig. 5d [Figure 7.52d], however, the ridge axis will move away from the trench, and they could never have met. Thus *J* lies to the south-east of *A*, and the junction will move down the boundary of *A*. This southward migration stops when the junction *RTF*(a) reaches the Murray fault (Fig. 6a [Figure 7.53a]) where it must change to *FFT* and move rapidly north-westward relative to the American plate, for it must then be fixed to *C* (Fig. 6b [Figure 7.53b]). The stability condition is again that the trench and transform fault between *A* and *C* form a straight line. The north-west motion then regenerates the trench on the western margin. During this period (Fig. 6b [Figure 7.53b]) the trench along the west coast continues to consume the two remaining pieces of the Farallon plate except between

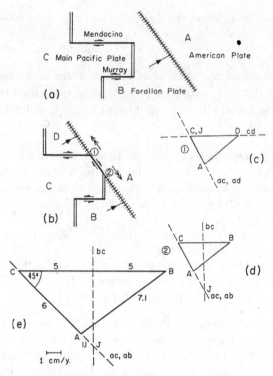

Figure 7.52 McKenzie and Morgan's Figure 5 (1969: 129). "(*a*) The geometry of the north-east Pacific at about the time of anomaly 13. All fracture zones except the Mendocino and the Murray have been omitted for simplicity. (*b*) Stable triple junctions at about the time of anomaly 9, formed when the east Pacific met the trench off western North America. The double headed arrows show the motion of the two junctions (1) and (2) relative to the American plate *A*. (*c*) is a sketch of the vector velocity diagram for junction (1) and shows it will move north-west with the Main Pacific plate. (*d*) is a similar diagram for (2). If the relative plate motions have not changed since at least the Middle Oligocene, the magnetic lineations and the present motion on the San Andreas may be used to draw the velocity diagrams to scale. (*e*) is such a drawing of (*d*), and shows that the triple junction *J* will slowly move to the south-east relative to *A*. The numbers are in cm/yr and the vector AB shows the direction and rate of consumption of the Farallon plate by the American."

the Mendocino and the Murray faults. Thus whether or not the oceanic transform faults possess continental extensions they can influence the tectonics of the continental margin. The geometry of the plate boundaries in Fig 6*b* [Figure 7.53b] changes back to Fig. 5*b* [Figure 7.52b] when the ridge south of the Murray migrates east to meet the trench. The resulting triple junction *RTF*(a) then continues the earlier slow movement to the south-east relative to the American plate.

(McKenzie and Morgan, 1969: 129; my bracketed additions)

Continuing to display the explanatory power of their analysis of triple junctions, they showed its quantitative strength.

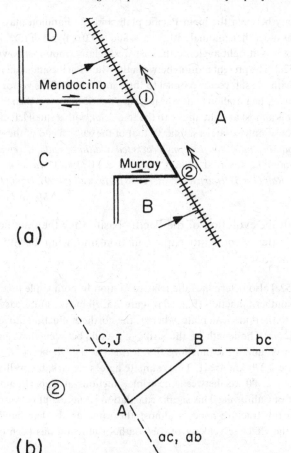

Figure 7.53 McKenzie and Morgan's Figure 7.53 (1969: 130). "When the southern junction (2) in Fig. 5*b* reaches the Murray, the right lateral offset of the ridge must change it from *RTF*(a) to *FFT*, and also cause the junction to move north-west, because it is then fixed to plate *C*. (*b*) gives the velocity triangle for junction (2) in (*a*)."

The complex series of events described is the inescapable consequence of the motion of rigid plates whose relative velocity remains constant. The geological history of the west coast of America and of the surrounding sea floor since the Cretaceous is in general compatible with the evolution of the plates and triple junctions outlined here. The magnetic lineations off the west coast[12, 22,27-31] show that the Farallon plate remained intact until the ridge first met the trench at the time of anomaly 10 (Figs. 4 [Figure 7.42] and 7 [not reproduced]) or in the Middle Oligocene[32,33]. As expected, the anomalies to the north and south show no change in spreading direction at the time. To the south the anomalies in contact with the continental margin become progressively younger to the present active ridge axis at the mouth of the Gulf of California, with probably a short interruption in the steady progression at the Murray fracture zone. If the motion of the Main Pacific and American plates has remained unchanged since at least the Oligocene, it is possible to determine the relative velocities of all three plates and their associated triple junctions.

The relative velocity between the main Pacific plate and the Farallon may be obtained from magnetic lineations older than anomaly 10[15, 32, 33], and is 5.0 cm/yr half rate. The fracture zones show that the ridge was at right angles to the east-west relative motion, shown to scale as *BC* in Fig. 5e [Figure 7.52e]. The present motion between the Main Pacific and American plates is close to 6 cm/yr[1, 7, 15], with the slip vector parallel to the San Andreas. This vector is shown as *AC* in Fig. 5e. The motion of the Farallon plate towards the American plate is then found to be 7 cm/yr. The motion vector is almost at right angles to the trench *ab*, with a small left-handed component of 1 cm/yr. This consumption rate is similar to that of the eastern end of the Aleutian arc.

(McKenzie and Morgan, 1969: 130; my bracketed additions; endnote references are to Sykes (1967b), Peter (1966b), Larson et al. (1968), Vacquier et al. (1961), Mason and Raff (1961), Raff and Mason (1961), Raff (1966), Heirtzler et al. (1968), Maxwell (1969), Hess (1962), Vine (1966), and Menard and Atwater (1968).

Not finished with the evolution of the Pacific basin, they then deduced the 2300 km distance between the Mendocino triple junction and what became known as the Riviera triple junction.

Fig. 5e [Figure 7.52e] also determines the relative motion of both triple junctions with respect to all plates. The southern junction (Fig. 5d [Figure 7.52d]) moves south-east with a velocity of 1.1 cm/yr relative to the American plate, whereas the northern junction moves north-west with a velocity of 6 cm/yr. The length of the strike slip fault between these junctions therefore increases at 7.1 cm/yr, and if the junctions first formed in the Oligocene 32 million years ago, they should now be 2,270 km apart. This estimate agrees remarkably well with the observed separation of about 2,300 km between the triple junctions at Cape Mendocino and at the mouth of the Gulf of California. This simple calculation is successful because the right handed offset on the Murray fracture zone is almost the same as the left handed offset on the Molokai[31]. Thus the effective velocity of the southern junction has been constant since the time it was formed.
(McKenzie and Morgan, 1969: 130; my bracketed additions; endnote reference is to Raff (1968)).

What a tour de force! Wegener, Daly, du Toit, and Holmes would have been pleased. McKenzie and Morgan also took a stab at explaining some of the tectonics of western North America.

I now turn to the second cause of plate evolution they promised to investigate which involved changes in plate motions. Here they demonstrated that "it is not possible for . . . three plates to rotate through finite angles about their instantaneous relative rotation axes," essentially an axiom of later attempts to unravel the past history of relative plate motions. Leaving a flat for a spherical Earth, they offered their proof.

Throughout our discussion, velocity, rather than angular velocity, triangles were used. This simplification is justified[3] because the behavior of a triple junction depends only on the relative motion of the three plates at the point where they meet. A quite different cause of plate evolution does, however, depend on the relative motions being rotations. This type of evolution produces real changes in the relative motion of three plates at a triple junction, and

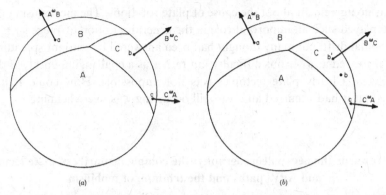

Figure 7.54 McKenzie and Morgan's Figure 10 (1969: 133). "(*a*) shows the angular velocity vectors of plates forming a triple junction. If $_A\omega_B$ and $_A\omega_C$ are taken to be fixed relative to *A* throughout the evolution, and their magnitudes also remain constant, then by (3) the direction and magnitude of $_B\omega_C$ are constant. The point *b* is fixed to *C* and is the point at which $_B\omega_C$ initially intersects *C*. As *B* and *C* rotate through finite angles with respect to *A* the axis of rotation $_B\omega_C$ moves with respect to both *B* and *C*, and does not continue to pass through the point *b* on *C*. (*b*) shows the geometry after finite rotations."

depends on the observation that finite rotations, unlike infinitesimal ones, do not add vectorially. Fig. 10a [Figure 7.54a] shows such a junction between three plates *A*, *B* and *C*, and the three axes of relative rotations $_A\omega_B$, $_B\omega_C$ and $_C\omega_A$ which satisfy:-

$$A\omega B + B\omega C + C\omega A = 0$$

By definition the points *a*, *b* and *c* where these axes intersect the plates *B* and *C* are fixed with respect to *A* and *B*, *B* and *C*, and *C* and *A* respectively. If finite rotations of *B* and *C* relative to *A* take place about the axes $_A\omega_B$ and $_C\omega_A$ and at a rotation rate given by the values of the two vectors, the orientation and magnitude of $_B\omega_C$ will remain constant and fixed relative to *A*, $_A\omega_B$, and $_C\omega_A$. The original point at which $_B\omega_C$ cuts plate *C*, *b*, is not, however, fixed to *A* but to *C*, which rotates about $_B\omega_C$ relative to *A*. Thus the final position of *b* after finite rotations of *B* and *C* relative to *A* will not be at the intersection of $_B\omega_C$ with *C* (Fig. 10*b* [Figure 7.54b]. Thus it is not possible for all three plates to rotate through finite angles about their instantaneous relative rotation axes. The only special case occurs if all ω vectors lie along the same axes, when finite rotations add as scalars and no changes need take place.

(McKenzie and Morgan, 1969: 132–133; bracketed additions mine)

With this proof, they now had established that plates do not continue rotating around the same instantaneous Euler poles once three plates are involved. The position of finite rotational poles cannot be inferred simply from positions of instantaneous Euler poles.

With the publication of this paper by McKenzie and Morgan on triple junctions, plate tectonics was essentially complete. Continental drift had morphed into seafloor spreading and it had morphed into plate tectonics. Plate tectonics was accepted immediately by those who had already accepted seafloor spreading, even though there was no agreement about the cause of plate rotations. The mechanism difficulty that had played such an important role in the general rejection of Wegener's version of continental drift, but many thought had been answered by seafloor spreading, was no longer treated as a serious difficulty but rather as a challenging unsolved theoretical problem. Clearly plate tectonics was not impossible. How could it be? The evidence that it had occurred and was still occurring was overwhelming.

7.22 Towards the new paleogeography: the complementarity of plate tectonics and APW paths and the triumph of mobilism

The subject matter of paleomagnetism is the natural remanent magnetization (NRM) of rocks as a record of the ancient geomagnetic field. Records from continents can be expressed as apparent polar wander paths, the first of which was constructed by Creer in 1954 (II, §3.6). But paleomagnetism embraces not only the study of the NRM of continental rocks but also the study of the NRM of oceanic rocks through the examination of marine magnetic anomalies, of dredge samples, and of deep-sea cores; the importance of marine paleomagnetism contrives to grow, bring an essential ingredient of the by now mandatory Vine–Matthews hypothesis (V–M). Of course, this was not always so. During the late 1950s and through the mid-1960s, workers studying marine magnetic anomalies at Lamont, Scripps, and before 1963 at Cambridge did not consider themselves to be working in paleomagnetism because they did not believe that it was the NRM of oceanic crust that caused the anomalies – they were not aware that they were looking at records of the ancient geomagnetic field.

Land-based and marine-based paleomagnetists originated and, by-and-large, have remained separate communities. Land-based paleomagnetism can be traced back to Alexander von Humboldt (1797). By the late nineteenth century, land-based paleo-magnetists had developed the clear objective of obtaining records of the history of Earth's magnetic field, but the field did not really come into its own until the 1950s. By the mid-twentieth century, paleomagnetists were busily estimating ancient field directions, compiling sequences of geomagnetic reversals and determining paleopoles (Volume II). Marine magnetics also began in earnest in the 1950s when magnetometers were towed behind ships or flown across oceans, but these workers did not consider themselves as paleomagnetists but as geophysicists or marine geophysicists, who happened to be studying the magnetism of the seafloor, which they thought was induced by the present geomagnetic field. Understandably their work was not considered a branch of paleomagnetism until the V-M was established and induced magnetization abandoned as a source for marine magnetic anomalies. For instance, Irving (1964), in the first book-length treatment of paleomagnetism, did not discuss marine magnetic anomalies. By-and-large the two communities still remain separate today, well after the acceptance of V-M, seafloor spreading, and plate tectonics.

Nowadays, students of marine geomagnetic anomalies, like paleomagnetists on land, have as their main subject matter the NRM of rocks and the history of the geomagnetic field.

Few workers have worked substantially in both communities. Cox and Doell (1962) at the USGS reported briefly on the NRM of basalt obtained in a deep-ocean core. Irving's group in Ottawa and colleagues wrote seven papers on ocean floor samples and the interpretation of marine magnetics (summaries in Irving, 1970; Irving *et al.*, 1970) but soon returned to continental work. At this time Neil Opdyke was the only permanent cross-over; after seven years of experience in three land-based groups, he migrated to Lamont where he worked on the paleomagnetism of deep-sea cores (§6.5) and as a result became the first to work extensively in both communities; his former student Dennis Kent, still affiliated with Lamont, is presently the most recognized member of both communities and Jeff Gee at Scripps also works in both. Results from land-based paleomagnetism (Volume II and the first two chapters of Volume III) were the reason the fortunes of mobilism greatly improved during the second half of the 1950s, and, to my mind, provided mobilism with its first difficulty-free solution. Having achieved this, as I explain below, land-based paleomagnetism did not dwindle, but branched out into many other applications notably establishing quantitatively the history of Earth's frame of reference – the positions of its rotational poles in the geological past. As researchers in marine magnetics unraveled the history of ocean basins they quickly sought to make comparisons between their reconstructions of seafloor spreading and the estimates of motions of continents determined from APW studies; they hoped for consilience (RS1). When attempting to reconstruct the Cenozoic history of the Atlantic Basin in terms of plate tectonics, Le Pichon appealed to land-based paleomagnetism (§7.17). Vine (1968b), studying the magnetic anomalies of the Great Magnetic Bight in the North Pacific, found a way to determine its latitude at the time it was formed (§7.20); it was, he thought, at least 25° *south* of its present latitude. This contrasted with the paleomagnetic pole for cratonic North America during the Cretaceous (as it was known at the time determined notable from Cretaceous plugs intruded through the Canadian Shield, Larochelle, 1968), which was at 70°N in the Chukchi Sea, indicating that northwestern North America was then 20° further *north* than at present. During the Cenozoic, there had been very considerable relative movement between the Pacific oceanic lithosphere and North America; vast slabs of oceanic lithosphere have been subsumed in an intervening trench system (Vine and Hess, 1971: 604).

McKenzie and Sclater (1971), building on Vine's work, turned to the Indian Ocean, hoping to reconstruct its Late Cretaceous and later history by means of plate tectonics and marine magnetic anomalies. McKenzie recalled the hectic times when he and Slater worked together.

I was at sea with John on the *Argo* on the Circe expedition, with John [Sclater] as chief scientist. He took over from Bob Fisher, who was chief scientist of the previous leg. I joined the ship at

Colombo (Sri Lanka) and left it at Mauritius. I am less sure of the dates, which must have been summer of 1968; I would guess July or August. I then went back to Cambridge after travelling all over East Africa on the buses. John and I decided to have a go at the Indian Ocean while we were at sea. I did most of the programming, partly at Cambridge, then I went to Caltech, probably around Christmas, then to Scripps, I would guess March-July 1969. I must have been back in Cambridge by the fall because that was when I met Indira again. She remembers the floor of my large room in King's (College) completely covered with maps of the Indian Ocean, with only odd patches of carpet showing through. I was making reconstructions. This was difficult, because I needed to use the 30 inch plotter, and, if anyone else's program crashed, the plot was ruined. So I had to do it in the very early morning, when the computer was running a huge program looking at the gravitational interaction between stars in a galaxy. I used to drop Indira off at night and then go into work. I think I went back to Scripps for about three months to finish off the paper, and perhaps to Caltech as well. I remember being there one very wet winter's night, and the house next door sliding down into the arroyo. But I lived like a gypsy for this period.

(McKenzie, January 30, 2012 email to author; my bracketed addition)

In order to check their reconstructions they made comparisons with land-based paleomagnetic results in a more rigorous manner than previously. Plate tectonics gave the relative motion between plates: land-based paleomagnetism gave the motion of each plate relative to Earth's rotational axis. They explained how to use Euler rotations to reconstruct the past positions of plates (which may or may not contain continents) relative to one another (1971: 438–439). They then explained how to generate a paleolatitude map of their reconstruction by rotating the paleopole to the present Earth's rotational pole about an axis in the equatorial plane. Thus:

Unlike magnetic lineations and fault plane solutions, palaeomagnetism does not determine the relative motion between two plates, but measures the motion of one plate relative to the Earth's rotational axis. Since rotations of a plate about the Earth's rotational pole have no effect, the instantaneous motion of a plate with respects to the Earth's rotational pole can always be described by an angular velocity vector in the equatorial plane. Similarly it is always possible to move the palaeomagnetic pole of any plate to the rotational pole by one finite rotation about some axis in the equatorial plane.

(McKenzie and Sclates, 1971: 438–439)

Like Vine, they found a way to determine the past latitude and orientation of ridges based on the magnetic anomalies produced at the ridge at the time the anomalies formed.

Though the magnetic anomaly patterns directly determine the relative motion between two plates, they can also be used to estimate the relative motion between the plate on which they now lie and the Earth's rotational axis. This is possible because the direction and magnitude of the magnetization ... depends on the latitude at which the anomalies are formed, and remains unchanged during later motions (Vine, 1968). Since the observed anomalies are dependent on the direction and intensity of the magnetization of the sea floor basalts, they can be used to estimate the original latitude and orientation of the ridge which produced this sea floor.

(McKenzie and Sclater, 1971; Vine, 1968 is equivalent to my Vine, 1968a)

They used the marine magnetic anomalies not only to determine the relative motion between plates, but also to determine a plate's motion relative to Earth's rotational axis.

The pattern of old anomalies in the sea floor south of Ceylon has several rather curious features ... Another striking feature of these anomalies is their amplitude ... The profile calculated on the assumption that the sea floor was formed by a ridge at the latitude and orientation that these anomalies now have is a poor fit. Not only is the calculated amplitude too small but the shape of the observed anomalies is quite different from the calculated ones. However, the calculated profile for the ridge at 8°S has a greater field over the negative blocks than over the positive. This disagreement in shape is more important than that in amplitude because the amplitude depends on the value of the susceptibility, which is probably variable and poorly known. The shape of anomalies, however, depends only on the Vine–Matthews hypothesis, the truth of which can no longer be doubted. This disagreement between the observed anomalies and those calculated at 8°S suggest the sea floor beneath this part of the Indian Ocean was not formed where it now is, but closer to the South Pole and has since drifted. A large number of calculated profiles were therefore produced, formed at many different latitudes and orientations by sea floor spreading, but all observed at the present latitude and orientation of the anomalies ...The resulting anomalies agreed both in shape and amplitude with those observed ... This result is in substantial agreement with the palaeomagnetic results from India.

(McKenzie and Sclater, 1971: 467–468)

There was, however, an ambiguity about the trend of the ridge, which they resolved.

The NW strike of the original ridge is uncertain by perhaps 30°, and its latitude by about 10° if the amplitude of the observed anomalies is not significant, but perhaps 5° if it is. The result is in substantial agreement with the palaeomagnetic results from India, and also with the reconstructions [of the Indian Ocean since the Upper Cretaceous based on analysis of all available marine magnetic anomalies and APW paths of continents surrounding the Indian Ocean]... In all determinations of the palaeolatitude and orientation of ridges there is a fundamental ambiguity about the strike of the ridge .... This arises because the magnetic anomalies are independent of the component of magnetization which is parallel to the lineations. In the case of the anomalies south of Ceylon the ridge at 50°S could strike either NW or NE. If, however, this piece of ocean floor has been part of the Indian plate since it was formed, the palaeomagnetic observations on Indian rocks remove this ambiguity. They are consistent with a NW and not a NE strike.

(McKenzie and Sclater, 1971: 468; my bracketed addition)

They then determined that their reconstruction (based on plate principles and marine magnetic anomalies) agreed with the equivalent paleopoles (based on land-based paleomagnetism) from any continents or continental fragments that were part of the same plate when the magnetic anomalies formed (RS1).

The proposed evolution [based on marine magnetic lineations, and evolution of the *RRR* triple junction where the South East, South West and Central Indian Ridges meet] agrees with the distribution of known continental fragments and with the Late Cretaceous paleomagnetic poles from surrounding continents and one obtained from the shape of the magnetic lineations south of India.

(McKenzie and Sclater, 1971: 437; my bracketed addition)

McKenzie and Sclater did not carry out a statistical analysis of the individual poles but chose the mean Cretaceous paleopole for Australia (Irving, 1964: 124 Table.2 based on the work of Robertson, 1963). It was a good choice. By rotating this pole along with the continents to the present geographic pole about an Euler pole in the equatorial plane they obtained a map (their Figure 45) with the paleogeographical grid appropriate to that time, the later Cretaceous. Although Figure 46 of McKenzie and Sclater is somewhat unpolished, their Figure 45, a fold out map of the Cretaceous Indian Ocean, was very polished – it was the shape of things to come. Other than the lack of a statistical analysis of the paleopoles (made later by Irving, 2004: supporting figs. 11a, 11b), the procedures used by them are essentially those used today in constructing paleogeographic maps. The spectacular improvement in the grouping of paleopoles obtained after assembly of the major continents into configurations determined by plate tectonics has since been demonstrated many times, most recently by Kent and Irving (2010), thus confirming that for the past 200 m.y. that the geomagnetic field has on average been a geocentric dipole and that plate-tectonically derived Euler rotation parameters and the determinations of APW paths have been determined correctly within statistical error.

McKenzie and Sclater's determination, using marine magnetic lineations, of the rate at which India moved northward away from Antarctica and their demonstration of its agreement with India's APW path from land-based paleomagnetic results from the Deccan Traps are analogous to Irving's (II, §3.4 and §5.16; II, Figure 5.18) and Clegg and company's (II, §3.13) determination of India's Deccan, early Eocene paleolatitude based on land-based paleomagnetic work and their demonstration that their results agreed with Köppen and Wegener's placement of India during the Eocene based on paleoclimatic evidence. The similarity of these two approaches makes clear that Irving and Clegg did in fact quantitatively confirm India's northward movement relative to Antarctica, and supported their belief in continental drift and their paleomagnetic work. Given the transitivity of consilience arguments, McKenzie and Sclater's results from magnetic anomalies are also consilient with Köppen and Wegener's paleoclimatic findings.

In December 1971, just as McKenzie and Sclater's regional study of the Indian Ocean and surrounding continents was appearing, there was a symposium organized in Cambridge, UK, by the Palaeontological Society, at which a sequence of global paleogeographic maps (Late Cambrian to early Cenozoic) constructed on principles similar to those just described was presented by Smith, Briden, and Drewry (published 1973).[29] Their work was done independently of McKenzie and Sclater. They presented the first cartographically rigorous "atlas" of past continental configurations back to the Cambrian/Ordovician. Alan Smith was the obvious person to produce the maps. He was on the organizing committee for the symposium; he had worked with Everett and Bullard (§3.4) using Euler's theorem to position the continents surrounding the Atlantic before it opened, and he had continued to work on continental reconstructions using Euler's theorem (Funnell and Smith, 1968; Smith and Hallam, 1970; Smith, 1971).Moreover, he was at the Sedgwick Museum along with Norman Hughes, who was then Secretary of the Paleontological Association and in charge of editing the symposium's planned volume. It also made perfect sense for Smith and Jim Briden to work together. Briden, a former Irving Ph.D. student at ANU (II, Chapter 1) was an expert in plotting land-based paleomagnetic paleopoles, and had previously worked with Smith (Briden, Smith, and Sallomy. 1970). Recalling what happened, they wrote:

Alan was on the planning committee for the symposium (so the preface to the book says!) but cannot recollect any more than having one or two meetings about the symposium. Alan and Jim both remember Stuart McKerrow (Oxford) as being a driving force in persuading us to compile the maps even though there were some important ages for which the databases we needed were worryingly sparse. We don't remember the specific contacts between us, but that is only natural because we all knew each other and had worked together previously. We had previously written a joint paper on the possibility of a Permian non-dipole field.

(Smith and Briden, February, 12, 2012 email to author)

Describing their mutual contributions and procedure in constructing the maps, they recalled (February 12, 2012, email to author):

Alan was responsible for repositioning the continents relative to one another in the Mesozoic using ocean-floor Euler rotations and both of us were responsible for re-positioning them in the Paleozoic. Jim was responsible for orienting the reassemblies using land-based paleomagnetic data to make a paleogeographic map with latitudes and relative longitudes. We only used marine magnetic anomalies to the extent that they provided Euler rotations. Our key collaboration was teasing out the reliability and statistical uncertainty in the data, and working out how to interpolate between data of slightly different ages.

They were not concerned with patterns of seafloor spreading. They were interested in getting snapshots of continents relative to each other and to Earth's rotational pole at a particular "moment" in time–the Eocene, Cretaceous, Jurassic, Permian, Lower Carboniferous, Lower Devonian or Cambrian/Ordoviciann – so paleontologists at the symposium could consider how well their fossil record accorded with them. As a result, they could and did skirt issues concerned with finite rotational poles, and simply determined net changes. Their aim was to produce maps that would suit the purposes of increasingly interested paleobiogeographers, paleontologists, and paleoclimatologists.

Successive rotations about different Euler poles and through different Euler angles can always be combined into a unique equivalent single rotation about a unique Euler pole and through a unique Euler angle. Thus the spreading motions on a mid-ocean ridge at a given instant of geological time may be described geometrically as though they occurred about a fixed Euler pole at a fixed rate of rotation. In this case the Euler is also the tectonic rotation pole. It may be that the same pole describes the spreading pattern for successive instants of geological time that may amount to several millions of years, but this is not necessarily so. As an ocean grows, tectonic rotation poles may change their position abruptly (e.g., Le Pichon and Fox, 1971), or migrate continuously. The net movement may be found by summing the sequential rotations and combining them into a unique Euler rotation. When the sequential rotations involve different tectonic rotation poles, then the rotation about the unique Euler pole does not trace out the actual spreading pattern; it merely describes the net changes that have taken place in the interval concerned, by means of a single rotation. Because plates are rigid, the Euler rotations describe the relative positions of every part of the corresponding plates. Thus the relative movement of two continents on either side of a spreading ridge is also described by the same Euler rotations as apply to the ridge itself.

(Smith et al., *1973: 4)*

With no marine magnetic anomalies to guide them once they got back to formation of Pangea, they used orogenic disjuncts as their guide.

All we did for the pre-Pangea maps was to cut up Pangea along the mountain belts created by the collision of the components that went into forming Pangea. We chose the configuration that implied the simplest sequence of relative motion, though longitudinal separations were somewhat arbitrary.

(Smith and Briden, February, 12, 2012 email to author)

Because extant oceanic crust is no older than Jurassic, marine magnetics cannot be used to determine pre-Mesozoic tectonics, and well-dated and constructed APW paths assume prime importance.

In this way APW paths began to be used in concert with magnetic lineations and plate tectonics to provide a global paleogeographic reference frame for mapping the distribution of continents and oceans. Before the development of land-based paleo-magnetism and plate tectonics, Wegener and Köppen used (as many workers must, perforce, still do for pre-Mesozoic work) jig-saw fits and other geological arguments to determine past positions of continents relative to each other, and then used paleoclimatology to fix their paleolatitudes. Now that plate tectonics is established as a means of determining relative motions of continents and APW paths are established as a means of determining their motions relative to the ancient geograph-ical pole, they are now used instead; they are the mainstays of Cenozoic, Mesozoic, and older paleogeography.

McKenzie also made a foray into paleobiogeography, teaming up with Nicholas Jardine. They co-authored a paper in which they examined the evolution and dispersal of marsupials (Jardine and McKenzie (1972). Wegener (I, §3.3) would had been pleased because he had himself placed much weight on his explanation of the dispersal of marsupials from South America to Australia across Antarctica. But, after G. G. Simpson sided against Wegener (I, §3.6), and most Australian paleozoologists followed suite (I, §9.4–§9.5). Wegener was left out in the cold. Needless to say, Jardine and McKenzie began with the assumption that marsupials arrive at Australia from Antarctica and not from Asia.

The narrative in this book began when the idea of fixed continents and oceans was paramount. It was challenged by the theory of continental drift. It was a century ago that Wegener proposed continental drift (1912) and Taylor introduced his narrower based version (1908). I have (I, §1.1) viewed the controversy in terms of two competing traditions, *mobilism* and *fixism*, terms introduced by Émile Argand. Fixists maintained that, except perhaps very early in Earth's history, continents remained fixed in the same place relative to one another. Fixists did not however always agree among themselves; some claimed that the axis of rotation remained fixed relative to the Earth as a whole, others that the axis remained fixed relative to the ecliptic but the Earth as a whole toppled (polar wander) so that although continents remained fixed relative to each other, they changed their positions relative to Earth's rotational poles. In contrast, most mobilists declared that continents have changed their position relative to one another and perforce relative to the rotational pole; over time, continents have undergone horizontal displacement, changed their latitude, and also their climate.

In the late 1960s plate tectonics brought in a conceptual change – that it is lithospheric plates not continents that move about as entities; most plates comprise oceanic and continental lithosphere, some oceanic lithosphere only. Plate boundaries are mainly within oceans or bordering continents. The plate tectonics revolution is a wonderful, romantic story. It was brought about through the acceptance of seafloor spreading as a result of the confirmation of the V-M hypothesis and ridge-ridge transform faults in the mid-1960s. Within a year, seafloor spreading and continental drift were recast into the mold of plate tectonics, largely by extending ideas of transform faults and rigid lithospheric plates. Plate tectonics was the geometrification of mobilism, a kinematic, not a dynamic theory. Within two more years, formal conditions for the stability of triple junctions were identified, and the relationship between instantaneous and finite Euler rotations explained. Meanwhile, plate tectonics was looped around the world, and its extensive seismological and paleomagnetic support (from the study of marine magnetic anomalies) repeatedly confirmed. Wegener and the host of deceased old-time mobilists would have been surprised, but they would have rejoiced.

Plate tectonics itself provides no general method for determining the latitude of plates; inter-plate motions are defined by it, but the general motions of plates relative to Earth's axis of rotation are not. Traditionally done from the evidences of past climate, during the 1950s and a decade and a half before the advent of plate tectonics, APW paths began to position continents relative to Earth's natural reference frame by an entirely different method. The term "natural" is appropriate, because both the axis of the time-averaged geomagnetic field and climate zones determined from geological record are controlled by Earth's rotation.

In the early 1970s the APW work of the 1950s and plate tectonics of the late 1960s came to complement one another, and Cenozoic and Mesozoic maps showing the distribution of the Earth's major surface features placed in their natural paleogeographic reference frame began to be constructed. Further back in time, in the Paleozoic and Precambrian, with drifting continents as a given and the idea of APW paths vindicated, paleomagnetists, paleoclimatologists, and biogeographers continued to propose mobilistic continental reconstructions. Tectonicists and paleomagnetists used paleomagnetism to determine paleolatitudes and rotations of crustal blocks within orogenic belts, and the idea of exotic terranes was born. The history of gross changes in Earth's climate zones have come under scrutiny – the glacial (icehouse), non-glacial (greenhouse), and pan-glacial climatic states. The latter embodies the idea of a snowball Earth based as it is on reconstructions of latest Precambrian paleogeography (Hoffman and Li, 2009). And so mobilism with its twin concepts of APW and plate tectonics has come to pervade the Earth sciences.

Notes

1 This brief account of Oliver's life is based on the two-part interview of him by Doel (March 8, 1996 and September 27, 1997), Niels Bohr Library & Archives, American Institute of Physics, www.aip.org/history/nbl/oralhistory.html), and three of his own works: *Shocks and Rocks* (1996); *Shakespeare Got It Wrong, It's Not "To Be," It's "To Do." The Autobiographical Memoirs of a Lucky Geophysicist* (1998), available online at ecommons.cornell.edu; and *The Incomplete Guide to the Art of Discovery* (2004), also available online at ecommons.cornell.edu.

2 Opdyke was not trained in paleomagnetism by Runcorn; he was trained by Irving at ANU. I believe Oliver's mistake is indicative of his ignorance, shared by most at Lamont, of the paleomagnetic work supportive of mobilism.

3 I thank Bryan Isacks for answering questions about his education, training, and work. I could not have written this section without his cooperation and straightforward answers, especially because, unlike most major figures at Lamont, it appears that he was not part of the interviewing project of the Niels Bohr Library and Archives and the Center for the History of Physics. As will become apparent, Isacks took the lead in his and Oliver's work on deep earthquakes.

4 Isacks did not mean to claim that Jeffreys denied the existence of deep foci earthquakes. However, their cause was a mystery.

5 Oliver and Isacks (1967: 4275) acknowledged the support of "the officers and staff of the Fiji Meteorological Office in Suva for their interest and assistance in many phases of the field program." They also thanked Liahona High School of Tonga. But there is no mention of the colonial Geology Survey of Fiji.

6 Chander and Brune's paper, as noted by Isacks, was published after he left for Fiji. Isacks, who knew Brune, believes that he saw a preprint of their paper before leaving (Isacks, October 3, 2011 email to author).

7 McKenzie's mother, the daughter of a laborer, was a forceful and talented person. She had more of an influence on him than his father. She earned a State Scholarship, applied to Newnham College, Cambridge University, but was not admitted because of her Yorkshire accent, emphasized by her being asked to read aloud. "She never forgave this snobbery" (McKenzie, May 11, 2007 interview by Alan Macfarlane, www.alanmacfarlane.com/DO/filmshow/mckenzie1tx.htm-/). She graduated from Royal Holloway College, University of London, with honors in English.

8 McKenzie was particularly disappointed with the right-wing politics at Cambridge. During his last year at Westminster he had been "very politically active in left wing groups."

He found the politics at Cambridge "half-baked and right wing so [he] had no interest in politics here." His parents were "absolutely old Labour because of the Welfare State which my father had joined as soon as it started." The modern welfare state of the United Kingdom is taken to have begun on July 5, 1948, the day that the National Health Service Act of the United Kingdom became effective (McKenzie, March 18, 1990 letter to author and May 11, 2007 interview by Alan Macfarlane).

9 Indeed, McKenzie was one of four graduate students then studying in the Department of Geodesy and Geophysics later elected to the Royal Society of London; Sclater and Vine began in fall 1962, and McKenzie and Parker, in 1963.

10 Both McKenzie and Morgan have pointed out how Lamont's digitizing of their data was an important advantage. Morgan recalled that he had heard Le Pichon talk about Ewing's insistence to digitize Lamont's data-sets.

Several times I've heard Le Pichon talk of the 'digital' database (getting all the ship records into a big database that could be accessed/plotted on a computer was pushed hard by Ewing). Such a thing didn't exist at the other oceanographic institutions – I remember looking at paper analog records at Woods Hole in the summer of 1967 to get a magnetic profile over a ridge.

(Morgan, February 8, 2012 email to author)

11 When referring to Morgan's extended outline of his talk at the 1967 AGU meeting, I shall refer directly to Morgan's extended outline as Morgan (1967b) even though it was not published until Le Pichon reproduced it in Le Pichon (1991).

12 I thank Jason Morgan for providing me with copies of documents relevant to publication of his paper. These include his letter of submission, editor's letter of receipt, editor's letter of acceptance, copies of the two reviewers' comments, and the submitted version of his paper to *JGR*.

13 Morgan (February 8, 2012 email to author) noted that he and Vine used the term "spreading rate" to refer to what we now call half-spreading rate, and referred to what we call "spreading rate" as "full-spreading rate."

14 McKenzie was not only the first to recognize the importance of using the slip component of earthquakes along the common boundary of two plates to determine their instantaneous Euler pole, he was the first to recognize that P (compressional) and T (tensional) stress axes of earthquake focal plane solutions are not important for plate tectonics. He wrote a paper about it which he has described as "essentially completing the argument" in McKenzie and Parker, 1967. P and T axes are only useful in understanding earthquakes that occur in the rock which is homogeneous and free of pre-existing faults. However, neither of these conditions applies to the interplate movements of adjacent plates given the theory of plate tectonics.

A good account of faulting is given by Anderson (1951) . . . Anderson suggested that the orientation of the stress axes could be obtained from that of the failure or fault plane. This simple argument cannot apply to most earthquakes for two reasons. The first is that when earthquakes produce surface displacements they almost always do so along preexisting faults . . . The other difficulty is that the shear stresses involved in shallow earthquakes are at least an order of magnitude too small to produce facture . . . These observations strongly suggest that a fault, once established is a plane of weakness, and that later movements are not simply related to the principal stress directions.

(McKenzie, 1969b: 591–592)

For movements of one plate relative to its adjacent plates, only slip directions are important, regardless of whether the earthquakes in question are strike-slip (transform) or dip-slip (normal or thrust earthquakes). McKenzie did note that stress axes are important in one place that is particularly relevant for plate tectonics, namely, for intermediate and deep earthquakes that occur *within* a downgoing lithosphere slab (see §7.16).

15 For works by Parker arising from his Ph.D. work see Parker (1966, 1968) and Bullard and Parker (1968).

16 McKenzie, who later came to regret not saying anything about Hawaii, recalled why he
   said nothing about it.

> My other regret [about McKenzie and Parker (1967)] concerns Hawaii. The projection Bob
> and I used suggested that the Hawaiian Ridge was formed by a volcano that was fixed
> to the North American plate, because it was parallel to the direction of movement between
> the two plates. This result was incomprehensible at the time, and furthermore confused the
> whole theory, which was constructed in terms of relative motions between plates, and
> explicitly rejected the whole concept of absolute motions. So I decided to say nothing.
> Jason noticed the same thing, and proposed an explanation which has been widely
> accepted.[22] He argued that Hawaii was on top of a host part of the mantle that is rising
> toward the surface, because its density is lower than that of the surrounding colder mantle.
> Such structures are called convective plumes.
>
> *(McKenzie (2001: 181); endnote 22 refers to Morgan (1971)*

17 I am not sure what triple junction they had in mind. Given its location at the entrance to the
   Gulf of California, it seems as if they had in mind what became known as the Rivera triple
   junction where the North America, Pacific and Cocos plates meet. Although it joins
   together a ridge, trench, and transform fault, the transform fault is not the San Andreas,
   which is at the northern end of the Gulf of California. However, McKenzie may have not
   realized this while writing the paper. He sorted it out by the time he and Morgan worked on
   the evolution of triple junctions when they discussed the evolution of the San Andreas
   Fault, and hypothesized that the Juan de Fuca and Cocos plates were originally part of the
   now subducted Farallon plate (McKenzie and Morgan, 1969). Moreover, McKenzie and
   Morgan also proposed that southern extension of the San Andreas Fault when it formed
   was the same type of triple junction as the Rivera triple junction, a ridge-trench-transform
   one (*RTF*(a) in Figures 7.52 and 7.53).

18 I shall often continue citing Le Pichon, Francheteau, and Bonnin's 1973 book *Plate
   Tectonics*. It was the first book-length treatment of plate tectonics. Just as Irving's 1964
   Paleomagnetism and Its Application to Geological and Geophysical Problems offered an
   excellent account of the current state of paleomagnetism, *Plate Tectonics* did the same for
   plate tectonics. Thus both are historic works.

19 Figuring out the nature of the linkage between the Tonga and New Hebrides trenches
   became a common pursuit among Lamont seismologists. Sykes *et al.* (1969: 1095) proposed
   that the "tectonics appears to be more complex than that of a single transform fault."
   Johnson and Molnar (1972) later reexamined the area. See Le Pichon *et al.* (1973: 89) for a
   brief summary.

20 Elsasser was quite pleased with the findings of the Lamont seismologists.

21 F. Suppe (1998a) analyzed the argumentative style employed by Morgan in his 1968b, the
   paper in which he introduced plate tectonics. He claimed that Morgan compares his model
   of plate tectonics linked through its corollaries to relevant observations, and argues that the
   agreement is good after explaining why some of the apparent observations are not relevant
   to the comparison. In this regard, Suppe is correct. He also claims that Morgan made no
   predictions that at the time were confirmable. He correctly claims that Morgan did not
   predict the latitude and longitude of any of the Euler poles he determined. However,
   Morgan did predict that determination of an Euler pole by means of independent data-sets,
   namely marine magnetic lineations and trends of active transform faults, will roughly agree
   within the bounds of data-reliability and geometrical restrictions as to the ability of either
   data set to fix both the latitude and longitude of the Euler pole. Indeed, this predicted and
   confirmed consilience provided strong support for plate tectonics. Suppe also claims that
   Morgan's mode of arguing threatens the philosophical model that scientists defend their
   views by arguing that it offers the best available explanation. I agree that Morgan did not
   argue that plate tectonics provided a better explanation of the data in question than any
   other competing theory. Nonetheless, his lack of such an argument did not threaten the
   claim that scientists often argue that their theory is the best available. Rapid Earth

expansion was the only theory other than plate tectonics that implied V-M and ridge-ridge transform faults. However, it faced serious well-known difficulties: support for subduction was strong, and paleomagnetic arguments had been brought to bear against it. Nonetheless, one may wonder why Morgan did not raise difficulties against it. There are, I believe, two reasons. First, he did not have anything new to add to the controversy based on data available to him. Second, Morgan had read a preprint of Le Pichon's (1968), and knew that he had raised a serious difficulty against Earth expansion based on his looping plate tectonics around the world (§7.17). Indeed, Morgan (February 8, 2012 email to author) retrospectively noted the importance of Le Pichon's argument against rapid Earth expansion. Morgan also knew of Oliver and Isacks' work on subduction. Leaving Morgan aside, we have shown that many papers written during the drift-mobilist controversy strongly support the idea that scientists when involved in a controversy often argue that their theory or solution is the best available. Suppe's paper led to several interesting responses (See Suppe, 1998a and b; Lipton, 1998; and Franklin and Howson, 1998).

22 I suspect Vine heard Peter speak at the 1967 spring AGU meeting (Elvers *et al.* 1967b).

23 Le Pichon *et al.* (1973: 148–150) offered a generally favorable assessment of Vine's determination of the paleolatitude of the Great Magnetic Bight.

24 Vine constructed the velocity diagram before McKenzie sent him an early draft of what would become McKenzie and Morgan (1969). McKenzie sent both Vine and Morgan copies of the draft in November 1968, and Vine sent McKenzie copies of what would become his Figure 8, with the velocity diagram on October 11, 1968 (see §7.21 for relevant excerpt from October 11 letter).

25 Although Vine and McKenzie do not remember, Morgan is pretty sure that McKenzie asked Vine if he wanted to be a co-author of the triple-junction paper, while they were discussing triple junctions at Princeton.

I remember the three of us (Dan [McKenzie], Fred [Vine], and myself) sitting at a table in the end room of the wing of Guyot Hall that was Dan's temporary office sketching/ discussing triple junctions. Dan had done a lot of work on the 'stability questions' and, I'm pretty sure by then, on the zipper effect along the California coast (drastic changes at a given site on the coast, yet no changes in the motions of the plates themselves – just the migration of the triple junction). This was when Fred was asked by Dan and Fred saying no

(Morgan, March 16, 2012 email to author; my bracketed additions).

Morgan (March 16, 2012 email to author) also confirmed that had he known how important his initial paper on the relative motion of plates would turn out to be, he would have asked Vine to be a co-author. Vine (March 15, 2012 email to author) however, would not have accepted. As now, he would not have considered it appropriate.

26 Once McKenzie left Caltech at the beginning of July 1967, and went to Scripps, where he wrote his first paper on plate tectonics with Parker, he became a peripatetic scholar. He recently reflected on how such a lifestyle affected him, and how it was "a strange way to begin an academic career."

Yes, I was all over the place, with no real base and no real position, neither of which bothered me. But I did not think I would get the same opportunity again, to have no teaching job, support from Kings [College] when I was in Cambridge, and the freedom to go where I wanted, not to be married and to be on the crest of the wave. Some of the time I was poor, most of it lonely, and I worked very hard. During this period I think I changed, and became thoroughly international. Thereafter I don't think I have changed much at all. After writing the paper with Bob [Parker] I ceased to worry about whether I would get a University job, or about the normal issues of advancing my career. But I thought the way I was living would only last a few years, and that it would then change. It did completely. I got married and accepted the most junior faculty position that existed at that time at Cambridge (called a Senior Assistant in Research, slightly below even an

Assistant Professorship in the US). I was offered full professorships at Manchester and at ETH in Zurich at the same time (I had become rather well known, even though I was still a post-doc). I thought I would be able to do more research at Cambridge and was happy living here (the post was in the Department of Geodesy and Geophysics, which did not teach undergraduates), so turned them both down. Indira [his wife] and I were a bit poor for some years, but I thought this also would change (and indeed it did). I have never regretted what I did, though it seemed to many people I knew at the time a strange way to begin an academic career. It seems to me stranger now than it did then.

(McKenzie, January 30, 2012 email to author; my bracketed additions)

27 The point where the three perpendicular bisectors meet is the circumcenter, not the centroid; the centroid is the point where the three medians of a triangle intersect. Various authors pointed out this mistake in McKenzie and Morgan (1969). The circumcenter and centroid of an equilateral triangle do coincide. The median of the vertex of an isosceles triangle opposite its base coincides with perpendicular bisector of its base, and therefore both pass through the centroid and circumpolar. Helpful discussions of triple junctions and their evolution appear in Le Pichon, Francheteau, and Bonnin, 1973: 38–39; Fowler, 2000: 19–24; Lowrie, 2007: 29–32; Kearney, Klepeis, and Vine, 2009: 113–120.

28 They also maintained that RRF triple junctions are always unstable but evolve into FFR triple junctions (see Figure 7.51). However, York (1973) found a stable geometry for FFR triple junctions where the two ridges are orthogonal.

29 See Irving (2004) for further discussion of McKenzie and Sclater (1971) and Smith *et al.* (1973). The appropriate sections may be found online at the website of the *Proceedings of the National Academy of Sciences*. The six Late Cretaceous poles used by McKenzie and Sclater with continents as they are today have a Fisher's estimate of precision $K = 11$, after reassembly of the southern continents $K = 54$, a vast improvement (See I, §1.13 on Fisher's statistics).

References

Abelson, P.H. 1998. Ross Gunn. *Biogr. Mem. Natl. Acad. Sci.*, **74**: 111–125.

Adams, D. R. 1963. Source characteristics of some deep New Zealand earthquakes. *N. Z. J. Geol. Geophys.*, **6**: 209–220.

Adams, D. R. and Christoffel, D. A. 1962. Total magnetic field surveys between New Zealand and Antarctica. *J. Geophys. Res.*, **67**: 805–813.

Ade-Hall, J.M. 1964. The magnetic properties of some submarine oceanic lavas. *Geophys. J.*, **9**: 85–92.

Aki, K. 1960. Study of earthquake mechanism by a method of phase equalization applied to Rayleigh and Love waves. *J. Geophys. Res.*, **65**: 729–740.

Aki, K. 1966a. Generation and propagation of G waves from the Niigata earthquake of June 16, 1964. Part I. A statistical analysis. *Tokyo Univ. Earthquake Res. Inst. Bull.*, **44**: 23–72.

Aki, K. 1966b. Earthquake generating stress in Japan for the years 1961 to 1963 obtained by smoothing the first motion radiation patterns. *Tokyo Univ. Earthquake Res. Inst. Bull.*, **44**: 447–471.

Allan, T.D. 1960. Magnetic measurements at sea. Ph.D. thesis (unpublished). University of Cambridge, Cambridge, 164 pp.

Allan, T.D. 1964. A preliminary magnetic survey in the Red Sea and Gulf of Aden. *Bull. Geofis. Teor. Appl.*, **6**: 199–214.

Allan, T.D. 1969. A review of marine geomagnetism. *Earth Sci. Rev.*, **5**: 217–254.

Allan, T. D. 2008. *Memories From a Life*. Privately published.

Allan T.D. and Pisani, M. 1966. Gravity and magnetic measurements in the Red Sea. In *The World Rift System*, Geological Survey of Canada, Special Paper 66-14: 62.

Allan, T.D., Charnock, H., and Morelli, C. 1964. Magnetic, gravity and depth surveys in the Mediterranean and Red Sea. *Nature*, **204**: 1245–1248.

Allègre, C. 1988. *The Behavior of the Earth*. Translated by van Dam, D. K. Harvard University Press, Cambridge, MA.

Allen, C. R. 1965. Transcurrent faults in continental areas. In Blackett, P. M. S., Bullard, E., and Runcorn, S. K., eds., *A Symposium on Continental Drift. Phil. Trans. Royal Soc. London A*, **258**: 12–26.

Almond, M., Clegg, J. A., and Jaeger, J. C. 1956. Remanent magnetism of some dolerites, basalts and volcanic tuffs from Tasmania. *Phil. Mag.*, **1**: 771–782.

Anderson, D. L. 1966. Earth's viscosity. *Science*, **151**: 321–322.

Anderson, D. L. 1967. Phase changes in the upper mantle. *Science*, **157**: 1165–1173.

Anderson, D. L. and Archambeau, C. B. 1964. The anelasticity of the earth. *J. Geophys. Res.*, **69**: 2071–2084.

Anderson, D.L., Ben-Menahem, A., and Archambeau, C.B. 1965. Attenuation of seismic energy in the upper mantle. *J. Geophys. Res.*, **70**: 1441–1448.

Anderson, E. M. 1951. *The Dynamics of Faulting* (2nd edition). Oliver and Boyd, Edinburgh.

Anonymous. 1967. Synopsis of the 1967 annual meeting – schedule. 1967. *Trans. Am. Geophys. Union*, **48**: 11–51.

Argand, E. 1924. La tectonique de l'Asie. *Proceedings of the 13th International Geological Congress*, **1**, Part 5: 171–372. Translated by Carozzi, A. V. 1977 as *The Tectonics of Asia*. Hafner Press, New York.

Argus, D.F., Gordon, R.C., and DeMetes, C. 2011. Geologically current motion of 56 plates relative to the no-net-rotation reference frame. *Geochem. Geophys. Geosyst.*, **12**: Q11001, doi:10.1029/2011GC003751.

Atwater, T. 1970. Implications of plate tectonics for the Cenozoic tectonic evolution of western North America. *Geol. Soc. Am. Bull.*, **81**: 3513–3536.

Atwater, T. 2001. When the plate tectonic revolution met western North America. In Oreskes, N., ed., with Le Grand, H., *Plate Tectonics*. Westview Press, Boulder, CO, 243–263.

Atwater, T. and Menard, H. W. 1971. Magnetic lineations in the northeast Pacific. *Earth Planet. Sci. Lett.*, **7**: 445–450.

Atwater, T. and Mudie, J. D. 1968. Block faulting on the Gorda Rise. *Science*, **159**: 729–731.

Atwater, T. and Molnar, P. 1973. Relative motion of the Pacific and North American plates deduced from sea-floor spreading in the Atlantic, Indian and South Pacific Oceans. In Kovach, R.L. and Nur, A. eds., *Proceedings of the Conference on Tectonic Problems of the San Andreas Fault*. Stanford University Publications in Geological Science, **13**: 136–148.

Backus, G. E. 1955. On the application of eigenfunction expansions to the problem of the thermal instability of a fluid sphere heated within. *Phil. Mag.*, **46**: 1310–1326.

Backus, G. E. 1956a. Non-existence of axisymmetric fluid dynamos. Ph.D. dissertation (unpublished). University of Chicago.

Backus, G. E. 1956b. The external electric field of a rotating magnet. *Astrophys. J.*, **123**: 508–512.

Backus, G. E. 1957a. The axisymmetric self-excited fluid dynamo. *Astrophys. J.*, **125**: 500–524.

Backus, G. E. 1957b. The existence and uniqueness of the velocity correlation derivative in Chandrasekhar's theory of turbulence. *J. Math. Mech.*, **6**: 215–234.

Backus, G. E. 1958. A class of self-sustaining dissipative spherical dynamos. *Ann. Phys.*, **4**: 372–447.

Backus, G. E. 1964. Magnetic anomalies over oceanic ridges. *Nature*, **201**: 591–592.

Backus, G. E. 1965. Possible forms of seismic anisotropy of the uppermost mantle under oceans. *Geophys. Res.*, **70**: 3429–3439.

Backus, G. E. and Chandrasekhar, S. 1956. On Cowlings theorem on the impossibility of self-maintained axisymmetric homogeneous dynamos. *Proc. Natl. Acad. Sci.*, **42**: 105–109.

Baker, P. E., Gass, I. G., Harris, P. G., and LeMaitre, R. W. 1964. Vulcanological report on the Royal Society expedition to Tristan de Cunha. *Phil. Trans. Roy. Soc. London*, **256**: 439–578.

Barazangi, M. and Dorman, M. 1969. World seismicity map of ESSA Coast and Geodetic Survey epicenter data for 1961–1967. *Bull. Seismol. Soc. Am.*, **59**: 369–380.

Barth, T.F.W. and Holmsen, P. 1939. Rocks from the Antarctandes and the Southern Antilles. *Sci. Res. Norwegian Antarctic Expeditions, 1927–1928*, No. 18. I Kommisjon Hos Jacob Dybwad, Oslo.

Benioff, H. 1949. Seismic evidence for the fault origin of oceanic deeps. *Geol. Soc. Am. Bull.*, **60**: 1837–1856.

Benioff, H. 1955. Orogenesis and deep crustal structure – additional evidence from seismology. *Geol. Soc. Am. Bull.*, **65**: 385–400.

Benioff, H. 1959. Circum-Pacific tectonics. *Publ. Dominion Obs.*, **20**: 395–402.

Benioff, H. 1962. Movements on major transcurrent faults. In Runcorn, S.K., ed., *Continental Drift*. Academic Press, New York, 103–134.

Bernal, J.D. 1961a. Continental and oceanic differentiation. *Nature*, **192**: 123–124.

Bernal, J.D. 1961b. Response to Dietz. *Nature*, **192**: 125.

Bernal, J.D. 1965. Discussion. In Blackett, P.M.S., Bullard, E., and Runcorn, S.K., eds., *A Symposium on Continental Drift*. *Phil. Trans. Roy. Soc. London A*, **258**: 318–321.

Betz, F. and Hess, H.H. 1940. Floor of the North Pacific Ocean. *Trans. Am. Geophys. Union*, 21st Annual Meeting: 348–349.

Betz, F. and Hess, H.H. 1942. The floor of the North Pacific Ocean. *Geogr. Rev.*, **32**: 99–116.

Blackett, P.M.S. 1956. *Lectures on Rock Magnetism*. Weizmann Science Press of Israel, Jerusalem.

Blackett, P.M.S. 1957. Introductory remarks. *Adv. Phys.*, **6**: 147–148.

Blackett, P.M.S. 1965. Introduction. In Blackett, P.M.S., Bullard, E., and Runcorn, S.K., eds., *A Symposium on Continental Drift*. *Phil. Trans. Roy. Soc. London A*, **258**: vii–x.

Blackett, P.M.S., Clegg, J.A., and Stubbs, P.H.S. 1960. An analysis of rock magnetic data. *Proc. Roy. Soc. A*, **256**: 291–322.

Bodvarsson, G. and Walker, G.P.L. 1964. Crustal drift in Iceland. *Geophys. J. Roy. Astron. Soc.*, **8**: 285–300.

Bolt, B.A. 1960. The revision of earthquake epicenters, focal depths and origin-times using a high-speed computer. *Geophys. J.*, **3**: 433–440.

Bolt, B.A. 2001. Locating earthquakes and plate boundaries. In Oreskes, N., ed., with Le Grand, H., *Plate Tectonics*. Westview Press, Boulder, CO, 148–154.

Bott, M.H.P. 1965. Discussion. In Blackett, P.M.S., Bullard, E., and Runcorn, S.K., eds., *A Symposium on Continental Drift*. *Phil. Trans. Roy. Soc. London A*, **258**: 212–213.

Bott, M.H.P. 1967. Solution of the linear inverse problem in magnetic interpretation: with application to oceanic magnetic anomalies. *Geophys. J. Roy. Soc.*, **13**: 313–323.

Boucot, A.J., Berry, W.B., and Johnson, J.G. 1968. The crust of the Earth from a Lower Paleozoic point of view. In Phinney, R.A., ed., *The History of the Earth's Crust*. Princeton University Press, Princeton, NJ, 208–228.

Bowin, C.O. 1960. Geology of central Dominican Republic. Ph.D. thesis (unpublished). Princeton University.

Bowin, C.O. 1966a. Geology of central Dominican Republic (a case history of part of an island arc). In Hess, H.H., ed., *Caribbean Geological Investigations*. *Geol. Soc. Am. Mem.*, **98**: 11–84.

Bowin, C.O. 1966b. Gravity over trenches and rifts. In *Continental Margins and Island Arcs*, Geological Survey of Canada, Special Paper 66-15: 430–439.

Bowin, C.O. 1987. Historical note on "Evolution of Ocean Basins" preprint by H. H. Hess [1906–1969]. *Geology*, **15**: 475–476.

Bowin, C.O. and Vogt, P.R. 1966. Magnetic lineation between Carlsberg Ridge and Seychelles Bank, Indian Ocean. *J. Geophys. Res.*, **71**: 2625–2630.

Bozorth, R.M. 1951. *Ferromagnetism*. D. Van Nostrand, New York.

Briden, J.C. 1968. Paleoclimatic evidence of a geocentric axial dipole field. In Phinney, R.A., ed., *The History of the Earth's Crust*. Princeton University Press, Princeton, NJ, 178–194.

Briden, J.C., Smith, A.G., and Sallomy, J.T. 1970. The geomagnetic field in Permo-Triassic time. *Geophys. J. Roy. Astron. Soc.*, **23**: 101–117.

Buddington, A.F. 1970. Harry Hammond Hess, 1906–1969. *Geol. Soc. Am. Proc.* **1969**: 1–9.

Bullard, E.C. 1951. Remarks on deformation of the Earth's crust. *Trans. Am. Geophys. Union*, **32**: 520.

Bullard, E.C. 1952. Heat flow through the floor of the Eastern Pacific Ocean. *Nature*, **170**: 199–200.

Bullard, E.C. 1954a. Introduction. *Proc. Roy. Soc. London*, **222A**: 287–289.

Bullard, E.C. 1954b. The flow of heat through the floor of the Atlantic Ocean. *Proc. Roy. Soc. London A*, **222**: 408–429.

Bullard, E.C. 1961. The automatic reduction of geophysical data. *Geophys. J.*, **3**: 237–243.

Bullard, E.C. 1964. Continental drift. *Q. J. Geol. Soc. London*, **120**: 1–26.

Bullard, E.C. 1965. Concluding remarks. In Blackett, P.M.S., Bullard, E., and Runcorn, S.K., eds., *A Symposium on Continental Drift. Phil. Trans. Roy. Soc. London A*, **258**: 322–323.

Bullard, E.C. 1967. Maurice Neville Hill. *Biogr. Mem. Fell. Roy. Soc.*, **13**: 193–203.

Bullard, E.C. 1968a. Reversals of the earth's magnetic field (the Bakerian Lecture for 1967). *Phil. Trans. Roy. Soc. A*, **263**: 481–524.

Bullard, E.C. 1968b. Conference on the history of the Earth's crust. In Phinney, R.A., ed., *The History of the Earth's Crust*. Princeton University Press, Princeton, NJ, 231–235.

Bullard, E.C. 1975a. The emergence of plate tectonics: a personal view. *Annu. Rev. Earth Planet. Sci.*, **3**: 1–30.

Bullard, E.C. 1975b. The effect of World War II on the development of knowledge in the physical sciences. *Proc. Roy. Soc. London A*, **343**: 519–536.

Bullard, E.C. 1980. William Maurice Ewing. *Biogr. Mem. Natl. Acad. Sci.*, **51**: 119–193.

Bullard, E.C. and Day, A. 1961. The flow of heat through the floor of the Atlantic Ocean. *Geophys. J.*, **4**: 282–292.

Bullard, E.C. and Mason, R.G. 1963. The magnetic field over the oceans. In Hill, M.H., ed., *The Sea, Volume III*. Interscience, New York, 175–217.

Bullard, E.C. and Parker, R.L. 1968. Electromagnetic induction in the oceans. In Maxwell, A.E., ed., *The Sea, Volume IV*. Interscience, New York, 695–730.

Bullard, E.C., Everett, J.E., and Smith, A.G., 1965. The fit of the continents around the Atlantic. In Blackett, P.M.S., Bullard, E., and Runcorn, S.K., eds., *A Symposium on Continental Drift. Phil. Trans. Roy. Soc. London A*, **258**: 41–51.

Bullard, E.C., Maxwell, A.E., and Revelle, R. 1956. Heat flow through the deep seafloor. *Adv. Geophys.*, **3**: 153–181.

Bunce, E. T. 1966. The Puerto Rico Trench. In *Continental Margins and Island Arcs*, Geological Survey of Canada, Special Paper 66-15: 165–176.

Burk, C. A. 1966. The Aleutian arc and Alaska continental margin. In *Continental Margins and Island Arcs*, Geological Survey of Canada, Special Paper 66-15: 206–214.

Cann, J. R. and Vine, F. J. 1966. An area on the crest of the Carlsberg Ridge: petrology and magnetic survey. *Phil. Trans. Roy. Soc. London A*, **259**: 198–217.

Carey, S. W. 1955a. The orocline concept in geotectonics. *Roy. Soc. Tasmania*, **89**: 255–288.

Carey, S. W. 1955b. Wegener's South America – Africa assembly, fit or misfit? *Geol. Mag.*, **XCII**: 196–200.

Carey, S. W. 1958. A tectonic approach to continental drift. In Carey, S. W., Convener, *Continental Drift: A Symposium*. University of Tasmania, Hobart, 177–355.

Carey, S. W. 1961. Paleomagnetic evidence relevant to a change in the Earth's radius. *Nature*, **190**: 35.

Carey, S. W. 1975. The expanding Earth – an essay review. *Earth Sci. Rev.*, **11**: 105–143.

Carey, S. W. 1976. *The Expanding Earth*. Elsevier, Amsterdam.

Carey, S. W. 1988. *Theories of the Earth and Universe: A History of Dogma in the Earth Sciences*. Stanford University Press, Stanford, CA.

Chamalaun, F. H. and McDougall, I. 1966. Dating geomagnetic polarity epochs in Réunion. *Nature*, **210**: 1212–1214.

Chander, R. and Brune, J. N. 1965. Radiation pattern of mantle Rayleigh waves and the source mechanism of the Hindu Kush earthquake of July 6, 1962. *Bull. Seismol. Soc. Am.*, **55**, 805–819.

Chandrasekhar, S. 1952. On the inhibition of convection by a magnetic field. *Phil. Mag.*, **43**: 501–532.

Chandrasekhar, S. 1953a. The onset of convection by thermal instability in spherical shells. *Phil. Mag.*, **44**: 233–241.

Chandrasekhar, S. 1953b. The instability of a layer of fluid heated below and subject to Coriolis forces. *Proc. Roy. Soc. London A.*, **217**: 306–327.

Chandrasekhar, S. 1954. The instability of a layer of fluid heated below and subject to the simultaneous action of a magnetic field and rotation. *Proc. Roy. Soc. London A*, **225**: 173–184.

Chandrasekhar, S. 1956. The instability of a layer of fluid heated below and subject to the simultaneous action of a magnetic field and rotation. II. *Proc. Roy. Soc. London A*, **237**: 476–484.

Chandrasekhar, S. 1957. Thermal convection (Rumford Medal Lecture). *Proc. Acad. Arts Sci.*, **86**: 323–339.

Christoffel, D. A. and Ross, D. I. 1961. Total magnetic field measurements between New Zealand and Antarctica. *Nature*, **190**: 776–778.

Christoffel, D. A. and Ross, D. I. 1965. Magnetic anomalies south of the New Zealand plateau. *J. Geophys. Res.*, **70**: 2857–2861.

Clegg, J. A., Almond, M., and Stubbs, P. H. S. 1954. The remanent magnetization of some sedimentary rocks in Britain. *Phil. Mag.*, **45**: 583–598.

Clegg, J. A., Deutsch, E. R., Everitt, C. W. R., and Stubbs, P. H. S. 1957. Some recent palaeomagnetic measurements made at Imperial College, London. *Adv. Phys.*, **6**: 219–230.

Clegg, J. A., Deutsch, E. R., and Griffiths, D. H. 1956. Rock magnetism in India. *Phil. Mag.*, **1**: 419–431.

Cloos, H. 1939. Hebung-Spaltung-Volcanismus. *Geologische Rundschau*, **30**: 506–510.

Coats, R. R. 1962. Magma type and crustal structure in the Aleutian arc. In MacDonald, G. A. and Kuno, H., eds., *The Crust of the Pacific Basin*. AGU Monograph 6, AGU, Washington, DC, 92–109.

Coode, A. M. 1965. A note on oceanic transcurrent faults. *Can. J. Earth Sci.*, **2**: 400–401.

Coode, A. M. 1966. An analysis of major tectonic features. *Geophys. J. Roy. Astron. Soc.*, **12**: 55–66.

Coode, A. M. 1968. The evidence of convection in the Earth's mantle from the Earth's surface and shape. Ph.D. dissertation. University of Newcastle upon Tyne.

Coode, A. M. and Runcorn, S. K. 1965. Satellite geoid and the structure of the Earth. *Nature*, **205**: 891.

Coode, A. M. and Tozer, D. C. 1965. Low-velocity layer as a source of the anomalous component of the geomagnetic variations at the coast. *Nature*, **205**: 164–165.

Cook, K. L. 1966a. Rift system in the Basin and Range Province. In *The World Rift System*, Geological Survey of Canada, Special Paper 66-14: 246–277.

Cook, K. L. 1966b. Comments. In *Continental Margins and Island Arcs*, Geological Survey of Canada, Special Paper 66-15: 313–315.

Cox, A. 1969. Geomagnetic reversals. *Science*, **163**: 237–245.

Cox, A. and Doell, R. R. 1960. Review of paleomagnetism. *Geol. Soc. Am. Bull.*, **71**: 645–768.

Cox, A. and Doell, R. R. 1961a. Palaeomagnetic evidence relevant to a change in the Earth's radius. *Nature*, **189**: 45–47.

Cox, A. and Doell, R. R. 1961b. Reply to Carey on palaeomagnetic evidence relevant to a change in the Earth's radius. *Nature*, **190**: 36–37.

Cox, A. and Doell, R. R. 1962. Magnetic properties of the basalt in hole EM7, Mohole project. *J. Geophys. Res.*, **678**: 3997–4007.

Cox, A., Doell, R. R., and Dalrymple, G. B. 1963a. Geomagnetic polarity epochs and Pleistocene geochronometry. *Nature*, **198**: 1049–1051.

Cox, A., Doell, R. R., and Dalrymple, G. B. 1963b. Geomagnetic polarity epochs: Sierra Nevada II. *Science*, **142**: 382–385.

Cox, A., Doell, R. R., and Dalrymple, G. B. 1964. Reversals of the Earth's magnetic field. *Science*, **144**: 1537–1543.

Cox, A., Doell, R. R. and Dalrymple, G. B. 1965. Quaternary paleomagnetic stratigraphy. In Wright, H. E. Jr. and Frey, D. G., eds., *The Quaternary of the United States*, Princeton University Press, Princeton, NJ, 817–830.

Cox, A., Doell, R. R., and Dalrymple, G. B. 1968. Time scale for geomagnetic reversals. In Phinney, R. A., ed., *The History of the Earth's Crust*. Princeton University Press, Princeton, NJ, 101–108.

Cox, A., Doell, R. R., and Thompson, G. 1964. *A Study of Serpentinite*, NAS-NRC Publication No. 1188, p. 49.

Creer, K. M. 1958. Preliminary palaeomagnetic measurements from South America. *Ann. Geóphys.*, **14**: 373–390.

Creer, K. M. 1965. Palaeomagnetic data from the Gondwanic continents. In Blackett, P. M. S., Bullard, E., and Runcorn, S. K., eds., *A Symposium on Continental Drift. Phil. Trans. Roy. Soc. London A*, **258**: 27–40.

Creer, K. M., Irving, E., Nairn, A. E. M. and Runcorn, S. K. 1958. Palaeomagnetic results from different continents and their relation to the problem of continental drift. *Ann. Geóphys.*, **14**: 492–501.

Creer, K. M., Irving, E., and Runcorn, S. K. 1960. The paleomagnetic poles for the lower Jurassic of Europe. *Geophys. J*, **3**: 367–370.

Crowell, J. C. 1960. The San Andreas fault in southern California. In *Report of the 21st International Geological Congress*, Copenhagen, part 18: 45–52.

Dalrymple, G. B., Cox, A., Doell, R. R. 1965. Potasium-argon age and paleomagnetism of the Bishop Tuff, California. *Geol. Soc. Am. Bull.*, **76**: 719–734.

Daly, R. A. 1940. *Strength and Structure of the Earth*, Prentice-Hall, New York.

Darwin, C. 1842. *Structure and Distribution of Coral Reefs*. Smith, Elder and Co., London.

David, P. 1904. Sur la stabilité de la direction d'aimantation dans quelques roches volcaniques. *C. R. Acad. Sci. Paris*, **188**: 41–42.

Davies, D. 1965. Discussion. In Blackett, P.M.S., Bullard, E., and Runcorn, S.K., eds., *A Symposium on Continental Drift. Phil. Trans. Roy. Soc. London A*, **258**: 140–141.

Davis, P. M. 2003. Azimuthal variation in seismic anisotropy of the southern California uppermost mantle. *J. Geophys. Res.*, **108**(B1): 1–24.

De Sitter, I. U. 1965. Discussion. In Blackett, P.M.S., Bullard, E., and Runcorn, S.K., eds., *A Symposium on Continental Drift. Phil. Trans. Roy. Soc. London A*, **258**: 205.

Deutsch, E. R. 1958. Recent palaeomagnetic evidence for northward movement of India. *J. Alberta Soc. Petrol. Geol.*, **6**: 155–162.

Deutsch, E. R. 1963. Polar wandering and continental drift: an evaluation of recent evidence. In Munyan, A.C., ed., *Polar Wandering and Continental Drift*. Society of Economic Paleontologists and Mineralogists, Special Publication No. 10: 4–46.

Deutsch, E. R., Radakrishnamurty, C., and Sahasrabudhe, P. W. 1958. The Remanent magnetism of some lavas in the Deccan Traps. *Phil. Mag.*, **3**: 170–184.

Dewey, J. and Kay, M. 1968. Appalachian and Caledonian evidence for drift in the North Atlantic. In Phinney, R. A., ed., *The History of the Earth's Crust*. Princeton University Press, Princeton, NJ, 161–167.

Dicke, R. H. 1957. Principle of equivalence and weak interactions. *Rev. Mod. Phys.*, **29**: 355–362.

Dicke, R. H. 1959. Gravitation – an enigma. *Am. Sci.*, **47**: 25–40.

Dickson, F. W. and Krauskopf, K. B. 1973. Memorial of Tom F. W. Barth, May 18, 1899–March 7, 1971. *Am. Minerol.*, **58**: 360–363.

Dickson, G. O. 1962a. Thermoremanent magnetization of igneous rocks. *J. Geophys. Res.*, **67**: 912–915.

Dickson, G. O. 1962b. The origin of small, randomly directed magnetic moments in demagnetized rock. *J. Geophys. Res.*, **67**: 4943–4945.

Dickson, G. O. 1963. The palaeomagnetism of Peat's Ridge Dolerite and Mt. Tomah Basalt. *J. Proc. Roy. Soc. NSW*, **96**: 129–132.

Dickson, G. O. and Foster, J. H. 1966. The magnetic stratigraphy of a deep sea core from the North Pacific Ocean. *Earth Planet. Sci. Lett.*, **1**: 458–462.

Dickson, G. O., Everitt, C. W. F., Parry, L. G., and Stacy, F. D. 1966. Origin of thermoremanent magnetization. *Earth Planet. Sci. Lett.*, **1**: 222–224.

Dickson, G. O., Pitman, W. C., III, Heirtzler, J.R. 1968. Magnetic anomalies in the South Atlantic and ocean floor spreading. *J. Geophys. Res.*, **70**: 3377–3406.

Dietz, R. S. 1946a. The meteoritic impact origin of the Moon's surface features. *J. Geol.*, **54**: 359–375.

Dietz, R.S. 1946b. Geological structures possibly related to lunar craters. *Pop. Astron.*, **54**: 455–467.

Dietz, R.S. 1947. Meteorite impact suggested by the orientation of shatter-cones at the Kentland, Indiana, disturbance. *Science*, **105**: 42–43.

Dietz, R.S. 1952a. The Pacific floor. *Sci. Am.*, **186**: 19–23.

Dietz, R.S. 1952b. Geomorphic evolution on continental terrace (continental shelf and slope). *Bull. Am. Assoc. Petrol. Geol.*, **36**: 1802–1819.

Dietz, R.S. 1954. Marine geology of northwestern Pacific: description of Japanese bathymetric chart 6901. *Geol. Soc. Am. Bull.*, **54**: 1199–1224.

Dietz, R.S. 1957. Office of Naval Research London, *European Scientific Notes*, No. 11-1.

Dietz, R.S. 1959a. Shatter cones in cryptoexplosion structures (meteorite impact?). *J. Geol.*, **67**: 496–505.

Dietz, R.S. 1959b. Point d'impact des astéroides comme origine des bassins océaniques: une hypothèse. *Coll. Int. CNRS, LXXXIII. La Topographie et la Geologie des Profondeurs Oceaniques*, 265–275.

Dietz, R.S. 1959c. Colloquium on the topography and geology of the deep sea floor. *Int. Geol. Rev.*, **1**: 113–122.

Dietz, R.S. 1959d. Drowned ancient islands of the Pacific. *New Scientist*, **5**: 14–17.

Dietz, R.S. 1960a. Meteorite impact suggested by shatter cones in rocks. *Science*, **131**: 1781–1784.

Dietz, R.S. 1960b. Vredefort ring structure; an astrobleme (meteorite impact structure). *Geol. Soc. Am. Bull.*, **71**: 2093.

Dietz, R.S. 1961a. Continent and ocean basin evolution by spreading of the sea floor. *Nature*, **190**: 854–857.

Dietz, R.S. 1961b. Astroblemes. *Sci. Am.*, **205**: 50–58.

Dietz, R.S. 1961c. Vredefort ring structure: meteorite impact scar? *J. Geol.*, **69**: 499–516.

Dietz, R.S. 1961d. The spreading ocean floor. *Saturday Evening Post*, **234**, No. 42: 34–35, 94, 96.

Dietz, R.S. 1961e. Response to Bernal. *Nature*, **192**: 124.

Dietz, R.S. 1962a. Ocean-basin evolution by sea-floor spreading. In Blackett, P.M.S., Bullard, E., and Runcorn, S.K., eds., *A Symposium on Continental Drift. Phil. Trans. Roy. Soc. London A*, **258**: 445–446.

Dietz, R.S. 1962b. Sudbury structure as an astrobleme. *Trans. Am. Geophys. Union*, **43**: 445–446.

Dietz, R.S. 1962c. Ocean basin evolution by sea floor spreading. *J. Oceanogr. Soc. Japan, 20th Century Volume*: 4–14.

Dietz, R.S. 1962d. Ocean-basin evolution by sea-floor spreading. In MacDonald, G.A. and Kuno, H., eds., *The Crust of the Pacific Basin*. American Geophysical Union, Washington, DC, 11–12.

Dietz, R.S. 1962e. *Continent and Ocean Basin Evolution by Sea Floor Spreading, Commotion in the ocean*. American Association of Petroleum Geologists, Distinguished Lecturer Series, 24 pages.

Dietz, R.S. 1963a. Reply to R. L. C. Gallant. *Nature*, **197**: 39–40.

Dietz, R.S. 1963b. Cryptoexplosion structures: a discussion. *Am. J. Sci.*, **261**: 650–664.

Dietz, R.S. 1963c. A theory of ocean basin origin. *Undersea Technology*, **4**: 26–29.

Dietz, R.S. 1964a. Sudbury structure as an astrobleme. *J. Geol.*, **72**: 412–434.

Dietz, R.S. 1964b. Commotion in the ocean: the growth of continents and ocean basins. In Yoshida, K., ed., *Studies in Oceanography: A collection of Papers Dedicated to Koji Hidaka*, University of Tokyo Press, Tokyo, Japan, 465–478.

Dietz, R.S. 1968. Reply. *J. Geophys. Res.*, **73**: 6567.

Dietz, R.S. 1989. Response by Robert S. Dietz for awarding of Penrose Medal. *Geol. Soc. Am. Bull.*, **101**: 987–989.

Dietz, R.S. 1994. Earth, sea and sky: life and times of a journeyman geologist. *Annu. Rev. Earth Planet. Sci.*, **22**: 1–32.

Dietz, R.S. and Butler, L.W. 1964. Shatter-cone orientation at Sudbury, Canada. *Nature*, **204**: 280–281.

Dietz, R.S. and Menard, H.W. 1951. Origin of the abrupt change in slope at the continental shelf margin. *Am. Assoc. Petrol. Geol. Bull.*, **35**: 1994–2016.

Dietz, R.S. and Menard, H.W. 1953. Hawaiian swell, deep, and arch, and subsidence of the Hawaiian Islands. *J. Geol.*, **61**: 99–113.

Dietz, R.S., Menard, H.W., and Hamilton, E.L. 1954. Echograms of the Mid-Pacific expedition. *Deep-Sea Res.*, **1**: 258–272.

Dirac, P.A.M. 1938. A new basis for cosmology. *Proc. Roy. Soc. London A*, **165**: 199–208.

Doel, R.E., Levin, T.J., and Marker, M.K. 2006. Extending modern cartography to the ocean depths: military patronage, Cold War priorities, and the Heezen–Tharp mapping project, 1952–1959. *J. Hist. Geogr.*, **32**: 605–626.

Doell, R.R. and Cox, A. 1963. The accuracy of the paleomagnetic method as evaluated from historic Hawaiian lava flows. *J. Geophys. Res.*, **68**: 1997–2009.

Doell, R.R. and Cox, A. 1965. Paleomagnetism of Hawaiian lava flows. *J. Geophys. Res.*, **70**: 3377–3406.

Doell, R.R. and Dalrymple, B.G. 1966. Geomagnetic polarity epochs: a new polarity event and the age of the Brunhes-Matuyama boundary. *Science*, **152**: 1060–1061.

Doell, R.R., Dalrymple, B.G., and Cox, A. 1966. Geomagnetic polarity epochs: Sierra Nevada data, 3. *J. Geophys. Res.*, **71**: 531–541.

Drake, C.L., Campbell, N.J., and Nafe, J.E. 1963. A Mid-Labrador Sea Ridge. *Nature*, **200**: 1085–1086.

Drake, C.L. and Girdler, R.W. 1964. A geophysical study of the Red Sea. *Geophys. J. Roy. Astron. Soc.*, **8**: 473–495.

Du Bois, P.M. 1957. Comparison of palaeomagnetic results for selected rocks of Great Britain and North America. *Adv. Phys.*, **6**: 177–186.

du Toit, A. 1937. *Our Wandering Continents: An Hypothesis of Continental Drifting*. Oliver and Boyd, Edinburgh.

Dymond, J.R., Watkins, N.D., and Nayudu, Y.R. 1968. Age of Cobb Seamount. *J. Geophys. Res.*, **73**: 3977–3979.

Egyed, L. 1963. The expanding Earth? *Nature*, **197**: 1059–1060.

Egyed, L. 1964. The satellite geoid and the structure of the Earth. *Nature*, **203**: 67–69.

Einarsson, T. 1957. Magneto-geological mapping in Iceland with the use of a compass. *Adv. Phys.*, **6**: 232–239.

Elsasser, W.M. 1963. Early history of the earth. In Geiss, J. and Goldberg, E.D., eds., *Earth Science and Meteoritics*. North-Holland, Amsterdam.

Elsasser, W.M. 1966. Thermal structure of the upper mantle and convection. In Hurley, P.M., ed., *Advances in Earth Science*. MIT Press, Cambridge, MA, 461–502.

Elsasser, W. M. 1967a. Sea-floor spreading as thermal convection. *J. Geophys. Res.*, **72**: 4768–4770.

Elsasser, W.M. 1967b. Convection and stress propagation in the upper mantle. Princeton University technical report 5, June 15, 1967.

Elsasser, W. M. 1968. The mechanics of continental drift. *Proc. Am. Phil. Soc.*, **112**: 344–353.

Elsasser, W. M. 1969a. Pattern of convective creep in the Earth's mantle. In Argon, A. S., ed., *Physics of Strength and Plasticity*. MIT Press, Cambridge, MA, 367–375.

Elsasser, W.M. 1969b. Convection and stress propagation in the upper mantle. In Runcorn, S. K., ed., *The Application of Modern Physics to the Earth and Planetary Interiors*. Interscience, New York, 223–246.

Elsasser, W. M. 1971. Sea-floor spreading as thermal convection. *J. Geophys. Res.*, **76**: 1101–1112.

Elvers, D., Mathewson, C., Kohler, R., and Moses, R. L. 1967a. Systematic ocean surveys by the USC and GSS Pioneer, 1961–1963, *Coast and Geodetic Survey Operational Data Report C and GSDR-1*, 19pp.

Elvers, D., Peter, G., and Moses, R. L. 1967b. Analysis of magnetic lineations in the North Pacific (abstract). *Trans. Am. Geophys. Union*, **48**: 89.

Elvers, D., Peter, G., and Yellin, M. 1964. *AGU Congress*.

Emiliani, C. 1981. Appendix I. In Emiliani, C., ed., *The Sea, Volume 7*. John Wiley & Sons, New York, 1717–1719.

Engdahl, E. R. and Flinn, E. A. 1969. Seismic waves reflected from discontinuities within earth's upper mantle. *Science*, **163**: 177–179.

Engel, A. E. J. and Engel, C.G. 1964. Composition of basalts from the Mid-Atlantic Ridge. *Science*, **144**: 1330–1333.

Engel, A. E. J., Engel, C.G., and Havens, R. G. 1965. Chemical characteristics of oceanic basalts and the upper mantle. *Geol. Soc. Am. Bull.*, **76**: 719–734.

Engel, C.G., Fisher, R. L., and Engel, A. E. J. 1965. Igneous rocks of the Indian ocean floor. *Science*, **150**: 605–609.

Ericson, D.B., Ewing, M., and Wollin, G. 1963. Pliocene-Pleistocene boundary in deep-sea sediments, *Science*, **139**: 727–737.

Everett, J. E. 1965. Magnetic variations. Ph.D. dissertation (unpublished). University of Cambridge.

Everett, J. E. and Smith, A. G. 2008. Genesis of a geophysical icon: the Bullard, Everett and Smith reconstruction of the circum-Atlantic continents. *Earth Sci. Hist.*, **27**: 1–11.

Ewing, J. and Ewing, M. 1967. Sediment distribution on the mid-ocean ridges with respect to spreading of the sea floor. *Science*, **156**: 1590–1592.

Ewing, J., Ewing, M., Aitken, T., and Ludwig, W. J. 1968. Upper mantle chemistry and evolution of Earth's crust. In Knopoff, L. *et al.*, eds., *The Crust and Upper Mantle of the Pacific Area*. AGU Geophysical Monograph 1, Washington, DC, 147–173.

Ewing, J., Worzel, J.L., Ewing, M., and Windisch, C. 1966. Ages of Horizon A and the oldest Atlantic sediments. *Science*, **154**: 1125–1131.

Ewing, M. 1963. Sediments of ocean basins. In *Man, Science, Learning, and Education: The Semicentennial Lectures at Rice University*. William Marsh Rice University and Chicago University Press, Chicago, 41–59.

Ewing, M. and Ewing, J. 1957. Seismic-refraction profiles in the Atlantic Ocean basins, in the Mediterranean Sea, on the Mid-Atlantic Ridge and in the Norwegian Sea. *Annual Meeting of the Seismological Society of America*.

Ewing, M. and Ewing, J. 1959. Seismic refraction profiles in the Atlantic Ocean basins, in the Mediterranean Sea, on the Mid-Atlantic Ridge and in the Norwegian Sea. *Geol. Soc. Am. Bull.*, **70**: 291–318.

Ewing, M. and Ewing, J. 1964. Distribution of ocean sediments. In Yoshida, K., ed., *Studies in Oceanography: A Collection of Papers Dedicated to Koji Hidaka*. University of Tokyo Press, Tokyo, Japan, 525–537.

Ewing, M. and Heezen, B. C. 1955. Puerto Rico trench topographic and geophysical data. *Geol. Soc. Am. Special Paper*, **62**: 255–268.

Ewing, M. and Heezen, B. C. 1956a. Some problems of Antarctic submarine geology. In *Antarctica in the I.G.Y.* AGU Geophysical Monograph, **1**: 75–81.

Ewing, M. and Heezen, B. C. 1956b. Mid-Atlantic ridge seismic belt. *Trans. AGU*, 337–343.

Ewing, M. and Heezen, B. C. 1960. Continuity of mid-oceanic ridge and rift valley in the southwestern Indian Ocean. *Science*, **131**: 1677–1679.

Ewing, M. and Landisman, M. 1961. Shape and structure of ocean basins. In Sears, M., ed. *Oceanography*, American Association for the Advancement of Science, Washington, DC, 3–38.

Ewing, M. and Worzel, J. L. 1954. Gravity anomalies and structure of the West Indies, part I. *Geol. Soc. Am. Bull.*, **65**: 165–174.

Ewing, M., Heezen, B. C., and Hirschman, J. 1957. Magnetic anomalies and seismicity in the Mid-Atlantic Ridge and its extensions (abstract). Communication No. 110 bis. *Assoc. Seismol. Ass. Gen. UGGI*, Toronto.

Ewing, M., Hirshman, J., and Heezen, B. C. 1959. *Magnetic anomalies of the Mid-Ocean Rift. International Ocean Congress preprints*. American Association for the Advancement of Science, Washington, DC, 24–25.

Ewing, M., Le Pichon, X., and Ewing, J. 1966. Crustal structure of the mid-ocean ridges. *J. Geophys. Rev.*, **71**: 1611–1636.

Ewing, M., Press, F., and Worzel, J. L. 1952. Further study of the T phase. *Bull. Seismol. Soc. Am.*, **42**: 37–51.

Ewing, M., Sutton, G. H., and Officer, C. B. 1954. Geophysical investigations in the emerged and submerged Atlantic coastal plain, Part VII. Continental shelf, continental slope and continental rise off Nova Scotia. *Geol. Soc. Am. Bull.*, **65**: 643–647.

Ewing, M., Tolstoy, I., and Press, F. 1950. Proposed use of the T phase in tsunami warning systems. *Bull. Seismol. Soc. Am.*, **40**: 53–58.

Fairbrother, N. 1954. *An English Year*. Knopf, New York.

Fairbrother, N. 1956. *Men and Gardens*. Hogarth Press, London.

Fairbrother, N. 1965. *The House in the Country*. Knopf, New York.

Fairbrother, N. 1970. *New Lives, New Landscapes*. Architectural Press, London.

Fairbrother, N. 1974. *The Nature of Landscape Design*. Architectural Press, London.

Feynman, R. 1999. *The Pleasure of Finding Things Out*. Perseus Publishing, Cambridge, MA.

Fisher, R. L. 1958. Downwind investigation of the Nasca Ridge. In Fisher, R. L., ed., *Preliminary Report on Expedition Downwind*, IGY General Report Series No. 2, National Research Council of National Academy of Science: 20–23.

Fisher, R. L. 1961. Middle America Trench: topography and structure. *Geol. Soc. Am. Bull.*, **72**: 703–720.

Fisher, R. L. 1965. Discussion. In Blackett, P. M. S., Bullard, E., and Runcorn, S. K., eds., *A Symposium on Continental Drift. Phil. Trans. Roy. Soc. London A*, **258**: 139–141.

Fisher, R. L. 1966. The median ridge in the south central Indian Ocean. In *The World Rift System*, Geological Survey of Canada, Special Paper 66-14: 135–147.

Fisher, R. L. and Goldberg, E. D. 1994. Henry William Menard. *Biogr. Mem. Natl. Acad. Sci.*, **64**: 267–276.

Fisher, R. L. and Hess, H. H. 1963. Trenches. In Hill, M. N., ed., *The Sea, Volume 3*. John Wiley & Sons, New York, 411–436.

Fisher, R. L., Sclater, J. G., and McKenzie, D. 1971. The evolution of the Central Indian Ridge. *Geol. Soc. Am. Bull.*, **82**: 553–562.

Fitch, F. J. 1965. Tectonics and continental drift. In Blackett, P. M. S., Bullard, E., and Runcorn, S. K., eds., *A Symposium on Continental Drift. Phil. Trans. Roy. Soc. London A*, **258**: 194–198.

Foster, J. H. 1966. A paleomagnetic spinner magnetometer using a fluxgate gradiometer. *Earth Planet. Sci. Lett.*, **1**: 463–466.

Fowler, C. M. R. 2000. *The Solid Earth*. Cambridge University Press, Cambridge.

Francheteau, J., Harrison, C. G. A., Sclater, J. G., and Richards, M. L. 1970. Magnetization of Pacific seamounts: a preliminary polar wander curve for the northeastern Pacific. *J. Geophys. Res.*, **75**: 2035–2061.

Francis, T. J. G. and Shor, G. G. Jr. 1966. Seismic refraction measurements from the northwest Indian Ocean, *J. Geophys. Res.*, **71**: 427–449.

Francis, T. J. G., Davis, D., and Hill, M. N., 1966. Crustal structure between Kenya and the Seychelles. *Phil. Trans. Roy. Soc. London A*, **259**: 240–261.

Frankel, H. 1979a. The career of continental drift theory: an application of Imre Lakatos' analysis of scientific growth to the rise of drift theory. *Stud. Hist. Phil. Sci.*, **10**: 21–66.

Frankel, H. 1979b. The reception and acceptance of continental drift theory as a rational episode in the history of science. In Mauskopf, S. H., ed., *The Reception of Unconventional Science*. American Association for the Advancement of Science, Washington, DC, 51–89.

Frankel, H. 1980. Hess's development of his seafloor spreading hypothesis. In Nickles, T., ed., *Scientific Discovery: Case Histories*. D. Reidel, Dordrecht, 345–366.

Frankel, H. 1982. The development, reception, and acceptance of the Vine–Matthews–Morley hypothesis. *Hist. Stud. Phys. Sci.*, **13**: 1–39.

Frankel, H. 1987. The continental drift controversy. In Englehart, H. T., Jr. and Caplan, A., eds., *Scientific Controversies*. Cambridge University Press, Cambridge, 203–248.

Franklin, A. and Howson, C. 1998. Comments on "The structure of a scientific paper. *Phil. Sci.*, **65**: 411–416.

Funnell, B. M. and Smith, A. G. 1968. The opening of the Atlantic Ocean. *Nature*, **219**: 1328–1333.

Garland, G. D. 1995. John Tuzo Wilson. *Biogr. Mem. Fell. Roy. Soc.*, **41**: 535–552.

Gartner, S., Jr. 1970. Sea-floor spreading, carbonate dissolution level, and the nature of Horizon A. *Science*, **169**: 1077–1079.

Gaskell, T. F. 1965. Discussion. In Blackett, P. M. S., Bullard, E., and Runcorn, S. K., eds., *A Symposium on Continental Drift. Phil. Trans. Roy. Soc. London A*, **258**: 139–141.

Gast, P. W. 1968. Upper mantle chemistry and evolution of Earth's crust. In Phinney, R. A., ed., *The History of the Earth's Crust*. Princeton University Press, Princeton, NJ, 15–27.

Gee, J.S. and Kent, D.V. 2007. Sources of oceanic magnetic anomalies and the geomagnetic polarity time-scale. In Kono, M., ed., *Treatise on Geophysics, 5 Geomagnetism*. Elsevier, Amsterdam, 455–507.

Gilliland, W. N. 1962. Possible continental continuation of the Mendocino Fracture Zone. *Science*, **137**: 685–686.

Gilliland, W. N. 1964. Extension of the theory of zonal rotation to explain global fracturing. *Nature*, **202**: 1276–1278.

Gilluly, J. 1963. The tectonic evolution of the western United States. *Q. J. Geol. Soc. London*, **119**: 133–174.

Gilluly, J. 1966. Continental drift: a reconsideration. *Science*, **152**: 946–950.

Girdler, R. W. 1958a. (1) Interpretation of gravity anomalies in the Red Sea area & the measurement of anisotropy of rocks and a study of the magnetic properties of some Jurassic rocks. (2) Some experiments on thermal convection in a rotating liquid. Ph.D. thesis (unpublished). University of Cambridge.

Girdler, R. W. 1958b. The relationship of the Red Sea to the East African rift system. *Q. J. Geol. Soc. London*, **114**: 79–105.

Girdler, R. W. 1958c. Some experiments on thermal convection in a rotating liquid, Ph.D. dissertation (unpublished). University of Cambridge.

Girdler, R. W. 1959a. Possible reversals of the Earth's magnetic field in the Jurassic period. *Nature*, **184**: 540–541.

Girdler, R. W. 1959b. A palaeomagnetic study of some Lower Jurassic rocks of N.W. Europe. *Geophys. J.*, **2**: 353–363.

Girdler, R. W. 1960. The case for an expanding Earth. *J. Durham Geol. Soc.*, **2**: 24–29.

Girdler, R. W. 1962. Initiation of continental drift. *Nature*, **194**: 521–524.

Girdler, R. W. 1963. Rift valleys, continental drift and convection in the Earth's mantle. *Nature*, **198**: 1037–1039.

Girdler, R. W. 1964. Geophysical studies of rift valleys. *Phys. Chem. Earth*, **5**: 122–154.

Girdler, R. W. 1965. The formation of new oceanic crust. In Blackett, P.M.S., Bullard, E., and Runcorn, S.K., eds., *A Symposium on Continental Drift. Phil. Trans. Roy. Soc. London A*, **258**: 123–136.

Girdler, R. W. 1966. The role of translational and rotational movements in the formation of the Red Sea and Gulf of Aden. In *The World Rift System*, Geological Survey of Canada, Special Paper 66-14: 65–76.

Girdler, R. W. 1967. Red Sea. In *International Dictionary of Geophysics*. Pergamon Press, Oxford, 1264–1268.

Girdler, R. W. 1968. Drifting and rifting of Africa. *Nature*, **217**: 1102–1106.

Girdler, R. W. and Peter, G. 1960. An example of the importance of natural remanent magnetization in the interpretation of magnetic anomalies. *Geophys. Prospecting*, **8**: 474–483.

Glen, W. 1982. *The Road to Jaramillo*. Stanford University Press, Stanford, CA.

Godby, E. A., Baker, R. C., Bower, M. E., and Hood, P. J. 1966. Aeromagnetic reconnaissance of the Labrador Sea. *J. Geophys. Res.*, **71**: 511–517.

Goguel, J. 1952. *Traité de Tectonique*. Masson, Paris.

Goguel, J. 1962. *Traité de Tectonique*. Translated from the 1952 French edition by Thalmann, H. E. W. H. Freeman, San Francisco.

Goguel, J. 1965. Tectonics and continental drift. In Blackett, P.M.S., Bullard, E., and Runcorn, S.K., eds., *A Symposium on Continental Drift. Phil. Trans. Roy. Soc. London A*, **258**: 194–198.

Gold, T. 1955. Instability of the Earth's axis of rotation. *Nature*, **175**: 526–529.

Gough, D. I. 1956. A study of the palaeomagnetism of the Pilansberg Dykes. *Mon. Not. Roy. Astron. Soc. Geophys. Suppl.*, **7**: 196–213.

Gough, D. I., Opdyke, N. D., and McElhinny, M. W. 1964. The significance of paleomagnetic results from Africa. *J. Geophys. Res.*, **69**: 2509–2519.

Graham, J. W. 1957. The role of magnetostriction in rock magnetism. *Adv. Phys.*, **6**: 362–363.

Graham, K. W. T. and Hales, A. L. 1957. Palaeomagnetic measurements on Karroo dolerites. *Adv. Phys.*, **6**: 149–161.

Graindor, M. J. 1965. Discussion. In Blackett, P. M. S., Bullard, E., and Runcorn, S. K., eds., *A Symposium on Continental Drift. Phil. Trans. Roy. Soc. London A*, **258**: 208.

Green, R. and Irving, E. 1958. The palaeomagnetism of the Cainozoic basalts from Australia. *Proc. Roy. Soc. Victoria*, **70**: 1–17.

Greene, M. T. 1998. Alfred Wegener and the origin of lunar craters. *Earth Sci. Hist.*, **7**: 111–138.

Griggs, D. 1954. Discussion, Verhoogen, 1954. *Trans. Am. Geophys. Union*, **35**: 93–96.

Griggs, D. T. 1967. Reflections on the earthquake mechanism. In Page, R., ed., *Proceedings of the Second United States–Japan Conference on Research Related to Earthquake Prediction Problems*. National Science Foundation, Washington, DC, 63.

Grommé, C. S. and Hay, R. L. 1963. Magnetization of basalt in Bed I, Olduvai Gorge, Tanganyika. *Nature*, **200**: 560–561.

Grow, J. A. and Atwater, T. 1970. Mid-Tertiary tectonic transition in the Aleutian Arc. *Geolog. Soc. Am. Bull.*, **81**: 3715–3722.

Gunn, R. 1947. Quantitative aspects of juxtaposed ocean deeps, mountain chains and volcanic ranges. *Geophysics*, **12**: 238–255.

Gutenberg, B. and Richter, C. F. 1954. *Seismicity of the Earth* (2nd edition). Princeton University Press, Princeton, NJ.

Hales, A. L. 1965. Discussion. In Blackett, P. M. S., Bullard, E., and Runcorn, S. K., eds., *A Symposium on Continental Drift. Phil. Trans. Roy. Soc. London A*, **258**: 58–59.

Hallam, A. 1973. *A Revolution in the Earth Sciences: From Continental Drift to Plate Tectonics*. Clarendon Press, New York.

Hamilton, E. L. 1966. Comments. In *Continental Margins and Island Arcs*, Geological Survey of Canada, Special Paper 66-15: 186–187.

Hamilton, R. M. and Gale, A. W. 1969. Thickness of the mantle seismic zone beneath the North Island of New Zealand. *J. Geophys. Res.*, **74**: 1608–1613.

Harland, W. B. 1965a. Discussion. In Blackett, P. M. S., Bullard, E., and Runcorn, S. K., eds., *A Symposium on Continental Drift. Phil. Trans. Roy. Soc. London A*, **258**: 59–76.

Harland, W. B. 1965b. Discussion. In Blackett, P. M. S., Bullard, E., and Runcorn, S. K., eds., *A Symposium on Continental Drift. Phil. Trans. Roy. Soc. London A*, **258**: 209–212.

Harrison, C. G. A. 1964. The magnetism of deep sea sediments. Ph.D. thesis (unpublished). University of Cambridge.

Harrison, C. G. A. and Funnell, B. M. 1964. Relationship of paleomagnetic reversals and micropalaeontology in two late Cenozoic cores from the Pacific Ocean. *Nature*, **204**: 566.

Hayes, D. E. and Heirtzler, J. 1968. Magnetic anomalies and their relation to the Aleutian Island arc. *J. Geophys. Res.*, **73**: 4637–4646.

Hays, J.D. 1965. Radiolaria and late Tertiary and Quaternary history of Antarctic Seas. *Antarct. Arct. Res., Am. Geophys. Union*, **5**: 125–184.

Hays, J.D. and Opdyke, N.D. 1967. *Science*, **158**: 1001–1011.

Heck, N.H. 1935. A new map of earthquake distribution. *Geogr. Rev.*, **25**: 125–130.

Heezen, B.C. 1955. Turbidity currents from the Magdalena River, Colombia. *Geol. Soc. Am. Bull.*, **66**: 1572.

Heezen, B.C. 1956. Outline of North Atlantic deep-sea geomorphology. *Geol. Soc. Am. Bull.*, **67**: 1703.

Heezen, B.C. 1957. Deep-sea physiographic provinces and crustal structure. *Trans. Am. Geophys. Union*, **38**: 394.

Heezen, B.C. 1959a. Géologie sous-marine et déplacements des continents. *Colloques Internationaux du Centre National de la Recherche Scientifique, LXXXIII. La Topographie et la Geologie des Profondeurs Oceaniques*, 295–304. Translation of paper and ensuing discussion by Annette Trefzer of the Lamont staff, 14 pages.

Heezen, B.C. 1959b. Paleomagnetism, continental displacements, and the origin of submarine topography. *International Ocean Congress preprints*. American Association for the Advancement of Science, Washington, DC, 26–28.

Heezen, B.C. 1959c. The tectonic evolution of the oceans. Unpublished paper presented on December 10, 1959 at Columbia University, 5 manuscript pages.

Heezen, B.C. 1960a. Comments. Unpublished transcription of meeting held in honor of Maurice Ewing's winning of the first Vetlesen Award held on March 25, 1960 at the Men's Faculty Club, Columbia University, 30–36a and 46.

Heezen, B.C. 1960b. The rift in the ocean floor. *Sci. Am.*, **203**: 98–110.

Heezen, B.C. 1960c. Tectonic evolution of the oceans. *Abstracts: International Union of Geodesy and Geophysics (IUGG)*, 12th General Assembly, Helsinki, Finland, July 7–August 6, 1960, 15.

Heezen, B.C. 1962. The deep-sea floor. In Runcorn, S.K., ed., *Continental Drift*. Academic Press, New York, 235–288.

Heezen, B.C. 1966a. Comments. In *Continental Margins and Island Arcs*, Geological Survey of Canada, Special Paper 66-15: 392–393.

Heezen, B.C. 1966b. Lecture at Conference, What's new on Earth. *Rutgers University*, New Brunswick, NJ, October (unpublished).

Heezen, B.C. 1974. Review of Hallam's *A Revolution in the Earth Sciences: From Continental Drift to Plate Tectonics. Science*, **183**: 504–505.

Heezen, B.C. and Ewing, M. 1952. Turbidity currents and submarine slumps, and the 1929 Grand Banks Earthquake. *Am. J. Sci.*, **250**: 849–873.

Heezen, B.C. and Ewing, M. 1961. The Mid-Oceanic Ridge and its extension through the Arctic Basin. In Raasch, G., ed., *Geology of the Arctic*. University of Toronto Press, Toronto, 622–642.

Heezen, B.C. and Ewing, M. 1963. The Mid-Ocean Ridge. In Hill, M.N., ed., *The Sea*, Volume 3, John Wiley & Sons, New York, 411–436.

Heezen, B.C. and Fox, P.J. 1966. Mid-Oceanic Ridge. In Fairbridge, R.W., ed., *The Encyclopedia of Oceanography, Encyclopedia of Earth Sciences Series*, **1**. Reinhold, New York, 506–517.

Heezen, B.C. and Tharp, M. 1963. The Mid-Ocean Ridge. In Hill, M.N., ed., *The Sea, Volume III*. John Wiley & Sons, New York, 411–436.

Heezen, B.C. and Tharp, M. 1965. Tectonic fabric of the Atlantic and India oceans. In Blackett, P.M.S., Bullard, E., and Runcorn, S.K., eds., *A Symposium on Continental Drift. Phil. Trans. Roy. Soc. London A*, **258**: 90–106.

Heezen, B. C., Bunce, E. T., Hersey, J. B. and Tharp, M. 1964. Chain and Romanche fracture zones. *Deep-Sea Res.*, **11**: 11–33.

Heezen, B. C., Ericson, D. B., and Ewing, M. 1954. Further evidence for a turbidity current following the 1929 Grand Banks earthquake. *Deep-Sea Res.*, **1**: 193–202.

Heezen, B. C., Ewing, M., and Ericson, D. B. 1954. Reconnaissance survey of abyssal plain south of Newfoundland. *Deep-Sea Res.*, **2**: 122–133.

Heezen, B. C., Ewing, M., and Miller, E. T. 1953. Trans-Atlantic profile of total magnetic intensity and topography, Dakar to Barbados. *Deep-Sea Res.*, **1**: 25–33.

Heezen, B. C. and Hollister, C. D. 1971. *The Face of the Deep*. Oxford University Press, New York.

Heezen, B. C., Tharp, M., and Ewing, M. 1959. The floors of the oceans, I, The North Atlantic. *Geol. Soc. Am. Special Paper*, **65**.

Heirtzler, J. R. 1961. *Vema* cruise No. 16 magnetic measurements. *Technical Report 2*, Contract CU-3-61 Nonr. Geology. Lamont Geological Observatory, Palisades, NY.

Heirtzler, J. R. 1968. Evidence for ocean floor spreading across the ocean basins. In Phinney, R. A., ed., *The History of the Earth's Crust*. Princeton University Press, Princeton, NJ, 90–100.

Heirtzler, J. R. and Le Pichon, X. 1965. Crustal structure of the mid-ocean ridges: 3 Magnetic anomalies over the Mid-Atlantic Ridge. *J. Geophys. Res.*, **79**: 4013–4033.

Heirtzler, J. R., Dickson, G. O., Herron, E. M., Pitman, W. C., and Le Pichon, X. 1968. Marine magnetic anomalies, geomagnetic field reversals and motions of the ocean floor and continents. *J. Geophys. Res.*, **73**: 2119–2136.

Heirtzler, J. R., Le Pichon, X., and Baron, J. G. 1966. Magnetic anomalies over the Reykjanes Ridge. *Deep-Sea Res.*, **13**: 427–443.

Henderson, P. 1965. Discussion. In Blackett, P. M. S., Bullard, E., and Runcorn, S. K., eds., *A Symposium on Continental Drift. Phil. Trans. Roy. Soc. London A*, **258**: 206.

Herron, E. M. and Heirtzler, J. R. 1967. Sea-floor spreading near the Galapagos. *Science*, **158**: 775–780.

Hersey, J. B. 1966. Marine geophysical investigations in the West. In *Continental Margins and Island Arcs*, Geological Survey of Canada, Special Paper 66-15: 151–164.

Hersey, J. B., Officer, C. B., Johnson, H. R., and Bergstom, S. 1957. Seismic refraction observations north of the Brownson Deep. *Bull. Seismol. Soc. Am.*, **42**: 291–306.

Hess, H. H. 1946. Drowned ancient islands of the Pacific Basin. *Am. J. Sci.*, **244**: 772–791.

Hess, H. H. 1954. Geological hypotheses and the Earth's crust under the oceans. *Proc. Roy. Soc. London A*, **222**: 341–348.

Hess, H. H. 1955a. Serpentines, orogeny, and epeirogeny. *Geol. Soc. Am. Special Paper*, **62**: 391–406.

Hess, H. H. 1955b. The oceanic crust. *J. Marine Res.*, **14**: 423–439.

Hess, H. H. 1959a. Nature of the great oceanic ridges. *International Oceanography Congress preprints*. American Association for the Advancement of Science, Washington, DC, 33–34.

Hess, H. H. 1959b. The AMSOC hole to the Earth's mantle. *Am. Geophys. Union*, **40**: 340–345.

Hess, H. H. 1960a. Scientific objectives of Mohole, and predicted section. *Am. Assoc. Petrol. Geol.*, **44**: 1250.

Hess, H. H. 1960b. Scientific objectives of Mohole, and predicted section. *Geol. Soc. Am. Bull.*, **71**: 2097. Reprint of (1960a).

Hess, H. H. 1960c. The AMSOC hole to the Earth's mantle. *Am. Sci.*, **47**: 254–263. Reprint of (1959b).

Hess, H. H. 1960d. Evolution of ocean basins. Preprint, 37 pages.

Hess, H. H. 1962. History of ocean basins. In *Petrologic Studies: A Volume to Honor A. F. Buddington.* Geological Society of America, New York, 599–620.

Hess, H. H. 1964a. Seismic anisotropy of the uppermost mantle under oceans. *Nature*, **203**: 629–631.

Hess, H. H. 1964b. The oceanic crust, the upper mantle and the Mayaguez serpentinized peridotite. In Burk, C. A., ed., *A Study of Serpentine: The AMSOC Core Hole Near Mayaguez, Puerto Rico.* Publication 1188, National Research Council, National Academy Sciences, Washington, DC, 169–175.

Hess, H. H. 1965. Mid-oceanic ridges and tectonics of the sea-floor. In Whittard, W. F. and Bradshaw, R., eds., *Submarine Geology and Geophysics.* Butterworths, London, 317–332.

Hess, H. H. 1966a. Comments on the Pacific basin. In *Continental Margins and Island Arcs*, Geological Survey of Canada, Special Paper 66-15: 311–312.

Hess, H. H. 1966b. Comments. In *Continental Margins and Island Arcs*, Geological Survey of Canada, Special Paper 66-15: 312–315.

Hess, H. H. 1968a. Response by Harry Hammond Hess, Penrose Medal. *Geol. Soc. Am. Proc.* for 1966: 85–86.

Hess, H. H. 1968b. Reply. *J. Geophys. Res.*, **73**: 65–69.

Hide, R. 1953a. Some experiments on thermal convection in a rotating liquid. *Q. J. Roy. Meteorol. Soc.*, **79**: 161.

Hide, R. 1953b. Fluid motions in the Earth's core, and some experiments on thermal convection in a rotating fluid. In Long, R. R., ed., *Fluid Models in Geophysics.* US Government Printing Office, Washington, DC, 101–116.

Hide, R. 1953c. Some experiments on thermal convection in a rotating liquid. Ph.D. dissertation (unpublished). University of Cambridge.

Hide, R. 1955. The hydrodynamics of the Earth's core. *Phys. Chem. Earth*, **1**, 94–137.

Hide, R. 1958. An experimental study of thermal convection in a rotating liquid. *Phil. Trans. Roy. Soc. London A*, **250**: 441–478.

Hide, R. 1997. Acceptance speech for Bowie Medal (1997). *Eos Trans. AGU*, **78**: 295.

Hide, R. 2006. Geomagnetism, 'vacillation', atmospheric predictability and deterministic chaos. *Pontif. Accad. Sci. Acta*, **18**: 257–274.

Hilgenberg, O. C. 1933. *Vom wachsenden Erdball.* Giessmann and Bartsch, Berlin.

Hill, M. L. 1960. A median valley of the Mid-Atlantic Ridge. *Deep-Sea Res.*, **6**: 193–205.

Hill, M. L. 1966. The San Andreas System, California and Mexico. In *The World Rift System*, Geological Survey of Canada, Special Paper 66-14: 239–245.

Hill, M. L. and Dibblee, T. W., Jr. 1963. San Andreas, Garlock, and Big Pine faults, California. *Geolog. Soc. Am. Bull.*, **64**: 443–458.

Hill, M. N. 1954. The topography of the Mid-Atlantic Ridge. Proces vertaux, General Assembly at Rome, September 1954: 269. Geofysisk Institutt, Bergen.

Hill, M. N. 1956. Notes on the bathymetric chart of the N. E. Atlantic. *Deep-Sea Res.*, **3**: 229–231.

Hill, M. N. 1957. Geophysical investigations on the floor of the Atlantic Ocean in *Discovery II*, 1956. *Nature*, **180**: 10–13.

Hill, M. N. 1966. Preface to *A Discussion Concerning the Floor of the Northwest Indian Ocean. Phil. Trans. Roy. Soc. London A*, **259**: 135–137.

Hirasawa, T. 1966. A least squares method for the focal mechanism determinations from S wave data (2). *Tokyo Univ. Earthquake Res. Inst. Bull.*, **44**: 919–938.

Hollingworth, S. E. 1965. Comments. In Blackett, P. M. S., Bullard, E., and Runcorn, S. K., eds., *A Symposium on Continental Drift. Phil. Trans. Roy. Soc. London A*, **258**: 141.

Holmes, A. 1929. A review of the continental drift hypothesis. *Mining Mag.*, **40**: 205–209; 286–288; 340–347.

Holmes, A. 1931. Radioactivity and earth movements. *Trans. Geol. Soc. Glasgow* (for 1928–9), **18**: 559–606.

Holmes, A. 1944. *Principles of Physical Geology* (1st edition). Thomas Nelson & Sons, Edinburgh.

Holmes, A. 1948. The oldest known minerals and rocks. *Trans. Edinburgh Geol. Soc.*, **14**: 176–194.

Holmes, A. 1953. The South Atlantic: land bridges or continental drift? *Nature*, **171**: 669–671.

Holmes, A. 1965. *Principles of Physical Geology*. Ronald Press, New York.

Honda, H. 1932. On the types of the seismograms and the mechanism of deep earthquakes. *Geophys. Mag.*, **5**: 301–324.

Honda, H., Masatsuka, A., and Emura, K. 1956. On the mechanisms of earthquakes and the stresses producing them in Japan and its vicinity (Second paper). *Tohoku Univ. Sci. Rep. Ser. 5. Geophys. Mag.*, **8**: 186–205.

Honda, H., Masatsuka, A., and Ichikawa, M. 1967. On the mechanisms of earthquakes and the stresses producing them in Japan and its vicinity (Third paper). *Tohoku Univ. Sci. Rep. Ser. 5. Geophys. Mag.*, **33**: 271–279.

Hood, P. J. and Godby, E. A. 1964. Magnetic anomalies over the Mid-Labrador Sea Ridge. *Nature*, **202**: 1099.

Hospers, J. 1953a. Reversals of the main geomagnetic field, part I. *Proc. Roy. Netherlands Acad. Sci. Amsterdam, Series B*, **56**: 467–476.

Hospers, J. 1953b. Reversals of the main geomagnetic field, part II. *Proc. Roy. Netherlands Acad. Sci. Amsterdam, Series B*, **56**: 477–491.

Hospers, J. 1954a. Reversals of the main geomagnetic field, part III. *Proc. Roy. Netherlands Acad. Sci. Amsterdam, Series B*, **57**: 112–121.

Hospers, J. 1954b. Magnetic correlation in volcanic districts. *Geol. Mag.*, **XCI**: 352–360.

Hurley, P. M. and Rand, J. R. 1968. Review of age data in West Africa and South America relative to a test of continental drift. In Phinney, R. A., ed., *The History of the Earth's Crust*. Princeton University Press, Princeton, NJ, 153–160.

Hurley, P. M., Almeida, F. F., Melcher, G. C., Cordani, U. G., Rand, J. R., Kawashita, K., Vandoros, P., Pinson, W. H., and Fairbairn, H. W. 1967. Test of drift by comparison of radiometric ages. *Science*, **157**: 495–500.

Ichikawa, M. 1961. On the mechanism of the earthquakes in and near Japan during the period from 1950 to 1957. *Geophys. Mag.*, **30**: 355–403.

Ichikawa, M. 1966. Mechanism of earthquakes in and near Japan, 1950–1962. *Pap. Meteorol. Geophys.*, **16**: 201–229.

Irving, E. 1956a. Palaeomagnetic and palaeoclimatological aspects of polar wandering. *Geofis. Pura Appl.*, **33**: 23–48.

Irving, E. 1956b. The magnetisation of the Mesozoic dolerites of Tasmania. *Pap. Proc. Roy. Soc. Tasmania*, **90**: 157–168.

Irving, E. 1957. Rock magnetism: a new approach to some palaeogeographic problems. *Adv. Phys.*, **6**: 194–218.

Irving, E. 1958a. Rock magnetism: a new approach to the problems of polar wandering and continental drift. In Carey, S.W., ed., *Continental Drift: A Symposium.* Geology Department, University of Tasmania, Hobart, 24–57.

Irving, E. 1958b. Palaeogeographic reconstruction from palaeomagnetism. *Geophys. J. Roy. Astron. Soc.*, **1**: 224–237.

Irving, E. 1959. Paleomagnetic pole positions: a survey and analysis. *Geophys. J. Roy. Astron. Soc.*, **2**: 51–59.

Irving, E. 1960. Palaeomagnetic directions and pole positions, part II. *Geophys. J.*, **3**: 444–449.

Irving, E. 1964. *Paleomagnetism and Its Application to Geological and Geophysical Problems.* John Wiley & Sons, New York.

Irving, E. 1966. The great Paleozoic reversal of the geomagnetic field (abstract). *Trans. Am. Geophys. Union*, **47**: 78.

Irving, E. 1970. The Mid-Atlantic Ridge at 45° XIV. Oxidation and magnetic properties of basalt; review and discussion. *Can. J. Earth Sci.*, **7**, 1528–1538.

Irving, E. 1984. Citation and response: J. Tuzo Wilson Medal. Geological Association of Canada, *Geology*, **3**: 31–32.

Irving, E. 1988. The paleomagnetic confirmation of continental drift. *Eos*, **69**: 994–1014.

Irving, E. 1991. Citation for Kenneth M. Creer for the John Adam Fleming Medal. *Eos*, **72**: 54–55.

Irving, E. 2000. Continental drift, organic evolution, and moral courage. *Eos*, **81**: 546.

Irving, E. 2004, The role of latitude in mobilism debates. *Proc. Natl. Acad. Sci.*, **102**, 1821–1828, also supporting text and figures on PNAS website.

Irving, E. 2008. Jan Hospers's key contributions to geomagnetism. *Eos Trans. AGU*, **89**: 457–468.

Irving, E. and Gaskell, T.F. 1962. The palaeogeographic latitude of oil fields. *Geophys. J.*, **7**: 54–63.

Irving, E. and Green, R. 1958. Polar movement relative to Australia. *Geophys. J. Roy. Astron. Soc.*, 164–172.

Irving, E., Park, J.K., Haggerty, S.E., Aumento, F., and Loncarovic, B. 1970. Magnetism and opaque mineralogy of basalts from the Mid-Atlantic Ridge at 45° N. *Nature*, **228**: 974–976.

Irving, E. and Parry, L.G. 1963. The magnetism of some Permian rocks from New South Wales. *Geophys. J.*, **7**: 395–411.

Irving, E. and Robertson, W.A. 1968. The distribution of continental crust and its relation to ice ages. In Phinney, R.A., ed., *The History of the Earth's Crust.* Princeton University Press, Princeton, NJ, 168–177.

Irving, E., Stott, P.M., and Ward, M.A. 1961. Demagnetization of igneous rocks by alternating magnetic fields. *Phil. Mag.*, **6**: 225–241.

Isacks, B. and Molnar, P. 1969. Mantle earthquake mechanisms and the sinking of the lithosphere. *Nature*, **223**: 1121–1124.

Isacks, B. and Molnar, P. 1971. Distribution of stresses in the descending lithosphere from a global survey of focal-mechanism solutions of mantle earthquakes. *Rev. Geophys. Space Phys.*, **9**: 103–174.

Isacks, B., Oliver, J., and Sykes, L. 1968. Seismology and the new global tectonics. *J. Geophys. Res.*, **73**: 5855–5899.

Isacks, B., Sykes, L. R., and Oliver, J. 1969. Focal mechanisms of deep and shallow earthquakes in the Tonga-Kermadec region and the tectonics of island arcs. *Geol. Soc. Am. Bull.*, **80**: 1443–1470.

Jacobs, J. A. and Atkinson, G. 1967. Planetary modulation of geomagnetic activity. In Hindmarsh, W. R., Lowes, F. J., Roberts, P. H., and Runcorn, S. K., eds., *Magnetism and the Cosmos: Proceedings of a NATO Advanced Study Institute on Planetary and Stellar Magnetism*. Oliver & Boyd, Edinburgh, 1967, 402–414.

Jacobs, J. A., Russell, R. D., and Wilson, J. T. 1959. *Physics and Geology*. McGraw-Hill, New York.

Jardine, N. and McKenzie, D. P. 1972. *Continental drift and the dispersal and evolution of organism. Nature*, **235**: 20–24.

Jeffreys, H. 1929. *The Earth: Its Origin, History and Physical Constitution* (2nd edition). Cambridge University Press, Cambridge.

Jeffreys, H. 1948. The origin of the solar system. *Mon. Not. Roy. Astron. Soc.*, **108**: 94–103.

Jeffreys, H. 1964. How soft is the Earth? *Q. J. Roy. Astron. Soc.*, **5**: 10–22.

Jeffreys, H. 1965. Discussion. In Blackett, P. M. S., Bullard, E., and Runcorn, S. K., eds., *A Symposium on Continental Drift, Phil. Trans. Roy. Soc. London A*, **258**: 314.

Johnson, T., and Molnar, P. 1972. Focal mechanism and plate tectonics of the southwest Pacific. *J. Geophys. Res.*, **77**: 5000–5032.

Kapp, R. O. 1960. *Towards a Unified Cosmology*. Hutchinson: London.

Katsumata, M. 1960. The effect of seismic zones upon the transmission of seismic waves. *Q. J. Seismol.*, **25**: 89–95 (in Japanese).

Katsumata, M. 1967. Seismic activities in and near Japan (3): Seismic activities versus depth. *J. Seismol. Soc. Japan (Zisin)*, **20**: 75–84 (in Japanese).

Katsumata, M. and Sykes, L. R. 1969. Seismicity and tectonics of the western Pacific: Izu-Mariana-Caroline and Ryuku-Taiwan regions. *J. Geophys. Res.*, **74**: 5923–5948.

Kay, M. 1951. *North American Geosynclines*. Geological Society of America Memoir, 48.

Kearney, P., Klepeis, K. A., and Vine, F. J. 2009. *Global Tectonics* (3rd edition). John Wiley & Sons, Oxford.

Kennedy, W. Q. 1946. The Great Glen Fault. *Q. J. Geol. Soc. London*, **102**: 41–76.

Kent, D. V. and Irving, E. 2010. Influence of inclination error in sedimentary rocks on the Triassic and Jurassic apparent pole wander path for North America and implications for Cordilleran tectonics, *J. Geophys. Res.*, **115**: B10103.

King, P. B. 1951. *The Tectonics of Middle North America*. Princeton University Press, Princeton, NJ.

King, P. B. 1959. *The Evolution of North America*. Princeton University Press, Princeton, NJ.

King, P. B., et al. 1944. *Tectonic Map of the United States*. American Association of Petroleum Geologists, Tulsa, OK.

Konigsberger, J. G. 1936. Residual magnetism and the measurement of geologic time. *International Geological Congress Reports, 16th Session, USA, 1933*, **i**: 225–231.

Krantz, W. 1924. Begleitworte zur geognostischen Spezialkarte von Wurttemberg. *Atlasblatt Heidenheim* (2nd edition). Stuttgart, 52–105.

Krause, D. C. 1966a. Equatorial shear zone. In *The World Rift System*, Geological Survey of Canada, Special Paper 66-14: 400–443.

Krause, D. C. 1966b. Comment. In *The World Rift System*, Geological Survey of Canada, Special Paper 66-14: 392.

Kropotkin, P. N. 1965. Discussion. In Blackett, P.M.S., Bullard, E., and Runcorn, S.K., eds., *A Symposium on Continental Drift. Phil. Trans. Roy. Soc. London A*, **258**: 316–318.

Kulp, J. L. 1961. Geologic time scale. *Science*, **133**: 1105–1114.

Langseth, M. 1966. Review of heat flow measurements along the mid-oceanic ridge system. In *The World Rift System*, Geological Survey of Canada, Special Paper 66-14: 349–362.

Langseth, M. and Grim, P. J. 1964. New heat-flow measurements in the Caribbean and western Atlantic. *J. Geophys. Res.*, **69**: 4916–4917.

Langseth, M., Grim, P. J., and Ewing, M. 1964. Heat-flow measurements in the East Pacific Ocean. *J. Geophys. Res.*, **70**: 367–380.

Langseth, M., Le Pichon, X., and Ewing, M. 1966. Crustal structure of the mid-ocean ridges. *J. Geophys. Res.*, **71**: 5321–5355.

Larochelle, A. 1968. Paleomagnetism of the Monteregian Hills: new results. *J. Geophys. Res.*, **73**: 3239–3246.

Larson, R. L., Menard, H. W., and Smith, S. M. 1968. Gulf of California: a result of ocean-floor spreading and transform faulting, *Science*, **161**: 781–784.

Laubscher, H. P. 1966. Comment. In *The World Rift System*, Geological Survey of Canada, Special Paper 66-14: 392.

Laughton, A. S. 1954. Laboratory measurements of seismic velocities in ocean sediments. *Proc. Roy. Soc. London*, **222**: 336–341.

Laughton, A. S. 1966a. The Gulf of Aden. *Phil. Trans. Roy. Soc. London A*, **259**: 150–171.

Laughton, A. S. 1966b. The Gulf of Aden, in relation to the Red Sea and the Afar Depression of Ethiopia. In *The World Rift System*, Geological Survey of Canada, Special Paper 66-14: 78–96.

Laughton, A. S. 1966c. Comments. In *Continental Margins and Island Arcs*, Geological Survey of Canada, Special Paper 66-15: 393.

Laughton, A.S., Hill, M. N., and Allan, T. D. 1960. Geophysical investigations of a seamount 150 miles north of Madeira. *Deep-Sea Res.*, **7**: 117–141.

Lear, J. 1967. Canada's unappreciated role as scientific innovator. *Sat. Rev.*, September 2: 45–50.

Lee, W. H. K. 1963. Heat flow data analysis. *Rev. Geophys.*, **1**: 449–479.

Leet, D.L., Linehan, D., and Berger, P.R. 1951. Investigation of the T phase. *Bull. Seismol. Soc. Am.*, **41**: 123–141.

Le Grand, H. E. 1988. *Drifting Continents and Shifting Theories*. Cambridge University Press, Cambridge.

Le Pichon, X. 1968a. Sea-floor spreading and continental drift. *J. Geophys. Res.*, **73**: 3661–3697.

Le Pichon, X. 1968b. Heat flow through the ocean floor and convection currents (abs). In Phinney, R. A., ed., *The History of the Earth's Crust*. Princeton University Press, Princeton, NJ, 119.

Le Pichon, X. 1986. The birth of plate tectonics. *Lamont Year Book, 1985–1986*, 53–61.

Le Pichon, X. 1991. Introduction to the publication of the extended outline of Jason Morgan's April 17, 1967 American Geophysical Union paper on "Rises, Trenches, Great Faults and Crustal Blocks." *Tectonophysics*, **187**: 1–6.

Le Pichon, X. 2001. My conversion to plate tectonics. In Oreskes, N., ed., with Le Grand, H., *Plate Tectonics*. Westview Press, Boulder, CO, 201–224.

Le Pichon, X. and Fox, P. J. 1971. Marginal offsets, fracture zones, and the early opening of the North Atlantic. *J. Geophys. Res.*, **76**: 6294–6308.

Le Pichon, X. and Heirtzler, J. R. 1966. Magnetic anomalies in the Indian Ocean and sea-floor spreading. *J. Geophys. Res.*, **73**: 2101–2117.

Le Pichon, X. and Langseth, M. G. 1969. Heat flow from the mid-ocean ridges and sea-floor spreading. *Tectonophysics*, **8**: 319–344.

Le Pichon, X. and Talwani, M. 1965. Gravity survey of a seamount near 35° N 46° W in the north Atlantic. *Marine Geol.*, **2**: 262–277.

Le Pichon, X., Francheteau, J., and Bonnin, J. 1973. *Plate Tectonics*. Elsevier Scientific Publishing Co., Amsterdam.

Linehan, D. 1940. Earthquakes in the West Indian region. *Trans. Am. Geophys. Union*: 229–232.

Lipton, P. 1998. The best explanation of a scientific paper. *Phil. Sci.*, **65**: 406–410.

Lister, C. R. B. 1962. Heat flow through the ocean floor. Ph.D. thesis (unpublished). University of Cambridge.

Lister, C. R. B. 1972. On the thermal balance of a mid-ocean ridge. *Geophys. J. Roy. Soc.*, **26**: 515–535.

Loncarevic, B. D. 1966. Comments. In *The World Rift System*, Geological Survey of Canada, Special Paper 66-14: 347.

Loncarevic, B. D., Mason, C. S., and Matthews, D. H. 1966a. Mid-Atlantic Ridge near 45° north. *Can. J. Earth Sci.*, **3**: 327–345.

Loncarevic, B. D., Mason, C. S., and Matthews, D. H. 1966b. Mid-Atlantic Ridge between 45°N and 46°N. In *The World Rift System*, Geological Survey of Canada, Special Paper 66-14: 212.

Longwell, C. R. 1945. The mechanics of orogeny. *Am. J. Sci.*, **243-A**: 417–447.

Lowrie, W. 2007. *Fundamentals of Geophysics* (2nd edition). Cambridge University Press, Cambridge.

Luskin, B., Heezen, B. C., Ewing, M., and Landisman, M. 1954. Precision measurement of ocean depth. *Deep Sea-Res.*, **1**: 131–140.

MacDonald, G. J. F. 1965. Continental structure and drift. In Blackett, P. M. S., Bullard, E., and Runcorn, S. K., eds., *A Symposium on Continental Drift. Phil. Trans. Roy. Soc. London A*, **258**: 215–227.

Mackey, R. 2007. Rhodes Fairbridge and the idea that the solar system regulates the Earth's climate. *J. Coastal Res.*, **50** (special issue): 955–968.

Martin, H. 1961. The hypothesis of continental drift in the light of recent advances of geological knowledge in Brazil and in South West Africa: Alex L. du Toit memorial lectures No. 7. *Trans. Proc. Geol. Soc. S. Afr.* (Annexure to), **64**: 1–47.

Marvin, U. B. 1966. Continental drift. In Lundquist, C. A. and Friedman, H. D., eds., *Scientific Horizons from Satellite Tracking*. Smithsonian Astrophysical Observatory Special Report 236, 31–74.

Mason, R. G. 1958. A magnetic survey off the west coast of the United States between Latitudes 32° and 36° N, Longitudes 121° and 128° W. *Geophys. J.*, **1**: 320–329.

Mason, R. G. 2001. Stripes on the sea floor. In Oreskes, N., ed., with Le Grand, H., *Plate Tectonics*. Westview Press, Boulder, CO, 31–45.

Mason, R. G. and Raff, A. D. 1961. Magnetic survey off the west coast of North America, 32° N. latitude to 42° N. latitude. *Geol. Soc. Am. Bull.*, **72**: 1259–1266.

Matthews, D. H. 1959. Aspects of the geology of the Scotia Arc. *Geol. Mag.*, **6**: 425–441.

Matthews, D. H. 1961a. Lavas from an abyssal hill on the floor of the north Atlantic Ocean. *Nature*, **190**: 158–159.

Matthews, D. H. 1961b. Rocks from the eastern north Atlantic. Ph.D. dissertation (unpublished). University of Cambridge.

Matthews, D. H. 1966a. The Owen fracture zone and the northern end of the Carlsberg Ridge. *Phil. Trans. Roy. Soc. London A*, **259**: 172–186.

Matthews, D. H. 1966b. The northern end of the Carlsberg Ridge. In *The World Rift System*, Geological Survey of Canada, Special Paper 66-14: 124–134.

Matthews, D. H. 1966c. Comments. In *Continental Margins and Island Arcs*, Geological Survey of Canada, Special Paper 66-15: 315–316.

Matthews, D. H. 1967. Mid-Ocean Ridges. In Runcorn, S. K., ed., *International Dictionary of Geophysics, Volume 2*. Pergamon Press, Oxford, 979–991.

Matthews, D. H. and Bath, J. 1967. Formation of magnetic anomaly pattern of mid-Atlantic ridge. *Geophys. J. Roy. Astron. Soc.*, **13**: 349–357.

Matthews, D. H. and Davies, D. 1966. Geophysical studies of the Seychelles Bank. *Phil. Trans. Roy. Soc. London A*, **259**: 227–239.

Matthews, D. H. and Maling, D. H. 1967. *The geology of the South Orkney Islands. I. Signey Island*. British Antarctic Survey, Scientific Reports No. 25.

Matthews, D. H., Vine, F. J., and Cann, J. R. 1965. Geology of an area of the Carlsberg Ridge. *Geol. Soc. Am. Bull.*, **76**: 675–682.

Matthews, D. H., Williams, C., and Laughton, A. S. 1967. Mid-Ocean ridge in the mouth of the Gulf of Aden. *Nature*, **215**: 1052–1053.

Matuyama, M. 1929a. On the direction of magnetization of basalt in Japan, Tyōsen and Manchuria. *Proceedings 4th Pacific Science Congress*: 1–3.

Matuyama, M. 1929b. On the direction of magnetization of basalt in Japan, Tyrōsen and Manchuria. *Proc. Imp. Acad. Japan*, **5**: 203–205.

Maxwell, A. E. 1969. Recent deep sea drilling results from the South Atlantic (abs.). *Trans. Am. Geophys. Union*, **50**: 113.

Maxwell, A. E., Von Herzen, R. P., Hsü, K. J., Andrews, J. E., Saito, T., Percival, S. F. P., Jr., Milow, E. D., and Boyce, R. E. 1970. Deep sea drilling in the South Atlantic. *Science*, **168**: 1047–1059.

McConnell, R. K., Jr. 1968. Viscosity of the Earth's mantle. In Phinney, R. A., ed., *The History of the Earth's Crust*. Princeton University Press, Princeton, NJ, 45–57.

McDougall, I. 1963. Potassium-argon ages from western Oahu, Hawaii. *Nature*, **197**: 344–345.

McDougall, I. 1964. Potassium-argon ages from lavas of the Hawaiian Islands, *Geol. Soc. Am. Bull.*, **75**: 107–128.

McDougall, I. and Chamalaun, F. H. 1966. Geomagnetic polarity scale of time. *Nature*, **212**: 1415–1418.

McDougall, I. and Tarling, D. H. 1963. Dating polarity zones in the Hawaiian Islands. *Nature*, **200**: 54–56.

McDougall, I. and Wensink, H. 1966. Paleomagnetism and geochronology of the Pliocene-Pleistocene lavas in Iceland. *Earth Planet. Sci. Lett.*, **1**: 232–236.

McDougall, I., Allsopp, H. L., and Chamalaun, F. H. 1966. Isotopic dating of the Newer Volcanics of Victoria, Australia, and geomagnetic polarity epochs. *J. Geophys. Res.*, **71**: 6107–6118.

McElhinny, M. W. 1970. Formation of the Indian Ocean. *Nature*, **228**: 977–979.

McElhinny, M. W. 1973. *Paleomagnetism and Plate Tectonics*. Cambridge University Press, Cambridge.

McElhinny, M.W. and McFadden, P.L. 2000. *Paleomagnetism: Continents and Oceans*. Academic Press, San Diego, CA.

McEvilly, T.V. 1966. Preliminary seismic data, June–July, 1966 in Parkfield earthquake of June 27–29, 1966, Monterey and San Louis Obispo counties, California: preliminary report. *Bull. Seismol. Soc. Am.*, **56**: 967–971.

McKenzie, D.P. 1966a. The viscosity of the lower mantle. *J. Geophys. Res.*, **71**: 3995–4010.

McKenzie, D.P. 1966b. The shape of the Earth. Ph.D. thesis (unpublished). University of Cambridge.

McKenzie, D.P. 1967a. Some remarks on heat flow and gravity anomalies. *J. Geophys. Res.*, **72**: 6261–6263.

McKenzie, D.P. 1967b. The viscocity of the mantle. *Geophys. J.*, **14**: 297–305.

McKenzie, D.P. 1968a. The geophysical importance of high-temperature creep. In Phinney, R.A., ed., *The History of the Earth's Crust*. Princeton University Press, Princeton, NJ, 28–44.

McKenzie, D.P. 1968b. The Earth's mantle. In Runcorn, S.K., ed., *International Dictionary of Geophysics*. Pergamon Press, Oxford.

McKenzie, D.P. 1969a. The oscillations of a viscous gravitating sphere. *Proc. Cambridge Phil. Soc.*, **65**: 123–137.

McKenzie, D.P. 1969b. Speculations on the consequences and causes of plate motions. *Geophys. J.*, **18**: 1–32.

McKenzie, D.P. 1969c. The relation between fault plane solutions for earthquakes and the directions of the principal stresses. *Bull. Seismol. Soc. Am.*, **59**: 591–601.

McKenzie, D.P. 1987. Edward Crisp Bullard. *Biogr. Mem. Fell. Roy. Soc.*, **33**: 65–98.

McKenzie, D.P. 2001. Plate tectonics: a surprising way to start a scientific career. In Oreskes, N., ed., with Le Grand, H., *Plate Tectonics*. Westview Press, Boulder, CO, 169–190.

McKenzie, D.P. and Bickle, M.J. 1988. The volume and composition of melt generated by extension of lithosphere. *J. Petrol.*, **29**: 625–679.

McKenzie, D.P. and Morgan, W.J. 1969. Evolution of triple junctions. *Nature*, **224**: 125–133.

McKenzie, D.P. and Parker, R.L. 1967. The North Pacific: an example of tectonics on a sphere. *Nature*, **216**: 1276–1280.

McKenzie, D.P. and Sclater, J.G. 1971. The evolution of the Indian Ocean in the Late Cretaceous. *Geophys. J. Roy. Astron. Soc.*, **25**: 437–528.

McKenzie, D., Molnar, P., and Davies, D. 1970. Plate tectonics of the Red Sea and East Africa. *Nature*, **226**: 239–243.

Menard, H.W. 1955a. Deformation of the northeastern Pacific Basin and the west coast of North America. *Geol. Soc. Am. Bull.*, **66**: 1149–1198.

Menard, H.W. 1955b. Deep-sea channels, topography, and sedimentation. *Bull. Am. Assoc. Petrol. Geol.*, **39**: 236–255.

Menard, H.W. 1958a. Development of median elevations in ocean basins. *Geol. Soc. Am. Bull.*, **69**: 1179–1186.

Menard, H.W. 1958b. Geology of the Pacific sea floor. *Experientia*, **15**: 205–213.

Menard, H.W. 1960. The East Pacific Rise. *Science*, **132**: 1737–1746.

Menard, H.W. 1961. The East Pacific Rise. *Sci. Am.*, **205**: 52–61.

Menard, H. W. 1964. *Marine Geology of the Pacific*. McGraw-Hill, New York.

Menard, H. W. 1965a. The world wide oceanic rise-ridge system. In Blackett, P. M. S., Bullard, E., and Runcorn, S. K., eds., *A Symposium on Continental Drift. Phil. Trans. Roy. Soc. London A*, **258**: 109–122.

Menard, H. W. 1965b. Discussion. In Blackett, P. M. S., Bullard, E., and Runcorn, S. K., eds., *A Symposium on Continental Drift. Phil. Trans. Roy. Soc. London A*, **258**: 206–207.

Menard, H. W. 1965c. Sea floor relief and mantle convection. *Physics Chem. Earth*, **6**: 315–364.

Menard, H. W. 1967. Extension of northeastern Pacific fracture zones. *Science*, **155**: 72–73.

Menard, H. W. 1968. Some remaining problems in sea-floor spreading. In Phinney, R. A. ed., *The History of the Earth's Crust*. Princeton University Press, Princeton, NJ, 109–118.

Menard, H. W. 1971. *Science, Growth and Change*. Harvard University Press, Cambridge, MA.

Menard, H. W. 1979. Very like a spear. In Schneer, C., ed., *Two Hundred Years of Geology in America*. University Press of New England, Hanover, NH, 19–30.

Menard, H. W. 1986. *The Ocean of Truth: A Personal History of Global Tectonics*. Princeton University Press, Princeton, NJ.

Menard, H. W. and Atwater, T. 1968. Changes in direction of sea floor spreading. *Nature*, **219**: 463–467.

Menard, H. W. and Atwater, T. 1969. Origin of fracture zone topography. *Nature*, **222**: 1037–1040.

Menard, H. W. and Dietz, R. S. 1951. Submarine geology of the Gulf of Alaska. *Geol. Soc. Am. Bull.*, **62**: 1263–1285.

Menard, H. W. and Dietz, R. S. 1952. Mendocino submarine escarpment. *J. Geol.*, **60**: 266–278.

Menard, H. W. and Vacquier, V. 1958. Magnetic survey of part of the deep sea floor off the coast of California. *United States Naval Research Office Research Reviews*, June issue: 1–6.

Menard, H. W., Allison, E. C., and Durham, J. W. 1962. A drowned Miocene terrace in the Hawaiian Islands. *Science*, **138**: 896–897.

Menard, H. W., Chase, T. E., and Smith, S. M. 1963. Galapagos Rise in the southeastern Pacific. *Deep-Sea Res.*, **11**: 223–242.

Mercanton, P. L. 1926a. Inversion de l'inclinasion magnetique terrestre aux ages geologiques. *Terr. Magn. Atmos. Elect.*, **31**: 187–190.

Mercanton, P. L. 1926b. Aimantation des basaltes groenlandais. *C. R. Acad. Sci. Paris*, **182**: 859–860.

Miller, J. A. 1964. Appendix II. Age determinations made on samples of basalt from the Tristan Da Cunha Group and other parts of the Mid-Atlantic Ridge. *Phil. Trans. Roy. Soc. London*, **256**: 556–569.

Miller, J. A. 1965a. Geochronology and continental drift – the North Atlantic. In Blackett, P. M. S., Bullard, E., and Runcorn, S. K., eds., *A Symposium on Continental Drift. Phil. Trans. Roy. Soc. London A*, **258**: 180–190.

Miller, J. A. 1965b. Discussion. In Blackett, P. M. S., Bullard, E., and Runcorn, S. K., eds., *A Symposium on Continental Drift. Phil. Trans. Roy. Soc. London A*, **258**: 213.

Minear, J. W. and Toksoz, M. N. 1970. Thermal regime of a downgoing slab and new global tectonics. *J. Geophys. Res.*, **75**: 1397–1419.

Mohr, O. 1928. *Abhandlungen aus dem Gebiete der technischen Mechanik* (3rd edition). Ernst & Shon, Berlin.

Molnar, P. 2001. From plate tectonics to continental tectonics. In Oreskes, N., ed., with Le Grand, H., *Plate Tectonics*. Westview Press, Boulder, CO, 288–328.

Molnar, P. and Oliver, J. 1969. Lateral variations of attenuation in the upper mantle and discontinuities in the lithosphere. *J. Geophys. Res.*, **74**: 2648–2682.

Molnar, P. and Sykes, L. R. 1969. Tectonics of the Caribbean and Middle America regions from focal mechanism and seismicity. *Geol. Soc. Am. Bull.*, **80**: 1639–1684.

Morgan, W.J. 1964. An astronomical and geophysical search for scalar gravitational waves. Ph.D. thesis (unpublished). Princeton University, Princeton, NJ.

Morgan, W.J. 1965a. Gravity anomalies and convection currents: 1. A sphere and cylinder sinking beneath the surface of a viscous fluid. *J. Geophys. Res.*, **70**: 6175–6187.

Morgan, W.J. 1965b. Gravity anomalies and convection currents: 2. The Puerto Rico Trench and the Mid-Atlantic Rise. *J. Geophys. Res.*, **70**: 6189–6204.

Morgan, W.J. 1967a. Convection in a viscous mantle and trenches. *Trans. Am. Geophys. Union*, **48**: 217–218.

Morgan, W.J. 1967b. Rises, trenches, and great faults and crustal bocks. Extended outline of 1967 AGU talk, originally unpublished, reproduced in Le Pichon, 1991.

Morgan, W. J. 1968a. Pliocene reconstruction of the Juan de Fuca ridge (abs.). *Trans. Am. Geophys. Union*, **49**: 327.

Morgan, W.J. 1968b. Rises, trenches, and great faults and crustal blocks. *J. Geophys. Res.*, **73**: 2131–2153.

Morgan, W.J. 1971. Convection plumes in the lower mantle. *Nature*, **230**: 42–43.

Morgan, W.J. 1991. Rises, trenches, great faults and crustal blocks. *Tectonophysics*, **187**: 7–21.

Morgan, W.J., Stoner, J.O., and Dicke, R. H. 1961. Periodicity of earthquakes and the invariance of the gravitational constant. *J. Geophys. Res.*, **66**: 3831–3843.

Morgan, W. J. Phipps 1988. 1987 Ewing Medal to W. Jason Morgan, Citation. *Eos*, **69**: 115.

Morgan, W. J., Vogt, P. R., and Falls, D. F. 1969. Magnetic anomalies and sea floor spreading on the Chile Rise. *Nature*, **222**: 137–142.

Morley, L. W. 1952. Correlation of the susceptibility and remanent magnetism with the petrology of rocks from some pre-Cambrian areas in Ontario. Ph.D. dissertation (unpublished). University of Toronto.

Morley, L. W. 1954. Comments. *Trans. Am. Geophys. Union*, **35**: 77–78.

Morley, L. W. 1981. An explanation of magnetic banding in ocean basins. In Emiliani, C., ed., *The Sea, Volume 7*. John Wiley & Sons, New York, 1717–1719.

Morley, L. W. 1986. Early work leading to the explanation of the banded geomagnetic imprinting of the ocean floor. *Eos*, **67**: 665–666.

Morley, L. W. 2001. The zebra pattern. In Oreskes, N., ed., with Le Grand, H., *Plate Tectonics*. Westview Press, Boulder, CO, 67–85.

Morley, L. W. and Larochelle, A. 1964. Paleomagnetism as a means of dating geological events. In Osborne, F.F., ed., *Geochronology in Canada*, Royal Society of Canada, Special Publication, no. 8, Toronto, 39–50.

Muir, I. D., Tilley, C. E., and Scoon, J. H. 1964. Basalts from the northern part of the Mid-Atlantic Ridge. *J. Petrol.*, **5**: 409–434.

Munk, W. H. and MacDonald, G. J. F. 1960a. *The Rotation of the Earth.* Cambridge University Press, Cambridge.

Murray, H. W. 1939. Submarine scarp off Mendocino, California. *Field Eng. Bull.,* **13**: 27–33.

Nafe, J. E., Hennion, J. F., and Peter, G. 1959. Geophysical measurements in the Gulf of Aden. *Preprints. International Oceanography Congress, Washington,* **42**.

Nagata, T., Akimoto, S., and Uyeda, S. 1951. Reverse thermo-remanent magnetization. *Proc. Imp. Acad. Japan,* **27**: 643–645.

Nairn, A. E. M. 1957. Palaeomagnetic collections from Britain and South Africa illustrating two problems of weathering. *Adv. Phys.,* **6**: 162–168.

Nairn, A. E. M. 1965. Discussion. In Blackett, P. M. S., Bullard, E., and Runcorn, S. K., eds., *A Symposium on Continental Drift. Phil. Trans. Roy. Soc. London A,* **258**: 59.

Néel, L. 1949. Théorie du traînage magnétique des ferromagnétiques en grains fins avec application aux terres cuites. *Ann. Géophys.,* **5**: 99–136.

Néel, L. 1951. Inversion de l'aimantation permenente des roches. *Ann. Geophys.,* **7**: 90–102.

Nichols, G. D. 1965. Petrological and geochemical evidence for convection in the Earth's mantle. In Blackett, P. M. S., Bullard, E., and Runcorn, S. K., eds., *A Symposium on Continental Drift. Phil. Trans. Roy. Soc. London A,* **258**: 168–179.

Ninkovich, D., Opdyke, N., Heezen, B. C., and Foster, J. H. 1966. Paleomagnetic stratigraphy, rates of deposition and tephrachronology in North Pacific deep-sea sediments. *Earth Planet. Sci. Lett.,* **1**: 476–492.

Northrop, J. and Heezen, B. C. 1951. An outcrop of Eocene sediment of the continental slope. *J. Geol.,* **59**: 396–399.

Officer, C. B. and Ewing, M. 1954. Geophysical investigations in the emerged and submerged Atlantic coastal plain, Part VII. Continental shelf, continental slope and continental rise off Nova Scotia. *Geol. Soc. Am. Bull.,* **65**: 643–674.

Officer, C. B., Ewing, J., Edwards, R. S., and Johnson, H. R. 1957. Geophysical investigations in the Eastern Caribbean, Trinidad Shelf, Tobago Trough, Barbados Ridge and Atlantic Ocean. *Geol. Soc. Am. Bull.,* **68**: 259–278.

Officer, C. B., Ewing, M., and Wuenschel, P. C. 1952. Seismic refraction measurements in the Atlantic Ocean, Part IV. Bermuda, Bermuda Rise and Nares Basin. *Geol. Soc. Am. Bull.,* **63**: 777–808.

O'Keefe, J. A. 1965. Discussion. In Blackett, P. M. S., Bullard, E., and Runcorn, S. K., eds., *A Symposium on Continental Drift. Phil. Trans. Roy. Soc. London A,* **258**: 272–275.

Oliver, J. 1996. *Shocks and Rocks.* American Geophysical Union, Washington, DC.

Oliver, J. 1998. *Shakespeare Got It Wrong, It's Not "To Be," It's "To Do." The Autobiographical Memoirs of a Lucky Geophysicist.* Available online at ecommons. cornell.edu.

Oliver, J. 2001. Earthquake seismology in the plate tectonics revolution. In Oreskes, N., ed., with Le Grand, H., *Plate Tectonics.* Westview Press, Boulder, CO, 155–166.

Oliver, J. 2004. *The Incomplete Guide to the Art of Discovery.* Available online at ecommons.cornell.edu.

Oliver, J. and Isacks, B. 1967. Deep earthquake zones, anomalous structures in the upper mantle, and the lithosphere. *J. Geophys. Res.,* **72**: 4259–4275.

Oliver, J. and Isacks, B. 1968. Structure and mobility of the crust and mantle in the vicinity of island arcs. *Can. J. Earth Sci.,* **5**: 985–991.

Oliver, J. and Murphy, L. 1971. WWNSS: seismology's global network of observing stations. *Science*, **174**: 254–261.

Oliver, J., Sykes, L., and Isacks, B. 1969. Seismology and the new global tectonics. *Tectonophysics*, **7**: 527–541.

Opdyke, N.D. 1964a. The paleomagnetism of the Permian red beds of southwest Tanganyika. *J. Geophys. Res.*, **69**: 2477–2487.

Opdyke, N.D. 1964b. The paleomagnetism of some Triassic red beds from Northern Rhodesia. *J. Geophys. Res.*, **69**: 2495–2497.

Opdyke, N.D. 1968. The paleomagnetism of oceanic cores. In Phinney, R.A., ed., *The History of the Earth's Crust*. Princeton University Press, Princeton, NJ, 61–72.

Opdyke, N.D. 1970. Paleomagnetism of the islands and sediments of the ocean basins. In Maxwell, A.E., ed., *The Sea, Volume IV*. Interscience, New York, 175–182.

Opdyke, N.D. 1985. Reversals of the Earth's magnetic field and the acceptance of crustal mobility in North America: a view from the trenches. *Eos, Trans. AGU*, **66**: 1177–1182.

Opdyke, N.D. 2001. The birth of plate tectonics. In Oreskes, N., ed., with Le Grand, H., *Plate Tectonics*. Westview Press, Boulder, CO, 95–107.

Opdyke, N.D. and Channell, J.E.T. 1996. *Magnetic Stratigraphy*. Academic Press, London and San Diego.

Opdyke, N.D. and Hekinian, R. 1967. Magnetic properties of some igneous rocks from the Mid-Atlantic Ridge. *J. Geophys. Res.*, **72**: 2257–2260.

Opdyke, N.D. and Wensink, H. 1966. Paleomagnetism of rocks from the White Mountain Plutonic-Volcanic Series in New Hampshire and Vermont. *J. Geophys. Res.*, **71**: 3045–3051.

Opdyke, N.D., Glass, B., Hays, J.D., and Foster, J. 1966. Paleomagnetic study of Antarctic deep-sea cores. *Science*, **154**: 349–357.

Oreskes, N. 1999. *The Rejection of Continental Drift*. Oxford University Press, New York.

Oreskes, N. 2000. *Laissez-tomber*: military patronage and women's work in mid-20th century oceanography. *Hist. Stud. Phys. Sci.*, **30**: 373–392.

Oreskes, N. 2001. From continental drift to plate tectonics. In Oreskes, N., ed., with Le Grand, H., *Plate Tectonics*. Westview Press, Boulder, CO, 3–27.

Orowan, E. 1964. Continental drift and the origin of mountains. *Science*, **146**: 1003–1008.

Orowan, E. 1965. Convection in a non-Newtonian mantle, continental drift, and mountain building. In Blackett, P.M.S., Bullard, E., and Runcorn, S.K., eds., *A Symposium on Continental Drift. Phil. Trans. Roy. Soc. London A*, **258**: 284–313.

Oxburgh, E.R. 1965. Discussion. In Blackett, P.M.S., Bullard, E., and Runcorn, S.K., eds., *A Symposium on Continental Drift. Phil. Trans. Roy. Soc. London A*, **258**: 142–144.

Parker, R.L. 1966. Reconnexion of lines of force in rotating spheres and cylinders. *Proc. Roy. Soc. London A*, **291**: 60–72.

Parker, R.L. 1968. Electromagnetic induction in a thin strip. *Geophys. J. Roy. Astron. Soc.*, **14**: 487–495.

Parker, R.L. 1994. *Geophysical Inverse Theory*. Princeton University Press, Princeton, NJ.

Parker, R. L. 2001. When plates were paving stones. In Oreskes, N., ed., with Le Grand, H., *Plate Tectonics*. Westview Press, Boulder, CO, 191–200.

Parsons B. and Sclater, J.G. 1977. An analysis of the variation of ocean floor bathymetry and heat flow with age. *J. Geophys. Res.*, **82**: 803–827.

Peter, G. 1966a. Preliminary results of a systematic geophysical survey south of the Alaska Peninsula. In *Continental Margins and Island Arcs*, Geological Survey of Canada, Special Paper 66-15: 223–237.

Peter, G. 1966b. Magnetic anomalies and fracture pattern in the Northeast Pacific Ocean. *J. Geophys. Res.*, **71**: 5365–5374.

Peter, G. and Stewart, H. B. 1965. Ocean surveys: the systematic approach. *Nature*, **206**: 1017–1018.

Phillips, J.D. 1967. Magnetic anomalies over the Mid-Atlantic Ridge near 27° N latitude. *Trans. Am. Geophys. Union*, **48**: 89.

Phillips, J.D. and Luyendyk, B.P. 1970. Central north Atlantic plate motions over the last 40 million years. *Science*, **170**: 727–729.

Piccard, J. and Dietz, R.S. 1961. *Seven Miles Down*. G. P. Putnam's Sons, New York.

Pitman, W. 2001. On board the Eltanin-19. In Oreskes, N., ed., with Le Grand, H., *Plate Tectonics*. Westview Press, Boulder, CO, 86–94.

Pitman, W. C. and Hayes, D. E. 1968. Sea-floor spreading in the Gulf of Alaska. *J. Geophys. Res.*, **73**: 6571–6580.

Pitman, W.C., III and Heirtzler, J.R. 1966. Magnetic anomalies over the Pacific-Antarctic ridge. *Science*, **154**: 1164–1171.

Pitman, W.C., III and Talwani, M. 1972. Sea-floor spreading and the North Atlantic. *Geol. Soc. Am. Bull.*, **83**: 619–646.

Pitman, W.C., III, Herron, E.M., and Heirtzler, J.R. 1968. Magnetic anomalies in the Pacific and sea floor spreading. *J. Geophys. Res.*, **73**: 2069–2085.

Poole, W.H. 1966. Editor's Note. In *The World Rift System*, Geological Survey of Canada, Special Paper 66-14: 391.

Press, F. and Ewing, M. 1948. A theory of microseisms with geological applications. *Trans. Am. Geophys. Union*, **29**: 163–174.

Press, F. and Ewing, M. 1952. Magnetic anomalies over oceanic structures. *Trans. Am. Geophys. Union*, **33**: 349–355.

Press, F., Ewing, M., and Tolstoy, I. 1950. The Airy phase of shallow-focus submarine earthquakes. *Bull. Seismol. Soc. Am.*, **40**: 111–148.

Quesnell, A.M. 1958. The structural and geomorphic evolution of the Dead Sea rift. *Geol. Soc. London Q. J.*, **114**: 1–24.

Raff, A.D. 1963. Magnetic anomaly over Mohole Drill Hole EM7. *J. Geophys. Res.*, **68**: 955–956.

Raff, A.D. 1966. Boundaries of an area of very long magnetic anomalies in the northeast Pacific. *J. Geophys. Res.*, **71**: 2631–2636.

Raff, A.D. 1968. Sea floor spreading: another rift. *J. Geophys. Res.*, **73**: 3699–3705.

Raff, A.D. and Mason, R.G. 1961. Magnetic survey off the west coast of North America, 40° N. Latitude to 52° N. Latitude. *Geol. Soc. Am. Bull.*, **72**: 1267–1270.

Rainger, R. 2000. Science at the crossroads: the navy, Bikini Atoll, and American oceanography in the 1940s. *Hist. Stud. Phys. Sci.*, **30**: 349–371.

Raitt, H. 1956. *Exploring the Deep Pacific*. W. W. Norton and Company, New York.

Raitt, R.W. 1956. Seismic refraction studies of the Pacific Ocean Basin. *Geol. Soc. Am. Bull.*, **67**: 1623–1640.

Raitt, R. W., Fisher, R. L., and Mason, R. G. 1954. Tonga Trench. *Symposium on the Crust of the Earth*, Columbia University, October 13–16, 1954. Abstract.

Raitt, R. W., Fisher, R. L., and Mason, R. G. 1955. Tonga Trench. *Geol. Soc. Am., Special Paper*, **62**: 237–254.

Raitt, R. W., Shor, G. G., Francis, T. J. G., and Morris, G. B. 1969. Anisotropy of the Pacific upper mantle. *J. Geophys. Res.*, **74**: 3095–3109.

Raitt, R. W., Shor, G. G., Morris, G. B., and Kirk, H. K. 1971. Mantle anisotropy in the Pacific Ocean. *Tectonophysics*, **12**: 173–186.

Raleigh, C. B. and Paterson, M. S. 1965. Experimental deformation of serpentinite and its tectonic implications. *J. Geophys. Res.*, **70**: 3965–3985.

Ramberg, R. 1963. Experimental study of gravity tectonics by means of centrifugal models. *Bull. Geol. Inst. Univ. Uppsala*, **62**: 1–97.

Revelle, R. and Maxwell, A. E. 1952. Heat flow through the floor of the eastern North Pacific Ocean. *Nature*, **170**: 199–202.

Ritsema, A. R. 1966. The fault-plane solutions of earthquakes of the Hindu Kush centre. *Tectonophysics*, **3**: 147–163.

Robertson, W. A. 1963. The paleomagnetism of some Mesozoic intrusives and tuffs from eastern Australia. *J. Geophys. Res.*, **68**: 2299–2312.

Roche, A. 1950. Sur les caractères magnétiques du système éruptif de Gergovie. *C. R. Acad. Sci. Paris*, **230**: 113–115.

Roche, A. 1951. Sur les inversions de l'aimantation rémanente des roches volcaniques dans les monts d'Auvergne. *C. R. Acad. Sci. Paris*, **233**: 1132–1134.

Roche, A. 1953. Sur l'origine des inversions d'aimantation constatées dans les roches d'Auvergne. *C. R. Acad. Sci. Paris*, **236**: 107–109.

Roche, A. 1957. Sur l'aimantation de laves Miocènes d'Auvergne. *C. R. Acad. Sci. Paris*, **250**: 377–379.

Rothé, J. P. 1954. La zone seismique mediane indo-Atlantique. *Proc. Roy. Soc. London A*, **222**: 387–397.

Rubey, W. W. 1968. Presentation of the 1966 Penrose Medal to Harry Hammond Hess. *GSA Proceedings 1968*: 83–84.

Runcorn, S. K. 1956. Palaeomagnetic comparisons between Europe and North America. *Proc. Geol. Assoc. Can.*, **8**: 77–85.

Runcorn, S. K. 1957. The sampling of rocks for palaeomagnetic comparisons between the continents. *Adv. Phys.*, **6**: 169–176.

Runcorn, S. K. 1962. Paleomagnetic evidence for continental drift and its geophysical cause. In Runcorn, S. K., ed., *Continental Drift*. Academic Press, New York, 1–39.

Runcorn, S. K. 1964. Changes in the Earth's moment of inertia. *Nature*, **204**: 824–825.

Runcorn, S. K. 1965a. Palaeomagnetic comparisons between Europe and North America. In Blackett, P. M. S., Bullard, E., and Runcorn, S. K., eds., *A Symposium on Continental Drift. Phil. Trans. Roy. Soc. London A*, **258**: 1–11.

Runcorn, S. K. 1965b. Changes in the convection pattern in the Earth's mantle and continental drift: evidence for a cold origin of the Earth. In Blackett, P. M. S., Bullard, E., and Runcorn, S. K., eds., *A Symposium on Continental Drift. Phil. Trans. Roy. Soc. London A*, **258**: 228–251.

Runcorn, S. K. 1966a. Satellite gravity observations and convection in the mantle. In *The World Rift System*, Geological Survey of Canada, Special Paper 66-14, 364–369.

Runcorn, S. K. 1966b. Comments. In *Continental Margins and Island Arcs*, Geological Survey of Canada, Special Paper 66-15: 186–187, 312, 396.

Runcorn, S. K. 1968. Review of *Principles of Physical Geology* by Arthur Holmes. *Phys. Earth Planet. Interiors*, **1**: 303–304.

Rutten, M. G. 1965a. Discussion. In Blackett, P. M. S., Bullard, E., and Runcorn, S. K., eds., *A Symposium on Continental Drift. Phil. Trans. Roy. Soc. London A*, **258**: 53–58.

Rutten, M. G. 1965b. Discussion. In Blackett, P. M. S., Bullard, E., and Runcorn, S. K., eds., *A Symposium on Continental Drift. Phil. Trans. Roy. Soc. London A*, **258**: 207–208.

Rutten, M. G. 1965c. Discussion. In Blackett, P. M. S., Bullard, E., and Runcorn, S. K., eds., *A Symposium on Continental Drift. Phil. Trans. Roy. Soc. London A*, **258**: 321.

Rutten, M. G. and Wensink, H. 1959. Geology of the Hvalfjördur-Skorradalur area (southwest Iceland). *Geol. Mijnbouw*, **21**: 172–181.

Rutten, M. G. and Wensink, H. 1960. Structure of the central graben of Iceland. *International Geological Congress, XXI Session, Norden*, Part XVIII, 81–88.

St. Amand, P. 1957. Geological and geophysical synthesis of the tectonics of portions of British Columbia, the Yukon Territory, and Alaska. *Geol. Soc. Am. Bull.*, **68**: 1343–1370.

Saito, T., Ewing, M., and Burckle, L. H. 1966a. Tertiary sediment from the Mid-Atlantic Ridge. *Science*, **151**: 1075–1079.

Saito, T., Ewing, M., and Burckle, L. H. 1966b. Lithology and paleontology of the reflective layer Horizon A. *Science*, **154**: 1173–1176.

Scheidegger, A. E. 1953a. Examination of the physics of theories of orogenesis. *Geol. Soc. Am. Bull.*, **64**: 127–150.

Scheidegger, A. E. 1953b. On some physical aspects of the theory of the origin of mountain belts and island arcs. *Can. J. Phys.*, **31**: 1148–1155.

Scheidegger, A. E. 1958. *Principles of Geodynamics*. Springer-Verlag, Berlin.

Scheidegger, A. E. 1963. *Principles of Geodynamics* (2nd edition). Springer-Verlag, Berlin.

Scheidegger, A. E. and Wilson, J. T. 1950. An investigation into possible methods of failure of the earth. *Proc. Geol. Assoc. Can.*, **3**: 167–190.

Schlee, S. 1973. *The Edge of an Unfamiliar World: A History of Oceanography*. E. P. Dutton, New York.

Schmidt, P. W. and Clark, D. A. 1980. The response of palaeomagnetic data to Earth expansion. *Geophys. J. Roy. Astron. Soc.*, **61**: 95–100.

Schuiling, R. D. 1966. Continental drift and oceanic heat-flow. *Nature*, **210**: 1027–1028.

Sclater, J. G. 1966. Heat flux through the ocean floor. Ph.D. thesis (unpublished). University of Cambridge.

Sclater, J. G. 2001. Heat flow under the oceans. In Oreskes, N., ed., with Le Grand, H., *Plate Tectonics*. Westview Press, Boulder, CO, 128–147.

Sclater, J. G. 2004. Variability of heat flux through the seafloor: discovery of hydrothermal circulation in the oceanic crust. In Davis, E. E. and Elderfield, H., eds., *Hydrology of the Oceanic Lithosphere*. Cambridge University Press, Cambridge, 3–27.

Sclater, J. G. and Crowe, J. 1979. a heat flow survey at Anomaly 13 on the Reykjanes Ridge: a critical test of the relation between heat flow and age. *J. Geophys. Res.*, **84B**: 1593–1602.

Sclater, J.G. and Francheteau, J. 1970. The implications of terrestrial heat observations on current tectonic and geochemical models of the crust and upper mantle of the earth. *Geophys. J. Roy. Soc.*, **20**: 509–537.

Scrutton, C.T. 1964. Periodicity in Devonian coral growth. *Palaeontology*, **7**: 552–558.

Shackleton, R.M. 1965. Discussion. In Blackett, P.M.S., Bullard, E., and Runcorn, S.K., eds., *A Symposium on Continental Drift. Phil. Trans. Roy. Soc. London A*, **258**: 59.

Shaler, N.S. 1903. A comparison of the features of the Earth and the Moon. *Smithsonian Contributions to Knowledge*, **34**.

Shand, S.J. 1949. Rocks of the Mid-Atlantic Ridge. *Geol. Soc. Am. Bull.*, **57**: 89–92.

Shepard, F.P. and Emery, K.O. 1941. Submarine topography off the California coast. *Geol. Soc. Am. Special Paper*, **31**.

Shor, G.G. 1966. Continental margins and island arcs of western North America. In *Continental Margins and Island Arcs*, Geological Survey of Canada, Special Paper 66-15: 216–221.

Shor, G.G. and Fisher, R.L. 1961. Middle America trench: seismic-refraction studies. *Geol. Soc. Am. Bull.*, **72**: 721–730.

Shumway, G. 1954. Carnegie Ridge and Cocos Ridge in the east equatorial Pacific. *J. Geol.*, **62**: 573–586.

Smith, A.G. 1963. The structure and stratigraphy of the Whitefish Range, Montana. Ph.D. dissertation (unpublished). Princeton University, Princeton, NJ.

Smith, A.G. 1971. Alpine deformation and the oceanic areas of the Tethys, Mediterranean and Atlantic. *Geolog. Soc. Am. Bull.*, **82**: 2039–2070.

Smith, A.G. and Hallam, A.G. 1970. The fit of the southern continents. *Nature*, **225**: 139–144.

Smith, A.G., Briden, J.C., and Drewry, G.E. 1973. Phanerozoic world maps. In Hughes, N.F., ed., *Organisms and Continents Through Time*. Special Papers in Palaeontology, No. 12, Palaeontological Association, London, 1–43.

Smitt, de, V.P. 1932. Earthquakes in the North Atlantic as related to submarine cables. *Trans. Am. Geophys. Union*, **13**: 103–109.

Stauder, W. 1960. The Alaska earthquake of July 10, 1958. *Bull. Seismol. Soc. Am.*, **50**: 293–322.

Stauder, W. 1968a. Mechanism of the Rat Island earthquake sequence of February 4, 1965 with relation to island arcs and sea-floor spreading. *J. Geophys. Res.*, **73**: 3847–3858.

Stauder, W. 1968b. Tensional character of earthquake foci beneath the Aleutian Trench with relation to sea-floor spreading. *J. Geophys. Res.*, **73**: 7693–7701.

Stauder, W. and Bollinger, G.A. 1964. The S wave project for focal mechanism studies: earthquakes of 1962. *Bull. Seismol. Soc. Am.*, **54**: 2199–2208.

Stauder, W. and Bollinger, G.A. 1966a. The S wave project for focal mechanism studies: earthquakes of 1963. *Bull. Seismol. Soc. Am.*, **56**: 1363–1371.

Stauder, W. and Bollinger, G.A. 1966b. The focal mechanism of the Alaska earthquake and its aftershocks. *J. Geophys. Res.*, **71**: 5283–5296.

Stauder, W. and Udías, A. 1963. S-wave studies of earthquakes of the north Pacific, Part II: Aleutian Islands. *Bull. Seismol. Soc. Am.*, **53**: 59–77.

Stehli, F.G. 1968. A paleoclimatic test of the hypothesis of an axial dipolar magnetic field. In Phinney, R.A., ed., *The History of the Earth's Crust*. Princeton University Press, Princeton, NJ, 195–207.

Sullivan, W. 1974. *Continents in Motion: The New Earth Debate*. McGraw-Hill Book Co., New York.

Suppe, F. 1998a. The structure of a scientific paper. *Phil. Sci.*, **65**: 381–405.

Suppe, F. 1998b. Reply to my critics. *Phil. Sci.*, **65**: 417–424.

Sutton, J. 1965. Comments. In Blackett, P.M.S., Bullard, E., and Runcorn, S.K., eds., *A Symposium on Continental Drift. Phil. Trans. Roy. Soc. London A*, **258**: 107–108.

Sykes, L. R. 1963. Seismicity of the South Pacific Ocean. *J. Geophys. Res.*, **68**: 5999–6006.

Sykes, L. R. 1964. Deep-focus earthquakes in the New Hebrides region. *J. Geophys. Res.*, **69**: 5353–5355.

Sykes, L. R. 1965. The seismicity of the Arctic. *Bull. Seismol. Soc. Am.*, **55**: 519–536.

Sykes, L. R. 1966. The seismicity and deep structure of island arcs. *J. Geophys. Res.*, **71**: 2981–3006.

Sykes, L. R. 1967a. The world-wide distribution of deep and shallow earthquakes. In Runcorn, S. K., ed., *Mantles of the Earth and Terrestrial Planets*. Interscience, London, 383–384.

Sykes, L. R. 1967b. Mechanism of earthquakes and nature of faulting on the Mid-Oceanic ridges. *J. Geophys. Res.*, **72**: 2131–2153.

Sykes, L. R. 1968. Seismological evidence for transform faults; sea floor spreading and continental drift. In Phinney, R. A., ed., *The History of the Earth's Crust*. Princeton University Press, Princeton, NJ, 120–150.

Sykes, L. R. and Ewing, M. 1965. The seismicity of the Caribbean region. *J. Geophys. Res.*, **70**: 5065–5074.

Sykes, L. R. and Landisman, M. 1964. The seismicity of East Africa, the GUIF of Aden and the Arabian and Red seas. *Bull. Seismol. Soc. Am.*, **54**: 1927–1940.

Sykes, L. R., Isacks B., and Oliver, J. 1969. Spatial distribution of deep and shallow earthquakes of small magnitudes in the Fiji–Tonga region. *Bull. Seismol. Soc. Am.*, **59**: 1093–1113.

Talwani, M. 1964. A review of marine geophysics. *Marine Geol.*, **2**: 29–80.

Talwani, M. 1966a. Gravity anomalies. In *Continental Margins and Island Arcs*, Geological Survey of Canada, Special Paper 66-15: 177.

Talwani, M. 1966b. Comments. In *The World Rift System*, Geological Survey of Canada, Special Paper 66-14: 346–347.

Talwani, M. 1966c. Comments. In *Continental Margins and Island Arcs*, Geological Survey of Canada, Special Paper 66-15: 313.

Talwani M. and Heirtzler, J. R. 1964. In Parks, A. G., ed., *Computers in the Mineral Industries*. Stanford University Press, Stanford, CA, 175–217.

Talwani, M., Heezen, B.C., and Worzel, J.L. 1961. Gravity anomalies, physiography and crustal structure of the Mid-Atlantic Ridge. *Trav. Scient. Sect. Seis. UGGI Ser. A*, **22**: 81–111.

Talwani, M., Le Pichon, X., and Ewing, J. 1965. Crustal structure of the Mid-Oceanic ridges, 2, Computed model from gravity and seismic refraction data. *J. Geophys. Res.*, **70**: 341–352.

Talwani, M., Le Pichon, X., and Heirtzler, J. R. 1965. East Pacific Rise: the magnetic pattern and the fracture zones. *Science*, **150**: 1109–1115.

Talwani, M., Le Pichon, X., and Heirtzler, J. R. 1966. Patterns of magnetic anomalies over the Mid-Oceanic Ridge. In *The World Rift System*, Geological Survey of Canada, Special Paper 66-14: 345–346.

Talwani, M., Sutton, G.H., and Worzel, J.L. 1959. A crustal section across the Puerto Rico Trench. *J. Geophys. Res.*, **64**: 1545–1555.

Talwani, M., Windisch, C. C., and Langseth, M. G. 1971. Reykjanes Ridge: a detailed geophysical study. *J. Geophys. Res.*, **76**: 478–517.

Tarling, D. H. 1963. Some aspects of rock magnetism. Ph.D. dissertation (unpublished). Australian National University.

Taylor, F. B. 1910. Bearing of the Tertiary mountain belt on the origin of the Earth's plan. *Geol. Soc. Am. Bull.*, **21**: 179–226.

Taylor, J. H. 1965. Discussion. In Blackett, P. M. S., Bullard, E., and Runcorn, S. K., eds., *A Symposium on Continental Drift. Phil. Trans. Roy. Soc. London A*, **258**: 52–53.

Tharp, M. 1982. Mapping the ocean floor – 1947 to 1997. In Scrutton, R. A. and Talwani, M., eds., *The Ocean Floor*, John Wiley & Sons, New York, 19–31.

Thompson, G. A. 1966a. The rift system of the western United States. In *The World Rift System*, Geological Survey of Canada, Special Paper 66–14: 280–289.

Thompson, G. A. 1966b. The rift system of the western United States. In *Continental Margins and Island Arcs*, Geological Survey of Canada, Special Paper 66–15: 393–394.

Tobin, D. G. and Sykes, L. R. 1968. Seismicity and tectonics of the Northeast Pacific Ocean. *J. Geophys. Res.*, **73**: 3821–3845.

Tocher, D. 1956. Earthquakes off the North Pacific coast of the United States. *Seismol. Soc. Am. Bull.*, **46**: 165–173.

Toksoz, N. M., Minear, J. W., and Julian, B. R. 1971. Temperature field and geophysical effects of a downgoing slab. *J. Geophys. Res.*, **76**: 1113–1138.

Tozer, D. C. 1965. Heat transfer and convection currents. In Blackett, P. M. S., Bullard, E., and Runcorn, S. K., eds., *A Symposium on Continental Drift. Phil. Trans. Roy. Soc. London A*, **258**: 252–271.

Turcotte, D. L. and Oxburgh, E. R. 1967. Finite amplitude convective cells and continental drift. *J. Fluid Mech.*, **28**: 29–42.

Udias, A. and Stauder, W. 1964. Application of numerical method for S-wave focal mechanism determinations to earthquakes of Kamchatka–Kurile Islands region. *Bull. Seismol. Soc. Am.*, **54**: 2049–2065.

Udintsev, G. B. 1966a. Results of upper mantle project studies in the Indian Ocean by the research vessel "Vityaz." In *The World Rift System*, Geological Survey of Canada, Special Paper 66-14: 148–171.

Udintsev, G. B. 1966b. Comments. In *Continental Margins and Island Arcs*, Geological Survey of Canada, Special Paper 66-15: 392.

Umbgrove, J. H. F. 1947. *The Pulse of the Earth* (2nd edition). Nijhoff, The Hague.

Urey, H. C. 1967. The origin of the Moon. In Runcorn, S. K., ed., *Mantles of the Earth and Terrestrial Planets*. Interscience, London, 251–260.

Utsu, T. 1966. Regional differences in absorption of seismic waves in the upper mantle as inferred from abnormal distributions of seismic intensities. *J. Fac. Sci. Hokkaido Univ. Japan, Ser. VII*, **2**(4): 359–374.

Utsu, T. 1967. Anomalies in seismic wave velocity and attenuation associated with a deep earthquake zone (I). *J. Fac. Sci. Hokkaido Univ. Japan, Ser. 7*, **3**: 1–25.

Utsu, T. 1971. Seismological evidence for the anomalous structure of island arcs with special reference to the Japanese region. *Rev. Geophys. Space Phys.*, **9**: 839–890.

Vacquier, V. 1938. Application of vertical variometer-measurements to the study of secular magnetic variations. *Trans. Am. Geophys. Union*, **19**: 206–210.

Vacquier, V. 1959. Measurement of horizontal displacement along faults in the ocean floor. *Nature*, **183**: 452–453.

Vacquier, V. 1962. Magnetic evidence for horizontal displacement in the floor of the Pacific Ocean. In Runcorn, S. K., ed., *Continental Drift*. Academic Press, New York, 135–144.

Vacquier, V. 1965. Transcurrent faulting in the ocean floor. In Blackett, P. M.S., Bullard, E., and Runcorn, S. K., eds., *A Symposium on Continental Drift*. *Phil. Trans. Roy. Soc. London A*, **258**: 77–81.

Vacquier, V. and Affleck, J. 1949. A computation of the average depth to the bottom of the Earth's magnetic crust based on a statistical study of local magnetic anomalies. *Trans. Am. Geophys. Union*, **22**: 446–450.

Vacquier, V. and Von Herzen, R. P. 1964. Evidence for connection between heat flow and the Mid-Atlantic Ridge magnetic anomaly. *J. Geophy. Res.*, **69**: 1093–1101.

Vacquier, V., Raff, A. D. and Warren, R. E. 1961. Horizontal displacements in the floor of the northeastern Pacific Ocean. *Bull. Geol. Soc. Am.*, **72**: 1251–1258.

Vacquier, V., Steenland, N. C., Roland, G., and Zietz, I. 1951. Interpretation of aeromagnetic maps. *Geol. Soc. Am. Mem.*, **47**.

Van Andel, S. I. and Hospers, J. 1967. Palaeomagnetism and the hypothesis of an expanding earth. *Tectonophysics*, **5**: 5–24.

Van Andel, S. I. and Hospers, J. 1968a. Palaeomagnetism and the hypothesis of an expanding earth: a new calculation method and its results. *Tectonophysics*, **5**: 273–285.

Van Andel, S. I. and Hospers, J. 1968b. A statistical analysis of ancient earth radii calculated from palaeomagnetic data. *Tectonophysics*, **6**: 491–497.

van der Gracht, W. A. J. M., *et al.* 1928. Remarks regarding the papers offered by the other contributors to the symposium. In van der Gracht, van Watershoot, ed., *Theory of Continental Drift: A Symposium on the Origin and Movement of Land Masses Both Inter-continental and Intra-continental, as Proposed by Alfred Wegener*. American Association of Petroleum Geologists, Tulsa, OK, 197–222.

van Hilten, D. 1963. Palaeomagnetic indications of an increase in the earth's radius. *Nature*, **200**: 1277–1279.

van Hilten, D. 1965. The ancient radius of the earth. *Geophys. J.*, **8**: 217–225.

van Hilten, D. 1968. Global expansion and paleomagnetic data. *Tectonophysics*, **5**: 191–210.

Vening Meinesz, F. A. 1948. Major tectonic phenomena and the hypothesis of convection currents in the earth. *Q. J. Geol. Soc. London*, **103**: 191–207.

Vening Meinesz, F. A. 1965. Discussion. In Blackett, P. M.S., Bullard, E., and Runcorn, S. K., eds., *A Symposium on Continental Drift*. *Phil. Trans. Roy. Soc. London A*, **258**: 314–316.

Verhoogen, J. 1965. Phase changes and convection in the Earth's mantle. In Blackett, P. M.S., Bullard, E., and Runcorn, S. K., eds., *A Symposium on Continental Drift*. *Phil. Trans. Roy. Soc. London A*, **258**: 276–283.

Vine, F. J. 1963a. Interpretation of magnetic measurements at sea. Unpublished literature review.

Vine, F. J. 1963b. Magnetic anomalies over the oceans. Draft of Vine and Matthews, 1963.

Vine, F. J. 1965. Interpretation of magnetic anomalies observed at sea. Ph.D. thesis. University of Cambridge.

Vine, F. J. 1966. Spreading of the ocean floor: new evidence. *Science*, **154**: 1405–1415.

Vine, F. J. 1968a. Magnetic anomalies associated with mid-ocean basins. In Phinney, R. A., ed., *The History of the Earth's Crust*. Princeton University Press, Princeton, NJ, 73–89.

Vine, F. J. 1968b. Paleomagnetic evidence for the northward movement of the North Pacific basin during the past 100 m.y. (abs.). *Trans. Am. Geophys. Union*, **49**: 156.

Vine, F. J. 2001. Reversals of fortune. In Oreskes, N., ed., with Le Grand, H., *Plate Tectonics*. Westview Press, Boulder, CO, 46–66.

Vine, F. J. 2003. Ophiolites, ocean crust formation, and magnetic studies: a personal view. In Dilek, Y. and Newcomb, S., eds., *Ophiolite Concept and the Evolution of Geological Thought. Geol. Soc. Am. Special Paper*, **373**: 65–75.

Vine, F. J. and Hess, H. H. 1971. Sea-floor spreading. In Maxwell, A. E., Bullard, E., and Worzel, J. L., eds., *The Sea, Volume IV, Part II*. John Wiley & Sons, New York, 587–622.

Vine, F. J. and Morgan, W. J. 1968. Simulation of mid-ocean ridge magnetic anomalies using a random injection model [abstract]. *Geol. Soc. Am. Special Paper*, **115**: 228.

Vine, F. J. and Matthews, D. H. 1963. Magnetic anomalies over the oceanic ridges. *Nature*, **199**: 947–949.

Vine, F. J. and Wilson, J. T. 1965. Magnetic anomalies over a young oceanic ridge of Vancouver Island. *Science*, **150**: 485–489.

Vogt, P. R. and Ostenso, N. A. 1966. Magnetic survey over the Mid-Atlantic Ridge between 42° N and 46° N. *J. Geophys. Res.*, **71**: 4389–4411.

Vogt, P. R. and Ostenso, N. A. 1967. Steady state crustal spreading. *Nature*, **215**: 810–817.

Von Herzen, R. P. 1959. Heat-flow values from the southeastern Pacific. *Nature*, **183**: 882–883.

Von Herzen, R. P. 1960. Pacific ocean floor heat-flow measurements, their interpretation and geophysical implications. Ph.D. dissertation (unpublished). University of California, Los Angeles.

Von Herzen, R. P. 1963. Heat flow through the Eastern Pacific Ocean Floor. *J. Geophys. Res.*, **68**: 4219–4250.

Von Herzen, R. P. and Langseth, M. G. 1965. Present status of oceanic heat-flow measurements. *Phys. Chem. Earth*, **6**: 367–407.

Von Herzen, R. P. and Maxwell, A. E. 1959. *J. Geophys. Res.*, **64**: 1557–1563.

Von Herzen, R. P. and Uyeda, S. 1963. Heat flow through the eastern Pacific ocean floor. *J. Geophys. Res.*, **68**: 4219–4250.

Von Herzen, R. P. and Vacquier, V. 1966. Heat flow and magnetic profiles on the Mid-Indian Ocean Ridge. *Trans. Roy. Soc. A*, **259**: 262–270.

Walker, G. P. L. 1965. Evidence of crustal drift from Icelandic geology. In Blackett, P. M. S., Bullard, E., and Runcorn, S. K., eds., *A Symposium on Continental Drift. Phil. Trans. Roy. Soc. London A*, **258**: 199–204.

Ward, M. A. 1963. On detecting changes in the Earth's radius. *Geophys. J.*, **8**: 217–225.

Wegener, A. 1924. *The Origin of Continents and Oceans*. Translated from the 3rd edition by Skerl, S. G. A. Methuen and Company, London.

Wegener, A. 1929/1966. *The Origin of Continents and Oceans*. Translated from the 4th revised German edition by Biram, J. Dover Publications, New York.

Wells, J. W. 1963. Coral growth and geochronometry. *Nature*, **197**: 948–950.

Wensink, H. 1964a. Paleomagnetic stratigraphy of younger basalt and intercalated Plio-Pleistocene tillites in Iceland. *Geologische Rundschau*, **54**: 364–384.

Wensink, H. 1964b. Secular variation of Earth magnetism in Plio-Pleistocene basalt of Eastern Iceland. *Geol. Mining*: **43**: 403–413.

Wertenbaker, W. 1974. *The Floor of the Sea: Maurice Ewing and the Search to Understand the Earth*. Little, Brown and Company, Boston, MA.

Westoll, T.S. 1965. Geological evidence bearing upon continental drift. In Blackett, P.M.S., Bullard, E., and Runcorn, S.K., eds., *A Symposium on Continental Drift. Phil. Trans. Roy. Soc. London A*, **258**: 12–26.

Wheeler, J.M. 1992. Applications of the EDSAC. *IEEE Ann. Hist. Comput.*, **14**: 27–33.

White, G.W. 1980. Permian–Triassic continental reconstruction of the Gulf of Mexico – Caribbean area. *Nature*, **283**: 823–826.

White, R.S. 1999. Drummond Hoyle Matthews. *Biogr. Mem. Roy. Soc.*, **45**: 272–294.

Whitten, C.A. 1948. *Trans. Am. Geophys. Union*, **29**: 318–323.

Wilkes, M.V. 1992. EDSAC 2. *IEEE Ann. Hist. Comput.*, **14**: 49–56.

Williams, C.A. and McKenzie, D. 1971. The evolution of the North-East Atlantic. *Nature*, **232**: 168–173.

Willis, B. 1932. Isthmian links. *Geol. Soc. Am. Bull.*, **43**: 917–952.

Wilson, J.Tuzo. 1949a. An extension of Lake's hypothesis concerning mountain and island arcs. *Nature*, **164**: 147–148.

Wilson, J.Tuzo. 1949b. Some major structures of the Canadian shield. *Can. Min. Metall. Trans.*, **52**: 231–242.

Wilson, J.Tuzo. 1949c. The origin of continents and Precambrian history. *Trans. Roy. Soc. Can.*, **43**: 157–184.

Wilson, J.Tuzo. 1950. An analysis of the pattern and possible cause of young mountain ranges and island arcs. *Proc. Geol. Assoc. Can.*, **3**: 141–166.

Wilson, J.Tuzo. 1951a. On the origin of continents, atmosphere and oceans. *Roy. Meteor. Soc. (London): Canadian Branch*, **2**: 1–9.

Wilson, J.Tuzo. 1951b. On the growths of continents. *Pap. Proc. Roy. Soc. Tasmania*, **85**: 85–111.

Wilson, J.Tuzo. 1952a. Some considerations regarding geochronology with special reference to Precambrian time. *Trans. Am. Geophys. Union*, **33**: 195–203.

Wilson, J.Tuzo. 1952b. Orogenesis as the fundamental geological process. *Trans. Am. Geophys. Union*, **33**: 444–449.

Wilson, J.Tuzo. 1953. The origin of continents as deduced from geological and geophysical evidence. *Proc. Geophys. Soc. Tulsa*, **1**: 66–69.

Wilson, J.Tuzo. 1954. The development and structure of the crust. In Kuiper, G.P., ed., *The Earth as a Planet*. Chicago University Press, Chicago, 138–257.

Wilson, J.Tuzo. 1959. Geophysics and continental growth. *Am. Sci.*, **47**: 1–24.

Wilson, J.Tuzo. 1960. Some consequences of expansion of the earth. *Nature*, **185**: 880–882.

Wilson, J.Tuzo. 1961. Discussion of R. S. Dietz: continent and ocean basin evolution by spreading of the sea floor. *Nature*, **192**: 123–128.

Wilson, J.Tuzo. 1962a. The effect of new orogenetic theories upon ideas of the tectonics of the Canadian Shield. In Stevenson, J.S., ed., *The Tectonics of the Canadian Shield*. Royal Society of Canada Special Publication, No. 4: 174–180. Toronto University Press and Royal Society of Canada.

Wilson, J.Tuzo. 1962b. Cabot Fault, an Appalachian equivalent of the San Andreas and Great Glen faults and some implications for continental displacement. *Nature*, **195**: 135–138.

Wilson, J.Tuzo. 1962c. Some further evidence in support of the Cabot Fault, a great Palaeozoic transcurrent fault zone in the Atlantic provinces and New England. *Trans. Roy. Soc. Can.*, **56**: 31–36.

Wilson, J.Tuzo. 1963a. Evidence from islands on the spreading of ocean floors. *Nature*, **197**: 536–538.

Wilson, J.Tuzo. 1963b. Pattern of uplifted islands in the main ocean basins. *Science*, **139**: 592–594.

Wilson, J.Tuzo. 1963c. A possible origin of the Hawaiian Islands. *Can. J. Phys.*, **41**: 863–870.

Wilson, J.Tuzo. 1963d. Continental drift. *Sci. Am.*, **208**: 86–100.

Wilson, J.Tuzo. 1963e. Hypothesis of Earth's behaviour. *Nature*, **198**: 925–929.

Wilson, J.Tuzo. 1965a. Evidence from ocean islands suggesting movements in the Earth. In Blackett, P.M.S., Bullard, E., and Runcorn, S.K., eds., *A Symposium on Continental Drift. Phil. Trans. Roy. Soc. London A*, **258**: 145–167.

Wilson, J.Tuzo. 1965b. A new class of faults and their bearing on continental drift. *Nature*, **207**: 343–347.

Wilson, J.Tuzo. 1965c. Transform faults, oceanic ridges, and magnetic anomalies southwest of Vancouver Island. *Science*, **150**: 482–485.

Wilson, J.Tuzo. 1965d. Submarine fracture zones, aseismic ridges and the International Council of Scientific Unions Line: proposed western margin of the East Pacific Rise. *Nature*, **207**: 907–914.

Wilson, J.Tuzo. 1966a. Patterns of growth of ocean basins and continents. In *Continental Margins and Island Arcs*, Geological Survey of Canada, Special Paper 66-15: 388–391.

Wilson, J.Tuzo. 1966b. Comments. In *Continental Margins and Island Arcs*, Geological Survey of Canada, Special Paper 66-15: 313, 392, 394–397.

Wilson, J. Tuzo. 1972. Walter H. Bucher medal to William Jason Morgan for original contributions to the basic knowledge of the earth's crust. *Eos*, **53**: 740.

Wilson, J.Tuzo. 1982. Early days in university geophysics. *Annu. Rev. Earth Planet Sci.*, **10**: 1–14.

Wilson, J.Tuzo. 1985. Development of ideas about the Canadian Shield: a personal account. In Drake, E.T. and Jordan, W.M., eds., *Geologists and Ideas: A History of North American Geology*, GSA Centennial Special Volume I. Geological Society America, Boulder, CO, 143–150.

Wilson, J.Tuzo. 1990. J. T. Wilson Killam Laureate, 1989. In Kenney-Wallace, G.A., MacLeod, M.G., and Stanton, R.G., eds., *Celebration of Canadian Scientists: A Decade of Killam Laureates*. Charles Babbage Research Centre, Winnipeg, 266–286.

Wilson, R. L. 1962. The palaeomagnetism of baked contact rocks and reversals of the Earth's magnetic field. *Geophys. J.*, **7**: 194–202.

Wise, D.U. 1963. An outrageous hypothesis for the tectonic pattern of the North American Cordillera. *Geol. Soc. Am. Bull.*, **74**: 357–362.

Wiseman, J.D.H. and Sewell, R. B. Seymour. 1937. The floor of the Arabian Sea. *Geol. Mag.*, **74**: 219–230.

Worzel, J.Lamar. 1965a. Discussion. In Blackett, P.M.S., Bullard, E., and Runcorn, S.K., eds., *A Symposium on Continental Drift. Phil. Trans. Roy. Soc. London A*, **258**: 137–139.

Worzel, J. Lamar. 1965b. Discussion. In Blackett, P.M.S., Bullard, E., and Runcorn, S.K., eds., *A Symposium on Continental Drift. Phil. Trans. Roy. Soc. London A*, **258**: 275.

Worzel, J. Lamar. 1965c. Deep structure of coastal margins and mid-ocean ridges. In Whittard, W. F. and Bradshaw, R., eds., *Submarine Geology and Geophysics*. Butterworths, London, 335–361.

Worzel, J. Lamar. 1966. Structure of continental margins and development of ocean trenches. In *Continental Margins and Island Arcs*, Geological Survey of Canada, Special Paper 66-15: 357–375.

Wyllie, P. J. 1976. *The Way the Earth Works*. John Wiley, New York.

York, D. 1973. Evolution of triple junctions. *Nature*, **244**: 341–342.

Index

Adams, R. D. 242
Ade-Hall, J. H. (formerly Hall) 102, 215–16, 234, 254–5, 315, 319
Adie, R. J. 88–9
aerial photography, use by Canadian Geological Survey 3–4
Africa
  fit with South America 170–86
  ocean ridges around 46, 60
African block
  movement relative to the Antarctic block 492–3
  movement relative to the South American block 486–90
Alaska 274–5
Aleutians 549–51
Allan, T. D. 69, 90, 171, 241
  early life and career 80–4
  magnetic anomalies and seamounts 78–9
  work on Atlantic magnetic anomalies 80–4
  work on Red Sea magnetic anomalies 82–4
Allen, C. R. 164, 165
Allsopp, H. L. 353, 355–7
Alpine Fault, New Zealand 30, 44, 595
American Miscellaneous Society (AMSOC) 214
Anderson, D. L. 413, 414, 453, 454, 458, 461–2
Anderson, O. 479, 481, 502
Anderson, Orson 561, 562
Antarctic block
  movement relative to the African block 492–3
  movement relative to the Pacific block 490–1
Antarctica
  implications of seafloor spreading 41–2
  Matthews time in 85–7
  oceanic ridges surrounding 41–2, 46, 57, 60, 275–7, 302–3
Appalachians 13–15
apparent polar wander (APW) paths 27–9, 186–7, 604–12
Arabian Sea, transform faults 272
Archambeau 454

Argand, E. 20, 25, 28, 40, 49, 177, 181, 611
Arkell, W. J. 87–8
astatic magnetometer (Cambridge) 91
asthenosphere 568
atolls in the Pacific 50–1
Atwater, T. 504, 507, 509, 576–84, 586, 592–3, 598–9
Australian National University (ANU) 144, 227, 228, 234
  improvements to the reversals timescale 351, 353–9

Backus, G. E. 461–2, 484, 510, 515
  analysis of marine magnetic anomalies 205–8, 210
  anisotropy in the upper mantle 213–14
  application of Euler's Point Theorem 205–8
  education and early career 202–4
  response to Everett's continental fit 204–5
  response to the Vine–Matthew hypothesis 204–9
Bailey, E. B. 45, 49
baked contact test 318–19
Baker, R. C. 242–3
Balsley, J. R. 128
Barazangi, M. 551, 563–4, 578
Baron, J. G. 310–16
basaltic nature of ocean rocks 89–90
Bath, J. 324–5
Beartooth Mountains, Montana 3
Beaumont, J. 139
Belshé, J. 91, 93–4, 101, 172, 360
Benioff, H. 160, 274, 275, 444, 552 see also Wadati–Benioff zone
Bernal, J. D. 35–6, 37, 39, 57, 58, 168
Bernard Price Institute of Geophysical Research, Johannesburg 355
Betz, F. 153–4
Bickle, M. J. 468
Bidgood, D. E. T. 93, 175, 185
Birch, F. 21, 458, 461–2
Bishop Tuff 352, 358

Black, B. 458–9
Black, M. 93–4
Blackett, P. M. S. 42, 72, 96, 163, 190, 219, 319, 460
Blanco Fracture Zone 297, 579
Blundell, D. J. 87
Bodvarsson, G. 239–40, 274, 284–5, 313–15
Bollinger, G. A. 393–4, 402–3, 575
Bolt, B. A. 389
Bonnin 538–9
Bonnin, J. 526
Bosco triple junction 524
Bott, M. H. P. 168, 319–20, 322–3
Boucot, J. 413
Bower, M. E. 242–3
Bowin, C. O. 244–5, 478
Brace, W. 393
Brant, A. 125, 126
Briden, J. C. 413, 416, 608–10
Britain, deep seismic profiling of the continental crust (BIRPS) 87
British Antarctic Survey 88
Brock, A. 142
Broeker, W. 361
Brown, J. (football coach) 438
Browne, B. 98
Bruckshaw, J.McG. 80, 107
Brune, J. N. 449
Brunhes epoch (Chron) 345–50
Brunhes–Matuyama boundary 352–3, 354–7
Brunhes normal series (deep-sea cores) 362–3
Bucher, W. H. 439–40
Buddington, A. F. 128, 279–80
Bullard, E. C. 3, 64–5, 78–9, 80–1, 83, 93–4, 96, 98, 105, 116, 121, 163, 164, 168, 169, 172, 187, 202, 219, 319, 402, 406, 457–8, 459, 460–1, 463, 495–6, 510–11, 512–13, 527, 558–9
    continental fit results presented in 1964 178–85, 206–7, 260–1
    fit of the continents around the Atlantic 170–86
    Goddard conference (November 1966) 412–13, 414, 415
    influence on Vine 99–101
    on marine magnetic anomalies 101–2
    on the mechanism of continental drift 192–3
    paleomagnetism's support for mobilism, 191–2
    reaction to Vine's draft paper 122–3
    reasons for lack of interest in Carey's fit 171
    Royal Society symposium (1964) presentation 170–86
    support for Vine's computer modeling 115–16
    testing Carey's fit of the continents 172–5
    use of Euler's Point Theorem 173–5

Cabot Fault 45, 46–50
Caledonides 45

Canada, support for the Vine–Matthews hypothesis 242–3
Canadian Geological Survey 3–4
Canadian Shield 6, 15–19, 30, 39
    basalts 135
    radiometric dating 15–16
    tectonics 42–6
Cann, J. R. 77, 97, 110, 111, 112, 113, 124, 279–80, 325
    work with Vine and Matthews 236, 237–9
Cape Verde Islands, marine magnetic anomalies 71
Carey, S. W. 28, 33, 49, 70, 93, 155, 170, 241, 259, 269, 329, 570
    continental fit 178–81, 185, 206–7, 260–1
    fit of Africa and South America 170–1
    rapid Earth expansion 224, 227, 228, 230, 231
Carlsberg Ridge 262
    computer modeling by Vine 114–24
    Matthews' magnetic survey (1962) 62, 64, 65, 91
    transform fault associated with 272
Chamalaun, F. H. 353–7, 358, 378
Chander, R. 449
Chandrasekhar, S. 202–4, 220, 221
Chase, C. G. 507–9
Chase, T. E. 194–5
Chile region 549–51
Christoffel, D. A. 241–2, 380
Clegg, J. A. 28–9, 31–2, 42, 460, 608
Coats 570
Cockroft, J. 3
Collet, L. 49
Collins, W. H. 3
computer modeling
    Carlsberg Ridge survey data (Vine) 114–24
    interpretation of magnetic anomalies 104–5, 107–8
    marine magnetic anomalies 64–5, 80–1
    Parker's map-projecting program 514
computers, Vine's use of 98–100
continental accretion, Wilson's defense of 5–18
Continental Drift (Runcorn ed. 1962) 106
continental drift
    attacks by Wilson 18–20
    debate about the mechanism 58–61
    implications of transform faults 267–8, 278
    mechanism difficulty 192–3
    Scheidegger's attack on 20–1
    see also Goddard conference; Royal Society symposium on continental drift (1964)
continental fit
    Bullard, Everett, and Smith 178–85, 260–1
    difficulty of Central America 185–6
    significance of (Hess) 260–1
    see also Carey
continental nuclei (Wilson) 15–18
continental reconstructions (Royal Society, 1964) 163–4, 166–7

continental shelves, formation and transformation (Wilson) 11, 18
continents
  development of (Wilson) 42–3
  implications of seafloor spreading 41
  mid-Mesozoic reconstruction (Wilson) 161
continuous seafloor spreading, evidence for 560
contractionism
  limitations 43–4
  rejection by Wilson (1960) 33–5
  Wilson's continued support for (1959) 21–31
  Wilson's defense of 5–18
Coode, A. M. 258
  challenging Wilson's ideas on faults 281–2
  differences from Wilson's account 288–9
  early life 280
  education and career 280–1
  idea of transform faults 280
  mention of transform fault idea to Wilson 262, 263, 264
  on mantle convection 287–8
  paper on types of oceanic faults 284–9
  publication of his paper 283–4
  rejection of his paper by *Nature* 282–3, 284, 289–90
  role as catalyst for Wilson 290–1
Cook, A. 512–13
Cook, K. L. 336
Cordillera mountain range 12, 15
Coriolis force, effect on mantle convection 220–3
Cowling, T. G. 202, 203
Cox, A. 56, 96, 102, 106–7, 123–4, 214, 234, 235, 298, 334, 357, 358, 402, 413, 416, 422, 605
  becomes a mobilist 374–5
  improvements in the reversals timescale 345–50, 351–3
  reaction to *Eltanin*-19 magnetic profile 374–5
Cox, J. 133–4
Creer, K. M. 29, 42, 79–80, 91, 163–4, 175, 281, 319, 460, 461, 604
Cretaceous Magnetic Quiet Zone 335
Cretaceous Normal Superchron 335
Crowe, J. 411, 468
crustal blocks/plates
  evolution and past rotations 524–6
  Hess 260
  Morgan identifies block boundaries 485
  Morgan's 1968 paper 485–94
  rigidity of 485, 493, 517
  structure of 494
  three types of block boundaries 485
  see also plate tectonics
Curie point 62, 68, 69, 71, 133, 135, 239
Curie point isotherm (Heirtzler and Le Pichon) 251–2

Dalrymple, G. B. 234, 357, 358, 413, 416
  improvements in the reversals timescale 345–50, 351–3

Daly, R. A. 155, 454, 455, 485
Dana, J. W. 6
Darwin Rise 50, 54, 56, 198, 260, 275, 301–2, 333–4
dating
  paleontological evidence 224–6
  radiometric dating 15–16
  radiometrically dated reversal timescale 298, 351
  relation between days in the year and geological time 224–6
  ridge crest basalts 343–4
  use of fossil corals 224–6, 230–1
Davies, D. 168, 462, 463
Davis, P. 570
Day, A. 144
days in the year
  change over geologic time 230–1
  measuring over geologic time 224–6
de Almeida, F. F. M. 184
De Geer Fault 269–71
De Geer line 31
De Sitter, L. U. 168
Dead Sea rift 272
Deccan Traps 608
deep-focus earthquakes
  explanations for 35–6
  Fiji–Tonga region 445–56
  Isacks and Oliver's study 443–6
deep-sea cores
  paleomagnetic analysis of the Lamont collection 359–63
  reversal timescale based on 359–63
Deer, W. A. 93–4
Department of Geodesy and Geophysics, Cambridge 79, 80, 87–8, 89–91, 96–101, 277, 406, 461
descending seismic zone length and slip rate 575
Deutsch, E. R. 28–9, 31–2
Dewey, John 413, 417
Dicke, R. W. 33, 474
Dickson, G. O. 63, 143–4, 363, 366–7
Dietz, R. S. 59, 134, 136, 153, 154, 241, 405, 416, 454
  discussion of seafloor spreading with Bernal 35–6
  Heezen's attack on seafloor spreading 56–8
  Menard's attack on seafloor spreading 200–2
  seafloor spreading 37, 39, 41, 53, 96, 106, 119, 131–2, 143, 155, 217, 218
Dill, R. F. 54
Dirac, P. A. M. 33, 227, 457
Doell, R. R. 56, 96, 102, 106–7, 123–4, 214, 234, 235, 298, 334, 357, 358, 413, 416, 422, 605
  becomes a mobilist 375
  improvements in the reversal timescale 345–50, 351–3
  reaction to *Eltanin*-19 magnetic profile 375
dolerites 20, 29

Dorman, H. J. 389, 551, 563–4, 578
Drake, C. L. 77, 247, 272, 273, 364, 386
Drewry 608
Du Bois, P. M. 129
du Toit, A. L. 38, 45, 88–9, 93, 155, 183
DuBridge, L. A. 495
Durham, J. W. 87–8

Earth expansion *see* expansionist view; rapid Earth
    expansion theories; slow Earth expansion
Earth's rate of rotation
    change over time 230–1
    evidence for slowing 224–6
earthquake seismology and global tectonics 562–75
earthquakes
    deep earthquake mechanisms 539–52
    epicenter distribution mapping 563–4
    first motion studies of earthquake mechanisms
        (Sykes) 389–96
    focal mechanism solutions 539–52
    locating hypocenters (Sykes) 388–91
    see also deep-focus earthquakes
    thrust-fault type 568–70
East Chile Ridge 277
East Pacific Rise 54, 274–5, 301–2, 331–2
    age of 50–1
    dimensions of 50–1
    extent of 52
    Hess (Ottawa, 1965) 333–4, 335, 336–7
    Lamont workers' interpretation 307–9
    origin of 56
    seafloor thinning 51
    Wilson 335, 336–7
Eckart, C. 204
EDSAC 2 computer (Cambridge) 64–5, 80–1
    Everett's use of 173–5, 178–9
    use by Vine 240
    Vine's use for modeling magnetic anomalies 114,
        115–16
Egyed, L., slow Earth expansion 33, 56, 216, 222,
    223–4, 227, 228, 230, 232
Einarsson, T. 133–4
Ellesmere Island 48–9
Elsasser, W. M. 202, 219, 328, 333, 474, 483, 485,
    493, 527, 540, 546–7, 549, 551
Eltanin-19 magnetic profile 366, 576
    influence at Lamont 441
    Opdyke's response 366–7
    use by Morgan 490–1
Eltanin-20 magnetic profile 365–6
Eltanin-21 magnetic profile 366
Eltanin survey magnetic profiles (Pitman) 363–74
Emiliani, C. 136–7, 140
Emura 575
Engel, A. E. J. 253
Engel, C. G. 253
episodic seafloor spreading 559–60, 572

eugeosynclines 11
Euler angles 525–6
Euler poles 558–9, 602–4
    instantaneous and finite 528–9
Euler poles, determining 518–22
Euler rotations 608–10
Euler's Point Theorem 173–5, 178–9, 205–8, 477,
    478, 481–5, 495–6, 510–12, 516, 518–22, 527,
    558–9, 568
Everett, J. E. 164, 417, 495–6, 510–11, 527
    continental fit results presented in 1964 178–85,
        260–1
    education and early career 171
    fit of the continents around the Atlantic 170–86
    Royal Society symposium (1964) presentation
        170–86
    testing Carey's fit of the continents 172–5, 177–8
    use of EDSAC 2 computer 173–5, 178–9
    use of Euler's Point Theorem 173–5, 178–9
    work on continental fit 204–5, 206–7
evolution and past rotations of the plates 524–6
Ewing, J. 1, 247, 341, 425, 428–31, 559–60, 572, 588
Ewing, M. 1, 3, 23, 41, 52, 59, 64, 72, 81, 109, 168,
    241, 246, 247, 270, 360, 389, 391–2, 396, 406,
    407–8, 413, 417, 455, 559–60, 560–1, 572, 584,
    588, 591–2
    influence of his anti-mobilist attitude 424–7
    influence on Oliver 438–40
    marine magnetic anomalies 70–1
    poor relationship with Hess 425–6
    reluctantly accepts discontinuous seafloor
        spreading 428–31
    seafloor sedimentation and seafloor spreading
        341
expansionist view 1, 557–8, 570–1, 552
    adoption by Wilson (1960) 33–5
    and seafloor spreading 41
    Heezen's defense of 56–8
    Menard's attack on 56
    views on (Royal Society, 1964) 169–70
    see also rapid Earth expansion theories; slow
        Earth expansion
Explorer Ridge 304, 538–9

Falkland Islands, Matthews' visit 88–9
Falkland Islands Dependency Survey (FIDS) 85,
    88–9
Farallon plate 586, 598–9
fault plane solutions 527–8
fault plane solutions, Sykes' 1967 paper 396–402
Ferraro, V. C. A. 514–15
Feynman, R. P., on open-mindedness 422
Field, R. M. 3
Fiji–Tonga region, deep earthquake studies
    445–6
finite rotational poles 602–4
finite rotations 558–9

finite rotations (cont.)
  describing 525–6
  distinction from instantaneous rotations
    528–9
first-motion studies 527–8
first motion studies of earthquake mechanisms
  (Sykes) 389–96
Fisher, R. A. 456
Fisher, R. L. 167–8, 259, 444, 605
Fisher statistics 228–9
Fitch, F. J. 165
Fitch, T. 570
fixism 1
fixism–mobilism debate, outcome 612
Flavill, L. 91, 219
fluxgate magnetometer 78–9, 104
fossil corals, as dating tool 224–6, 230–1
Foster, J. H. 360–3
Foster, M. 457–8
Fowler, W. 457–8
fracture zones
  Hess's explanation for 260, 333–4
  northeastern Pacific 275
  Pacific Ocean floor 52, 53, 333–4
Francheteau 526, 538–9
Francis, T. J. G. 236
Fuller, M. 91
Funnell, B. 97, 360

GAD hypothesis 187
Galapagos Rift Zone 590–1, 597
Galapagos rise 194–5
Gartner, S. 431
Gaskell, T. G. 167–8
Gast, P. 395–6, 412–13
Gauss normal epoch 346–9, 358–9
Gauss normal series (deep-sea cores) 362–3
Gee, D. 457
Gee, Jeff 605
Geikie, A. 456–7, 458–9
Gelletich 142, 143, 319
geologic time, change in days in the year 230–1
geological periods, number of days in the year
  224–6
Geological Society of America (GSA)
  fall 1966 meeting 417
  November 1965 meeting 345–50
Geological Survey of Canada (GSC) 2
geological timescale, history of development
  350–1
geomagnetic field reversals
  and marine magnetic anomalies 105–7
  and the Vine–Matthews hypothesis 62, 234
  record in the seafloor 134
  status in the early 1960s 317–19
  timescale 133–4
  see also reversal timescale

geomagnetic timescales 558–60
geosynclinal theory 5–6, 11, 12
Gilbert, F. 209, 461–2, 515
Gilbert reversed epoch 368–9
Gilbert reversed series (deep-sea cores) 362–3
Gilchrist, L. 3, 125
Gill, A. 463
Gilliland, W. N. 263, 264–5, 286
Gilluly, J. 162, 185–6, 193
Girdler, R. W. 71–8, 81, 91, 101, 106, 164, 165, 215,
  241, 272, 273, 380
Glass, B. 361–3
Glisá event 357–8
global kinematic model (Le Pichon) 552–62
Glossopteris flora distribution 155
Godby, E. A. 242–3, 386
Goddard conference (November 1966) 395–6, 402,
  404, 449, 463
  Bullard 412–13, 414, 415
  cancellations and withdrawals 413
  continental drift and seafloor spreading prevail
    412–18
  Le Pichon 413
  MacDonald 412–13, 415, 417
  McKenzie 463
  Menard 413, 414–15, 417
  Opdyke 412–13
  sessions and participants 413–17
  Vine 414
Goguel, J. 165
Gold, T. 28, 87–8
Gondwana 329
Gorda Ridge 297, 304, 534, 538–9, 581
Gough, D. I. 63, 141–3, 319
Graham, K. W. T. 29
Graham, J. W. 126–7, 246
Graindor, M. J. 168
Great Glen Fault, Scotland 30, 44–5, 46–50
Great Magnetic Bight 597, 598, 605
  origin of 584–91
Great Rift Valley 3
Green, C. 461–2
Green, R. 29, 31, 355, 356
Greenland 31, 48–9
greenstone rocks 17
Griffiths, D. H. 28, 31–2
Griggs, D. T. 19, 202, 375, 402
Grommé, C. S. 348, 351
Gross, H. 38
Grossling, B. 107
group-think, at Lamont 421–4, 439–40
Gulf of Aden
  magnetic anomalies 71, 72–7
  transform faults 272–3
Gutenberg, B. 160, 444, 551, 563–4, 572
Gutenberg fault zones 546–7
guyots 50–1, 54, 134–5

Hales, A. L. 29, 166–7, 413
Hall, J. H. (later Ade-Hall) 101, 103–4, 105, 107, 116, 135
Hallam, A. 257
Hamilton, E. L. 201, 337–8
Hamilton, W. 21–2, 50, 593–4
Harland, W. B. 91, 92, 93–4, 166–7, 175–6, 185, 270, 458–9, 460
Harrison, C. G. A. 360
Haruna dacite 100–1
Hawaiian Islands, origins of 152–5
Hay, R. L. 351
Hayes, D. 584–6
Hays, J. D. 361
heat flow measurement
  difficulty for seafloor spreading 405–12
  history of measurement at sea 406
Heezen, B. C. 23, 25, 33, 41, 52, 59, 164–5, 167, 169–70, 185, 239, 270, 328, 330–1, 361, 413, 425, 484, 486
  attack on seafloor spreading 56–8
  attack on seafloor thinning 56–8
  criticism of proposed convection patterns 57–8
  Menard's attack on Earth expansion 56
  Mid-Atlantic Ridge magnetic anomalies 70–1
  renounces rapid Earth expansion 405
  support for rapid Earth expansion 56–8, 224, 227, 228, 231
Heirtzler, J. R. 143, 235, 236, 247, 360, 363–4, 375–6, 378, 392, 413, 414–15, 416, 484, 552–3, 554–6, 576, 584, 590–1
  accepts seafloor spreading 372–3
  accepts the Vine–Matthews hypothesis 372–3
  alternative to the Vine–Matthews hypothesis 249–55
  arguments against seafloor spreading 304–10
  axial and flank magnetic anomalies 320
  becomes a mobilist 367–73
  education and career 246
  extends the reversal timescale 471
  geomagnetic timescale 558–60
  interpretation of Reykjanes Ridge symmetrical anomalies 310–16
  motivation to prove Hess wrong 418–27
  work in Ridge and Trough Province 304–10
  work on marine magnetic anomalies 245–6
  work on the *Eltanin* magnetic profiles 367–73
  work on the Mid-Atlantic Ridge 248–9, 249–55
Hekinian, R. 254–5, 360
Henderson, P. 168
Herron, E. M. 365–6, 590–1
Hersey, J. B. 386
Herzenberg, A. 203
Hess, H. H. 3, 25, 59, 72, 93–6, 136, 138–9, 142, 143, 149, 152, 153–4, 164, 167, 176–7, 238, 243, 275, 279–80, 294, 297, 316, 328, 329–30, 386, 416, 454, 455, 459, 463, 474, 495, 591
  at Madingley rise (1965) 255–7, 258
  award of the Penrose medal 418
  caution from Irving (1961) 58–61
  changes to account of mantle convection 259–60
  correspondence with Menard about seafloor spreading 50–3
  Darwin Rise 333–4
  discussion of "crustal plates" 260
  East Pacific Rise 333–4, 335, 336–7
  explanation for fracture zones 260
  fine tunes and extends seafloor spreading (1965) 259–61
  fracture zones in the Pacific 333–4
  Heezen's attack on seafloor spreading 56–8
  Lamont workers' prejudice against seafloor spreading 418–27
  letter from Vine about Princeton 255–7
  mantle convection 321–3
  meeting with Holmes (1965) 257
  Menard's attack on seafloor spreading 200–2
  nature of Layer 2 oceanic crust 214–15
  on Lamont Reykjanes Ridge magnetic data 326
  on Lamont seafloor sediments argument 344–5
  on Wilson's transform faults idea 261–2
  origin of the Great Magnetic Bight 586–90
  poor relationship with Ewing 425–6
  presentation and discussion (Ottawa meeting, 1965) 333–7
  response to Lamont heat flow difficulty 408–12
  response to the Vine–Matthews hypothesis 210–15
  seafloor spreading 37, 39, 41, 119, 155
  seismic anisotropy in the upper mantle 210–14
  significance of circum-Atlantic continental fits 260–1
  source of marine magnetic anomalies in the crust 299
  support for the Vine–Matthews hypothesis 334
  UK lecture tour (1965) 255–7
  work with Vine and Wilson 255, 259
Hey, D. 507
Hide, R. 49, 79–80, 127, 154, 203, 219–23, 413
Hilgenberg, O. C. 230
Hill, M. N. 69, 71, 77, 78–9, 80–1, 82, 89, 90, 96–7, 99, 100, 109, 123, 140, 176, 178, 236–7, 319–20, 322–3, 457–8
Hirschman, J. 71
Hodgson, J. H. 22, 160, 392, 393, 449
Hodych, J. 38
Hollingworth, S. E. 168
Holmes, A. 16, 18, 20, 49, 59, 81, 93, 94, 162, 241, 328, 458–9
  meeting with Hess (1965) 257
  on mantle convection 216, 217–23
  on seafloor spreading 216–17
  Principles of Physical Geology (1965) 216–27, 230
  rejection of rapid Earth expansion 227

Holmes, A. (cont.)
  speculation on the cause of expansion 227
  support for slow Earth expansion 216, 222,
    223–6, 227, 230
  support for Wilson's work 216
Holmes, D. (Reynolds) 257
Holtedahl, O. 45
Hood, P. J. 242–3
Honda 575
horizontal displacements in the Earth's crust
  (Royal Society, 1964) 164–5, 167–8
Hospers, J. 72, 74, 79–80, 81, 89, 98, 100, 122, 127,
  133–4, 230, 319, 321, 461
hot spots within the mantle 152–5
Hoyle, F. 457–8
Hughes, N. F. 92, 93–4, 458–9
Humboldt, Alexander von 604
Hunkins, K. 442
Hurley, P. M. 184, 413, 417

IBM 650 computer 563–4
Ice Station Alpha (T3, Fletcher's Ice Island) 441–2
Iceland 239–40
Icelandic lavas, dating of polarity reversals 357–8
Ichikawa 575
ICSU (International Council of Scientific Unions)
  line 301–2
Imbre, J. 413
Imperial College group 107–8, 114
India, northward drift 31–2
India, northward movement 608
Indian Ocean paleogeography (McKenzie and
  Sclater) 605–9
Indian Ocean, transform faults 272–4
instantaneous rotational poles 602–4
instantaneous rotations 558–9
  describing 525–6
  distinction from finite rotations 528–9
Institute of Geophysics and Planetary Physics
  (IGPP) 204–5
International Union of Geodesy and Geophysics
  (IUGG) 4
International Upper Mantle Committee meeting,
  Ottawa 1965 *see* Ottawa meeting (September
  1965)
inverse theory 515
Irving, E. 29, 31, 42, 79–80, 89, 98, 101, 127, 129,
  144, 167, 175, 189, 227, 228, 272, 319, 328, 355,
  356, 413, 416, 423, 460, 461, 605, 608
  caution to Hess (1961) 58–61
  concerns about continental drift mechanism
    debate 58–61
  influence on Wilson 45–6
  on J. Tuzo Wilson 26–7, 38–9
  *Paleomagnetism and Its Application to Geological
    and Geophysical Problems* (1964) 187–8,
    215–16

response to the Vine–Matthews hypothesis
  215–16
Isacks, B. 386, 388–9, 395, 402
  becomes a mobilist 441
  early life and education 440
  earthquake focal mechanism solutions 539–52
  explanation of mantle subduction 446–56
  Ph.D. thesis on high-frequency earthquake
    waves 442–3
  seismology and global tectonics 562–75
  study of deep earthquakes 443–6
  work at Lamont 440–3
  work on Ice Station Alpha (T3, Fletcher's Ice
    Island) 441–2
  work with Oliver 441–3
island arcs
  and seafloor spreading 43–4
  earthquake focal mechanisms 539–52
  evolution into mountain belts 5–18
  fault mechanisms beneath 22
  length of seismic zones 575
  lithosphere bending beneath 568–70
  origins and evolution (Wilson) 41
island chains, creation of 152–5
Izu-Bonin 549–51

Jacobs, J. A. 27–31, 281, 283–4
Jaeger, J. C. 228
Jaramillo normal event (Jaramillo Subchron)
  345–50, 351, 353, 357, 358, 362–3, 367–8, 375,
  376, 380
Jardetsky, W. S. 264
Jardine 610
Jastrow, R. 412–13
Jeffreys, H. 18, 21, 33, 59, 87–8, 93, 96, 163, 166,
  168–9, 235
  contraction theory 6–7, 15
  dismissal of Carey's fit 170–1, 172–3
  influence on Smith 176–7, 184–5
  influence on Wilson 3, 25
  on Hide's thermal convection model 219–20
  on the evidence for continental drift 190–1, 193
  opposition to continental fit 184–5, 205
  opposition to mobilism 185
Joint Oceanographic Institutions for Deep Earth
  Sampling (JOIDES) 560
Jones, A. G. 262, 263, 274
Jones, D. L. 142
Jones, O. T. 93–4
Juan de Fuca plate 528, 534, 538–9, 586
Juan de Fuca region 530, 533
Juan de Fuca Ridge 274–5, 294–5, 534, 538–9, 581,
  592
  bilateral symmetry of magnetic anomalies 295–7,
    299–301
  contrasting interpretations of discoveries 309–10
  evidence for seafloor spreading 295–7

evidence for the Vine–Matthews hypothesis 297–9
Lamont workers' interpretation 304–10
similarity of the *Eltanin* magnetic profiles 365, 366–8
spreading rate 298–9, 345–50, 367–8

Kaena event 358–9, 378
Kalgoorlie Series, Western Australia 17
Kapp, R. O. 227
Karig, D. E. 507–9
Karroo dolerites 29
Katsumata 575
Kay, M. 5–6, 9, 11, 13–14, 15, 413, 417, 439
Keevil, N. 125
Kent, Dennis 605
Kermadec region earthquake focal mechanisms 539–52
Khramov, A. N. 169
Kiaman Reversed Superchron 318, 335
King, L. 4, 37–8, 155, 329, 439–40
King, P. B. 5–6, 9, 11, 13–14
Koenigsberger, J. G. 321
Koenigsberger ratio 104, 135, 215–16, 315, 316, 319
Köppen, W. 608, 610
Krause, D. C. 102
Kropotkin, P. N. 169
Kula plate 586
Kunaratnam, K. 65, 107–8, 114, 115–16, 117
Kuriles 549–51

Lamont Geological Observatory 51, 63, 64, 71, 72–8, 79, 109, 143–4, 168, 202, 234, 265, 406, 530–1, 533, 547, 560–1, 591–2, 605
alternative to the Vine–Matthews hypothesis 249–55
Atwater's visit 578
disagreement with Morgan 505
discovery of support for Vine–Matthews 248–9
group-think and bias 421–4, 439–40
heat flow over the Mid-Atlantic ridge and seafloor spreading 405–12
influence of Ewing's anti-mobilist attitude 424–7
insular research philosophy 444
integrating seismology and global tectonics 562–75
lack of interest in Morgan's work 552–3
predominance of fixist ideas 438–43
reasons for opposition to Vine–Matthews 418–27
reasons for prejudice against mobilism 418–27
rejection of seafloor spreading 304–16
rejection of the Vine–Matthews hypothesis 245–6, 249, 304–16
reversal timescale based on deep-sea cores 359–63
sedimentation argument against seafloor spreading 341

seismologists 562–75
symmetrical anomalies on Reykjanes Ridge 310–16
trenches viewed as regions of tension 570
view of mid-ocean ridges 248–9, 304–16
work in the northeast Pacific 304–10
work of Heirtzler 246
work of Le Pichon 247
work of Talwani 246
work on the Mid-Atlantic Ridge 245–6, 248–55
work on the Reykjanes Ridge 246
Landisman, M. 74, 388, 389
Langseth, M. 330, 405–12, 467
Larochelle, A. 129, 130, 137, 138
Latham, G. 442
Laughton, A. S. 69, 83, 90, 102, 105, 236, 272–3, 319, 331
early life and career 79–80
magnetic anomalies and seamounts 78–9, 81–2
Le Pichon, X. 235, 365–6, 378, 413, 414, 417, 467, 468–9, 478, 484, 494, 503, 505, 507, 526, 529, 530–1, 533, 536–9, 544, 570, 605
accepts seafloor spreading 374
alternative to the Vine–Matthews hypothesis 249–55
arguments against seafloor spreading 304–10
axial and flank magnetic anomalies 320
education and career 247–8
episodic seafloor spreading 572
global kinematic model 552–62
heat flow difficulty for seafloor spreading 405–6, 407–8
interpretation of Reykjanes Ridge symmetrical anomalies 310–16
motivation to prove Hess wrong 418–27
seafloor sedimentation and seafloor spreading 341
six-plate model 553–62, 564–6, 570–1
solitary follow-up on Morgan's work 552–3
testing Morgan's version of plate tectonics 552–62
work at Lamont 247
work in Ridge and Trough Province 304–10
work on marine magnetic anomalies 245–6
work on the Mid-Atlantic Ridge 248–55
Lenches, Laszlo 593, 594
Lister, C. R. B. 411, 468
lithosphere 568
bending beneath island arcs 568–70
Loncarevic, B. D. 323–4, 327
Lyell, C. 456–7, 458–9

Ma, T. Y. H. 226
MacDonald, G. J. F. 21, 53, 59, 163, 166, 177, 185, 193, 201, 204, 235, 408, 412–13, 415, 417, 458, 459, 461–2

Mackenzie, C. J. 4
Madden, T. 171, 172
Maddox, J. 504
magnetic anomalies, question of self-reversal
    318–19 *see also* marine magnetic anomalies
magnetic field reversal *see* geomagnetic field
    reversals; reversal timescale
magnetometers 91
    fluxgate magnetometer 78–9, 104
    proton precession magnetometer 78–9, 104, 109
Mammoth event 352, 358, 362–3, 378
mantle convection 1
    and continental drift (Royal Society, 1964) 165,
        168
    and seafloor spreading 46, 302–3
    Coode 287–8
    debate over role in seafloor spreading 58–61
    disagreement between Runcorn and Hamilton
        337–8
    disconnection from ocean ridges (Morgan) 494
    effect of the Coriolis force 220–3
    Hess 321–3
    Hess changes his account of 259–60
    Hide's model 219–23
    Holmes' view (1965) 216, 217–23
    Matthews' support for 240
    patterns of convection cells 49, 57–8
    physics of convection currents (Royal Society,
        1964) 166, 168
    Scheidegger's difficulties with 20–1
    separation from ridge formation 277
    Vine 321–3
    Wilson's dismissal of 18–20
mantle subduction, work of Isacks and Oliver
    446–56
map-projecting program (Parker) 514
Marianas region 549–51
Marine Geology of the Pacific (Menard, 1964) 53–6
marine magnetic anomalies 527–8
    application of Euler's Point Theorem 205–8
    Backus' analysis 205–8, 210
    bilateral symmetry about spreading ridges
        299–301
    computer modeling 64–5
    development of interpretative models 64–5
    development of the Vine–Matthews hypothesis
        114–24
    Dickson's hypothesis 143–4
    distribution of the source in the crust 299
    dominance of remanent magnetization 319
    early attempts to explain 63–5
    Eltanin survey magnetic profiles 363–74
    Gough, McElhinny, and Opdyke's hypothesis
        141–3
    greater amplitude of the central anomaly 323–5
    Hess (Ottawa, 1965) 333–4
    interpretation challenges 104–8

interpretation of ridges (prior to Vine) 70–8
interpretation of seamounts (prior to Vine) 78–84
Matthews' graduate work at Cambridge 89–91
Morley's hypothesis 130–6
nature of the magnetic body 64
northeast Pacific interpretation (prior to Vine)
    65–70
Pitman's "magic" profile 363–74
potential causes 63–4
relation to topography 64
reversed magnetization caused by field reversal
    105–7
significance of remanent magnetization 105
size of the central anomaly 299
use of computer simulations 104–5, 107–8
Vine's review of literature 104–8
Martin, H. 183
Marvin, U. 405
Masatsuka 575
Mason, R. G. 78–9, 80, 90, 96, 105, 107, 108, 114,
    116, 117, 121, 207, 215, 241, 294–5, 297, 304,
    382, 403, 416
    northeast Pacific magnetic anomalies 65–9
    northeast Pacific magnetic survey 130–1, 132,
        134–6, 142
    on marine magnetic anomalies 101–2, 103–4
Matthews, D. H. 62, 105, 313, 329–30, 416, 457–8, 463
    basaltic nature of ocean rocks 89–90
    becomes sympathetic to continental drift 88–9
    becomes Vine's supervisor 97–9
    Carlsberg Ridge survey (1962) 64, 65, 99, 109–14
    data from Carlsberg Ridge survey 114
    defense of the Vine–Matthews hypothesis 236–40
    early life and career 84–7
    graduate work at Cambridge 89–91
    greater amplitude of the central anomaly 323–5
    importance of spirituality 85–7
    influence on Vine 99–101
    on Wilson's transform faults idea 261, 262
    Owen Fracture Zone 261, 262, 272, 273–4
    remanent magnetization of ocean basalts 90–1
    response to Lamont magnetic anomaly profiles
        312–13
    response to Vine's draft paper 122–3
    shared views with Vine 91
    support for mantle convection 240
    sympathetic view of mobilism 89
    time in Antarctica 85–7
    view of drift while an undergraduate 87–8
    views on seafloor spreading 91
    visits the Falkland Islands 88–9
    see also Vine–Matthews hypothesis
Matuyama–Brunhes boundary 352–3, 354–7
Matuyama epoch (Chron) 345–50
    Glisá event 357–8
Matuyama reversed series (deep-sea cores) 362–3
Maxwell, A. E. 560

Maxwell, J. 176
McConnell, R. K. 413
McDougall, I. 234, 352, 353–8, 358, 378
McElhinny, M. W. 63, 141–3, 318, 354, 610
McFadden, P. L. 318
McKenzie, D. P. 277, 302, 337–8, 385, 402, 403,
    413, 423–4, 539, 554–6, 573–5, 577, 578
  early life and education 456–8
  early response to Vine–Matthews 459–60
  evolution of triple junctions 591–604
  heat flow data in terms of seafloor spreading
    463–9
  independent discovery of plate tectonics 494–510
  Indian Ocean paleogeography 605–9
  keys to his discovery of plate tectonics 510–15
  learning of Morgan's plate tectonics work 503–4,
    505–10
  letters from Morgan 529–39
  model of mid-ocean ridges 463–9
  on Bullard 175
  on paleomagnetic support for mobilism 460–1
  on the shape of the Earth 462–3
  Ph.D. thesis 462–3
  rotation of plates about triple juctions 517–18,
    522–6
  time in the US (1965) 461–2
  use of slip vectors to determine Euler poles
    518–22
  version of plate tectonics 516–26
  version of plate tectonics compared to Morgan's
    527–39
McKenzie, N. (née Fairbrother) 456, 458–9
Melanesian Rise 54
Menard, H. W. 25, 59, 65, 152, 153, 154, 162, 164,
    165, 208, 209, 241, 275, 333, 386, 413, 414–15,
    417, 425, 461–2, 476–8, 479, 480, 494, 504–5,
    507–9, 568, 577–84, 586, 598–9
  accepts seafloor spreading 403–5
  active oceanic rises , 54–5
  attack on Earth expansion (1962) 56
  attack on seafloor spreading 200–2
  attack on Wilson's oceanic islands work
    198–200
  correspondence with Hess about seafloor
    spreading 50–3
  extent of mid-ocean ridge system 52
  fracture zones in the Pacific 52, 53
  growing doubts about paleomagnetic evidence
    52–3
  Heezen's attack on seafloor thinning 56–8
  loss of enthusiasm for mobilism 55
  Mesozoic Mid-Pacific Ridge 50–1, 54
  on Lamont seafloor sediments argument 344–5
  on Wilson 148–9
  origins of ocean ridges 53–6
  patterns in distribution of ocean ridges and rises
    193–8

re-embraces fixism 52–3, 193–8
seafloor thinning theory 51, 53–6
width of ocean ridges 51
Mendocino Fracture Zone 45, 65, 68, 263, 264, 275,
    577–84
  basalts 103–4, 105
  relation to the San Andreas Fault 599–602
Mendocino triple junction 524, 525, 578
mesosphere 568
Mesozoic Mid-Pacific Ridge 50–1, 54
Meyer, B. 176
Mid-Atlantic Ridge 40, 54–5
  equatorial fracture zones 271–2
  magnetometer surveys 70–1
  origins of 57
  rift valley 70–1
  termination in the North Atlantic 269–71
  transform faults associated with 269–72
Middle America region 549–51
Mid-Indian Ocean Ridge 54–5
mid-ocean ridges
  age of 50–1
  and Earth expansion (Wilson) 33–5
  discovery of 52
  origin of 44
  width of 50–1
  Wilson's explanation (1959) 23–5
  see also oceanic ridges
Mid-Pacific Ridge (Mesozoic) 50–1, 54
Miller, J. A. 165, 177, 183, 343
miogeosynclines 12
mobilism 1
  and uniformitarianism 41
  explanation for mountain belt formation 44–5
  Wilson's rejection of (1959) 21–31
mobilism, support from paleomagnetism 604–12
mobilism–fixism debate, outcome 612
Moho discontinuity 65
Mohole project 102–3, 106–7, 123–4
Mohr's theory of fracture 7, 12, 22
Moine Thrust, Newfoundland 49
Moon, origins of 18
Molnar, P. 539, 540, 547–52
Morgan, W. J. 208, 247, 333, 337, 385, 578, 579
  disconnecting mantle convection and ocean
    ridges 494
  discovery of plate tectonics 476–8
  discussions with Vine 474
  early life and education 474
  evolution of triple junctions 591–604
  factors affecting location of ocean ridges 494
  identifies crustal block boundaries 485
  Le Pichon's test of plate tectonics 552–62
  Le Pichon's understanding of his work 552–3
  letters to McKenzie 529–39
  move to Princeton 474
  paper on crustal blocks (1968) 485–94

Morgan, W. J. (cont.)
    preparation of his paper on plate tectonics
        478–81
    presentation of plate tectonics to the AGU (April
        1967) 478–9, 481–5
    relative movement of African and South
        American blocks 486–90
    relative movement of Antarctic and African
        blocks 492–3
    relative movement of Pacific and Antarctic
        blocks 490–1
    relative movement of Pacific and North
        American blocks 490
    reviews of his paper on plate tectonics 479–81
    rigidity of crustal blocks 485, 493
    structure of crustal blocks 494
    structure of ocean ridges 494
    three types of block boundaries 485
    transfer of transform faults to a sphere 485
    use of Euler's Point Theorem 485
    version of plate tectonics 553
    version of plate tectonics compared to
        McKenzie's 527–39
    work on mantle convection 476
Morley, L. W.
    accepts continental drift 129
    accepts geomagnetic field reversals 127–9
    aeromagnetic survey work 126, 127, 129
    education 125–7
    influence of seafloor spreading 131–2
    interpretation of northeast Pacific magnetic
        survey 130–1, 132
    magnetization of seamounts 134–5
    using paleomagnetism to test continental drift
        126–7
Morley's hypothesis 62–3, 124–5, 130–6
    appearance of Vine and Matthews' paper 137–8
    attempts to get his paper published 136–9
    differences from Vine and Matthews 134–6
    induced magnetization 134–6
    interpretation of NE Pacific magnetic survey 134–6
    pattern of magnetic anomalies 134
    published versions of his paper 132–6
    reasons for rejection of his paper 139–41
    rejections of his paper 132, 136–9
    timescale for field reversals 133–4
Mount Hague 3
mountain belt formation
    and mobilism 39, 44–5
    contraction theory of evolution 5–18
mountain building, difficulties with continental
    drift 18–20, 20–1
Mudie, J. 495, 496, 507, 509
Mudie, John 577, 578
Munk, W. H. 53, 142, 204, 413, 414, 458, 461–2
Murray escarpment/fracture zone 44–5
Murray Fracture Zone 65, 69, 275, 578–84

Nafe, J. E. 364
Nairn, A. E. M. 29, 166–7, 175
National Institute of Oceanography (NIO), UK 79
National Research Council (NRC) of Canada 4
natural remanent magnetism (NRM) 604–12
Nature
    acceptance of Vine and Matthews paper 139–41
    rejection of Morley's paper 136–7, 139–41
Néel, L. 89, 321
Nel, H. J. 38
Nettleton 65
new global tectonics 562–75
New Hebrides 549–51
New Zealand 549–51
New Zealand, support for Vine–Matthews
    hypothesis 241–2
Newcastle conference (1965) 391–2
Newer Volcanics of Victoria, dating of polarity
    reversals 355–7
Nicholls, G. D. 165
Ninkovich, D. 361
North American block, movement relative to the
    Pacific block 490
North American Geosynclines (Kay) 5–6
North America, tectonics of its western margin
    598–602
North Honshu region 549–51
north Pacific, origin of the Great Magnetic Bight
    584–91
northeast Pacific
    challenges of unraveling its Cenozoic history
        338–9
    evolution of 598–602
    explanations for anomalies 528
    fracture zones 275
    magnetic anomalies, interpretations prior to
        Vine 65–70
    magnetic survey 130–1, 132, 134–6, 142
    work at Lamont 304–10
    work of Wilson and Vine 294–304
northeast Pacific fracture zones, application of
    plate tectonics 576–84

O'Keefe, J. A. 168
oceanic basalts 89–90
    remanent magnetization 90–1
oceanic islands, Wilson's study of 148–62
oceanic ridge crest basalts, dating by Lamont
    workers 343–4
oceanic ridge formation, separation from mantle
    convection 277
oceanic ridges
    factors affecting location of 494
    magnetic anomaly interpretation (prior to Vine)
        70–8
    migration theory (Wilson) 41–2, 46, 57, 60
    origins of (Menard) 53–6

structure of (Morgan) 494
  surrounding Antarctica 41–2, 46, 57, 60, 275–7,
    302–3
  see also mid-ocean ridges
Odell, N. 2–3
Olduvai normal event 346–9, 351–2, 355, 357–8,
    362–3
Oliver, J. 386, 387, 388–9, 392, 395, 402, 426–7,
    540–8
  early life and education 438–40
  explanation of mantle subduction 446–56
  influence of Ewing 438–40
  seismology and global tectonics 562–75
  study of deep earthquakes 443–6
  work at Lamont 438–40
Opdyke, N. D. 63, 141–3, 234, 246, 254–5, 310, 315,
    363, 364, 368–9, 372, 375–6, 413, 416, 423, 425,
    440, 605
  analysis of Lamont's deep-sea cores 359–63
  arrival at Lamont 359–60
  influence on Pitman 366–7
  response to the *Eltanin*-19 magnetic profile 366–7
  reversal timescale based on deep-sea cores
    359–63
ophiolites 9, 11
Ornach-Nal Fault 262, 272
orogenesis and mobilism 39
orogeny and slow Earth expansion (Wilson) 33–5
orogneic belts, Caledonides 45
Orowan, E. 166, 260, 328
Osemeikhian, J. 172
Ottawa meeting (September 1965) 325–38
  disagreement on mantle convection 337–8
  Hess's presentation and discussion 333–7
  range of topics of disagreement 337–8
  Talwani's presentation 326–8
  Wilson's presentation and discussion 328–33
Owen Fracture Zone 261, 262, 272, 273–4, 331
Oxburgh, E. R. 168, 466–7

Pacific–Antarctic Ridge 54–5
  Pitman's "magic" profile 363–74
Pacific block
  movement relative to the Antarctic block 490–1
  movement relative to the North American block
    490
Pacific Ocean
  great fracture zones 301–2
  seafloor downwarping (Wilson) 155
  transform faults 274–7
  see also northeast Pacific
paleoclimatic evidence 31–2
  support for mobilism 45–6
paleogeography
  contributions of plate tectonics and APW paths
    604–12
  extending further back in time 604–12

Indian Ocean (McKenzie and Sclater) 605–9
  mapping (Smith and Briden) 608–10
paleomagnetism
  support for the mobilist argument 604–12
  history of study 604–12
*Paleomagnetism and Its Application to Geological
    and Geophysical Problems* (Irving, 1964) 187–8,
    215–16
paleomagnetism's support for mobilism 25–7,
    27–31
  attack by Wilson 34–5
  Blackett's view 190
  Bullard's view 191–2
  challenge of persuading doubters 186–7
  comparison with other support 189
  difficulty-free status 191–2
  GAD hypothesis 187
  influence on Wilson 38–9
  Irving's 1964 monograph 187–8
  Jeffreys' view 190–1
  Menard's growing doubts 52–3
  Rutten's perspective 189
  solution to divergent APW paths 186–7
  status of the debate by 1964 186–92
paleontological evidence, as a dating tool 224–6
Pangea 40
Parker, R. L. 323–4, 337–8, 459, 464, 527, 554–6,
    577
  education and career 512–15
  map-projecting program 514
  response to Vine–Matthews hypothesis 513
  version of plate tectonics produced with
    McKenzie 516–26
  work with McKenzie on plate tectonics 494, 495,
    496, 497–502, 503–4, 512, 514, 515
Parsons, B. 468
paving stone theory, McKenzie and Parker
    516–26
Peter, G. 72–8, 81, 106, 235, 243–4, 273, 380, 587
Peters, L. 126
Petterson, H. 406
Philippines 549–51
Phinney, R. 591
Philipps, J. D. 478, 484
Physics and Geology (Jacobs *et al.*, 1959) 27–31
Pilansberg dykes, South Africa 142
Pioneer Fracture Zone 65, 579, 583, 584
Pitman, W. C. 236, 375–6
  accepts seafloor spreading 372–3
  accepts the Vine–Matthews hypothesis 372–3
  becomes a mobilist 366–7
  early life and education 363–4
  Eltanin-19 profile 363–74, 392, 490–1
  Eltanin survey 338
  Eltanin survey magnetic profiles 363–74
  "magic" profile over the Pacific–Antarctic ridge
    363–74, 392, 490–1

Pitman, W. C. (cont.)
  move to Lamont 363–4
  work on the *Eltanin* magnetic profiles 367–73
plate evolution
  and stability of triple junctions 595–8
  changes in plate motions 602–4
plate rigidity 537
  evidence for 556–7
plate tectonics 60–1, 208
  application to paleogeography 611–12
  application to NE Pacific fracture zones
    576–84
  as a kinematic theory 277
  comparison of Morgan and McKenzie's versions
    527–39
  conceptualization 528–9
  discovery by Morgan 476–8
  explanations for NE Pacific anomalies 528
  impacts of the concept 611–12
  keys to McKenzie's discovery 510–15
  impacts of the concept 611–12
  kinematic methods of testing 527–8
  Le Pichon's test of Morgan's work 552–62
  mantle convection and ridges 586
  McKenzie and Parker's version 516–26
  McKenzie's independent discovery 494–510
  Morgan's explanation (AGU 1967) 481–5
  Morgan's first talk on 247
  Morgan's preparation of his paper 478–81
  Morgan's presentation to the AGU (April 1967)
    478–9, 481–5
  testing by Morgan and McKenzie 528
  work of seismologists 562–75
plates on the Earth's surface
  evolution and past rotations 524–6
  types of plate borders (Wilson) 265–6
  Wilson 265–6
  see also crustal blocks/plates
Plumstead, E. P. 4, 37–8
polar wandering, view of Jacobs, Wilson, and
    Russell 25–9
polarity reversals *see* geomagnetic field reversals
Poldervaart, A. 440
Prague, A. 456
Precambrian, continental evolution (Wilson)
    15–18
Press, F. 64, 71, 72, 81, 413, 414, 439
Princeton 591–2
Principles of Geodynamics (Scheidegger) 31–2
Principles of Physical Geology (Holmes, 1965)
    216–27, 230, 230
proto-Atlantic (Argand) 40, 49
proton precession magnetometer 78–9, 104, 109

quaternions 525–6
Queen Charlotte Islands Fault 274–5
Quesnell, A. M. 272

radiometric dating, Canadian provinces 15–16
radiometrically dated reversal timescale 298, 351
Raff, A. D. 65, 96, 123–4, 207, 241, 294–5, 304, 382,
    403, 416, 590–1
  northeast Pacific magnetic anomalies 67–9
  northeast Pacific magnetic survey 130–1, 132,
    134–6, 142
  on marine magnetic anomalies 102–3
Raitt, R. W. 51, 78–9, 167, 210–14, 304, 552
Ramberg, R. 195–7
Rand, J. R. 413
rapid Earth expansion 570–1, 552
rapid Earth expansion theories 1, 33
  attacks on (1964–1964) 227–32
  Carey 224, 227, 228, 230, 231
  Heezen 224, 227, 228, 231, 328
  Heezen renounces 405
  rejection by Holmes 227
  Runcorn's attack on 230
  Ward's attack on 227–30
Rat Islands 530, 533, 535
Raven, J. 456
Red Sea
  magnetic anomalies 71–2, 82–4
  transform faults 272–3
remanent magnetization of ocean basalts 90–1, 234
  dominance in marine magnetic anomalies 319
  Koenigsberger ratio 104
  Mendocino Fracture Zone basalts 103–4
  significance in magnetic anomalies 105
Research Strategy 1 (RS1)
  agreement between anomaly patterns from
    different locations 380
  agreement between determinations of Atlantic
    opening 431
  agreement between independent dating methods
    363
  agreement between independent reversal dating
    methods 380
  agreement between reversal dating methods 298
  agreement between Runcorn and Ward's
    calculations 231
  agreement between seafloor spreading rates 381
  agreement between sediment and magnetic
    findings 431
  agreement of paleogeographic data 607
  aspects of continental fit 183
  calculating the rate of seafloor spreading 239
  consilience between data used for
    reconstructions 605
  consilient instantaneous relative plate motion
    determinations 594
  consilience throughout evidence for seafloor
    spreading 431
  counterexample to support ridge activity 386
  delineation of the Juan de Fuca Ridge 297
  explanation of east Pacific magnetic anomalies 239

intensity of remanent magnetization of basalts 321
Juan de Fuca as a spreading ridge 294–5
matching magnetic profiles across different
  ridges 431
McKenzie's solution to seafloor spreading
  difficulties 464
Morley's explanation of marine magnetic
  anomalies 133
past movement of Africa 151
removal of difficulties for the Vine–Matthews
  hypothesis 297
removal of plate tectonics difficulties (McKenzie)
  522–6
response to difficulties with seafloor spreading 35
unrecognised problems in seafloor spreading 149
Vine–Matthews hypothesis as difficulty-free 377
Wilson on continental evolution 15
Wilson on continental nuclei 17
Wilson on continental shelf formation 18
Wilson on fractured arcs 12
Wilson on island arc formation 7
Wilson on mountain belt formation 7, 12
Wilson's explanation of the Wadati–Benioff
  zone 7
Wilson's recognition of Juan de Fuca Ridge
  294–5
Wilson's secondary mountain belts 12
Research Strategy 2 (RS2), Carey's view of trenches
  570
Research Strategy 2 (RS2)
  continental drift and mountain building 18
  differences between axial and flank magnetic
    anomalies 320
  difficulties with constant seafloor spreading 428
  difficulties with explanations for the Scotia Arc 88
  difficulties with explanations of magnetic
    anomalies 133
  difficulties with McKenzie and Parker's plate
    tectonics 522–6
  difficulties with mobilism (Jacobs *et al.*) 30
  difficulties with patterns of convection cells 57
  difficulties with seafloor spreading 35, 52, 57
  difficulties with the Vine–Matthews hypothesis
    209–10, 249, 431
  difficulty with continental fit 182, 185
  difficulty with contractionism 34
  difficulty with Earth expansion 56
  difficulty with mantle convection 18, 221
  difficulty with mid-oceanic ridges 57
  difficulty with origin of marine magnetic
    anomalies 67
  difficulty with seafloor subduction results 391
  distribution of guyots and atolls in the Pacific 50
  Ewing's difficulty with seafloor spreading 342
  explanations for the central magnetic anomaly
    239
  Hess's response to heat flow difficulty 408

importance of raising difficulties in science 418
island age and distance from the ridge 150, 199
lack of seismic detection of magnetic anomalies
  68
magnetic anomaly correlation with topography
  68
Menard on Wilson's origin of island chains 199
mobilism and uniformitarianism 41
Morley's difficulty with marine magnetic
  anomalies 133
origin of transcurrent faults (Wilson) 34
poor fit in magnetic profiles with normal
  magnetization only 118
Scheidegger on mantle convection 20, 21
Scheidegger's difficulties with continental
  drift 20
support for reversed magnetization 118
Wilson's difficulties with mantle convection
  19, 20
Research Strategy 3 (RS3)
  alternative explanation for island arcs 41
  alternative theory to seafloor spreading 52
  alternatives to fixist solutions 149
  contractionism preferred to mantle convection
    18, 20
  difficulty with contractionism 21
  features explained by mantle convection 34
  Morley's hypothesis compared to other
    explanations 133
  seafloor thinning as alternative to expansion 56
  slow expansion as alternative to contractionism
    34
  Vine's explanation of marine magnetic anomalies
    121
Réunion volcanics, dating of polarity reversals
  353–5
Revelle, R. 204, 406
reversal timescale 60
  accurate dating by USGS 345–50
  based on deep-sea cores 359–63
  contribution to the mobilism debate 351
  extension by Heirtzler 471
  history of development 350–1
  improvements during 1966 350–9
  radiometric dating of reversals 298, 351
  work of the ANU group 351, 353–9
  work of the USGS Menlo Park group 351–3
  work on *Eltanin* magnetic profiles 368–9
Reykjanes Ridge 246, 248, 251–2
  Lamont discovery of symmetrical anomalies
    310–16
  similarity to *Eltanin* magnetic profiles 367–8
  spreading rate 367–8
Richter, C. F. 160, 444, 551, 563–4, 572
Ridge and Trough Province 274–5, 294–5
ridge-ridge-ridge (*RRR*) triple junctions 584–91,
  595–8 *see also* triple junctions

ridge-ridge transform faults *see* transform faults
ridges *see* mid-ocean ridges; oceanic ridges
Riviera triple junction 602
Roche, A. 128–9, 317
Ross, D. I. 241–2, 380
Rothé, J. P. 247, 563–4
Royal Society 459
Royal Society symposium on continental drift
     (1964) 162, 459
  balance of viewpoints of participants 163–70
  balance of viewpoints on mobilism 169
  Continental reconstructions (session 1) 163–4,
     166–7
  Convection currents and continental drift
     (session 3) 165, 168
  Everett, Bullard, and Smith's fit of the continents
     170–86
  Gilluly's review 193
  Horizontal displacements in the Earth's crust
     (session 2) 164–5, 167–8
  influence of paleomagnetism on participants'
     views 169
  Menard re-embraces fixism 193–8
  Menard's attack on Wilson's work 198–200
  Menard's attack on seafloor spreading
     200–2
  organization by Runcorn 162–3
  participants 163–70
  Physics of convection currents in the Earth's
     mantle (session 4) 166, 168
  reactions to the continental fit model 185–6
  status of paleomagnetism's support for mobilism
     186–92
  status of the mechanism difficulty 192–3
  structure of the sessions 163–70
  Vacquier's difficulties with the Vine–Matthews
     hypothesis 209–10
  views on Earth expansion 169–70
Rubey, W. W. 202, 418
Runcorn, S. K. 27, 31, 76–7, 79–80, 87–8, 101,
     127–9, 138–9, 163–4, 166, 175, 178, 202, 203,
     219, 221, 257–8, 281, 329, 332–3, 335, 460, 461,
     463
  attack on rapid Earth expansion 230
  Continental Drift (1962) 106
  mantle convection disagreement with Hamilton
     337–8
  organization of the Royal Society symposium
     (1964) 162–3
Russell, R. D. 27–31
Rutten, M. G. 166–7, 189
Ryukyus 549–51

Saito, T. 343–4
San Andreas Fault 30, 44, 274–5, 295–7, 298, 309,
     490, 518, 524, 525, 538–9, 577–8
  origin of 599–602

relation to the Mendocino Fracture Zone
     599–602
Scheidegger, A. E. 11, 12, 15, 22, 23, 25, 26, 33, 39,
     40, 156, 160
  acknowledges paleomagnetic support for
     mobilism 31–2
  attack on continental drift 20–1
  attack on mantle conversion 20–1
  development of Jeffreys' contraction theory 6–7
Sclater, J. G. 140, 173, 236–7, 295, 410–12, 459,
     468, 494, 507–9, 593, 605–9
Scotia Arc, origins 88–9
Scripps Institution of Oceanography 51, 53, 54,
     108, 109, 141, 142, 167–8, 204, 209, 244, 253,
     259, 360, 372, 406, 414–15, 416, 444, 461–2,
     463, 515, 554–6, 576–84, 605
Scrutton, C. T. 231
seafloor basalts 89–90
seafloor evolution, competing theories 1
seafloor sedimentation
  and seafloor spreading 341, 343–5
  lack of Pre-Cretaceous seafloor sediments 43–4
seafloor spreading 131–2, 527
  and Earth expansion 41
  and fracture zones 52, 53
  and heat flow over the Mid-Atlantic ridge 405–8,
     408–12
  and island arcs (Wilson) 43–4
  and mantle convection 46, 302–3
  and *RRR* triple junctions 584–91
  and sedimentation 341, 343–4, 344–5
  and the Vine–Matthews hypothesis 62
  bilateral symmetry of magnetic anomalies
     299–301
  Dietz 37, 39, 41, 53, 96, 106, 119, 131–2, 143, 155,
     217, 218
  discussion between Bernal and Dietz 35–6
  Eltanin survey magnetic profiles 363–74
  episodic 559–60, 572
  evidence for continuous spreading 560
  first published exchange about 35–6
  fossil evidence against 343–4
  Heezen's attack on 56–8
  Heirtzler accepts 372–3
  Hess 37, 39, 41, 119, 155
  Hess fine tunes and extends his account 259–61
  Holmes' view (1965) 216–17
  ideas which developed from 59–61
  increasing continental thickness 41
  Juan de Fuca Ridge evidence 295–7, 299–301
  Le Pichon accepts 374
  McKenzie's solution to difficulties 463–9
  mechanism difficulty 58–61
  Menard accepts 403–5
  Menard's attack on 200–2
  Menard's correspondence with Hess 50–3
  ocean ridges around Antarctica 41–2

origin of the Great Magnetic Bight 584–91
  Pitman accepts 372–3
  Pitman's "magic" profile 363–74
  premise of the Vine–Matthews hypothesis 234–6
  reasons for eventual general acceptance 431–4
  rejection at Lamont 304–10, 310–16
  Talwani accepts 373–4
  Wilson's comments on consequences 39–42
  Wilson's search for evidence 148–62
  Zed patterns 581–4
seafloor spreading rate
  based on the Vine–Matthews hypothesis
    298–9
  calculating the rate of spread 239
  Juan de Fuca ridge 345–50
  Vine and Wilson's determinations 298–9
seafloor subduction, Sykes' evidence for 388–91
seafloor thinning theory 1
  and width of ocean ridges 51
  Heezen's attack on 56–8
  Menard's explanation 51
seamounts 50–1
  Kunaratnam's computer program 107–8
  magnetic anomalies 101–2, 105–6, 134–5
  magnetic anomaly interpretation (prior to Vine)
    78–84
  magnetization 68–9
sedimentary rock, conversion to metamorphic rock
    43–4
seismic anisotropy in the upper mantle (Hess)
    210–14
seismic profiling of the continental crust around
    Britain (BIRPS) 87
seismology and plate tectonics 562–75
self-reversal of rock magnetization 318–19
self-reversing Haruna dacite 100–1
serpentinized peridotite 70, 136
Serra Geral basalts, South America 29
Shackleton, R. M. 166–7, 185
Shire, E. 79
six-plate model (Le Pichon) 553–62, 564–6, 570–1
Shor, G. G. 51, 167, 210–14, 332
Sigurgeirsson, T. 133–4
slip rate and length of descending seismic zone
    575
slow Earth expansion
  adoption by Wilson (1960) 33–5
  and seafloor spreading 41
  Egyed. 33, 56, 216, 222, 223–4, 227, 228,
    230, 232
  Holmes' speculation on the cause 227
  Holmes' support for 216, 222, 223–6, 227, 230
Smith, A. G. 164, 174, 417, 495–6, 510–11, 527,
    608–10
  continental fit results presented in 1964 178–85,
    206–7, 260–1
  contribution to the fit of continents 177

education and early career 175–7
fit of the continents around the Atlantic 170–86
influence of Jeffreys 176–7, 184–5
Royal Society symposium (1964) presentation
    170–86
Smith, S. M. 194–5
South African dolerites 29
South America, fit with Africa 170–86
South American block, movement relative to the
    African block 486–90
Spain, rotation relative to France 28
spinner magnetometer 127
Sprague 108
stable triple junctions 595–8
Stacey, F. 144, 366, 367
Stauder, W. V. 393–4, 402–3, 449, 511–12, 530, 533,
    534–5, 536–7, 570, 575
Steenland, N. C. 65
Stehli, F. G. 413
Steinhart, John 176
Stewart, H. B. 235, 243–4
Stokoe, Austin 456
Stubbs, Peter 42
subduction rate calculation 571–2
subduction, work of Isacks and Oliver 446–56
Suess, F. E. 45, 49
Sunda region 549–51
Sutton, J. 167–8
Sykes, L. R. 265, 275, 284–5, 286–7, 304, 307, 413,
    416, 443, 444, 449, 455, 478, 487, 530, 533, 534,
    540–8
  accepts the Vine–Matthews hypothesis 392
  computer program to process earthquake data
    389
  confirms ridge-ridge transform faults 386–403
  contact with Wilson 396
  early life and education 386–7
  early skepticism about continental drift 387–8,
    391–2
  evidence for seafloor subduction 388–91
  fault plane solutions paper (1967) 396–402
  first motion studies of earthquake mechanisms
    389–96
  locating hypocenters of earthquakes 388–91
  move to Lamont 387
  Ph.D. work on shallow earthquakes 388
  response to Pitman's "magic" profile 392
  seismology and global tectonics 562–75
  talk at the Goddard conference (1966) 402
  tests Wilson's transform fault proposition
    389–96

Talwani, M. 71, 238, 239, 247, 253, 315–16,
    329–30, 484
  accepts seafloor spreading 373–4
  arguments against seafloor spreading 304–10
  education and career 246

Talwani, M. (cont.)
  motivation to prove Hess wrong 418–27
  presentation (Ottawa meeting, 1965) 326–8
  rejection of seafloor spreading 326–8
  rejection of the Vine–Matthews hypothesis
    326–8
  response to the Vine–Matthews hypothesis 202
  work at Lamont 246
  work in Ridge and Trough Province 304–10
  work on marine magnetic anomalies 245–6
Tarling, D. H. 234, 272, 281, 352, 355, 358–9
Taylor, F. B. 36, 611
Taylor, Isabel 577
Taylor, J. H. 166–7, 185
  Cabot Fault–Great Glen Fault 46–50
tectonic theory, requirements (Wilson) 42–6
Tharp, M. 71, 164–5, 169–70, 185, 330–1, 486
The Evolution of North America (King) 5–6
The Pulse of the Earth (Umbgrove) 5–6
thrust-fault earthquakes 568–70
Thom, T. 3
Thompson, G. A. 214, 331–2
Tilley, C. E. 215
Tobin, D. G. 389, 530, 533, 534
Tonga–Fiji region earthquake focal mechanisms
    539–52
Tozer, D. C. 166, 169–70, 227, 281, 328, 333, 459,
    462
transcurrent faults 22, 30–1, 47
  and expansion (Wilson) 34
  differences from transform faults 268
  direction of motion 268
  Wilson's global perspective 264
  Wilson's paper model 268
transform fault trends 527–8
transform faults 44, 50, 59–61, 110, 111, 513–14, 527
  and aseismic ridges 301–2
  Arabian Sea 272
  associated with Carlsberg Ridge 272
  associated with Mid-Atlantic Ridge, 269–72
  Coode's independent idea 280
  Coode's mention of the idea to Wilson 262,
    263, 264
  development of Wilson's idea 261–8
  differences between Coode and Wilson's
    accounts 288–9
  differences from transcurrent faults 268
  direction of motion 268
  first motion studies of earthquake mechanisms
    (Sykes) 389–96
  Gulf of Aden 272–3
  identification worldwide by Wilson 269–78
  implications for continental drift 267–8, 278
  Indian Ocean 272–4
  Lamont workers' view 308–9
  Pacific Ocean 274–7
  presentation by Wilson (Ottawa, 1965) 328

  projection onto a sphere 511, 517
  Red Sea 272–3
  Sykes confirms ridge-ridge transform faults
    386–403
  Sykes tests Wilson's proposition 389–96
  transfer to a spherical surface (Morgan) 485
  types of (Wilson) 266–8
  Vine's independent proposal 278–80
  Vine's mention of the idea to Wilson 262, 263, 264
  Wilson's 1965 paper 265–8
  Wilson's description 265–6
  Wilson's paper model 268
transform (half-shear) junctions (Wilson) 266
Transvaal, Africa 17
trenches, viewed as regions of tension 570
triple junctions 385, 528
  evolution of 591–604
  origin of the Great Magnetic Bight 584–91
  rotation of plates about 517–18, 522–6
  stability of different types 595–8
  types of 595–8
tunnel vision, among Lamont workers 421–4
Turcotte, D. L. 466–7

Udintsev, G. B. 329
Umbgrove, J. H. F. 5, 11, 15
uniformitarianism 12, 15–18, 22, 27, 39, 41,
    42–3, 45
United States Coast and Geodetic Survey
    (USCGS) 388
United States Geological Survey (USGS)
    5, 234
  improvements in the reversal timescale 345–50,
    351–3
United States Office of Naval Research 109
United States Upper Mantle project 443
Utsu, T. 444, 575
Uyeda, S., self-reversing Haruna dacite 100–1

Vacquier, V. 64, 65, 69–70, 96, 107, 108, 126, 130–1,
    142, 164, 165, 209–10, 235, 275, 284–5, 307–9,
    416, 466, 484, 513
van Andel, S. I. 230
Van Bemmelen, R. W. 189
van der Gracht, W. A. J.M. 155
van Hilten, D. 230
Veldkamp, J. 189
Vening Meinesz, F. A. 19, 168–9, 221
Verhoogen, J. 166
Verkhoyansk Mountains 269–71
Vine, F. J. 62, 76, 77, 243, 252, 325, 413, 416, 425–6,
    458–9, 478, 485, 529, 572, 591, 597, 605
  adaptation of Kunaratnam's seamount program
    107–8, 114, 115–16, 117
  awareness of significance of remanence 101
  awareness of work on reversals 100–1
  begins research at Cambridge (1962) 96–101

challenges of magnetic anomaly interpretation 104–8
computer modeling of the Carlsberg Ridge survey data 114–24
computer simulations of magnetic anomalies 104–5, 107–8
defense of the Vine–Matthews hypothesis 236–40
development of the Vine–Matthews hypothesis (1963) 114–24
discussions with Morgan 474
early interest in continental drift 91–3
enthusiasm for Hess' ideas 93–6
fully accepts seafloor spreading 348–50
independently proposes ridge-ridge transform faults 278–80
influence of Bullard 99–101
influence of Matthews 99–101
internal reactions to his draft paper 122–3
interpretation of marine magnetic anomalies 96
learns of corrections to reversal timescale 345–50
letter to Hess about Princeton (1965) 255–7
magnetic anomalies and seamounts 105–6
Matthews becomes his supervisor 97–9
McKenzie letter about plate tectonics discovery 502–3
mention of transform fault idea to Wilson 262, 263, 264
on Vacquier's criticisms 210
origin of the Great Magnetic Bight 586–90
overview of evidence supporting Vine–Matthews (1966) 375–86
presentation at the Goddard conference 414
rejection of self-reversal in magnetic anomalies 105–7
reversed magnetization caused by field reversal 105–7
review of literature on marine magnetic anomalies 104–8
role as catalyst for Wilson 290–1
shared views with Matthews 91
significance of remanent magnetization 105
turns Vine–Matthews into a difficulty-free solution 375–86
undergraduate years at Cambridge 93–6
use of computers 98–100
use of interpretative modeling 65
view of paleomagnetic support for mobilism 96
work on Juan de Fuca spreading rate 345–50
work with Wilson and Hess 255, 259
work with Wilson in the northeast Pacific 245–6, 294–304
Vine–Matthews hypothesis 59–61, 62, 91, 604
bilateral symmetry of magnetic anomalies 295–7, 301
defense and development (1964–1965) 236–40
development by Vine (1963) 114–24
differences from Morley's hypothesis 134–6

difficulties raised 243–5
difficulty-free status (1966) 375–86
early responses to 202–16
empirical difficulties 235–6
evidence from the Juan de Fuca Ridge 297–9
geomagnetic field reversal premise 234
Heirtzler accepts 372–3
Heirtzler and Le Pichon's alternative 249–55
Hess's response 210–15
hypotheses which were similar to 62–3
initial difficulties 234–6
Irving's response 215–16
McKenzie's early response 459–60
mild support and criticism (early 1965) 241–5
Parker's response 513
Pitman accepts 372–3
premise of geomagnetic field reversals 317–19
publication in *Nature* (1963) 124, 137–8
reasons for acceptance for publication 139–41
reasons for opposition at Lamont 418–27
rejection at Lamont 202, 249, 304–16
rejection by Heirtzler, Le Pichon and Talwani 245–6
remanence of ocean floor basalts 234
researchers who ignored it 241
response of Talwani 202
seafloor spreading premise 234–6
seafloor spreading rates based on 298–9
support from Canada 242–3
support from Hess 334
support from New Zealand 241–2
Sykes accepts 392
three premises 234–6
Vacquier's criticism 209–10
Vine's overview of supporting evidence (1966) 375–86
Vine's Ph.D. thesis (1965) 319–23
axial and flank magnetic anomalies 320
remanent magnetization of basalts 321
speculations on mantle convection 321–3
Vogt, P. 244–5
volcanic arcs 568–70
volcanoes, formation of 568–70
Von Herzen, R. P. 209, 304, 406–7, 466, 484

Wadati, K. 444
Wadati–Benioff zone 7, 9, 36, 443–6, 546–7
Walker, G. P. L. 165, 239–40, 274, 284–5, 313–15
Ward, M. A. 227–30, 231
Wasserberg, J. C. 413, 414, 461–2
Wegener, A. 18, 36, 37, 38, 45, 59, 88, 93, 155, 162, 270, 608, 610, 611
Wegener Fault 31, 48–9, 156, 157, 269–71
Wegmann, C. E. 31, 45, 270
Weiss, N. 499, 513, 514–15
Wells, J. W. 224–6, 230–1

Wensink, H. 254–5, 313, 353, 357–8, 360
Westoll, T. S. 163–4, 169–70, 227
White, G. W. 186
Whittard, W. F. 257
Williams, C. A. 459
Wilson, H. (Tuzo) 2
Wilson, J. A. 2
Wilson, J. Tuzo 1, 59, 125, 126–7, 165, 167, 216,
     243, 386, 404, 405, 416, 463, 478, 485, 527, 576,
     591
  ability to generate new ideas 161–2
  adoption of slow Earth expansion (1960) 33–5
  aseismic ridges that branch off active ridges 151
  at Madingley rise (1965) 258, 259
  attack on paleomagnetic support for mobilism
     34–5
  attacks on continental drift 18–20
  attacks on mantle convection 18–20
  becomes a mobilist (1961) 36–46
  becoming a globalist 4
  comments on seafloor spreading concept 35,
     39–42, 57
  consequences of seafloor spreading 39–42
  contact with Sykes 396
  Coode's ideas on faults 281–2
  creation of island chains 152–5
  defense of continental accretion 5–18
  defense of contractionism 5–18, 21–31
  description of transform faults 265–6
  developing the idea of transform faults 261–8
  development of continents 42–3
  differences between transform faults and
     transcurrent faults 268
  difficulties with seafloor spreading 41–2
  difficulty of the circum-Antarctic ridge 275–7
  direction of motion in transcurrent faults 268
  direction of motion in transform faults 268
  discussion of the Pacific and seafloor spreading
     160–1
  early life and career 2–5
  East Pacific Rise 335, 336–7
  explanation for large horizontal movements
     44–5
  explanation of mid-ocean ridges 23–5
  explanation of oceanic seafloor features 156–7
  first paper on transform faults 245–6
  formation of continental shelves 11
  global perspective on transcurrent faults 264
  honors and awards 5
  hot spots within the mantle 152–5
  identification of transform faults around the
     world 269–78
  influence of Jeffreys 3, 18, 25
  influence of paleomagnetic findings 38–9
  island age and distance from ridge 149–51
  island arcs and seafloor spreading 43–4
  island arcs origin and evolution 41

  lack of Pre-Cretaceous seafloor sediments 43–4
  lecture invitation from Runcorn 257–8
  limitations of contraction theory 43–4
  mantle convection and seafloor spreading 46
  mantle convection patterns 49
  matches the Cabot and Great Glen faults 46–50
  Menard's attack on his oceanic islands work
     198–200
  migrating ocean ridges theory 41–2, 46, 57, 60
  mobilism and mountain belt formation 44–5
  models of transform and transcurrent faults 268
  on paleomagnetism's support for mobilism 25–7,
     27–31
  on polar wandering 25–9
  on the paleomagnetic case for mobilism 45–6
  origin of mid-ocean ridges 44
  origin of the Hawaiian Islands 152–5
  orogeny and slow Earth expansion 33–5
  paper on transform faults (1965) 265–8
  plates on the Earth's surface 265–6
  predicted ridge in the Labrador Sea 242, 243
  presentation and discussion (Ottawa meeting,
     1965), 328–33
  presentation of transform faults (Ottawa, 1965)
     328
  proposed down-warping of the Pacific seafloor
     155
  recognition of his work 5
  reconciling orogenesis and mobilism 39
  reconstruction of continents in the mid-Mesozoic
     161
  rejection of contractionism (1960) 33–5
  rejection of mobilism (1959) 21–31
  requirements for a tectonic theory 42–6
  role of Vine and Coode as catalysts 290–1
  search for evidence of seafloor spreading 148–62
  search for further support for mobilism 148–62
  sedimentary rock conversion to metamorphic
     rocks 43–4
  speculation about the ocean ridge system
     157–9
  study of oceanic islands 148–62
  summary papers 155–6
  tectonic disjuncts 45
  theory of island arc formation 5–18
  theory of mountain formation 5–18
  transcurrent faults 30–1
  transcurrent faults and expansion 34
  transform (half-shear) junctions 266
  transform faults 44, 50, 265–8, 308–9, 511,
     513–14, 517
  transform faults and continental drift 267–8
  travels in Europe (1965) 257–9
  types of plate border 265–6
  types of transform fault 266–8
  uniformitarianism 12, 15–18, 22, 27, 39, 41,
     42–3, 45

Vine's mention of transform fault idea 279–80
work with Vine and Hess 255, 259
work with Vine in the northeast Pacific 245–6,
   294–304
Wilson, R. L. 133–4, 318–19
Wise, D. U. 382
Woods Hole Oceanographic Institute (WHOI) 214,
   244, 386
Woollard, G. P. 3
Wordie, J. 3

World Wide Standardized Seismograph Network
   (WWSSN) 449
Worzel, J. Lamar 71, 164, 168, 185, 239, 246,
   414–15, 422, 484, 570

Yamaska Mountains, Canada 129

Zed patterns 581–4
Zeitz, I. 130
Zijderveld, J. D. A. 189

Printed in the United States
By Bookmasters